Int
the
of

Introduction to the Mathematics of Medical Imaging

Second Edition

Charles L. Epstein

University of Pennsylvania
Philadelphia, Pennsylvania

siam

Society for Industrial and Applied Mathematics • Philadelphia

This SIAM edition is a second edition of the work first published by Pearson Education, Inc., 2003.

10 9 8 7 6 5 4 3 2 1

Library of Congress Cataloging-in-Publication Data

Epstein, Charles L.

Introduction to the mathematics of medical imaging / Charles L. Epstein. -- 2nd ed.

p. cm.

ISBN 978-0-89871-642-9 (alk. paper)

1. Diagnostic imaging--Mathematics. 2. Imaging systems in medicine--Mathematics. 3. Medical physics--Mathematics. I. Title.

RC78.7.D53E676 2008

616.07'54--dc22 2007061801

This book is dedicated to my wife, Jane,
and our children, Leo and Sylvia.

They make it all worthwhile.

Contents

Preface to the second edition

It seems like only yesterday that I was sending the "camera ready" pdf file of this book off to Prentice Hall. Despite a very positive response from the mathematics and engineering communities, Pearson decided, last year, to let the book go out of print. I would like to thank George Lobell, my editor at Prentice Hall, for making it so easy to reacquire the publication rights. Secondly I would like to thank SIAM, and my editors Sarah Granlund and Ann Manning Allen, for making it so easy to prepare this second edition. I would be very remiss if I did not thank Sergei Gelfand, editor at the AMS, who prodded me to get the rights back, so I could prepare a second edition.

The main differences between this edition and the Prentice Hall edition are: 1. A revised section on the relationship between the continuum and discrete Fourier transforms, Section 10.2.2 (reflecting my improved understanding of this problem); 2. A short section on Grangreat's formula, Section 10.2.2, which forms the basis of most of the recent work on cone-beam reconstruction algorithms; 3. A better description of the gridding method, Section 11.8 (many thanks to Leslie Greengard and Jeremy Magland for helping me to understand this properly); 4. A chapter on magnetic resonance imaging, Chapter 14; 5. A short section on noise analysis in MR-imaging, Section 16.3. For the last two items I would like to express my deep gratitude to Felix Wehrli, for allowing me to adapt an article we wrote together for the Elsevier Encyclopedia on Mathematical Physics, and for his enormous hospitality, welcoming me into his research group, the Laboratory for Structural NMR Imaging, at the Hospital of the University of Pennsylvania.

With a bit more experience teaching the course and using the book, I now feel that it is essential for students to have taken at least one semester of undergraduate analysis, beyond calculus, and a semester of linear algebra. Without this level of sophistication, it is difficult to appreciate what all the fuss is about.

I have received a lot of encouragement to prepare this second edition from the many people who used the book, either as a course textbook or for self study. I would like to thank Rafe Mazzeo, Petra Bonfert-Taylor, Ed Taylor, Doug Cochran, John Schotland, Larry Shepp, and Leslie Greengard for their kind words and advice. Finally, I thank my wife, Jane, and our children, Leo and Sylvia, for their forebearance during the endless preparation of the first edition, and their encouragement to produce this second edition.

Charles L. Epstein
May 16, 2007

Preface

Over the past several decades, advanced mathematics has quietly insinuated itself into many facets of our day-to-day life. Mathematics is at the heart of technologies from cellular telephones and satellite positioning systems to online banking and metal detectors. Arguably no technology has had a more positive and profound effect on our lives than medical imaging, and in no technology is the role of mathematics more pronounced or less appreciated. X-ray tomography, ultrasound, positron emission tomography, and magnetic resonance imaging have fundamentally altered the practice of medicine. At the core of each modality is a mathematical model to interpret the measurements and a numerical algorithm to reconstruct an image. While each modality operates on a different physical principle and probes a different aspect of our anatomy or physiology, there is a large overlap in the mathematics used to model the measurements, design reconstruction algorithms, and analyze the effects of noise. In this text we provide a tool kit, with detailed operating instructions, to work on the sorts of mathematical problems that arise in medical imaging. Our treatment steers a course midway between a complete, rigorous mathematical discussion and a cookbook engineering approach.

The target audience for this book is junior or senior math undergraduates with a firm command of multi-variable calculus, linear algebra over the real and complex numbers, and the basic facts of mathematical analysis. Some familiarity with basic physics would also be useful. The book is written in the language of mathematics, which, as I have learned, is quite distinct from the language of physics or the language of engineering. Nonetheless, the discussion of every topic begins at an elementary level and the book should, with a little translation, be usable by advanced science and engineering students with some mathematical sophistication. A large part of the mathematical background material is provided in two appendices.

X-ray tomography is employed as a *pedagogical machine*, similar in spirit to the elaborate devices used to illustrate the principles of Newtonian mechanics. The *physical principles* used in x-ray tomography are simple to describe and require little formal background in physics to understand. This is not the case in any of the other modalities listed nor in less developed modalities like infrared imaging or impedance tomography. The *mathematical* problems that arise in x-ray tomography and the tools used to solve them have a great deal in common with those used in the other imaging modalities. This is why our title is *Introduction to the Mathematics of Medical Imaging* instead of *Introduction to the Mathematics of X-Ray Tomography.* A student with a thorough understanding of the material in this book should be mathematically prepared for further investigations in most subfields of medical imaging.

Very good treatments of the physical principles underlying the other modalities can be found in *Radiological Imaging* by Harrison H. Barrett and William Swindell, [6], *Principles of Computerized Tomographic Imaging* by Avinash C. Kak and Malcolm Slaney, [76], *Foundations of Medical Imaging* by Cho, Jones, Singh, [22], *Image Reconstruction from Projections* by Gabor T. Herman, [52], and *Magnetic Resonance Imaging* by E. Mark Haacke, Robert W. Brown, Michael R. Thompson, Ramesh Venkatesan, [50]. Indeed these books were invaluable sources as I learned the subject myself. My treatment of many topics owes a great deal to these books as well as to the papers of Larry Shepp and Peter Joseph and their collaborators. More advanced treatments of the mathematics and algorithms introduced here can be found in *The Mathematics of Computerized Tomography* by Frank Natterer, [95], and *Mathematical Methods in Image Reconstruction* by Frank Natterer and Frank Wübbelling, [96].

The order and presentation of topics is somewhat nonstandard. The organizing principle of this book is the evolutionary development of an accurate and complete model for x-ray tomography. We start with a highly idealized mathematical model for x-ray tomography and work toward more realistic models of the actual data collected and the algorithms used to reconstruct images. After some preliminary material we describe a continuum, complete data model phrased in terms of the Radon transform. The Fourier transform is introduced as a tool, first to invert the Radon transform and subsequently for image processing. The first refinement of this model is to take account of the fact that real data are always sampled. This entails the introduction of Fourier series, sampling theory, and the finite Fourier transform. After introducing terminology and concepts from filtering theory, we give a detailed synthesis of the foregoing ideas by describing how continuum, shift invariant, linear filters are approximately implemented on finitely sampled data. With these preliminaries in hand, we return to the study of x-ray tomography, per se. Several designs for x-ray computed tomography machines are described, after which we derive the corresponding implementations of the filtered back-projection algorithm. At first we assume that the x-ray beam is one dimensional and monochromatic. Subsequently we analyze the effects of a finite width beam and various sorts of measurement and modeling errors. The last part of the book is concerned with noise analysis. The basic concepts of probability theory are reviewed and applied to problems in imaging. The notion of signal-to-noise ratio (SNR) is introduced and used to analyze the effects of quantum noise on images reconstructed using filtered back-projection. A maximum likelihood algorithm for image reconstruction in positron emission tomography is described. The final chapter introduces the idea of a random process. We describe the random processes commonly encountered in imaging and an elementary example of an optimal filter. We conclude with a brief analysis of noise in the continuum model of filtered back-projection.

The book begins with an introduction to the idea of using a mathematical model as a tool to extract the physical state of system from feasible measurements. In medical imaging, the "state of the system" in question is the anatomy and physiology of a *living* human being. To probe it nondestructively requires considerable ingenuity and sophisticated mathematics. After considering a variety of examples, each a toy problem for some aspect of medical imaging, we turn to a description of x-ray tomography. This leads us to our first mathematical topic, *integral transforms*. The transform of immediate interest is the Radon transform, though we are quickly led to the Abel transform, Hilbert transform, and Fourier transform. Our study of the Fourier transform is dictated by the applications we have in mind, with a strong emphasis on the connection between the smoothness of a function and the decay of its Fourier transform and vice versa. Many of the basic ideas of functional analysis appear as we consider these examples. The concept of a weak derivative, which is ubiquitous in the engineering literature and essential to a precise understanding of the Radon inversion formula, is introduced. This part of the book culminates in a study of the Radon inversion formula. A theme in these chapters is the difference between finite- and infinite-dimensional linear algebra.

The next topics we consider are Fourier series, sampling, and filtering theory. These form the basis for applying the mathematics of the Fourier transform to real-world problems. Chapter 8 is on sampling theory; we discuss the Nyquist theorem, the Shannon–Whittaker interpolation formula, the Poisson summation formula, and the consequences of undersampling. In Chapter 9, on filtering theory, we recast Fourier analysis as a tool for image and signal processing. The chapter concludes with an overview of image processing and a linear systems analysis of some basic imaging hardware. We then discuss the mathematics of approximating continuous time, linear shift invariant filters on finitely sampled data, using the finite Fourier transform.

In Chapters 11 and 12 the mathematical tools are applied to the problem of image reconstruction in x-ray tomography. These chapters are largely devoted to the filtered back-projection algorithm, though other methods are briefly considered. After deriving the reconstruction algorithms, we analyze the point spread function and modulation transfer function of the full measurement and reconstruction process. We use this formalism to analyze a variety of imaging artifacts. Chapter 13 contains a brief description of "algebraic reconstruction techniques," which are essentially methods for solving large, sparse systems of linear equations.

The final topic is noise in the filtered back-projection algorithm. This part of the book begins with an introduction to probability theory. Our presentation uses the language and ideas of measure theory, in a metaphoric rather than a technical way. Chapter 15 concludes with a study of specific probability distributions that are important in imaging. In Chapter 16 we apply probability theory to a variety of problems in medical imaging. This chapter includes the famous resolution-dosage fourth power relation, which shows that to double the resolution in a CT image, keeping the SNR constant, the radiation dosage must be increased by a factor of 16! The chapter ends with an introduction to positron emission tomography and the maximum likelihood algorithm. Chapter 17 introduces the ideas of random processes and their role in signal and image processing. Again the focus is on those processes needed to analyze noise in x-ray imaging. A student with a good grasp of Riemann integration should not have difficulty with the material in these chapters.

Acknowledgments

Perhaps the best reward for writing a book of this type is the opportunity it affords for thanking the many people who contributed to it in one way or another. There are a lot of people to thank, and I address them in roughly chronological order.

First I would like to thank my parents, Jean and Herbert Epstein, for their encouragement to follow my dreams and the very high standards they set for me from earliest childhood. I would also like to thank my father and Robert M. Goodman for imparting the idea that observation and careful thought go a long way toward understanding the world in which we live. Both emphasized the importance of expressing ideas simply but carefully.

My years as an undergraduate at the Massachusetts Institute of Technology not only provided me with a solid background in mathematics, physics, and engineering but also a belief in the unity of scientific enquiry. I am especially grateful for the time and attention Jerry Lettvin lavished on me. My interest in the intricacies of physical measurement surely grew out of our many conversations. I was fortunate to be a graduate student at the Courant Institute, one of the few places where "pure" and "applied" mathematics lived together in harmony. In both word and deed, my thesis advisor, Peter Lax, placed mathematics and its applications on an absolutely equal footing. It was a privilege to be his student. I am very grateful for the enthusiasm that he and his late wife, Anneli, showed for turning my lecture notes into a book.

I would like to acknowledge my friends and earliest collaborators in the enterprise of becoming a scientist—Robert Indik and Carlos Tomei. I would also thank my friends and current collaborators, Gennadi Henkin and Richard Melrose, for the vast wealth of culture and knowledge they have shared with me and their forbearance, while I have been "missing in action."

Coming closer to the present day, I would like to thank Dennis Deturck for his unflagging support, both material and emotional, for the development of my course on medical imaging and this book. Without the financial support he and Jim Gee provided for the spring of 2002, I could not have finished the manuscript. The development of the original course was supported in part by National Science Foundation grant DUE95-52464. I would like to thank Dr. Ben Mann, my program director at the National Science Foundation, for providing both moral and financial support. I am very grateful to Hyunsuk Kang, who transcribed the first version of these notes from my lectures in the spring of 1999. Her hard work provided the solid foundation on which the rest was built. I would also like to thank the students who attended Math 584 in 1999, 2000, 2001, and 2002 for their invaluable input on the content of the course and their numerous corrections to earlier versions of this book. I would also like to thank Paula Airey for flawlessly producing what must have seemed like endless copies of preliminary versions of the book.

John D'Angelo read the entire text and provided me with an extremely useful critique as well as a lot of encouragement. Chris Croke, my colleague at the University of Pennsylvania, also carefully read much of the manuscript while teaching the course and provided many corrections. I would like to thank Phil Nelson for his help with typesetting, publishing, and the writing process and Fred Villars for sharing with me his insights on medicine, imaging, and a host of other topics.

The confidence my editor, George Lobell, expressed in the importance of this project was a strong impetus for me to write this book. Jeanne Audino, my production editor, and Patricia M. Daly, my copy editor, provided the endless lists of corrections that carried my manuscript from its larval state as lecture notes to the polished book that lies before you. I am most appreciative of their efforts

I am grateful to my colleagues in the Radiology Department—Dr. Gabor Herman, Dr. Peter Joseph, Dr. David Hackney, Dr. Felix Wehrli, Dr. Jim Gee, and Brian Avants—for sharing with me their profound, first-hand knowledge of medical imaging. Gabor Herman's computer program, *SNARK93*®, introduced me to the practical side of image reconstruction and was used to make some of the images in the book. I used Dr. Kevin Rosenberg's program *ctsim* to make many other images. I am very grateful for the effort he expended to produce a version of his marvelous program that would run on my computer. David Hackney provided beautiful, state-of-the-art images of brain sections. Felix Wehrli provided the image illustrating aliasing in magnetic resonance imaging and the x-ray CT micrograph of trabecular bone. I am very grateful to Peter Joseph for sharing his encyclopedic first-hand knowledge of x-ray tomography and its development as well as his treasure trove of old, artifact-ridden CT images. Jim Gee and Brian Avants showed me how to use MATLAB® for image processing. Rob Lewitt provided some very useful suggestions and references to the literature.

I would also like the acknowledge my World Wide Web colleagues. I am most appreciative for the x-ray spectrum provided by Dr. Andrew Karellas of the University of Massachusetts, the chest x-ray provided by Drs. David S. Feigen and James Smirniotopoulos of the Uniformed Services University, and the nuclear magnatic resonance spectrum (NMR) provided by Dr. Walter Bauer of the Erlangen-Nürnberg Univerisity. Dr. Bergman of the Virtual Hospital at the University of Iowa (www.vh.org) provided an image of an anatomical section of the brain. The World Wide Web is a resource for imaging science of unparalleled depth, breadth, and openness.

Finally and most of all I would like to thank my wife, Jane, and our children, Leo and Sylvia, for their love, constant support, and daily encouragement through the many, many, moody months.

Charles L. Epstein
Philadelphia, PA
January 8, 2003

How to Use This Book

This chapter is strongly recommended for all readers.

The structure of this book: It is unlikely that you will want to read this book like a novel, from beginning to end. The book is organized somewhat like a hypertext document. Each chapter builds from elementary material to more advanced material, as is also the case with the longer sections. The elementary material in each chapter (or section) depends only on the elementary material in previous chapters. The more advanced material can be regarded as "links" that you may or may not want to follow. Each chapter is divided into sections, labeled $c.s$, where c is the chapter number and s, the section number. Many sections are divided into subsections, labeled $c.s.ss$, with ss the subsection number. Sections, subsections, examples, exercises, and so on that are *essential* for later parts of the book are marked with a star: $\star$. These should not be skipped.

Sections, subsections, exercises, and so on that require more prerequisites than are listed in the preface are marked with an asterisk: $*$. These sections assume a familiarity with the elementary parts of functions of a complex variable or more advanced real analysis. The asterisk is inherited by the subparts of a marked part. The sections without an asterisk do not depend on the sections with an asterisk. All sections with an asterisk can be omitted without any loss in continuity. The book has two appendices. Appendix A contains background material on a variety of topics ranging from very elementary things like numbers and their representation in computers to more advanced topics like spaces of function and approximation theory. This material serves several purposes: It compiles definitions and results used in the book, and provides a larger context for some material and references for further study. Appendix B is a quick review of the definitions and theorems usually encountered in an undergraduate course in mathematical analysis. Many sections begin with a box:

See: A.1, B.3, B.4, B.5.

containing a list of recommended background readings drawn from the appendices and earlier sections.

For the student: This book is wide ranging, and it is expected that students with a variety of backgrounds will want to use it. Nonetheless, choices needed to be made regarding notation and terminology. In most cases I use notation and terminology that is standard in the world of mathematics. If you come on a concept or term you do not understand, you should first look in *List of Notations* at the beginning of the book and then in the Index at the end. There are many exercises and examples scattered throughout the text. Doing the exercises and working through the details of the examples will teach you a lot more than simply reading the text. The book does not contain many problems involving machine computation; nonetheless I strongly recommend implementing as many things as you can using MATLAB®, Maple®, and so on.

The World Wide Web is a fantastic resource for medical imaging, with many sites devoted to images and the physics behind imaging. At www.ctsim.org you will find an excellent program, written by Dr. Kevin Rosenberg, that simulates the data collection, processing and postprocessing done by variety of x-ray CT machines. The pictures you will produce with *ctsim* are surely worth thousands of words! A list of additional imaging Web sites can be found at the end of Chapter 11. I hope you enjoy this book and get something out of it. I would be most grateful to hear about which parts of the book you liked and which parts you feel need improvement. Contact me at: cle@math.upenn.edu.

For the instructor: Teaching from this book requires careful navigation. I have taught several courses from the notes that grew into this book. Sometimes the students had a mathematical bent, sometimes a physical bent, and sometimes an engineering bent. I learned that the material covered *and* its presentation must be tailored to the audience. For example, proofs or outlines of proofs are provided for most of the mathematical results. These should be covered in detail for math students and perhaps assigned as reading or skipped for engineering students. How far you go into each topic should depend on how much time you have and the composition of the class. If you are unfamiliar with the practicalities of medical imaging I would strongly recommend that you read the fundamental articles [113], [114], and [55] as well as the parts of [76] devoted to x-ray tomography.

In the following section I have provided suggestions for one- and two-semester courses with either a mathematical or an engineering flavor (or student body). These guidelines should make it easier to fashion coherent courses out of the wide range of material in the text. Exercises are collected at the ends of the sections and subsections. Most develop ideas presented in the text; only a few are of a standard, computational character. I would recommend using computer demonstrations to illustrate as many of the ideas as possible and assigning exercises that ask the students to implement ideas in the book as computer programs. I would be most grateful if you would share your exercises with me, for inclusion in future editions. As remarked previously, Dr. Kevin Rosenberg's program *ctsim*, which is freely available at www.ctsim.org, can be used to provide students with irreplaceable hands-on experience in reconstructing images.

Some suggestions for courses: In the following charts material shaded in *dark gray* forms the basis for classroom presentation; that shaded in *medium gray* can be assigned as reading or used as additional material in class. The *light gray* topics are more advanced enrichment material; if the students have adequate background, some of this material can be selected for presentation in class. Students should be encouraged to read the sections in the appendices listed in the boxes at the beginnings of the sections. I have not recommended spending a lot of class time on the higher-dimensional generalizations of the Fourier transform, Fourier series, and so on. One is tempted to say that "things go just the same." I would instead give a brief outline in class and assign the appropriate sections in the book, *including some exercises*, for home study.

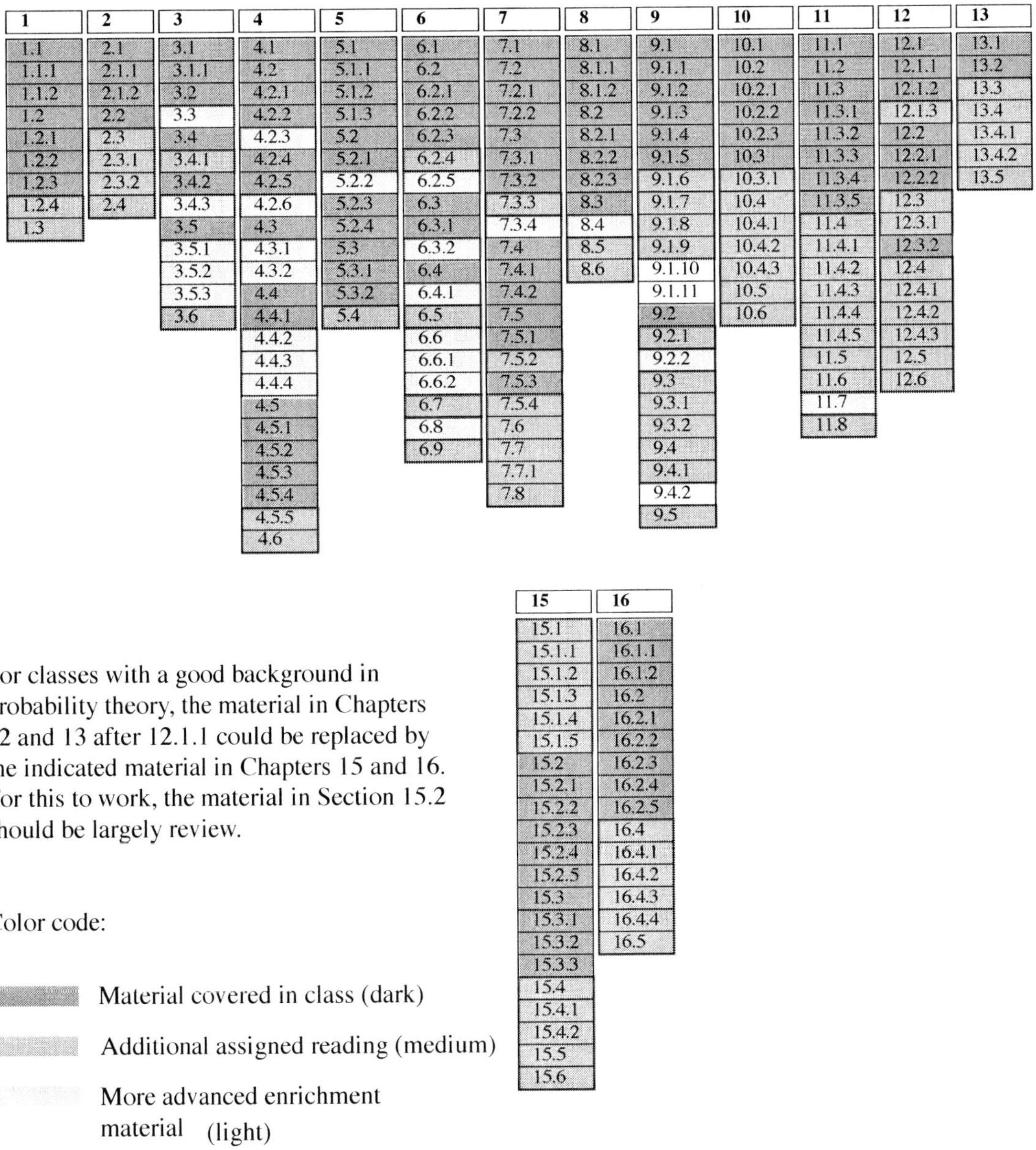

Figure 1. An outline for a one-semester course with a mathematical emphasis.

The chart in Figure 1 outlines a one semester course with an emphasis on the mathematical aspects of the subject. The dark gray sections assume a background in undergraduate analysis and linear algebra. If students have better preparation, you may want to select some of the light gray topics for classroom presentation.

If at all possible, I would recommend going through the section on generalized functions, A.4.5, when introducing the concept of weak derivatives in Section 4.3. Much of the subsequent red material assumes a familiarity with these ideas. As noted, if students have a good background in measure and integration or probability, you may want to skip the material after 12.1.1 and go instead to the analysis of noise, beginning in Chapter 14.

The chart in Figure 2 outlines a one semester course with an emphasis on the engineering aspects of the subject. For such a class the material in this book could be supplemented with more applied topics taken from, for example, [76] or [6]. The proofs of mathematical results can be assigned as reading or skipped.

1	2	3	4	5	6	7	8	9	10	11	12
1.1	2.1	3.1	4.1	5.1	6.1	7.1	8.1	9.1	10.1	11.1	12.1
1.1.1	2.1.1	3.1.1	4.2	5.1.1	6.2	7.2	8.1.1	9.1.1	10.2	11.2	12.1.1
1.1.2	2.1.2	3.2	4.2.1	5.1.2	6.2.1	7.2.1	8.1.2	9.1.2	10.2.1	11.3	12.1.2
1.2	2.2	3.3	4.2.2	5.1.3	6.2.2	7.2.2	8.2	9.1.3	10.2.2	11.3.1	12.1.3
1.2.1	2.3	3.4	4.2.3	5.2	6.2.3	7.3	8.2.1	9.1.4	10.2.3	11.3.2	12.2
1.2.2	2.3.1	3.4.1	4.2.4	5.2.1	6.2.4	7.3.1	8.2.2	9.1.5	10.3	11.3.3	12.2.1
1.2.3	2.3.2	3.4.2	4.2.5	5.2.2	6.2.5	7.3.2	8.2.3	9.1.6	10.3.1	11.3.4	12.2.2
1.2.4	2.4	3.4.3	4.2.6	5.2.3	6.3	7.3.3	8.3	9.1.7	10.4	11.3.5	12.3
1.3		3.5	4.3	5.2.4	6.3.1	7.3.4	8.4	9.1.8	10.4.1	11.4	12.3.1
		3.5.1	4.3.1	5.3	6.3.2	7.4	8.5	9.1.9	10.4.2	11.4.1	12.3.2
		3.5.2	4.3.2	5.3.1	6.4	7.4.1	8.6	9.1.10	10.4.3	11.4.2	12.4
		3.5.3	4.4	5.3.2	6.4.1	7.4.2		9.1.11	10.5	11.4.3	12.4.1
		3.6	4.4.1	5.4	6.5	7.5		9.2	10.6	11.4.4	12.4.2
			4.4.2		6.6	7.5.1		9.2.1		11.4.5	12.4.3
			4.4.3		6.6.1	7.5.2		9.2.2		11.5	12.5
			4.4.4		6.6.2	7.5.3		9.3		11.6	12.6
			4.5		6.7	7.5.4		9.3.1		11.7	
			4.5.1		6.8	7.6		9.3.2		11.8	
			4.5.2		6.9	7.7		9.4			
			4.5.3			7.7.1		9.4.1			
			4.5.4			7.8		9.4.2			
			4.5.5					9.5			
			4.6								

Due to time constraints, it may be necessary to choose one of 9.4.1 or 9.4.2 and leave the material not covered in class as a reading assignment.

Figure 2. An outline for a one-semester course with an engineering emphasis.

The chart in Figure 2 gives suggestions for a full-year course with a mathematical emphasis. A full year should afford enough time to introduce generalized functions in Section A.4.5. This should be done along with the material in Section 4.3. This allows inclusion of Section 4.4.4, which covers the Fourier transform on generalized functions.

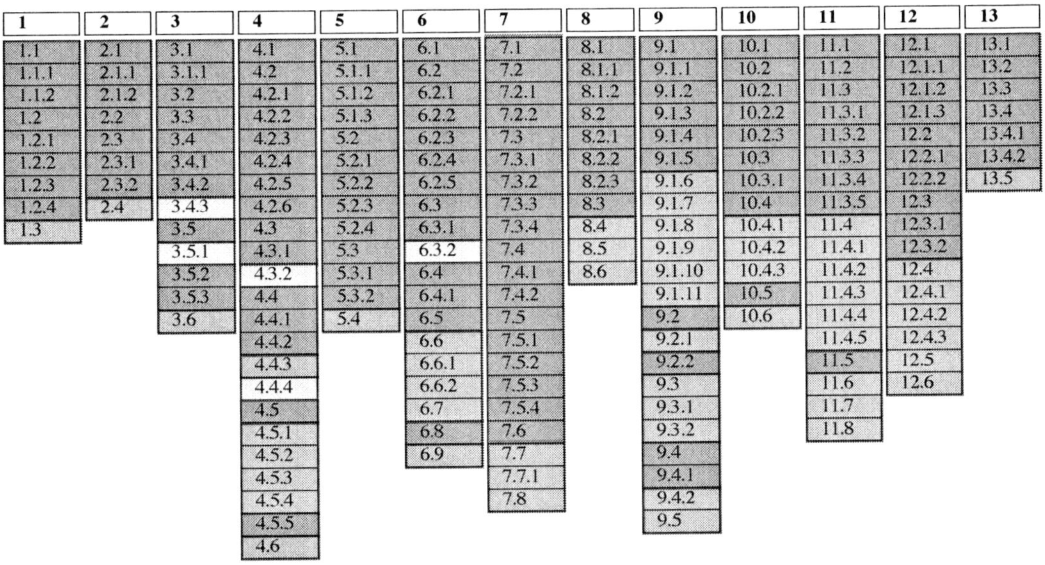
1
1.1 1.1.1 1.1.2 1.2 1.2.1 1.2.2 1.2.3 1.2.4 1.3
2
2.1 2.1.1 2.1.2 2.2 2.3 2.3.1 2.3.2 2.4
3
3.1 3.1.1 3.2 3.3 3.4 3.4.1 3.4.2 3.4.3 3.5 3.5.1 3.5.2 3.5.3 3.6
4
4.1 4.2 4.2.1 4.2.2 4.2.3 4.2.4 4.2.5 4.2.6 4.3 4.3.1 4.3.2 4.4 4.4.1 4.4.2 4.4.3 4.4.4 4.5 4.5.1 4.5.2 4.5.3 4.5.4 4.5.5 4.6
5
5.1 5.1.1 5.1.2 5.1.3 5.2 5.2.1 5.2.2 5.2.3 5.2.4 5.3 5.3.1 5.3.2 5.4
6
6.1 6.2 6.2.1 6.2.2 6.2.3 6.2.4 6.2.5 6.3 6.3.1 6.3.2 6.4 6.4.1 6.5 6.6 6.6.1 6.6.2 6.7 6.8 6.9
7
7.1 7.2 7.2.1 7.2.2 7.3 7.3.1 7.3.2 7.3.3 7.3.4 7.4 7.4.1 7.4.2 7.5 7.5.1 7.5.2 7.5.3 7.5.4 7.6 7.7 7.7.1 7.8
8
8.1 8.1.1 8.1.2 8.2 8.2.1 8.2.2 8.2.3 8.3 8.4 8.5 8.6
9
9.1 9.1.1 9.1.2 9.1.3 9.1.4 9.1.5 9.1.6 9.1.7 9.1.8 9.1.9 9.1.10 9.1.11 9.2 9.2.1 9.2.2 9.3 9.3.1 9.3.2 9.4 9.4.1 9.4.2 9.5
10
10.1 10.2 10.2.1 10.2.2 10.2.3 10.3 10.3.1 10.4 10.4.1 10.4.2 10.4.3 10.5 10.6
11
11.1 11.2 11.3 11.3.1 11.3.2 11.3.3 11.3.4 11.3.5 11.4 11.4.1 11.4.2 11.4.3 11.4.4 11.4.5 11.5 11.6 11.7 11.8
12
12.1 12.1.1 12.1.2 12.1.3 12.2 12.2.1 12.2.2 12.3 12.3.1 12.3.2 12.4 12.4.1 12.4.2 12.4.3 12.5 12.6
13
13.1 13.2 13.3 13.4 13.4.1 13.4.2 13.5

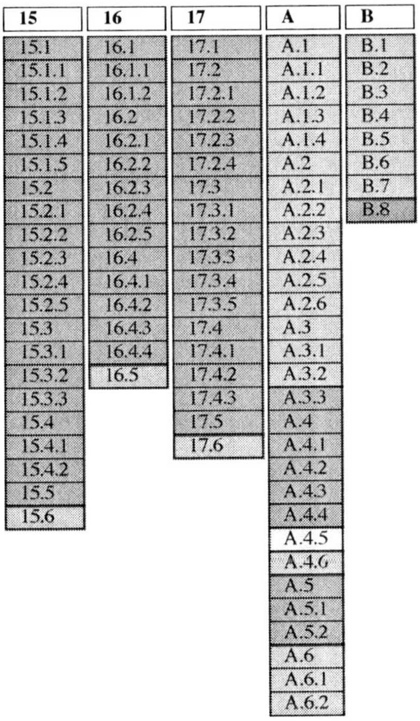
15
15.1 15.1.1 15.1.2 15.1.3 15.1.4 15.1.5 15.2 15.2.1 15.2.2 15.2.3 15.2.4 15.2.5 15.3 15.3.1 15.3.2 15.3.3 15.4 15.4.1 15.4.2 15.5 15.6
16
16.1 16.1.1 16.1.2 16.2 16.2.1 16.2.2 16.2.3 16.2.4 16.2.5 16.4 16.4.1 16.4.2 16.4.3 16.4.4 16.5
17
17.1 17.2 17.2.1 17.2.2 17.2.3 17.2.4 17.3 17.3.1 17.3.2 17.3.3 17.3.4 17.3.5 17.4 17.4.1 17.4.2 17.4.3 17.5 17.6
A
A.1 A.1.1 A.1.2 A.1.3 A.1.4 A.2 A.2.1 A.2.2 A.2.3 A.2.4 A.2.5 A.2.6 A.3 A.3.1 A.3.2 A.3.3 A.4 A.4.1 A.4.2 A.4.3 A.4.4 A.4.5 A.4.6 A.5 A.5.1 A.5.2 A.6 A.6.1 A.6.2
B
B.1 B.2 B.3 B.4 B.5 B.6 B.7 B.8

The sections in the appendices, indicated in dark gray, should be presented in class when they are referred to in the boxes at the starts of sections. If at all possible, Section A.4.5, on generalized functions, should be done while doing Section 4.3, thus allowing the inclusion of 4.4.4.

Figure 3. An outline for a one-year course with a mathematical emphasis.

The chart in Figure 3 gives suggestions for a full-year course with an engineering emphasis. As before, this material should be supplemented with more applied material coming from, for example [76] or [6].

1	2	3	4	5	6	7	8	9	10	11	12	13
1.1	2.1	3.1	4.1	5.1	6.1	7.1	8.1	9.1	10.1	11.1	12.1	13.1
1.1.1	2.1.1	3.1.1	4.2	5.1.1	6.2	7.2	8.1.1	9.1.1	10.2	11.2	12.1.1	13.2
1.1.2	2.1.2	3.2	4.2.1	5.1.2	6.2.1	7.2.1	8.1.2	9.1.2	10.2.1	11.3	12.1.2	13.3
1.2	2.2	3.3	4.2.2	5.1.3	6.2.2	7.2.2	8.2	9.1.3	10.2.2	11.3.1	12.1.3	13.4
1.2.1	2.3	3.4	4.2.3	5.2	6.2.3	7.3	8.2.1	9.1.4	10.2.3	11.3.2	12.2	13.4.1
1.2.2	2.3.1	3.4.1	4.2.4	5.2.1	6.2.4	7.3.1	8.2.2	9.1.5	10.3	11.3.3	12.2.1	13.4.2
1.2.3	2.3.2	3.4.2	4.2.5	5.2.2	6.2.5	7.3.2	8.2.3	9.1.6	10.3.1	11.3.4	12.2.2	13.5
1.2.4	2.4	3.4.3	4.2.6	5.2.3	6.3	7.3.3	8.3	9.1.7	10.4	11.3.5	12.3	
1.3		3.5	4.3	5.2.4	6.3.1	7.3.4	8.4	9.1.8	10.4.1	11.4	12.3.1	
		3.5.1	4.3.1	5.3	6.3.2	7.4	8.5	9.1.9	10.4.2	11.4.1	12.3.2	
		3.5.2	4.3.2	5.3.1	6.4	7.4.1	8.6	9.1.10	10.4.3	11.4.2	12.4	
		3.5.3	4.4	5.3.2	6.4.1	7.4.2		9.1.11	10.5	11.4.3	12.4.1	
		3.6	4.4.1	5.4	6.5	7.5		9.2	10.6	11.4.4	12.4.2	
			4.4.2		6.6	7.5.1		9.2.1		11.4.5	12.4.3	
			4.4.3		6.6.1	7.5.2		9.2.2		11.5	12.5	
			4.4.4		6.6.2	7.5.3		9.3		11.6	12.6	
			4.5		6.7	7.5.4		9.3.1		11.7		
			4.5.1		6.8	7.6		9.3.2		11.8		
			4.5.2		6.9	7.7		9.4				
			4.5.3			7.7.1		9.4.1				
			4.5.4			7.8		9.4.2				
			4.5.5					9.5				
			4.6									

15	16	17	A
15.1	16.1	17.1	A.1
15.1.1	16.1.1	17.2	A.1.1
15.1.2	16.1.2	17.2.1	A.1.2
15.1.3	16.2	17.2.2	A.1.3
15.1.4	16.2.1	17.2.3	A.1.4
15.1.5	16.2.2	17.2.4	A.2
15.2	16.2.3	17.3	A.2.1
15.2.1	16.2.4	17.3.1	A.2.2
15.2.2	16.2.5	17.3.2	A.2.3
15.2.3	16.4	17.3.3	A.2.4
15.2.4	16.4.1	17.3.4	A.2.5
15.2.5	16.4.2	17.3.5	A.2.6
15.3	16.4.3	17.4	A.3
15.3.1	16.4.4	17.4.1	A.3.1
15.3.2	16.5	17.4.2	A.3.2
15.3.3		17.4.3	A.3.3
15.4		17.5	A.4
15.4.1		17.6	A.4.1
15.4.2			A.4.2
15.5			A.4.3
15.6			A.4.4
			A.4.5
			A.4.6
			A.5
			A.5.1
			A.5.2
			A.6
			A.6.1
			A.6.2

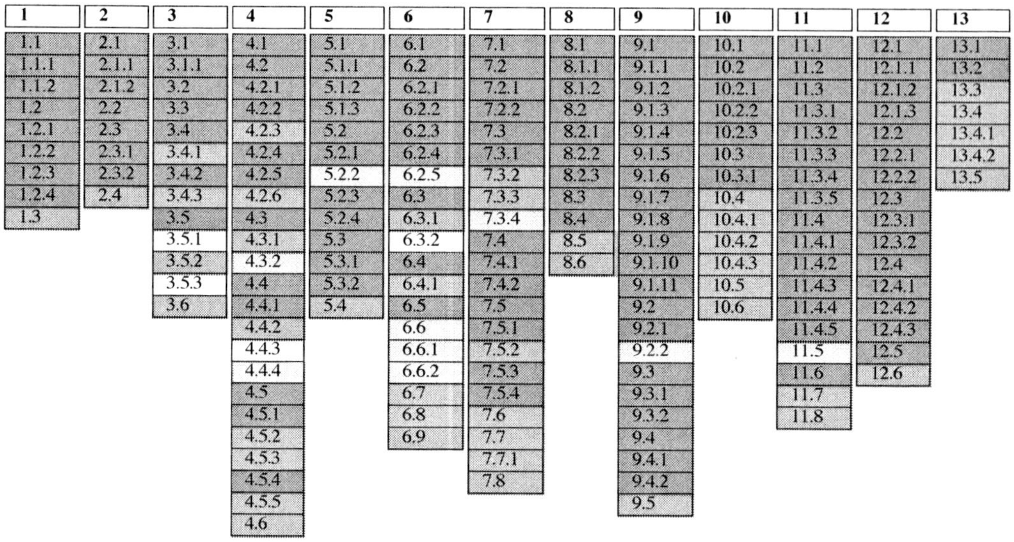

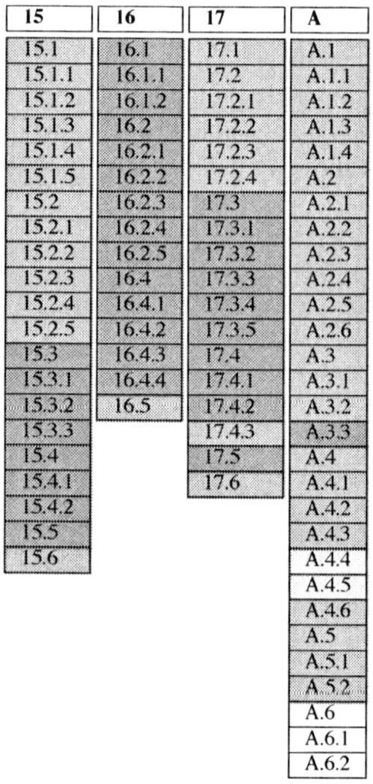

Either 9.4.1 or 9.4.2 should be presented in class with the other assigned as reading. If at all possible, the material in A.4.4 and A.4.5 should be presented in class. The medium gra sections of Chapters 15 and 17 should be outlined in class and assigned as reading. Section A.3.3 should be presented when the material is called for in the text.

Figure 4. An outline for a one-year course with an engineering emphasis.

Notational Conventions

DEFINITIONS: $A \stackrel{d}{=} B$ the expression appearing on the right defines the symbol on the left.

SETS: $\{A : P\} \stackrel{d}{=}$ the set of elements A satisfying property P.

CARTESIAN PRODUCT: $A \times B \stackrel{d}{=}$ the set of *ordered* pairs (a, b), where a belongs to the set A and b to the set B.

REPEATED CARTESIAN PRODUCT: $A^n \stackrel{d}{=}$ the set of ordered n-tuples $(a_1, \ldots, a_n)$, where the a_i belong to the set A.

THE NATURAL NUMBERS: $\mathbb{N}$.

THE INTEGERS: $\mathbb{Z}$.

THE RATIONAL NUMBERS: $\mathbb{Q}$.

THE REAL NUMBERS: $\mathbb{R}$.

THE COMPLEX NUMBERS: $\mathbb{C}$.

MULTI-INDEX NOTATION: If $\boldsymbol{j} = (j_1, \ldots, j_n)$ is an n-vector of integers and $\boldsymbol{x} = (x_1, \ldots, x_n)$ is an n-vector, then

$$\boldsymbol{x}^{\boldsymbol{j}} \stackrel{d}{=} x_1^{j_1} \cdots x_n^{j_n}.$$

THE GREATEST INTEGER FUNCTION: For a real number x, $[x] \stackrel{d}{=}$ the largest integer smaller than or equal to x.

INTERVALS: If a and b are real numbers with $a \leq b$, then

$(a, b) \stackrel{d}{=} \{x \in \mathbb{R} : a < x < b\}$	an open interval,
$[a, b) \stackrel{d}{=} \{x \in \mathbb{R} : a \leq x < b\}$	a half-open interval,
$(a, b] \stackrel{d}{=} \{x \in \mathbb{R} : a < x \leq b\}$	a half-open interval,
$[a, b] \stackrel{d}{=} \{x \in \mathbb{R} : a \leq x \leq b\}$	a closed interval.

$$(-\infty,\infty) \stackrel{d}{=} \mathbb{R},$$

$(a,\infty) \stackrel{d}{=} \{x \in \mathbb{R} : a < x < \infty\}$ an open positive half-ray,

$[a,\infty) \stackrel{d}{=} \{x \in \mathbb{R} : a \leq x < \infty\}$ a closed positive half-ray,

$(-\infty,b) \stackrel{d}{=} \{x \in \mathbb{R} : -\infty < x < b\}$ an open negative half-ray,

$(-\infty,b] \stackrel{d}{=} \{x \in \mathbb{R} : -\infty < x \leq b\}$ a closed negative half-ray.

THE EUCLIDEAN NORM: $\|(x_1,\ldots,x_n)\|_2 \stackrel{d}{=} \sqrt{\sum_{j=1}^n x_j^2}.$

THE n-DIMENSIONAL UNIT SPHERE: $S^n \stackrel{d}{=} \{\boldsymbol{x} \in \mathbb{R}^{n+1} : \|\boldsymbol{x}\|_2 = 1\}.$

BALLS IN $\mathbb{R}^n$: If $\boldsymbol{a} \in \mathbb{R}^n$ and r is a positive number then

$$B_r(\boldsymbol{a}) = \{\boldsymbol{x} \in \mathbb{R}^n : \|\boldsymbol{x} - \boldsymbol{a}\| < r\}.$$

The ball of radius r centered at $(0,\ldots,0)$ is often denoted B_r.

SEQUENCES: $< \cdot >$ angle brackets enclose the elements of a sequence, for example, $< x_n >$.

INNER PRODUCTS: $\langle \boldsymbol{x}, \boldsymbol{y} \rangle$ is the inner product of the vectors $\boldsymbol{x}$ and $\boldsymbol{y}$.

KRONECKER DELTA: $\delta_{ij} \stackrel{d}{=}$ a square matrix with $\delta_{ii} = 1$ and $\delta_{ij} = 0$ if $i \neq j$.

CONTINUOUS FUNCTIONS: $\mathscr{C}^0(A) \stackrel{d}{=}$ the set of continuous functions defined on the set A.

DIFFERENTIABLE FUNCTIONS: $\mathscr{C}^k(A) \stackrel{d}{=}$ the set of k-times continuously differentiable functions defined on the set A.

SMOOTH FUNCTIONS: $\mathscr{C}^\infty(A) \stackrel{d}{=}$ the set of infinitely differentiable functions defined on the set A.

DERIVATIVES: If f is a function of the variable x then the first derivative is denoted, variously by

$$f', \quad \partial_x f, \text{ or } \quad \frac{df}{dx}.$$

The jth derivative is denoted by

$$f^{[j]}, \quad \partial_x^j f, \text{ or } \quad \frac{d^j f}{dx^j}.$$

INTEGRABLE FUNCTIONS: $L^1(A) \stackrel{d}{=}$ the set of absolutely integrable functions defined on the set A.

SQUARE-INTEGRABLE FUNCTIONS: $L^2(A) \stackrel{d}{=}$ the set of functions, defined on the set A, whose square is absolutely integrable.

"BIG OH" NOTATION: A function f is $O(g(x))$ for x near to x_0 if there is an $\epsilon > 0$ and a constant M so that

$$|f(x)| \leq Mg(x) \quad \text{if } |x - x_0| < \epsilon.$$

"LITTLE OH" NOTATION: A function f is $o(g(x))$ for x near to x_0 if

$$\lim_{x \to x_0} \frac{|f(x)|}{g(x)} = 0.$$

Chapter 1

Measurements and Modeling

quantitative model of a physical system is expressed in the language of mathematics. A qualitative model often precedes a quantitative model. For many years clinicians used medical x-ray images without employing a precise quantitative model. X-rays were thought of as high frequency 'light' with three very useful properties:

1. If x-rays are incident on a human body, some fraction of the incident radiation is absorbed or scattered, though a sizable fraction is transmitted. The fraction absorbed or scattered is proportional to the total density of the material encountered. The overall decrease in the intensity of the x-ray beam is called *attenuation*.

2. A beam of x-ray light travels in a straight line.

3. X-rays darken photographic film. The opacity of the film is a monotone function of the incident energy.

Taken together, these properties mean that using x-rays one can "see through" a human body to obtain a shadow or projection of the internal anatomy on a sheet of film [Figure 1.1(a)].

This model was adequate given the available technology. In their time, x-ray images led to a revolution in the practice of medicine because they opened the door to non-invasive examination of internal anatomy. They are still useful for locating bone fractures, dental caries, and foreign objects, but their ability to visualize soft tissues and more detailed anatomic structure is limited. There are several reasons for this. The principal difficulty is that an x-ray image is a two-dimensional representation of a three-dimensional object. In Figure 1.1(b), the opacity of the film at a point on the film plane is inversely proportional to an average of the density of the object, measured along the line joining the point to the x-ray source. This renders it impossible to deduce the spatial ordering in the missing third dimension.

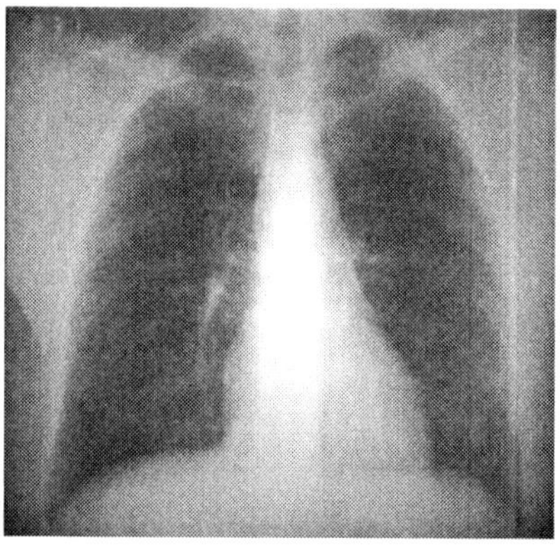

(a) A old-fashioned chest x-ray image. (Image provided courtesy of Dr. David S. Feigin, ENS Sherri Rudinsky, and Dr. James G. Smirniotopoulos of the Uniformed Services University of the Health Sciences, Dept. of Radiology, Bethesda, MD.)

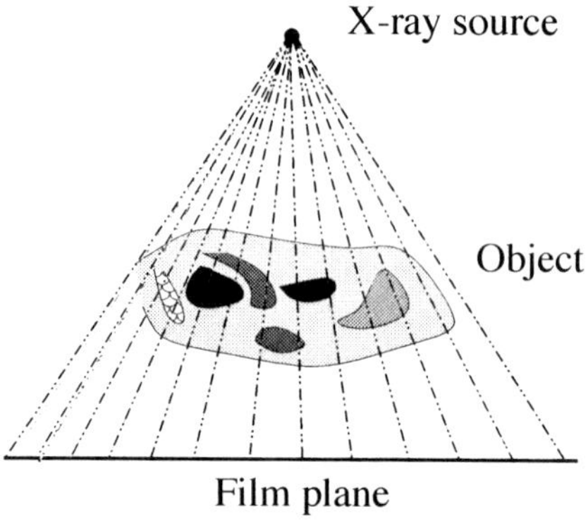

(b) Depth information is lost in a projection.

Figure 1.1. The world of old fashioned x-rays imaging.

A second problem is connected to the "detector" used in traditional x-ray imaging. Photographic film is used to record the total energy in the x rays that are transmitted through the object. Unfortunately film is rather insensitive to x rays. To get a usable image, a light emitting phosphor is sandwiched with the film. This increases the sensitivity of the overall "detector," but even so, large differences in the intensity of the incident x-ray beam produce small differences in the opacity of film. This means that the contrast between different soft tissues is poor. Beyond this there are other problems caused by the scattering of x rays and noise. Because of these limitations a qualitative theory was adequate for the interpretation of traditional x-ray images.

A desire to improve upon this situation led Alan Cormack, [24], and Godfrey Hounsfield, [64], to independently develop x-ray *tomography* or slice imaging. The first step in their work was to use a quantitative theory for the attenuation of x-rays. Such a theory already existed and is little more than a quantitative restatement of (1) and (2). It is not needed for old fashioned x-ray imaging because traditional x-ray images are read "by eye," and no further processing is done after the film is developed. Both Cormack and Hounsfield realized that mathematics could be used to infer three-dimensional anatomic structure from a large collection of *different* two-dimensional projections. The possibility for making this idea work relied on two technological advances:

1. The availability of scintillation crystals to use as detectors
2. Powerful, digital computers to process the tens of thousands of measurements needed to form a usable image

A detector using a scintillation crystal is about a hundred times more sensitive than photographic film. Increasing the dynamic range in the basic measurements makes possible much finer distinctions. As millions of arithmetic operations are needed for each image, fast computers are a necessity for reconstructing an image from the available measurements. It is an interesting historical note that the mathematics underlying x-ray tomography was done in 1917 by Johan Radon, [105]. It had been largely forgotten, and both Hounsfield and Cormack worked out solutions to the problem of reconstructing an image from its projections. Indeed, this problem had arisen and been solved in contexts as diverse as radio astronomy and statistics.

This book is a detailed exploration of the mathematics that underpins the reconstruction of images in x-ray tomography. While our emphasis is on understanding these mathematical foundations, we constantly return to the practicalities of x-ray tomography. Of particular interest is the relationship between the mathematical treatment of a problem and the realities of numerical computation and physical measurement. There are many different imaging *modalities* in common use today, such as x-ray computed tomography (CT), magnetic resonance imaging (MRI), positron emission tomography (PET), ultrasound, optical imaging, and electrical impedance imaging. Because each relies on a different physical principle, each provides different information. In every case the mathematics needed to process and interpret the data has a large overlap with that used in x-ray CT. We concentrate on x-ray CT because of the simplicity of the physical principles underlying the measurement process. Detailed descriptions of the other modalities can be found in [91], [76], or [6].

Mathematics is the language in which any quantitative theory or model is eventually expressed. In this introductory chapter we consider a variety of examples of physical systems, measurement processes, and the mathematical models used to describe them. These models illustrate different aspects of more complicated models used in medical imaging. We define the notion of degrees of freedom and relate it to the dimension of a vector space. The chapter concludes by analyzing the problem of reconstructing a region in the plane from measurements of the shadows it casts.

1.1 Mathematical Modeling

The first step in giving a mathematical description of a system is to isolate that system from the universe in which it sits. While it is no doubt true that a butterfly flapping its wings in Siberia in midsummer will affect the amount of rainfall in the Amazon rain forest a decade hence, it is surely a tiny effect, impossible to accurately quantify. A practical model includes the system of interest and the *major* influences on it. Small effects are ignored, though they may come back, as measurement error and noise, to haunt the model. After the system is isolated, we need to find a collection of numerical parameters that describe its state. In this generality these parameters are called *state variables*. In the idealized world of an isolated system the exact measurement of the state variables uniquely determines the state of the system. It may happen that the parameters that give a convenient description of the system are not directly measurable. The mathematical model then describes relations among the state variables. Using these relations, the state of the system can often be determined from feasible measurements. A simple example should help clarify these abstract-sounding concepts.

***Example* 1.1.1.** Suppose the system is a ball on a rod. For simplicity we assume that the ball has radius zero. The state of the system is described by (x, y), the coordinates of the ball. These are the state variables. If the rod is of length r and one end of it is fixed at the point $(0, 0)$, then the state variables satisfy the relation

$$x^2 + y^2 = r^2. \tag{1.1}$$

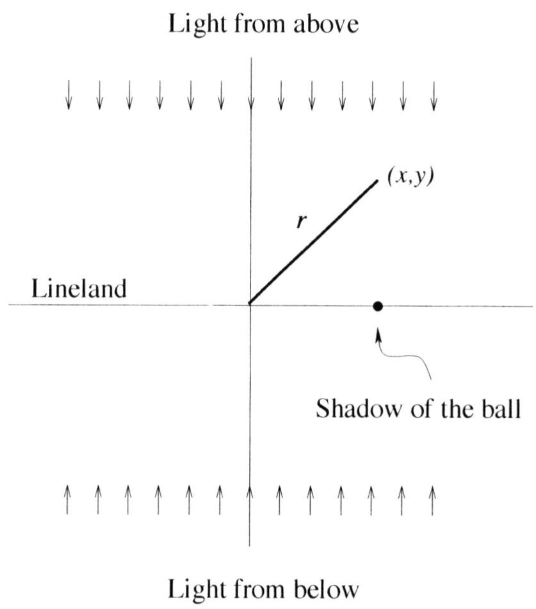

Figure 1.2. A rod of length r casts a shadow on lineland.

Imagine now that one-dimensional creatures, living on the x-axis $\{y = 0\}$, can observe a shadow of the ball, cast by very distant light sources so that the rays of light are perpendicular to the x-axis (Figure 1.2). The line creatures want to predict whether or not the ball is about to collide with their world. Locating the shadow determines the x-coordinate of the ball and using equation (1.1) gives

$$y = \pm\sqrt{r^2 - x^2}.$$

To determine the sign of the y-coordinate requires additional information not available in the model. On the other hand, this information is adequate if one only wants to predict if the ball is about to collide with the x-axis. If the x-axis is illuminated by red light from above and blue light from below, then a ball approaching from below would cast of red shadow while a ball approaching from above would cast a blue shadow. With these additional data, the location of the ball is completely determined.

Ordered pairs of real numbers, $\{(x, y)\}$, are the state variables for the system in Example 1.1.1. Because of the constraint (1.1), not every pair defines a state of this system. Generally we define the *state space* to be values of state variables which correspond to actual states of the system. The state space in Example 1.1.1 is the circle of radius r centered at $(0, 0)$.

Exercises

***Exercise* 1.1.1.** Suppose that in Example 1.1.1 light sources are located at $(0, \pm R)$. What is the relationship between the x-coordinate and the shadow?

***Exercise* 1.1.2.** Suppose that in Example 1.1.1 the ball is tethered to $(0, 0)$ by a string of length r. What relations do the state variables (x, y) satisfy? Is there a measurement the line creatures can make to determine the location of the ball? What is the state space for this system?

***Exercise* 1.1.3.** Suppose that the ball is untethered but is constrained to lie in the region $\{(x, y) : 0 \leq y < R\}$. Assume that the points $\{(x_1, y_1), (x_2, y_2), (x_3, y_3)\}$ do not lie on a line and have $y_j > R$. Show that the shadows cast on the line $y = 0$ by light sources located at these three points determine the location of the ball. Find a formula for (x, y) in terms of the shadow locations. Why are three sources needed?

1.1.1 Finitely Many Degrees of Freedom

See: A.1, B.3, B.4, B.5.

The collection of ordered n-tuples of real numbers

$$\{(x_1, \ldots, x_n) : x_j \in \mathbb{R}, j = 1, \ldots, n\}$$

is called Euclidean n-space and is denoted by $\mathbb{R}^n$. We often use boldface letters $\boldsymbol{x}$, $\boldsymbol{y}$ to denote points in $\mathbb{R}^n$, which we sometimes call vectors. Recall that if $\boldsymbol{x} = (x_1, \ldots, x_n)$ and $\boldsymbol{y} = (y_1, \ldots, y_n)$, then their sum $\boldsymbol{x} + \boldsymbol{y}$ is defined by

$$\boldsymbol{x} + \boldsymbol{y} = (x_1 + y_1, \ldots, x_n + y_n), \tag{1.2}$$

and if $a \in \mathbb{R}$, then $a\boldsymbol{x}$ is defined by

$$a\boldsymbol{x} = (ax_1, \ldots, ax_n). \tag{1.3}$$

These two operations make $\mathbb{R}^n$ into a real vector space.

***Definition* 1.1.1.** If the state of a system is described by a finite collection of real numbers, then the system has *finitely many degrees of freedom*.

Euclidean n-space is the simplest state space for a system with n degrees of freedom. Most systems encountered in elementary physics and engineering have finitely many degrees of freedom. Suppose that the state of a system is specified by a point $\boldsymbol{x} \in \mathbb{R}^n$. Then the mathematical model is expressed as relations that these variables satisfy. These often take the form of functional relations,

$$\begin{aligned} f_1(x_1, \ldots, x_n) &= 0 \\ \vdots \quad & \quad \vdots \\ f_m(x_1, \ldots, x_n) &= 0. \end{aligned} \tag{1.4}$$

In addition to conditions like those in (1.4) the parameters describing the state of a system might also satisfy inequalities of the form

$$\begin{aligned} g_1(x_1, \ldots, x_n) &\geq 0 \\ \vdots \quad & \quad \vdots \\ g_l(x_1, \ldots, x_n) &\geq 0. \end{aligned} \tag{1.5}$$

The state space for the system is then the subset of $\mathbb{R}^n$ consisting of solutions to (1.4) which also satisfy (1.5).

***Definition* 1.1.2.** An equation or inequality that must be satisfied by a point belonging to the state space of a system is called a *constraint*.

Example 1.1.1 considers a system with one degree of freedom. The state space for this system is the subset of $\mathbb{R}^2$ consisting of points satisfying (1.1). If the state variables satisfy constraints, then this generally reduces the number of degrees of freedom.

A function $f : \mathbb{R}^n \to \mathbb{R}$ is linear if it satisfies the conditions

$$\begin{aligned} f(\boldsymbol{x} + \boldsymbol{y}) &= f(\boldsymbol{x}) + f(\boldsymbol{y}) \text{ for all } \boldsymbol{x}, \boldsymbol{y} \in \mathbb{R}^n \text{ and} \\ f(a\boldsymbol{x}) &= af(\boldsymbol{x}) \text{ for all } a \in \mathbb{R} \text{ and } \boldsymbol{x} \in \mathbb{R}^n. \end{aligned} \tag{1.6}$$

Recall that the dot or inner product is the map from $\mathbb{R}^n \times \mathbb{R}^n \to \mathbb{R}$ defined by

$$\langle \boldsymbol{x}, \boldsymbol{y} \rangle = \sum_{j=1}^{n} x_j y_j. \tag{1.7}$$

Sometimes it is denoted by $\boldsymbol{x} \cdot \boldsymbol{y}$. The Euclidean length of $\boldsymbol{x} \in \mathbb{R}^n$ is defined to be

$$\|\boldsymbol{x}\| = \sqrt{\langle \boldsymbol{x}, \boldsymbol{x} \rangle} = \left[\sum_{j=1}^{n} x_j^2 \right]^{\frac{1}{2}}. \tag{1.8}$$

From the definition it is easy to establish that

$$\begin{aligned} \langle \boldsymbol{x}, \boldsymbol{y} \rangle &= \langle \boldsymbol{y}, \boldsymbol{x} \rangle \text{ for all } \boldsymbol{x}, \boldsymbol{y} \in \mathbb{R}^n, \\ \langle a\boldsymbol{x}, \boldsymbol{y} \rangle &= a\langle \boldsymbol{x}, \boldsymbol{y} \rangle \text{ for all } a \in \mathbb{R} \text{ and } \boldsymbol{x} \in \mathbb{R}^n, \\ \langle \boldsymbol{x}_1 + \boldsymbol{x}_2, \boldsymbol{y} \rangle &= \langle \boldsymbol{x}_1, \boldsymbol{y} \rangle + \langle \boldsymbol{x}_2, \boldsymbol{y} \rangle \text{ for all } \boldsymbol{x}_1, \boldsymbol{x}_2, \boldsymbol{y} \in \mathbb{R}^n. \\ \|c\boldsymbol{x}\| &= |c|\|\boldsymbol{x}\| \text{ for all } c \in \mathbb{R} \text{ and } \boldsymbol{x} \in \mathbb{R}^n. \end{aligned} \tag{1.9}$$

For y a point in $\mathbb{R}^n$, define the function $f_y(x) = \langle x, y \rangle$. The second and third relations in (1.9) show that f_y is linear. Indeed, every linear function has a such a representation.

Proposition 1.1.1. *If $f : \mathbb{R}^n \to \mathbb{R}$ is a linear function, then there is a unique vector y_f such that $f(x) = \langle x, y_f \rangle$.*

This fact is proved in Exercise 1.1.5. The inner product satisfies a basic inequality called the Cauchy-Schwarz inequality.

Proposition 1.1.2 (Cauchy-Schwarz inequality). *If $x, y \in \mathbb{R}^n$, then*

$$|\langle x, y \rangle| \le \|x\| \|y\|. \tag{1.10}$$

A proof of this result is outlined in Exercise 1.1.6. The Cauchy-Schwarz inequality shows that if neither x nor y is zero, then

$$-1 \le \frac{\langle x, y \rangle}{\|x\| \|y\|} \le 1;$$

this in turn allows us the define the angle between two vectors.

***Definition* 1.1.3.** If $x, y \in \mathbb{R}^n$ are both nonvanishing, then the angle $\theta \in [0, \pi]$, between x and y is defined by

$$\cos\theta = \frac{\langle x, y \rangle}{\|x\| \|y\|}. \tag{1.11}$$

In particular, two vector are *orthogonal* if $\langle x, y \rangle = 0$.

The Cauchy-Schwarz inequality implies that the Euclidean length satisfies the *triangle inequality*.

Proposition 1.1.3. *For $x, y \in \mathbb{R}^n$, the following inequality holds:*

$$\|x + y\| \le \|x\| + \|y\|. \tag{1.12}$$

This is called the triangle inequality.

***Remark* 1.1.1.** The Euclidean length is an example of a *norm* on $\mathbb{R}^n$. A real-valued function N defined on $\mathbb{R}^n$ is a norm provided it satisfies the following conditions:

NON-DEGENERACY:
$N(x) = 0$ if and only if $x = 0$,

HOMOGENEITY:
$N(ax) = |a| N(x)$ for all $a \in \mathbb{R}$ and $x \in \mathbb{R}^n$,

THE TRIANGLE INEQUALITY:
$N(x + y) \le N(x) + N(y)$ for all $x, y \in \mathbb{R}^n$.

Any norm provides a way to measure distances. The distance between $\boldsymbol{x}$ and $\boldsymbol{y}$ is defined to be

$$d_N(\boldsymbol{x}, \boldsymbol{y}) \stackrel{d}{=} N(\boldsymbol{x} - \boldsymbol{y}).$$

Suppose that the state of a system is specified by a point in $\mathbb{R}^n$ subject to the constraints in (1.4). If all the functions $\{f_1, \ldots, f_m\}$ are linear, then we say that this is a *linear model.* This is the simplest type of model and also the most common in applications. In this case the set of solutions to (1.4) is a *subspace* of $\mathbb{R}^n$. We recall the definition.

***Definition* 1.1.4.** A subset $S \subset \mathbb{R}^n$ is a subspace if

1. the zero vector belongs to S,
2. $\boldsymbol{x}_1, \boldsymbol{x}_2 \in S$, then $\boldsymbol{x}_1 + \boldsymbol{x}_2 \in S$,
3. if $c \in \mathbb{R}$ and $\boldsymbol{x} \in S$, then $c\boldsymbol{x} \in S$.

For a linear model it is a simple matter to determine the number of degrees of freedom. Suppose the state space consists of vectors satisfying a single linear equation. In light of Proposition 1.1.1, it can be expressed in the form

$$\langle \boldsymbol{a}_1, \boldsymbol{x} \rangle = 0, \tag{1.13}$$

with $\boldsymbol{a}_1$ a nonzero vector. This is the equation of a *hyperplane* in $\mathbb{R}^n$. The solutions to (1.13) are the vectors in $\mathbb{R}^n$ orthogonal to $\boldsymbol{a}_1$. Recall the following definition:

***Definition* 1.1.5.** The vectors $\{\boldsymbol{v}_1, \ldots, \boldsymbol{v}_k\}$ are *linearly independent* if the only linear combination, $c_1\boldsymbol{v}_1 + \cdots + c_k\boldsymbol{v}_k$, that vanishes has all its coefficients, $\{c_i\}$, equal to zero. Otherwise the vectors are *linearly dependent.*

The dimension of a subspace of $\mathbb{R}^n$ can now be defined.

***Definition* 1.1.6.** Let $S \subset \mathbb{R}^n$ be a subspace. If there is a set of k linearly independent vectors contained in S but any set with $k + 1$ or more vectors is linearly dependent, then the dimension of S equals k. In this case we write $\dim S = k$.

There is a collection of $(n - 1)$ linearly independent n-vectors $\{\boldsymbol{v}_1, \ldots, \boldsymbol{v}_{n-1}\}$ so that $\langle \boldsymbol{a}_1, \boldsymbol{x} \rangle = 0$ if and only if

$$\boldsymbol{x} = \sum_{i=1}^{n-1} c_i \boldsymbol{v}_i.$$

The hyperplane has dimension $n - 1$, and therefore a system described by a single linear equation has $n - 1$ degrees of freedom. The general case is not much harder. Suppose that the state space is the solution set of the system of linear equations

$$\begin{aligned} \langle \boldsymbol{a}_1, \boldsymbol{x} \rangle &= 0 \\ \vdots \quad &\ \ \vdots \\ \langle \boldsymbol{a}_m, \boldsymbol{x} \rangle &= 0. \end{aligned} \tag{1.14}$$

Suppose that $k \leq m$ is the largest number of linearly independent vectors in the collection $\{\boldsymbol{a}_1, \ldots, \boldsymbol{a}_m\}$. By renumbering, we can assume that $\{\boldsymbol{a}_1, \ldots, \boldsymbol{a}_k\}$ are linearly independent, and for any $l > k$ the vector $\boldsymbol{a}_l$ is a linear combination of these vectors. Hence if $\boldsymbol{x}$ satisfies

$$\langle \boldsymbol{a}_i, \boldsymbol{x} \rangle = 0 \text{ for } 1 \leq i \leq k$$

then it also satisfies $\langle \boldsymbol{a}_l, \boldsymbol{x} \rangle = 0$ for any l greater than k. The argument in the previous paragraph can be applied recursively to conclude that there is a collection of $n - k$ linearly independent vectors $\{\boldsymbol{u}_1, \ldots, \boldsymbol{u}_{n-k}\}$ so that $\boldsymbol{x}$ solves (1.14) if and only if

$$\boldsymbol{x} = \sum_{i=1}^{n-k} c_i \boldsymbol{u}_i.$$

Thus the system has $n - k$ degrees of freedom.

A nonlinear model can often be approximated by a linear model. If f is a differentiable function, then the gradient of f at $\boldsymbol{x}$ is defined to be

$$\nabla f(\boldsymbol{x}) = \left(\frac{\partial f}{\partial x_1}(\boldsymbol{x}), \ldots, \frac{\partial f}{\partial x_n}(\boldsymbol{x}) \right).$$

From the definition of the derivative it follows that

$$f(\boldsymbol{x}_0 + \boldsymbol{x}_1) = f(\boldsymbol{x}_0) + \langle \boldsymbol{x}_1, \nabla f(\boldsymbol{x}_0) \rangle + e(\boldsymbol{x}_1), \tag{1.15}$$

where the *error* $e(\boldsymbol{x}_1)$ satisfies

$$\lim_{\boldsymbol{x}_1 \to 0} \frac{|e(\boldsymbol{x}_1)|}{\|\boldsymbol{x}_1\|} = 0.$$

In this case we write

$$f(\boldsymbol{x}_0 + \boldsymbol{x}_1) \approx f(\boldsymbol{x}_0) + \langle \boldsymbol{x}_1, \nabla f(\boldsymbol{x}_0) \rangle. \tag{1.16}$$

Suppose that the functions in (1.4) are differentiable and $f_j(\boldsymbol{x}_0) = 0$ for $j = 1, \ldots, m$. Then

$$f_j(\boldsymbol{x}_0 + \boldsymbol{x}_1) \approx \langle \boldsymbol{x}_1, \nabla f_j(\boldsymbol{x}_0) \rangle.$$

For small values of $\boldsymbol{x}_1$ the system of equations (1.4) can be approximated, near to $\boldsymbol{x}_0$, by a system of linear equations,

$$\begin{aligned} \langle \boldsymbol{x}_1, \nabla f_1(\boldsymbol{x}_0) \rangle &= 0 \\ \vdots \quad &\quad \vdots \\ \langle \boldsymbol{x}_1, \nabla f_m(\boldsymbol{x}_0) \rangle &= 0. \end{aligned} \tag{1.17}$$

This provides a linear model that approximates the non-linear model. The accuracy of this approximation depends, in a subtle way, on the collection of vectors $\{\nabla f_j(\boldsymbol{x})\}$, for $\boldsymbol{x}$ near to $\boldsymbol{x}_0$. The simplest situation is when these vectors are linearly independent at $\boldsymbol{x}_0$. In this case the solutions to

$$f_j(\boldsymbol{x}_0 + \boldsymbol{x}_1) = 0, \quad j = 1, \ldots, m,$$

are well approximated, for small x_1, by the solutions of (1.17). This is a consequence of the implicit function theorem; see [119].

Often the state variables for a system are divided into two sets, the *input variables*, $(w_1, \ldots, w_k)$, and *output variables*, $(z_1, \ldots, z_m)$, with constraints rewritten in the form

$$\begin{aligned} F_1(w_1, \ldots, w_k) &= z_1 \\ \vdots \quad & \quad \vdots \\ F_m(w_1, \ldots, w_k) &= z_m. \end{aligned} \tag{1.18}$$

The output variables are thought of as being measured; the remaining variables must then be determined by solving this system of equations. For a linear model this amounts to solving a system of linear equations. We now consider some examples of physical systems and their mathematical models.

***Example* 1.1.2.** We would like to find the height of a mountain without climbing it. To that end, the distance x between the point P and the base of the mountain, as well as the angle θ, are measured (Figure 1.3). If x and θ are measured exactly, then the height, h, of the mountain is given by

$$h(x, \theta) = x \tan \theta. \tag{1.19}$$

Measurements are never exact; using the model and elementary calculus, we can relate the error in the measurement θ to the error in the computed value of h. Suppose that x is measured exactly but there is an uncertainty $\Delta\theta$ in the value of θ. Equation (1.16) gives the linear approximation

$$h(x, \theta + \Delta\theta) - h(x, \theta) \approx \frac{\partial h}{\partial \theta}(x, \theta)\Delta\theta.$$

As $\partial_\theta h = x \sec^2 \theta$, the height, h_m, predicted from the measurement of the angle is given by

$$h_m = x \tan(\theta + \Delta\theta) \approx x(\tan \theta + \sec^2 \theta \Delta\theta).$$

The approximate value of the *absolute error* is

$$h_m - h \approx x \frac{\Delta\theta}{\cos^2 \theta}.$$

The absolute error is a number with the same units as h; in general, it is not an interesting quantity. If, for example, the true measurement were 10,000 m, then an error of size 1 m would not be too significant. If the true measurement were 2 m, then this error would be significant. To avoid this obvious pitfall, we normally consider the *relative error*. In this problem the relative error is

$$\frac{h_m - h}{h} = \frac{\Delta\theta}{\cos^2 \theta \tan \theta} = \frac{\Delta\theta}{\sin \theta \cos \theta}.$$

Generally the relative error is the absolute error divided by the correct value. It is a dimensionless quantity that gives a quantitative assessment of the accuracy of the measurement.

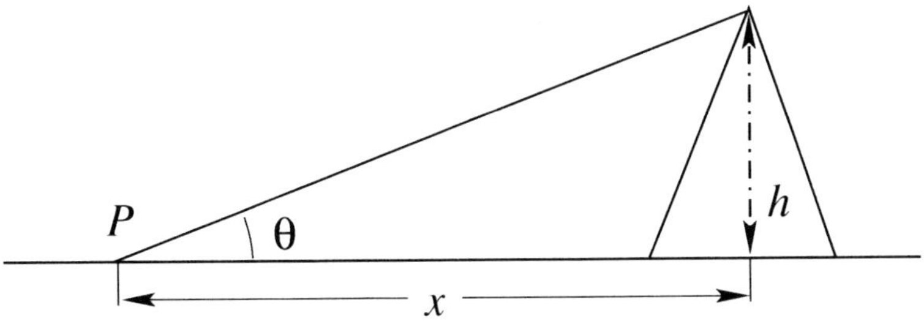

Figure 1.3. Using trigonometry to find the height of a mountain.

If the angle θ is measured from a point too near to or too far from the mountain (i.e., θ is very close to 0 or $\pi/2$), then small measurement errors result in a substantial loss of accuracy. A useful feature of a precise mathematical model is the possibility of estimating how errors in measurement affect the accuracy of the parameters we wish to determine. In Exercise 1.1.13 we consider how to estimate the error entailed in using a linear approximation.

***Example* 1.1.3.** In a real situation we cannot measure the distance to the base of the mountain. Suppose that we measure the angles, θ_1 and θ_2, from two different points, P_1 and P_2, as well as the distance $x_2 - x_1$ between the two points, as shown in Figure 1.4.

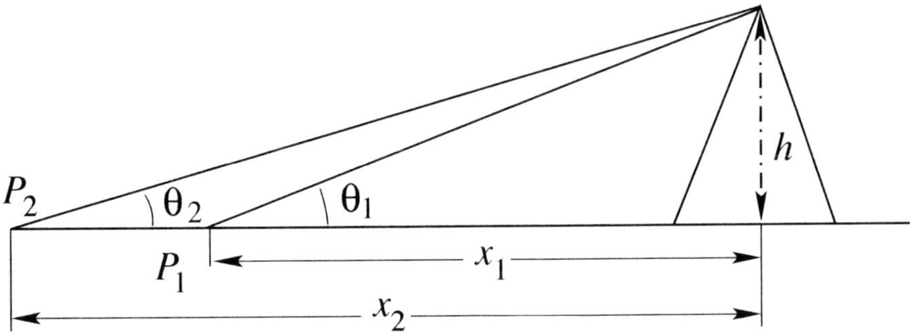

Figure 1.4. A more realistic measurement.

From the previous example we know that

$$\begin{aligned} h &= x_1 \tan\theta_1, \\ h &= x_2 \tan\theta_2. \end{aligned} \tag{1.20}$$

Using these equations and elementary trigonometry, we deduce that

$$x_1 = \frac{x_2 - x_1}{\left[\frac{\tan\theta_1}{\tan\theta_2} - 1\right]}, \tag{1.21}$$

which implies that

$$h = x_1 \tan\theta_1$$
$$= (x_2 - x_1)\frac{\sin\theta_1 \sin\theta_2}{\sin(\theta_1 - \theta_2)}. \tag{1.22}$$

Thus h can be determined from θ_1, θ_2 and $x_2 - x_1$. With $d = x_2 - x_1$, equation (1.22) expresses h as a function of (d, θ_1, θ_2). At the beginning of this example, $(x_1, \theta_1, x_2, \theta_2, h)$ were the state variables describing our system; by the end we used $(d, \theta_1, \theta_2, h)$. The first three are directly measurable, and the last is an explicit function of the others. The models in this and the previous example, as expressed by equations (1.22) and (1.19), respectively, are nonlinear models.

In this example there are many different ways that the model may fail to capture important features of the physical situation. We now consider a few potential problems.

1. If the shape of a mountain looks like that in Figure 1.5 and we measure the distance and angle at the point P, we are certainly not finding the real height of the mountain. Some a priori information is always incorporated in a mathematical model.

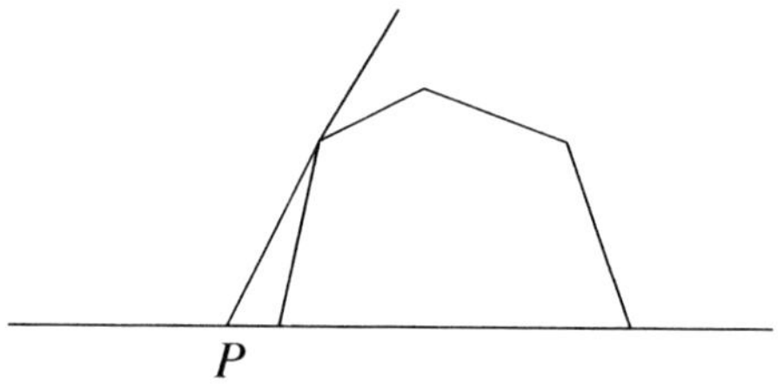

Figure 1.5. Not exactly what we predicted!

2. The curvature of the earth is ignored. A more sophisticated geometric model is needed to correct for such errors. This becomes a significant problem as soon as the distances, x, x_1, x_2, are large compared to the distance to the horizon (about 25 km for a 2-meter-tall person). The approximations used in the model must be adapted to the actual physical conditions of the measurements.

3. The geometry of the underlying measurements could be quite different from the simple Euclidean geometry used in the model. To measure the angles θ_1, θ_2, we would normally use a transit to sight the peak of the mountain. If the mountain is far away, then the light traveling from the mountain to the transit passes through air of varying density. The light is refracted by the air and therefore the ray path is not a straight line, as assumed in the model. To include this effect would vastly complicate the model. This is an important consideration in the similar problem of creating a map of the sky from earth based observations of stars and planets.

Analogous problems arise in medical imaging. If the wavelength of the energy used to probe the human anatomy is very small compared to the size of the structures that are present, then it is reasonable to assume that the waves are not refracted. For example, x-rays can be assumed to travel along straight lines. For energies with wavelengths comparable to the size of structures present in the human anatomy, this assumption is simply wrong. The waves are then bent and diffracted by the medium, and the difficulty of modeling the ray paths is considerable. This is an important issue in ultrasound imaging that remains largely unresolved.

***Example* 1.1.4.** Refraction provides another example of a simple physical system. Suppose that we have two fluids in a tank, as shown in Figure 1.6, and would like to determine the height of the interface between them. Suppose that the refractive indices of the fluids are known. Let n_1 be the refractive index of the upper fluid and n_2 the refractive index of the lower one. Snell's law states that

$$\frac{\sin(\theta_1)}{\sin(\theta_2)} = \frac{n_2}{n_1}.$$

Let h denote the total height of the fluid; then

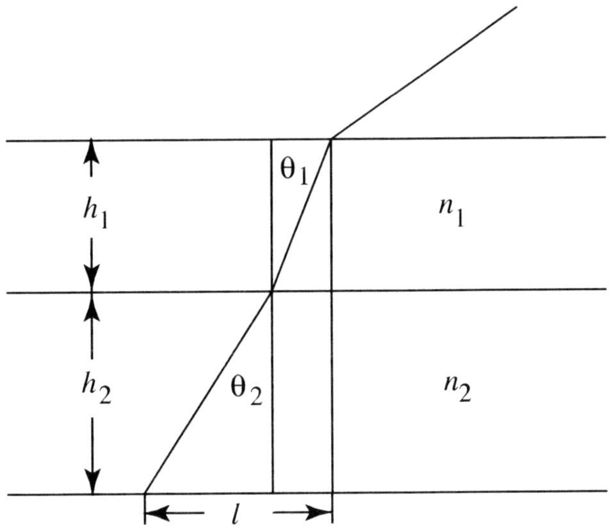

Figure 1.6. Using refraction to determine the height of an interface.

$$h_1 + h_2 = h.$$

The measurement we make is the total displacement, l, of the light ray as it passes through the fluids. It satisfies the relationship

$$h_1 \tan(\theta_1) + h_2 \tan(\theta_2) = l.$$

The heights h_1 and h_2 are easily determined from these three formulæ. The assumption that we know n_1 implies, by Snell's law, that we can determine θ_1 from a measurement of the angle of the light ray above the fluid. If n_2 is also known, then using these observations we can determine θ_2 as well:

$$\sin(\theta_2) = \frac{n_1}{n_2}\sin(\theta_1).$$

The pair (h_1, h_2) satisfies the 2×2 linear system

$$\begin{pmatrix} 1 & 1 \\ \tan(\theta_1) & \tan(\theta_2) \end{pmatrix} \begin{pmatrix} h_1 \\ h_2 \end{pmatrix} = \begin{pmatrix} h \\ l \end{pmatrix}. \tag{1.23}$$

In Example 2.1.1 we consider a slightly more realistic situation where the refractive index of the lower fluid in not known. By using more measurements, n_2 can also be determined.

Exercises

***Exercise* 1.1.4.** Prove the formulæ in (1.9).

***Exercise* 1.1.5.** Let $\boldsymbol{e}_j \in \mathbb{R}^n$, $j = 1, \ldots, n$ denote the vector with a 1 in the jth place and otherwise zero,

$$\boldsymbol{e}_1 = (1, 0, 0, \ldots, 0),\ \boldsymbol{e}_2 = (0, 1, 0, \ldots, 0), \ldots, \boldsymbol{e}_n = (0, \ldots, 0, 1).$$

1. Show that if $\boldsymbol{x} = (x_1, \ldots, x_n)$, then
$$\boldsymbol{x} = \sum_{j=1}^{n} x_j \boldsymbol{e}_j.$$

2. Use the previous part to prove the existence statement in Proposition 1.1.1; that is, show that there is a vector $\boldsymbol{y}_f$ so that $f(\boldsymbol{x}) = \langle \boldsymbol{x}, \boldsymbol{y}_f \rangle$. Give a formula for $\boldsymbol{y}_f$.

3. Show that the uniqueness part of the proposition is equivalent to the statement "If $\boldsymbol{y} \in \mathbb{R}^n$ satisfies
$$\langle \boldsymbol{x}, \boldsymbol{y} \rangle = 0 \text{ for all } \boldsymbol{x} \in \mathbb{R}^n,$$
then $\boldsymbol{y} = 0$." Prove this statement.

***Exercise* 1.1.6.** In this exercise we use calculus to prove the Cauchy-Schwarz inequality. Let $\boldsymbol{x}, \boldsymbol{y} \in \mathbb{R}^n$ be nonzero vectors. Define the function

$$F(t) = \langle \boldsymbol{x} + t\boldsymbol{y}, \boldsymbol{x} + t\boldsymbol{y} \rangle.$$

Use calculus to find the value of t, where F assumes its minimum value. By using the fact that $F(t) \geq 0$ for all t, deduce the Cauchy-Schwarz inequality.

***Exercise* 1.1.7.** Show that (1.12) is a consequence of the Cauchy-Schwarz inequality. *Hint*: Consider $\|\boldsymbol{x} + \boldsymbol{y}\|^2$.

***Exercise* 1.1.8.** Define a real-valued function on $\mathbb{R}^n$ by setting

$$N(\boldsymbol{x}) = \max\{|x_1|, \ldots, |x_n|\}.$$

Show that N defines a norm.

***Exercise* 1.1.9.** Let N be a norm on $\mathbb{R}^n$ and define $d(\boldsymbol{x}, \boldsymbol{y}) = N(\boldsymbol{x} - \boldsymbol{y})$. Show that for any triple of points $\boldsymbol{x}_1, \boldsymbol{x}_2, \boldsymbol{x}_3$, the following estimate holds:

$$d(\boldsymbol{x}_1, \boldsymbol{x}_3) \leq d(\boldsymbol{x}_1, \boldsymbol{x}_2) + d(\boldsymbol{x}_2, \boldsymbol{x}_3).$$

Explain why this is also called the triangle inequality.

***Exercise* 1.1.10.** Let $S \subset \mathbb{R}^n$ be a subspace of dimension k. Show that there exists a collection of vectors $\{\boldsymbol{v}_1, \ldots, \boldsymbol{v}_k\} \subset S$ such that every vector $\boldsymbol{x} \in S$ has a *unique* representation of the form

$$\boldsymbol{x} = c_1\boldsymbol{v}_1 + \cdots + c_k\boldsymbol{v}_k.$$

***Exercise* 1.1.11.** Let $\boldsymbol{a}$ be a nonzero n-vector. Show that there is a collection of $n-1$ linearly independent n-vectors, $\{\boldsymbol{v}_1, \ldots, \boldsymbol{v}_{n-1}\}$, so that $\boldsymbol{x}$ solves $\langle \boldsymbol{a}, \boldsymbol{x}\rangle = 0$ if and only if

$$\boldsymbol{x} = \sum_{i=1}^{n-1} c_i \boldsymbol{v}_i$$

for some real constants $\{c_1, \ldots, c_{n-1}\}$.

***Exercise* 1.1.12.** Let $\{\boldsymbol{a}_1, \ldots, \boldsymbol{a}_k\}$ be linearly independent n-vectors. Show that there is a collection of $n-k$ linearly independent n-vectors, $\{\boldsymbol{v}_1, \ldots, \boldsymbol{v}_{n-k}\}$, so that $\boldsymbol{x}$ solves

$$\langle \boldsymbol{a}_j, \boldsymbol{x}\rangle = 0 \text{ for } j = 1, \ldots, k$$

if and only if

$$\boldsymbol{x} = \sum_{i=1}^{n-k} c_i \boldsymbol{v}_i$$

for some real constants $\{c_1, \ldots, c_{n-k}\}$. *Hint*: Use the previous exercise and an induction argument.

***Exercise* 1.1.13.** If a function f has two derivatives, then Taylor's theorem gives a formula for the error $e(y) = f(x+y) - [f(x) + f'(x)y]$. There exists a z between 0 and y such that

$$e(z) = \frac{f''(z)y^2}{2};$$

see (B.13). Use this formula to bound the error made in replacing $h(x, \theta + \Delta\theta)$ with $h(x, \theta) + \partial_\theta h(x, \theta)\Delta\theta$. *Hint*: Find the value of z between 0 and $\Delta\theta$ that maximizes the error term.

***Exercise* 1.1.14.** In Example 1.1.3 compute the gradient of h to determine how the absolute and relative errors depend on θ_1, θ_2, and d.

1.1.2 Infinitely Many Degrees of Freedom

See: A.3, A.5.

In the previous section we examined some simple physical systems with finitely many degrees of freedom. In these examples, the problem of determining the state of the system from feasible measurements reduces to solving systems of finitely many equations in finitely many unknowns. In imaging applications the state of a system is usually described by a function or functions of continuous variables. These systems have infinitely many degrees of freedom. In this section we consider several examples.

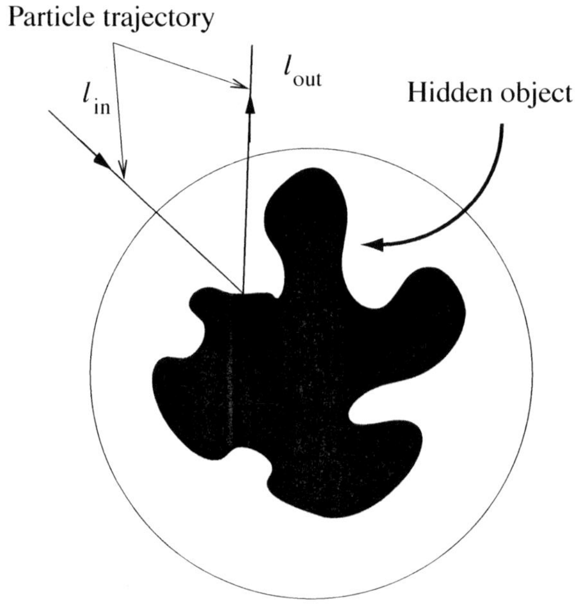

Figure 1.7. Particle scattering can be used to explore the boundary of an unknown region.

***Example* 1.1.5.** Suppose that we would like to determine the shape of a planar region, D, that cannot be seen. The object is lying inside a disk and we can fire particles at the object. Assume that the particles bounce off according to a simple scattering process. Each particle strikes the object once and is then scattered along a straight line off to infinity (Figure 1.7). The outline of the object can be determined by knowing the correspondence between incoming lines, l_{in}, and outgoing lines, l_{out}. Each intersection point $l_{\text{in}} \cap l_{\text{out}}$ lies on the boundary of the object. Measuring $\{l^j_{\text{out}}\}$ for finitely many incoming directions $\{l^j_{\text{in}}\}$ determines finitely many points $\{l^j_{\text{in}} \cap l^j_{\text{out}}\}$ on the boundary of D. In order to use this finite collection of points to make any assertions about the rest of the boundary of D, more information is required. If we know that D consists of a single piece or component, then these points would lie on a single closed curve, though it might be difficult to decide in what order they should appear on the curve. On the other hand, these measurements provide a lot of information about *convex regions.*

Definition* 1.1.7. A region D in the plane is *convex* if it has the following property: For each pair of points p and q lying in D, the line segment $\overline{pq}$ is also contained in D. See Figure 1.8.

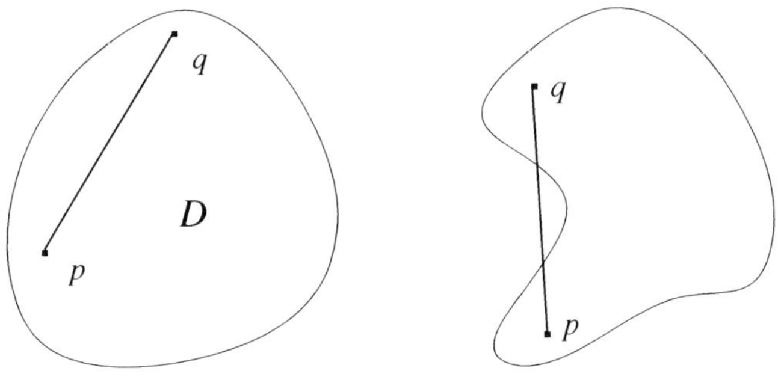

(a) A convex region (b) A non-convex region

Figure 1.8. Convex and non-convex regions.

Convex regions have many special properties. If p and q are on the boundary of D, then the line segment $\overline{pq}$ lies inside of D. From this observation we can show that if $\{p_1, \ldots, p_N\}$ are points on the boundary of a convex region, then the smallest polygon with these points as vertices lies entirely within D [Figure 1.9(a)]. Convexity can also be defined by a property of the boundary of D: For each point p on the boundary of D there is a line l_p that passes through p but is otherwise disjoint from D. This line is called a *support line* through p. If the boundary is smooth at p, then the tangent line to the boundary is the unique support line. A line divides the plane into two half-planes. Let l_p be a support line to D at p. Since the interior of D does not meet l_p it must lie entirely in one of the half-planes determined by this line [see Figure 1.9(b)]. If each support line meets the boundary of D at exactly one point, then the region is *strictly* convex.

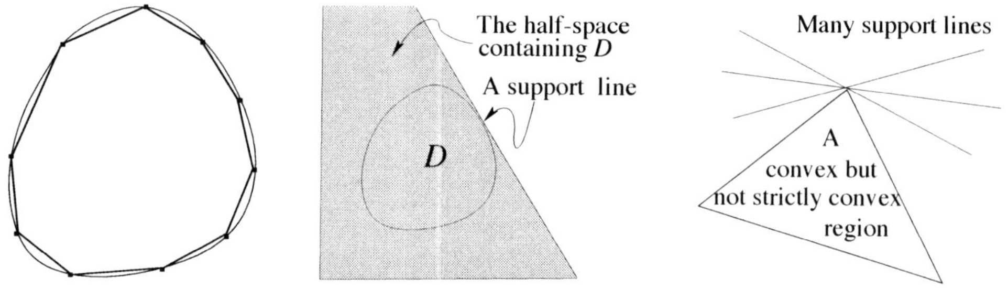

(a) Joining the vertices lying on the boundary of a convex region defines an inscribed polygon.

(b) A support line and half-space.

(c) A triangle bounds a convex region that is not strictly convex.

Figure 1.9. Inscribed polygons and support lines.

***Example* 1.1.6.** A triangle is the boundary of a convex region, with each edge of the triangle a support line. As infinitely many points of the boundary belong to each edge, the region bounded by a triangle in not strictly convex. On the other hand, through each vertex of the triangle, there are infinitely many support lines. These observations are illustrated in Figure 1.9(c).

Suppose that the object is convex and more is known about the scattering process: for example, that the angle of incidence is equal to the angle of reflection. From a finite number of incoming and outgoing pairs, $\{(l^i_{\text{in}}, l^i_{\text{out}}) : i = 1, \ldots, N\}$, we can now determine an approximation to D with an estimate for the error. The intersection points, $p_i = l^i_{\text{in}} \cap l^i_{\text{out}}$ lie on the boundary of the convex region, D. If we use these points as the vertices of a polygon P^{in}_N, then, as remarked previously, P^{in}_N is completely contained within D. On the other hand, as the angle of incidence equals the angle of reflection, we can also determine the tangent lines $\{l_{p_i}\}$ to the boundary of D at the points $\{p_i\}$. These lines are support lines for D. Hence by intersecting the half-planes that contain D, defined by these tangent lines, we obtain another convex polygon, P^{out}_N, that contains D. Thus with these N-measurements we obtain the both an *inner* and *outer* approximation to D:

$$P^{\text{in}}_N \subset D \subset P^{\text{out}}_N.$$

An example is shown in Figure 1.10.

A convex region is determined by its boundary, and each point on the boundary is, in effect, a state variable. Therefore, the collection of convex regions is a system with *infinitely many degrees of freedom*. A nice description for the state space of smooth convex regions is developed in Section 1.2.2. As we have seen, a convex region can be approximated by polygons. Once the number of sides is fixed, then we are again considering a system with finitely many degrees of freedom. In all practical problems, a system with infinitely many degrees of freedom must eventually be approximated by a system with finitely many degrees of freedom.

***Remark* 1.1.1.** For a non-convex body, the preceding method does not work as the correspondence between incoming and outgoing lines can be complicated: Some incoming lines may undergo multiple reflections before escaping, and in fact some lines might become permanently trapped.

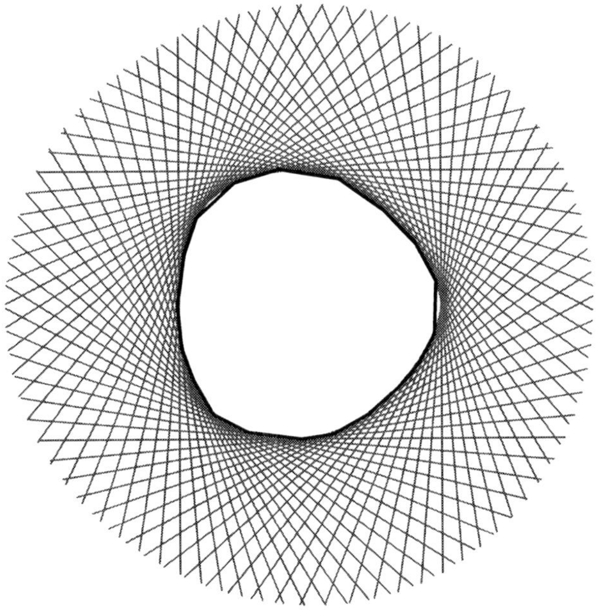

Figure 1.10. An inner and an outer approximation to a convex region.

Exercises

***Exercise* 1.1.15.** Find state variables to describe the set of polygons with n-vertices in the plane. For the case of triangles, find the relations satisfied by your variables. Extra credit: Find a condition, in terms of your parameters, implying that the polygon is convex.

***Exercise* 1.1.16.** Suppose that D_1 and D_2 are convex regions in the plane. Show that their intersection $D_1 \cap D_2$ is also a convex region.

Exercise* 1.1.17. Suppose that D is a possibly non-convex region in the plane. Define a new region D' as the intersection of all the half-planes that contain D. Show that $D = D'$ if and only if D is convex.

***Exercise* 1.1.18.** Find an example of a planar region such that at least one particle trajectory is trapped forever.

1.2 A Simple Model Problem for Image Reconstruction

The problem of image reconstruction in x-ray tomography is sometimes described as reconstructing an object from its "projections." Of course, these are projections under the illumination of x-ray "light." In this section we consider the analogous but simpler problem of determining the outline of a convex object from its shadows. As is also the case in medical applications, we consider a two-dimensional problem. Let D be a convex region in the plane. Imagine that a light source is placed very far from D. Since the light source is very far away, the rays of light are all traveling in essentially the same direction. We can think of them as a collection of parallel lines. We want to measure the shadow that D casts for each position of the light source. To describe the measurements imagine that a screen is placed on the "other side" of D perpendicular to the direction of the light rays (Figure 1.11). In a real apparatus sensors would be placed on the screen, allowing us to determine where the shadow begins and ends.

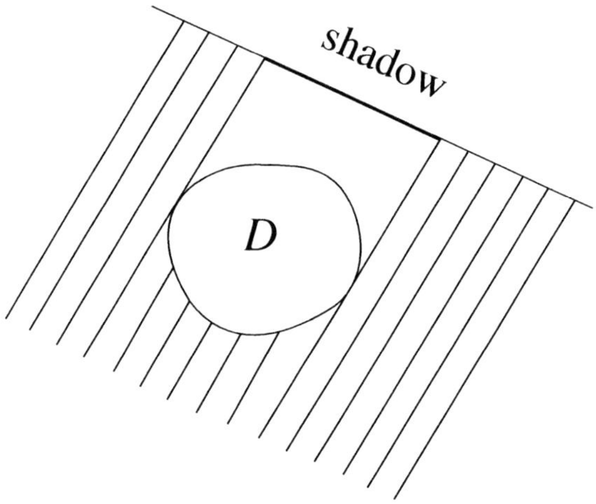

Figure 1.11. The shadow of a convex region.

The region, D, blocks a certain collection of light rays and allows the rest to pass. Locating the shadow amounts to determining the "first" and "last" lines in this family of parallel lines to intersect D. To describe the object completely, we need to rotate the source and detector through π radians, measuring, at each angle, where the shadow begins and ends.

The first and last lines to intersect a region just meet it along its boundary. These lines are therefore tangent to the boundary of D. The problem of reconstructing a region from its shadows is mathematically the same as the problem of reconstructing a region from a knowledge of the tangent lines to its boundary. As a first step in this direction we need a good way to organize our measurements. To that end we give a description for the *space of lines in the plane.*

1.2.1 The Space of Lines in the Plane*

A line in the plane is a set of points that satisfies an equation of the form

$$ax + by = c,$$

where $a^2 + b^2 \neq 0$. We could use (a, b, c) to parameterize the set of lines, but note that we get the same set of points if we replace this equation by

$$\frac{a}{\sqrt{a^2 + b^2}} x + \frac{b}{\sqrt{a^2 + b^2}} y = \frac{c}{\sqrt{a^2 + b^2}}.$$

The coefficients $(\frac{a}{\sqrt{a^2+b^2}}, \frac{b}{\sqrt{a^2+b^2}})$ define a point $\boldsymbol{\omega}$ on the unit circle, $S^1 \subset \mathbb{R}^2$, and the constant $\frac{c}{\sqrt{a^2+b^2}}$ can be any number. The lines in the plane are parameterized by a pair consisting of a unit vector

$$\boldsymbol{\omega} = (\omega_1, \omega_2)$$

and a real number t. The line $l_{t,\boldsymbol{\omega}}$ is the set of points in $\mathbb{R}^2$ satisfying the equation

$$\langle (x, y), \boldsymbol{\omega} \rangle = t. \tag{1.24}$$

The vector $\boldsymbol{\omega}$ is perpendicular to this line (Figure 1.12).

It is often convenient to parameterize the points on the unit circle by a real number; to that end we set

$$\boldsymbol{\omega}(\theta) = (\cos(\theta), \sin(\theta)). \tag{1.25}$$

Since cos and sin are 2π-periodic, it clear that $\boldsymbol{\omega}(\theta)$ and $\boldsymbol{\omega}(\theta + 2\pi)$ are the same point on the unit circle. Using this parameterization for points on the circle, the line $l_{t,\theta} \overset{d}{=} l_{t,\boldsymbol{\omega}(\theta)}$ is the set of solutions to the equation

$$\langle (x, y), (\cos(\theta), \sin(\theta)) \rangle = t.$$

Both notations for lines and points on the circle are used in the sequel.

While the parameterization provided by $(t, \boldsymbol{\omega})$ is much more efficient than that provided by (a, b, c), note that the set of points satisfying (1.24) is unchanged if $(t, \boldsymbol{\omega})$ is replaced by $(-t, -\boldsymbol{\omega})$. Thus, as sets,

$$l_{t,\boldsymbol{\omega}} = l_{-t,-\boldsymbol{\omega}}. \tag{1.26}$$

It is not difficult to show that if $l_{t_1,\boldsymbol{\omega}_1} = l_{t_2,\boldsymbol{\omega}_2}$ then either $t_1 = t_2$ and $\boldsymbol{\omega}_1 = \boldsymbol{\omega}_2$ or $t_1 = -t_2$ and $\boldsymbol{\omega}_1 = -\boldsymbol{\omega}_2$.

The pair $(t, \boldsymbol{\omega})$ actually specifies an *oriented line*. That is, we can use these data to define the positive direction along the line. The vector

$$\hat{\boldsymbol{\omega}} = (-\omega_2, \omega_1)$$

is perpendicular to $\boldsymbol{\omega}$ and is therefore parallel to $l_{t,\boldsymbol{\omega}}$. In fact, $\hat{\boldsymbol{\omega}}$ and $-\hat{\boldsymbol{\omega}}$ are both unit vectors that are parallel to $l_{t,\boldsymbol{\omega}}$. The vector $\hat{\boldsymbol{\omega}}$ is selected by using the condition that the 2×2 matrix,

$$\begin{pmatrix} \omega_1 & -\omega_2 \\ \omega_2 & \omega_1 \end{pmatrix},$$

has determinant $+1$. The vector $\hat{\boldsymbol{\omega}}$ *defines* the positive direction or *orientation* of the line $l_{t,\boldsymbol{\omega}}$. This explains how the pair $(t, \boldsymbol{\omega})$ determines an *oriented line*. We summarize these computations in a proposition.

Proposition 1.2.1. *The pairs* $(t, \boldsymbol{\omega}) \in \mathbb{R} \times S^1$ *are in one-to-one correspondence with the set of oriented lines in the plane.*

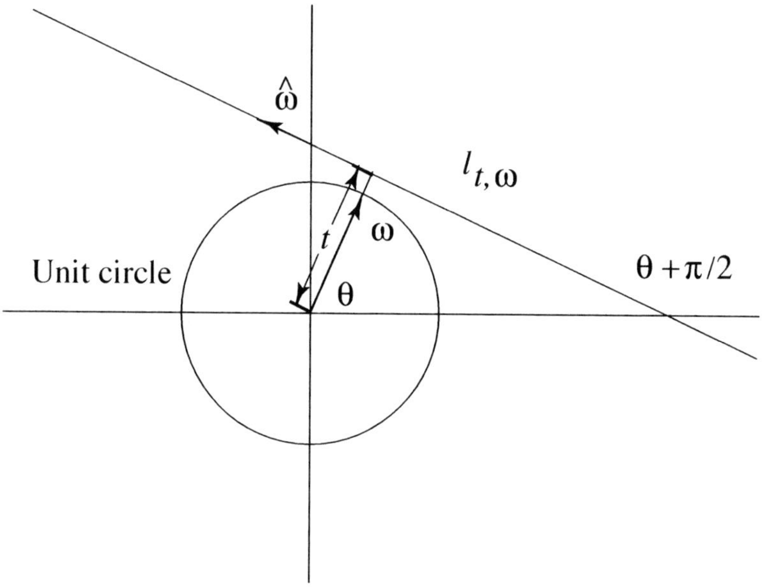

Figure 1.12. Parameterization of oriented lines in the plane.

The vector $\boldsymbol{\omega}$ is the direction orthogonal to the line and the number t is called the *affine parameter* of the line; $|t|$ is the distance from the line to the origin of the coordinate system. The pair $(t, \boldsymbol{\omega})$ defines two half-planes

$$H^+_{t,\boldsymbol{\omega}} = \{\boldsymbol{x} \in \mathbb{R}^2 \mid \langle \boldsymbol{x}, \boldsymbol{\omega} \rangle > t\} \text{ and } H^-_{t,\boldsymbol{\omega}} = \{\boldsymbol{x} \in \mathbb{R}^2 \mid \langle \boldsymbol{x}, \boldsymbol{\omega} \rangle < t\}; \tag{1.27}$$

the line $l_{t,\boldsymbol{\omega}}$ is the common boundary of these half-planes. Facing along the line $l_{t,\boldsymbol{\omega}}$ in the direction specified by $\hat{\boldsymbol{\omega}}$, the half-plane $H^-_{t,\boldsymbol{\omega}}$ lies to the left.

Exercises

Exercise* 1.2.1. Show that $l_{t,\boldsymbol{\omega}}$ is given parametrically as the set of points

$$l_{t,\boldsymbol{\omega}} = \{t\boldsymbol{\omega} + s\hat{\boldsymbol{\omega}} \ : \ s \in (-\infty, \infty)\}.$$

Exercise* 1.2.2. Show that if $\boldsymbol{\omega} = (\cos(\theta), \sin(\theta))$, then $\hat{\boldsymbol{\omega}} = (-\sin(\theta), \cos(\theta))$, and as a function of θ :

$$\hat{\boldsymbol{\omega}}(\theta) = \partial_\theta \boldsymbol{\omega}(\theta).$$

***Exercise* 1.2.3.** Suppose that $(t, \boldsymbol{\omega})$ and $(t_1, \boldsymbol{\omega}_1)$ are different points in $\mathbb{R} \times S^1$ such that $l_{t_1,\boldsymbol{\omega}_1} = l_{t_2,\boldsymbol{\omega}_2}$. Show that $(t_1, \boldsymbol{\omega}_1) = (-t_2, -\boldsymbol{\omega}_2)$.

***Exercise* 1.2.4.** Show that

$$|t| = \min\{\sqrt{x^2 + y^2} \ : \ (x, y) \in l_{t,\omega}\}.$$

***Exercise* 1.2.5.** Show that if ω is fixed, then the lines in the family $\{l_{t,\omega} \ : \ t \in \mathbb{R}\}$ are parallel.

***Exercise* 1.2.6.** Show that every line in the family $\{l_{t,\hat{\omega}} \ : \ t \in \mathbb{R}\}$ is orthogonal to every line in the family $\{l_{t,\omega} \ : \ t \in \mathbb{R}\}$.

***Exercise* 1.2.7.** Each choice of direction ω defines a coordinate system on $\mathbb{R}^2$,

$$(x, y) = t\omega + s\hat{\omega}.$$

Find the inverse, expressing (t, s) as functions of (x, y). Show that the area element in the plane satisfies

$$dx\,dy = dt\,ds.$$

1.2.2 Reconstructing an Object from Its Shadows

Now we can quantitatively describe the shadow. Because there are two lines in each family of parallel lines that are tangent to the boundary of D, we need a way to select one of them. To do this we choose an orientation for the boundary of D; this operation is familiar from Green's theorem in the plane. The positive direction on the boundary is selected so that, when facing in this direction the region lies to the left; the counterclockwise direction is, by convention, the positive direction (Figure 1.13).

Fix a source position $\omega(\theta)$. In the family of parallel lines $\{l_{t,\omega(\theta)} : t \in \mathbb{R}\}$ there are two values of t, $t_0 < t_1$, such that the lines $l_{t_0,\omega(\theta)}$ and $l_{t_1,\omega(\theta)}$ are tangent to the boundary of D (Figure 1.13). Examining the diagram, it is clear that the orientation of the boundary at the point of tangency and that of the oriented line agree for $l_{t_1,\omega}$, and are opposite for $l_{t_0,\omega}$. Define h_D, the *shadow function* of D, by setting

$$h_D(\theta) = t_1 \text{ and } h_D(\theta + \pi) = -t_0. \tag{1.28}$$

The shadow function is completely determined by values of θ belonging to an interval of length π. Because $\omega(\theta) = \omega(\theta+2\pi)$, the shadow function can be regarded as a 2π-periodic function defined on the whole real line. The mathematical formulation of reconstruction problem is as follows: Can the boundary of the region D be determined from h_D?

As $\omega(\theta) = (\cos(\theta), \sin(\theta))$, the line $l_{h_D(\theta),\omega(\theta)}$ is given parametrically by

$$\{h_D(\theta)(\cos(\theta), \sin(\theta)) + s(-\sin(\theta), \cos(\theta)) \mid s \in (-\infty, \infty)\}.$$

To determine the boundary of D, it would suffice to determine the point of tangency of $l_{h_D(\theta),\omega(\theta)}$ with the boundary of D; in other words, we would like to find the function $s(\theta)$ so that for each θ,

$$(x(\theta), y(\theta)) = h_D(\theta)(\cos(\theta), \sin(\theta)) + s(\theta)(-\sin(\theta), \cos(\theta)) \tag{1.29}$$

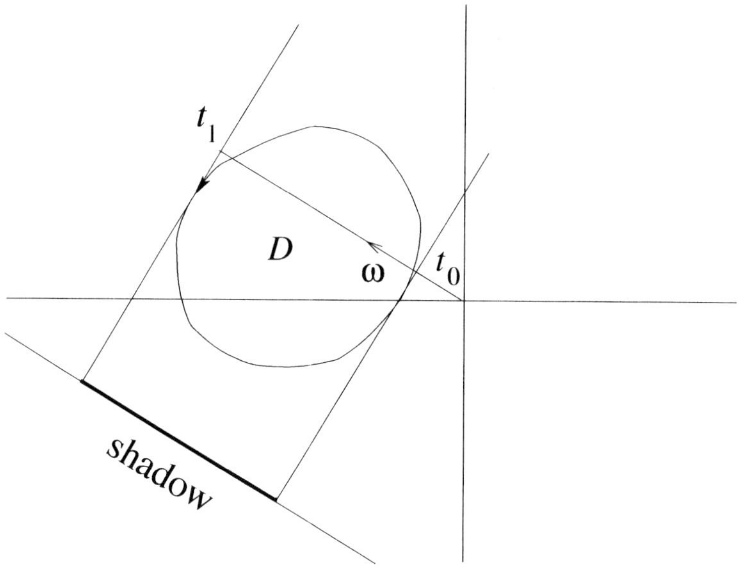

Figure 1.13. The measurement of the shadow.

is a point on the boundary of D. For the remainder of this section we suppose that s is differentiable.

The function s is found by recalling that, at the point of tangency, the direction of the tangent line to D is $\hat{\omega}(\theta)$. For a curve in the plane given parametrically by differentiable functions $(x(\theta), y(\theta))$, the direction of the tangent line is found by differentiating. At a parameter value θ_0 the direction of the tangent line is the same as that of the vector $(x'(\theta_0), y'(\theta_0))$. Differentiating the expression given in (1.29) and using the fact that $\partial_\theta \boldsymbol{\omega} = \hat{\boldsymbol{\omega}}$, we find that

$$(x'(\theta), y'(\theta)) = (h'_D(\theta) - s(\theta))\boldsymbol{\omega}(\theta) + (h_D(\theta) + s'(\theta))\hat{\boldsymbol{\omega}}(\theta). \tag{1.30}$$

Since the tangent line at $(x(\theta), y(\theta))$ is parallel to $\hat{\boldsymbol{\omega}}(\theta)$ it follows from (1.30) that

$$h'_D(\theta) - s(\theta) = 0. \tag{1.31}$$

This gives a parametric representation for the boundary of a convex region in terms of its shadow function: If the shadow function is $h_D(\theta)$, then the boundary of D is given parametrically by

$$(x(\theta), y(\theta)) = h_D(\theta)\boldsymbol{\omega}(\theta) + h'_D(\theta)\hat{\boldsymbol{\omega}}(\theta). \tag{1.32}$$

Note that we have assumed that D is strictly convex and the $h_D(\theta)$ is a differentiable function. This is not always true; for example, if the region D is a polygon, then neither assumption holds.

Let D denote a convex region and h_D its shadow function. We can think of $D \mapsto h_D$ as a mapping from convex regions in the plane to 2π-periodic functions. It is reasonable to enquire if every 2π-periodic function is the shadow function of a convex region. The answer to this question is no. For strictly convex regions with smooth boundaries, we are able to characterize the range of this mapping. If h is twice differentiable, then the tangent vector to the curve defined by

$$(x(\theta), y(\theta)) = h(\theta)\omega(\theta) + h'(\theta)\hat{\omega}(\theta) \tag{1.33}$$

is given by

$$(x'(\theta), y'(\theta)) = (h''(\theta) + h(\theta))\hat{\omega}(\theta).$$

In our construction of the shadow function, we observed that the tangent vector to the curve at $(x(\theta), y(\theta))$ and the vector $\hat{\omega}(\theta)$ point in the same direction. From our formula for the tangent vector, we see that this implies that

$$h''(\theta) + h(\theta) > 0 \text{ for all } \theta \in [0, 2\pi]. \tag{1.34}$$

This gives a necessary condition for a twice differentiable function h to be the shadow function for a strictly convex region with a smooth boundary. Mathematically we are determining the range of the map that takes a convex body $D \subset \mathbb{R}^2$ to its shadow function h_D, under the assumption that h_D is twice differentiable. This is a convenient mathematical assumption, though in an applied context it is likely to be overly restrictive. The state space of the "system" which consists of strictly convex regions with smooth boundaries is parameterized by the set of smooth, 2π-periodic functions satisfying the inequality (1.34). This is an example of a system where the constraint defining the state space is an inequality rather than an equality.

Exercises

***Exercise* 1.2.8.** Justify the definition of $h_D(\theta+\pi)$ in (1.28) by showing that the orientation of the boundary at the point of tangency with $l_{t_0,\omega(\theta)}$ agrees with that of $l_{-t_0,\omega(\theta+\pi)}$.

***Exercise* 1.2.9.** Suppose that D_n is a regular n-gon. Determine the shadow function $h_{D_h}(\theta)$.

***Exercise* 1.2.10.** Suppose that D is a bounded, convex planar region. Show that the shadow function h_D is a continuous function of θ.

***Exercise* 1.2.11.** Suppose that h is a 2π-periodic, twice differentiable function that satisfies (1.34). Show that the curve given by (1.33) is the boundary of a strictly convex region.

***Exercise* 1.2.12.** How is the assumption that D is strictly convex used in the derivation of (1.31)?

***Exercise* 1.2.13.** If h is a differentiable function, then equation (1.33) defines a curve. By plotting examples, determine what happens if the condition (1.34) is not satisfied.

***Exercise* 1.2.14.** Suppose that h is a function satisfying (1.34). Show that the area of D_h is given by the

$$\text{Area}(D_h) = \frac{1}{2}\int_0^{2\pi} [(h(\theta))^2 - (h'(\theta))^2]\, d\theta.$$

Explain why this implies that a function satisfying (1.34) also satisfies the estimate

$$\int_0^{2\pi} (h'(\theta))^2 d\theta < \int_0^{2\pi} (h(\theta))^2 \, d\theta.$$

***Exercise* 1.2.15.** Let h be a smooth 2π-periodic function that satisfies (1.34). Prove that the curvature of the boundary of the region with this shadow function, at the point $h(\theta)\boldsymbol{\omega}(\theta) + h'(\theta)\hat{\boldsymbol{\omega}}(\theta)$, is given by

$$\kappa(\theta) = \frac{1}{h(\theta) + h''(\theta)}. \tag{1.35}$$

***Exercise* 1.2.16.** Suppose that h is a function satisfying (1.34). Show that another parametric representation for the boundary of the region with this shadow function is

$$\theta \mapsto \left(-\int_0^{\theta} (h(s) + h''(s))\sin(s)\, ds, \int_0^{\theta} (h(s) + h''(s))\cos(s)\, ds \right).$$

***Exercise* 1.2.17.** In this exercise we determine which positive functions κ defined on S^1 are the curvatures of closed strictly convex curves. Prove the following result: A positive function κ on S^1 is the curvature of a closed, strictly convex curve (parameterized by its tangent direction) if and only if

$$\int_0^{\infty} \frac{\sin(s)\, ds}{\kappa(s)} = 0 = \int_0^{\infty} \frac{\cos(s)\, ds}{\kappa(s)}.$$

***Exercise* 1.2.18.** Let D be a convex region with shadow function h_D. For a vector $\boldsymbol{v} \in \mathbb{R}^2$, define the translated region

$$D^{\boldsymbol{v}} = \{\boldsymbol{x} + \boldsymbol{v} \, : \, \boldsymbol{x} \in D\}.$$

Find the relation between h_D and $h_{D^{\boldsymbol{v}}}$. Explain why this answer is inevitable in light of the formula (1.35) for the curvature.

***Exercise* 1.2.19.** Let D be a convex region with shadow function h_D. For a rotation $A = \begin{pmatrix} \cos\phi & -\sin\phi \\ \sin\phi & \cos\phi \end{pmatrix}$, define the rotated region

$$D^A = \{A\boldsymbol{x} \, : \, \boldsymbol{x} \in D\}.$$

Find the relation between h_D and h_{D^A}.

Exercise* 1.2.20. If h_1 and h_2 are 2π-periodic functions satisfying (1.34) then they are the shadow functions of convex regions D_1 and D_2. The sum $h_1 + h_2$ also satisfies (1.34) and so is the shadow function of a convex region, D_3. Describe geometrically how D_3 is determined by D_1 and D_2.

Exercise* 1.2.21. Suppose that D is non-convex planar region. The shadow function h_D is defined as before. What information about D is encoded in h_D?

1.2.3 Approximate Reconstructions

See: A.6.2.

In a realistic situation the shadow function is measured at a finite set of angles

$$\{\theta_1, \ldots, \theta_m\}.$$

How can the data, $\{h_D(\theta_1), \ldots, h_D(\theta_m)\}$, be used to construct an approximation to the region D? We consider two different strategies; each relies on the special geometric properties of convex regions. Recall that a convex region always lies in one of the half-planes determined by the support line at any point of its boundary. Since the boundary of D and $l_{h(\theta),\omega(\theta)}$ have the same orientation at the point of contact, it follows that D lies in each of the half-planes

$$H^-_{h(\theta_j),\omega(\theta_j)}, \quad j = 1, \ldots, m;$$

see (1.27). As D lies in each of these half-planes, it also lies in their intersection. This defines a convex polygon

$$P_m = \bigcap_{j=1}^{m} H^-_{h(\theta_j),\omega(\theta_j)}$$

that contains D. This polygon provides one sort of approximation for D from the measurement of a finite set of shadows. It is a stable approximation to D because small errors in the measurements of either the angles θ_j or the corresponding affine parameters $h(\theta_j)$ lead to small changes in the approximating polygon.

The difficulty with using the exact reconstruction formula (1.32) is that h is only known at finitely many values, $\{\theta_j\}$. From this information it is not possible to compute the exact values of the derivatives, $h'(\theta_j)$. We could use a finite difference approximation for the derivative to determine a finite set of points that approximate points on the boundary of D:

$$(x_j, y_j) = h(\theta_j)\omega(\theta_j) + \frac{h(\theta_j) - h(\theta_{j+1})}{\theta_j - \theta_{j+1}}\hat{\omega}(\theta_j).$$

If the measurements were perfect, the boundary of D smooth and the numbers $\{|\theta_j - \theta_{j+1}|\}$ small, then the finite difference approximations to $h'(\theta_j)$ would be accurate and these points would lie close to points on the boundary of D. Joining these points in the given order gives a polygon, P', that approximates D. If $\{h'(\theta_j\}$ could be computed exactly, then P' would be contained in D. With approximate values this cannot be asserted with certainty, though P' should be largely contained within D.

This gives a different way to reconstruct an approximation to D from a finite set of measurements. This method is not as robust as the first technique because it requires the measured data to be differentiated. In order for the finite difference $\frac{h(\theta_j)-h(\theta_{j+1})}{\theta_j-\theta_{j+1}}$ to be a good approximation to $h'(\theta_j)$, it is generally necessary for $|\theta_j - \theta_{j+1}|$ to be small. Moreover, the errors in the measurements of $h(\theta_j)$ and $h(\theta_{j+1})$ must also be small *compared to* $|\theta_j - \theta_{j+1}|$. This difficulty arises in solution of the reconstruction problem in x-ray CT; the exact reconstruction formula calls for the measured data to be differentiated.

In general, measured data are corrupted by noise, and noise is usually non-differentiable. This means that the measurements cannot be used directly to approximate the derivatives of a putative underlying smooth function. This calls for finding a way to improve the accuracy of the measurements. If the errors in individual measurements are random then repeating the same measurement many times and averaging the results should give a good approximation to the true value. This is the approach taken in magnetic resonance imaging. Another possibility is to make a large number of measurements at closely spaced angles $\{(h_j, j\Delta\theta) : j = 1, \dots, N\}$, which are then averaged to give less noisy approximations on a coarser grid. There are many ways to do the averaging. One way is to find a differentiable function, H, belonging to a family of functions of dimension $M < N$ that minimizes the *square error*

$$e(H) = \sum_{j=1}^{N} (h_j - H(j\Delta\theta))^2.$$

For example, H could be taken to be a polynomial of degree $M - 1$, or a continuously differentiable, piecewise cubic function. The reconstruction formula can be applied to H to obtain a different approximation to D. The use of averaging reduces the effects of noise but fine structure in the boundary is also blurred by any such procedure.

Exercises

***Exercise* 1.2.22.** Suppose that the angles $\{\theta_j\}$ can be measured exactly but there is an uncertainty of size ϵ in the measurement of the affine parameters, $h(\theta_j)$. Find a polygon $P_{m,\epsilon}$ that gives the best possible approximation to D and certainly contains D.

***Exercise* 1.2.23.** Suppose that we know that $|h''(\theta)| < M$, and the measurement errors are bounded by $\epsilon > 0$. For what angle spacing $\Delta\theta$ is the error, using a finite difference approximation for h', due to the uncertainty in the measurements equal to that caused by the nonlinearity of h itself?

1.2.4 Can an Object Be Reconstructed from Its Width?

To measure the location of the shadow requires an expensive detector that can accurately locate a transition from light to dark. It would be much cheaper to build a device, similar to the exposure meter in a camera, to measure the length of the shadow region without determining its precise location. It is therefore an interesting question whether or not the boundary of a region can be reconstructed from measurements of the *widths* of its shadows. Let $w_D(\theta)$ denote the width of the shadow in direction θ. A moment's consideration shows that

$$w_D(\theta) = h_D(\theta) + h_D(\theta + \pi). \tag{1.36}$$

Using this formula and Exercise 1.2.11, it is easy to show that w_D does *not* determine D. From Exercise 1.2.11 we know that if h_D has two derivatives such that $h_D'' + h_D > 0$, then h_D is the shadow function of a strictly convex region. Let e be an *odd* smooth function [i.e., $e(\theta) + e(\theta + \pi) \equiv 0$] such that

$$h_D'' + h_D + e'' + e > 0.$$

If $e \not\equiv 0$, then $h_D + e$ is the shadow function for D', a different strictly convex region. Observe that D' has the same *width* of shadow for each direction as D; that is,

$$w_D(\theta) = (h_D(\theta) + e(\theta)) + (h_D(\theta + \pi) + e(\theta + \pi)) = w_{D'}(\theta).$$

To complete this discussion, note that any function expressible as a series of the form

$$e(\theta) = \sum_{j=0}^{\infty} [a_j \sin(2j + 1)\theta + b_j \cos(2j + 1)\theta]$$

is an odd function. This is an infinite-dimensional space of functions. This implies that if w_D is the width of the shadow function for a convex region D, then there is an infinite-dimensional set of regions with the same width of the shadow function. Consequently, the simpler measurement is inadequate to reconstruct the boundary of a convex region. Figure 1.14 shows the unit disk and another region that has constant shadow width equal to 2.

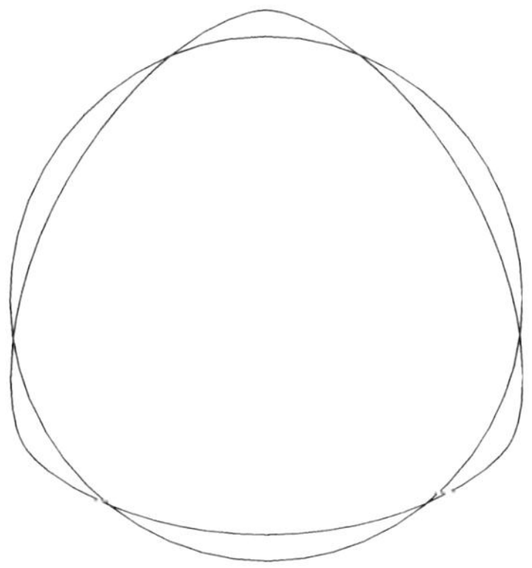

Figure 1.14. Two regions of constant width 2.

Exercises

***Exercise* 1.2.24.** Justify the formula (1.36) for the shadow width.

***Exercise* 1.2.25.** Show that the width function satisfies $w_D'' + w_D > 0$.

***Exercise* 1.2.26.** Is it true that every twice differentiable, π-periodic function, w satisfying $w'' + w > 0$ is the width function of a convex domain?

***Exercise* 1.2.27.** We considered whether or not a convex body is determined by the width of its shadows in order use a less expensive detector. The cheaper detector can only measure the width of the covered region. Can you find a way to use a detector that only measures the length of an illuminated region to locate the edge of the shadow? *Hint*: Cover only half of the detector with photosensitive material.

1.3 Conclusion

By considering examples, we have seen how physical systems can be described using mathematical models. The problem of determining the state of the system from measurements is replaced by that of solving equations or systems of equations. It is important to keep in mind that mathematical models are just models, indeed often toy models. A good model must satisfy two opposing requirements: The model should accurately depict the system under study while at the same time being simple enough to be usable. In addition, it must also have accurate, finite-dimensional approximations.

In mathematics, problems of determining the state of a physical system from feasible measurements are gathered under the rubric of *inverse problems*. The division of problems into inverse problems and *direct problems* is often a matter of history. Usually a physical theory that models how the state of the system determines feasible measurements preceded a description of the inverse process: how to use measurements to determine the state of the system. While many of the problems that arise in medical imaging are considered to be inverse problems, we do not give a systematic development of this subject. The curious reader is referred to the article by Joe Keller, [80], which contains analyses of many classical inverse problems or the book *Introduction to Inverse Problems in Imaging*, [10].

The models used in medical imaging usually involve infinitely many degrees of freedom. The state of the system is described by a function of continuous variables. Ultimately only a finite number of measurements can be made and only a finite amount of time is available to process them. Our analysis of the reconstruction process in x-ray CT passes through several stages. We begin with a description of the complete, perfect data situation. The measurement is described by a function on the space of lines. By finding an explicit inversion formula, we show that the state of the system can be determined from these measurements. The main tool in this analysis is the Fourier transform. We next consider the consequences of having only discrete samples of these measurements. This leads us to sampling theory and Fourier series. In the next chapter we quickly review linear algebra and the theory of linear equations, recasting this material in the language of measurement. The chapter ends with a brief introduction to the issues that arise in the extension of linear algebra to infinite-dimensional spaces.

Chapter 2

Linear Models and Linear Equations

Using measurements to determine the state of a system eventually reduces to the problem of solving systems of equations. In almost all practical applications, we are eventually reduced to solving systems of linear equations. This is true even for physical systems that are described by non-linear equations. As we saw in Section 1.1.1, nonlinear equations may be approximated by linear equations. A nonlinear equation is usually solved iteratively, where the iteration step involves the solution of the approximating linear system.

Linear algebra is the mathematical tool most often used in applied subjects. There are many reasons why linear equations and linear models are ubiquitous. As we will see in this chapter, there is a complete mathematical theory for systems of linear equations. Indeed, there are necessary and sufficient conditions for linear equations to have solutions and a description of the space of solutions when they exist. More pragmatically, there are efficient, generally applicable algorithms for solving systems of linear equations. This is not true even for the simplest systems of nonlinear equations. On a more conceptual level, for systems with some sort of intrinsic smoothness, a linear model often suffices to describe small deviations from a known state.

In this chapter we review some basic concepts from linear algebra, in particular the theory of systems of linear equations. This is not intended to serve as a text on finite-dimensional linear algebra but rather to situate these familiar concepts in the context of measurement problems. A more complete introduction to the theory of vector spaces is presented in Appendix A.2. As the models that arise in medical imaging are not usually finite dimensional, we briefly consider linear algebra on infinite-dimensional spaces. As a prelude to our discussion of the Fourier transform, the chapter concludes with an introduction to complex numbers and complex vector spaces.

2.1 Linear Equations and Linear Maps

See: A.2.

Suppose that the state of a system is described by the variables $(x_1, \ldots, x_n; y_1, \ldots, y_m)$ or, more concisely, by $(\boldsymbol{x}; \boldsymbol{y})$ with $\boldsymbol{x} \in \mathbb{R}^n$ and $\boldsymbol{y} \in \mathbb{R}^m$. Here $\boldsymbol{x}$ is regarded as an input variable and $\boldsymbol{y}$ is regarded as an output variable. The model for the system is expressed by the system of linear equations,

$$\begin{aligned} a_{11}x_1 + a_{12}x_2 + &\cdots + a_{1n}x_n &= y_1 \\ a_{21}x_1 + a_{22}x_2 + &\cdots + a_{2n}x_n &= y_2 \\ \vdots \qquad\quad & \qquad \vdots & \vdots \\ a_{m1}x_1 + a_{m2}x_2 + &\cdots + a_{mn}x_n &= y_m. \end{aligned} \tag{2.1}$$

Before proceeding with our analysis, we first need to simplify the notation. It is very cumbersome to have to work with complicated expressions like (2.1); instead we use standard matrix and vector notation. Let $\boldsymbol{a}$ denote the $m \times n$ array of numbers $(a_{ij})_{i=1\ldots m, j=1\ldots n}$, $\boldsymbol{x}$ an n-vector and $\boldsymbol{y}$ an m-vector. The system of equations (2.1) is concisely expressed as

$$\boldsymbol{a}\boldsymbol{x} = \boldsymbol{y}, \tag{2.2}$$

where $\boldsymbol{a}\boldsymbol{x}$ denotes the matrix product. We briefly recall the properties of matrix multiplication. Let $\boldsymbol{x}_1$ and $\boldsymbol{x}_2$ be n-vectors; then

$$\boldsymbol{a}(\boldsymbol{x}_1 + \boldsymbol{x}_2) = \boldsymbol{a}\boldsymbol{x}_1 + \boldsymbol{a}\boldsymbol{x}_2$$

and for any number c

$$\boldsymbol{a}(c\boldsymbol{x}_1) = c(\boldsymbol{a}\boldsymbol{x}_1).$$

In other words, the map from $\mathbb{R}^n$ to $\mathbb{R}^m$ defined by $\boldsymbol{x} \mapsto \boldsymbol{a}\boldsymbol{x}$ is linear. The state space for the system is the graph of this linear transformation,

$$\{(\boldsymbol{x}; \boldsymbol{a}\boldsymbol{x}) : \boldsymbol{x} \in \mathbb{R}^n\}.$$

This is a linear subspace of $\mathbb{R}^n \times \mathbb{R}^m$.

Let $\boldsymbol{a}_i = (a_{i1}, \ldots, a_{in})$ denote the ith row of $\boldsymbol{a}$. The equations in (2.1) can also be rewritten

$$\begin{aligned} \langle \boldsymbol{a}_1, \boldsymbol{x} \rangle &= y_1 \\ &\vdots \\ \langle \boldsymbol{a}_m, \boldsymbol{x} \rangle &= y_m. \end{aligned} \tag{2.3}$$

The inner product $\langle \boldsymbol{a}_i, \boldsymbol{x} \rangle$ is interpreted as the outcome of the ith measurement when the system is in the state described by $\boldsymbol{x}$. The matrix $\boldsymbol{a}$ is called the *measurement matrix*. The state of the system is determined by $\boldsymbol{x}$, which can take any value in $\mathbb{R}^n$. The question of principal interest is the extent to which the measurements, $\boldsymbol{y}$, determine the state, $\boldsymbol{x}$. If the rows of $\boldsymbol{a}$ are linearly dependent, then there is some index l and constants $\{c_i\}$ so that

$$\boldsymbol{a}_l = \sum_{i \neq l} c_i \boldsymbol{a}_i.$$

By linearity this means that

$$\langle \boldsymbol{a}_l, \boldsymbol{x} \rangle = \sum_{i \neq l} c_i \langle \boldsymbol{a}_i, \boldsymbol{x} \rangle.$$

In other words, the outcome of the lth measurement is already determined by the others. It is not an independent measurement and, in pure mathematics, would be regarded as redundant information. In applications, measurement error and noise make it useful to repeat experiments, so such a measurement might very well be retained.

There is a final way to rewrite (2.1). If

$$\boldsymbol{a}^j = \begin{pmatrix} a_{1j} \\ \vdots \\ a_{mj} \end{pmatrix}$$

denotes the jth column of $\boldsymbol{a}$, then (2.1) is equivalent to

$$x_1 \boldsymbol{a}^1 + \cdots + x_n \boldsymbol{a}^n = \boldsymbol{y}.$$

In this form it is clear that the set of possible outcomes of these measurements is the subspace of $\mathbb{R}^m$ spanned by the columns of $\boldsymbol{a}$. This subspace is called the range or *image* of the linear transformation $\boldsymbol{a}$. It is denoted by $\operatorname{Im} \boldsymbol{a}$.

There are three questions which require answers:

Q1: EXISTENCE: For a given m-vector $\boldsymbol{y}$, does there exist an n-vector $\boldsymbol{x}$ that satisfies the equations in (2.2)? In other words, is $\boldsymbol{y}$ in the image of $\boldsymbol{a}$?

Q2: UNIQUENESS: When a solution exists is it unique? More generally, describe the space of solutions.

Q3: STABILITY: How sensitive is the solution to small variations in the measurement matrix $\boldsymbol{a}$ or the measured values $\boldsymbol{y}$?

It is a somewhat unexpected, but important, fact that these issues are rather independent of one another. For applications it is also necessary to have an algorithm to find approximate solutions of (2.1) and criteria to select a solution when there is more than one.

2.1.1 Solving Linear Equations

Suppose that $\boldsymbol{x}_0$ is a solution of the equation $\boldsymbol{ax} = 0$ and $\boldsymbol{x}_1$ is a solution of the equation $\boldsymbol{ax}_1 = \boldsymbol{y}$. By linearity it follows that

$$\boldsymbol{a}(c\boldsymbol{x}_0 + \boldsymbol{x}_1) = c\boldsymbol{ax}_0 + \boldsymbol{ax}_1 = \boldsymbol{ax}_1 = \boldsymbol{y} \text{ for any } c \in \mathbb{R}.$$

From the case $\boldsymbol{y} = 0$ we conclude that the set of solutions to the equation

$$\boldsymbol{ax} = 0$$

is a linear subspace; that is, if $\boldsymbol{x}_0$ and $\boldsymbol{x}_1$ solve this equation, then so does $\boldsymbol{x}_0 + \boldsymbol{x}_1$ as well as $c\boldsymbol{x}_0$, for any number c. This subspace is called the *null space or kernel* of $\boldsymbol{a}$. It is denoted by $\ker(\boldsymbol{a})$ and always contains at least the zero vector $\boldsymbol{0} = (0, \ldots, 0)$. If $\ker(\boldsymbol{a})$ is not just the zero vector, then the available measurements are inadequate to distinguish the states of the system. These observations answer question Q2.

Theorem 2.1.1. *Let $\boldsymbol{a}$ be an $m \times n$ matrix. Suppose that $\boldsymbol{x}_1$ satisfies $\boldsymbol{ax}_1 = \boldsymbol{y}$. Then every other solution to this equation is of the form $\boldsymbol{x}_1 + \boldsymbol{x}_0$, where $\boldsymbol{x}_0 \in \ker(\boldsymbol{a})$. Moreover, every vector of this form solves the equation $\boldsymbol{ax} = \boldsymbol{y}$.*

The solution of the equation $\boldsymbol{ax} = \boldsymbol{y}$ is unique only if the null space of $\boldsymbol{a}$ contains only the zero vector or, in other words, the columns of $\boldsymbol{a}$ are linearly independent.

We now turn to question Q1. Suppose that $\boldsymbol{a}$ is an $m \times n$ matrix, $\boldsymbol{x}$ is an n-vector, and $\boldsymbol{y}$ is an m-vector. Then $\boldsymbol{ax}$ is an m-vector and the inner product is given by

$$\langle \boldsymbol{ax}, \boldsymbol{y} \rangle = \sum_{i=1}^{m} \sum_{j=1}^{n} a_{ij} x_j y_i.$$

The transpose of the matrix $\boldsymbol{a}$ is the $n \times m$ matrix $\boldsymbol{a}^t$ whose ij-entry is a_{ji}. From the previous formula it follows that

$$\langle \boldsymbol{ax}, \boldsymbol{y} \rangle = \langle \boldsymbol{x}, \boldsymbol{a}^t \boldsymbol{y} \rangle.$$

Suppose that $\boldsymbol{b}$ is a nonzero vector in the null space of the transpose, $\boldsymbol{a}^t$ and the equation, $\boldsymbol{ax} = \boldsymbol{y}$ has a solution. Using the preceding calculations, we see that

$$\langle \boldsymbol{y}, \boldsymbol{b} \rangle = \langle \boldsymbol{ax}, \boldsymbol{b} \rangle = \langle \boldsymbol{x}, \boldsymbol{a}^t \boldsymbol{b} \rangle = 0.$$

The last equality follows from the fact that $\boldsymbol{a}^t \boldsymbol{b} = 0$. This gives a necessary condition for existence of a solution to the equation $\boldsymbol{ax} = \boldsymbol{y}$. The vector $\boldsymbol{y}$ must satisfy the conditions

$$\langle \boldsymbol{b}, \boldsymbol{y} \rangle = 0,$$

for every solution of the homogeneous equation $\boldsymbol{a}^t \boldsymbol{b} = 0$. This also turns out to be sufficient.

Theorem 2.1.2. *Let $\boldsymbol{a}$ be an $m \times n$ matrix and $\boldsymbol{y}$ an m-vector. The equation $\boldsymbol{ax} = \boldsymbol{y}$ has a solution if and only if*

$$\langle \boldsymbol{b}, \boldsymbol{y} \rangle = 0$$

for every vector $\boldsymbol{b} \in \ker(\boldsymbol{a}^t)$.

The equation $a^t b = 0$ has nontrivial solutions if and only if the rows of $\boldsymbol{a}$ are linearly dependent. This means that the outcomes of the measurements $\{\langle \boldsymbol{a}_i, \boldsymbol{x}\rangle : i = 1, \dots, m\}$ are not independent of one another. The condition for the solvability of $\boldsymbol{a}\boldsymbol{x} = \boldsymbol{y}$ is simply that entries of $\boldsymbol{y}$ should satisfy the same relations as the measurements themselves.

***Example* 2.1.1.** In the refraction problem considered in Example 1.1.4, we remarked that the refractive index of the lower fluid n_2 could be determined by an additional measurement. Suppose that we shine a beam of light in at a different angle so that the upper angle is ϕ_1 and the lower angle is ϕ_2. This light beam is displaced by l_2 as it passes through the fluid. We now have three equations for the two unknowns:

$$\begin{pmatrix} 1 & 1 \\ \tan(\theta_1) & \tan(\theta_2) \\ \tan(\phi_1) & \tan(\phi_2) \end{pmatrix} \begin{pmatrix} h_1 \\ h_2 \end{pmatrix} = \begin{pmatrix} h \\ l_1 \\ l_2 \end{pmatrix}. \tag{2.4}$$

In order for this equation to have a solution, the measurements (h, l_1, l_2) must satisfy the condition

$$\begin{pmatrix} 1 \\ \tan(\theta_1) \\ \tan(\phi_1) \end{pmatrix} \times \begin{pmatrix} 1 \\ \tan(\theta_2) \\ \tan(\phi_2) \end{pmatrix} \cdot \begin{pmatrix} h \\ l_1 \\ l_2 \end{pmatrix} = 0.$$

Here $\times$ is the vector cross product:

$$(x_1, y_1, z_1) \times (x_2, y_2, z_2) = (y_1 z_2 - z_1 y_2, z_1 x_2 - x_1 z_2, x_1 y_2 - y_1 x_2).$$

Since

$$\frac{\sin(\theta_1)}{\sin(\theta_2)} = \frac{\sin(\phi_1)}{\sin(\phi_2)} = \frac{n_2}{n_1}$$

and the angles θ_1 and ϕ_1 as well as (h, l_1, l_2) are assumed known, this solvability condition gives a nonlinear equation that allows the determination of $\frac{n_2}{n_1}$ from the measured data.

***Example* 2.1.2.** Suppose we have a collection of photon sources, labeled by $1 \le i \le n$ and an array of detectors, labeled by $1 \le j \le m$. The matrix $\boldsymbol{P}$ has entries $0 \le p_{ij} \le 1$ with the ij-entry the probability that a particle emitted from source i is detected by detector j. Since a given photon can be detected by at most one detector, it follows that

$$\sum_{j=1}^{m} p_{ij} \le 1 \text{ for } i = 1, \dots n.$$

If d_j, $j = 1, \dots, m$ is the number of photons detected at detector j and s_i, $i = 1, \dots, n$ is the number of photons emitted by source i then our model predicts that

$$\boldsymbol{P}\boldsymbol{s} = \boldsymbol{d}. \tag{2.5}$$

If $m = n$ and the measurements are independent, then $\boldsymbol{P}$ is an invertible matrix. The measurements $\boldsymbol{d}$ determine a unique vector $\boldsymbol{s}$, satisfying (2.5). Since the model is probabilistic, this should be regarded as an average value for the distribution of sources. If $m > n$, then we have more measurements than unknowns, so any measurement errors or flaws in the model could make it impossible to find a vector $\boldsymbol{s}$ so that $\boldsymbol{P}\boldsymbol{s} = \boldsymbol{d}$. This is a frequent situation in image reconstruction problems. We choose a way to measure the error, usually a non-negative function of the difference $\boldsymbol{P}\boldsymbol{s} - \boldsymbol{d}$, and seek a vector $\boldsymbol{s}$ that minimizes this error function. Finally, we may have more sources than detectors. The measurements are then inadequate, in principle to determine their distribution. This is also a common circumstance in image reconstruction problems and is resolved by making some a priori assumptions about the allowable distribution of sources to obtain a determined (or even over-determined) problem.

It is a consequence of Theorem 2.1.2 that there are essentially three types of linear models for systems with finitely many degrees of freedom.

DETERMINED: The simplest case arises when the numbers of *independent* measurements and parameters describing the state of the system are the same. This implies that $n = m$ and that the measurements uniquely determine the state of the system. Mathematically we say that the matrix $\boldsymbol{a}$ is invertible. For a square matrix this is equivalent to the statement that the homogeneous equation, $\boldsymbol{a}\boldsymbol{x} = 0$, has only the trivial solution, $\boldsymbol{x} = 0$. The inverse matrix is denoted by $\boldsymbol{a}^{-1}$; it is both a left and a right inverse to $\boldsymbol{a}$,

$$\boldsymbol{a}^{-1}\boldsymbol{a} = \mathrm{Id}_n = \boldsymbol{a}\boldsymbol{a}^{-1}.$$

Here Id_n denotes the $n \times n$ identity matrix, that in

$$(\mathrm{Id}_n)_{ij} = \begin{cases} 1 \text{ if } i = j, \\ 0 \text{ if } i \neq j. \end{cases}$$

From the mathematical point of view, the unique solution is obtained by setting

$$\boldsymbol{x} = \boldsymbol{a}^{-1}\boldsymbol{y}.$$

In practical applications it is unusual to compute the inverse matrix explicitly.

OVER-DETERMINED: In this case we have more measurements than parameters (i.e., $m > n$). If the model and measurements were perfect, then there would be a unique $\boldsymbol{x}$ with $\boldsymbol{a}\boldsymbol{x} = \boldsymbol{y}$. In general, neither is true and there does not exist any $\boldsymbol{x}$ exactly satisfying this equation. Having more measurements than parameters can be used to advantage in several different ways. In Example 2.1.1 we explain how to use the conditions for solvability given in Theorem 2.1.2 to determine physical parameters. Often, measurements are noisy. A model for the noise in the measurements can be used to select a criterion for a best approximate solution. The most common way to measure the error is to use the norm defined in (1.8), setting

$$e(\boldsymbol{x}) = \|\boldsymbol{a}\boldsymbol{x} - \boldsymbol{y}\|^2.$$

There are two reasons why this measure of the error is often employed: (1) It is a natural choice if the noise is normally distributed; (2) the problem of minimizing $e(\boldsymbol{x})$ can be reduced to the problem of solving a system of linear equations.

UNDERDETERMINED: Most problems in image reconstruction are underdetermined; that is, we do not have enough data to uniquely determine a solution. In mathematical tomography, a perfect reconstruction would require an infinite number of exact measurements. These are, of course, never available. In a finite-dimensional linear algebra problem, this is the case where $m < n$. When the measurements $\boldsymbol{y}$ do not uniquely determine the state $\boldsymbol{x}$, additional criteria are needed to determine which solution to use. For example, we might use the solution to $\boldsymbol{ax} = \boldsymbol{y}$, which is of smallest norm. Another approach is to assume that $\boldsymbol{x}$ belongs to a subspace whose dimension is at most the number of independent measurements. Both of these approaches are used in medical imaging.

The problem of stability, Q3, is considered in the next section.

Exercises

***Exercise* 2.1.1.** Let $\boldsymbol{a}$ be an $m \times n$ matrix. Show that if $\ker \boldsymbol{a} = \ker \boldsymbol{a}^t = \boldsymbol{0}$, then $n = m$. Is the converse true?

***Exercise* 2.1.2.** Let $\boldsymbol{a}$ be an $m \times n$ matrix. Prove that

$$n = \dim \operatorname{Im} \boldsymbol{a} + \dim \ker \boldsymbol{a}.$$

Use the results in this section to conclude that

$$\dim \operatorname{Im} \boldsymbol{a} = \dim \operatorname{Im} \boldsymbol{a}^t.$$

In linear algebra texts, this is the statement that the "row rank" of a matrix equals its "column rank." Conclude that a square matrix is invertible if and only if its null space equals the zero vector.

***Exercise* 2.1.3.** Let $\boldsymbol{a} : \mathbb{R}^n \to \mathbb{R}^n$ be an invertible linear transformation. Show that $\boldsymbol{a}^{-1}$ is also linear.

***Exercise* 2.1.4.** Suppose that the state of a system is described by the vector $\boldsymbol{x}$. The measurements are modeled as inner products $\{\langle \boldsymbol{a}_i, \boldsymbol{x} \rangle : i - 1, \dots, m\}$. However, the measurements are noisy and each is repeated m_i times leading to measured values $\{y_i^1, \dots, y_j^{m_i}\}$. Define an error function by

$$e(\boldsymbol{x}) = \sum_{i=1}^{m} \sum_{k=1}^{m_i} (\langle \boldsymbol{a}_i, \boldsymbol{x} \rangle - y_i^k)^2.$$

Show that $e(\boldsymbol{x})$ is minimized by a vector $\boldsymbol{x}$ that satisfies the averaged equations

$$\langle \boldsymbol{a}_i, \boldsymbol{x} \rangle = \frac{1}{m_i} \sum_{k=1}^{m_i} y_i^k.$$

2.1.2 Stability of Solutions

Suppose that $\boldsymbol{a}$ is an $n \times n$ invertible matrix that models a measurement process. If $\boldsymbol{x}_1$ and $\boldsymbol{x}_2$ are two states of our system, then, because the model is linear, the difference in the measurements can easily be computed:

$$\boldsymbol{y}_1 - \boldsymbol{y}_2 = \boldsymbol{a}\boldsymbol{x}_1 - \boldsymbol{a}\boldsymbol{x}_2 = \boldsymbol{a}(\boldsymbol{x}_1 - \boldsymbol{x}_2).$$

When studying stability it is useful to have a measure for the "size" of a linear transformation.

***Definition* 2.1.1.** Let $\boldsymbol{a}$ be an $m \times n$ matrix and let $\|\cdot\|$ denote the Euclidean norm. The Euclidean *operator norm* of $\boldsymbol{a}$ is defined to be

$$|||\boldsymbol{a}||| = \max\{\|\boldsymbol{a}\boldsymbol{x}\| \ : \ \boldsymbol{x} \in \mathbb{R}^n \text{ with } \|\boldsymbol{x}\| = 1\}.$$

This method for measuring the size of linear transformations has many useful properties. Let $\boldsymbol{a}_1$ and $\boldsymbol{a}_2$ by $m \times n$-matrices and c a real number; then

$$\begin{aligned} |||c\boldsymbol{a}_1||| &= |c|\,|||\boldsymbol{a}_1|||, \\ |||\boldsymbol{a}_1 + \boldsymbol{a}_2||| &\leq |||\boldsymbol{a}_1||| + |||\boldsymbol{a}_2|||, \\ \|\boldsymbol{a}_1\boldsymbol{x}\| &\leq |||\boldsymbol{a}_1|||\,\|\boldsymbol{x}\| \text{ for all } \boldsymbol{x} \in \mathbb{R}^n. \end{aligned} \tag{2.6}$$

Using the linearity of $\boldsymbol{a}$ and the last line in (2.6), we obtain the estimate

$$\|\boldsymbol{y}_1 - \boldsymbol{y}_2\| \leq |||\boldsymbol{a}|||\,\|\boldsymbol{x}_1 - \boldsymbol{x}_2\|.$$

Therefore, nearby states lead to nearby measurements. The reverse may not be true. There may exist states $\boldsymbol{x}_1$ and $\boldsymbol{x}_2$ that are not nearby, in the sense that $\|\boldsymbol{x}_1 - \boldsymbol{x}_2\|$ is large but $\|\boldsymbol{a}(\boldsymbol{x}_1 - \boldsymbol{x}_2)\|$ is small. Physically, the measurements performed are not sufficiently independent to distinguish such pairs of states. In numerical analysis this is known as an *ill conditioned* problem. Briefly, a small error in the measurement process can be magnified by applying $\boldsymbol{a}^{-1}$ to the measurement vector. For an ill conditioned problem even a good algorithm for solving linear equations can produce meaningless results.

***Example* 2.1.3.** Consider the 2×2 system with

$$\boldsymbol{a} = \begin{pmatrix} 1 & 0 \\ 1 & 10^{-5} \end{pmatrix}.$$

The measurement matrix is invertible and

$$\boldsymbol{a}^{-1} = \begin{pmatrix} 1 & 0 \\ -10^5 & 10^5 \end{pmatrix}.$$

If the actual data are $\boldsymbol{y} = (1, 1)$ but we make an error in measurement and measure, $\boldsymbol{y}_m = (1, 1 + \epsilon)$, then the relative error in the predicted state is

$$\frac{\|\boldsymbol{a}^{-1}\boldsymbol{y}_m - \boldsymbol{a}^{-1}\boldsymbol{y}\|}{\|\boldsymbol{a}^{-1}\boldsymbol{y}\|} = \epsilon 10^5.$$

Even though the measurements uniquely determine the state of the system, a small error in measurement is vastly amplified.

As above $\boldsymbol{a}$ is an $n \times n$ matrix that models a measurement process. Let $\delta\boldsymbol{a}$ be an $n \times n$ matrix representing an aggregation of the errors in this model. The uncertainty in the model can be captured by using $\boldsymbol{a} + \delta\boldsymbol{a}$ for the measurement matrix. The measurements may also be subject to error and should be considered to have the form $\boldsymbol{y} + \delta\boldsymbol{y}$. A more realistic problem is therefore to solve the system(s) of equations

$$(\boldsymbol{a} + \delta\boldsymbol{a})\boldsymbol{x} = \boldsymbol{y} + \delta\boldsymbol{y}. \tag{2.7}$$

But what does this mean?

We consider only the simplest case where $\boldsymbol{a}$ is an $n \times n$, invertible matrix. Suppose that we can bound the uncertainty in both the model and the measurements in the sense that we have constants $\epsilon > 0$ and $\eta > 0$ such that

$$\|\delta\boldsymbol{y}\| < \epsilon \text{ and } |||\delta\boldsymbol{a}||| < \eta.$$

In the absence of more detailed information about the systematic errors, "the solution" to (2.7) could be defined as the set of vectors

$$\{\boldsymbol{x} \, : \, (\boldsymbol{a} + \delta\boldsymbol{a})\boldsymbol{x} = \boldsymbol{y} + \delta\boldsymbol{y} \text{ for some choice of } \delta\boldsymbol{a}, \delta\boldsymbol{y} \text{ with } \|\delta\boldsymbol{y}\| < \epsilon, |||\delta\boldsymbol{a}||| < \eta\}.$$

This is a little cumbersome. In practice, we find a vector that satisfies

$$\boldsymbol{a}\boldsymbol{x} = \boldsymbol{y}$$

and a bound for the error made in asserting that the actual state of the system is $\boldsymbol{x}$.

To proceed with this analysis, we assume that all the possible model matrices, $\boldsymbol{a} + \delta\boldsymbol{a}$, are invertible. If $|||\delta\boldsymbol{a}|||$ is sufficiently small, then this condition is satisfied. As $\boldsymbol{a}$ is invertible, the number

$$\mu = \min_{x \neq 0} \frac{\|\boldsymbol{a}\boldsymbol{x}\|}{\|\boldsymbol{x}\|} \tag{2.8}$$

is a positive. If $|||\delta\boldsymbol{a}||| < \mu$, then $\boldsymbol{a} + \delta\boldsymbol{a}$ is also invertible. The proof of this statement is given in Exercise 2.1.10. In the remainder of this discussion we assume that η, the bound on the uncertainty in the model, is smaller than μ.

An estimate on the error in $\boldsymbol{x}$ is found in two steps. First, fix the model and consider only errors in measurement. Suppose that $\boldsymbol{a}\boldsymbol{x} = \boldsymbol{y}$ and $\boldsymbol{a}(\boldsymbol{x} + \delta\boldsymbol{x}) = \boldsymbol{y} + \delta\boldsymbol{y}$. Taking the difference of these two equations gives

$$\boldsymbol{a}\delta\boldsymbol{x} = \delta\boldsymbol{y}$$

and therefore $\delta\boldsymbol{x} = \boldsymbol{a}^{-1}\delta\boldsymbol{y}$. Using (2.6), we see that

$$\|\delta\boldsymbol{x}\| \leq |||\boldsymbol{a}^{-1}||| \, \|\delta\boldsymbol{y}\|.$$

This is a bound on the absolute error; it is more meaningful to bound the relative error $\|\delta\boldsymbol{x}\|/\|\boldsymbol{x}\|$. To that end observe that

$$\|\boldsymbol{y}\| \leq |||\boldsymbol{a}||| \, \|\boldsymbol{x}\|$$

and therefore

$$\frac{\|\delta x\|}{\|x\|} \leq |||a||| \, |||a^{-1}||| \frac{\|\delta y\|}{\|y\|}. \tag{2.9}$$

This is a useful inequality: It estimates the relative uncertainty in the state in terms of the relative uncertainty in the measurements. The coefficient

$$c_a = |||a||| \, |||a^{-1}||| \tag{2.10}$$

is called the *condition number* of the matrix a. It is useful measure of the stability of a model of this type.

To complete our analysis we need to incorporate errors in the model. Suppose that $x + \delta x$ solves

$$(a + \delta a)(x + \delta x) = y + \delta y.$$

Subtracting this from $ax = y$ gives

$$(a + \delta a)\delta x = \delta y - \delta a x.$$

Proceeding as before, we see that

$$\frac{\|\delta x\|}{\|x\|} \leq |||(a + \delta a)^{-1}||| \, |||a||| \frac{\|\delta y\|}{\|y\|} + |||(a + \delta a)^{-1}||| \, |||\delta a|||. \tag{2.11}$$

If δa is very small (relative to μ), then

$$\begin{aligned} (a + \delta a)^{-1} &= a^{-1}(\mathrm{Id} + \delta a a^{-1})^{-1} \\ &= a^{-1} - a^{-1}\delta a a^{-1} + O([\delta a]^2). \end{aligned} \tag{2.12}$$

Here $O([\delta a]^2)$ is a linear transformation whose operator norm is bounded by a constant times $|||\delta a|||^2$. The second estimate in (2.6) implies that

$$|||(a + \delta a)^{-1}||| \lessapprox |||a^{-1}||| + |||a^{-1}\delta a a^{-1}|||.$$

Ignoring quadratic error terms in (2.12), this gives the estimate

$$\frac{\|\delta x\|}{\|x\|} \lessapprox c_a \left[\frac{\|\delta y\|}{\|y\|} + \frac{|||\delta a|||}{|||a|||} \right]. \tag{2.13}$$

Once again, the condition number of a relates the relative error in the predicted state to the relative errors in the model and measurements.

This analysis considers a special case, but it indicates how gross features of the model constrain the accuracy of its predictions. We have discussed neither the effects of using a particular algorithm to solve the system of equations, nor round-off error. A similar analysis applies to study these problems. Our discussion of this material is adapted from [127]. In image reconstruction the practical problems of solving systems of linear equations are considerable. It is not uncommon to have 10,000 equations in 10,000-unknowns. These huge systems arise as finite-dimensional approximations to linear equations for functions of continuous variables. The next section contains a short discussion of linear algebra in infinite-dimensional spaces, a theme occupying a large part of this book.

Exercises

***Exercise* 2.1.5.** Show that μ, defined in (2.8), equals $|||\boldsymbol{a}^{-1}|||$.

***Exercise* 2.1.6.** Show that the condition number is given by the following ratio:

$$c_a = \frac{\max_{x \neq 0} \frac{\|\boldsymbol{a}x\|}{\|x\|}}{\min_{x \neq 0} \frac{\|\boldsymbol{a}x\|}{\|x\|}}. \tag{2.14}$$

This shows that the condition number of any matrix is at least one. Which matrices have condition number equal to one?

***Exercise* 2.1.7.** Suppose that $\boldsymbol{a}$ is a positive definite, symmetric matrix with eigenvalues $0 < \lambda_1 \leq \cdots \leq \lambda_n$. Show that

$$c_a = \frac{\lambda_n}{\lambda_1}.$$

If $\boldsymbol{a}$ is an arbitrary matrix with $\ker(\boldsymbol{a}) = \mathbf{0}$, show that

$$c_a = \sqrt{c_{a^*a}}.$$

***Exercise* 2.1.8.** Show that $|||\boldsymbol{a}||| = 0$ if and only if $\boldsymbol{a}$ is the zero matrix. It therefore follows from (2.6) that $|||\cdot|||$ defines a norm on the space of $n \times n$-matrices. Let $< \boldsymbol{a}_k >$ be a sequence of matrices with entries $(a_k)_{ij}$. Show that

$$\lim_{k\to\infty} |||\boldsymbol{a}_k - \boldsymbol{b}||| = 0$$

if and only if

$$\lim_{k\to\infty} (a_k)_{ij} = b_{ij} \text{ for all } 1 \leq i, j \leq n.$$

***Exercise* 2.1.9.** Let $\boldsymbol{a}_1$ and $\boldsymbol{a}_2$ be $n \times n$ matrices. Show that

$$|||\boldsymbol{a}_1\boldsymbol{a}_2||| \leq |||\boldsymbol{a}_1|||\,|||\boldsymbol{a}_2|||. \tag{2.15}$$

***Exercise* 2.1.10.** Use (2.15) to show that if $\boldsymbol{a}$ is an $n \times n$ matrix with $|||\boldsymbol{a}||| < 1$, then the sequence of matrices

$$\boldsymbol{a}_k = \mathrm{Id} + \sum_{j=1}^{k} (-1)^j \boldsymbol{a}^j$$

converges to a matrix $\boldsymbol{b}$ in the sense that

$$\lim_{k\to\infty} |||\boldsymbol{a}_k - \boldsymbol{b}||| = 0.$$

Conclude that $\mathrm{Id} + \boldsymbol{a}$ is invertible by showing that $\boldsymbol{b} = (\mathrm{Id} + \boldsymbol{a})^{-1}$. Justify equation (2.12).

2.2 Infinite-dimensional Linear Algebra

See: A.4.

In the foregoing sections we considered the spaces $\mathbb{R}^n$ and $\mathbb{C}^n$. These are examples of finite-dimensional vector spaces. Recall that a set V is vector space if it has two operations defined on it, addition and scalar multiplication that satisfy several conditions. Let $\boldsymbol{v}_1$ and $\boldsymbol{v}_2$ denote two points in V. Then their sum is denoted by $\boldsymbol{v}_1 + \boldsymbol{v}_2$; if a is a scalar, then the product of a and $\boldsymbol{v} \in V$ is denoted $a\boldsymbol{v}$. In our applications the scalars are either real or complex numbers. The basic properties of complex numbers are reviewed in the next section. In addition, there must be a *zero vector* $\mathbf{0} \in V$ such that $\boldsymbol{v} + \mathbf{0} = \boldsymbol{v}$ for all $\boldsymbol{v} \in V$ and

$$a(\boldsymbol{v}_1 + \boldsymbol{v}_2) = a\boldsymbol{v}_1 + a\boldsymbol{v}_2, \quad (a_1 + a_2)\boldsymbol{v} = a_1\boldsymbol{v} + a_2\boldsymbol{v}, \quad 0\boldsymbol{v} = \mathbf{0}. \tag{2.16}$$

The vector space V is finite dimensional if there is a finite set of elements $\{\boldsymbol{v}_1, \ldots, \boldsymbol{v}_n\} \subset V$ so that every element is a *linear combination* of these elements. In other words, for each vector $\boldsymbol{v}$ there are scalars $\{c_1, \ldots, c_n\}$ so that

$$\boldsymbol{v} = \sum_{i=1}^{n} c_i \boldsymbol{v}_i.$$

Vector spaces that are not finite dimensional are infinite-dimensional. Many familiar spaces are infinite dimensional vector spaces.

***Example* 2.2.1.** Let $\mathcal{P}$ denote polynomials in the variable x with real coefficients. An element of $\mathcal{P}$ can be expressed as

$$p = \sum_{i=0}^{n} c_i x^i,$$

where the $\{c_i\}$ are real numbers. The usual rules for addition and scalar multiplication of polynomials make $\mathcal{P}$ into a vector space. Since n is arbitrary, the dimension of $\mathcal{P}$ is not finite.

***Example* 2.2.2.** Let $\mathcal{C}^0(\mathbb{R})$ be the set of continuous, real-valued functions defined on $\mathbb{R}$. Since sums and scalar multiples of continuous functions are continuous, this is also a vector space. As it contains $\mathcal{P}$ as a subspace, it is not finite dimensional.

***Example* 2.2.3.** Let $L^1([0, 1])$ denote the set of real-valued functions defined on $[0, 1]$ whose absolute value is integrable. If $f, g \in L^1([0, 1])$, then we define $f + g$ by $(f + g)(x) = f(x) + g(x)$. If $a \in \mathbb{R}$, then we define af by $(af)(x) = af(x)$. Clearly, if $f \in L^1([0, 1])$, then so is af. The zero function is the zero vector. If $f, g \in L^1([0, 1])$, then

$$\int_0^1 |f(x) + g(x)|\, dx \leq \int_0^1 |f(x)|\, dx + \int_0^1 |g(x)|\, dx.$$

This shows that $L^1([0, 1])$ is a vector space.

As with finite-dimensional vector spaces, it is important to be able to measure lengths and distances in infinite-dimensional vector spaces. This is usually done using a norm. A norm on an infinite-dimensional vector space, V, is a function $N : V \to [0, \infty)$ satisfying the properties enumerated in Remark 1.1.1, with the understanding that $\boldsymbol{x}$ and $\boldsymbol{y}$ are arbitrary elements of V.

Example* 2.2.4. Let $L^1(\mathbb{R})$ denote the set of absolutely integrable functions defined on $\mathbb{R}$. The L^1-norm is defined by

$$\|f\|_{L^1} = \int\limits_{-\infty}^{\infty} |f(x)|\, dx.$$

Example* 2.2.5. Let X be a set and $F_b(X)$ denote the vector space of bounded functions defined on X. The *sup-norm* is defined by

$$\|f\|_\infty = \sup\{|f(\boldsymbol{x})| \;:\; \boldsymbol{x} \in X\}.$$

The properties above are easily established for the sup norm.

***Example* 2.2.6.** Let $\mathscr{C}_b^0(\mathbb{R}^n)$ denote the set of bounded, continuous functions on $\mathbb{R}^n$. The sup norm also defines a norm on $\mathscr{C}_b^0(\mathbb{R}^n)$.

***Example* 2.2.7.** If $\mathscr{C}^0([0, 1])$ denote the set of continuous functions on $[0, 1]$, the function

$$\|f\|_1 = \int\limits_0^1 |f(x)|\, dx$$

defines a norm on this vector space.

The exact state of a system in medical imaging is usually described by a point in an infinite-dimensional vector space. Let f describe the state of the system. Suppose that f is real-valued function, defined on the real line and integrable over any finite interval. A linear measurement of the state is often represented as an integral:

$$\mathcal{M}(f)(x) = \int\limits_{-\infty}^{\infty} m(x, y) f(y)\, dy.$$

Here $m(x, y)$ is a function on $\mathbb{R} \times \mathbb{R}$ that provides a model for the measurement process. It can be thought of as an infinite matrix with indices x and y. A linear transformation of an infinite-dimensional vector space is called a *linear operator*. A linear transformation that can be expressed as an integral is called an *integral operator*.

Suppose that the function g is the output of the measurement process; to reconstruct f means solving the linear equation

$$\mathcal{M}f = g.$$

This is a concise way to write a system of infinitely many equations in infinitely many unknowns. Theorems 2.1.1 and 2.1.2 contain the complete theory for the existence and uniqueness of solutions to linear equations in finitely many variables. In these theorems no mention is made of norms; indeed, they are entirely algebraic in character. No such theory exists for equations in infinitely many variables. It is usually a difficult problem to describe either the domain or range of a linear transformation of an infinite-dimensional space. We consider a few illustrative examples.

***Example* 2.2.8.** A simple example of a linear operator is the indefinite integral

$$\mathcal{I}(f)(x) = \int_0^x f(y)\,dy.$$

If we use the vector space of continuous functions on $\mathbb{R}$ as the domain of $\mathcal{I}$, then every function in the range is continuously differentiable. Moreover the null space of $\mathcal{I}$ is the zero function. Observe that the domain and range of $\mathcal{I}$ are fundamentally different spaces. Because $\mathcal{I}(f)(0) = 0$, not every continuously differentiable function is in the range of $\mathcal{I}$. The derivative is a left inverse to $\mathcal{I}$. The fundamental theorem of calculus states that if f is continuous, then

$$\frac{d}{dx} \circ \mathcal{I}(f)(x) = f(x).$$

On the other hand, it is not quite a right inverse because

$$\mathcal{I}(\frac{df}{dx})(x) = f(x) - f(0).$$

The domain of $\mathcal{I}$ can be enlarged to include all *locally integrable* functions. These are functions such that

$$\int_0^x |f(y)|\,dy < \infty$$

for every $x \in \mathbb{R}$; this is again a vector space. Enlarging the domain also enlarges the range. For example, the function $|x|$ lies in the enlarged range of $\mathcal{I}$,

$$|x| = \int_0^x \operatorname{sgn}(y)\,dy,$$

where $\operatorname{sgn}(y) = 1$ if $y \geq 0$ and -1 if $y < 0$. Even though $|x|$ is not differentiable at $x = 0$, it is still the indefinite integral of a locally integrable function. The formula

$$\frac{d|x|}{dx} = \operatorname{sgn}(x)$$

does not make sense at $x = 0$.

***Example* 2.2.9.** Changing the lower limit of integration to $-\infty$ leads to a different sort of linear transformation. Initially $\mathcal{I}_\infty$ is defined for continuous functions f, vanishing for sufficiently negative x by

$$\mathcal{I}_\infty(f)(x) = \int_{-\infty}^{x} f(y)\,dy.$$

Once again the null space of $\mathcal{I}_\infty$ consists of the zero function alone. The domain can be enlarged to include locally integrable functions such that

$$\lim_{R\to\infty} \int_{-R}^{0} |f(y)|\,dy < \infty. \tag{2.17}$$

If f is continuous, then we can apply the fundamental theorem of calculus to obtain

$$\frac{d}{dx} \circ \mathcal{I}_\infty(f) = f.$$

If a function g belongs to the range of $\mathcal{I}_\infty$, then

$$\lim_{x\to-\infty} g(x) = 0. \tag{2.18}$$

There are once differentiable functions satisfying this condition that do not belong to the range of $\mathcal{I}_\infty$. For example,

$$f(x) = \frac{x\cos x - \sin x}{x^2} = \frac{d}{dx}\frac{\sin x}{x}$$

satisfies (2.18) but f does not satisfy (2.17). With the domain defined by (2.17), the precise range of $\mathcal{I}_\infty$ is rather difficult to describe.

This example illustrates how a integral operator may have a simple definition on a certain domain, which by a limiting process can be extended to a larger domain. The domain of such an operator is often characterized by a size condition like (2.17).

***Example* 2.2.10.** A real physical measurement is always some sort of an average. If the state of the system is described by a function f of a single variable x, then the average of f over an interval of length 2δ is

$$\mathcal{M}_\delta(f)(x) = \frac{1}{2\delta}\int_{x-\delta}^{x+\delta} f(y)\,dy.$$

A natural domain for $\mathcal{M}_\delta$ is all locally integrable functions. To what extent is f determined by $\mathcal{M}_\delta(f)$? Suppose that f and g are two states; then, because the integral is linear,

$$\mathcal{M}_\delta(f) - \mathcal{M}_\delta(g) = \mathcal{M}_\delta(f - g).$$

The extent to which $\mathcal{M}_\delta(f)$ determines f is characterized by the null space of $\mathcal{M}_\delta$,

$$\mathcal{N}_\delta = \{f \,:\, \mathcal{M}_\delta(f) = 0\}.$$

Proceeding formally, we can differentiate $\mathcal{M}_\delta(f)$ to obtain

$$\frac{d\mathcal{M}_\delta(f)}{dx} = f(x+\delta) - f(x-\delta) \tag{2.19}$$

If $f \in \mathcal{N}_\delta$, then $\mathcal{M}_\delta(f)$ is surely constant and therefore

$$f \in \mathcal{N}_\delta \Rightarrow f(x+\delta) - f(x-\delta) = 0, \text{ for all } x \in \mathbb{R}.$$

In other words, f is periodic with periodic 2δ. The functions

$$\{\cos\left(\frac{\pi jx}{\delta}\right), \sin\left(\frac{\pi jx}{\delta}\right) \,:\, j \in \{1, 2, 3, \dots\}\} \tag{2.20}$$

are linearly independent and belong to $\mathcal{N}_\delta$. This shows that the null space of $\mathcal{M}_\delta$ is infinite dimensional.

In applications we often have additional information about the state of the system. For example, we might know that

$$\lim_{|x|\to\infty} f(x) = 0. \tag{2.21}$$

A periodic function that tends to zero at infinity must be identically zero, so among such functions the measurements $\mathcal{M}_\delta(f)$ would appear to determine f completely. To *prove* this statement we need to know somewhat more about f than (2.21). With a more quantitative condition like

$$\|f\|_p = \left[\int_{-\infty}^{\infty} |f(y)|^p \, dy\right]^{\frac{1}{p}} < \infty, \tag{2.22}$$

for a p between 1 and 2, it is possible to show that $\mathcal{M}_\delta(f) = 0$ implies that $f = 0$ and therefore the measurement $\mathcal{M}_\delta(f)$ uniquely determines f. As we shall see, f cannot be *stably* reconstructed from $\mathcal{M}_\delta(f)$. A small error in measurement can lead to a very large error in the reconstructed state.

For each p between 1 and infinity, the function $f \mapsto \|f\|_p$ defines a norm on the space of functions that satisfy (2.22). This is called the L^p-norm. The first step in analyzing linear transformations of infinite-dimensional spaces is the introduction of norms on the domain and range. This was not necessary in finite dimensions but is absolutely essential in the infinite-dimensional case. Using the the L^1-norm on the domain, we can control the sup norm in the output of the measurement process modeled by $f \mapsto \mathcal{M}_\delta(f)$.

***Example* 2.2.11.** Let f be a function we would like to determine and model the measurement as $\mathcal{M}_\delta(f)$. To control the dependence in the sup norm of $\mathcal{M}_\delta f$, on f it suffices to control the L^1-norm of f,

$$|\mathcal{M}_\delta f(t)| \le \frac{1}{2\delta} \int_{t-\delta}^{t+\delta} |f(y)| \, dy \le \frac{1}{2\delta} \int_{-\infty}^{\infty} |f(y)| \, dy.$$

Because the measurement is linear, we see that if f_1 and f_2 are two functions, then

$$|\mathcal{M}_\delta f_1(t) - \mathcal{M}_\delta f_2(t)| \leq \frac{1}{2\delta} \int_{-\infty}^{\infty} |f_1(y) - f_2(y)| \, dy.$$

The length of the interval over which we average is an indication of the *resolution* in the measurement. The smaller δ is, the more sensitive the measurement is to the size of the difference, $f_1 - f_2$, measured in the L^1-norm.

In medical image reconstruction there is a short list of important linear transformations: the Fourier transform, the Radon transform, the Hilbert transform, and the Abel transform. Our study of infinite dimensional linear algebra is directed toward understanding these linear operators.

Exercises

***Exercise* 2.2.1.** Show that the sup norm defined in Example 2.2.5 satisfies the properties of a norm.

***Exercise* 2.2.2.** Let $\mathscr{C}^0([0, 1])$ denote the continuous functions on $[0, 1]$. Show that $f \mapsto \|f\|_\infty$ defines a norm on this space. Suppose that $< f_k >$ is a sequence of functions in $\mathscr{C}^0([0, 1]$ and there exists a function f such that $\lim_{k\to\infty} \|f - f_k\|_\infty = 0$. Show that $f \in \mathscr{C}^0([0, 1])$ as well.

***Exercise* 2.2.3.** Is the conclusion of the previous exercise still correct if we use $f \mapsto \|f\|_1$, defined in Example 2.2.7, to define the norm? Give a proof or a counterexample.

***Exercise* 2.2.4.** Prove that the null space of $\mathcal{I}$ acting on $\mathscr{C}^0(\mathbb{R})$ is the zero function.

***Exercise* 2.2.5.** Show that the functions listed in (2.20) belong to $\ker \mathcal{M}_\delta$.

***Exercise* 2.2.6.** Show that the functions listed in (2.20) are linearly independent.

2.3 Complex Numbers and Vector Spaces*

Thus far we have only considered real numbers and vector spaces over the real numbers. We quickly introduce the complex numbers and the basic features of $\mathbb{C}^n$. Complex numbers are essential to define the Fourier transform.

2.3.1 Complex Numbers

There is no real number that solves the algebraic equation

$$x^2 = -1.$$

To remedy this we simply introduce a new symbol i, which is *defined* to be the square root of -1, that is,

$$i^2 = -1.$$

This is called the *imaginary unit*.

***Definition* 2.3.1.** The *complex numbers* are defined to be the collection of symbols

$$\{x + iy \ : \ x, y \in \mathbb{R}\}$$

with the addition operation defined by

$$(x_1 + iy_1) + (x_2 + iy_2) \stackrel{d}{=} (x_1 + x_2) + i(y_1 + y_2)$$

and multiplication defined by

$$(x_1 + iy_1)(x_2 + iy_2) \stackrel{d}{=} (x_1x_2 - y_1y_2) + i(x_1y_2 + x_2y_1).$$

The set of complex numbers is denoted by $\mathbb{C}$.

Note that addition and multiplication are defined in terms of the addition and multiplication operations on real numbers. The complex number $0 = 0+i0$ satisfies $0+(x+iy) = x+iy$, and the complex number $1 = 1 + i0$ satisfies $(1 + i0)(x + iy) = x + iy$. We often use the letters z or w to denote complex numbers. The sum is denoted by $z + w$ and the product by zw.

Proposition 2.3.1. *A nonzero complex number has a multiplicative inverse. The addition and multiplication defined for complex numbers satisfy the commutative, associative, and distributive laws. That is, if* z_1, z_2, z_3 *are three complex numbers, then*

$$\begin{aligned} z_1 + z_2 = z_2 + z_1, \quad (z_1 + z_2) + z_3 &= z_1 + (z_2 + z_3), \\ z_1z_2 = z_2z_1, \quad (z_1z_2)z_3 &= z_1(z_2z_3), \\ z_1(z_2 + z_3) &= z_1z_2 + z_1z_3. \end{aligned} \tag{2.23}$$

***Definition* 2.3.2.** If $z = x + iy$ then the *real part* of z is x and the *imaginary part* of z is y. These functions are written symbolically as

$$\operatorname{Re} z = x, \quad \operatorname{Im} z = y.$$

The set underlying the complex numbers is $\mathbb{R}^2$,

$$x + iy \leftrightarrow (x, y);$$

addition of complex numbers is the same as vector addition. Often the set of complex numbers is called the *complex plane*. The Euclidean norm is used to define the *absolute value* or norm of a complex number,

$$|x + iy| = \sqrt{x^2 + y^2}.$$

There is a useful operation defined on complex numbers called *complex conjugation*. If $z = x + iy$, then its complex conjugate is $\bar{z} = x - iy$. In engineering texts the complex conjugate of z is often denoted z^*.

Exercise 2.3.2 shows that the multiplication defined previously has all of the expected properties of a product. The real numbers sit inside the complex numbers as the set $\{z \in \mathbb{C} : \operatorname{Im} z = 0\}$. It is easy to see that the two definitions of addition and multiplication agree on this subset. Complex conjugation is simply reflection across the real axis.

It is useful to understand the multiplication of complex numbers geometrically. For this purpose we represent points in the complex plane using polar coordinates. The radial coordinate is simply $r = |z|$. The ratio $\omega = z|z|^{-1}$ is a number of unit length and therefore has a representation as $\omega = \cos\theta + i\sin\theta$. The polar representation of z is

$$z = r(\cos\theta + i\sin\theta). \tag{2.24}$$

The angle θ is called the *argument* of z, which is denoted by $\arg(z)$. It is only determined up to multiples of 2π. If z and w are two complex numbers, then they can be expressed in polar form as

$$z = r(\cos\theta + i\sin\theta), \qquad w = R(\cos\phi + i\sin\phi).$$

Computing their product we find that

$$\begin{aligned} zw =& rR([\cos\theta\cos\phi - \sin\theta\sin\phi] + i(\cos\theta\sin\phi + \sin\theta\cos\phi]) \\ =& rR(\cos(\theta+\phi) + i\sin(\theta+\phi)). \end{aligned} \tag{2.25}$$

This is shown in Figure 2.1. In the second line we used the sum formulæ for the sine and cosine. This shows us that complex multiplication of w by z can be understood geometrically as scaling the length of w by $|z|$ and rotating it in the plane through an angle $\arg(z)$.

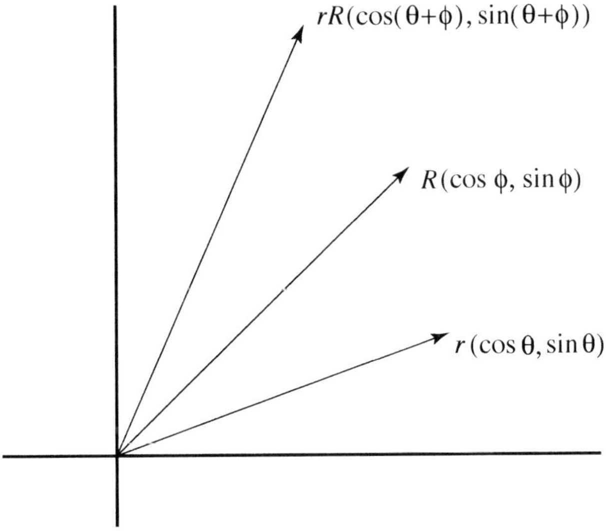

Figure 2.1. Multiplication of complex numbers.

Using the notion of distance on $\mathbb{C}$ defined previously we can define the concept of convergence for sequences of complex numbers.

***Definition* 2.3.3.** Let $< z_n >$ be a sequence of complex numbers. The sequence converges to z_0 if

$$\lim_{n\to\infty} |z_n - z_0| = 0.$$

In this case z_0 is called the limit of $< z_n >$ and we write

$$\lim_{n\to\infty} z_n = z_0.$$

Exercises

***Exercise* 2.3.1.** Prove Proposition 2.3.1.

***Exercise* 2.3.2.** Let z and w be complex numbers. Show that

$$\begin{aligned} \overline{(z+w)} = \bar{z} + \bar{w}, \quad \overline{zw} = \bar{z}\bar{w}, \\ z\bar{z} = |z|^2. \end{aligned} \tag{2.26}$$

Using the second condition, show that if $z \neq 0$, then the complex number defined by

$$z^{-1} = \frac{\bar{z}}{|z|^2}$$

satisfies $zz^{-1} = 1 = z^{-1}z$.

***Exercise* 2.3.3.** Let $< z_n >$ be a sequence of complex numbers. Show that z_n converges to z_0 if and only if $\operatorname{Re} z_n$ converges to $\operatorname{Re} z_0$ and $\operatorname{Im} z_n$ converges to $\operatorname{Im} z_0$.

2.3.2 Complex Vector Spaces

The collection of ordered n-tuples of complex numbers is denoted by $\mathbb{C}^n$. It is a vector space with

$$(z_1, \ldots, z_n) + (w_1, \ldots, w_n) = (z_1 + w_1, \ldots, z_n + w_n).$$

Since the entries of the vectors are now complex numbers, we can define scalar multiplication for $w \in \mathbb{C}$ by

$$w \cdot (z_1, \ldots, z_n) = (wz_1, \ldots, wz_n).$$

For this reason $\mathbb{C}^n$ is called a *complex vector space*.The boldface letters $\boldsymbol{w}$ and $\boldsymbol{z}$ are often used to denote vectors in $\mathbb{C}^n$.

***Definition* 2.3.4.** A subset S of $\mathbb{C}^n$ is a *complex* subspace provided that

1. the zero vector belongs to S,
2. if $\boldsymbol{z}, \boldsymbol{w} \in S$, then $\boldsymbol{z} + \boldsymbol{w} \in S$,
3. if $w \in \mathbb{C}$ and $\boldsymbol{z} \in S$, then $w\boldsymbol{z} \in S$.

***Example* 2.3.1.** The subset $\{(z, 0) : z \in \mathbb{C}\}$ is a complex subspace of $\mathbb{C}^2$. The subset $\{(x, y) : x, y \in \mathbb{R}\}$ is *not* a complex subspace of $\mathbb{C}^2$.

The Euclidean norm on $\mathbb{C}^n$ is defined by

$$\|(z_1, \ldots, z_n)\|_2 = \sqrt{\sum_{j=1}^{n} |z_j|^2}.$$

An inner product is defined by setting

$$\langle (z_1, \ldots, z_n), (w_1, \ldots, w_n) \rangle = \sum_{j=1}^{n} z_j \bar{w}_j.$$

The Euclidean norm can again be expressed in terms of the inner product:

$$\|\boldsymbol{z}\|_2^2 = \langle \boldsymbol{z}, \boldsymbol{z} \rangle.$$

The inner product has the following properties:

Proposition 2.3.2. *If $w \in \mathbb{C}$ and $\boldsymbol{z} \in \mathbb{C}^n$, then $\|w\boldsymbol{z}\| = |w|\|\boldsymbol{z}\|$. Moreover, if $\boldsymbol{z}, \boldsymbol{w} \in \mathbb{C}^n$, then*

$$\langle \boldsymbol{z}, \boldsymbol{w} \rangle = \overline{\langle \boldsymbol{w}, \boldsymbol{z} \rangle} \tag{2.27}$$

and if $w \in \mathbb{C}$, then

$$\langle w\boldsymbol{z}, \boldsymbol{w} \rangle = w \langle \boldsymbol{z}, \boldsymbol{w} \rangle \text{ and } \langle \boldsymbol{z}, w\boldsymbol{w} \rangle = \bar{w} \langle \boldsymbol{z}, \boldsymbol{w} \rangle. \tag{2.28}$$

It also satisfies the Cauchy-Schwarz inequality

$$|\langle \boldsymbol{z}, \boldsymbol{w} \rangle| \leq \|\boldsymbol{z}\|_2 \|\boldsymbol{w}\|_2. \tag{2.29}$$

The Euclidean inner product on $\mathbb{C}^n$ is not symmetric, but rather *Hermitian or conjugate symmetric*. It is often referred to as the *Hermitian inner product.*

Proof. We prove the Cauchy-Schwarz inequality; the proofs of the other parts of the proposition are left as exercises. We may assume that $s = \langle \boldsymbol{z}, \boldsymbol{w} \rangle$ is not zero, for otherwise the statement is trivial. The complex number $\rho = s|s|^{-1}$ has norm one and

$$\langle \boldsymbol{z}, \rho\boldsymbol{w} \rangle - |\langle \boldsymbol{z}, \boldsymbol{w} \rangle|. \tag{2.30}$$

Define a real-valued function of $t \in \mathbb{R}$ by setting

$$F(t) = \|\boldsymbol{z} + t\rho\boldsymbol{w}\|^2 = \|\boldsymbol{z}\|^2 + 2t|\langle \boldsymbol{z}, \boldsymbol{w} \rangle| + t^2 \|\boldsymbol{w}\|^2.$$

This is a nonnegative, quadratic function that assumes its minimum value at

$$t_0 = -\frac{|\langle \boldsymbol{z}, \boldsymbol{w} \rangle|}{\|\boldsymbol{w}\|^2}.$$

Substituting, we find that

$$F(t_0) = \frac{\|\boldsymbol{z}\|^2 \|\boldsymbol{w}\|^2 - |\langle \boldsymbol{z}, \boldsymbol{w} \rangle|}{\|\boldsymbol{w}\|^2}.$$

The Cauchy-Schwarz inequality follows from the fact that $F(t_0) \geq 0$. □

The theory of complex vector spaces is very similar to that of real vector spaces. The concepts of linear functions, transformations, bases, matrices, and matrix multiplication all carry over without change. We simply allow the scalars to be complex numbers. For example, a function f from $\mathbb{C}^n$ to $\mathbb{C}$ is linear if

$$\begin{aligned} f(z_1 + z_2) &= f(z_1) + f(z_2) \text{ for all } z_1, z_2 \in \mathbb{C}^n, \\ f(wz) &= wf(z) \text{ for all } w \in \mathbb{C} \text{ and } z \in \mathbb{C}^n. \end{aligned} \tag{2.31}$$

In this case every linear function has a representation in terms of the Hermitian inner product,

$$f(z) = \langle z, \boldsymbol{w}_f \rangle$$

for a unique vector $\boldsymbol{w}_f$. As is evident from (2.27) and (2.28), some small differences arise when dealing with inner products on complex vector spaces.

***Example* 2.3.2.** The space $\mathbb{C}^n$ can be regarded as either a real on complex vector space. If it is thought of as a complex vector space, then the vectors $\{\boldsymbol{e}_1, \ldots, \boldsymbol{e}_n\}$ are a basis. Thinking of it as a real vector space means that we only allow scalar multiplication by real numbers. In this case n vectors do *not* suffice to define a basis. Instead we could use

$$\{\boldsymbol{e}_1, \ldots, \boldsymbol{e}_n, i\boldsymbol{e}_1, \ldots, i\boldsymbol{e}_n\}.$$

As a real vector space, $\mathbb{C}^n$ has dimension $2n$.

Exercises

***Exercise* 2.3.4.** Prove Proposition 2.3.2.

***Exercise* 2.3.5.** Prove that every linear function f on $\mathbb{C}^n$ has a representation as $\langle z, \boldsymbol{w}_f \rangle$ for a unique $\boldsymbol{w}_f \in \mathbb{C}^n$. Explain how to find $\boldsymbol{w}_f$.

***Exercise* 2.3.6.** By restricting the scalar multiplication to real numbers, $\mathbb{C}^n$ can be regarded as a real vector space. Show that, as a real vector space, $\mathbb{C}^n$ is isomorphic to $\mathbb{R}^{2n}$.

2.4 Conclusion

Finite-dimensional linear algebra is the most often applied part of mathematics. As we have seen, there is a complete theory for the solvability of linear equations. The text by Peter D. Lax, [82] provides an excellent introduction to the theoretical aspects of this subject. There is also a huge literature on practical methods for solving linear equations. Trefethen and Bau, [127], is a very nice treatise on the numerical side of the subject. Infinite-dimensional linear algebra, which usually goes under the name of *functional analysis* or *operator theory*, is a vast subject. An introduction that emphasizes the theory of integral equations can be found in [131]. We return to problems associated with infinite-dimensional linear algebra in Chapter 4. In the next chapter we give a mathematical model for the measurements made in x-ray tomography. The system in question has infinitely many degrees of freedom and its state is specified by a nonnegative function defined in $\mathbb{R}^3$. The measurements are modeled as a linear transformation, called the Radon transform, applied to this function.

Chapter 3

A Basic Model for Tomography

We begin our study of medical imaging with a mathematical model of the measurement process used in x-ray tomography. The model begins with a quantitative description of the interaction of x-rays with matter called Beer's law. In this formulation the physical properties of an object are encoded in a function μ, called the *attenuation coefficient*. This function quantifies the tendency of an object to absorb or scatter x-rays. Using this model, idealized measurements are described as certain averages of μ. In mathematical terms, the averages of μ define a linear transformation, called the *Radon transform*. These measurements are akin to the determination of the shadow function of a convex region described in Section 1.2. In Chapter 6 we show that if the Radon transform of μ could be measured exactly, then μ could be completely determined from this data. In reality, the data collected are a limited part of the mathematically "necessary" data, and the measurements are subject to a variety of errors. In later chapters we refine the measurement model and reconstruction algorithm to reflect more realistic models of the physical properties of x-rays and the data that are actually collected.

This chapter begins with a model for the interaction of x-rays with matter. To explicate the model we then apply it to describe several simple physical situations. After a short excursion into the realm of x-ray physics, we give a formal definition of the Radon transform and consider some of its simpler properties. The last part of the chapter is more mathematical, covering the inversion of the Radon transform on radial functions, the Abel transform, and a discussion of Volterra integral equations.

3.1 Tomography

Literally, tomography means *slice imaging*. Today this term is applied to many methods used to reconstruct the internal structure of a solid object from external measurements. In x-ray tomography the literal meaning persists in that we reconstruct a three-dimensional object from its two-dimensional slices. Objects of interest in x-ray imaging are described by a real-valued function defined on $\mathbb{R}^3$, called the *attenuation coefficient*. The attenuation coefficient quantifies the tendency of an object to absorb or scatter x-rays of a given energy. This function varies from point-to-point within the object and is usually taken to vanish

Material	**Attenuation coefficient in Hounsfield units**
water	0
air	−1000
bone	1086
blood	53
fat	−61
brain white/gray	−4
breast tissue	9
muscle	41
soft tissue	51

Table 3.1. Attenuation coefficients of human tissues for 100-keV x-rays, adapted from [67].

outside it. The attenuation coefficient, like density, is nonnegative. It is useful for medical imaging because different anatomical structures have different attenuation coefficients. Bone has a much higher attenuation coefficient than soft tissue and different soft tissues have slightly different coefficients. For medical applications it is crucial that normal and cancerous tissues also have slightly different attenuation coefficients.

While in this book we work almost exclusively with the attenuation coefficient, as described previously, it is rarely used by radiologists. Instead the attenuation coefficient is compared to the attenuation coefficient water and quoted in terms of a dimensionless quantity called a *Hounsfield unit*. The normalized attenuation coefficient, in Hounsfield units is defined by

$$H_{\text{tissue}} = \frac{\mu_{\text{tissue}} - \mu_{\text{water}}}{\mu_{\text{water}}} \times 1000.$$

Unlike the actual attenuation coefficient, this relative measure assumes both positive and negative values. Table 3.1 lists typical normalized attenuation coefficients for different parts of the body.

The attenuation coefficients of air (−1000) and bone (1100) define the range of values present in a typical clinical situation. This means that the *dynamic range* of a clinical CT measurement is about 2000 Hounsfield units. From the table it is apparent that the variation in the attenuation coefficients of soft tissues is about 2% of this range. For x-ray CT to be clinically useful this means that the reconstruction of the attenuation coefficient must be accurate to about 10 Hounsfield units or less than a half a percent of its dynamic range.

Let μ be a function defined on $\mathbb{R}^3$. To define the slices of μ we need to fix a coordinate system $\boldsymbol{x} = (x_1, x_2, x_3)$. For each fixed value c of x_3, the x_3-slice of μ is the function of two variables, f_c, defined by

$$f_c(x_1, x_2) = \mu(x_1, x_2, c).$$

A knowledge of the collection of functions $\{f_c : c \in [a, b]\}$ is equivalent to a knowledge of μ for all points in the slab

$$\{(x_1, x_2, x_3) : -\infty < x_1 < \infty, \ -\infty < x_2 < \infty, \ a \leq x_3 \leq b\}.$$

While the choice of coordinates is arbitrary, having a fixed frame of reference is a crucial element of any tomographic method. By convention the slices are defined by fixing the last coordinate. In general, different coordinate systems lead to different collections of slices. In actual practice the x-ray CT machine fixes the frame of reference.

***Definition* 3.1.1.** Let D be a subset of $\mathbb{R}^n$. The *characteristic function* of D is defined by

$$\chi_D(\boldsymbol{x}) = \begin{cases} 1 & \text{if } \boldsymbol{x} \in D, \\ 0 & \text{if } \boldsymbol{x} \notin D. \end{cases}$$

***Remark* 3.1.1.** Evaluation of χ_D at $\boldsymbol{x}$ is a simple mathematical model for an experiment that tries to decide whether or not the point $\boldsymbol{x}$ lies in the set D.

***Example* 3.1.1.** Let D be a subset of $\mathbb{R}^3$; the function χ_D models an object with constant attenuation coefficient. In this case the object is determined by its intersection with the planes

$$H_c = \{(x_1, x_2, c) : x_1, x_2 \in \mathbb{R}\}.$$

For each c we let $D_c = D \cap H_c$. Figure 3.1 shows a two-dimensional slice of a three-dimensional object.

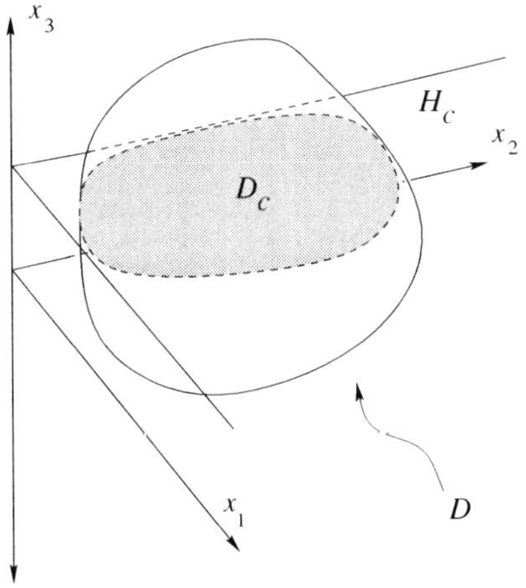

Figure 3.1. A 2D slice of a 3D object.

***Example* 3.1.2.** Suppose that the object is contained in the ball of radius 1 and its attenuation coefficient is

$$\mu(\boldsymbol{x}) = \begin{cases} 1 - \|\boldsymbol{x}\| & \text{if } \|\boldsymbol{x}\| \leq 1, \\ 0 & \text{if } \|\boldsymbol{x}\| > 1. \end{cases}$$

The slices of μ are the functions

$$f_c(x_1, x_2) = \begin{cases} 1 - \sqrt{x_1^2 + x_2^2 + c^2} & \text{if } \sqrt{x_1^2 + x_2^2 + c^2} \leq 1, \\ 0 & \text{if } \sqrt{x_1^2 + x_2^2 + c^2} > 1. \end{cases}$$

Note in particular that if $|c| > 1$, then $f_c \equiv 0$.

As these examples show, an important feature of a function is the set of points where it is nonzero. From the point of view of measurement it is difficult to distinguish points where a function is nonzero from points that are "arbitrarily close" to such points. Indeed, to get a useful mathematical concept we need to add points that are "arbitrarily close" to points where the function is nonzero. As the definition is the same in all dimensions we give it for functions defined on $\mathbb{R}^n$.

***Definition* 3.1.2.** Let f be a function defined on $\mathbb{R}^n$. A point $\boldsymbol{x}$ belongs to the *support* of f if there is a sequence of points $< \boldsymbol{x}_n >$ such that

1. $f(\boldsymbol{x}_n) \neq 0$,
2. $\lim_{n\to\infty} \boldsymbol{x}_n = \boldsymbol{x}$.

This set is denoted by $\operatorname{supp}(f)$.

***Example* 3.1.3.** The support of the function $f(x) = x$ is the whole real line, even though $f(0) = 0$. The support of the function $f(x, y) = xy$ is the whole plane, even though $f(0, y) = f(x, 0) = 0$. The support of the function $\chi_{(0,1)}(x)$ is $[0, 1]$.

***Example* 3.1.4.** If D is a subset of $\mathbb{R}^n$ then the support of χ_D is the *closure* of D. This is the collection of all points which are limits of sequences contained in D. A set that contains all such limit points is called a *closed set*.

***Definition* 3.1.3.** A function f defined in $\mathbb{R}^n$ is said to have *bounded support* if there is an R so that $f(\boldsymbol{x}) = 0$ if $\|\boldsymbol{x}\| > R$. In this case we say that the support of f is contained in the ball of radius R.

For the purposes of medical imaging, air is usually assumed to be transparent to x-rays. This means that the attenuation coefficient is set to zero at points outside of the patient. The support of μ can therefore be determined by non-invasive measurements.

3.1.1 Beer's Law and X-ray Tomography

We now turn our attention to a simple quantitative model for the interaction of x-rays with matter. Sometimes x-rays are thought of as a flux of very high-energy, electromagnetic radiation. The x-ray beam is described by a vector valued function $\boldsymbol{I}(\boldsymbol{x})$. The direction of $\boldsymbol{I}$ at $\boldsymbol{x}$ is the direction of the flux at $\boldsymbol{x}$ and its magnitude,

$$I(\boldsymbol{x}) = \|\boldsymbol{I}(\boldsymbol{x})\|$$

is the intensity of the beam. If dS is an infinitesimal surface element at $\boldsymbol{x}$ of area $|dS|$, placed at right angles to $\boldsymbol{I}(\boldsymbol{x})$, then the energy-per-unit-time passing through dS is $I(\boldsymbol{x})|dS|$.

More generally, the intensity of the beam passing through the infinitesimal area element dS, located at $\boldsymbol{x}$, with unit normal vector $\boldsymbol{n}(\boldsymbol{x})$, is

$$\langle \boldsymbol{I}(\boldsymbol{x}), \boldsymbol{n}(\boldsymbol{x})\rangle \, |dS|.$$

The total intensity passing through a surface S is the surface integral

$$\int_S \langle \boldsymbol{I}(\boldsymbol{x}), \boldsymbol{n}(\boldsymbol{x})\rangle \, dA_S,$$

where dA_S is the area element on S.

In other circumstances it is more useful to describe an x-ray beam as being composed of a stream of discrete particles called *photons*. Each photon has a well defined energy, often quoted in units of *electron-volts*. The energy E is related to electromagnetic frequency ν by de Broglie's Law:

$$E = h\nu;$$

h is Planck's constant. While the photon description is more correct, quantum mechanically, the two descriptions only differ significantly when the number of photons (or the intensity of the beam) is small.

Our model for the interaction of x-rays with matter is phrased in terms of the continuum model; it rests on three basic assumptions:

(1) NO REFRACTION OR DIFFRACTION: X-ray beams travel along straight lines that are not "bent" by the objects they pass through.

(2) THE X-RAYS USED ARE MONOCHROMATIC: The waves making up the x-ray beam are all of the same frequency.

(3) BEER'S LAW: Each material encountered has a characteristic linear attenuation coefficient μ for x-rays of a given energy. The intensity, I of the x-ray beam satisfies Beer's law

$$\frac{dI}{ds} = -\mu(\boldsymbol{x})I. \tag{3.1}$$

Here s is the arc-length along the straight-line trajectory of the x-ray beam.

Because x-rays have very high energies, and therefore very short wavelengths, assumption (1) is a good approximation to the truth. Ordinary light is also electromagnetic radiation, we experience the frequency of this radiation as the color of the light. The second assumption is that the x-ray beam is "all of one color." This is not a realistic assumption, but it is needed to construct a *linear* model for the measurements. The implications of the failure of this assumption are discussed later in this chapter. If we think of the x-ray beam as a stream of photons, then the beam is monochromatic if all the photons in the beam have the same energy E. The intensity $I(\boldsymbol{x})$ then equals $EN(\boldsymbol{x})$. Here $N(\boldsymbol{x})$ is the *photon flux*, or the number of photons per unit time crossing a surface at right angles to $\boldsymbol{I}(\boldsymbol{x})$. If the beam is not monochromatic, then the relationship between the intensity and photon flux is described by the spectral function which is introduced in Section 3.3.

Beer's law requires some explanation. Suppose that an x-ray beam encounters an object. Beer's law describes how the presence of the object affects the intensity of the beam. For the moment suppose that we live in a one-dimensional world, with s a coordinate in our world. Let $I(s)$ be the intensity of the x-ray beam at s. For a one-dimensional problem I would be quoted in units of electron-volts/s. Beer's law predicts the change in intensity due to the material lying between s and $s + \Delta s$:

$$I(s + \Delta s) - I(s) \approx -\mu(s) I(s)\, \Delta s. \tag{3.2}$$

Think now of the beam as being composed of a large number, $N(s)$ photons/second, each of the given energy E traveling along the same straight line. Then $I(s) = EN(s)$ and equation (3.2) implies that

$$N(s + \Delta s) - N(s) \approx -\mu(s) N(s)\, \Delta s.$$

This formulation suggests a probabilistic interpretation for Beer's law: $\mu(s)\Delta s$ can be regarded as giving the probability that a photon incident on the material at coordinate s is absorbed. We return to this in Section 16.1.2.

Using the assumption that x-rays travel along straight lines, Beer's law is easily applied to two- and three-dimensional problems. The two- or three-dimensional x-ray flux is modeled as a collection of noninteracting, one-dimensional beams. As noted previously, the x-ray flux at a point $\boldsymbol{x}$ is described by a vector $\boldsymbol{I}(\boldsymbol{x})$. In light of assumption (1) the direction of $\boldsymbol{I}$ at $\boldsymbol{x}$ gives the direction of the *straight line* along which the beam is traveling. Suppose that one of the x-ray beams is traveling along the line given parametrically by

$$\{\boldsymbol{x}_0 + s\boldsymbol{v} \;:\; s \in \mathbb{R}\}.$$

Here $\boldsymbol{v}$ is a unit vector in $\mathbb{R}^3$. The function

$$i(s) = I(\boldsymbol{x}_0 + s\boldsymbol{v})$$

gives the intensity of the beam at points along this line, and

$$m(s) = \mu(\boldsymbol{x}_0 + s\boldsymbol{v})$$

gives the attenuation coefficient. Beer's law states that

$$\frac{di}{ds} = -m(s) i(s) \text{ or } \frac{d(\log i)}{ds} = -m(s).$$

Integrating this equation from $s = a$ to $s = b$ gives

$$\log\left[\frac{i(b)}{i(a)}\right] = -\int_a^b m(s)\, ds.$$

Between the points $\boldsymbol{x}_0 + a\boldsymbol{v}$ and $\boldsymbol{x}_0 + b\boldsymbol{v}$, the intensity of the beam is attenuated by

$$\exp\left[-\int_a^b m(s)\,ds\right].$$

Typical units for the intensity of an x-ray beam are electron-volts/(s$\times$cm^2) in three-dimensions and electron-volts/(s$\times$cm) in two-dimensions. Beer's law describes how the material attenuates each of these one-dimensional beams; it implicitly asserts that the attenuation of x-rays is an isotropic process: It does not depend on the direction of the line along which the x-ray travels. An application of Beer's law to a two-dimensional situation is considered next.

***Example* 3.1.5.** Assume that we have a point source of x-rays of intensity I_0 in the plane [see Figure 3.2(a)]. The x-ray source is isotropic which means the outgoing flux is the same in all directions.

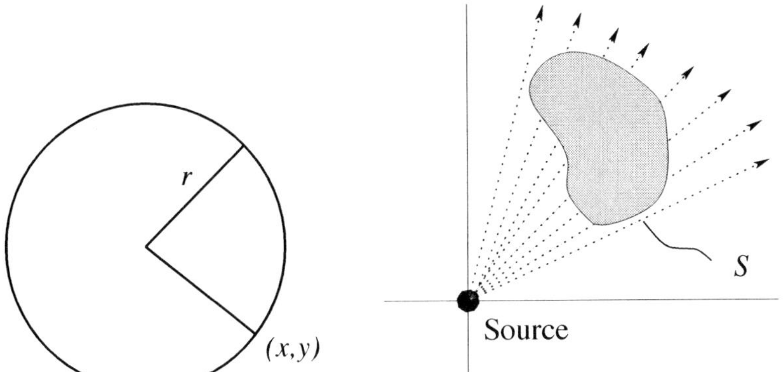

(a) A point source of x-rays

(b) The flux through a curve not enclosing the source is zero.

Figure 3.2. Analysis of an isotropic point source.

Because the source is isotropic, the intensity of the beam is only a function of the distance to the source. Let $I(r)$ denote the intensity of the flux at distance r from the source. By conservation of energy,

$$I_0 = \int_{x^2+y^2=r^2} I(r)\,ds = 2\pi r I(r). \tag{3.3}$$

The intensity of the beam at distance r from the source is therefore

$$I(r) = \frac{I_0}{2\pi r}. \tag{3.4}$$

The x-ray flux is modeled as beams traveling along the rays that pass through the source. If I_0 is measured in units of electron-volts/second, then $I(r)$ has units $\frac{\text{electron-volts}}{\text{cm}\times\text{second}}$.

Fixing coordinates so that the source is placed at $(0, 0)$, the x-ray flux at a point (x, y) travels along the line from (x, y) to $(0, 0)$ and is given by

$$\begin{aligned} \boldsymbol{I}(x, y) = I(r)\frac{(x, y)}{\sqrt{x^2 + y^2}} &= I_0\frac{\hat{r}}{2\pi r}, \\ \text{where } \hat{r} &= \frac{(x, y)}{r}, \quad r = \sqrt{x^2 + y^2}. \end{aligned} \tag{3.5}$$

If a curve S does not enclose the source, then conservation of energy implies that

$$\int_S \boldsymbol{I}(x, y)\hat{r} \cdot \hat{n}\, ds = 0; \tag{3.6}$$

here $\hat{n}$ is the outward normal vector to S. As the curve encloses no sources or sinks, the line integral of the flux is zero: everything that comes into this surface has to go out.

For a point source the intensity of the rays diminish as you move away from the source; this is called beam spreading. Beer's law can be used to model this effect. Let μ_s denote the attenuation coefficient that accounts for the spreading of the beam. As a guess we let $\mu_s = 1/r$ and see that

$$\frac{dI}{dr} = -\frac{1}{r}I \Rightarrow \frac{d\log I}{dr} = -\frac{1}{r}. \tag{3.7}$$

Integrating equation (3.7) from an $r_0 > 0$ to r gives

$$I(r) = I(r_0)\frac{r_0}{r}.$$

This agrees with (3.4). That we cannot integrate down to $r = 0$ reflects the non-physical nature of a point source. In x-ray tomography it is often assumed that the attenuation of the beam due to beam spreading is sufficiently small, compared to the attenuation due to the object that it can be ignored. This is called a *non-diverging source* of x-rays.

In a real measurement, the x-ray source is turned on for a known period of time. The total energy, I_i, incident on the object along a given line, l is known. The total energy, I_o, emerging from the object along l is then measured by an x-ray detector. Integrating Beer's law we obtain

$$\log\frac{I_o}{I_i} = -\int_l \mu\, ds. \tag{3.8}$$

Here ds is the arc length parameter along the straight line path l. A perfect measurement of the ratio I_o/I_i would therefore furnish the line integral of the attenuation coefficient along the line l. Indeed we model the measurements made in x-ray tomography as precisely these line integrals.

An ordinary x-ray image is formed by sending a beam of x-rays through an object, with the detector a sheet of photographic film. Suppose that the x-rays travel along parallel lines, passing through an object before arriving on a photographic plate, as shown in Figure 3.3. By measuring the density of the exposed film, we can determine the intensity of the x-ray beam at the surface of the film. More absorbent parts of an object result in fewer x-rays photons at the surface of the film. If the intensity of the incident beam is known, then the density of the film can be used to determine the integrals of the attenuation coefficient along this family of parallel lines.

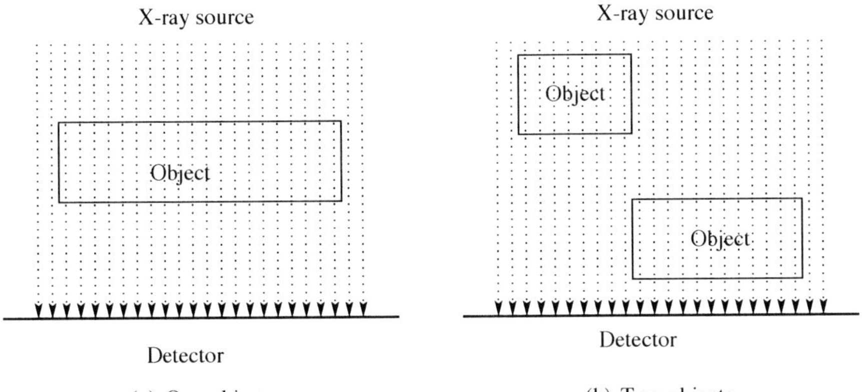

(a) One object. (b) Two objects.

Figure 3.3. The failure of ordinary x-ray images to distinguish objects.

The result is a "projection" or shadow of the object. The shadows of the objects in Figures 3.3(a) and (b) are the same, so it is not possible to distinguish between them using this projection. Placing the x-ray source at a different angle gives a different measurement. The measurement in Figure 3.4 distinguishes between the objects in Figures 3.3(a) and (b).

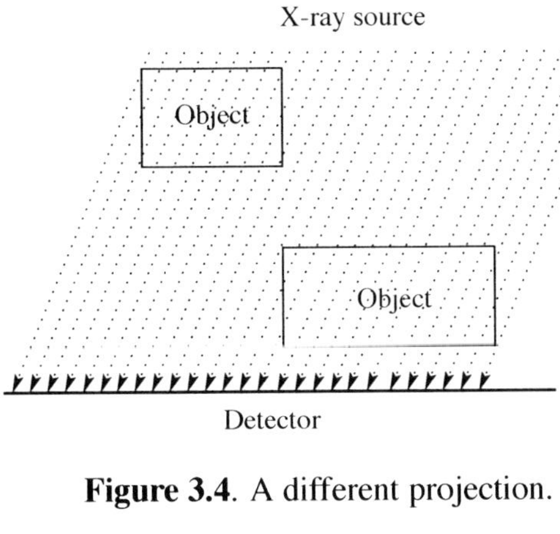

Figure 3.4. A different projection.

The principle is clear: The more directions from which you make measurements, the more arrangements of objects you can distinguish. The goal of x-ray tomography is much more ambitious; we would like to use these projections to reconstruct a picture of the slice. This problem is similar to that considered in Example 1.1.5. However, it is much more challenging to reconstruct a function from its averages along lines than to reconstruct the outline of an object from its shadows. To accomplish this in principle and in practice requires a great deal more mathematics.

Exercises

***Exercise* 3.1.1.** Suppose that we have an isotropic point source of x-rays in three-dimensions of intensity I_0. Find the formula for the intensity of the beams at a distance r from the source. What are the units of $I(r)$?

***Exercise* 3.1.2.** Verify (3.6) by direct computation.

***Exercise* 3.1.3.** Describe an apparatus that would produce a uniform, non-divergent source of x-rays.

3.2 Analysis of a Point Source Device

In this section we use Beer's law to study a simple two-dimensional apparatus and analyze what it measures. Figure 3.5 shows an apparatus with a point source of x-rays, an attenuating body and a photographic plate. We derive an expression for the flux at a point P on the photographic plate in terms of the attenuation of the beam caused by the object as well as beam spreading. The final expression involves the line integral of the attenuation coefficient.

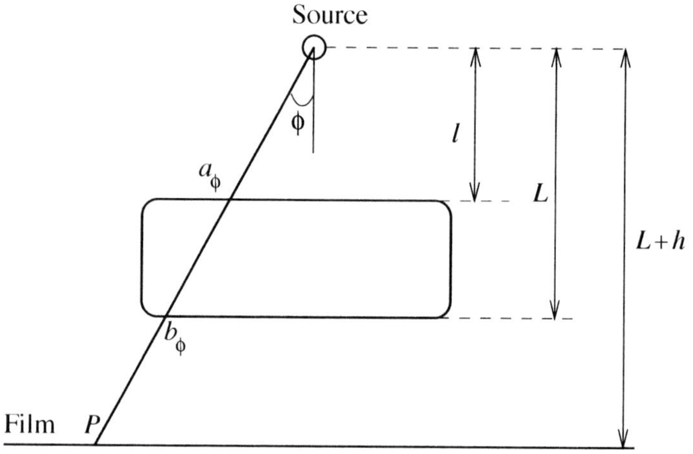

Figure 3.5. A point source device for measuring line integrals of the attenuation coefficient.

The geometry of our apparatus suggests the use of polar coordinates to label points in the plane. Let r denote the distance from the source and ϕ the angle indicated in the diagram. The attenuation coefficient for the absorbing body in Figure 3.5 is then a function of (r, ϕ), denoted by $\mu_a(r, \phi)$. The effect of beam spreading on the intensity of the flux is analyzed in Example 3.1.5. The total attenuation coefficient is obtained by adding $\mu_s(r) = r^{-1}$ to μ_a. For the beam of x-rays traveling along the line through the source, at angle ϕ, the differential equation describing the attenuation of the x-ray beam is

$$\frac{dI}{dr} = -\left(\mu_a(r,\phi) + \frac{1}{r}\right)I. \tag{3.9}$$

The sum, $\mu_a + r^{-1}$, is an *effective* attenuation coefficient as it captures both the attenuation due to the object and that due to beam spreading. The film is exposed by turning the source on for a known period of time. In order to avoid introducing more notation, I is also used to denote the total energy per unit length resulting from this exposure. The units for I in this case would be electron-volts/cm.

Label the radius of the first point of intersection of the line at angle ϕ with the absorbing body by a_ϕ, and the last by b_ϕ. The other distances describing the apparatus are labeled by h, L, and $L+h$, respectively. Integrating equation (3.9) from $r = r_0$ to the film plane $r = r_\phi$ gives

$$\log \frac{I(r_\phi,\phi)}{I(r_0,\phi)} = \log \frac{r_0}{r_\phi} - \int_{a_\phi}^{b_\phi} \mu_a(s,\phi)\, ds.$$

Using

$$a_\phi = \frac{l}{\cos\phi}, \quad b_\phi = \frac{L}{\cos\phi}, \quad r_\phi = \frac{L+h}{\cos\phi},$$

we get

$$I(r_\phi,\phi) = I_0 \frac{\cos\phi}{2\pi(L+h)} \exp\left[-\int_{a_\phi}^{b_\phi} \mu_a(s,\phi)\, ds\right].$$

The density of the developed film at a point is proportional to the logarithm of the total energy incident at that point; that is,

$$\text{density of the film} = \gamma \times \log(\text{total energy intensity,}) \tag{3.10}$$

where γ is a constant. We now compute this energy. As the film plane is not perpendicular to the direction of the beam of x-rays, we need to determine the flux across the part of the film subtended by the angle $\Delta\phi$. It is given by

$$\Delta F = \int_{\phi}^{\phi+\Delta\phi} I(r_\phi,\phi)\hat{r}\cdot\hat{n}\, d\sigma, \quad \hat{r} = -(\sin\phi, \cos\phi).$$

Here $\hat{n} = (0,-1)$ is the outward, unit normal vector to the film plane and $d\sigma$ is the arc length element along the film plane. In polar coordinates it is given by

$$d\sigma = \frac{L+h}{\cos^2\phi} d\phi.$$

Since $\Delta\phi$ is small, we can approximate the integral by

$$\Delta F \approx \int_{\phi}^{\phi+\Delta\phi} I(r_\phi, \phi)\hat{r}\cdot\hat{n}\frac{L+h}{\cos^2\phi}d\phi \approx I_0\frac{\cos^2\phi}{2\pi(L+h)}\exp\left[-\int_{a_\phi}^{b_\phi}\mu_a(s,\phi)\,ds\right]\frac{L+h}{\cos^2\phi}\Delta\phi. \tag{3.11}$$

The length of film subtended by the angle $\Delta\phi$ is approximately

$$\Delta\sigma = \frac{L+h}{\cos^2\phi}\Delta\phi.$$

The energy density at the point P_ϕ, where the line making angle ϕ with the source, meets the film plane, is ΔF divided by this length. Indeed, letting $\Delta\phi$ tend to zero gives

$$\frac{dF}{d\sigma} = \frac{I_0\cos^2\phi}{2\pi(L+h)}\exp\left[-\int_{a_\phi}^{b_\phi}\mu_a(s,\phi)\,ds\right].$$

According to (3.10), the density of the film at P_ϕ is therefore

$$\gamma\log\frac{dF}{d\sigma} = \gamma\left[\log\frac{I_0\cos^2\phi}{2\pi(L+h)} - \int_{a_\phi}^{b_\phi}\mu_a(s,\phi)\,ds\right].$$

The first term comes from the attenuation due to beam spreading. Subtracting it from the measurement gives the line integral of the attenuation coefficient of the absorbing body along the ray at angle ϕ. Let $\delta(\phi)$ denote the density of the film at P_ϕ, this formula can be rewritten

$$-\int_{a_\phi}^{b_\phi}\mu_a(s,\phi)\,ds = \gamma^{-1}\delta(\phi) - \log\left[\frac{I_0\cos^2\phi}{2\pi(L+h)}\right].$$

On the right-hand side are quantities determined by measurement; the left-hand side is the line integral of the attenuation coefficient. This formula expresses the measurements as a *linear function* of the attenuation coefficient.

By varying the position of the source, we can measure the line integrals of the attenuation coefficient along another family of lines. If we move the source and film plane together, around a circle enclosing the absorbent material, making the measurements described previously for each position of the source, then we can measure the line integrals of the attenuation coefficient for all lines that intercept the object (Figure 3.6). This brings us to an essentially mathematical problem: Can a function be recovered from a knowledge of its line integrals along **all** lines? We shall see that this can in principle be done. That it can also be done in practice forms the basis for image reconstruction in an x-ray CT machine.

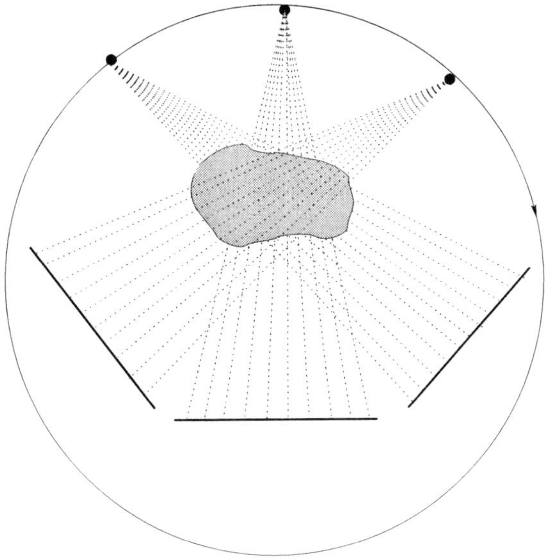

Source positions

Figure 3.6. Collecting data from many views.

We now describe a idealized model for what is measured in x-ray tomography. As remarked above the CT machine determines a coordinate system (x_1, x_2, x_3) for $\mathbb{R}^3$. Let μ denote the attenuation coefficient of an object. Suppose the support of μ lies in the cube $[-a, a] \times [-a, a] \times [-a, a]$. For each c between $\pm a$ and each pair $(t, \boldsymbol{\omega}) \in \mathbb{R} \times S^1$ we measure integral of μ along the line

$$\{(x_1, x_2, x_3) : x_3 = c \text{ and } \langle (x_1, x_2), \boldsymbol{\omega} \rangle = t\},$$

that is,

$$\int_{-\infty}^{\infty} \mu(t\boldsymbol{\omega} + s\hat{\boldsymbol{\omega}}, c)\, ds.$$

We measure the line integrals of μ along all lines that lie in planes where x_3 is constant.

Figure 3.7(a) is a density plot of a function that models the attenuation coefficient of a slice of the human head. The densities in this plot are scaled so that the highest attenuation (the skull) is white and the lowest (surrounding air) is black. A model of this sort is called a *mathematical phantom*. Such models were introduced into medical image reconstruction by Larry Shepp. These are discussed in greater detail in Section 11.2. The example shown in Figure 3.7(a) is adapted from a paper by Shepp and Logan, [114]. Figure 3.7(b) depicts the line integrals of this function. The horizontal axis is the angular variable, with θ corresponding to $\boldsymbol{\omega}(\theta) = (\cos(\theta), \sin(\theta))$. The angle varies between 0 and 2π. The vertical axis is the affine parameter. Again white corresponds to high densities and black to low densities.

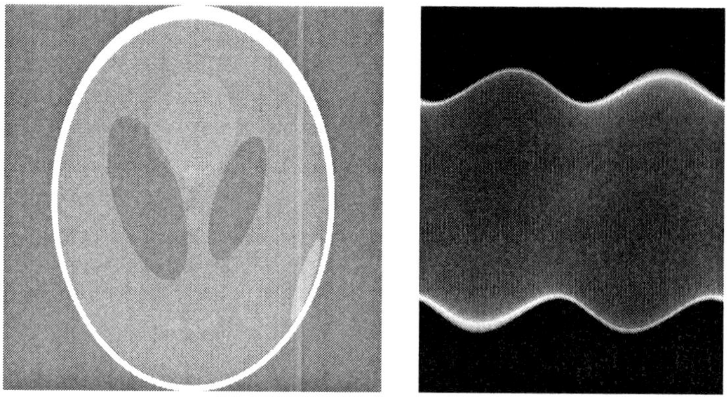

(a) A density plot of the Shepp-Logan phantom.

(b) The line integrals for the function shown in (a).

Figure 3.7. The line integrals of the Shepp-Logan phantom.

In Chapter 6 we give a formula for determining a function defined in $\mathbb{R}^2$ from its integrals on *all* lines in the plane. Of course, it is not possible to make an infinite number of measurements. This means that we need a method for reconstructing an approximation to a function from a finite collection of line integrals. In Chapter 11, beginning with the exact reconstruction formula, we obtain algorithms for use with finitely many measurements.

Exercises

***Exercise* 3.2.1.** Explain why the top and bottom of Figure 3.7(b) are identical.

***Exercise* 3.2.2.** Why is Figure 3.7(b) black near the left and right edges?

***Exercise* 3.2.3.** Explain the significance of the white bands in Figure 3.7(b).

3.3 Some Physical Considerations

Before proceeding with the mathematical development, we briefly revisit the assumptions underlying our model for the attenuation of x-rays. This discussion previews topics considered in later chapters and is not essential to the remainder of this chapter. The x-ray source is assumed to be monochromatic. In fact, the beam of x-rays is made up of photons having a wide range of energies. The distribution of photon energies is described by its *spectral function*, S. If $\mathscr{E}_1$ and $\mathscr{E}_2$ are nearby energies, then the energy in the beam due to photons with energies lying in the interval $[\mathscr{E}_1, \mathscr{E}_2]$ is about $S(\mathscr{E}_1)(\mathscr{E}_2 - \mathscr{E}_1)$, or more precisely

$$\int_{\mathscr{E}_1}^{\mathscr{E}_2} S(\mathscr{E})\, d\mathscr{E}.$$

The graph of a typical spectral function is shown in Figure 3.8. The total energy output of the source is given by

$$\Psi_i = \int_0^\infty S(\mathscr{E})d\mathscr{E}.$$

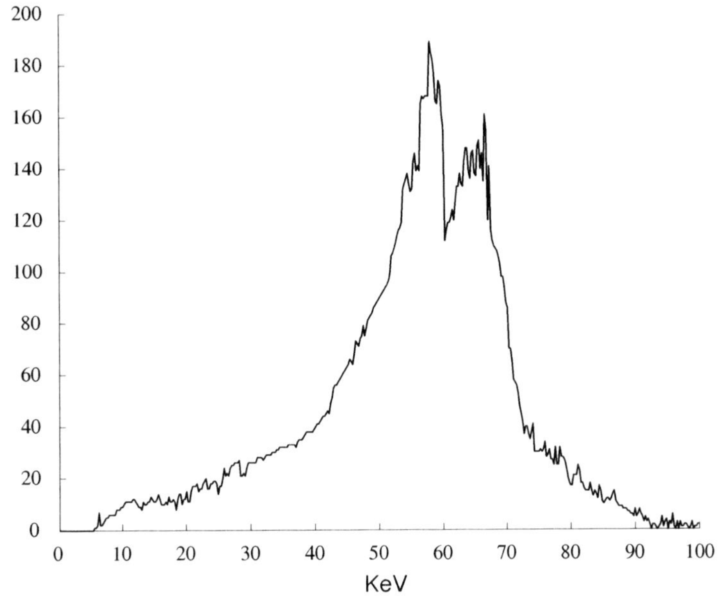

Figure 3.8. A typical x-ray source spectral function. (Spectral data provided by Dr. Andrew Karellas, University of Massachusetts Medical School, Department of Radiology, Worcester, Massachusetts.)

The attenuation coefficient of a given material is a complicated function of the energy, monotonely decreasing as the energy increases. The attenuation coefficient is the sum total of the results of several physical processes that x-rays undergo. A discussion of the physics behind the attenuation coefficient can be found in [6]. Let $\mu(\boldsymbol{x}, \mathscr{E})$ denote the attenuation coefficient of the object at $\boldsymbol{x}$ for photons of energy $\mathscr{E}$. Beer's law, applied to the photons of energy $\mathscr{E}$, traveling along a line l states that the ratio $I_o(\mathscr{E})/I_i(\mathscr{E})$ of emitted flux to incident flux at this energy is

$$\frac{I_o(\mathscr{E})}{I_i(\mathscr{E})} = \exp\left[-\int_l \mu(\boldsymbol{x}, \mathscr{E})\,ds\right].$$

The incident flux at energy $\mathscr{E}$ is $S(\mathscr{E})\,d\mathscr{E}$ and therefore

$$I_o(\mathscr{E}) = S(\mathscr{E})\,d\mathscr{E}\exp\left[-\int_l \mu(\boldsymbol{x}, \mathscr{E})\,ds\right].$$

Because low-energy (or soft) x-rays are attenuated more efficiently than high-energy (or hard) x-rays, the distribution of energies in the output beam is skewed toward higher energies. In medical imaging, this is called *beam hardening*. Along a given line the spectral function at energy $\mathcal{E}$ of the output beam is

$$S_{\text{out}}(\mathcal{E}) = S(\mathcal{E}) \exp\left[-\int_l \mu(x, \mathcal{E}) ds\right].$$

Integrating S_{out} over the energy gives the measured output

$$\Psi_o = \int_0^\infty S(\mathcal{E}) \exp\left[-\int_l \mu(\boldsymbol{x}, \mathcal{E})\, ds\right] d\mathcal{E}.$$

As before, we would like to reconstruct μ or perhaps some average of this function over energies. Mathematically this is a *very* difficult problem as the measurement, Ψ_o, is a non-linear function of μ. We have avoided this problem by assuming that the x-ray beam used to make the measurements is monochromatic. This provides the much simpler *linear* relationship (3.8) between the measurements and the attenuation coefficient. In Chapter 12 we briefly consider the artifacts that result from using polychromatic x-rays and methods used to ameliorate them.

The fact that the x-ray "beam" is not a continuous flux but is composed of discrete particles produces random errors in the measurements. This type of error is called Poisson noise, quantum noise, or photon noise. In Chapter 16 we analyze this effect, showing that the available information in the data is proportional to the square root of the *number* of photons used to make the measurement. The accuracy of the measurements is the ultimate limitation on the number of significant digits in the reconstructed attenuation coefficient. Table 3.1 lists the attenuation coefficients of different structures encountered in medical CT. The attenuation coefficients of air (-1000) and bone (1086) define the range of values present in a typical clinical situation. The dynamic range of a clinical CT measurement is about 2000 Hounsfield units. From the table it is apparent that the variation in the attenuation coefficients of soft tissues is about 2% of this range. For x-ray CT to be clinically useful this means that the reconstruction of the attenuation coefficient needs to be accurate to less than a half a percent.

An obvious solution to this problem would be to increase the number of photons. Since each x-ray photon carries a very large amount of energy, considerations of patient safety preclude this solution. The number of x-ray photons involved in forming a CT image is approximately $10^7/\text{cm}^2$. This should be compared with the 10^{11} to $10^{12}/\text{cm}^2$ photons, needed to make a usable photographic image. In ordinary photography, quantum noise is not a serious problem because the number of photons involved is very large. In x-ray tomography, patient safety and quantum noise place definite limits on the contrast and resolution of a CT image.

3.4 The Definition of the Radon Transform

See: A.3, A.4.1.

The first step in determining a function from its integrals along all straight lines is to organize this information in a usable fashion. We use the parameterization for the set of oriented lines in the plane described in Section 1.2.1. Recall that for $(t, \omega) \in \mathbb{R} \times S^1$, $l_{t,\omega}$ is the set

$$\{x \in \mathbb{R}^2 \mid \langle x, \omega \rangle = t\},$$

with orientation determined by $\hat{\omega}$ (Figure 1.12).

***Definition* 3.4.1.** Suppose that f is a function defined in the plane, which, for simplicity, we assume is continuous with bounded support. The integral of f along the line $l_{t,\omega}$ is denoted by

$$\begin{aligned}\mathcal{R}f(t,\omega) &= \int_{l_{t,\omega}} f\,ds \\ &= \int_{-\infty}^{\infty} f(s\hat{\omega} + t\omega)\,ds \\ &= \int_{-\infty}^{\infty} f(t\omega_1 - s\omega_2, t\omega_2 + s\omega_1)\,ds.\end{aligned} \tag{3.12}$$

The collection of integrals of f along the lines in the plane defines a function on $\mathbb{R} \times S^1$, called the *Radon transform* of f.

***Example* 3.4.1.** In this example we give a graphical description for the computation of the Radon transform. Figure 3.9(a) shows the graph of a function f defined in the plane along with some lines in the family

$$\{\langle x, (\cos(\frac{\pi}{4}), \sin(\frac{\pi}{4}))\rangle = t\}.$$

In Figure 3.9(b) we show graphs of f restricted to these lines. Behind the slices is the graph of $\mathcal{R}f(t, (\cos(\frac{\pi}{4}), \sin(\frac{\pi}{4})))$. The height of this curve at each point is just the area under the corresponding slice of the graph of f.

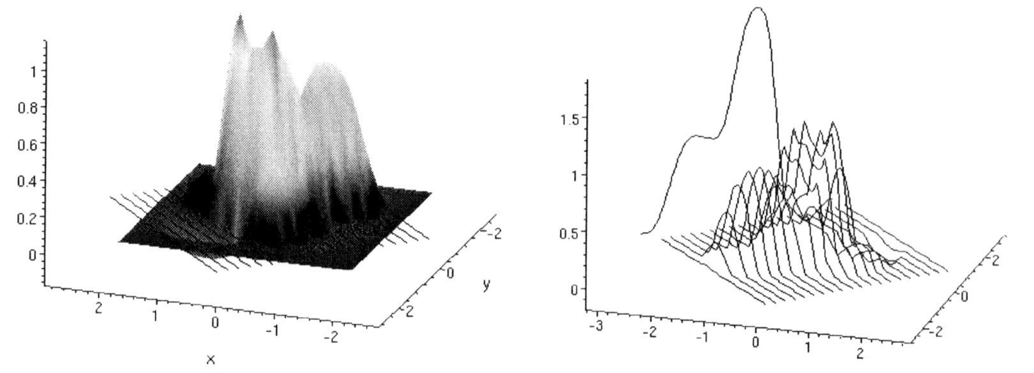

(a) The graph of f and a family of parallel lines in the plane.

(b) Slices of f and its Radon transform for this family of lines.

Figure 3.9. Graphical depiction of the Radon transform.

It is not necessary for f to be either continuous or of bounded support. The Radon transform can be defined, a priori for a function, f whose restriction to each line is *locally integrable* and

$$\int_{-\infty}^{\infty} |f(t\omega + s\hat{\omega})\, ds| < \infty \text{ for all } (t, \omega) \in \mathbb{R} \times S^1. \tag{3.13}$$

With these conditions the improper integrals in (3.12) are unambiguously defined. Functions that satisfy (3.13) are in the *natural domain* of the Radon transform. This is really two different conditions:

1. The function is regular enough so that restricting it to any line gives a locally integrable function.

2. The function goes to zero rapidly enough for the improper integrals to converge.

The function $f \equiv 1$ is not in the natural domain of the Radon transform because it does not decay at infinity. The function $f(x, y) = (x^2 + y^2)^{-1}$ is not in the natural domain of $\mathcal{R}$ because the integrals in (3.13) diverge if $t = 0$. While an understanding of the domain of Radon transform is a important part of its mathematical analysis. In applications to medical imaging, functions of interest are usually piecewise continuous and zero outside of some disk and therefore belong to the natural domain of $\mathcal{R}$.

Functions in the range of the Radon transform have a special symmetry property.

***Definition* 3.4.2.** A function h on $\mathbb{R} \times S^1$ is an *even function* if

$$h(t, \omega) = h(-t, -\omega). \tag{3.14}$$

We summarize the elementary properties of the Radon transform in a proposition.

Proposition 3.4.1. *The Radon transform is linear; that is, if f and g are functions in the natural domain of the Radon transform, then*

$$\begin{aligned} \mathcal{R}(af) &= a\mathcal{R}f \text{ for all } a \in \mathbb{R} \text{ and} \\ \mathcal{R}(f+g) &= \mathcal{R}f + \mathcal{R}g. \end{aligned} \tag{3.15}$$

The Radon transform of f is an even function; that is,

$$\mathcal{R}f(t,\omega) = \mathcal{R}f(-t,-\omega). \tag{3.16}$$

The Radon transform is monotone: if f is a non-negative function in the natural domain of the Radon transform then

$$\mathcal{R}f(t,\omega) \geq 0 \quad \text{for every } (t,\omega). \tag{3.17}$$

Proof. The linearity is a consequence of the linearity of integration over a line. The second statement follows from the fact that, as sets, $l_{t,\omega} = l_{-t,-\omega}$. The last statement follows from the analogous property for the integral. □

We now define some simple classes of functions and compute their Radon transforms.

***Example* 3.4.2.** Let E be a subset of $\mathbb{R}^2$; the Radon transform of χ_E has a simple geometric description.

$$\mathcal{R}\chi_E(t,\omega) = \text{ the length of the intersection } l_{t,\omega} \cap E.$$

If E is a closed, bounded subset, then χ_E belongs to the natural domain of the Radon transform. These functions model objects with constant attenuation coefficient.

In some cases it is possible to give a more explicit formula for $\mathcal{R}\chi_E$.

***Definition* 3.4.3.** The ball in $\mathbb{R}^n$ of radius r centered at $\boldsymbol{a}$ is denoted

$$B_r(\boldsymbol{a}) = \{\boldsymbol{x} \, : \, \|\boldsymbol{x}-\boldsymbol{a}\| < r\}.$$

Often $B_r(\mathbf{0})$ is denoted by B_r. Balls in $\mathbb{R}^2$ are often called *disks*.

***Example* 3.4.3.** Consider the disk of radius 1 in $\mathbb{R}^2$. The function χ_{B_1} is a special case of the general class considered in the previous example. The formula for the Radon transform of χ_{B_1} is

$$\mathcal{R}\chi_{B_1}(t,\omega) = \begin{cases} 2\sqrt{1-t^2} & \text{if } |t| \leq 1, \\ 0 & \text{if } |t| > 1. \end{cases}$$

Note that $|t| > 1$ corresponds to lines $l_{t,\omega}$ that do not intersect B_1.

***Definition* 3.4.4.** A function, f defined on $\mathbb{R}^n$ is *radial* if its value only depends on the distance to the origin. In this case there exists a function, F, of a single variable so that

$$f(\boldsymbol{x}) = F(\|\boldsymbol{x}\|).$$

***Example* 3.4.4.** The Radon transform of a radial function takes a simpler form. From geometric considerations it is clear that $\mathcal{R} f(t, \omega)$ does not depend on ω. Fixing a convenient direction—for example, $\omega = (1, 0)$—we obtain

$$\begin{aligned} \mathcal{R} f(t, \omega) &= \int_{-\infty}^{\infty} f(t, s)\, ds \\ &= \int_{-\infty}^{\infty} F(\sqrt{t^2 + s^2})\, ds. \end{aligned} \tag{3.18}$$

Using the change of variable, $r^2 = t^2 + s^2$, $2r\, dr = 2s\, ds$, we obtain

$$\mathcal{R} f(t, \omega) = 2 \int_{t}^{\infty} \frac{F(r) r\, dr}{\sqrt{r^2 - t^2}}. \tag{3.19}$$

Formula (3.19) expresses the Radon transform of a radial function as a one-dimensional *integral transform*.

Our goal is the recovery of a function, f, from a knowledge of its Radon transform, $\mathcal{R} f$. Since $\mathcal{R}$ is a linear map, we might hope that there is a linear map $\mathcal{R}^{-1}$ from functions on $\mathbb{R} \times S^1$ to functions on $\mathbb{R}^2$ satisfying

$$\mathcal{R}^{-1} \circ \mathcal{R} f = f.$$

The inverse map should also be given by an integral formula. This turns out to be the case, but the derivation and analysis of this formula are rather involved. Because these spaces of functions are infinite dimensional, finding the inverse is not just a problem in linear algebra. The domain of $\mathcal{R}^{-1}$ is the range of $\mathcal{R}$, and neither the domain of $\mathcal{R}$ nor of $\mathcal{R}^{-1}$ is easy to describe explicitly. These issues are studied in Chapter 6. The remainder of this section is devoted to further properties of the Radon transform and its inverse.

Naively, we would expect that in order for $\mathcal{R}^{-1}$ to exist it would be necessary that $\mathcal{R} f(t, \omega) = 0$ for all pairs (t, ω) only if $f \equiv 0$. In fact, it is easy to construct examples of functions that are not zero but have zero Radon transform.

***Example* 3.4.5.** Define the function

$$f(x, y) = \begin{cases} 1 & \text{if } (x, y) = (0, 0), \\ 0 & \text{if } (x, y) \neq (0, 0). \end{cases}$$

Clearly, $\mathcal{R} f(t, \omega) = 0$ for all (t, ω).

From the point of view of measurement, this is a very trivial example. The next example is somewhat more interesting.

***Example* 3.4.6.** Define a function f by setting $f(x, y) = 1$ if $x \in [-1, 1]$ and $y = 0$ and zero otherwise. Then $\mathcal{R}f(t, \boldsymbol{\omega}) = 0$ if $\boldsymbol{\omega} \neq (0, \pm 1)$ and $\mathcal{R}f(0, (0, \pm 1)) = 2$. In this case the Radon transform is usually zero, but for certain special lines it is not. Observe that if we replace f by a function $\tilde{f}$ that is 1 on some other subset of $\mathbb{R} \times 0$ of total length 2, then $\mathcal{R}f = \mathcal{R}\tilde{f}$. This gives examples, which are not entirely trivial, where the Radon transform does not contain enough information to distinguish between two functions.

The concept of a *set of measure zero* helps clarify these examples. We give a precise definition.

***Definition* 3.4.5.** A subset $E \subset \mathbb{R}^n$ is said to be of *n-dimensional measure zero* if for any $\epsilon > 0$ there is a collection of balls $B_{r_i}(\boldsymbol{x}_i)$ so that

$$E \subset \bigcup_{i=1}^{\infty} B_{r_i}(\boldsymbol{x}_i)$$

and

$$\sum_{i=1}^{\infty} r_i^n < \epsilon.$$

Such a set carries no n-dimensional mass.

For example, a point is set of measure zero in the line, a line is a set of measure zero in the plane, a plane is a set of measure zero in $\mathbb{R}^3$, and so on. From our perspective the basic fact about sets of measure zero is the following: If f is a function defined in $\mathbb{R}^n$ and the set of points where $f \neq 0$ is a set of measure zero, then [1]

$$\int_{\mathbb{R}^n} |f(\boldsymbol{x})| \, d\boldsymbol{x} = 0.$$

Indeed if φ is any function in $L^1(\mathbb{R}^n)$, then

$$\int_{\mathbb{R}^n} f(\boldsymbol{x})\varphi(\boldsymbol{x}) \, d\boldsymbol{x} = 0.$$

As most linear models for realistic measurements are given by expressions of this sort, it follows that no practical measurement can distinguish such a function from the zero function.

Using the map $(t, \theta) \mapsto (t, \boldsymbol{\omega}(\theta))$, we can identify $\mathbb{R} \times [0, 2\pi)$ with $\mathbb{R} \times S^1$. A set in $\mathbb{R} \times S^1$ has measure zero if its preimage under this map does. With these concepts we can state a basic result about the Radon transform.

Proposition 3.4.2. *If f is a function defined in the plane supported on a set of measure zero, then the set of values $(t, \boldsymbol{\omega}) \in \mathbb{R} \times S^1$ for which $\mathcal{R}f(t, \boldsymbol{\omega}) \neq 0$ is itself a set of measure zero.*

[1] Strictly speaking, we should use the Lebesgue integral when discussing the integral of a function which is only nonzero on a set a measure zero. Such a function may fail to be Riemann integrable. See [43].

As Example 3.4.6 shows, a function supported on a set of measure zero cannot, in general, be reconstructed from its Radon transform. Since the Radon transform is linear, it cannot distinguish functions which differ only on a set of measure zero. This is a feature common to any measurement process defined by integrals. While it is important to keep in mind, it does not lead to serious difficulties in medical imaging.

The support properties of f are reflected in the support properties of $\mathcal{R}f$.

Proposition 3.4.3. *Suppose that f is a function defined in the plane with $f(x, y) = 0$ if $x^2 + y^2 > R^2$. Then*

$$\mathcal{R}f(t, \omega) = 0 \qquad \text{if } |t| > R. \tag{3.20}$$

Proof. Any line $l_{t,\omega}$ with $|t| > R$ lies entirely outside of the support of f. From the definition it follows that $\mathcal{R}f(t, \omega) = 0$ if $|t| > R$. □

If f is *known* to vanish outside a certain disk, then we do not need to compute its Radon transform for lines that are disjoint from the disk. It would be tempting to assert that the converse statement is also true, that is, "If $\mathcal{R}f(t, \omega) = 0$ for $|t| > R$, then $f(x, y) = 0$ if $x^2 + y^2 > R^2$." As the next set of examples shows, this is false. We return to this question in Chapter 6.

***Example* 3.4.7.** For each integer $n > 1$ define a function, in polar coordinates, by setting

$$f_n(r, \theta) = r^{-n} \cos(n\theta).$$

These functions all blow up at $r = 0$ faster than r^{-1} and therefore do not belong to the natural domain of $\mathcal{R}$. This is because f_n cannot be integrated along any line that passes through $(0, 0)$. On the other hand, since f_n goes to zero as $r \to \infty$ like r^{-n} and $n > 1$, the integrals defining $\mathcal{R}f_n(t, \omega)$ converge absolutely for any $t \neq 0$. We use the following result:

***Lemma* 3.4.1.** *For each $n > 1$ and $t \neq 0, \omega \in S^1$ the integrals*

$$\int_{l_{t,\omega}} f_n(t\omega + s\hat{\omega})\, ds$$

converge absolutely and equal zero.

The proof of the lemma is at the end of this section. It already indicates the difficulty of inverting the Radon transform. These functions are not in the natural domain of the Radon transform because

$$\int_{-\infty}^{\infty} |f_n(-s \sin\theta, s \cos\theta)|\, ds = \infty$$

for any value of θ. On the other hand, $\mathcal{R}f_n(t, \omega) = 0$ for all $t \neq 0$. So in some sense, $\mathcal{R}f_n$ is supported on the set of measure zero $\{0\} \times S^1$.

For each n we modify f_n to obtain a function F_n in the natural domain of $\mathcal{R}$ such that $\mathcal{R}F_n(t, \omega) = 0$ for all (t, ω), with $|t| > 1$. On the hand, the functions F_n do not vanish outside the disk or radius 1. The modified functions are defined by

$$F_n(r, \theta) = \begin{cases} f_n(r, \theta) & \text{for } r > 1, \\ 0 & \text{for } r \leq 1. \end{cases}$$

A line $l_{t,\omega}$ with $|t| > 1$ lies entirely outside the unit disk. On such a line, the lemma applies to show that

$$\mathcal{R}F_n(t, \omega) = \int\limits_{l_{t,\omega}} f_n \, ds = 0.$$

On the other hand, F_n is bounded in a neighborhood of $(0, 0)$ and therefore $\mathcal{R}F_n(t, \omega)$ is defined for all $(t, \omega) \in \mathbb{R} \times S^1$. This shows that the Radon transform of a function may vanish for all t with $|t| > r$ without the function being zero outside disk of radius r.

Exercises

***Exercise* 3.4.1.** Provide a detailed proof for Proposition 3.4.1.

***Exercise* 3.4.2.** Find an explicit formula for the Radon transform of $\chi_{B_1(\boldsymbol{a})}$, for any $\boldsymbol{a} \in \mathbb{R}^2$.

***Exercise* 3.4.3.** Compute the Radon transform of $f(x, y) = xy\chi_{[-1,1]}(x)\chi_{[-1,1]}(y)$.

***Exercise* 3.4.4.** Let $f(x, y) = 1$ if $x^2 + y^2 = 1$ and zero otherwise. Show that $\mathcal{R}f(t, \omega) = 0$ for all $(t, \omega) \in \mathbb{R} \times S^1$.

***Exercise* 3.4.5.** Suppose that f and g are functions in the natural domain of the Radon transform. Show that if $f(\boldsymbol{x}) \geq g(\boldsymbol{x})$ for every $\boldsymbol{x} \in \mathbb{R}^2$, then

$$\mathcal{R}f(t, \omega) \geq \mathcal{R}g(t, \omega) \text{ for every } (t, \omega) \in S^1 \times \mathbb{R}.$$

***Exercise* 3.4.6.** Suppose that f is a function defined on $\mathbb{R}^2$ such that the set of points where f is non-zero has measure zero. Show that if φ is a continuous function with bounded support, then

$$\int\limits_{-\infty}^{\infty} \mathcal{R}f(s, \omega)\varphi(t - s) \, ds = 0$$

for every t. Explain the relevance of this fact to medical imaging.

***Exercise* 3.4.7.** * Show that the set of points where $\chi_{\mathbb{Q}}$ is non-zero is a set of measure zero. Show that $\chi_{\mathbb{Q}}$ is *not* a Riemann integrable function.

***Exercise* 3.4.8.** Suppose that $\{\boldsymbol{x}_1, \ldots, \boldsymbol{x}_n\}$ are n distinct points on the unit circle. For $i \neq j$, let l_{ij} denote the line segment joining $\boldsymbol{x}_i$ to $\boldsymbol{x}_j$ and r_{ij} denote a real number. Show that if $r_{ij} = r_{ji}$ for all $i \neq j$, then there is function f supported on the line segments $\{l_{ij}\}$ such that

$$\int\limits_{l_{ij}} f \, ds = r_{ij} \text{ for all } i \neq j.$$

***Exercise* 3.4.9.** Show that a line segment has measure zero as a subset of the plane.

***Exercise* 3.4.10.** Show that the x-axis has measure zero as a subset of the plane.

***Exercise* 3.4.11.** Show that the set $\{(0, \boldsymbol{\omega}) : \boldsymbol{\omega} \in S^1\}$ has measure zero as a subset of $\mathbb{R} \times S^1$.

3.4.1 Appendix: Proof of Lemma 3.4.1*

The proof of the theorem makes use of the elementary theory of complex variables and is not needed for the subsequent development of the book.

Proof. Let $z = x + iy$ and observe that by Euler's formula it follows that

$$f_n = \operatorname{Re} z^{-n}.$$

This means that for $t \neq 0$

$$\mathcal{R} f_n(t, \boldsymbol{\omega}) = \operatorname{Re} \int\limits_{l_{t,\omega}} z^{-n}\, ds,$$

where ds is the arc element along the line. The line $(t, \boldsymbol{\omega}(\theta))$ can be represented as

$$z = (t + is)e^{i\theta}, \quad t \in \mathbb{R}.$$

Using this complex parameterization, the Radon transform of f_n can be reexpressed as a complex contour integral:

$$\mathcal{R} f_n(t, \theta) = \int\limits_{-\infty}^{\infty} f_n((t + is)e^{i\theta})\, ds = \operatorname{Re}\left[-ie^{-i\theta} \int\limits_{\operatorname{Re}(e^{-i\theta} z) = t} z^{-n}\, dz \right], \tag{3.21}$$

where the arc length element, along the line, is written in complex notation as

$$ds = -ie^{-i\theta} dz.$$

For $n > 1$ $\int z^{-n} = (1 - n)^{-1} z^{1-n}$; hence the theorem follows from (3.21). □

3.4.2 The Back-Projection Formula

Even though the line integrals of a function are concrete data, it is difficult to use these data directly to reconstruct the function. An obvious thing to try is averaging the values of the $\mathcal{R} f$ over the lines that pass through a point. For a direction $\boldsymbol{\omega}$, the line in the family $\{l_{t,\omega} : t \in \mathbb{R}\}$, passing through a point $\boldsymbol{x}$ is given by $t = \langle \boldsymbol{x}, \boldsymbol{\omega} \rangle$. Thus we could try setting

$$\tilde{f}(\boldsymbol{x}) = \frac{1}{2\pi} \int\limits_{0}^{2\pi} \mathcal{R} f(\langle \boldsymbol{x}, \boldsymbol{\omega}(\theta) \rangle, \theta)\, d\theta. \tag{3.22}$$

This is called the *back-projection formula*. While it is a reasonable guess, it does not give the correct answer. Figure 3.10 shows the result of using back-projection to reconstruct a simple black and white image. The object is recognizable but blurry.

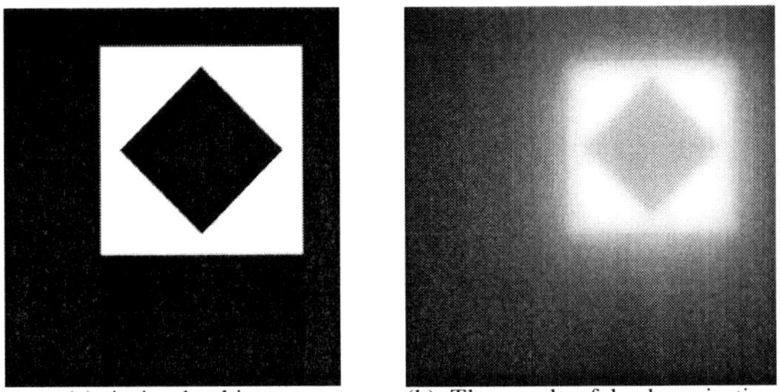

(a) A simple object.

(b) The result of back-projecting the object's Radon transform.

Figure 3.10. Back-projection does not work.

To find the true inverse of the Radon transform requires an indirect approach passing through the Fourier transform. The Fourier transform, while perhaps more familiar, is a less transparent integral transform than the Radon transform. On the other hand, the inverse of the Fourier transform is easier to obtain. In the next chapter we consider the Fourier transform in some detail as it is of fundamental importance in the theory of image reconstruction and signal processing.

3.4.3 Continuity of the Radon Transform*

See: A.4.2, A.4.4.

The Radon transform is a linear transformation from functions defined in the plane to even functions on the space of lines. In medical applications, the function $\mathcal{R}f$ is an idealization for what is measured. Generally speaking, f is taken to be a bounded, though possibly discontinuous, function that vanishes outside of the patient. Suppose that f vanishes outside the disk of radius L and $f \leq M$. The lengths of the intersections of $l_{t,\omega}$ with the support of f are bounded above $2L$, giving a crude estimate for $\mathcal{R}f$:

$$|\mathcal{R}f(t,\omega)| = \left| \int\limits_{l_{t,\omega}} f\, ds \right| \leq 2ML. \tag{3.23}$$

How sensitive are the measurements to errors? This is a question about the continuity properties of the map $f \mapsto \mathcal{R}f$. The answer to this question is important, but in the final analysis, it is not the correct question. What we really want to know is how sensitive the *reconstruction method* is to measurement errors. In other words, we want to understand the continuity properties of $\mathcal{R}^{-1}$. Since we have not yet constructed $\mathcal{R}^{-1}$, we consider the somewhat easier question of the continuity of $\mathcal{R}$. That is, how sensitive are the measurements to the data? We need to choose a way to measure the size of the errors in both the data and the measurements. For the present we make the following choices: on the measurements we use the maximum of the integrals in the affine parameter

$$\|\mathcal{R}f(t,\boldsymbol{\omega})\|_{1,\infty} = \max_{\boldsymbol{\omega}\in S^1} \int_{-\infty}^{\infty} |\mathcal{R}f(t,\boldsymbol{\omega})|\,dt; \tag{3.24}$$

as a norm on the data we use the standard L^1-norm

$$\|f\|_1 = \int_{\mathbb{R}^2} |f(\boldsymbol{x})|\,d\boldsymbol{x}.$$

A function for which $\|f\|_1 < \infty$ is called an *absolutely integrable* or L^1-function. The set of such functions is denoted by $L^1(\mathbb{R}^2)$.

Proposition 3.4.4. *Suppose that f is an L^1-function in the natural domain of the Radon transform. Then*

$$\|\mathcal{R}f\|_{1,\infty} \leq \|f\|_1. \tag{3.25}$$

Proof. This proposition is a consequence of the triangle inequality, the change-of-variables formula, and the Fubini theorem. For each $\boldsymbol{\omega} \in S^1$ the triangle inequality implies that

$$\begin{aligned}\int_{-\infty}^{\infty} |\mathcal{R}f(t,\boldsymbol{\omega})|\,dt &\leq \int_{-\infty}^{\infty}\int_{-\infty}^{\infty} |f(t\boldsymbol{\omega}+s\hat{\boldsymbol{\omega}})|\,ds\,dt \\ &= \int_{\mathbb{R}^2} |f(x,y)|\,dx\,dy.\end{aligned} \tag{3.26}$$

In the second line we use the fact that $(s,t) \mapsto t\boldsymbol{\omega}+s\hat{\boldsymbol{\omega}}$ is an orthogonal change of variables. Since the last line is independent of $\boldsymbol{\omega}$, this proves the proposition. □

Because the Radon transform is linear this estimate implies that for any pair of functions, f, g, in the natural domain of $\mathcal{R}$, we have the estimate

$$\|\mathcal{R}f - \mathcal{R}g\|_{1,\infty} \leq \|f - g\|_1. \tag{3.27}$$

The fact that it takes time to make measurements means that, in the course of acquiring a full set of samples, the patient often moves. For a vector $\boldsymbol{\tau} = (\tau_1, \tau_2)$, let $f_{\boldsymbol{\tau}}$ denote the function f translated by the vector $\boldsymbol{\tau}$:

$$f_{\boldsymbol{\tau}}(x,y) = f(x-\tau_1, y-\tau_2).$$

Suppose that we attempt to measure the $\mathcal{R}f(t,\boldsymbol{\omega})$ but the patient moves, a little, during the measurement process. A better model for what is measured is $\{\mathcal{R}f_{\boldsymbol{\tau}(\boldsymbol{\omega})}(t,\boldsymbol{\omega})\}$, where, as indicated, $\{\boldsymbol{\tau}(\boldsymbol{\omega})\}$ are vectors in $\mathbb{R}^2$ describing the position of the patient as a function of $\boldsymbol{\omega}$. The estimate (3.27) implies that

$$\| \mathscr{R} f - \mathscr{R} f_{\boldsymbol{\tau}} \|_{1,\infty} \leq \max_{\boldsymbol{\omega} \in S^1} \| f - f_{\boldsymbol{\tau}(\boldsymbol{\omega})} \|_1 .$$

If *on average* f does not vary too quickly and the motions that arise are not too large then this estimate shows that the "actual" measurements $\{\mathscr{R} f_{\boldsymbol{\tau}(\boldsymbol{\omega})}(t, \boldsymbol{\omega})\}$ are close to the model measurements $\{\mathscr{R} f(t, \boldsymbol{\omega})\}$. Since the functions that arise in imaging are not continuous, it is important that while the average variation needs to be controlled, the pointwise variation does not. This is illustrated by an example.

***Example* 3.4.8.** Let $f(x, y) = \chi_{[-1,1]}(x)\chi_{[-1,1]}(y)$. If $\tau \neq 0$, then

$$\max_{\boldsymbol{x}} |f(\boldsymbol{x}) - f_{(\tau,0)}(\boldsymbol{x})| = 1.$$

On the other hand, for $|\tau| < 1$, it is also true that

$$\int_{\mathbb{R}^2} |f(\boldsymbol{x}) - f_{(0,\tau)}(\boldsymbol{x})| \, d\boldsymbol{x} = 4\tau.$$

Because of the averaging that occurs in the measurement process, it is often sufficient to keep the measurement errors small in a norm like $\| \cdot \|_{1,\infty}$. This fact is explained in Example 2.2.11.

As an application of functional analytic methods to the study of the Radon transform we show that the estimate (3.25) allows the extension of the Radon transform beyond its natural domain to $L^1(\mathbb{R}^2)$. To define this extension we observe that continuous functions with bounded support are dense in the space $L^1(\mathbb{R}^2)$. This means that for $f \in L^1(\mathbb{R}^2)$, we can choose a sequence $< f_n >$ of continuous functions, with bounded support so that

$$\lim_{n \to \infty} \| f - f_n \|_{L^1} = 0.$$

The Radon transform of f is the function on $\mathbb{R} \times S^1$ *defined* as the limit of the sequence of functions $< \mathscr{R} f_n >$ with respect to the L^1-norm. We let $\mathscr{R} f$ denote the limit. It also has the property that, for all $\boldsymbol{\omega} \in S^1$,

$$\int_{-\infty}^{\infty} |\mathscr{R} f(t, \boldsymbol{\omega})| \, dt \leq \| f \|_{L^1}.$$

On the other hand, $\mathscr{R} f$ is no longer given by the formula (3.12) as it is not known, a priori that these integrals converge. Fubini's theorem implies that these integrals are finite for almost every t.

Exercises

***Exercise* 3.4.12.** Prove that the sequence of functions $< \mathcal{R} f_n >$ has a limit.

***Exercise* 3.4.13.** Compute the Radon transform of

$$f = \frac{\chi_{[0,1]}(x^2 + y^2)}{\sqrt{x^2 + y^2}}.$$

Is f in the natural domain of $\mathcal{R}$? What is $\| \mathcal{R} f \|_1$?

To close this chapter we study the Radon transform acting on radial functions. Inverting the transform in this case is simpler, reducing to a special case of the Abel transform.

3.5 The Radon Transform on Radial Functions

The Radon transform of a radial function does not depend on $\boldsymbol{\omega}$. It is given in (3.19) as an integral transform of a function of one variable. In this section, we suppress the dependence of the Radon transform on $\boldsymbol{\omega}$. After changing variables, we see that $\mathcal{R} f$ is a special case of an Abel transform. For $0 < \alpha \leq 1$, the *α-Abel transform* of g is defined by

$$\mathrm{A}_\alpha\, g(t) = \frac{1}{\Gamma(\alpha)} \int_t^\infty \frac{g(s)\, ds}{(s - t)^{1-\alpha}};$$

the coefficient $\Gamma(\alpha)$ is the *Gamma* function, defined in Section A.3.3. The theory of the Abel transform is outlined in Section 3.5.2. Among other things we show that

$$\mathrm{A}_{\frac{1}{2}}^{-1} = -\partial_t [\mathrm{A}_{\frac{1}{2}}]. \tag{3.28}$$

Suppose that $f(x, y) = F(\sqrt{x^2 + y^2})$; then changing variables shows that

$$\mathcal{R} f(t) = \sqrt{\pi} (\mathrm{A}_{\frac{1}{2}} (F_{\surd}) (t^2). \tag{3.29}$$

Here $F_{\surd}$ denotes the function $r \mapsto F(\sqrt{r})$. Using the formula for the inverse of the Abel transform (3.28) that is derived in Section 3.5.2 and a change of variables, we can solve equation (3.19) to obtain

$$F(r) = -\frac{1}{\pi r} \partial_r \left[\int_r^\infty \frac{\mathcal{R} f(t) t\, dt}{(t^2 - r^2)^{1/2}} \right]. \tag{3.30}$$

The inversion formula involves the Radon transform itself followed by a derivative. Differentiation is an unbounded operation, and this is where the subtlety in approximating the inverse of Radon transform lies. It is a consequence of (3.30) and Exercise 3.5.10 that a radial function f vanishes outside the disk of radius L if and only if $\mathcal{R} f(t) = 0$ for $|t| \geq L$.

***Example* 3.5.1.** The characteristic function of an annular region,

$$\chi_{A_{ab}}(\boldsymbol{x}) = \chi_{[a^2,b^2]}(\|\boldsymbol{x}\|^2),$$

is a simple model for the sorts of functions encountered in medical imaging. It is piecewise differentiable with jump discontinuities. Using formula (3.19), we easily compute $\mathcal{R}\chi_{A_{ab}}$:

$$\mathcal{R}\chi_{A_{ab}}(t) = \begin{cases} 2[\sqrt{b^2 - t^2} - \sqrt{a^2 - t^2}] & \text{for } |t| \le a, \\ 2\sqrt{b^2 - t^2} & \text{for } a < |t| \le b, \\ 0 & \text{for } b < |t|. \end{cases} \tag{3.31}$$

Exercises

***Exercise* 3.5.1.** Prove formula (3.29).

***Exercise* 3.5.2.** Use formula (3.30) to compute the integrals

$$\int_r^1 \sqrt{\frac{1-t^2}{t^2-r^2}}\, t\,dt, \quad \text{for } 0 \le r \le 1.$$

***Exercise* 3.5.3.** Prove the formulæ in (3.31).

3.5.1 The Range of the Radial Radon Transform*

The remainder of this chapter is of a more advanced character, assuming some familiarity with elementary functional analysis. We study the Abel transforms and the range of the Radon transform on radial functions. The chapter concludes with an analysis of Volterra operators of the first kind. This is a class of operators which appear frequently in the study of inverse problems. The following notation is standard in functional analysis.

***Definition* 3.5.1.** The set of functions defined on $\mathbb{R}^n$ with k continuous derivatives is denoted $\mathcal{C}^k(\mathbb{R}^n)$.

Due to the simple structure of the inversion formula for radial functions, there are simple sufficient conditions for a function, ψ, to be the Radon transform of a bounded continuous, radial function. The next proposition is an example of such a result.

Proposition 3.5.1. *Let $\psi \in \mathcal{C}^2(\mathbb{R})$ satisfy the following conditions:*

1. $\psi(t) = \psi(-t)$.
2. *There is a constant M so that*

$$|\psi(t)| \le M \text{ and } |\psi'(t)| \le M.$$

3. *Both ψ and ψ' are absolutely integrable.*

Then there is a bounded, continuous function $g(\boldsymbol{x}) = G(\|\boldsymbol{x}\|)$, in the natural domain of the Radon transform, such that

$$\mathcal{R}g = \psi.$$

Proof. For a function satisfying the preceding conditions we can integrate by parts to show that, for $r > 0$,

$$\tilde{g}(r) = \int_r^\infty \frac{\psi(t)t\,dt}{\sqrt{t^2 - r^2}} = -\int_r^\infty \psi'(t)\sqrt{t^2 - r^2}\,dt. \tag{3.32}$$

As both integrals have the same limit as r tends to 0, this identity holds for $r \geq 0$. It is not difficult to prove that $\tilde{g}$ is differentiable. Set

$$G(r) = -\frac{1}{\pi r}\partial_r \tilde{g}(r) = -\frac{1}{\pi}\int_r^\infty \frac{\psi'(t)\,dt}{\sqrt{t^2 - r^2}} \tag{3.33}$$

and $g(x, y) = G(\sqrt{x^2 + y^2})$.

To show that G is bounded, we split the integral into two parts. If $r \geq 1$, then

$$\begin{aligned} |G(r)| &\leq \frac{1}{\pi}\left[\int_r^{\sqrt{2}r} \frac{M\,dt}{\sqrt{t^2 - r^2}} + \frac{1}{r}\int_{\sqrt{2}r}^\infty |\psi'(t)|\,dt\right] \\ &\leq C\left[M + \int_0^\infty |\psi'(t)|\,dt\right]. \end{aligned} \tag{3.34}$$

If $r < 1$, a different argument is required. Because ψ is twice differentiable and even, there is a constant M' so that

$$|\psi'(t)| \leq M'|t|.$$

We then have the estimate

$$\begin{aligned} |G(r)| &\leq \frac{1}{\pi}\left[\int_r^1 \frac{M'|t|\,dt}{\sqrt{t^2 - r^2}} + \int_1^\infty |\psi'(t)|\,dt\right] \\ &\leq C'\left[M' + \int_0^\infty |\psi'(t)|\,dt\right]. \end{aligned} \tag{3.35}$$

The continuity of G is left as an exercise.

To show that g is absolutely integrable on lines, we interchange order of the integrations to obtain

$$\begin{aligned} \int_0^\infty |G(r)|\,dr &\leq \frac{1}{\pi}\int_0^\infty \int_0^t \frac{|\psi'(t)|\,dr\,dt}{\sqrt{t^2 - r^2}} \\ &= \frac{1}{2}\int_0^\infty |\psi'(t)|\,dt. \end{aligned} \tag{3.36}$$

As $G(r)$ is bounded and absolutely integrable it follows that the integrals in (3.19), defining $\mathcal{R}g$, converge absolutely. It is now an elementary calculation to show that $\mathcal{R}G = \psi$. □

***Remark* 3.5.1.** Formula (3.33) gives an alternate form for the inverse of the Radon transform if $\mathcal{R}f$ satisfies the hypotheses of the proposition. A characterization of the range of the Radon transform similar to that in Proposition 3.5.1, though without the assumption of spherical symmetry, was already given by Radon.

Exercises

***Exercise* 3.5.4.** Prove that $\tilde{g}$, defined in (3.32), is a differentiable function.

***Exercise* 3.5.5.** Prove that $G(r)$ is a continuous function.

***Exercise* 3.5.6.** Prove the fact, used in (3.36), that

$$\int_0^t \frac{dr}{\sqrt{t^2 - r^2}} = \frac{\pi}{2}.$$

***Exercise* 3.5.7.** Give complete justifications for the statements that g is in the natural domain of the Radon transform and $\mathcal{R}g = \psi$.

3.5.2 The Abel Transform*

The Abel transform is a familiar feature of many problems involving measurement. It is also an important example of a nontrivial integral transform that nonetheless admits an explicit analysis. Formally, the inverse of the α-Abel transform is

$$\mathrm{A}_\alpha^{-1} = -\partial_x \mathrm{A}_{1-\alpha}. \tag{3.37}$$

This formula is a consequence of the identity

$$\int_x^s \frac{dt}{(t-x)^\alpha (s-t)^{1-\alpha}} = \Gamma(\alpha)\Gamma(1-\alpha), \tag{3.38}$$

which holds for $0 < \alpha < 1$. To derive the Abel inversion formula, we let

$$g(t) = \mathrm{A}_\alpha f = \frac{1}{\Gamma(\alpha)} \int_t^\infty \frac{f(s)\,ds}{(s-t)^{1-\alpha}}. \tag{3.39}$$

Changing the order of the integration and using the preceding identity gives

$$\begin{aligned}\frac{1}{\Gamma(1-\alpha)}\int_x^\infty \frac{g(t)\,dt}{(t-x)^\alpha} &= \frac{1}{\Gamma(1-\alpha)\Gamma(\alpha)}\int_x^\infty\int_t^\infty \frac{f(s)\,ds\,dt}{(s-t)^{1-\alpha}(t-x)^\alpha}\\ &= \frac{1}{\Gamma(1-\alpha)\Gamma(\alpha)}\int_x^\infty\int_x^s \frac{f(s)\,dt\,ds}{(s-t)^{1-\alpha}(t-x)^\alpha}\\ &= \int_x^\infty f(s)\,ds.\end{aligned} \tag{3.40}$$

Taking the derivative, we obtain

$$f(x) = -\partial_x\left[\frac{1}{\Gamma(1-\alpha)}\int_x^\infty \frac{g(t)\,dt}{(t-x)^\alpha}\right]. \tag{3.41}$$

In other words,

$$I = -\partial_x \mathrm{A}_{1-\alpha} \circ \mathrm{A}_\alpha. \tag{3.42}$$

The operator $-\partial_x \mathrm{A}_{1-\alpha}$ is a *left inverse* to A_α.

Our derivation of the inversion formula is a formal computation, assuming that the various manipulations make sense for the given function f. The main points are the interchange of the order of integrations in the second line of (3.40) and the application of the fundamental theorem of calculus in (3.41). If f is continuous and absolutely integrable, then these steps are easily justified. For an L^1-function f, it follows that

$$\mathrm{A}_{1-\alpha} \circ \mathrm{A}_\alpha f = \int_x^\infty f(s)\,ds.$$

If f is also continuous, then this indefinite integral is differentiable and therefore

$$f = \partial_x \mathrm{A}_{1-\alpha} \circ \mathrm{A}_\alpha f.$$

It is not reasonable to use continuous functions to model the data of interest in medical imaging. Piecewise continuous functions provide a more accurate description. A piecewise continuous function of one variable can be represented as a sum

$$f(x) = f_c(x) + \sum_{j=1}^N \alpha_j \chi_{[a_j,b_j]}(x),$$

where $f_c(x)$ belongs to $\mathscr{C}^0(\mathbb{R})$, and the other term collects the jumps in f. As noted,

$$\mathrm{A}_{1-\alpha} \circ \mathrm{A}_\alpha \chi_{[a,b]}(x) = \int_x^\infty \chi_{[a,b]}(s)\,ds.$$

If $x \neq a$ or b, then this function is differentiable with derivative 0 or 1. In order to interpret the formula at the exceptional points, we need to extend our notion of differentiability.

***Definition* 3.5.2.** A locally integrable function f has a *weak derivative* if there is a locally integrable function f_1 such that, for every continuously differentiable function, with bounded support g, the following identity holds:

$$\int_{-\infty}^{\infty} f(x)g'(x)\,dx = -\int_{-\infty}^{\infty} f_1(x)g(x)\,dx. \tag{3.43}$$

In this case f_1 is called the weak derivative of f.

If f is a differentiable function, then formula (3.43), with $f_1 = f'$, is just the usual integration by parts formula. The weak derivative of the indefinite integral of a piecewise continuous function is the function itself. This shows that the inversion formula, properly understood, is also valid for the sort of data that arise in imaging applications.

***Remark* 3.5.2.** Often the α-Abel transform is defined by

$$\mathscr{A}_\alpha f(t) = \frac{1}{\Gamma(\alpha)} \int_0^t \frac{f(s)\,ds}{(t-s)^{1-\alpha}}, \qquad \text{for } t \geq 0.$$

As before, α is in the interval (0, 1]. Using (3.38), we can show, at least formally, that

$$\mathscr{A}_\alpha^{-1} = \partial_x \mathscr{A}_{1-\alpha}. \tag{3.44}$$

The derivation of (3.44) is left as an exercise.

Exercises

***Exercise* 3.5.8.** Prove (3.38) by using the change of variables

$$t = \lambda x + (1-\lambda)s$$

and the classical formula

$$\int_0^1 \frac{d\lambda}{\lambda^\alpha (1-\lambda)^{1-\alpha}} = \Gamma(\alpha)\Gamma(1-\alpha);$$

see [130].

***Exercise* 3.5.9.** Let f be a piecewise continuous, L^1-function and $0 < \alpha \leq 1$. Show that if $\mathrm{A}_\alpha f(x) = 0$ for $x > R$, then $f(x) = 0$ for $x > R$ as well.

***Exercise* 3.5.10.** Use Exercise 3.5.9 to prove the following uniqueness result for the Radon transform. If f is a piecewise continuous, radial function in the natural domain of the Radon transform and $\mathscr{R}f(t) = 0$ for $|t| > R$, then $f(r) = 0$ if $r > R$.

***Exercise* 3.5.11.** Generalize the argument given in (3.39)– (3.40) to prove that

$$\mathrm{A}_\alpha \circ \mathrm{A}_\beta = \mathrm{A}_{\alpha+\beta}\,.$$

For what range of α and β does this formula make sense?

***Exercise* 3.5.12.** For $0 < a < b$ compute $g_{a,b} = \mathrm{A}_\alpha(\chi_{[a,b]})$ and verify by explicit calculation that $\chi_{[a,b]}$ is the weak derivative of $-\mathrm{A}_{1-\alpha}(g_{a,b})$.

***Exercise* 3.5.13.** Provide the detailed justification for the derivation of (3.42) for f a continuous L^1-function.

***Exercise* 3.5.14.** Suppose that $f \in \mathscr{C}^1(\mathbb{R})$ and that f and f' are absolutely integrable. Show that

$$\mathrm{A}_\alpha\left[-\partial_x \mathrm{A}_{1-\alpha}\right] f = f.$$

***Exercise* 3.5.15.** Suppose that f is an L^1-function. Show that f is the weak derivative of

$$F(x) = -\int\limits_x^\infty f(s)\,ds.$$

***Exercise* 3.5.16.** Derive the inversion formula (3.44) for the operator $\mathscr{A}_\alpha$. What hypotheses on f are needed to conclude that

$$\partial_x \mathscr{A}_{1-\alpha}(\mathscr{A}_\alpha f) = f?$$

How about

$$\mathscr{A}_\alpha(\partial_x \mathscr{A}_{1-\alpha} f) = f?$$

3.5.3 Volterra Equations of the First Kind*

See: A.2.6, A.6.2.

The Abel transforms are examples of a class of integral operators called Volterra operators. These operators are infinite-dimensional generalizations of upper triangular matrices. A linear transformation K is a Volterra operator of the first kind if it can be represented in the form

$$Kf(x) = \int\limits_0^x k(x,y) f(y)\,dy.$$

This differs a little from the form of the Abel transform as the integral there extends from x to infinity, rather than 0 to x. The function $k(x, y)$ is called the *kernel function* of the integral operator K. The kernel functions for the Abel transforms are singular where $x = y$. In this section we restrict ourselves to kernel functions that satisfy an estimate of the form

$$|k(x, y)| \leq M,$$

and analyze Volterra operators acting on functions defined on the interval [0, 1].

Volterra operators often appear in applications where one is required to solve an equation of the form

$$g = f + Kf = (\mathrm{Id} + K)f.$$

Such equations turn out to be easy to solve. Formally, we would write

$$f = (\mathrm{Id} + K)^{-1} g.$$

Still proceeding formally, we can express $(\mathrm{Id} + K)^{-1}$ as an infinite series:

$$(\mathrm{Id} + K)^{-1} f = \sum_{j=0}^{\infty} (-1)^j K^j f. \tag{3.45}$$

This is called the *Neumann series* for $(\mathrm{Id} + K)^{-1}$; it is obtained from the Taylor expansion of the function $(1 + x)^{-1}$ about $x = 0$:

$$(1 + x)^{-1} = \sum_{j=0}^{\infty} (-1)^j x^j.$$

Here $K^j f$ means the j-fold composition of K with itself. The sum on the right-hand side of (3.45) is an infinite sum of functions, and we need to understand in what sense it converges. That is, we need to choose a norm to measure the sizes of the terms in this sum. A useful property of Volterra operators is that this series converges for almost any reasonable choice of norm.

The basic estimates are summarized in the proposition.

Proposition 3.5.2. *Let* $1 \leq p \leq \infty$. *Suppose that* $|k(x, y)| \leq M$ *and* $f \in L^p([0, 1])$. *Then for* $x \in [0, 1]$ *and* $j \geq 1$

$$|K^j f(x)| \leq \frac{M^j x^{j-1}}{(j-1)!} \|f\|_{L^p}. \tag{3.46}$$

Proof. If $f \in L^p([0, 1])$ for a $p \geq 1$, then $f \in L^1([0, 1])$ and Hölder's inequality implies that

$$\|f\|_{L^1} \leq \|f\|_{L^p}.$$

In light of this, it suffices to prove (3.46) with $p = 1$. The proof is by induction on j. First consider $j = 1$:

$$\begin{aligned} |Kf(x)| &= \left| \int_0^x k(x, y) f(y)\, dy \right| \\ &\leq \int_0^x M |f(y)|\, dy \\ &\leq M \|f\|_{L^1}. \end{aligned} \tag{3.47}$$

This verifies (3.46) for $j = 1$; assume it has been proved for j. Then

$$\begin{aligned} |K^{j+1} f(x)| &= \left| \int_0^x k(x,y) K^j f(y)\, dy \right| \\ &\le \int_0^x M \frac{M^j y^{j-1}}{(j-1)!} \|f\|_{L^1}\, dy \\ &\le \frac{M^{j+1} x^j}{j!} \|f\|_{L^1}. \end{aligned} \tag{3.48}$$

This completes the induction step and thereby the proof of the proposition. □

The proposition implies that $(\mathrm{Id} + K)^{-1} f - f$ converges pointwise uniformly, even if f is only in $L^p([0,1])$. Indeed we have the pointwise estimate

$$|(\mathrm{Id} + K)^{-1} f(x) - f(x)| \le M \|f\|_{L^1} \sum_{j=0}^{\infty} \frac{M^j x^j}{j!} = M \|f\|_{L^1} e^{Mx}. \tag{3.49}$$

Proposition 3.5.3. *If* $f \in L^p([0,1])$, *then the equation* $f = (\mathrm{Id} + K) g$ *has a unique solution of the form* $g = f + f_0$, *where* f_0 *is a continuous function on* $[0,1]$ *that satisfies the estimate*

$$|f_0(x)| \le M \|f\|_{L^p} e^{Mx}.$$

In applications sometimes we encounter equations of the form

$$f = Kg, \tag{3.50}$$

where K is a Volterra operator of the first kind. A similar equation arose in the previous section. Provided that $k(x,y)$ is differentiable and $k(x,x)$ does not vanish, this sort of equation can be reduced to the type of equation considered in the last proposition. If (3.50) is solvable, then f must, in some sense, be differentiable. We formally differentiate equation (3.50) to obtain

$$f'(x) = k(x,x) g(x) + \int_0^x k_x(x,y) g(y)\, dy.$$

If K' denotes the Volterra operator with kernel function $k_x(x,y)/k(x,x)$, then this equation can be rewritten

$$\frac{f'(x)}{k(x,x)} = (\mathrm{Id} + K') g.$$

Applying our earlier result, we obtain the solution of the original equation in the form

$$g = (\mathrm{Id} + K')^{-1} \left(\frac{f'(x)}{k(x,x)} \right) = \left(\frac{f'(x)}{k(x,x)} \right) + \sum_{j=1}^{\infty} (-1)^j (K')^j \left(\frac{f'(x)}{k(x,x)} \right). \tag{3.51}$$

In applications K describes a measurement process and f represents measurements. In this context it can be quite difficult to approximate f' accurately. As a result, it is often stated that a problem that involves solving an equation of the form (3.50) is *ill posed.* Small errors in measurement can lead to substantial errors in the solution of this type of equation. While it is reasonable to expect that we can control measurement errors in the sup norm, it is usually not possible to control errors in the derivatives of measurements, even in an L^p-norm. The inverse problem is ill posed because K^{-1} is not continuous as a map from $\mathscr{C}^0([0, 1])$ to itself.

***Remark* 3.5.3.** *Tikhonov regularization* provides a general method for finding approximate solutions to equation like (3.50), which is more stable than the exact solution. An exposition of this method can be found in [49]. The material in this section is a small sample from the highly developed field of integral equations. A good introductory treatment can be found in [131] or [106].

Exercises

***Exercise* 3.5.17.** Suppose that instead of assuming that $k(x, y)$ is uniformly bounded we assume that

$$\int_0^x |k(x, y)|^q \, dy \leq M$$

for a $1 < q < \infty$ and all $x \in [0, 1]$. Show that estimates analogous to (3.46) hold for $f \in L^p([0, 1])$, where $p = q(q-1)^{-1}$.

***Exercise* 3.5.18.** Using the previous exercise, show that the equation $g = (\text{Id} + K)f$ is solvable for $g \in L^p([0, 1])$.

***Exercise* 3.5.19.** Volterra operators of the first kind are infinite-dimensional generalizations of strictly upper triangular matrices. These are matrices a_{ij} such that $a_{ij} = 0$ if $i \leq j$. Suppose that A is an $n \times n$ strictly upper triangular matrix. Show that $A^n = 0$. Prove that $I + A$ is always invertible, and give a formula for its inverse.

3.6 Conclusion

The interaction of x-rays with matter is described, via Beer's law, in terms of the attenuation coefficient. The attenuation coefficient is a nonnegative function whose values reflect the internal structure of an object. In this chapter we have introduced a simple model for the measurements made in x-ray CT as the Radon transforms of the two-dimensional slices of the attenuation coefficient. As is typical of realistic measurement processes, the Radon transform is defined by an integral. As such it is not possible to distinguish functions that differ on very small sets, called sets of measure zero.

In the final sections we showed how to invert the Radon transform on radial functions. The inverse is the composition of an integral transform, essentially the Radon transform itself, *followed by* a derivative. While the inversion formula for radial functions is quite useful for working out simple examples and getting a feel for the Radon transform as mapping of infinite-dimensional spaces, it is not used in practical imaging applications. In the next several chapters we develop the mathematical tools needed to invert the Radon transform and process images in medical applications. The workhorse, throughout the subject, is the Fourier transform. In Chapter 6 we use the Fourier transform to derive a formula for the inverse of the Radon transform. This formula is the starting point for deriving the practical algorithms used in medical image reconstruction.

Chapter 4

Introduction to the Fourier Transform

In this chapter we introduce the Fourier transform and review some of its basic properties. The Fourier transform is the "Swiss army knife" of mathematical analysis; it is a sturdy general-purpose tool with many useful special features. The Fourier transform makes precise the concept of decomposing a function into its harmonic components. In engineering it is used to define the power spectrum and describe many filtering operations. Because it has efficient numerical approximations, it forms the foundation for most image and signal processing algorithms.

Using its properties as a linear transformation of infinite-dimensional, normed vector spaces, we define the Fourier transform and its inverse for several different spaces of functions. Among other things we establish the Parseval formula relating the energy content of a function to that of its Fourier transform. We also study the connections between the smoothness and decay of a function and that of its Fourier transform.

In marked contrast to the Radon transform, the theory of the Fourier transform is largely independent of the dimension: The theory of the Fourier transform for functions of one variable is formally the same as the theory for functions of 2, 3, or n variables. For simplicity we begin with a discussion of the basic concepts for functions of a single variable, though in some definitions, where there is no additional difficulty, we treat the general case from the outset. The chapter ends with a recapitulation of the main results for functions of n variables.

4.1 The Complex Exponential Function

See: 2.3, A.3.1.

The building block for the Fourier transform is the complex exponential function, e^{ix}.

We also use the notation $\exp(ix)$ for this function. The basic facts about the exponential function can be found in Section A.3.1. Recall that the polar coordinates (r, θ) correspond to the point with rectangular coordinates $(r\cos\theta, r\sin\theta)$. As a complex number this is

$$r(\cos\theta + i\sin\theta) = re^{i\theta}.$$

Multiplication of complex numbers is easy in the polar representation. If $z = re^{i\theta}$ and $w = \rho e^{i\phi}$, then

$$zw = re^{i\theta}\rho e^{i\phi} = r\rho e^{i(\theta+\phi)}.$$

A positive number r has a real logarithm, $s = \log r$, so that a complex number can also be expressed in the form

$$z = e^{s+i\theta}.$$

The logarithm of z is the complex number defined by

$$\log z \stackrel{d}{=} s + i\theta = \log|z| + i\tan^{-1}\left(\frac{\operatorname{Im} z}{\operatorname{Re} z}\right).$$

As $\exp(2\pi i) = 1$, the imaginary part of the $\log z$ is only determined up to integer multiplies of 2π.

Using the complex exponential we build a family of functions, $\{e^{ix\xi} : \xi \in \mathbb{R}\}$. Sometimes we think of x as the variable and ξ as a parameter and sometimes their roles are interchanged. Thinking of ξ as a parameter, we see that $e^{ix\xi}$ is a $\frac{2\pi}{\xi}$-periodic function, that is,

$$\exp(i(x + \frac{2\pi}{\xi})\xi) = \exp(ix\xi).$$

In physical applications $e^{ix\xi}$ describes an oscillatory state with *frequency* $\frac{\xi}{2\pi}$ and *wavelength* $\frac{2\pi}{\xi}$. The goal of Fourier analysis is to represent "arbitrary" functions as linear combinations of these oscillatory states. The units associated to the frequency and wavelength are dictated by the intended application. If we are studying a function of time, then the frequency is expressed in terms of cycles-per-unit time and the wavelength is a length of time (e.g., measured in seconds). If we are studying a function of a spatial variable, then the *spatial frequency* is expressed in terms of cycles-per-unit length and the wavelength is an ordinary length (e.g., measured in centimeters).

An important feature of the exponential is that it satisfies an ordinary differential equation:

$$\partial_x e^{ix\xi} = i\xi e^{ix\xi}. \tag{4.1}$$

This follows easily from (A.40). Loosely speaking, this formula says that $e^{ix\xi}$ is an *eigenvector* with *eigenvalue* $i\xi$ for the linear operator ∂_x. A linear differential operator with constant coefficients is defined by

$$Df = \sum_{j=0}^{m} a_j \partial_x^j f,$$

where the coefficients, $\{a_0, \ldots, a_m\}$, are complex numbers. It is a simple consequence of (4.1) that for each complex number ξ

$$De^{ix\xi} = \left[\sum_{j=0}^{m} a_j (i\xi)^j\right] e^{ix\xi}. \tag{4.2}$$

This observation explains the centrality of the exponential function in the analysis of physical models: Many physical systems are described by linear, constant coefficient, differential operators, and the complex exponentials "diagonalize" all such operators!

Exercises

***Exercise* 4.1.1.** If a is a real number then it is a consequence of the fundamental theorem of calculus that

$$\int_0^x e^{ay} dy = \frac{e^{ax} - 1}{a}. \tag{4.3}$$

Use the power series for the exponential to prove that this formula remains correct, even if a is a complex number.

***Exercise* 4.1.2.** Use the power series for the exponential to prove that (4.1) continues to hold for ξ any complex number.

***Exercise* 4.1.3.** Derive equation (4.2).

***Exercise* 4.1.4.** Use the differential equation satisfied by e^x to show that $e^x e^{-x} = 1$. *Hint:* Use the uniqueness theorem for solutions of ODEs.

***Exercise* 4.1.5.** If $\operatorname{Re} a < 0$, then the improper integral is absolutely convergent:

$$\int_0^\infty e^{ax} dx = \frac{-1}{a}.$$

Using the triangle inequality (not the explicit formula), show that

$$\left|\int_0^\infty e^{ax} dx\right| \leq \frac{1}{|\operatorname{Re} a|}.$$

***Exercise* 4.1.6.** Which complex numbers have purely imaginary logarithms?

4.2 The Fourier Transform for Functions of a Single Variable

We now turn our attention to the Fourier transform for functions of a single real variable. As the complex exponential itself assumes complex values, it is natural to consider complex-valued functions from the outset. The theory for functions of several variables is quite similar and is treated later in the chapter.

4.2.1 Absolutely Integrable Functions

See: 2.3.2, A.4.1.

Let f be a function defined on $\mathbb{R}^n$. We say that f is *locally absolutely integrable* if

$$\int_{\|x\|<R} |f(x)|\,dx$$

is defined and finite for any R, and f is an absolutely integrable or L^1-function if

$$\|f\|_1 = \int_{\mathbb{R}^n} |f(x)|\,dx < \infty.$$

The set of L^1-functions is a vector space. It would be natural to use $\|\cdot\|_1$ to define a norm on this vector space, but there is a small difficulty: If f is supported on a set of measure zero, then $\|f\|_1 = 0$. In other words, there are nonzero L^1-functions with "norm" zero. As real measurements are usually expressible as integrals, two functions that differ on a set of measure zero cannot be distinguished by any practical measurement. For example, the functions $\chi_{[0,1]}$ and $\chi_{[0,1)}$ are indistinguishable from the point of view of measurements and of course

$$\|\chi_{[0,1]} - \chi_{[0,1)}\|_1 = 0.$$

The way to circumvent this technical problem is to *declare* that two L^1-functions, f_1 and f_2, are the *same* whenever $f_1 - f_2$ is supported on a set of measure zero. In other words, we identify two states that cannot be distinguished by any realistic measurement. This defines a equivalence relation on the set of integrable functions. The normed vector space $L^1(\mathbb{R}^n)$ is defined to be the set of L^1-functions modulo this equivalence relation with norm defined by $\|\cdot\|_1$. This is a complete, normed linear space.

This issue arises whenever an integral is used to define a norm. Students unfamiliar with this concept need not worry: As it plays very little role in imaging, we will usually be sloppy and ignore this point, acting as if the elements of $L^1(\mathbb{R}^n)$ and similar spaces are ordinary functions rather than equivalence classes of functions.

4.2.2 The Fourier Transform for Integrable Functions*

The natural domain for the Fourier transform is the space of L^1-functions.

***Definition* 4.2.1.** The Fourier transform of an L^1-function f, defined on $\mathbb{R}$, is the function $\hat{f}$ defined on $\mathbb{R}$ by the integral

$$\hat{f}(\xi) = \int_{-\infty}^{\infty} f(x) e^{-ix\xi}\,dx. \tag{4.4}$$

The utility of the Fourier transform stems from the fact that f can be "reconstructed" from $\hat{f}$. A result that suffices for most of our applications is the following:

Theorem 4.2.1 (Fourier inversion formula). *Suppose that f is an L^1-function such that $\hat{f}$ is also in $L^1(\mathbb{R})$. Then*

$$f(x) = \frac{1}{2\pi}\int_{-\infty}^{\infty} \hat{f}(\xi)e^{ix\xi}d\xi. \tag{4.5}$$

***Remark* 4.2.1.** Formula (4.5) is called the *Fourier inversion formula.* It is *the* prototype of all reconstruction formulæ used in medical imaging.

Proof. We give a proof of the inversion formula under the additional assumption that f is continuous, this assumption is removed in Section 5.1. We need to show that

$$f(x) = \frac{1}{2\pi}\int_{-\infty}^{\infty} \hat{f}(\xi)e^{i\xi x}d\xi.$$

Because $\hat{f}$ is in $L^1(\mathbb{R})$, it is not difficult to show that

$$\begin{aligned}\frac{1}{2\pi}\int_{-\infty}^{\infty} \hat{f}(\xi)e^{i\xi x}d\xi &= \lim_{\epsilon\to 0^+}\frac{1}{2\pi}\int_{-\infty}^{\infty} \hat{f}(\xi)e^{-\epsilon\xi^2}e^{i\xi x}d\xi \\ &= \lim_{\epsilon\to 0^+}\frac{1}{2\pi}\int_{-\infty}^{\infty}\int_{-\infty}^{\infty} f(y)e^{-\epsilon\xi^2}e^{i\xi(x-y)}dyd\xi.\end{aligned} \tag{4.6}$$

For each positive ϵ, Fubini's theorem implies that we can interchange the order of the integrations in the last formula. Using Example 4.2.4 to compute the Fourier transform of the Gaussian, we get

$$\begin{aligned}\lim_{\epsilon\to 0^+}\frac{1}{2\pi}\int_{-\infty}^{\infty}\int_{-\infty}^{\infty} f(y)e^{-\epsilon\xi^2}e^{i\xi(x-y)}d\xi dy &= \lim_{\epsilon\to 0^+}\frac{1}{2\sqrt{\epsilon\pi}}\int_{-\infty}^{\infty} f(y)e^{-\frac{(x-y)^2}{4\epsilon}}dy \\ &= \lim_{\epsilon\to 0^+}\frac{1}{\sqrt{\pi}}\int_{-\infty}^{\infty} f(x-2\sqrt{\epsilon}t)e^{-t^2}dt.\end{aligned} \tag{4.7}$$

As f is continuous and integrable, it follows that the limit in the last line is

$$\frac{f(x)}{\sqrt{\pi}}\int_{-\infty}^{\infty} e^{-t^2}dt.$$

The proof is completed by observing that the integral or e^{-t^2} equals $\sqrt{\pi}$. □

***Remark* 4.2.2.** The Fourier transform and its inverse are integral transforms that are frequently thought of as mappings. In this context it is customary to use the notation

$$\begin{aligned}\mathcal{F}(f) &= \int_{-\infty}^{\infty} f(x)e^{-ix\xi}\,dx,\\ \mathcal{F}^{-1}(f) &= \frac{1}{2\pi}\int_{-\infty}^{\infty} f(\xi)e^{ix\xi}\,d\xi.\end{aligned} \tag{4.8}$$

Observe that the operation performed to recover f from $\hat{f}$ is almost the same as the operation performed to obtain $\hat{f}$ from f. Indeed, if $f_r(x) \stackrel{d}{=} f(-x)$, then

$$\mathcal{F}^{-1}(f) = \frac{1}{2\pi}\mathcal{F}(f_r). \tag{4.9}$$

This symmetry accounts for many of the Fourier transform's remarkable properties. As the following example shows, the assumption that f is in $L^1(\mathbb{R})$ does not imply that $\hat{f}$ is as well.

***Example* 4.2.1.** Recall that $\chi_{[a,b)}(x)$ equals 1 for $a \le x < b$ and is otherwise zero. Its Fourier transform is given by

$$\hat{\chi}_{[a,b)}(\xi) = \frac{e^{-ia\xi} - e^{-ib\xi}}{i\xi}. \tag{4.10}$$

***Example* 4.2.2.** Define the function

$$r_1(x) = \begin{cases} 1 \text{ for } -1 < x < 1,\\ 0 \text{ for } 1 < |x|. \end{cases} \tag{4.11}$$

The Fourier transform of r_1 is

$$\widehat{r_1}(\xi) = \int_{-1}^{1} e^{-i\xi x}\,dx = \left.\frac{e^{-i\xi x}}{-i\xi}\right|_{-1}^{1} = \frac{2\sin\xi}{\xi},$$

and

$$\int_{-\infty}^{\infty} |\widehat{r_1}(\xi)|\,d\xi = 2\int_{-\infty}^{\infty} \frac{|\sin\xi|}{|\xi|}$$

diverges. So while r_1 is absolutely integrable, its Fourier transform $\hat{r}_1$ is not. The Fourier transform of $\frac{1}{2}r_1$ is such an important function in image processing, it is called the "sinc" function[1]:

$$\operatorname{sinc}(x) \stackrel{d}{=} \frac{\sin(x)}{x}.$$

[1] This differs slightly from the definition used in engineering texts where $\operatorname{sinc}(x)$ is defined to be $\frac{\sin(\pi x)}{\pi x}$.

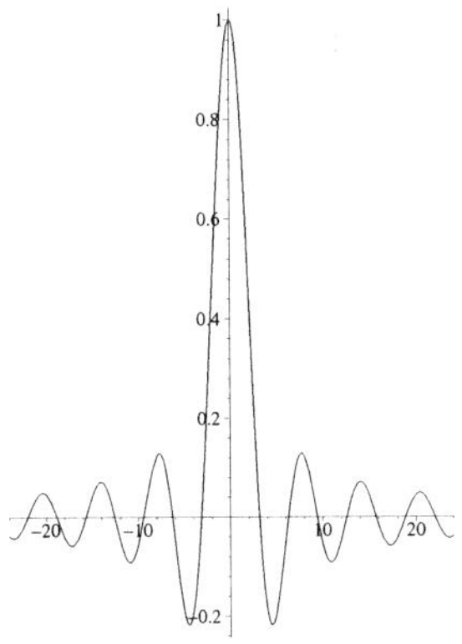

Figure 4.1. The sinc function.

A graph of the sinc function is shown in Figure 4.1.

***Example* 4.2.3.** A family of functions arising in magnetic resonance imaging consists of those of the form

$$f(x) = \chi_{[0,\infty)}(x)e^{iax}e^{-\beta x}, \ \alpha \in \mathbb{R} \text{ and } \beta > 0.$$

By simply computing the integral, we find that

$$\hat{f}(\xi) = \frac{1}{\beta + i(\xi - \alpha)}.$$

Using the fact that $e^{iax} = \cos(\alpha x) + i\sin(\alpha x)$, it is not difficult to show that

$$\begin{aligned} \mathcal{F}(\cos(\alpha x)e^{-\beta x}\chi_{[0,\infty)}(x)) &= \frac{\beta + i\xi}{\beta^2 + \alpha^2 - \xi^2 + 2i\xi\beta} \\ &\text{and} \\ \mathcal{F}(\sin(\alpha x)e^{-\beta x}\chi_{[0,\infty)}(x)) &= \frac{\alpha}{\beta^2 + \alpha^2 - \xi^2 + 2i\xi\beta}. \end{aligned} \tag{4.12}$$

***Example* 4.2.4.** The *Gaussian*, e^{-x^2}, is a function of considerable importance in image processing and mathematics. Its Fourier transform was already used in the proof of the inversion formula. For later reference we record its Fourier transform:

$$\begin{aligned} \mathcal{F}(e^{-x^2})(\xi) &= \int_{-\infty}^{\infty} e^{-x^2}e^{-i\xi x}\,dx \\ &= \sqrt{\pi}e^{-\frac{\xi^2}{4}}, \end{aligned} \tag{4.13}$$

or, more generally,

$$\mathcal{F}(e^{-ax^2})(\xi) = \sqrt{\frac{\pi}{a}}e^{-\frac{\xi^2}{4a}}. \tag{4.14}$$

This is derived in Section 4.2.3. Note that $e^{-\frac{x^2}{2}}$ is an eigenvector of the Fourier transform, with eigenvalue $\sqrt{2\pi}$. The Gaussian and its Fourier transform are shown in figure 4.2.

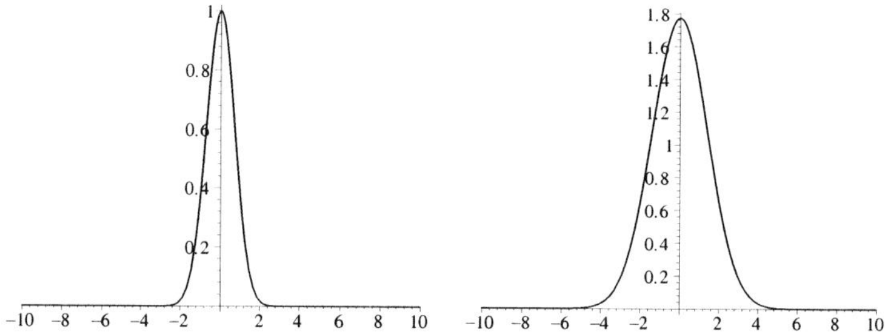

(a) The Gaussian e^{-x^2}. (b) Its Fourier transform $\sqrt{\pi}e^{-\frac{\xi^2}{4}}$.

Figure 4.2. The Gaussian and its Fourier transform.

Table 4.1 shows graphs of several functions and their Fourier transforms.

4.2.3 Appendix: The Fourier Transform of a Gaussian*

For completeness we include a derivation of the Fourier transform of the Gaussian e^{-x^2}. It uses the Cauchy integral formula for analytic functions of a complex variable. The Fourier transform is given by

$$\begin{aligned}\mathcal{F}(e^{-x^2})(\xi) &= \int_{-\infty}^{\infty} e^{-(x^2+ix\xi)}dx \\ &= e^{-\frac{\xi^2}{4}} \int_{-\infty}^{\infty} e^{-(x+i\xi/2)^2}dx.\end{aligned} \tag{4.15}$$

The second integral is the complex contour integral of the analytic function e^{-z^2} along the contour $\operatorname{Im} z = \xi/2$. Because e^{-z^2} decays rapidly to zero as $|\operatorname{Re} z|$ tends to infinity, Cauchy's theorem implies that the contour can be shifted to the real axis without changing the value of the integral; that is,

$$\int_{-\infty}^{\infty} e^{-(x+i\xi/2)^2}dx = \int_{-\infty}^{\infty} e^{-x^2}dx. \tag{4.16}$$

The function	f	$\hat{f}$
Tent	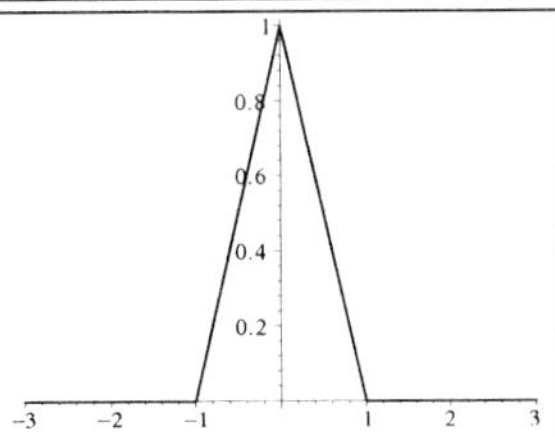	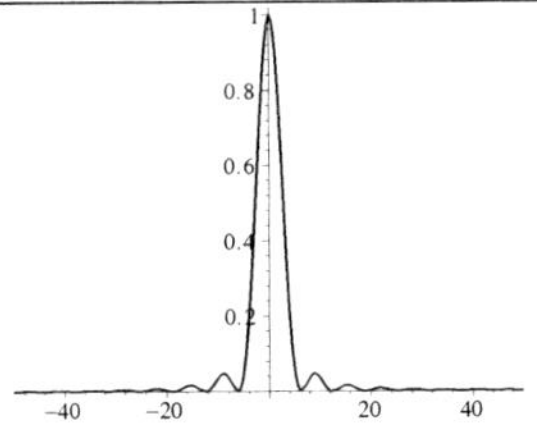
Half-tent	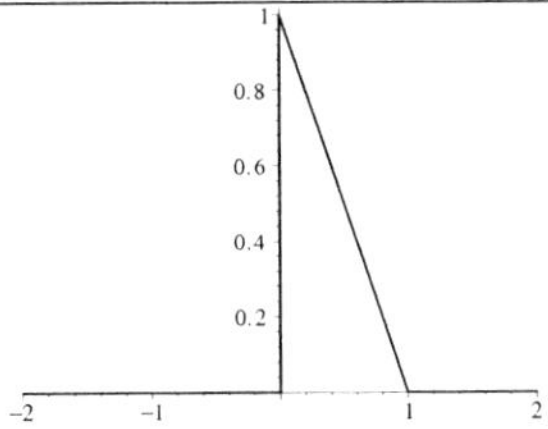	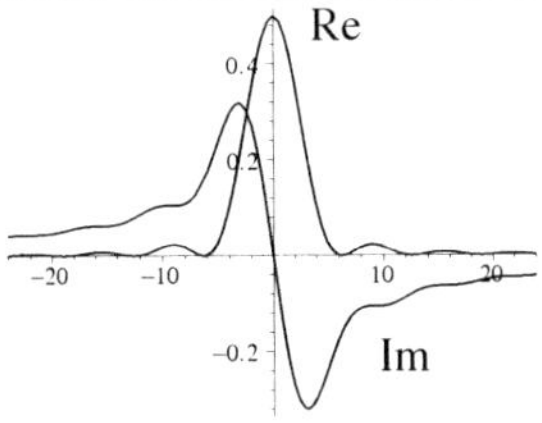
$\chi_{[-1,1]}(x)\sqrt{1-x^2}$	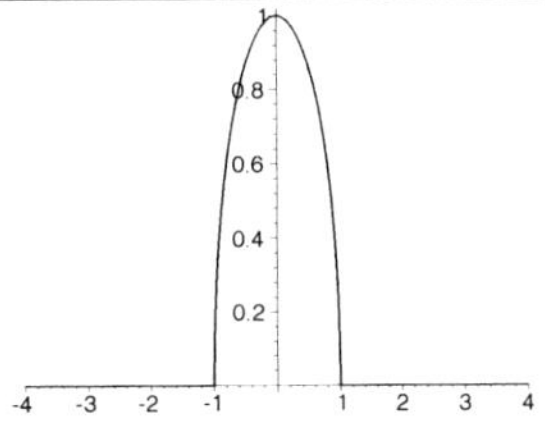	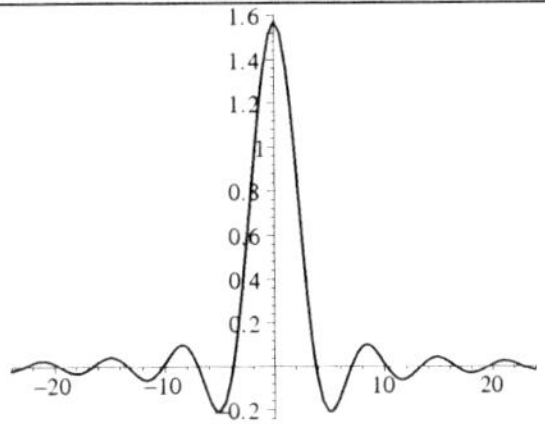
$\frac{\chi_{[-1,1]}(x)}{\sqrt{1-x^2}}$	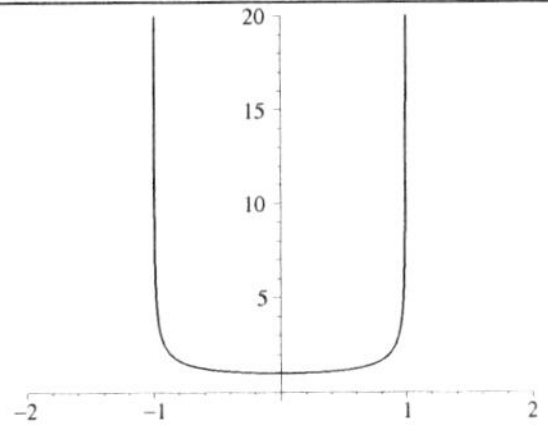	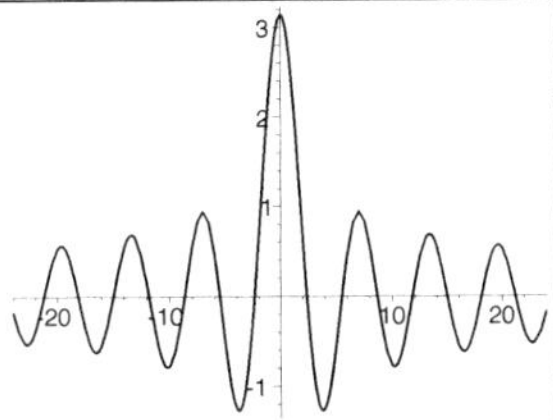
$\frac{1}{1+x^2}$	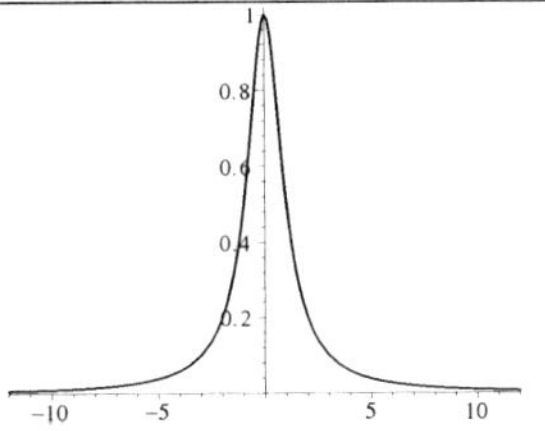	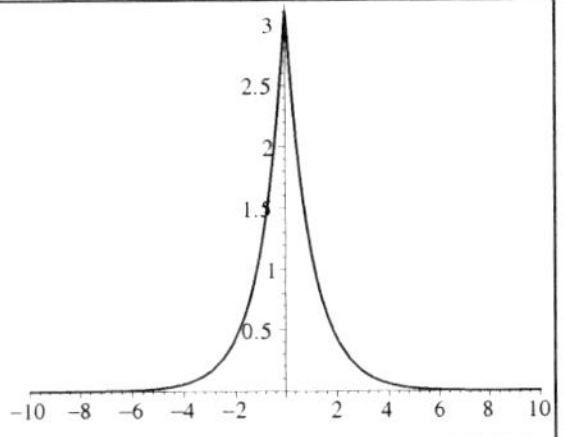
$\frac{x}{1+x^2}$	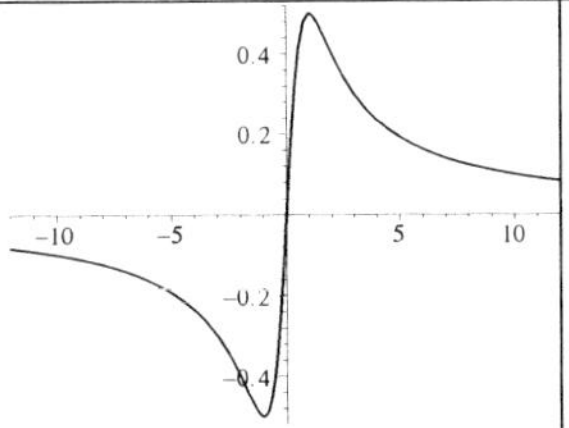	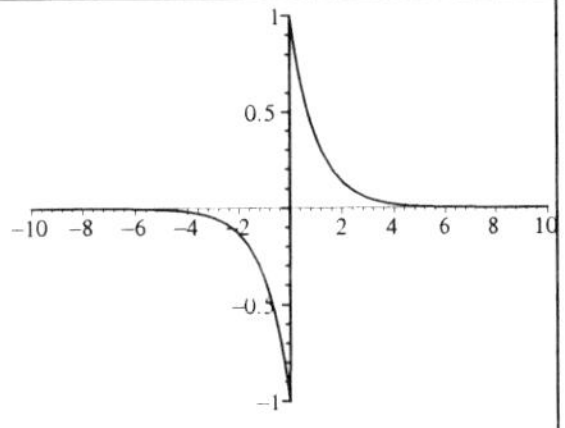
$\chi_{[0,\infty)}(x)e^{-(3-5i)x}$	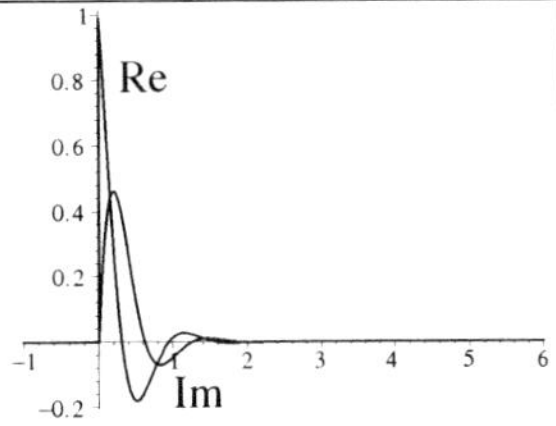	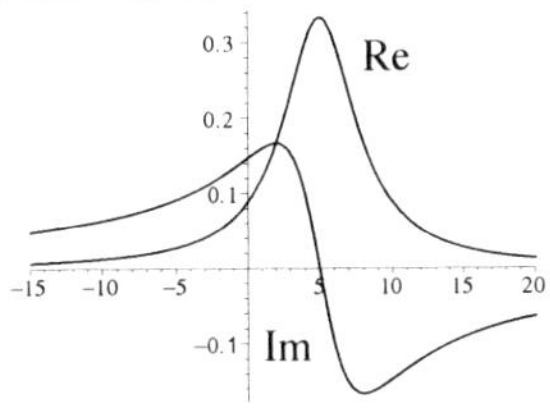

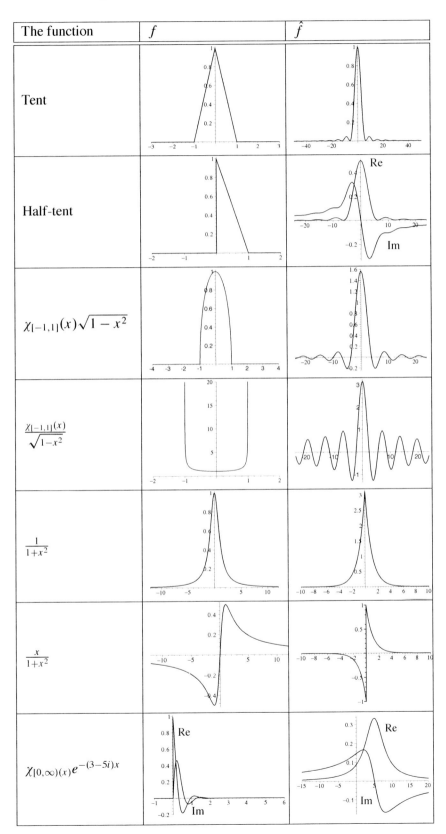

Table 4.1. Several Fourier transform pairs. Two plots on the same graph are the real and imaginary parts of the function.

To compute the last integral, observe that

$$\begin{aligned}\left[\int_{-\infty}^{\infty} e^{-x^2}dx\right]^2 &= \int_{-\infty}^{\infty} e^{-x^2}dx \int_{-\infty}^{\infty} e^{-y^2}dy \\ &= \int_0^{2\pi}\int_0^{\infty} e^{-r^2} r\,dr\,d\theta \\ &= 2\pi \int_0^{\infty} e^{-s}\frac{ds}{2} \\ &= \pi .\end{aligned} \tag{4.17}$$

Polar coordinates are used in the second line; in the second to last line we let $s = r^2$. Combining these formulæ gives

$$\mathcal{F}(e^{-x^2}) = \sqrt{\pi}e^{-\frac{\xi^2}{4}}.$$

Exercises

***Exercise* 4.2.1.** Show that if f is a continuous L^1-function then

$$\lim_{\epsilon\to 0^+} \frac{1}{\sqrt{\pi}}\int_{-\infty}^{\infty} f(x-2\sqrt{\epsilon}t)e^{-t^2}dt = f(x).$$

***Exercise* 4.2.2.** Suppose that f is absolutely integrable. Show that $\hat{f}$ is a bounded, continuous function.

***Exercise* 4.2.3.** Prove the identity (4.9).

***Exercise* 4.2.4.** Prove formula (4.10). Show that for any numbers $a < b$ there is a constant M so that

$$|\hat{\chi}_{[a,b)}(\xi)| \leq \frac{M}{1+|\xi|}.$$

***Exercise* 4.2.5.** Prove the formulæ in (4.12) and show that

$$\mathcal{F}(e^{-\beta|x|}e^{iax}) = \frac{2\beta}{\beta^2+(\xi-\alpha)^2}.$$

***Exercise* 4.2.6.** Derive the formula for $\mathcal{F}(e^{-ax^2})$ from the formula for $\mathcal{F}(e^{-x^2})$.

***Exercise* 4.2.7.** Show that for any $k \in \mathbb{N}\cup\{0\}$ the function $h_k = (\partial_x - x)^k e^{-\frac{x^2}{2}}$ is an eigenfunction of the Fourier transform. That is, $\mathcal{F}(h_k) = \lambda_k h_k$. Find λ_k. *Hint*: Integrate by parts and use induction. Find formulæ for h_1, h_2, h_3.

Exercise* 4.2.8. Give a detailed justification for (4.16).

4.2.4 Regularity and Decay*

See: A.4.1, A.5.1, B.8.

It is a general principle that the *regularity* properties of a function, f, defined on $\mathbb{R}^n$ are reflected in the *decay* properties of its Fourier transform $\hat{f}$ and, similarly, the regularity of the Fourier transform is a reflection of the decay properties of f. Without any regularity, beyond absolute integrability we have the following fundamental result:

Theorem 4.2.2 (The Riemann-Lebesgue lemma). *If f is an L^1-function, then its Fourier transform $\hat{f}$ is a continuous function that goes to zero at infinity. That is, for $\eta \in \mathbb{R}$,*

$$\lim_{\xi \to \eta} \hat{f}(\xi) = \hat{f}(\eta) \text{ and } \lim_{\xi \to \pm\infty} \hat{f}(\xi) = 0. \tag{4.18}$$

Proof. The second statement is a consequence of the basic approximation theorem for L^1-functions, Theorem A.5.2. According to this theorem, given $\epsilon > 0$ there is a step function F, given by

$$F(x) = \sum_{j=1}^{N} c_j \chi_{[a_j,b_j)}(x)$$

so that

$$\int_{-\infty}^{\infty} |f(x) - F(x)| < \epsilon.$$

Estimating the difference of their Fourier transforms gives

$$\begin{aligned} |\hat{F}(\xi) - \hat{f}(\xi)| &= \left| \int_{-\infty}^{\infty} (F(x) - f(x))e^{-ix\xi} dx \right| \\ &\leq \int_{-\infty}^{\infty} |F(x) - f(x)|\, dx \\ &\leq \epsilon. \end{aligned} \tag{4.19}$$

Since ϵ is an arbitrary positive number, it therefore suffices to show that $\lim_{|\xi|\to\infty} \hat{F}(\xi) = 0$. The Fourier transform of F is

$$\begin{aligned} \hat{F}(\xi) &= \sum_{j=1}^{N} c_j \hat{\chi}_{[a_j,b_j)}(\xi) \\ &= \sum_{j=1}^{N} c_j \frac{e^{-ib_j\xi} - e^{-ia_j\xi}}{i\xi}. \end{aligned} \tag{4.20}$$

The second line shows that there is a constant C so that

$$|\hat{F}(\xi)| \leq \frac{C}{1+|\xi|}.$$

The continuity of $\hat{f}(\xi)$ is left as an exercise. □

To go beyond (4.18), we need to introduce quantitative measures of regularity and decay. A simple way to measure regularity is through differentiation: The more derivatives a function has, the more regular it is.

***Definition* 4.2.2.** For $k \in \mathbb{N} \cup \{0\}$, the set of functions on $\mathbb{R}$ with k continuous derivatives is denoted by $\mathscr{C}^k(\mathbb{R})$. The set of infinitely differentiable functions is denoted by $\mathscr{C}^\infty(\mathbb{R})$.

Since the Fourier transform involves integration over the whole real, line it is important to assume that these derivatives are also integrable. To quantify rates of decay, we compare a function f to a simpler function such as a power of $\|\boldsymbol{x}\|$.

***Definition* 4.2.3.** A function, f, defined on $\mathbb{R}^n$, decays like $\|\boldsymbol{x}\|^{-\alpha}$ if there are constants C and R so that

$$|f(\boldsymbol{x})| \leq \frac{C}{\|\boldsymbol{x}\|^\alpha} \text{ for } \|\boldsymbol{x}\| > R.$$

This is sometimes denoted by "$f = O(\|\boldsymbol{x}\|^{-\alpha})$ as $\|\boldsymbol{x}\|$ tends to infinity."

Recall the integration by parts formula: Let f and g be differentiable functions on the interval $[a, b]$. Then

$$\int_a^b f'(x)g(x)\,dx = f(x)g(x)\Big|_{x=a}^{x=b} - \int_a^b f(x)g'(x)\,dx. \tag{4.21}$$

To use integration by parts in Fourier analysis, we need an extension of this formula with $a = -\infty$ and $b = \infty$. For our purposes it suffices to assume that fg, $f'g$ and fg' are absolutely integrable. The integration by parts formula then becomes

$$\int_{-\infty}^{\infty} f'(x)g(x)dx = -\int_{-\infty}^{\infty} f(x)g'(x)\,dx. \tag{4.22}$$

This formula follows by taking letting a and b tend to infinity in (4.21). That the integrals converge is an immediate consequence of the assumption that $f'g$ and fg' are absolutely integrable. The assumption that fg is also absolutely integrable implies the existence of sequences $< a_n >$ and $< b_n >$ so that

$$\begin{aligned} \lim_{n\to\infty} a_n = -\infty \text{ and } \lim_{n\to\infty} b_n = \infty, \\ \lim_{n\to\infty} fg(a_n) = 0 \text{ and } \lim_{n\to\infty} fg(b_n) = 0. \end{aligned} \tag{4.23}$$

Taking the limits in (4.21) along these sequences gives (4.22).

Suppose that f is an L^1-function with an absolutely integrable first derivative; that is,

$$\int_{-\infty}^{\infty} [|f(x)| + |f'(x)|]\,dx < \infty.$$

Provided $\xi \neq 0$, we can use (4.22) to obtain a formula for $\hat{f}$:

$$\begin{aligned}\hat{f}(\xi) &= \int_{-\infty}^{\infty} f(x)e^{-ix\xi}\,dx \\ &= \int_{-\infty}^{\infty} f'(x)\frac{e^{-ix\xi}}{i\xi}\,dx.\end{aligned} \tag{4.24}$$

That is,

$$\hat{f}(\xi) = \frac{\widehat{f'}(\xi)}{i\xi}.$$

Because f' is absolutely integrable, the Riemann-Lebesgue lemma implies that $\widehat{f'}$ tends to zero as $|\xi|$ tends to ∞. Combining our formula for $\hat{f}$ with this observation, we see that $\hat{f}$ goes to zero more rapidly than $|\xi|^{-1}$. This should be contrasted with the computation of the Fourier transform of r_1. The function $\hat{r}_1$ tends to zero as $|\xi|$ tends to infinity exactly like $|\xi|^{-1}$. This is a reflection of the fact that r_1 is not everywhere differentiable, having jump discontinuities at ± 1.

If f has j integrable derivatives then, by repeatedly integrating by parts, we get additional formulæ for $\hat{f}$:

$$\hat{f}(\xi) = \left[\frac{1}{i\xi}\right]^j \widehat{f^{[j]}}(\xi).$$

Again, because $f^{[j]}$ is absolutely integrable, $\widehat{f^{[j]}}$ tends to zero as $|\xi| \to \infty$. We state the result of these computations as a proposition.

Proposition 4.2.1. *Let j be a positive integer. If f has j integrable derivatives, then there is a constant C so $\hat{f}$ satisfies the estimate*

$$|\hat{f}(\xi)| \leq \frac{C}{(1+|\xi|)^j}.$$

Moreover, for $1 \leq l \leq j$, the Fourier transform of $f^{[l]}$ is given by

$$\widehat{f^{[l]}}(\xi) = (i\xi)^l \hat{f}(\xi). \tag{4.25}$$

The rate of decay in $\hat{f}$ is also reflected in the smoothness of f.

Proposition 4.2.2. *Suppose that f is absolutely integrable and j is a nonnegative integer. If $\hat{f}(\xi)(1 + |\xi|)^j$ is absolutely integrable, then f is continuous and has j continuous derivatives.*

Proof. The hypotheses of the proposition show that we may use the Fourier inversion formula to obtain

$$f(x) = \frac{1}{2\pi} \int_{-\infty}^{\infty} \hat{f}(\xi) e^{ix\xi} d\xi.$$

In light of the hypothesis on $\hat{f}$, we may differentiate this formula up to j times obtaining formulæ for derivatives of f as absolutely convergent integrals:

$$f^{[l]}(x) = \frac{1}{2\pi} \int_{-\infty}^{\infty} \hat{f}(\xi)[i\xi]^l e^{ix\xi} d\xi \qquad \text{for } 0 \leq l \leq j.$$

As $[i\xi]^l \hat{f}(\xi)$ is absolutely integrable for $l \leq j$, this formula implies that f has j continuous derivatives. □

***Remark* 4.2.3.** Note that if f has j integrable derivatives, then $\hat{f}$ decays faster than $|\xi|^{-j}$. The exact rate of decay depends on how continuous $f^{[j]}$ is. On the other hand, the rate of decay that corresponds to the hypothesis that $\hat{f}(\xi)(1+|\xi|)^j$ be integrable is $|\xi|^{-(1+j+\epsilon)}$ for any positive ϵ. So we appear to "lose" one order of differentiability when inverting the Fourier transform. Both results are actually correct. The function r_1 provides an example showing the second result is sharp. It has a jump discontinuity, and its Fourier transform, $2\,\text{sinc}(\xi)$, decays like $|\xi|^{-1}$. To construct an example showing that the first result is sharp, we begin with the case $j = 0$. By integrating the following examples, we obtain integrable functions with integrable derivatives whose Fourier transforms decay slower than $|\xi|^{-(1+\epsilon)}$, for any fixed positive $\epsilon > 0$.

***Example* 4.2.5.** Let φ be a smooth, rapidly decaying odd function with Fourier transform $\hat{\varphi}$ that satisfies the following conditions:

1. $-1 \leq \hat{\varphi}(\xi) \leq 1$ for all ξ,
2. $\hat{\varphi}(0) = 0$,
3. $\hat{\varphi}(\xi) = 0$ if $|\xi| > 1$.

For example, we could take

$$\hat{\varphi}(\xi) = \begin{cases} i\xi e^{-\frac{1}{1-\xi^2}} & \text{if } |\xi| < 1, \\ 0 & \text{if } |\xi| \geq 1. \end{cases}$$

In fact, the details of this function are not important; only the listed properties are needed to construct the examples. For each $k \in \mathbb{N}$ define the function

$$\hat{f}_k(\xi) = \sum_{n=1}^{\infty} \frac{\hat{\varphi}(\xi - n^k) + \hat{\varphi}(\xi + n^k)}{n^2}.$$

For a given ξ, at most one term in the sum is nonzero. If $k > 1$, then $\hat{f}_k$ is zero "most of the time." On the other hand, the best *rate of decay* that holds for all ξ is

$$|\hat{f}_k(\xi)| \le \frac{C}{|\xi|^{\frac{2}{k}}}.$$

By using a large k, we can make this function decay as slowly as we like. Because $\sum n^{-2} < \infty$, the Fourier inversion formula applies to give

$$f_k(x) = \varphi(x) \sum_{n=1}^{\infty} \frac{2\cos(n^k x)}{n^2}.$$

The infinite sum converges absolutely and uniformly in x, and therefore f_k is a continuous function. Because φ decays rapidly at infinity, so does f_k. This means that f_k is an continuous L^1-function whose Fourier transform goes to zero like $|\xi|^{-\frac{2}{k}}$. These examples show that the rate of decay of the Fourier transform of a continuous L^1-function can be as slow as one likes. The function, f_k, is the smooth function φ, *modulated* by noise. The graphs in figure 4.3 show the real parts of these functions; they are very "fuzzy." The fact that these functions are not differentiable is visible in figure 4.4. These graphs show f_{12} at smaller and smaller scales. Observe that f_{12} does not appear smoother at small scales than at large scales.

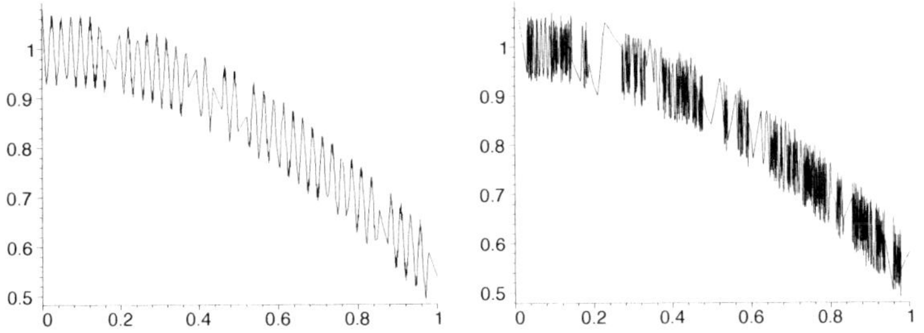

(a) $k = 8$ (b) $k = 12$

Figure 4.3. Fuzzy functions.

The function

$$F_k(x) = \int_{-\infty}^{x} f_k(y)dy$$

is continuously differentiable with an absolutely integrable derivative. It Fourier transform, given by

$$\widehat{F}_k(\xi) = \frac{\hat{f}_k(\xi)}{i\xi},$$

decays at the rate $|\xi|^{-(1+\frac{2}{k})}$.

These examples demonstrate that there are two different phenomena governing the rate of decay of the Fourier transform. The function r_1 is very smooth, except where it has a jump. This kind of very localized failure of smoothness produces a characteristic $|\xi|^{-1}$ rate of decay in the Fourier transform. In the L^1-sense the function r_1 is very close to being a continuous function. In fact, by using linear interpolation we can find piecewise differentiable functions very close to r_1. These sort of functions frequently arise in medical imaging. Each function f_k is continuous but very fuzzy. The larger k is, the higher the amplitude of the high-frequency components producing the fuzz and the slower the rate of decay for $\hat{f}_k$. These functions are not close to differentiable functions in the L^1-sense. Such functions are typical of random processes used to model noise.

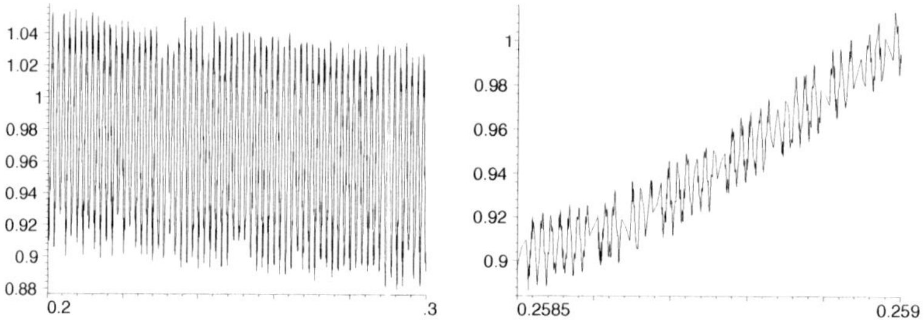

(a) (b)

Figure 4.4. The function f_{12} at smaller scales.

The foregoing results establish the connection between the regularity of f and the decay of its Fourier transform. If, on the other hand, we know that f itself decays, then this is reflected in increased regularity of its Fourier transform.

Proposition 4.2.3. *Suppose that j is a positive integer and*

$$\int_{-\infty}^{\infty} |f(x)|(1+|x|)^j dx < \infty,$$

then $\hat{f}$ has j-continuous derivatives that tend to zero as $|\xi|$ tends to infinity. In fact, for $0 \leq k \leq j$,

$$\partial_\xi^k \hat{f}(\xi) = \int_{-\infty}^{\infty} (-ix)^k f(x) e^{-ix\xi} dx. \tag{4.26}$$

Of course, (4.26) gives a formula for the Fourier transform of $x^k f(x)$ in terms of the Fourier transform of f:

$$\mathcal{F}(x^k f)(\xi) = i^k \partial_\xi^k \hat{f}(\xi). \tag{4.27}$$

A special case of this proposition arises if f vanishes outside a bounded interval. In this case $x^k f(x)$ is absolutely integrable for any positive integer k, and therefore $\hat{f}$ is an infinitely differentiable function. The derivatives tend to zero as $|\xi|$ tends to infinity, but the *rate* of decay may be the same for all the derivatives; for example,

$$\hat{r}_1(\xi) = \frac{2\sin\xi}{\xi}.$$

Differentiating this function repeatedly gives a sum of terms, one of which tends to zero exactly like $|\xi|^{-1}$. This further confirms our principle that the rate of decay of the Fourier transform is a reflection of the smoothness of the function.

***Example* 4.2.6.** An important application of the Fourier transform is to study ordinary differential equations with constant coefficients. Suppose that $\{a_0, \dots, a_n\}$ are complex numbers. We would like to study the solutions of the differential equation

$$Df \stackrel{d}{=} \sum_{j=0}^{n} a_j \partial_x^j f = g.$$

Proceeding formally, taking the Fourier transform of both sides of (4.25) gives relation

$$\left[\sum_{j=0}^{n} a_j (i\xi)^j\right] \hat{f}(\xi) = \hat{g}(\xi). \tag{4.28}$$

The polynomial

$$P_D(\xi) = \sum_{j=0}^{n} a_j (i\xi)^j$$

is called the *characteristic polynomial* for the differential operator D. If a complex number ξ_0 is a root of this equation [i.e., $P_D(\xi_0) = 0$], then the exponential function $v_0 = \exp(i\xi_0 x)$ is a solution of the homogeneous equation $Dv_0 = 0$.

If, on the other hand, P_D has no real roots and g is absolutely integrable, then we can divide in (4.28) to obtain

$$\hat{f}(\xi) = \frac{\hat{g}(\xi)}{P_D(\xi)}.$$

Using the Fourier inversion formula, we obtain a particular solution to the equation $Df = g$,

$$f_p(x) = \frac{1}{2\pi} \int_{-\infty}^{\infty} \frac{\hat{g}(\xi) e^{i\xi x} d\xi}{P_D(\xi)}. \tag{4.29}$$

The general solution is of the form $f_p + f_0$, where $Df_0 = 0$. If P_D has real roots, then a more careful analysis is required; see [12].

The Parseval Formula*

In the foregoing discussion we considered L^1-functions. The Fourier transform is then defined in terms of an absolutely convergent integral. As we observed, this does not imply that the Fourier transform is itself absolutely integrable. In fact, it is quite difficult to describe the range of $\mathcal{F}$ when the domain is $L^1(\mathbb{R})$. Using the L^1-norm, there are also discrepancies in the quantitative relationships between the smoothness of a function and the rate of decay of its Fourier transform. A more natural condition when working with the Fourier transform is *square integrability*.

***Definition* 4.2.4.** A complex-valued function f, defined on $\mathbb{R}^n$, is L^2 or square integrable if

$$\|f\|_{L^2}^2 = \int_{\mathbb{R}^n} |f(\mathbf{x})|^2 d\mathbf{x} < \infty.$$

The set of such functions, with norm defined by $\|\cdot\|_{L^2}$, is denoted $L^2(\mathbb{R}^n)$. With this norm $L^2(\mathbb{R}^n)$ is a complete, normed linear space.

The norm on $L^2(\mathbb{R}^n)$ is defined by an inner product,

$$\langle f, g\rangle_{L^2} = \int_{\mathbb{R}^n} f(\mathbf{x})\overline{g(\mathbf{x})}\, d\mathbf{x}.$$

This inner product satisfies the usual Cauchy-Schwarz inequality.

Proposition 4.2.4. *If $f, g \in L^2(\mathbb{R}^n)$, then*

$$|\langle f, g\rangle_{L^2}| \leq \|f\|_{L^2}\|g\|_{L^2}. \tag{4.30}$$

Proof. The proof of the Cauchy-Schwarz inequality for $L^2(\mathbb{R}^n)$ is formally identical to the proof for $\mathbb{C}^n$ given in the proof of Proposition 2.3.2. The verification of this fact is left to the Exercises. □

Recall that a normed linear space is complete if every Cauchy sequence has a limit. The completeness of L^2 is quite important for what follows.

***Example* 4.2.7.** The function $f(x) = (1 + |x|)^{-\frac{3}{4}}$ is not absolutely integrable, but it is square integrable. On the other hand, the function

$$g(x) = \frac{\chi_{[-1,1]}(x)}{\sqrt{|x|}}$$

is absolutely integrable but not square integrable.

An L^2-function is always locally absolutely integrable. For a function of one variable this means that for any finite interval, $[a, b]$, the integral of $|f|$ over $[a, b]$ is finite. To prove this we use the Cauchy-Schwarz inequality with $g = 1$:

$$\int_a^b |f(x)|\, dx \leq \sqrt{|b-a|}\sqrt{\int_a^b |f(x)|^2\, dx} \leq \sqrt{|b-a|}\|f\|_{L^2}.$$

The reason square integrability is a natural condition is contained in the following theorem.

Theorem 4.2.3 (Parseval formula). *If f is absolutely integrable and also square integrable, then $\hat{f}$ is square integrable and*

$$\int_{-\infty}^{\infty} |f(x)|^2\,dx = \int_{-\infty}^{\infty} |\hat{f}(\xi)|^2 \frac{d\xi}{2\pi}. \tag{4.31}$$

Though very typical of arguments in this subject, the proof of this result is rather abstract. It can safely be skipped as nothing in the sequel relies upon it.

Proof. To prove (4.31), we use the Fourier inversion formula, Propositions 4.2.1 and 4.2.3, and the following lemma.

Lemma 4.2.1. *Suppose that f and g are integrable functions that are $O(|x|^{-2})$ as $|x|$ tend to infinity. Then we have the identity*

$$\int_{-\infty}^{\infty} f(x)\hat{g}(x)\,dx = \int_{-\infty}^{\infty} \hat{f}(x)g(x)\,dx. \tag{4.32}$$

The proof of the lemma is left as an exercise.

Suppose for the moment that f is an infinitely differentiable function with bounded support. Proposition 4.2.1 shows that, for any positive k, $\hat{f}$ is $O(|\xi|^{-k})$ as $|\xi|$ tends to infinity while Proposition 4.2.3 shows that $\hat{f}$ is smooth and similar estimates hold for its derivatives. Let $g = [2\pi]^{-1}\overline{\hat{f}}$; the Fourier inversion formula implies that $\hat{g} = \bar{f}$. The identity (4.32) applies to this pair, giving

$$\int_{-\infty}^{\infty} |f(x)|^2\,dx = \frac{1}{2\pi}\int_{-\infty}^{\infty} |\hat{f}(\xi)|^2\,d\xi,$$

thus verifying the Parseval formula for the special case of smooth functions with bounded support.

In Chapter 5 we show that, for any function satisfying the hypotheses of the theorem, there is a sequence $< f_n >$ of smooth functions, with bounded support, such that

$$\lim_{n\to\infty} \|f - f_n\|_{L^1} = 0 \text{ and } \lim_{n\to\infty} \|f - f_n\|_{L^2} = 0.$$

The argument given so far applies to the differences to show that

$$\int_{-\infty}^{\infty} |f_n(x) - f_m(x)|^2\,dx = \frac{1}{2\pi}\int_{-\infty}^{\infty} |\hat{f}_n(\xi) - \hat{f}_m(\xi)|^2\,d\xi. \tag{4.33}$$

The left-hand side tends to zero as m, n tend to infinity. Therefore, $< \hat{f}_n >$ is also an L^2-Cauchy sequence converging to a limit in $L^2(\mathbb{R})$. As $< f_n >$ converges to f in $L^1(\mathbb{R})$, the sequence $< \hat{f}_n >$

converges pointwise to $\hat{f}$. This implies that $<\hat{f}_n>$ tends to $\hat{f}$ in $L^2(\mathbb{R})$ as well. This completes the proof of the theorem as

$$\int_{-\infty}^{\infty}|f(x)|^2\,dx=\lim_{n\to\infty}\int_{-\infty}^{\infty}|f_n(x)|^2\,dx=\lim_{n\to\infty}\frac{1}{2\pi}\int_{-\infty}^{\infty}|\hat{f}_n(\xi)|^2d\xi=\frac{1}{2\pi}\int_{-\infty}^{\infty}|\hat{f}(\xi)|^2\,d\xi.$$

□

In many physical applications the square integral of a function is interpreted as the total energy. Up to the factor of 2π, Parseval's formula says that the total energy in f is the same as that in $\hat{f}$. Often the variable $\frac{\xi}{2\pi}$ is thought of as a frequency. Following the quantum mechanical practice, higher frequencies correspond to higher energies. In this context $[2\pi]^{-1}|\hat{f}(\xi)|^2$ is interpreted as the *power spectral density* or energy *density* of f at frequency $\frac{\xi}{2\pi}$. As we shall see, "noise" is essentially a high-frequency phenomenon, and a noisy signal has a lot of energy at high frequencies. Figure 4.5(a) shows the real and imaginary parts of the Fourier transform of $e^{-(3-5i)x}\chi_{[0,\infty)}$ while Figure 4.5(b) shows the power spectral density for this function.

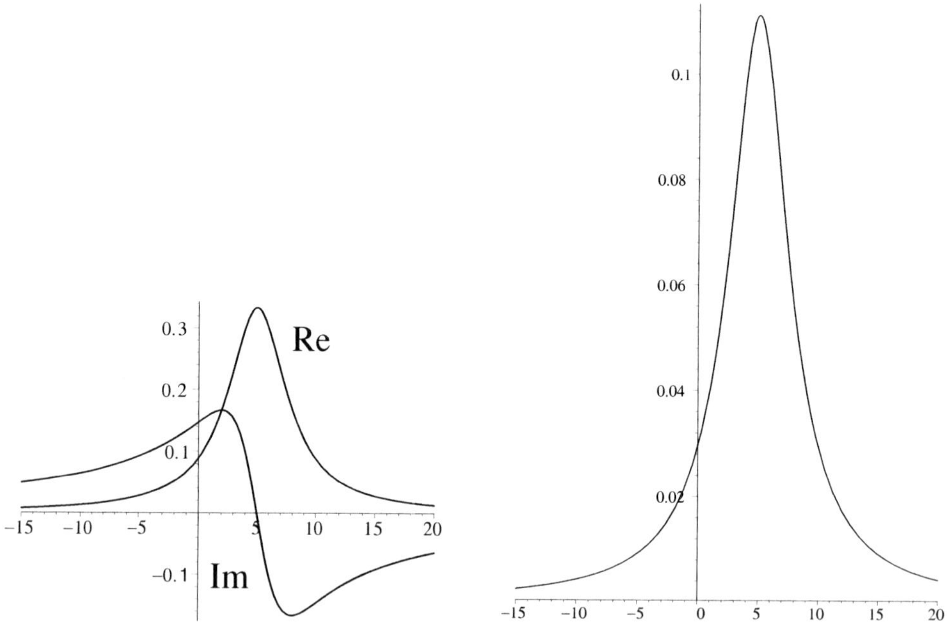

(a) The Fourier transform $\mathcal{F}(e^{-(3-5i)x}\chi_{[0,\infty)})$.

(b) The power spectral density $|\mathcal{F}(e^{-(3-5i)x}\chi_{[0,\infty)})|^2$.

Figure 4.5. The Fourier transform of $e^{-(3-5i)x}\chi_{[0,\infty)}$ and its power spectral density.

Exercises

***Exercise* 4.2.9.*** Let f be an L^1-function. Show that $\hat{f}$ is a continuous function. Extra credit: Show that $\hat{f}$ is uniformly continuous on the whole real line.

***Exercise* 4.2.10.** If fg is absolutely integrable, show that sequences exist satisfying (4.23).

***Exercise* 4.2.11.** Suppose that $f'g$ and fg' are absolutely integrable. Show that the limits

$$\lim_{x\to\infty} fg(x) \text{ and } \lim_{x\to-\infty} fg(x)$$

both exist. Does (4.22) hold even if fg is not assumed to be absolutely integrable?

***Exercise* 4.2.12.** Prove that for any number j the jth-derivative $\partial_\xi^j \hat{r}_1$ has a term that decays exactly like $|\xi|^{-1}$.

***Exercise* 4.2.13.** Show that f_p, defined in Example 4.29 and its first n derivatives tend to zero as $|x|$ tends to infinity. *Hint*: The first $(n-1)$ derivatives are easy; a different argument is needed for the nth derivative.

***Exercise* 4.2.14.** Show that the function φ defined in Example 4.2.5 is infinitely differentiable.

***Exercise* 4.2.15.** Give a detailed proof of Proposition 4.2.4. Explain the following statement: "The Cauchy-Schwarz inequality is a statement about the two-dimensional subspaces of a vector space."

4.2.5 Fourier Transform on $L^2(\mathbb{R})$

See: A.2.4, A.2.5, A.4.2.

The Parseval formula shows that the L^2-norm is intimately connected to the Fourier transform. When the L^2-norm is used in both the domain and range, Parseval's formula says that $\mathcal{F}$ is a continuous linear transformation. This result indicates that it should be possible to extend the Fourier transform to all functions in $L^2(\mathbb{R})$. This is indeed the case. Let $f \in L^2(\mathbb{R})$. For each $R > 0$, define

$$\hat{f}_R(\xi) = \int_{-R}^{R} f(x)e^{-ix\xi}dx. \tag{4.34}$$

From Parseval's formula it follows that, if $R_1 < R_2$, then

$$\|\hat{f}_{R_1} - \hat{f}_{R_2}\|_{L^2}^2 = 2\pi \int_{R_1\le|x|\le R_2} |f(x)|^2\, dx.$$

Because f is square integrable, the right-hand side of this formula goes to zero as R_1 and R_2 tend to infinity. Hence if we measure the distance in the L^2-norm, then the functions

$< \hat{f}_R >$ are clustering closer and closer together as $R \to \infty$. In other words, $< \hat{f}_R >$ is an L^2-Cauchy sequence. Because $L^2(\mathbb{R})$ is a complete, normed vector space, this implies that $< \hat{f}_R >$ converges to a limit as $R \to \infty$; this limit *defines* $\hat{f}$. The limit of a sequence in the L^2-norm is called a *limit in the mean*; it is denoted by the symbol LIM.

***Definition* 4.2.5.** If f is a function in $L^2(\mathbb{R})$, then its Fourier transform is defined to be

$$\hat{f} = \operatorname*{LIM}_{R\to\infty} \hat{f}_R,$$

where $\hat{f}_R$ is defined in (4.34).

We summarize these observations in a proposition.

Proposition 4.2.5. *The Fourier transform extends to define a continuous map from $L^2(\mathbb{R})$ to itself. If $f \in L^2(\mathbb{R})$, then*

$$\int_{-\infty}^{\infty} |f(x)|^2\,dx = \frac{1}{2\pi}\int_{-\infty}^{\infty} |\hat{f}(\xi)|^2\,d\xi.$$

Proof. The continuity statement follows from the Parseval formula. That the Parseval formula holds for all $f \in L^2(\mathbb{R})$ is a consequence of the definition of $\hat{f}$ and the fact that

$$\int_{-R}^{R} |f(x)|^2\,dx = \frac{1}{2\pi}\int_{-\infty}^{\infty} |\hat{f}_R(\xi)|^2\,d\xi \qquad \text{for } R > 0.$$

□

While the Fourier transform extends to define a continuous map from $L^2(\mathbb{R})$ to itself, there is a price to pay. The Fourier transform of a function in $L^2(\mathbb{R})$ cannot be directly defined by a simple formula like (4.4). For a function like f in Example 4.2.7, the integral defining $\hat{f}$ is not absolutely convergent.

***Example* 4.2.8.** The function

$$f(x) = \frac{1}{\sqrt{1+x^2}}$$

is square integrable but not absolutely integrable. We use integration by parts to compute $\hat{f}_R(\xi)$:

$$\hat{f}_R(\xi) = \frac{2\operatorname{sinc}(R\xi)}{\sqrt{1+R^2}} - \int_{-R}^{R} \frac{xe^{-ix\xi}}{i\xi(1+x^2)^{\frac{3}{2}}}.$$

It is now a simple matter to obtain the pointwise limit as R tends to infinity:

$$\hat{f}(\xi) = \int_{-\infty}^{\infty} \frac{xe^{-ix\xi}}{i\xi(1+x^2)^{\frac{3}{2}}}. \tag{4.35}$$

In the Exercise 4.2.16 you are asked to show that $\hat{f}_R$ converges to $\hat{f}$ in the L^2-norm.

A consequence of Parseval's formula is the identity

$$\int_{-\infty}^{\infty} f(x)\overline{g(x)}\,dx = \int_{-\infty}^{\infty} \hat{f}(\xi)\overline{\hat{g}(\xi)}\frac{d\xi}{2\pi}. \tag{4.36}$$

This is proved by applying (4.31) to $f + tg$ and comparing the coefficients of powers t on the left- and right-hand sides. Up to the factor of 2π, the Fourier transform preserves the inner product. Recall that this is also a property of rotations of Euclidean space. Such transformations of complex vector spaces are called *unitary*. Another consequence of the Parseval formula is a uniqueness statement: A function in L^2 is determined by its Fourier transform.

Corollary 4.2.1. *If* $f \in L^2(\mathbb{R})$ *and* $\hat{f} = 0$, *then* $f \equiv 0$.

***Remark* 4.2.4.** As noted in Section 4.2.1, it would be more accurate to say that the set of points for which $f \neq 0$ has measure 0.

The Fourier transform of an L^2-function is generally not absolutely integrable, so the inversion formula, proved previously, does not directly apply. The inverse is defined in much the same way as the Fourier transform itself.

Proposition 4.2.6 (Fourier inversion for $L^2(\mathbb{R})$). *For* $f \in L^2(\mathbb{R})$ *define*

$$f_R(x) = \frac{1}{2\pi}\int_{-R}^{R} \hat{f}(\xi)e^{ix\xi}\,d\xi;$$

then $f = \underset{R\to\infty}{\mathrm{LIM}}\, f_R$.

The proof is left to the Exercises.

Symmetry Properties of the Fourier Transform*

We conclude this section by summarizing the basic properties of the Fourier transform that hold for integrable or L^2-functions. These properties are consequences of elementary properties of the integral.

1. LINEARITY: The Fourier transform is a linear operation:

$$\mathcal{F}(f + g) = \mathcal{F}(f) + \mathcal{F}(g), \quad \mathcal{F}(\alpha f) = \alpha\mathcal{F}(f),\ \alpha \in \mathbb{C}.$$

2. SCALING: The Fourier transform of $f(ax)$, the function f dilated by $a \in \mathbb{R}$, is given by

$$\begin{aligned}\int_{-\infty}^{\infty} f(ax)e^{-i\xi x}\,dx &= \int_{-\infty}^{\infty} f(y)e^{-\frac{i\xi y}{a}}\frac{dy}{a} \\ &= \frac{1}{a}\hat{f}\left(\frac{\xi}{a}\right).\end{aligned} \tag{4.37}$$

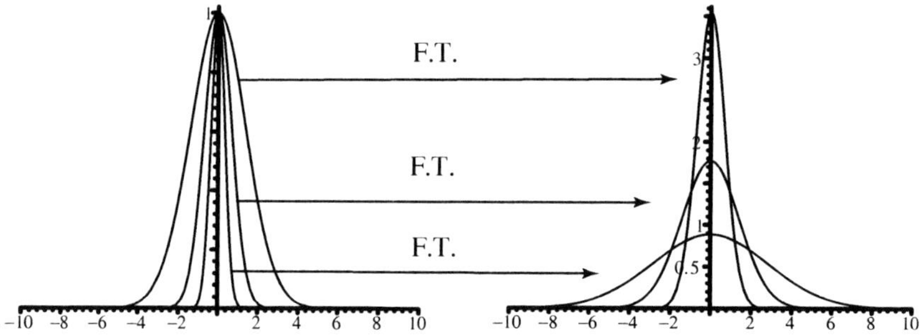

Figure 4.6. The effect of scaling on the Fourier transform.

Figure 4.6 shows the functions e^{-ax^2}, for various values of a, along with their Fourier transforms.

3. TRANSLATION: Let f_t be the function f shifted by t [i.e., $f_t(x) = f(x-t)$]. The Fourier transform of f_t is given by

$$\begin{aligned} \widehat{f_t}(\xi) &= \int_{-\infty}^{\infty} f(x-t)e^{-i\xi x}\,dx \\ &= \int f(y)e^{-i\xi(y+t)}\,dy \\ &= e^{-i\xi t}\hat{f}(\xi). \end{aligned} \tag{4.38}$$

Figure 4.7 shows the functions $e^{-(x-t)^2}$, for various values of t, along with their Fourier transforms.

4. REALITY: If f is a real-valued function, then its Fourier transform satisfies $\hat{f}(\xi) = \overline{\hat{f}(-\xi)}$. This shows that the Fourier transform of a real-valued function is completely determined by its values for positive (or negative) frequencies.

Recall the following definitions.

***Definition* 4.2.6.** A function f defined on $\mathbb{R}^n$ is *even* if $f(\boldsymbol{x}) = f(-\boldsymbol{x})$. A function f defined on $\mathbb{R}^n$ is *odd* if $f(\boldsymbol{x}) = -f(-\boldsymbol{x})$.

5. EVENNESS: If f is real-valued and even, then $\hat{f}$ is real-valued. If f is real-valued and odd, then $\hat{f}$ takes purely imaginary values. If f is even, then its Fourier transform is also even. It is given by the formula

$$\hat{f}(\xi) = 2\int_0^{\infty} f(x)\cos(\xi x)\,dx. \tag{4.39}$$

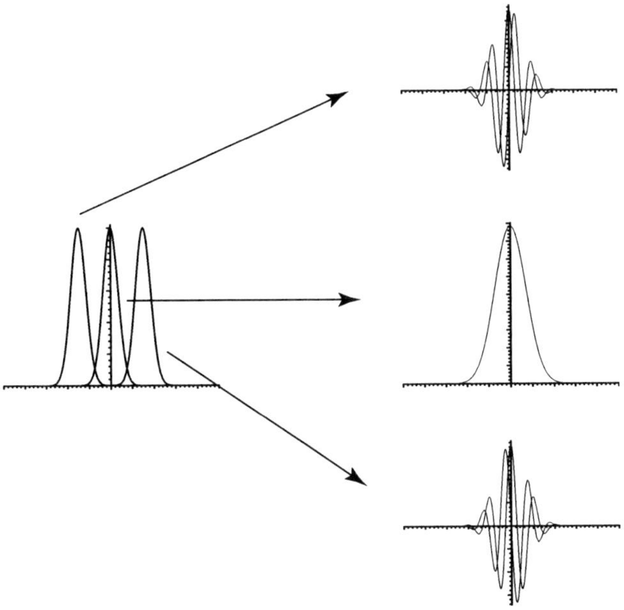

Figure 4.7. The effect of translation on the Fourier transform. The top and bottom graphs on the right show the real and imaginary parts.

Exercises

Exercise **4.2.16.** Prove that $\hat{f}_R$, defined in (4.35), converges to $\hat{f}$ in the L^2-norm.

Exercise **4.2.17.** Let $f(x) = \chi_{[1,\infty)}(x)x^{-1}$. Using the method of Example 4.2.8, compute the Fourier transform of f. Verify the convergence, in the L^2-norm, of $\hat{f}_R$ to $\hat{f}$.

Exercise **4.2.18.*** Prove Proposition 4.2.6. *Hint:* Use the Parseval formula to estimate the difference $\|f - f_R\|_{L^2}^2$.

Exercise **4.2.19.** Let $f, g \in L^2(\mathbb{R})$ by considering the functions $f + tg$, where $t \in \mathbb{C}$ show that the Parseval formula implies (4.36).

Exercise **4.2.20.*** Prove Lemma 4.2.1 and show that (4.32) holds for any pair of functions in $L^2(\mathbb{R})$.

Exercise **4.2.21.** Verify the statement that if $g = \overline{\hat{f}}$, then $\hat{g} = f$.

Exercise **4.2.22.** Show that a function $f \in L^2(\mathbb{R})$ is zero if and only if $\langle f, g\rangle = 0$ for every $g \in L^2(\mathbb{R})$. Use this fact and the formula in Exercise 4.2.20 to show that $\mathcal{F}(L^2(\mathbb{R})) = L^2(\mathbb{R})$. *Hint:* If this were false, then there would exist a nonzero function $g \in L^2(\mathbb{R})$ such that $\langle g, \hat{f}\rangle = 0$ for every $f \in L^2(\mathbb{R})$.

Exercise **4.2.23.** Verify properties (4) and (5).

Exercise **4.2.24.** Find a formula like (4.39) for the Fourier transform of an odd function.

Exercise **4.2.25.*** Suppose that m is a bounded, locally integrable function. Show that the map from $L^2(\mathbb{R})$ to itself defined by $\mathcal{A}_m(f) = \mathcal{F}^{-1}(m\hat{f})$ is continuous.

4.2.6 A General Principle in Functional Analysis*

In the previous section we extended the definition of the Fourier transform to $L^2(\mathbb{R})$ by using the Plancherel formula *and* the completeness of $L^2(\mathbb{R})$. This is an example of a general principle in functional analysis and explains, in part, why *completeness* is such an important property for a normed linear space. As we will encounter this situation again, we pause for a moment to enunciate this principle explicitly. Recall the following definition.

***Definition* 4.2.7.** Let $(V, \|\cdot\|)$ be a normed linear space. A subspace S of V is *dense* if for every $\boldsymbol{v} \in V$ there is a sequence $< \boldsymbol{v}_k > \subset S$ such that

$$\lim_{k\to\infty} \|\boldsymbol{v} - \boldsymbol{v}_k\| = 0.$$

The general principle is that a bounded linear map, defined on a dense subset, extends to the whole space.

Theorem 4.2.4. *Let $(V_1, \|\cdot\|_1)$ and $(V_2, \|\cdot\|_2)$ be normed, linear spaces and assume that V_2 is complete. Suppose that S_1 is a dense subspace of V_1 and A is a linear map from S_1 to V_2. If there exists a constant M such that*

$$\|A\boldsymbol{v}\|_2 \leq M\|\boldsymbol{v}\|_1, \tag{4.40}$$

for all $\boldsymbol{v}$ in S_1, then A extends to define a linear map from V_1 to V_2, satisfying the same estimate.

Proof. Let $\boldsymbol{v}$ be an arbitrary point in V_1 and let $< \boldsymbol{v}_k >$ be a sequence contained in S_1 converging to $\boldsymbol{v}$. Because A is linear and S_1 is a subspace, (4.40) implies that

$$\|A(\boldsymbol{v}_j - \boldsymbol{v}_k)\|_2 \leq M\|\boldsymbol{v}_j - \boldsymbol{v}_k\|_1.$$

This estimate shows that $< A\boldsymbol{v}_k >$ is a Cauchy sequence in V_2. From the completeness of V_2 we conclude that this sequence has a limit $\boldsymbol{u}$. Provisionally define $A\boldsymbol{v} = \boldsymbol{u}$. To show that $A\boldsymbol{v}$ is well defined, we need to show that if $< \boldsymbol{v}'_k > \subset S_1$ is another sequence converging to $\boldsymbol{v}$, then $< A\boldsymbol{v}'_k >$ also converges to $\boldsymbol{u}$. Since the two sequences have the same limit, the difference $\|\boldsymbol{v}_k - \boldsymbol{v}'_k\|_1$ converges to zero. The estimate (4.40) implies that

$$\|A\boldsymbol{v}_k - A\boldsymbol{v}'_k\|_2 \leq M\|\boldsymbol{v}_k - \boldsymbol{v}'_k\|_1,$$

showing that the limit is well defined. The fact that the extended map is linear is left as an exercise. □

Exercises

***Exercise* 4.2.26.** Show that the extension of A defined in the proof of Theorem 4.2.4 is linear.

***Exercise* 4.2.27.** Show that the only dense subspace of a finite-dimensional normed linear space is the whole space.

4.3 Functions With Weak Derivatives

See: A.4.5, B.4.

In Section 4.2.5 we extended the definition of the Fourier transform to square integrable functions. For an L^2-function, f, the integral defining the Fourier transform, (4.4) may fail to converge and so an indirect method was needed to define $\hat{f}$. This was accomplished by thinking of the Fourier transform as a linear mapping from one space of functions to another and using the continuity of this map, as expressed by the Parseval formula. In elementary calculus, a function f is said to be differentiable at x if: (1) f is defined in an interval containing x and (2) the limit of the difference quotients,

$$f'(x) = \lim_{h \to 0} \frac{f(x+h) - f(x)}{h},$$

exists. For clarity, we sometimes say that f has a *classical derivative* at x. The function is differentiable on an interval (a, b) if its derivative exists for each x in (a, b), and continuously differentiable on (a, b) if f' is a continuous function on (a, b). When working with measurements and with integral transforms, this definition proves to be inadequate. In this section we extend the notion of differentiation in a manner consistent with our linear model for a realistic measurement as a weighted average:

$$m_g(f) = \int_{-\infty}^{\infty} f(x)g(x)dx.$$

With this model, a measurement of the derivative of f would be $m_g(f')$.

If f and g are differentiable functions that vanish outside a bounded interval, then the integration by parts formula states that

$$\int_{-\infty}^{\infty} f'(x)\overline{g(x)}\,dx = -\int_{-\infty}^{\infty} f(x)\overline{g'(x)}\,dx. \tag{4.41}$$

This formula says that, if g is differentiable, then $m_g(f') = -m_{g'}(f)$, which in turn suggests a way to extend the notion of differentiation to some functions that do not have a classical derivative. Suppose that f is a locally integrable function and there exists another locally integrable function f_1 such that, for *every* $\mathcal{C}^1$-function g that vanishes outside a bounded interval, we have the identity

$$\int_{-\infty}^{\infty} f_1(x)\overline{g(x)}\,dx = -\int_{-\infty}^{\infty} f(x)\overline{g'(x)}\,dx. \tag{4.42}$$

From the perspective of any measurement defined by a $\mathcal{C}^1$-function with bounded support, the function f_1 *looks like* the derivative of f. If this condition holds, then we say that f has a *weak derivative* and write $f' = f_1$. In this context the function g is called a *$\mathcal{C}^1$-test function.* It is clear that a function, differentiable in the ordinary sense, is weakly differentiable and the two definitions of the derivative agree.

It is easy to see from examples that a weak derivative can exist even when f does not have a classical derivative.

***Example* 4.3.1.** The function

$$f(x) = \begin{cases} 0 & \text{if } |x| > 1, \\ |x+1| & \text{if } -1 \le x \le 0, \\ |x-1| & \text{if } 0 < x \le 1 \end{cases}$$

does not have a classical derivative at $x = -1, 0$, and 1. Nonetheless the function

$$f_1(x) = \begin{cases} 0 & \text{if } |x| > 1, \\ 1 & \text{if } -1 \le x \le 0, \\ -1 & \text{if } 0 < x \le 1 \end{cases}$$

is the weak derivative of f.

A classically differentiable function is continuous. Properly understood this is also true of a function with a weak derivative. The subtlety is that if f is a function with a weak derivative and h is a function that equals f, except possibly on a set of measure zero, then h also has a weak derivative. In fact their weak derivatives are equal. The precise continuity statement is as follows: If f has a weak derivative, then, after modification on a set of measure zero, f is a continuous function. In the context of weak derivatives, we always assume that the function has been so modified. With this understood, weak derivatives satisfy the fundamental theorem of calculus. If f_1 is the weak derivative of f, then for any $a < b$ we have that

$$f(b) - f(a) = \int_a^b f_1(x)\,dx. \tag{4.43}$$

The proof of this statement is a bit more involved than we might expect. The basic idea is to use a sequence of test functions in (4.41) that converge to $\chi_{[a,b]}$; see [43].

We also note that (4.42) does not uniquely determine the value of $f_1(x)$ at every x. Suppose that f_1 satisfies (4.42) for every $\mathcal{C}^1$-test function. If the set of points where $f_1(x) - \tilde{f}_1(x) \neq 0$ is a set of measure zero then

$$\int_{-\infty}^{\infty} f_1(x)g(x)dx = \int_{-\infty}^{\infty} \tilde{f}_1(x)g(x)dx, \tag{4.44}$$

for every $\mathcal{C}^1$-test function. Thus it would be equally correct to say that $\tilde{f}_1$ is the weak derivative of f. From (4.44), it is clear that, from the point of view of measurement, this ambiguity in the definition of the weak derivative does not lead to problems. It should also come as no surprise that the weak derivative does not share the properties of the classical derivative that depend on pointwise evaluation. For example, the weak derivative does *not* satisfy the mean value theorem: If a function f has a weak derivative f_1, then there may or may *not* exist a point $c \in (a, b)$ so that $f(b) - f(a) = f'(c)(b - a)$. For instance, if f is the function in Example 4.3.1, then f_1 takes only the values ± 1 in the interval $(-1, 1)$. On the other hand $f(\frac{1}{2}) - f(-\frac{1}{2}) = 0$. In most of the applications we consider the functions are classically differentiable except at a finite set of points.

An especially useful condition is for the weak derivative to belong to L^2.

***Definition* 4.3.1.** Let $f \in L^2(\mathbb{R})$ we say that f has an *L^2-derivative* if f has a weak derivative that also belongs to $L^2(\mathbb{R})$.

A function in $L^2(\mathbb{R})$ that is differentiable in the ordinary sense and whose classical derivative belongs to $L^2(\mathbb{R})$ is also differentiable in the L^2-sense. Its classical and L^2-derivatives are equal. If a function f has a classical derivative that is bounded, then the mean value theorem implies that, for some constant M, we have the estimate:

$$\frac{|f(b) - f(a)|}{|b - a|} \leq M.$$

Such an estimate is generally false for a function with a weak derivative. Using (4.43), it is not difficult to show that a function with an L^2-derivative satisfies a similar estimate. Applying the Cauchy-Schwarz inequality to the right-hand side of (4.43) gives the estimate

$$|f(b) - f(a)| \leq \sqrt{|b - a|}\|f_1\|_{L^2}.$$

In other words,

$$\frac{|f(b) - f(a)|}{\sqrt{|b - a|}} \leq \|f_1\|_{L^2}.$$

A function for which this ratio is bounded is called a *Hölder-$\frac{1}{2}$* function. Such a function is said to have a *half a classical derivative.*

The definition of *weak derivative* can be applied recursively to define higher-order weak derivatives. Suppose a locally integrable function f has a weak derivative f'. If f' also has a weak derivative, then we say that f has two weak derivatives with $f^{[2]} \stackrel{d}{=} (f')'$. More generally, if f has j weak derivatives, $\{f', \dots, f^{[j]}\}$, and $f^{[j]}$ has a weak derivative, then we say that f has $j + 1$ weak derivatives. The usual notations are also used for weak derivatives (i.e., f', $f^{[j+1]}$, $\partial_x^j f$, etc.).

Weak differentiability is well adapted to the Fourier transform. Suppose that f is an L^1-function with a weak derivative that is also absolutely integrable. The functions $fe^{-ix\xi}$ and $f'e^{-ix\xi}$ are in $L^1(\mathbb{R})$, and therefore formula (4.22) applies to show that

$$\int_{-\infty}^{\infty} f(x)e^{-ix\xi}dx = \frac{1}{i\xi}\int_{-\infty}^{\infty} f'(x)e^{-ix\xi}dx.$$

Thus the Fourier transform of the weak derivative is related to that of the original function precisely as in the classically differentiable case.

The notion of weak derivative extends the concept of differentiability to a larger class of functions. This approach can be used to define derivatives of objects more general than functions called *generalized functions*. This topic is discussed in Appendix A.4.5; the reader is urged to look over this section.

Exercises

***Exercise* 4.3.1.** In Example 4.3.1, prove that f_1 is the weak derivative of f.

***Exercise* 4.3.2.** Suppose that f has a weak derivative and g equals f on the complement of a set of measure zero. Show that g also has a weak derivative that equals that of f.

***Exercise* 4.3.3.** Suppose that f has a continuous classical derivative and g has a weak derivative. Show that fg has a weak derivative which is given by the usual product formula:

$$(fg)' = f'g + fg'.$$

***Exercise* 4.3.4.** Show that if f has a weak derivative, then (after possible modification on a set of measure zero) it is continuous.

***Exercise* 4.3.5.** Give an example to show that, if f is a weakly differentiable function with weak derivative f', and f assumes a local maximum value at x_0, then $f'(x_0)$ need *not* equal zero.

***Exercise* 4.3.6.** Suppose that f is a locally integrable function defined on $\mathbb{R}$. What should it mean for such a function to be weakly differentiable on the interval (a, b)? Does it make sense to say that a function is weakly differentiable at a point?

***Exercise* 4.3.7.** Show that if f has two weak derivatives, then it has one classical derivative.

***Exercise* 4.3.8.** Show that the definition for higher-order weak derivatives is consistent: If f has j weak derivatives and $f^{[j]}$ has k weak derivatives, then f has $k+j$ weak derivatives and

$$f^{[k+j]} = (f^{[j]})^{[k]}.$$

***Exercise* 4.3.9.** Show that Proposition 4.2.1 remains true if f is assumed to have j absolutely integrable, *weak* derivatives.

4.3.1 Functions With L^2-Derivatives*

If $f \in L^2(\mathbb{R})$ has an L^2-derivative, then the Fourier transforms of f and f' are related just as they would be if f had a classical derivative

$$\widehat{f'}(\xi) = i\xi \hat{f}(\xi).$$

Moreover, the Parseval identity carries over to give

$$\int_{-\infty}^{\infty} |f'(x)|^2\, dx = \frac{1}{2\pi} \int_{-\infty}^{\infty} |\xi|^2 |\hat{f}(\xi)|^2\, d\xi.$$

On the other hand, if $\xi\hat{f}(\xi)$ is square integrable, then we can show that f has an L^2-derivative and its Fourier transform is $i\xi\hat{f}(\xi)$. This is what was meant by the statement that the relationship between the smoothness of a function and the decay of the Fourier transform is very close when these concepts are defined with respect to the L^2-norm.

The higher L^2-derivatives are defined exactly as in the classical case. If $f \in L^2(\mathbb{R})$ has an L^2-derivative, and $f' \in L^2$ also has an L^2-derivative, then we say that f has two L^2-derivatives. This can be repeated to define all higher derivatives. A simple condition for a function $f \in L^2(\mathbb{R})$ to have j L^2-derivatives is that there are functions $\{f_1, \dots, f_j\} \subset L^2(\mathbb{R})$ so that for every j-times differentiable function φ, vanishing outside a bounded interval and $1 \leq l \leq j$, we have that

$$\langle f, \varphi^{[l]}\rangle_{L^2} = (-1)^l \langle f_l, \varphi\rangle_{L^2}.$$

The function f_l is then the lth L^2-derivative of f. Standard notations are also used for the higher L^2-derivatives (e.g., $f^{[l]}$, $\partial_x^l f$, etc.).

The basic result about L^2-derivatives is as follows.

Theorem 4.3.1. *A function $f \in L^2(\mathbb{R})$ has j L^2-derivatives if and only if $\xi^j\hat{f}(\xi)$ is in $L^2(\mathbb{R})$. In this case*

$$\widehat{f^{[l]}}(\xi) = (i\xi)^l \hat{f}(\xi); \tag{4.45}$$

moreover,

$$\int_{-\infty}^{\infty} |f^{[l]}(x)|^2\,dx = \frac{1}{2\pi}\int_{-\infty}^{\infty} |\xi|^{2l}|\hat{f}(\xi)|^2\,d\xi. \tag{4.46}$$

Exercises

***Exercise* 4.3.10.** Suppose that $f \in L^2(\mathbb{R})$ has an $L^2(\mathbb{R})$-derivative f'. Show that if f vanishes for $|x| > R$, then so does f'.

***Exercise* 4.3.11.** Prove that if $f \in L^2(\mathbb{R})$ has an L^2-derivative, then $\widehat{f'}(\xi) = i\xi\hat{f}(\xi)$. *Hint:* Use (4.36).

***Exercise* 4.3.12.** Show that if f has an L^2-derivative, then $\hat{f}$ is absolutely integrable. Conclude that f is a continuous function.

***Exercise* 4.3.13.** Use the result of Exercise 4.3.12 to prove that (4.43) holds under the assumption that f and f' are in $L^2(\mathbb{R})$.

4.3.2 Fractional Derivatives and L^2-Derivatives*

See: A.4.5 .

In the previous section we extended the notion of differentiability to functions that do not have a classical derivative. In the study of the Radon transform it turns out to be useful to have other generalizations of differentiability. We begin with a generalization of the classical notion of differentiability.

The basic observation is the following: A function f has a derivative if the difference quotients

$$\frac{f(x+h)-f(x)}{h}$$

have a limit as $h \to 0$. In order for this limit to exist, it is clearly necessary that the ratios

$$\frac{|f(x+h)-f(x)|}{|h|}$$

be uniformly bounded, for small h. Thus the basic estimate satisfied by a continuously differentiable function is that the ratios

$$\frac{|f(x)-f(y)|}{|x-y|}$$

are locally, uniformly bounded. The function $f(x) = |x|$ shows that these ratios can be bounded without the function being differentiable. However, from the point of view of measurements, such a distinction is very hard to make.

***Definition* 4.3.2.** Let $0 \leq \alpha < 1$. We say that a function f, defined in an interval $[a, b]$, has an *αth-classical derivative* if there is a constant M so that

$$\frac{|f(x)-f(y)|}{|x-y|^{\alpha}} \leq M, \tag{4.47}$$

for all $x, y \in [a, b]$. Such a function is also said to be *α-Hölder continuous*.

The same idea can be applied to functions with L^2-derivatives. Recall that an L^2-function has an L^2-derivative if and only if $\xi \hat{f}(\xi) \in L^2(\mathbb{R})$. This is just the estimate

$$\int_{-\infty}^{\infty} |\xi|^2 |\hat{f}(\xi)|^2 \, d\xi < \infty.$$

By analogy to the classical case, we make the following definition.

***Definition* 4.3.3.** A function $f \in L^2(\mathbb{R})$ has an *αth L^2-derivative* if

$$\int_{-\infty}^{\infty} |\xi|^{2\alpha} |\hat{f}(\xi)|^2 d\xi < \infty. \tag{4.48}$$

For example, an L^2-function, f has half an L^2-derivative if

$$\int_{-\infty}^{\infty} |\hat{f}(\xi)|^2 |\xi| \, d\xi < \infty.$$

There is no canonical way to define the "αth-L^2-derivative operator." The following definition is sometimes useful. For $\alpha \in (0, 1)$, define the αth-L^2-derivative to be

$$D_\alpha f = \underset{R\to\infty}{\mathrm{LIM}} \frac{1}{2\pi} \int_{-R}^{R} |\xi|^\alpha \hat{f}(\xi) e^{i\xi x} d\xi.$$

This operation is defined precisely for those functions satisfying (4.48). Note that with $\alpha = 1$ this definition does *not* give the classical answer.

The relationship between these two notions of fractional differentiability is somewhat complicated. As shown in the previous section, a function with one L^2-derivative is Hölder-$\frac{1}{2}$. On the other hand, the function $f(x) = \sqrt{x}$ is Holder-$\frac{1}{2}$. That $(\sqrt{x})^{-1}$ is not square integrable shows that having half a classical derivative does not imply that a function has one L^2-derivative.

Exercise

***Exercise* 4.3.14.** Suppose that f satisfies the estimate in (4.47) with an $\alpha > 1$. Show that f is constant.

4.4 Some Refined Properties of the Fourier Transform

See: B.2, A.3.1.

In this section we consider some properties of the Fourier transform that are somewhat less elementary than those considered so far. Several results in this section use elementary facts from the theory of analytic functions of a complex variable. The first question we consider concerns the *pointwise* convergence of the inverse Fourier transform.

4.4.1 Localization Principle

Let f be a function in either $L^1(\mathbb{R})$ or $L^2(\mathbb{R})$; for each $R > 0$ define

$$f_R(x) = \mathcal{F}^{-1}(\chi_{[-R,R]}\hat{f})(x) = \frac{1}{2\pi} \int_{-R}^{R} \hat{f}(\xi) e^{ix\xi} d\xi.$$

The function f_R can be expressed directly in terms of f by the formula

$$f_R(x) = \int_{-\infty}^{\infty} f(y) \frac{\sin(R(x-y))}{\pi(x-y)} dy. \tag{4.49}$$

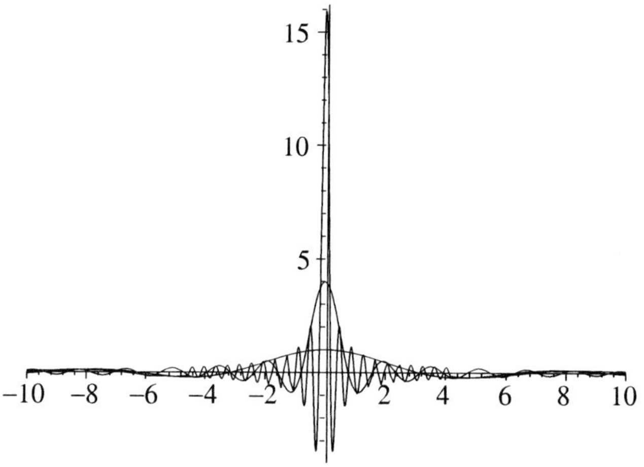

Figure 4.8. The function $\frac{\sin(Rx)}{x}$ for several values of R.

If f is in $L^1(\mathbb{R})$, then (4.49) follows by inserting the definition of $\hat{f}$ in the integral defining f_R and interchanging the order of the integrations. If f is $L^2(\mathbb{R})$, then (4.49) follows from the formula in Exercise 4.2.20. Figure 4.8 shows $\frac{\sin(Rx)}{x}$ for several values of R. As R increases the central peak becomes taller and narrower, while the tails become more oscillatory.

If $\hat{f}$ is absolutely integrable, then Theorem 4.2.1 shows that $f(x)$ is the limit, as $R \to \infty$ of $f_R(x)$. In fact, if f is well enough behaved *near to* x, then this is always the case, whether or not $\hat{f}$ (or for that matter f) is absolutely integrable. This is Riemann's famous localization principle for the Fourier transform.

Theorem 4.4.1 (Localization principle). *Suppose that f belongs to either $L^1(\mathbb{R})$ or $L^2(\mathbb{R})$. If f vanishes in an open interval containing x_0, then*

$$\lim_{R\to\infty} f_R(x_0) = 0.$$

Proof. The proof of this result is not difficult. The same proof works if f is in L^1 or L^2. From (4.49) we obtain

$$\begin{aligned} f_R(x_0) &= \int_{-\infty}^{\infty} f(y)\frac{\sin(R(x_0-y))}{\pi(x_0-y)}dy \\ &= \int_{-\infty}^{\infty} [e^{iR(x_0-y)} - e^{-iR(x_0-y)}]\frac{f(y)}{2\pi i(x_0-y)}dy. \end{aligned} \tag{4.50}$$

Because f vanishes in an interval containing x_0, it follows that $f(y)(x_0-y)^{-1}$ is an L^1-function. The conclusion of the theorem is therefore a consequence of the Riemann-Lebesgue lemma. □

***Remark* 4.4.1.** In light of the linearity of the Fourier transform, this result holds for any function f that can be written as a sum $f = f_1 + f_2$, where $f_p \in L^p(\mathbb{R})$. The set of such functions is denoted $L^1(\mathbb{R}) + L^2(\mathbb{R})$. This set is clearly a vector space.

This result has a simple corollary that makes clearer why it is called the localization principle. Suppose that f and g are functions in $L^1(\mathbb{R}) + L^2(\mathbb{R})$ such that

1. $\lim_{R\to\infty} g_R(x_0) = g(x_0)$, and
2. $f(x) = g(x)$ for x in an interval containing x_0.

The second condition implies that $f(x) - g(x) = 0$ in an interval containing x_0 and therefore

$$\begin{aligned} f(x_0) = g(x_0) &= \lim_{R\to\infty} g_R(x_0) \\ &= \lim_{R\to\infty} g_R(x_0) + \lim_{R\to\infty} (f_R(x_0) - g_R(x_0)) \\ &= \lim_{R\to\infty} f_R(x_0). \end{aligned} \tag{4.51}$$

In the second line we use Theorem 4.4.1. The Fourier inversion process is very sensitive to the local behavior of f. It is important to note that this result is special to one dimension. The analogous result is *false* for the Fourier transform in $\mathbb{R}^n$ if $n \geq 2$. This phenomenon is carefully analyzed in [103]; see also Section 4.5.5.

Exercises

***Exercise* 4.4.1.** Give a complete derivation for (4.49) with f either integrable or square integrable.

***Exercise* 4.4.2.** Suppose that f is a square integrable function which is continuously differentiable for $x \in (a, b)$. Show that for every x in this interval $\lim_{R\to\infty} f_R(x) = f(x)$.

4.4.2 The Heisenberg Uncertainty Principle*

In this section we study relationships between the supp f and supp $\hat{f}$. The simplest such result states that if a function has bounded support, then its Fourier transform cannot.

Proposition 4.4.1. *Suppose* supp f *is contained the bounded interval* $(-R, R)$. *If* $\hat{f}$ *also has bounded support then* $f \equiv 0$.

Proof. The radius of convergence of the series $\sum_0^\infty (-ix\xi)^j/j!$ is infinity, and it converges to $e^{-ix\xi}$, uniformly on bounded intervals. Combining this with the fact that f has bounded support, we conclude that we may interchange the integration with the summation to obtain

$$\begin{aligned} \hat{f}(\xi) &= \int_{-\infty}^{\infty} f(x)e^{-i\xi x}dx \\ &= \int_{-R}^{R} \sum_{j=0}^{\infty} f(x)\frac{(-ix\xi)^j}{j!}dx \\ &= \sum_{j=0}^{\infty} \frac{1}{j!}(-i\xi)^j \int_{-R}^{R} f(x)x^j dx. \end{aligned} \tag{4.52}$$

Since

$$\left|\int_{-R}^{R} f(x)x^j dx\right| \leq R^j \int_{-R}^{R} |f(x)|\, dx,$$

the terms of the series representing $\hat{f}(\xi)$ are bounded by the terms of a series having an infinite radius of convergence; the jth term is bounded by

$$\frac{(R|\xi|)^j}{j!} \int_{-R}^{R} |f(x)|\, dx.$$

Therefore, the series expansion for $\hat{f}(\xi)$ also has an infinite radius of convergence. This argument can be repeated to obtain the Taylor expansion of $\hat{f}(\xi)$ about an arbitrary ξ_0:

$$\begin{aligned}
\hat{f}(\xi) &= \int_{-R}^{R} e^{-i(\xi-\xi_0)x} f(x) e^{i\xi_0 x} dx \\
&= \int_{-R}^{R} \sum_{j=0}^{\infty} \frac{[-i(\xi-\xi_0)x]^j}{j!} f(x) e^{i\xi_0 x} dx \\
&= \sum_{j=0}^{\infty} \int_{-R}^{R} \frac{[-i(\xi-\xi_0)x]^j}{j!} f(x) e^{i\xi_0 x} dx \\
&= \sum_{j=0}^{\infty} \frac{[-i(\xi-\xi_0)]^j}{j!} \int_{-R}^{R} f(x) x^j e^{i\xi_0 x} dx.
\end{aligned} \tag{4.53}$$

If we let $a_j^{\xi_0} = \int f(x)x^j e^{i\xi_0 x} dx$, then

$$\hat{f}(\xi) = \sum_{0}^{\infty} a_j^{\xi_0} \frac{[-i(\xi-\xi_0)]^j}{j!}.$$

As previously, this expansion is valid for all ξ.

Suppose there exists ξ_0 such that $\partial_\xi^j \hat{f}(\xi_0) = 0$ for all $j = 0, 1, \ldots$. Then $\hat{f}(\xi) \equiv 0$ since all the coefficients, $a_j^{\xi_0} = \partial_\xi^j \hat{f}(\xi_0)$, equal zero. This proves the proposition. □

***Remark* 4.4.2.** The proof actually shows that if f is supported on a finite interval and all the derivatives of $\hat{f}$ vanishcs at a single point, then $f \equiv 0$.

This result indicates that we cannot obtain arbitrarily good localization of a function in both x-space and ξ-space simultaneously. A famous quantitative version of this statement is the Heisenberg uncertainty principle, which we now briefly discuss using physical terms coming from quantum mechanics and probability theory. The latter subject is discussed in Chapter 15. In this context an L^2-function f describes the state of a particle. The *probability* of finding the particle in the interval $[a, b]$ is defined to be $\int_a^b |f(x)|^2\,dx$. We normalize so that the total probability is 1. By the Parseval formula,

$$\int_{-\infty}^{\infty} |f(x)|^2\,dx = \int_{-\infty}^{\infty} |\hat{f}(\xi)|^2 \frac{d\xi}{2\pi} = 1.$$

The expected value of the position of a particle is given by

$$E(x) = \int_{-\infty}^{\infty} x|f(x)|^2\,dx.$$

By translating in x, we can normalize f to make $E(x)$ zero. In physics, the Fourier transform of f describes the momentum of a particle. The expected value of the momentum is

$$E(\xi) = \int \xi|\hat{f}(\xi)|^2 \frac{d\xi}{2\pi}.$$

By replacing f by $e^{i\xi_0 x} f$ for an appropriate choice of ξ_0, we can also make $E(\xi) = 0$. With these normalizations, the variance of the position and the momentum, $(\Delta x)^2$ and $(\Delta\xi)^2$, are given by

$$\begin{aligned}(\Delta x)^2 &= \int_{-\infty}^{\infty} x^2|f(x)|^2\,dx,\\ (\Delta\xi)^2 &= \int_{-\infty}^{\infty} \xi^2|\hat{f}(\xi)|^2 \frac{d\xi}{2\pi}.\end{aligned}$$

The Parseval formula implies that

$$(\Delta\xi)^2 = \int_{-\infty}^{\infty} |\partial_x f(x)|^2\,dx.$$

The basic result is as follows

Theorem 4.4.2 (The Heisenberg uncertainty principle). *If f and $\partial_x f$ belong to $L^2(\mathbb{R})$ then*

$$\int_{-\infty}^{\infty} |x|^2|f(x)|^2\,dx \int_{-\infty}^{\infty} |\xi|^2|\hat{f}(\xi)|^2 \frac{d\xi}{2\pi} \ge \frac{1}{4}\left[\int_{-\infty}^{\infty} |f(x)|^2\right]^2 dx. \tag{4.54}$$

Because the product of the variances has a lower bound, this means that we cannot localize the position and the momentum of a particle, arbitrarily well *at the same time*. The proof of this theorem is a simple integration by parts followed by an application of the Cauchy-Schwarz inequality.

Proof. If f decays sufficiently rapidly, we can integration by parts to obtain that

$$\begin{aligned}\int_{-\infty}^{\infty} x f f_x \, dx &= \frac{1}{2}(x f^2)\Big|_{-\infty}^{\infty} - \int_{-\infty}^{\infty} \frac{1}{2} f^2 \, dx \\ &= -\frac{1}{2}\int_{-\infty}^{\infty} f^2.\end{aligned} \tag{4.55}$$

The Cauchy-Schwarz inequality implies that

$$\left|\int_{-\infty}^{\infty} x f f_x \, dx\right| \leq \left[\int_{-\infty}^{\infty} x^2 |f|^2 \, dx\right]^{\frac{1}{2}} \left[\int_{-\infty}^{\infty} |f_x|^2 \, dx\right]^{\frac{1}{2}}.$$

Using (4.55), the Parseval formula and this estimate we obtain

$$\frac{1}{2}\int_{-\infty}^{\infty} |f|^2 \, dx \leq \left[\int_{-\infty}^{\infty} x^2 |f|^2 \, dx\right]^{\frac{1}{2}} \left[\frac{1}{2\pi}\int_{-\infty}^{\infty} \xi^2 |\hat{f}|^2 \, dx\right]^{\frac{1}{2}}. \tag{4.56}$$

□

With the expected position and momentum normalized to be zero, the variance in the position and momentum are given by

$$\Delta x = \left(\int_{-\infty}^{\infty} x^2 f^2\right)^{1/2} \text{ and } \Delta\xi = \left(\int_{-\infty}^{\infty} f_x^2\right)^{1/2}.$$

The estimate (4.56) is equivalent to $\Delta x \cdot \Delta\xi \geq \frac{1}{2}$. If a, b are nonnegative numbers, then the arithmetic-geometric mean inequality states that

$$ab \leq \frac{a^2 + b^2}{2}.$$

Combining this with the Heisenberg uncertainty principle shows that $1 \leq (\Delta x)^2 + (\Delta\xi)^2$. That is,

$$\int_{-\infty}^{\infty} f^2 dx \leq \int_{-\infty}^{\infty} [x^2 f^2 + f_x^2] \, dx. \tag{4.57}$$

The inequality (4.57) becomes an *equality* if we use the Gaussian function $f(x) = e^{-\frac{x^2}{2}}$. A reason why the Gaussian is often used to smooth measured data is that it provides the optimal resolution (in the L^2-norm) for a given amount of de-noising.

Exercise

***Exercise* 4.4.3.** Show that both (4.54) and (4.57) are *equalities* if $f = e^{-\frac{x^2}{2}}$. Can you show that the only functions for which this is true are multiples of f?

4.4.3 The Paley-Wiener Theorem*

In imaging applications we usually work with functions of bounded support. The question naturally arises whether it is possible to recognize such a function from its Fourier transform. There is a variety of theorems that relate the support of a function to properties of its Fourier transform. They go collectively by the name of Paley-Wiener theorems.

Theorem 4.4.3 (Paley-Wiener Theorem I). *An L^2-function f satisfies $f(x) = 0$ for $|x| > L$ if and only if its Fourier transform $\hat{f}$ extends to be an analytic function in the whole complex plane that satisfies*

$$\int_{-\infty}^{\infty} |\hat{f}(\xi + i\tau)|^2 d\xi \leq Me^{2L|\tau|} \quad \text{for all } \tau \text{ and}$$
$$|\hat{f}(\xi + i\tau)| \leq \frac{Me^{L|\tau|}}{\sqrt{|\tau|}} \tag{4.58}$$

Proof. The proof of the forward implication is elementary. The Fourier transform of f is given by an integral over a finite interval,

$$\hat{f}(\xi) = \int_{-L}^{L} f(x)e^{-ix\xi} dx. \tag{4.59}$$

The expression clearly makes sense if ξ is replaced by $\xi + i\tau$, and differentiating under the integral shows that $\hat{f}(\xi + i\tau)$ is a analytic function. The first estimate follows from the Parseval formula as $\hat{f}(\xi + i\tau)$ is the Fourier transform of the L^2-function $f(x)e^{-\tau x}$. Using the Cauchy-Schwartz inequality, we obtain

$$\begin{aligned} |\hat{f}(\xi + i\tau)| &= \left| \int_{-L}^{L} f(x)e^{-ix\xi - x\tau} dx \right| \\ &\leq \frac{e^{L|\tau|}}{\sqrt{|\tau|}} \sqrt{\int_{-L}^{L} |f(x)|^2 dx}; \end{aligned} \tag{4.60}$$

from which the estimate is immediate.

The proof of the converse statement is a little more involved; it uses the Fourier inversion formula and a change of contour. We outline of this argument, the complete justification for the change of contour can be found in [79]. Let $x > L > 0$. The Fourier inversion formula states that

$$f(x) = \frac{1}{2\pi} \int_{-\infty}^{\infty} \hat{f}(\xi)e^{ix\xi} d\xi.$$

Since $\hat{f}(z)e^{ixz}$ is an analytic function, satisfying appropriate estimates, we can shift the integration to the line $\xi + i\tau$ for any $\tau > 0$,

$$f(x) = \frac{1}{2\pi} \int_{-\infty}^{\infty} \hat{f}(\xi + i\tau)e^{-x\tau}e^{ix\xi} d\xi.$$

In light of the first estimate in (4.58), we obtain the bound

$$|f(x)| \leq Me^{(L-x)\tau}.$$

Letting τ tend to infinity shows that $f(x) = 0$ for $x > L$. A similar argument using $\tau < 0$ shows that $f(x) = 0$ if $x < -L$. □

For later applications we state a variant of this result whose proof can be found in [79].

Theorem 4.4.4 (Paley-Wiener II). *A function $f \in L^2(\mathbb{R})$ has an analytic extension $F(x + iy)$ to the upper half-plane ($y > 0$) satisfying*

$$\begin{aligned} &\int_{-\infty}^{\infty} |F(x+iy)|^2 \, dx \leq M, \\ &\lim_{y \downarrow 0} \int_{-\infty}^{\infty} |F(x+iy) - f(x)|^2 = 0 \end{aligned} \tag{4.61}$$

if and only if $\hat{f}(\xi) = 0$ for $\xi < 0$.

4.4.4 The Fourier Transform of Generalized Functions*

See: A.4.5.

Initially the Fourier transform is defined for L^1-functions, by an explicit formula (4.4). It is then extended, in Definition 4.2.5, to L^2-functions by using its *continuity* properties. The Parseval formula implies that the Fourier transform is a continuous map from $L^2(\mathbb{R})$ to itself; indeed it is an invertible, isometry. For an L^2-function, the Fourier transform may *not* defined by an integral; nonetheless the Fourier transform on $L^2(\mathbb{R})$ shares all the important properties of the Fourier transform defined earlier for L^1-functions.

It is reasonable to seek the largest class of functions to which the Fourier transform can be extended. In turns out that the answer is *not* a class of functions, but rather the generalized functions (or tempered distributions) defined in Section A.4.5. In the discussion that follows we assume a familiarity with this section. The definition of the Fourier transform on generalized functions closely follows the pattern of the definition of the derivative of a generalized function, with the result again a generalized function. To accomplish this extension we need to revisit the definition of a generalized function. In Section A.4.5 we gave the following definition:

Let $\mathscr{C}_c^\infty(\mathbb{R})$ denote infinitely differentiable functions defined on $\mathbb{R}$ that vanish outside of bounded sets. These are called *test functions.*

***Definition* 4.4.1.** A generalized function on $\mathbb{R}$ is a linear function, l, defined on the set of test functions such that there is a constant C and an integer k so that, for every $f \in \mathscr{C}_c^\infty(\mathbb{R})$, we have the estimate

$$|l(f)| \leq C \sup_{x\in\mathbb{R}} \left[(1+|x|)^k \sum_{j=0}^{k} |\partial_x^j f(x)| \right]. \tag{4.62}$$

These are linear functions on $\mathscr{C}_c^\infty(\mathbb{R})$ that are, in a certain sense, continuous. The constants C and k in (4.62) depend on l but not on f. The expression on the right-hand side defines a norm on $\mathscr{C}_c^\infty(\mathbb{R})$, for convenience we let

$$\|f\|_k = \sup_{x\in\mathbb{R}} \left[(1+|x|)^k \sum_{j=0}^{k} |\partial_x^j f(x)| \right].$$

The observation that we make is the following: If a generalized function satisfies the estimate

$$|l(f)| \leq C\|f\|_k$$

then it can be extended, by continuity, to any function f that is the limit of a sequence $< f_n > \subset \mathscr{C}_c^\infty(\mathbb{R})$ in the sense that

$$\lim_{n\to\infty} \|f - f_n\|_k = 0.$$

Clearly, $f \in \mathscr{C}^k(\mathbb{R})$ and $\|f\|_k < \infty$. This motivates the following definition:

***Definition* 4.4.2.** A function $f \in \mathscr{C}^\infty(\mathbb{R})$ belongs to *Schwartz class* if $\|f\|_k < \infty$ for every $k \in \mathbb{N}$. The set of such functions is a vector space denoted by $\mathscr{S}(\mathbb{R})$.

From the definition it is clear that

$$\mathscr{C}_c^\infty(\mathbb{R}) \subset \mathscr{S}(\mathbb{R}). \tag{4.63}$$

Schwartz class does not have a norm with respect to which it is a complete normed linear space; instead a sequence $< f_n > \subset \mathscr{S}(\mathbb{R})$ converges to $f \in \mathscr{S}(\mathbb{R})$ if and only if

$$\lim_{n\to\infty} \|f - f_n\|_k = 0 \qquad \text{for every } k \in \mathbb{N}.$$

With this notion of convergence, Schwartz class becomes a complete metric space, and the distance is defined by

$$d_{\mathscr{S}}(f, g) = \sum_{j=0}^{\infty} 2^{-j} \frac{\|f - g\|_j}{1 + \|f - g\|_j}.$$

***Remark* 4.4.3.** Of course, each $\|\cdot\|_k$ satisfies all the axioms for a norm. Nonetheless, in this context they are called *semi-norms* because each one alone does not define the topology on $\mathscr{S}(\mathbb{R})$.

Let $\varphi(x) \in \mathscr{C}_c^\infty(\mathbb{R})$ be a nonnegative function with the following properties:

1. $\varphi(x) = 1$ if $x \in [-1, 1]$,
2. $\varphi(x) = 0$ if $|x| > 2$.

Define $\varphi_n(x) = \varphi(n^{-1}x)$. It is not difficult to prove the following proposition.

Proposition 4.4.2. *If $f \in \mathscr{S}(\mathbb{R})$, then $f_n = \varphi_n f \in \mathscr{C}_c^\infty(\mathbb{R})$ converges to f in $\mathscr{S}(\mathbb{R})$. That is*

$$\lim_{n\to\infty} \|f_n - f\|_k = 0 \qquad \textit{for every } k. \tag{4.64}$$

The proof is left as an exercise.

From the previous discussion it therefore follows that *every* generalized function can be extended to $\mathscr{S}(\mathbb{R})$. Because (4.64) holds for every k, if l is a generalized function and $f \in \mathscr{S}(\mathbb{R})$, then $l(f)$ is defined as

$$l(f) = \lim_{n\to\infty} l(\varphi_n f).$$

To show that this makes sense, it is only necessary to prove that if $< g_n > \subset \mathscr{C}_c^\infty(\mathbb{R})$, which converges to f in Schwartz class, then

$$\lim_{n\to\infty} l(g_n - \varphi_n f) = 0. \tag{4.65}$$

This is an immediate consequence of the triangle inequality and the estimate that l satisfies: There is a C and k so that

$$\begin{aligned} |l(g_n - \varphi_n f)| &\leq C\|g_n - \varphi_n f\|_k \\ &\leq C[\|g_n - f\|_k + \|f - \varphi_n f\|_k]. \end{aligned} \tag{4.66}$$

Since both terms on the right-hand side of the second line tend to zero as $n \to \infty$, equation (4.65) is proved. In fact, the generalized functions are exactly the set of continuous linear functions on $\mathscr{S}(\mathbb{R})$. For this reason the set of generalized functions is usually denoted by $\mathscr{S}'(\mathbb{R})$.

Why did we go to all this trouble? How will this help extend the Fourier transform to $\mathscr{S}'(\mathbb{R})$? The integration by parts formula was the "trick" used to extend the notion of derivative to generalized functions. The reason it works is that if $f \in \mathscr{S}(\mathbb{R})$, then $\partial_x f \in \mathscr{S}(\mathbb{R})$ as well. This implies that $l(\partial_x f)$ is a generalized function whenever l itself is. The Schwartz class has a similar property *vis á vis* the Fourier transform.

Theorem 4.4.5. *The Fourier transform is an isomorphism of $\mathscr{S}(\mathbb{R})$ onto itself; that is, if $f \in \mathscr{S}(\mathbb{R})$, then both $\mathscr{F}(f)$ and $\mathscr{F}^{-1}(f)$ also belong to $\mathscr{S}(\mathbb{R})$. Moreover, for each k there is a k' and constant C_k so that*

$$\|\mathscr{F}(f)\|_k \leq C_k\|f\|_{k'} \qquad \textit{for all } f \in \mathscr{S}(\mathbb{R}). \tag{4.67}$$

The proof of this theorem is an easy consequence of results in Section 4.2.4. We give the proof for $\mathscr{F}$; the proof for $\mathscr{F}^{-1}$ is essentially identical.

Proof. Since $f \in \mathcal{S}(\mathbb{R})$ for any $j, k \in \mathbb{N} \cup \{0\}$, we have the estimates

$$|\partial_x^j f(x)| \leq \frac{\|f\|_k}{(1+|x|)^k}. \tag{4.68}$$

From Propositions 4.2.1 and 4.2.3, it follows that $\hat{f}$ is infinitely differentiable and that, for any k, j,

$$\sup_{\xi \in \mathbb{R}} |\xi|^k |\partial_\xi^j \hat{f}(\xi)| < \infty.$$

To prove this we use the formula

$$\xi^k \partial_\xi^j \hat{f}(\xi) = \int_{-\infty}^{\infty} (i\partial_x)^k \left[(-ix)^j f(x)\right] e^{-ix\xi} dx.$$

Because $f \in \mathcal{S}(\mathbb{R})$ the integrand is absolutely integrable and, in fact, if $m = \max\{j, k\}$, then

$$|\xi^k \partial_\xi^j \hat{f}(\xi)| \leq C_{k,l} \|f\|_{m+2}; \tag{4.69}$$

here $C_{k,l}$ depends only on k and l. This completes the proof. □

Instead of integration by parts, we now use this theorem and the identity

$$\int_{-\infty}^{\infty} f(x)\hat{g}(x)\, dx = \int_{-\infty}^{\infty} \hat{f}(x) g(x)\, dx, \tag{4.70}$$

to extend the Fourier transform to generalized functions. The identity follows by a simple change in the order of integrations, which is easily justified if $f, g \in \mathcal{S}(\mathbb{R})$. It is now clear how we should define the Fourier transform of a generalized function.

***Definition* 4.4.3.** If $l \in \mathcal{S}'(\mathbb{R})$, then the Fourier transform of l is the generalized function $\hat{l}$ *defined* by

$$\hat{l}(f) = l(\hat{f}) \qquad \text{for all } f \in \mathcal{S}(\mathbb{R}). \tag{4.71}$$

Theorem 4.4.5 implies that $\hat{f} \in \mathcal{S}(\mathbb{R})$ so that the right-hand side in (4.71) defines a generalized function.

But why did we need to extend the definition of generalized functions from $\mathcal{C}_c^\infty(\mathbb{R})$ to $\mathcal{S}(\mathbb{R})$? The answer is simple: If $0 \neq f \in \mathcal{C}_c^\infty(\mathbb{R})$, then Proposition 4.4.1 implies that $\hat{f} \notin \mathcal{C}_c^\infty(\mathbb{R})$. This would prevent using (4.71) to define $\hat{l}$ because we would not know that $l(\hat{f})$ made sense! This appears to be a rather abstract definition, and it is not at all clear that it can be used to compute the Fourier transform of a generalized function. In fact, there are many distributions whose Fourier transforms can be explicitly computed.

***Example* 4.4.1.** If φ is an L^1-function, then

$$\widehat{l_\varphi} = l_{\hat{\varphi}}.$$

If $f \in \mathcal{S}(\mathbb{R})$, then the identity in (4.70) holds with $g = \hat{\varphi}$, as a simple interchange of integrations shows. Hence, for all $f \in \mathcal{S}(\mathbb{R})$,

$$l_{\varphi}(\hat{f}) = \int_{-\infty}^{\infty} f(x)\hat{\varphi}(x)\,dx = l_{\hat{\varphi}}(f).$$

This shows that the Fourier transform for generalized functions is indeed an extension of the ordinary transform: If a generalized function l is *represented* by an integrable function in the sense that $l = l_{\varphi}$, then the definition of the Fourier transform of l is consistent with the earlier definition of the Fourier transform of φ.

***Example* 4.4.2.** If $f \in \mathcal{S}(\mathbb{R})$, then

$$\hat{f}(0) = \int_{-\infty}^{\infty} f(x)\,dx.$$

This shows that $\hat{\delta} = l_1$, which is *represented* by an ordinary function equal to the constant 1.

***Example* 4.4.3.** On the other hand, the Fourier inversion formula implies that

$$\int_{-\infty}^{\infty} \hat{f}(\xi)\,d\xi = 2\pi f(0)$$

and therefore $\widehat{l_1} = 2\pi\delta$. This is an example of an ordinary function that does not have a Fourier transform, in the usual sense, and whose Fourier transform, as a generalized function, is *not* an ordinary function.

Recall that a sequence $< l_n > \subset \mathcal{S}'(\mathbb{R})$ converges to l in $\mathcal{S}'(\mathbb{R})$ provided that

$$l(g) = \lim_{n\to\infty} l_n(g) \qquad \text{for all } g \in \mathcal{S}(\mathbb{R}). \tag{4.72}$$

This is very useful for computing Fourier transforms because the Fourier transform is continuous with respect to the limit in (4.72). It follows from the definition that

$$\widehat{l_n}(g) = l_n(\hat{g}) \tag{4.73}$$

and therefore

$$\lim_{n\to\infty} \widehat{l_n}(g) = \lim_{n\to\infty} l_n(\hat{g}) = l(\hat{g}) = \hat{l}(g). \tag{4.74}$$

***Example* 4.4.4.** The generalized function $l_{\chi_{[0,\infty)}}$ can be defined as a limit by

$$l_{\chi_{[0,\infty)}}(f) = \lim_{\epsilon\downarrow 0} \int_{0}^{\infty} e^{-\epsilon x} f(x)\,dx.$$

The Fourier transform of $l_{e^{-\epsilon x}\chi_{[0,\infty)}}$ is easily computed using Example 4.4.1; it is

$$\mathcal{F}(l_{e^{-\epsilon x}\chi_{[0,\infty)}})(f) = \int_{-\infty}^{\infty} \frac{f(x)\,dx}{ix+\epsilon}.$$

This shows that

$$\mathcal{F}(l_{\chi_{[0,\infty)}})(f) = \lim_{\epsilon \downarrow 0} \int_{-\infty}^{\infty} \frac{f(x)\,dx}{ix+\epsilon}. \tag{4.75}$$

In fact, it proves that the limit on the right-hand side exists!

We close this discussion by verifying that the Fourier transform on generalized functions has many of the properties of the ordinary Fourier transform. Recall that if l is a generalized function and f is an infinitely differentiable function that satisfies estimates

$$|\partial_x^j f(x)| \leq C_j(1+|x|)^k,$$

for a *fixed* k, then the product $f \cdot l$ is defined by

$$f \cdot l(g) = l(fg).$$

If $l \in \mathcal{S}'(\mathbb{R})$, then so are all of its derivatives. Using the definition, it is not difficult to find formulæ for $\mathcal{F}(l^{[j]})$:

$$\mathcal{F}(l^{[j]})(f) = l^{[j]}(\hat{f}) = (-1)^j l(\partial_x^j \hat{f}) = l(\widehat{(ix)^j f}). \tag{4.76}$$

This shows that

$$\mathcal{F}(l^{[j]}) = (ix)^j \cdot \hat{l}. \tag{4.77}$$

A similar calculation shows that

$$\mathcal{F}((-ix)^j \cdot l) = \hat{l}^{[j]}. \tag{4.78}$$

Exercises

***Exercise* 4.4.4.** Prove (4.63).

***Exercise* 4.4.5.** Prove that $d_{\mathcal{S}}$ defines a metric. Show that a sequence $< f_n >$ converges in $\mathcal{S}(\mathbb{R})$ to f if and only if

$$\lim_{n\to\infty} d_{\mathcal{S}}(f_n, f) = 0.$$

***Exercise* 4.4.6.** Prove Proposition 4.4.2.

***Exercise* 4.4.7.** Prove (4.70). What is the "minimal" hypothesis on f and g so this formula makes sense, as absolutely convergent integrals?

***Exercise* 4.4.8.** Give a detailed proof of (4.69).

***Exercise* 4.4.9.** Prove, by direct computation, that the limit on the right-hand side of (4.75) exists for any $f \in \mathcal{S}(\mathbb{R})$.

***Exercise* 4.4.10.** If $l_{1/x}$ is the Cauchy principal value integral

$$l_{1/x}(f) = \text{P.V.} \int_{-\infty}^{\infty} \frac{f(x)\,dx}{x}$$

then show that $\mathcal{F}(l_{1/x}) = l_{\operatorname{sgn} x}$.

***Exercise* 4.4.11.** Prove (4.78).

***Exercise* 4.4.12.** The inverse Fourier transform of a generalized function is defined by

$$[\mathcal{F}^{-1}(l)](g) = l(\mathcal{F}^{-1}(g)).$$

Show that $\mathcal{F}^{-1}(\hat{l}) = l = \widehat{\mathcal{F}^{-1}(l)}$.

4.5 The Fourier Transform for Functions of Several Variables.

See: B.8.

The Fourier transform can also be defined for functions of several variables. This section presents the definition and some of the elementary properties of the Fourier transform for functions in $L^1(\mathbb{R}^n)$ and $L^2(\mathbb{R}^n)$. In most respects the higher-dimensional theory is quite similar to the one-dimensional theory. A notable difference is discussed in Section 4.5.5.

Recall that we use lowercase, bold Roman letters $\boldsymbol{x}$, $\boldsymbol{y}$, and so on to denote points in $\mathbb{R}^n$; that is,

$$\boldsymbol{x} = (x_1, \dots, x_n) \text{ or } \boldsymbol{y} = (y_1, \dots, y_n).$$

In this case x_j is called the jth-coordinate of $\boldsymbol{x}$. The Fourier transform of a function of n-variables is also a function of n-variables. It is customary to use the lowercase, bold Greek letters, $\boldsymbol{\xi}$ or $\boldsymbol{\eta}$, for points on the Fourier transform space with

$$\boldsymbol{\xi} = (\xi_1, \dots, \xi_n) \text{ or } \boldsymbol{\eta} = (\eta_1, \dots, \eta_n).$$

The volume form on Fourier space is denoted $d\boldsymbol{\xi} = d\xi_1 \dots d\xi_n$.

4.5.1 L^1-case

As before, we begin with the technically simpler case of L^1-functions.

***Definition* 4.5.1.** If f belongs to $L^1(\mathbb{R}^n)$, then the Fourier transform, $\hat{f}$ of f, is defined by

$$\hat{f}(\boldsymbol{\xi}) = \int_{\mathbb{R}^n} f(\boldsymbol{x}) e^{-i\langle\boldsymbol{\xi},\boldsymbol{x}\rangle} d\boldsymbol{x} \quad \text{for} \quad \boldsymbol{\xi} \in \mathbb{R}^n. \tag{4.79}$$

Since f is absolutely integrable over $\mathbb{R}^n$, the integral can be computed as an iterated integral

$$\int_{\mathbb{R}^n} f(\boldsymbol{x}) e^{-i\langle\boldsymbol{\xi},\boldsymbol{x}\rangle} d\boldsymbol{x} = \int_{-\infty}^{\infty} \cdots \int_{-\infty}^{\infty} f(x_1, \ldots, x_n) e^{-ix_1\xi_1} dx_1 \cdots e^{-ix_n\xi_n} dx_n; \tag{4.80}$$

changing the order of the one-dimensional integrals does not affect the result. When thought of as a linear mapping, it is customary to use $\mathcal{F}(f)$ to denote the Fourier transform of f.

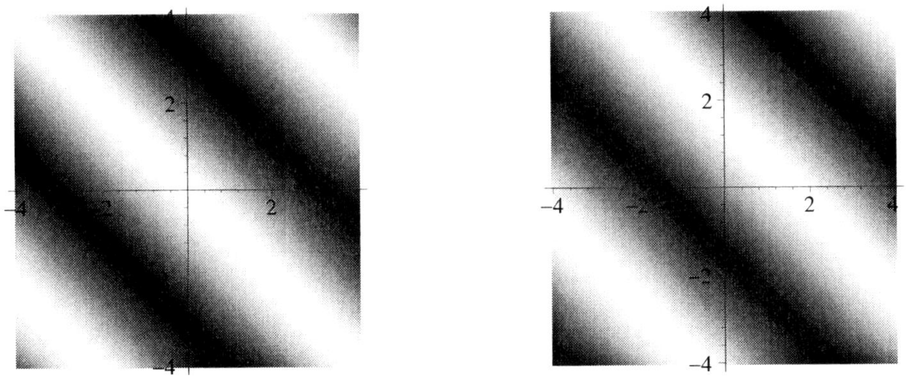

(a) $\operatorname{Re} e^{i\langle(x,y),(1,1)\rangle}$ (b) $\operatorname{Im} e^{i\langle(x,y),(1,1)\rangle}$

Figure 4.9. Real and imaginary parts of $e^{i\langle(x,y),(1,1)\rangle}$.

Using a geometric picture for the inner product leads to a better understanding of the functions $e^{i\langle\boldsymbol{\xi},\boldsymbol{x}\rangle}$. To that end we write $\boldsymbol{\xi}$ in polar form as $\boldsymbol{\xi} = r\boldsymbol{\omega}$. Here $r = \|\boldsymbol{\xi}\|$ is the length of $\boldsymbol{\xi}$ and $\boldsymbol{\omega}$ its direction. Write $\boldsymbol{x} = \boldsymbol{x}' + t\boldsymbol{\omega}$, where $\boldsymbol{x}'$ is orthogonal to $\boldsymbol{\omega}$, (i.e., $\langle\boldsymbol{x}', \boldsymbol{\omega}\rangle = 0$). As $\langle\boldsymbol{x}, \boldsymbol{\omega}\rangle = t$, the function $\langle\boldsymbol{x}, \boldsymbol{\omega}\rangle$ depends only on t. Thus

$$e^{i\langle\boldsymbol{x},\boldsymbol{\xi}\rangle} = e^{irt}.$$

This function oscillates in the $\boldsymbol{\omega}$-direction with wavelength $\frac{2\pi}{r}$. To illustrate this we give a density plot in the plane of the real and imaginary parts of

$$e^{i\langle\boldsymbol{x},\boldsymbol{\xi}\rangle} = \cos\langle\boldsymbol{x}, \boldsymbol{\xi}\rangle + i\sin\langle\boldsymbol{x}, \boldsymbol{\xi}\rangle$$

for several choices of $\boldsymbol{\xi}$. In Figures 4.9 and 4.10 white corresponds to $+1$ and black corresponds to -1. When using the Fourier transform to analyze functions of spatial variables, the vector $[2\pi]^{-1}\boldsymbol{\xi}$ is called the *spatial frequency* and its components have units of cycles-per-unit length.

The Fourier transform at $\boldsymbol{\xi} = r\boldsymbol{\omega}$ can be reexpressed as

$$\hat{f}(r\boldsymbol{\omega}) = \int_{-\infty}^{\infty} \int_{L} f(\boldsymbol{x}' + t\boldsymbol{\omega}) e^{-irt} d\boldsymbol{x}' \, dt. \tag{4.81}$$

Here L is the $(n-1)$-dimensional subspace of $\mathbb{R}^n$ orthogonal to $\boldsymbol{\omega}$:

$$L = \{x' \in \mathbb{R}^n \,:\, \langle \boldsymbol{x}', \boldsymbol{\omega} \rangle = 0\}$$

and $d\boldsymbol{x}'$ is the $(n-1)$-dimensional Euclidean measure on L.

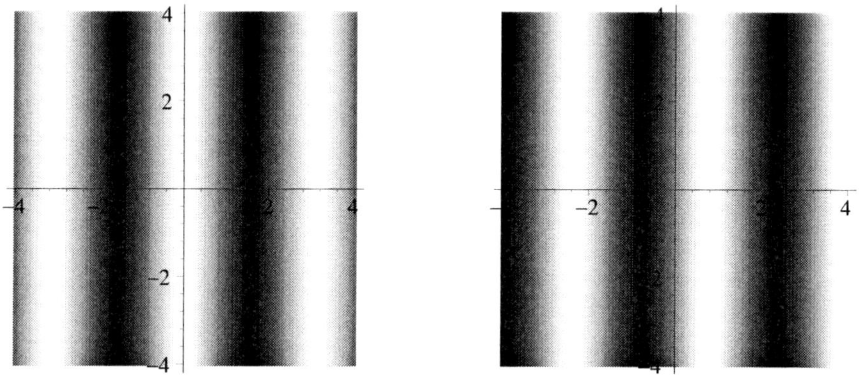

(a) $\operatorname{Re} e^{i\langle (x,y),(2,0)\rangle}$ (b) $\operatorname{Im} e^{i\langle (x,y),(2,0)\rangle}$

Figure 4.10. Real and imaginary parts of $e^{i\langle (x,y),(2,0)\rangle}$.

The Fourier transform is invertible; under appropriate hypotheses there is an explicit formula for the inverse.

Theorem 4.5.1 (Fourier Inversion Formula). *Suppose that f is an L^1-function defined on $\mathbb{R}^n$. If $\hat{f}$ also belongs to $L^1(\mathbb{R}^n)$, then*

$$f(\boldsymbol{x}) = \frac{1}{[2\pi]^n} \int_{\mathbb{R}^n} \hat{f}(\boldsymbol{\xi}) e^{i\boldsymbol{x}\cdot\boldsymbol{\xi}} d\boldsymbol{\xi}. \tag{4.82}$$

Proof. The proof is formally identical to the proof of the one dimensional result. As before, we begin by assuming that f is continuous. The basic fact used is that the Fourier transform of a Gaussian can be computed explicitly:

$$\mathcal{F}(e^{-\epsilon\|x\|^2}) = \left[\frac{\pi}{\epsilon}\right]^{\frac{n}{2}} e^{-\frac{\|x\|^2}{4\epsilon}}. \tag{4.83}$$

Because $\hat{f}$ is absolutely integrable,

$$\begin{aligned}\frac{1}{[2\pi]^n}\int_{\mathbb{R}^n}\hat{f}(\boldsymbol{\xi})e^{i\boldsymbol{x}\cdot\boldsymbol{\xi}}d\boldsymbol{\xi} &= \lim_{\epsilon\downarrow 0}\frac{1}{[2\pi]^n}\int_{\mathbb{R}^n}\hat{f}(\boldsymbol{\xi})e^{i\boldsymbol{x}\cdot\boldsymbol{\xi}}e^{-\epsilon\|\boldsymbol{\xi}\|^2}d\boldsymbol{\xi}\\ &= \lim_{\epsilon\downarrow 0}\frac{1}{[2\pi]^n}\int_{\mathbb{R}^n}\int_{\mathbb{R}^n} f(\boldsymbol{y})e^{-i\boldsymbol{y}\cdot\boldsymbol{\xi}}d\boldsymbol{y}e^{i\boldsymbol{x}\cdot\boldsymbol{\xi}}e^{-\epsilon\|\boldsymbol{\xi}\|^2}d\boldsymbol{\xi}.\end{aligned} \tag{4.84}$$

The order of the integrations in the last line can be interchanged; using (4.83) gives

$$\begin{aligned}\frac{1}{[2\pi]^n}\int_{\mathbb{R}^n}\hat{f}(\boldsymbol{\xi})e^{i\boldsymbol{x}\cdot\boldsymbol{\xi}}d\boldsymbol{\xi} &= \lim_{\epsilon\downarrow 0}\frac{1}{[2\pi]^n}\int_{\mathbb{R}^n} f(\boldsymbol{y})\left[\frac{\pi}{\epsilon}\right]^{\frac{n}{2}} e^{-\frac{\|\boldsymbol{x}-\boldsymbol{y}\|^2}{4\epsilon}}d\boldsymbol{y}\\ &= \lim_{\epsilon\downarrow 0}\frac{1}{[2\pi]^{\frac{n}{2}}}\int_{\mathbb{R}^n} f(\boldsymbol{x}-2\sqrt{\epsilon}\boldsymbol{t})e^{-\|\boldsymbol{t}\|^2}d\boldsymbol{t}.\end{aligned} \tag{4.85}$$

In the last line we use the change of variables $\boldsymbol{y} = \boldsymbol{x} - 2\sqrt{\epsilon}\boldsymbol{t}$. As f is continuous and absolutely integrable, this converges to

$$\frac{f(\boldsymbol{x})}{[2\pi]^{\frac{n}{2}}}\int_{\mathbb{R}^n} e^{-\|\boldsymbol{t}\|^2}d\boldsymbol{t}.$$

Since

$$\int_{\mathbb{R}^n} e^{-\|\boldsymbol{t}\|^2}d\boldsymbol{t} = [2\pi]^{\frac{n}{2}},$$

this completes the proof of the theorem for continuous functions. As in the one-dimensional case, an approximation argument is used to remove the additional hypothesis. The details are left to the reader. □

Exercises

***Exercise* 4.5.1.** Prove formula (4.81).

***Exercise* 4.5.2.** If $g_1(x), \dots, g_n(x)$ belong to $L^1(\mathbb{R})$, show that

$$f(x_1, \dots, x_n) = g_1(x_1)\cdots g_n(x_n) \in L^1(\mathbb{R}^n).$$

Show that

$$\hat{f}(\xi_1, \dots, \xi_n) = \hat{g}_1(\xi_1)\cdots \hat{g}_n(\xi_n).$$

Use this to compute the Fourier transform of $e^{-\|x\|^2}$.

4.5.2 Regularity and Decay

See: B.8.

There is once again a close connection between the smoothness of a function and the decay of its Fourier transform and vice versa. A convenient way to quantify the smoothness of a function on $\mathbb{R}^n$ is in terms of the existence of partial derivatives. Formulæ in several variables that involve derivatives can rapidly become cumbersome and unreadable. Fortunately there is a compact notation, called *multi-index* notation, giving n-variable formulæ with the simplicity and readability of the one-variable case.

***Definition* 4.5.2.** A *multi-index* is an ordered n-tuple of nonnegative integers usually denoted by a bold, lowercase Greek letter. For $\boldsymbol{\alpha} = (\alpha_1, \dots, \alpha_n)$, a multi-index, set

$$\boldsymbol{\alpha}! = \alpha_1! \cdots \alpha_n! \qquad \text{and} \qquad |\boldsymbol{\alpha}| = \alpha_1 + \cdots + \alpha_n.$$

The function $|\boldsymbol{\alpha}|$ is called the length of $\boldsymbol{\alpha}$. The following conventions are useful:

$$\boldsymbol{x}^{\boldsymbol{\alpha}} = x_1^{\alpha_1} x_2^{\alpha_2} \cdots x_n^{\alpha_n} \text{ and } \partial_{\boldsymbol{x}}^{\boldsymbol{\alpha}} = \partial_{x_1}^{\alpha_1} \partial_{x_2}^{\alpha_2} \cdots \partial_{x_n}^{\alpha_n}.$$

***Example* 4.5.1.** The binomial formula has an n-dimensional analogue:

$$(x_1 + \cdots + x_n)^k = k! \sum_{\{\boldsymbol{\alpha}\,:\,|\boldsymbol{\alpha}|=k\}} \frac{\boldsymbol{x}^{\boldsymbol{\alpha}}}{\boldsymbol{\alpha}!}.$$

***Example* 4.5.2.** If f is a k-times differentiable function on $\mathbb{R}^n$, then there is an n-dimensional analogue of Taylor's formula:

$$f(\boldsymbol{x}) = \sum_{\{\boldsymbol{\alpha}\,:\,|\boldsymbol{\alpha}|\leq k\}} \frac{\partial_{\boldsymbol{x}}^{\boldsymbol{\alpha}} f(0) \boldsymbol{x}^{\boldsymbol{\alpha}}}{\boldsymbol{\alpha}!} + R_k(\boldsymbol{x}). \tag{4.86}$$

Here R_k is the remainder term; it satisfies

$$\lim_{\|\boldsymbol{x}\|\to 0} \frac{|R_k(\boldsymbol{x})|}{\|\boldsymbol{x}\|^k} = 0.$$

As in the one-dimensional case, the most general decay result is the Riemann-Lebesgue lemma.

Proposition 4.5.1 (Riemann-Lebesgue lemma). *Let f be an L^1-function on $\mathbb{R}^n$, then $\hat{f}$ is a continuous function and* $\lim_{|\boldsymbol{\xi}|\to\infty} \hat{f}(\boldsymbol{\xi}) = 0$.

The proof is very similar to the one-dimensional case and is left to the reader.

The smoothness of f is reflected in the decay properties of its Fourier transform. Suppose that f is continuous and has a continuous partial derivative in the x_j-direction that is integrable, that is,

$$\int_{\mathbb{R}^n} |\partial_{x_j} f|\, d\boldsymbol{x} < \infty.$$

For notational convenience we suppose that $j = 1$ and set $\boldsymbol{x} = (x_1, \boldsymbol{x}')$. For any finite limits R, R_1, R_2 we can integrate by parts in the x_1-variable to obtain

$$\begin{split}\int_{-R_1}^{R_2} \int_{\|\boldsymbol{x}'\|<R} f(x_1, \boldsymbol{x}')e^{-i\langle \boldsymbol{x},\boldsymbol{\xi}\rangle} d\boldsymbol{x}'\, dx_1 = \\ \left[\frac{1}{-i\xi_1}\right]\Bigg[\int_{\|\boldsymbol{x}'\|<R} [f(R_2, \boldsymbol{x}')e^{-i\langle (R_2,\boldsymbol{x}'),\boldsymbol{\xi}\rangle} - f(-R_1, \boldsymbol{x}')e^{-i\langle (-R_1,\boldsymbol{x}'),\boldsymbol{\xi}\rangle}]\, d\boldsymbol{x}' - \\ \int_{-R_1}^{R_2} \int_{\|\boldsymbol{x}'\|<R} \partial_{x_1} f(x_1, \boldsymbol{x}')e^{-i\langle \boldsymbol{x},\boldsymbol{\xi}\rangle} d\boldsymbol{x}'\, dx_1\Bigg].\end{split} \tag{4.87}$$

Because $\partial_{x_1} f$ is integrable, the second integral on the right-hand side of (4.87) tends to $\mathcal{F}(\partial_{x_1} f)$ as R, R_1, R_2 tend to ∞. The boundary terms are bounded by

$$\int_{\mathbb{R}^{n-1}} [|f(R_2, \boldsymbol{x}')| + |f(-R_1, \boldsymbol{x}')|]\, d\boldsymbol{x}'.$$

As f is absolutely integrable, there exist sequences $< R_1^k >, < R_2^k >$ tending to infinity so that these $(n-1)$-dimensional integrals tend to zero. This shows that $\mathcal{F}(\partial_{x_1} f) = i\xi_1 \mathcal{F}(f)$. The same argument applies to any coordinate, proving the following proposition.

Proposition 4.5.2. *If f is a continuous L^1-function with an absolutely integrable, continuous jth-partial derivative, then*

$$\mathcal{F}(\partial_{x_j} f)(\boldsymbol{\xi}) = i\xi_j \mathcal{F}(f)(\boldsymbol{\xi}).$$

There is a constant C_n such that if f has a continuous, integrable gradient, then $\hat{f}$ satisfies the estimate:

$$|\hat{f}(\boldsymbol{\xi})| \leq \frac{C_n \int_{\mathbb{R}^n} [|f| + \|\nabla f\|]\, d\boldsymbol{x}}{(1 + \|\boldsymbol{\xi}\|)}.$$

The integration by parts argument can be iterated to obtain formulæ for the Fourier transform of $\partial_{\boldsymbol{x}}^{\boldsymbol{\alpha}} f$ for any multi-index $\boldsymbol{\alpha}$.

Proposition 4.5.3. *Suppose that f is a continuous L^1-function with a continuous, absolutely integrable $\boldsymbol{\alpha}$th-partial derivative for any $\boldsymbol{\alpha}$ with $|\boldsymbol{\alpha}| \leq k$. Then there exists a constant C so that*

$$|\hat{f}(\boldsymbol{\xi})| \leq \frac{C}{(1 + \|\boldsymbol{\xi}\|)^k}, \tag{4.88}$$

and for each such $\boldsymbol{\alpha}$ we have

$$\widehat{\partial_{\boldsymbol{x}}^{\boldsymbol{\alpha}} f}(\boldsymbol{\xi}) = (i\boldsymbol{\xi})^{\alpha} \hat{f}(\boldsymbol{\xi}).$$

The proposition relates the rate of decay of the Fourier transform to the smoothness of f. As in the one-dimensional case, it has a partial converse.

Proposition 4.5.4. *Suppose that f is an L^1-function on $\mathbb{R}^n$ such that, for a nonnegative integer k, $|\hat{f}(\boldsymbol{\xi})|(1 + \|\boldsymbol{\xi}\|)^k$ is integrable. Then $\partial_x^{\boldsymbol{\alpha}} f$ exists, is continuous and tends to zero at infinity, for any $\boldsymbol{\alpha}$ with $|\boldsymbol{\alpha}| \leq k$.*

Proof. The proof is a consequence of the Fourier inversion formula. The decay hypothesis implies that

$$f(\boldsymbol{x}) = \frac{1}{[2\pi]^n} \int\limits_{\mathbb{R}^n} \hat{f}(\boldsymbol{\xi}) e^{i\boldsymbol{x}\cdot\boldsymbol{\xi}} d\boldsymbol{\xi}.$$

The estimate satisfied by $\hat{f}$ implies that this expression can be differentiated up to k times. Hence the Fourier transform of $\partial_{\boldsymbol{x}}^{\boldsymbol{\alpha}} f$ is $(i\boldsymbol{\xi})^{\alpha} \hat{f}(\boldsymbol{\xi})$. Because $\mathscr{F}(\partial_{\boldsymbol{x}}^{\boldsymbol{\alpha}} f)$ is an L^1-function, the last statement follows from the Riemann-Lebesgue lemma. □

***Remark* 4.5.1.** An estimate estimate similar to (4.88) that implies the hypothesis of the proposition is

$$|\hat{f}(\boldsymbol{\xi})| = O(\|\boldsymbol{\xi}\|^{-(n+k+\epsilon)})$$

for a positive ϵ. It is apparent that the discrepancy between this estimate and that in Proposition 4.5.3 grows as the dimension increases. As in the one-dimensional case, more natural and precise results are obtained by using weak derivatives and the L^2-norm. As these results are not needed in the rest of the book, we will not pursue this direction. The interested reader should consult [43].

In order to understand how decay at infinity for f is reflected in properties of $\hat{f}$, we first suppose that f vanishes outside the ball of radius R. It can be shown without difficulty that $\hat{f}$ is a differentiable function, and its partial derivatives are given by

$$\partial_{\xi_j} \hat{f}(\boldsymbol{\xi}) = \int\limits_{B_R} \partial_{\xi_j}[f(\boldsymbol{x})e^{-i\boldsymbol{\xi}\cdot\boldsymbol{x}}]d\boldsymbol{x} = \int\limits_{B_R} f(\boldsymbol{x})(-ix_j)e^{-i\boldsymbol{\xi}\cdot\boldsymbol{x}} d\boldsymbol{x} = \mathscr{F}(-ix_j f)(\boldsymbol{\xi}). \tag{4.89}$$

Iterating (4.89) gives

$$\partial_{\boldsymbol{\xi}}^{\boldsymbol{\alpha}} \hat{f}(\boldsymbol{\xi}) = (-i)^{|\alpha|} \int\limits_{\mathbb{R}^n} \boldsymbol{x}^{\alpha} f(\boldsymbol{x}) e^{-i\boldsymbol{\xi}\cdot\boldsymbol{x}} d\boldsymbol{x} = (-i)^{|\alpha|}\mathscr{F}(\boldsymbol{x}^{\alpha} f)(\boldsymbol{\xi}). \tag{4.90}$$

If, instead of assuming that f has bounded support we assume that $(1 + \|x\|)^k f$ is integrable, then a standard limiting argument shows that $\hat{f}$ is k times differentiable and the $\boldsymbol{\alpha}$th derivative is given by the right-hand side of (4.90).

Summarizing these computations, we have

Proposition 4.5.5. *If $(1 + \|x\|)^k f$ is absolutely integrable for a positive integer k, then the Fourier transform of f has k continuous derivatives. The partial derivatives of $\hat{f}$ are given by*

$$\partial_{\xi}^{\alpha} \hat{f}(\xi) = (-i)^{|\alpha|} \mathcal{F}(x^{\alpha} f)(\xi).$$

They satisfy the estimates

$$|\partial_{\xi}^{\alpha} \hat{f}(\xi)| \leq \int_{\mathbb{R}^n} \|x\|^{|\alpha|} |f(x)|\, dx,$$

and tend to zero as $\|\xi\|$ tends to infinity.

Exercises

***Exercise* 4.5.3.** Suppose that f is in $L^1(\mathbb{R}^n)$. Show that there exist sequences $< a_n >$ tending to $\pm\infty$ so that

$$\lim_{n \to \infty} \int_{\mathbb{R}^{n-1}} |f(a_n, x')|\, dx'.$$

***Exercise* 4.5.4.** Suppose that f is an integrable function that vanishes outside the ball of radius R. Show that $\hat{f}$ is a differentiable function and justify the interchange of the derivative and the integral in (4.89).

***Exercise* 4.5.5.** Suppose that f is an integrable function that vanishes outside the ball of radius R. Show that $\hat{f}$ is an infinitely differentiable function.

***Exercise* 4.5.6.** Give the details of the limiting argument used to pass from (4.90) with f of bounded support to the conclusion of Proposition 4.5.5.

***Exercise* 4.5.7.** Prove the n-variable binomial formula.

***Exercise* 4.5.8.** Explain the dependence on the dimension in the hypothesis of Proposition 4.5.4.

***Exercise* 4.5.9.** Find a function f of n-variables so that

$$|\hat{f}(\xi)| \leq \frac{C}{(1 + \|\xi\|)^n}$$

but f is *not* continuous.

4.5.3 L^2-Theory

See: A.4.2.

As in the one-dimensional case, the n-dimensional Fourier transform extends to L^2-functions. The basic result is the Parseval formula.

Theorem 4.5.2 (Parseval formula). *If f is absolutely integrable and square integrable, then*

$$\int_{\mathbb{R}^n} |f(x)|^2\, dx = \frac{1}{[2\pi]^n} \int_{\mathbb{R}^n} |\hat{f}(\xi)|^2\, d\xi. \tag{4.91}$$

***Remark* 4.5.2.** Again $|\hat{f}(\xi)|^2$ is called the power spectral density of f at ξ.

The proof is quite similar to the one-dimensional case. It uses an approximation argument and the identity, valid for L^1-functions with L^1 Fourier transforms:

$$\int_{\mathbb{R}^n} f(x)\hat{g}(x)\, dx = \int_{\mathbb{R}^n} f(x) \int_{\mathbb{R}^n} e^{-ix\cdot y} g(y)\, dy\, dx = \int_{\mathbb{R}^n} \hat{f}(y) g(y)\, dy. \tag{4.92}$$

The details are left to reader.

As in the single-variable case, the Fourier transform is extended to $L^2(\mathbb{R}^n)$ by continuity. If we set

$$\hat{f}_R(\xi) = \int_{\|x\|<R} f(x) e^{-i\xi\cdot x}\, dx,$$

then Parseval's formula implies that

$$\|\hat{f}_R\|_{L^2} \leq \|\chi_{B_R} f\|_{L^2}.$$

Because $L^2(\mathbb{R}^n)$ is complete and L^2-functions with bounded support are dense in L^2, it follows from Theorem 4.2.4 that the Fourier transform of f can be defined as the L^2-limit

$$\hat{f} = \operatorname*{LIM}_{R\to\infty} \hat{f}_R.$$

Moreover, the Parseval formula extends to all functions in $L^2(\mathbb{R}^n)$. This shows that the Fourier transform is a continuous mapping of $L^2(\mathbb{R}^n)$ to itself: If $< f_n >$ is a sequence with $\operatorname*{LIM}_{n\to\infty} f_n = f$, then

$$\operatorname*{LIM}_{n\to\infty} \hat{f}_n = \hat{f}.$$

The L^2-inversion formula is also a consequence of the Parseval formula.

Proposition 4.5.6 (**L^2-inversion formula**). *Let $f \in L^2(\mathbb{R}^n)$ and define*

$$f_R(x) = \frac{1}{[2\pi]^n} \int\limits_{\|\xi\|<R} \hat{f}(\xi) e^{ix\cdot\xi} d\xi;$$

then $f = \underset{R\to\infty}{\mathrm{LIM}} f_R$.

Proof. We need to show that $\lim_{R\to\infty} \|f_R - f\|_{L^2} = 0$. Because the norm is defined by an inner product, we have

$$\|f_R - f\|_{L^2}^2 = \|f_R\|_{L^2}^2 - 2\,\mathrm{Re}\langle f_R, f\rangle_{L^2} + \|f\|_{L^2}.$$

The Parseval formula implies that

$$\|f_R\|_{L^2}^2 = \frac{1}{[2\pi]^n} \int\limits_{\|\xi\|<R} |\hat{f}(\xi)|^2 d\xi \text{ and } \|f\|_{L^2}^2 = \frac{1}{[2\pi]^n} \int\limits_{\mathbb{R}^n} |\hat{f}(\xi)|^2 d\xi.$$

The proof is completed by using the following lemma.

Lemma 4.5.1. *Let $g \in L^2(\mathbb{R}^n)$, then*

$$\langle f_R, g\rangle = \frac{1}{[2\pi]^n} \int\limits_{\|\xi\|<R} \hat{f}(\xi)\overline{\hat{g}(\xi)}\, d\xi. \tag{4.93}$$

The proof of the lemma is a consequence of the Parseval formula; it is left as an exercise for the reader. Using (4.93) gives

$$\|f_R - f\|_{L^2}^2 = \frac{1}{[2\pi]^n} \int\limits_{\|\xi\|\geq R} |\hat{f}(\xi)|^2 d\xi.$$

This implies that $\underset{R\to\infty}{\mathrm{LIM}} f_R = f$. □

***Remark* 4.5.3.** The extension of the Fourier transform to functions in $L^2(\mathbb{R}^n)$ has many nice properties. In particular, the range of the Fourier transform on $L^2(\mathbb{R}^n)$ is exactly $L^2(\mathbb{R}^n)$. However, the formula for the Fourier transform as an integral is *purely symbolic*. The Fourier transform itself is only defined as a LIM; for a given ξ the pointwise limit

$$\lim_{R\to\infty} \int\limits_{\|x\|<R} f(x)^{-ix\cdot\xi} dx$$

may or may not exist.

We conclude this section with an enumeration of the elementary properties of the Fourier transform for functions of n-variables. As before, these hold for L^1- or L^2-functions and follow from elementary properties of the integral.

1. LINEARITY: The Fourier transform is a linear operation. If $\alpha \in \mathbb{C}$, then

$$\mathcal{F}(f+g) = \mathcal{F}(f) + \mathcal{F}(g), \quad \mathcal{F}(\alpha f) = \alpha\mathcal{F}(f).$$

2. SCALING: The Fourier transform of $f(ax)$, a function dilated by $a \in \mathbb{R}$, is given by

$$\int_{\mathbb{R}^n} f(ax)e^{-i\xi\cdot x}dx = \int_{\mathbb{R}^n} f(y)e^{-i\frac{\xi\cdot y}{a}}\frac{dy}{a^n} = \frac{1}{a^n}\hat{f}(\frac{\xi}{a}). \tag{4.94}$$

3. TRANSLATION: Let f_t be the function f shifted by the vector t, $f_t(x) = f(x - t)$. The Fourier transform of f_t is given by

$$\begin{aligned}\widehat{f_t}(\xi) &= \int_{\mathbb{R}^n} f(x - t)e^{-i\xi\cdot x}\,dx \\ &= \int_{\mathbb{R}^n} f(y)e^{-i\xi\cdot(y+t)}\,dy \\ &= e^{-i\xi\cdot t}\hat{f}(\xi).\end{aligned} \tag{4.95}$$

4. REALITY: If $f(x)$ is real valued, then $\hat{f}(\xi) = \overline{\hat{f}(-\xi)}$.

5. EVENNESS: If f is even and real valued, then $\hat{f}$ is real valued, if f is odd and real valued, then $\hat{f}$ is purely imaginary valued.

Exercises

***Exercise* 4.5.10.** Give the details of the proof of the n-dimensional Parseval formula.

***Exercise* 4.5.11.** Show that (4.91) implies that

$$\int_{\mathbb{R}^n} f(x)\overline{g(x)}\,dx = \int_{\mathbb{R}^n} \hat{f}(\xi)\overline{\hat{g}(\xi)}\frac{d\xi}{[2\pi]^n}.$$

***Exercise* 4.5.12.** Prove Lemma 4.5.1.

***Exercise* 4.5.13.** Verify properties (4) and (5).

***Exercise* 4.5.14.** Prove that the Fourier transform of a radial function is also a radial function and formula (4.96).

4.5.4 The Fourier Transform on Radial Functions

See: A.3.3.

Recall that a function that only depends on $\|x\|$ is said to be *radial*. The Fourier transform of a radial function is also radial and can be given by a one-dimensional integral transform.

Theorem 4.5.3. *Suppose that $f(\boldsymbol{x}) = F(\|\boldsymbol{x}\|)$ is an integrable function; then the Fourier transform of f is given by the one-dimensional integral transform*

$$\hat{f}(\boldsymbol{\xi}) = \frac{c_n}{\|\boldsymbol{\xi}\|^{\frac{n-2}{2}}} \int_0^\infty J_{\frac{n-2}{2}}(r\|\boldsymbol{\xi}\|)F(r)r^{\frac{n}{2}}\,dr. \tag{4.96}$$

Here c_n is a constant depending on the dimension.

If $\operatorname{Re}(\nu) > -\frac{1}{2}$, then $J_\nu(z)$, the order ν Bessel function is defined by the integral

$$J_\nu(z) = \frac{\left(\frac{z}{2}\right)^\nu}{\Gamma\left(\nu+\frac{1}{2}\right)\Gamma\left(\frac{1}{2}\right)} \int_0^\pi e^{iz\cos(\theta)}\sin^{2\nu}(\theta)\,d\theta.$$

Proof. The derivation of (4.96) uses polar coordinates on $\mathbb{R}^n$. Let $\boldsymbol{x} = r\boldsymbol{\omega}$, where r is a nonnegative number and $\boldsymbol{\omega}$ belongs to the unit $(n-1)$-sphere, S^{n-1}. In these coordinates the volume form on $\mathbb{R}^n$ is

$$d\boldsymbol{x} = r^{n-1}dr\,dV_{S^{n-1}};$$

here $dV_{S^{n-1}}$ is the volume form on S^{n-1}. In polar coordinates, the Fourier transform of f is given by

$$\hat{f}(\boldsymbol{\xi}) = \int_0^\infty \int_{S^{n-1}} F(r)e^{-ir\langle\boldsymbol{\omega},\boldsymbol{\xi}\rangle}\,dV_{S^{n-1}}r^{n-1}\,dr. \tag{4.97}$$

It not difficult to show that the integral over S^{n-1} only depends on $\|\boldsymbol{\xi}\|$, and therefore it suffices to evaluate it for $\boldsymbol{\xi} = (0, \dots, 0, \|\boldsymbol{\xi}\|)$. Points on the $(n-1)$-sphere can be expressed in the form

$$\boldsymbol{\omega} = \sin\theta(\boldsymbol{\omega}', 0) + (0, \dots, 0, \cos\theta),$$

where $\boldsymbol{\omega}'$ is a point on the unit $(n-2)$-sphere and $\theta \in [0, \pi]$. Using this parameterization for S^{n-1}, we obtain a formula for the volume form,

$$dV_{S^{n-1}} = \sin^{n-2}\theta\,dV_{S^{n-2}}. \tag{4.98}$$

Using these observations, the spherical integral in (4.97) becomes

$$\begin{aligned}\int_{S^{n-1}} e^{-ir\langle\boldsymbol{\omega},\boldsymbol{\xi}\rangle}\,dV_{S^{n-1}} &= \int_0^\pi \int_{S^{n-2}} e^{-ir\|\boldsymbol{\xi}\|\cos\theta}\sin^{n-2}\theta\,dV_{S^{n-2}}d\theta \\ &= \sigma_{n-2}\int_0^\pi e^{-ir\|\boldsymbol{\xi}\|\cos\theta}\sin^{n-2}\theta d\theta.\end{aligned} \tag{4.99}$$

The coefficient σ_{n-2} is the $(n-2)$-dimensional volume of S^{n-2}. Comparing this integral with the definition of the Bessel function gives (4.96). □

***Example* 4.5.3.** The Fourier transform of the characteristic function of the unit ball $B_1 \subset \mathbb{R}^n$ is given by the radial integral

$$\widehat{\chi_{B_1}}(\xi) = \frac{c_n}{\|\xi\|^{\frac{n-2}{2}}} \int_0^1 J_{\frac{n-2}{2}}(r\|\xi\|) r^{\frac{n}{2}}\, dr.$$

Using formula 6.561.5 in [45] gives

$$\widehat{\chi_{B_1}}(\xi) = \frac{c_n}{\|\xi\|^{\frac{n}{2}}} J_{\frac{n}{2}}(\|\xi\|).$$

As $\|\xi\|$ tends to infinity, the Bessel function is a oscillatory term times $[\sqrt{\|\xi\|}]^{-1}$. Overall we have the estimate

$$\widehat{\chi_{B_1}}(\xi) \leq \frac{C}{(1+\|\xi\|)^{\frac{n+1}{2}}}.$$

Exercises

***Exercise* 4.5.15.** Prove that the spherical integral in (4.97) only depends on $\|\xi\|$.

***Exercise* 4.5.16.** Verify the parameterization of the $(n-1)$-sphere used to obtain (4.98) as well as this formula.

***Exercise* 4.5.17.** Determine the constant c_n in (4.96).

***Exercise* 4.5.18.** Using (4.98) show that σ_n the n-volume of S^n is given by

$$\sigma_n = \frac{2\pi^{\frac{n+1}{2}}}{\Gamma\left(\frac{n+1}{2}\right)}.$$

***Exercise* 4.5.19.** Using the connection between the one-dimensional integral transform defined in (4.96) and the n-dimensional Fourier transform, find a formula for the inverse of this transform. *Hint:* Use symmetry; this does not require any computation!

4.5.5 The Failure of Localization in Higher Dimensions

The localization principle is a remarkable feature of the 1-dimensional Fourier transform. Suppose that f is an integrable function defined on $\mathbb{R}$. According to the localization principle, the convergence of the partial inverse

$$f_R(x) = \frac{1}{2\pi} \int_{-R}^{R} \hat{f}(\xi) e^{ix\xi} d\xi$$

to $f(x)$ only depends on the behavior of f in an interval about x. This is a uniquely one-dimensional phenomenon. In this section we give an example due to Pinsky showing the failure of the localization principle in three dimensions. A complete discussion of this phenomenon can be found in [103].

Pinsky's example is very simple; it concerns $f = \chi_{B_1}$, the characteristic function of the unit ball. The Fourier transform of f was computed in Example 4.5.3; it is

$$\hat{f}(\xi) = \frac{cJ_{\frac{3}{2}}(\|\xi\|)}{\|\xi\|^{\frac{3}{2}}}.$$

In this example c denotes various positive constants. Using formula 8.464.3 in [45], this can be reexpressed in terms of elementary functions by

$$\hat{f}(\xi) = \frac{c[\|\xi\|\cos(\|\xi\|) - \sin(\|\xi\|)]}{\|\xi\|^3}.$$

Using polar coordinates, we compute the partial inverse:

$$\begin{aligned} f_R(0) &= \frac{c}{[2\pi]^3}\int_0^R \left[\cos(r) - \frac{\sin(r)}{r}\right] dr \\ &= c\left[\sin(R) - \int_0^R \frac{\sin(r)}{r} dr\right]. \end{aligned} \tag{4.100}$$

In the second line the integral has a limit as $R \to \infty$; however, $\sin(R)$ does not! Thus $f_R(0)$ remains bounded as R tends to infinity but does not converge.

4.6 Conclusion

In this chapter we introduced the Fourier transform and considered some of its basic properties. Of particular importance in applications to imaging are the connections between the smoothness of a function and the decay of its Fourier transform. These connections are usually expressed as estimates satisfied by the Fourier transform. The Fourier transform is an example of an invertible linear operator. We used the Parseval formula to extend the domain of definition of the Fourier transform beyond $L^1(\mathbb{R}^n)$, obtaining a bounded linear map of $L^2(\mathbb{R}^n)$ to itself. The map is well defined even though the integral expression defining $\hat{f}(\xi)$ may not be meaningful. Many linear operators encountered in imaging applications are defined in terms of the Fourier transform and computed using a fast approximate implementation of it called the fast Fourier transform.

The other basic concept introduced in this chapter is that of a weak derivative. This is an extension of the notion of differentiability defined using the integration by parts formula. It provides a systematic extension of the theory of differentiation to functions that do not have classical pointwise derivatives. As we shall see, such a theory is needed in order to study the inverse of the Radon transform for the sort of data that arise in medical imaging. The notion of weak derivative is also well adapted to measurements defined by averages.

From this point of view it is not possible to distinguish between a function with classical derivatives and a function with weak derivatives. In the next chapter we introduce the *convolution product* as a generalization of a moving average. In part because of its intimate connection to the Fourier transform, this operation is the tool used to defined most filters employed in signal processing and medical imaging.

A text covering much of the material in this chapter and the next, with a similarly applied bent, is *The Fourier Transform and Its Applications* by R. N. Bracewell, [15]. The first edition of Bracewell's book, published in 1965, strongly influenced the early development of algorithms in x-ray tomography. The reader interested in a more complete mathematical treatment of the Fourier transform is directed to [79], for the one-dimensional case, and to [120], for higher dimensions.

Chapter 5

Convolution

Suppose we would like to study a smooth function of one variable, but the available data are contaminated by noise. For the purposes of the present discussion, this means that the measured signal is of the form $f = s + \epsilon n$. Here ϵ is a (small) number and n is function which models the noise. An example is shown in Figure 5.1.

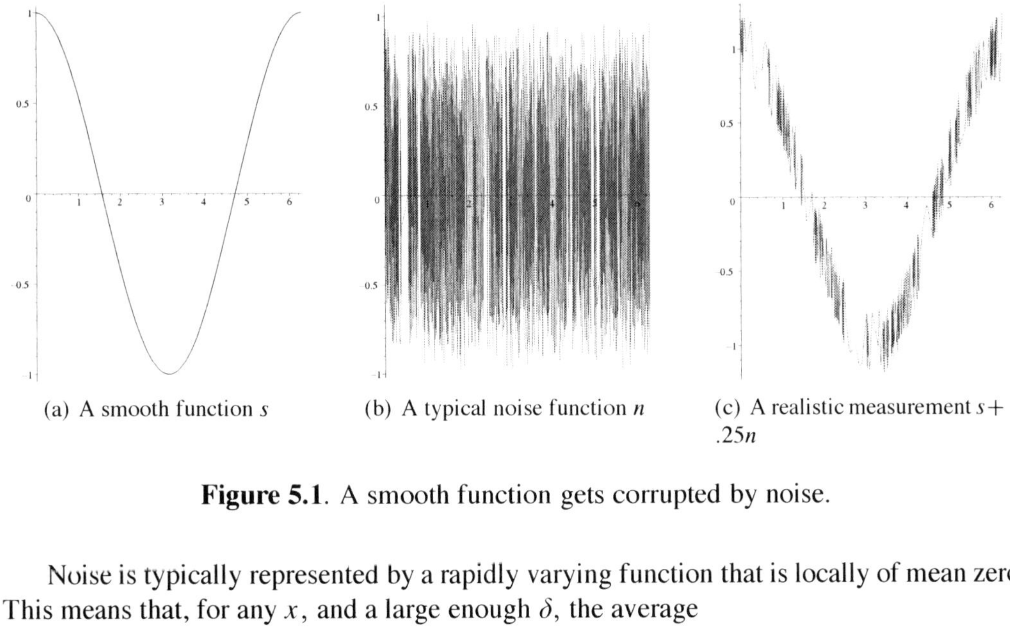

(a) A smooth function s

(b) A typical noise function n

(c) A realistic measurement $s + .25n$

Figure 5.1. A smooth function gets corrupted by noise.

Noise is typically represented by a rapidly varying function that is locally of mean zero. This means that, for any x, and a large enough δ, the average

$$\frac{1}{\delta}\int_{x}^{x+\delta} n(y)\,dy$$

is small compared to the size of n. The more random the noise, the smaller δ can be taken. On the other hand, since s is a smooth function, the analogous average of s should be close to $s(x)$. The moving average of f is defined to be

$$\mathcal{M}_\delta(f)(x) = \frac{1}{\delta}\int_x^{x+\delta} f(y)\,dy. \tag{5.1}$$

If the noise is very random, so that δ can be taken small, then $\mathcal{M}_\delta(f)$ should be close to s. The results of applying this averaging process to the function shown in Figure 5.1(c) are shown in Figure 5.2.

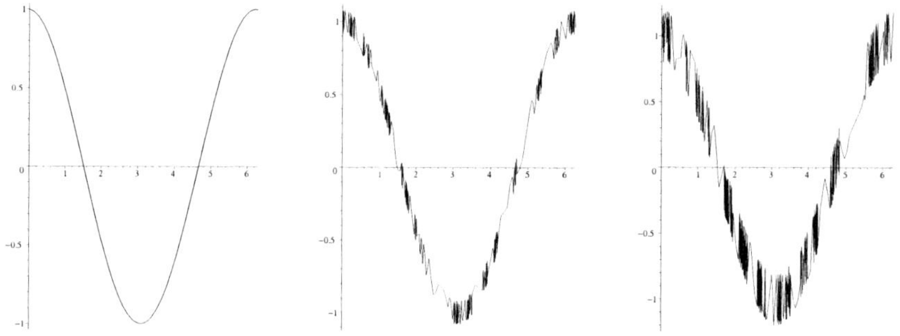

(a) $\delta = .1$ (b) $\delta = .001$ (c) $\delta = .00001$

Figure 5.2. The moving average $\mathcal{M}_\delta(f)$ for various values of δ.

There is a somewhat neater and more flexible way to express the operation defined in (5.1). Define the weight function

$$m_\delta(x) = \begin{cases} \frac{1}{\delta} & \text{for } x \in [-\delta, 0], \\ 0 & \text{otherwise.} \end{cases}$$

The moving average then becomes

$$\mathcal{M}_\delta(f)(x) = \int_{-\infty}^{\infty} f(y) m_\delta(x-y)\,dy. \tag{5.2}$$

In this formulation we see that the value of $\mathcal{M}_\delta(f)(x)$ is obtained by translating the weight function along the axis, multiplying it by f, and integrating. To be a little more precise, the weight function is first reflected around the vertical axis [i.e., $m_\delta(y)$ is replaced by $m_\delta(-y)$ and then translated to give $m_\delta(-(y-x)) = m_\delta(x-y)$]. At this stage it is difficult to motivate the reflection step, beyond saying that in the end it leads to a simpler theory.

The weight function in (5.2) is just one possible choice. Depending on the properties of the noise (or the signal), it might be advantageous to use a different weight function. For w an integrable function, define the w-weighted moving average by

$$\mathcal{M}_w(f) = \int_{-\infty}^{\infty} f(y) w(x-y)\,dy. \tag{5.3}$$

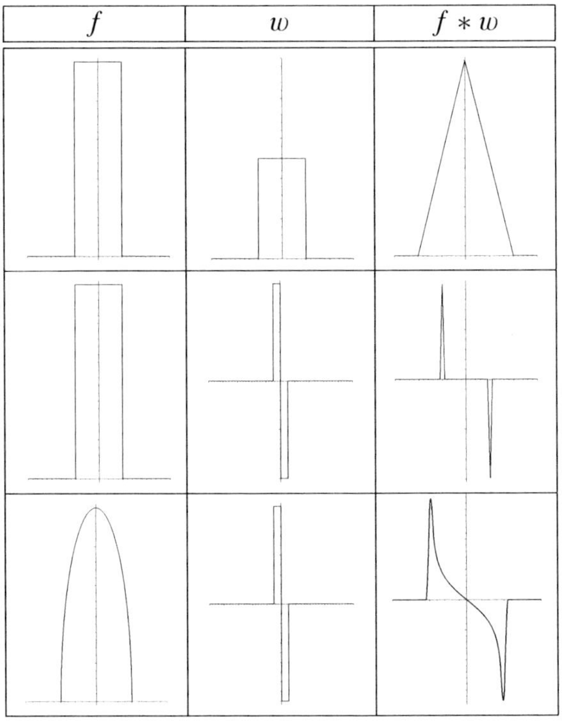

Table 5.1. Examples of the convolution product

For this to be an "average" in the usual sense, w should be nonnegative with total integral equal to one, though the operation $f \mapsto \mathcal{M}_w(f)$ is defined under much weaker conditions. The main features of this operation are as follows: (1) It is linear in f, and (2) the weight assigned to $f(y)$ in the output $\mathcal{M}_w(f)(x)$ depends only on the difference $x - y$. Many operations of this type appear in mathematics and image processing.

It turns out that the simplest theory results from thinking of this as a bilinear operation in the two functions f and w. The result, denoted by $f * w$, is called the *convolution product*. Several examples are shown in Tables 5.1 and 5.2. As we show in this chapter, this operation has many of the properties of ordinary pointwise multiplication, with one important addition: Convolution is intimately connected to the Fourier transform. Because there are very efficient algorithms for approximating the Fourier transform and its inverse, convolution lies at the heart of many practical filters. After defining the convolution product for functions on $\mathbb{R}^n$ and establishing its basic properties, we briefly turn our attention to filtering theory.

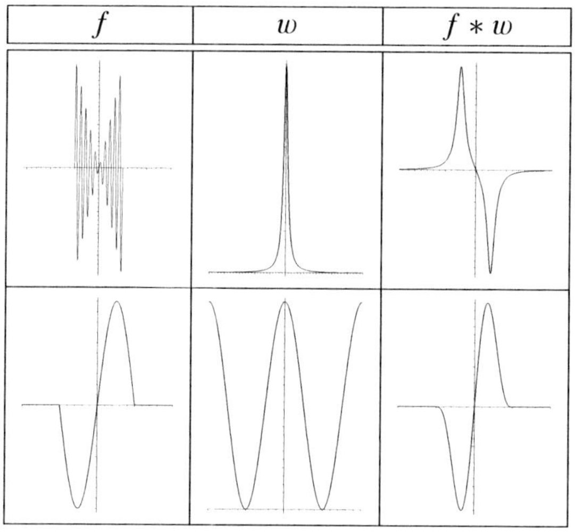

Table 5.2. More examples of the convolution product

5.1 Convolution

See: A.5.1.

For applications to medical imaging we use convolution in one, two and three dimensions. As the definition and formal properties of this operation do not depend on the dimension, we define it and consider its properties, once and for all, for functions defined on $\mathbb{R}^n$.

***Definition* 5.1.1.** If f is an L^1-function defined on $\mathbb{R}^n$ and g is a bounded, locally integrable function then, the *convolution product* of f and g is the function on $\mathbb{R}^n$ defined by the integral

$$f * g(x) = \int_{\mathbb{R}^n} f(x - y)g(y)\,dy. \tag{5.4}$$

***Remark* 5.1.1.** There are many different conditions under which this operation is defined. If the product $f(y)g(x - y)$ is an integrable function of y, then $f * g(x)$ is defined by an absolutely convergent integral. For example, if g is bounded with bounded support, then it is only necessary that f be locally integrable in order for $f * g$ to be defined. In this chapter we use functional analytic methods to extend the definition of convolution to situations where these integrals are not absolutely convergent. This closely follows the pattern established to extend the Fourier transform to $L^2(\mathbb{R}^n)$.

We consider a couple of additional examples:

***Example* 5.1.1.** Let $g(x) = c_n r^{-n} \chi_{B_r}(\|x\|)$; here B_r is the ball of radius r in $\mathbb{R}^n$ and c_n^{-1} is the volume of B_1. For any locally integrable function f the value of $f * g(x)$ is given by

$$\begin{aligned} f * g(x) &= \int_{\mathbb{R}^n} f(x-y)g(y)\,dy \\ &= \frac{c_n}{r^n} \int_{B_r} f(x-y)\,dy. \end{aligned} \tag{5.5}$$

This is the ordinary average of the values of f over points in $B_r(x)$.

Convolution also appears in the partial inverse of the Fourier transform. In this case the weighting function assumes both positive and negative values.

***Example* 5.1.2.** Let f belong to either $L^1(\mathbb{R})$ or $L^2(\mathbb{R})$. In Section 4.4.1 we defined the partial inverse of the Fourier transform

$$S_R(f)(x) = \frac{1}{2\pi} \int_{-R}^{R} \hat{f}(\xi) e^{ix\xi}\,d\xi. \tag{5.6}$$

This can be represented as a convolution,

$$S_R(f) = f * D_R,$$

where

$$D_R(x) = \frac{R\,\mathrm{sinc}(Rx)}{\pi}.$$

For functions in either L^1 or L^2 this convolution is given by an absolutely convergent integral.

5.1.1 Basic Properties of the Convolution Product

The convolution product satisfies many estimates, the simplest is a consequence of the triangle inequality for integrals:

$$\|f * g\|_\infty \le \|f\|_{L^1} \|g\|_\infty. \tag{5.7}$$

We now establish another estimate that, via Theorem 4.2.4, extends the domain of the convolution product.

Proposition 5.1.1. *Suppose that f and g are integrable and g is bounded. Then $f * g$ is absolutely integrable and*

$$\|f * g\|_{L^1} \le \|f\|_{L^1} \|g\|_{L^1}. \tag{5.8}$$

Proof. It follows from the triangle inequality that

$$\begin{aligned}\int_{\mathbb{R}^n} |f * g(x)|\,dx &\leq \int_{\mathbb{R}^n}\int_{\mathbb{R}^n} |f(x-y)g(y)|\,dy\,dx \\ &= \int_{\mathbb{R}^n}\int_{\mathbb{R}^n} |f(x-y)g(y)|\,dx\,dy.\end{aligned} \tag{5.9}$$

Going from the first to the second lines, we interchanged the order of the integrations. This is allowed by Fubini's theorem, since $f(y)g(x-y)$ is absolutely integrable over $\mathbb{R}^n \times \mathbb{R}^n$. Changing variables in the x-integral by setting $t = x - y$, we get

$$\|f * g\|_{L^1} \leq \int_{\mathbb{R}^n}\int_{\mathbb{R}^n} |f(t)||g(y)|\,dt\,dy = \|f\|_{L^1}\|g\|_{L^1}.$$

□

For a fixed f in $L^1(\mathbb{R}^n)$ the map from bounded, integrable functions to $L^1(\mathbb{R}^n)$ defined by $C_f(g) = f * g$ is linear and satisfies (5.8). As bounded integrable functions are dense in $L^1(\mathbb{R}^n)$, Theorem 4.2.4 applies to show that C_f extends to define a map from $L^1(\mathbb{R}^n)$ to itself. Because f is an arbitrary integrable function, convolution extends as a map from $L^1(\mathbb{R}^n) \times L^1(\mathbb{R}^n)$ to $L^1(\mathbb{R}^n)$. Indeed, the linearity of the integral implies that

$$C_f(g_1 + g_2) = C_f(g_1) + C_f(g_2) \text{ and } C_{f_1+f_2}(g) = C_{f_1}(g) + C_{f_2}(g).$$

The following proposition summarizes these observations.

Proposition 5.1.2. *The convolution product extends to define a continuous bilinear map from $L^1(\mathbb{R}^n) \times L^1(\mathbb{R}^n)$ to $L^1(\mathbb{R}^n)$ that satisfies* (5.8).

***Remark* 5.1.2.** If f and g are both in $L^1(\mathbb{R}^n)$, then the integral defining $f * g(x)$ may not converge for every x. The fact that $f(y)g(x-y)$ is integrable over $\mathbb{R}^n \times \mathbb{R}^n$ implies that

$$\int_{\mathbb{R}^n} f(y)g(x-y)\,dy$$

might diverge, but only for x belonging to a set of measure zero. An inequality analogous to (5.8) holds for any $1 \leq p \leq \infty$. That is, if $f \in L^p(\mathbb{R}^n)$ and $g \in L^1(\mathbb{R}^n)$, then $f * g$ is defined as an element of $L^p(\mathbb{R}^n)$, satisfying the estimate

$$\|f * g\|_{L^p} \leq \|f\|_{L^p}\|g\|_{L^1}. \tag{5.10}$$

The proof of this statement is left to the exercises.

***Example* 5.1.3.** Some decay conditions are required for $f * g$ to be defined. If $f(x) = [\sqrt{1+|x|}]^{-1}$, then

$$f * f(x) = \int_{-\infty}^{\infty} \frac{1}{\sqrt{1+|y|}} \frac{1}{\sqrt{1+|x-y|}}\,dy = \infty \qquad \text{for all } x.$$

If, for any positive ϵ, we let $g = [\sqrt{1+|x|}]^{-(1+\epsilon)}$; then $f * g$ is defined.

The basic properties of integration lead to certain algebraic properties for the convolution product.

Proposition 5.1.3. *The convolution product is commutative, distributive, and associative; that is, if f_1, f_2, f_3 belong to $L^1(\mathbb{R}^n)$, then the following identities hold:*

$$\begin{aligned} f_1 * f_2 &= f_2 * f_1, \\ f_1 * (f_2 + f_3) &= f_1 * f_2 + f_1 * f_3, \\ f_1 * (f_2 * f_3) &= (f_1 * f_2) * f_3. \end{aligned} \tag{5.11}$$

***Remark* 5.1.3.** If convolution were defined without "reflecting" the argument of the second function through the origin [i.e., if instead $f * g(\boldsymbol{x}) = \int f(\boldsymbol{y})g(\boldsymbol{y} - \boldsymbol{x})\,d\boldsymbol{y}$], then the convolution product would not be commutative but would satisfy the identity $f * g(\boldsymbol{x}) = g * f(-\boldsymbol{x})$.

Proof. We prove the first assertion; it suffices to assume that f_2 is bounded, the general case, then follows by taking limits and using (5.8). The definition states that

$$f_1 * f_2(\boldsymbol{x}) = \int_{\mathbb{R}^n} f_1(\boldsymbol{y}) f_2(\boldsymbol{x} - \boldsymbol{y})\,d\boldsymbol{y}.$$

Letting $\boldsymbol{t} = \boldsymbol{x} - \boldsymbol{y}$, this integral becomes

$$\int_{\mathbb{R}^n} f_1(\boldsymbol{x} - \boldsymbol{t}) f_2(\boldsymbol{t})\,d\boldsymbol{t} = f_2 * f_1(\boldsymbol{x}).$$

The proofs of the remaining parts are left to the Exercises. □

Convolution defines a multiplication on $L^1(\mathbb{R}^n)$ that is commutative, distributive, and associative. The only thing missing is a multiplicative unit, which is a function $i \in L^1(\mathbb{R}^n)$, so that $f * i = f$ for every f in $L^1(\mathbb{R}^n)$. It is not hard to see that such a function cannot exist. For if

$$f(\boldsymbol{x}) = \int_{\mathbb{R}^n} f(\boldsymbol{x} - \boldsymbol{y}) i(\boldsymbol{y})\,d\boldsymbol{y},$$

for every point $\boldsymbol{x}$ and every function $f \in L^1(\mathbb{R}^n)$, then $i(x)$ must vanish for $\boldsymbol{x} \neq 0$. But in this case $f * i \equiv 0$ for any function $f \in L^1(\mathbb{R}^n)$. In Section 5.3 we return to this point.

A reason that the convolution product is so important in applications is that the Fourier transform converts convolution into ordinary pointwise multiplication.

Theorem 5.1.1. *Suppose that f and g are L^1-functions. Then*

$$\mathscr{F}(f * g) = \mathscr{F}(f)\mathscr{F}(g). \tag{5.12}$$

Proof. The convolution, $f * g$ is an L^1-function and therefore has a Fourier transform. Because $f(\boldsymbol{x}-\boldsymbol{y})g(\boldsymbol{y})$ is an L^1-function of $(\boldsymbol{x}, \boldsymbol{y})$, the following manipulations are easily justified:

$$\begin{aligned}\mathcal{F}(f * g)(\boldsymbol{\xi}) &= \int_{\mathbb{R}^n} (f * g)(\boldsymbol{x}) e^{-i\langle \boldsymbol{\xi}, \boldsymbol{x}\rangle} d\boldsymbol{x} \\ &= \int_{\mathbb{R}^n}\int_{\mathbb{R}^n} f(\boldsymbol{x}-\boldsymbol{y})g(\boldsymbol{y}) e^{-i\langle \boldsymbol{\xi}, \boldsymbol{x}\rangle} d\boldsymbol{y}\, d\boldsymbol{x} \\ &= \int_{\mathbb{R}^n}\int_{\mathbb{R}^n} f(\boldsymbol{t})g(\boldsymbol{y}) e^{-i\langle \boldsymbol{\xi}, (\boldsymbol{y}+\boldsymbol{t})\rangle} d\boldsymbol{t}\, d\boldsymbol{y} \\ &= \hat{f}(\boldsymbol{\xi})\hat{g}(\boldsymbol{\xi}).\end{aligned} \tag{5.13}$$

□

***Remark* 5.1.4.** The conclusion of Theorem 5.1.1 remains true if $f \in L^2(\mathbb{R}^n)$ and $g \in L^1(\mathbb{R}^n)$. In this case $f * g$ also belongs to $L^2(\mathbb{R}^n)$. Note that $\hat{g}$ is a bounded function, so that $\hat{f}\hat{g}$ belongs to $L^2(\mathbb{R}^n)$ as well.

***Example* 5.1.4.** Let $f = \chi_{[-1,1]}$. Formula (5.12) simplifies the computation of the Fourier transform for $f * f$ or even the j-fold convolution of f with itself:

$$f *_j f \stackrel{d}{=} \underbrace{f * \cdots * f}_{j\text{-times}}.$$

In this case

$$\mathcal{F}(f *_j f)(\xi) = [2\operatorname{sinc}(\xi)]^j.$$

***Example* 5.1.5.** A partial inverse for the Fourier transform in n dimensions is defined by

$$S_R^n(f) = \frac{1}{[2\pi]^n} \int_{-R}^{R} \cdots \int_{-R}^{R} \hat{f}(\boldsymbol{\xi}) e^{i\langle \boldsymbol{x}, \boldsymbol{\xi}\rangle} d\boldsymbol{\xi}.$$

The Fourier transform of the function

$$D_R^n(\boldsymbol{x}) = \left[\frac{R}{\pi}\right]^n \prod_{j=1}^{n} \operatorname{sinc}(Rx_j)$$

is $\chi_{[-R,R]}(\xi_1)\cdots\chi_{[-R,R]}(\xi_n)$, and therefore Theorem 5.1.1 implies that

$$S_R^n(f) = D_R^n * f.$$

Exercises

***Exercise* 5.1.1.** For $f \in L^1(\mathbb{R})$ define

$$f_B(x) = \begin{cases} f(x) & \text{if } |f(x)| \leq B, \\ 0 & \text{if } |f(x)| > B. \end{cases}$$

Show that $\lim_{B \to \infty} \|f - f_B\|_{L^1} = 0$. Use this fact and the inequality (5.8) to show that the sequence $< f_B * g >$ has a limit in $L^1(\mathbb{R})$.

***Exercise* 5.1.2.** Prove the remaining parts of Proposition 5.1.3. Explain why it suffices to prove these identities for *bounded* integrable functions.

***Exercise* 5.1.3.** Compute $\chi_{[-1,1]} *_j \chi_{[-1,1]}$ for $j = 2, 3, 4$ and plot these functions on a single graph.

***Exercise* 5.1.4.** Prove that $\|f * g\|_{L^2} \leq \|f\|_{L^2} \|g\|_{L^1}$. *Hint:* Use the Cauchy-Schwarz inequality.

***Exercise* 5.1.5.** * For $1 < p < \infty$ use Hölder's inequality to show that $\|f * g\|_{L^p} \leq \|f\|_{L^p} \|g\|_{L^1}$.

***Exercise* 5.1.6.** Show that $\mathcal{F}(D_R^n)(\boldsymbol{\xi}) = \chi_{[-R,R]}(\xi_1) \cdots \chi_{[-R,R]}(\xi_n)$.

***Exercise* 5.1.7.** Prove that the conclusion of Theorem 5.1.1 remains true if $f \in L^2(\mathbb{R}^n)$ and $g \in L^1(\mathbb{R}^n)$. *Hint:* Use the estimate $\|f * g\|_{L^2} \leq \|f\|_{L^2} \|g\|_{L^1}$ to reduce to a simpler case.

***Exercise* 5.1.8.** Suppose that the convolution product were defined by

$$f * g(x) = \int f(y) g(y - x) \, dy.$$

Show that (5.12) would not hold. What would replace it?

***Exercise* 5.1.9.** Show that there does not exist an integrable function i so that $i * f = f$ for every integrable function f. *Hint:* Use Theorem 5.1.1 and the Riemann-Lebesgue Lemma.

***Exercise* 5.1.10.** A different partial inverse for the n-dimensional Fourier transform is defined by

$$\Sigma_R(f) = \frac{1}{[2\pi]^n} \int\limits_{\|\boldsymbol{\xi}\| \leq R} \hat{f}(\boldsymbol{\xi}) e^{i\langle x, \boldsymbol{\xi} \rangle} d\boldsymbol{\xi}.$$

This can also be expressed as the convolution of f with a function F_R^n. Find an explicit formula for F_R^n.

***Exercise* 5.1.11.** Use the Fourier inversion formula to prove that

$$\widehat{fg}(\xi) = \frac{1}{2\pi} \hat{f} * \hat{g}(\xi). \tag{5.14}$$

What assumptions are needed for $\hat{f} * \hat{g}$ to make sense?

5.1.2 Shift Invariant Filters*

In engineering essentially any operation that maps inputs to outputs is called a *filter*. Since most inputs and outputs are represented by functions, a filter is usually a map from one space of functions to another. The filter is a *linear filter* if this map of function spaces is linear. In practice many filtering operations are given by convolution with a fixed function. If $\psi \in L^1(\mathbb{R}^n)$, then

$$C_\psi(g) = \psi * g$$

defines such a filter. A filter that takes bounded inputs to bounded outputs is called a *stable filter*. The estimate (5.7) shows that any filter defined by convolution with an L^1-function is stable. Indeed the estimates in (5.10) show that such filters act continuously on many function spaces.

Filters defined by convolution have an important physical property: They are *shift invariant*.

***Definition* 5.1.2.** For $\boldsymbol{\tau} \in \mathbb{R}^n$ the *shift of f by $\boldsymbol{\tau}$* is the function $f_{\boldsymbol{\tau}}$, defined by

$$f_{\boldsymbol{\tau}}(\boldsymbol{x}) = f(\boldsymbol{x} - \boldsymbol{\tau}).$$

A filter, $\mathcal{A}$, mapping functions defined on $\mathbb{R}^n$ to functions defined on $\mathbb{R}^n$ is *shift invariant* if

$$\mathcal{A}(f_{\boldsymbol{\tau}}) = (\mathcal{A}f)_{\boldsymbol{\tau}}.$$

If $n = 1$ and the input is a function of time, then a filter is shift invariant if the action of the filter does not depend on *when* the input arrives. If the input is a function of spatial variables, then a filter is shift invariant if its action does not depend on *where* the input is located.

***Example* 5.1.6.** Suppose $\boldsymbol{\tau}$ is a point in $\mathbb{R}^n$; the shift operation $f \mapsto f_{\boldsymbol{\tau}}$ defines a shift invariant filter.

Proposition 5.1.4. *A filter defined by convolution is shift invariant.*

Proof. The proof is a simple change of variables:

$$\begin{aligned} C_\psi(f_{\boldsymbol{\tau}})(\boldsymbol{x}) &= \int\limits_{\mathbb{R}^n} \psi(\boldsymbol{x} - \boldsymbol{y}) f(\boldsymbol{y} - \boldsymbol{\tau}) \, d\boldsymbol{y} \\ &= \int\limits_{\mathbb{R}^n} \psi(\boldsymbol{x} - \boldsymbol{\tau} - \boldsymbol{w}) f(\boldsymbol{w}) \, d\boldsymbol{w} \\ &= C_\psi(f)(\boldsymbol{x} - \boldsymbol{\tau}). \end{aligned} \tag{5.15}$$

In going from the first to the second line, we used the change of variable $\boldsymbol{w} = \boldsymbol{y} - \boldsymbol{\tau}$. □

In a certain sense the converse is also true: Any shift invariant, linear filter can be represented by convolution. What makes this a little complicated is that the function ψ may need to be replaced by a generalized function.

Beyond the evident simplicity of shift invariance, this class of filters is important for another reason: Theorem 5.1.1 shows that the output of such a filter can be computed using the Fourier transform and its inverse, explicitly

$$C_\psi(f) = \mathcal{F}^{-1}(\hat{\psi}\hat{f}). \tag{5.16}$$

This is significant because, as noted previously, the Fourier transform has a very efficient, approximate numerical implementation.

***Example* 5.1.7.** Let $\psi = \frac{1}{2}\chi_{[-1,1]}$, the convolution $\psi * f$ is the moving average of f over intervals of length 2. It can be computed using the Fourier transform by

$$\psi * f(x) = \frac{1}{2\pi}\int_{-\infty}^{\infty} \operatorname{sinc}(\xi)\hat{f}(\xi)e^{ix\xi}d\xi.$$

Exercises

***Exercise* 5.1.12.** For each of the following filters, decide if it is shift invariant or non-shift invariant. Justify your answers.

1. Translation: $\mathcal{A}_\tau(f)(x) \stackrel{d}{=} f(x-\tau)$
2. Scaling: $\mathcal{A}_\epsilon(f)(x) \stackrel{d}{=} \frac{1}{\epsilon^n} f\left(\frac{x}{\epsilon}\right)$
3. Multiplication by a function: $\mathcal{A}_\psi(f) \stackrel{d}{=} \psi f$
4. Indefinite integral from 0: $\mathcal{I}_0(f)(x) \stackrel{d}{=} \int_0^x f(y)\,dy$
5. Indefinite integral from $-\infty$: $\mathcal{I}_{-\infty}(f)(x) \stackrel{d}{=} \int_{-\infty}^x f(y)\,dy$
6. Time reversal: $\mathcal{T}_r(f)(x) \stackrel{d}{=} f(-x)$
7. An integral filter: $f \mapsto \int_{-\infty}^{\infty} xyf(y)\,dy$
8. Differentiation: $\mathcal{D}(f)(x) = f'(x)$

***Exercise* 5.1.13.** Suppose that $\mathcal{A}$ and $\mathcal{B}$ are shift invariant. Show that their composition $\mathcal{A} \circ \mathcal{B}(f) \stackrel{d}{=} \mathcal{A}(\mathcal{B}(f))$ is also shift invariant.

5.1.3 Convolution Equations

Convolution provides a model for many measurement and filtering processes. For example, suppose that f is the state of a system and, for a fixed function ψ, the result of measuring f is modeled by the convolution $g = \psi * f$. To recover the state of the system from this measurement, we must therefore *solve* this equation for f as a function of g. Formally this equation is easy to solve; (5.13) implies that

$$\hat{f}(\xi) = \frac{\hat{g}(\xi)}{\hat{\psi}(\xi)} \qquad \text{so } f = \mathcal{F}^{-1}\left(\frac{\hat{g}}{\hat{\psi}}\right).$$

There are several problems with this approach. The most obvious problem is that $\hat{\psi}$ may vanish for some values of ξ. If the model were perfect, then, of course, $\hat{g}(\xi)$ would also have to vanish at the same points. In real applications this leads to serious problems with stability. A second problem is that, if ψ is absolutely integrable, then the Riemann-Lebesgue lemma implies that $\hat{\psi}(\xi)$ tends to 0 as $\|\xi\|$ goes to infinity. Unless the measurement g is smooth and noise free, it is not possible to exactly determine f by applying the inverse Fourier transform to this ratio. In Chapter 9 we discuss how these issues are handled in practice.

***Example* 5.1.8.** The rectangle function defines a simple weight, $\psi_\epsilon = (2\epsilon)^{-1}\chi_{[-\epsilon,\epsilon]}$. Its Fourier transform is given by

$$\hat{\psi}_\epsilon(\xi) = \operatorname{sinc}(\epsilon\xi).$$

This function has zeros at $\xi = \pm(\epsilon^{-1}m\pi)$, where m is any positive integer. These zeros are isolated, so it seems reasonable that an integrable function f should be uniquely specified by the averages $\psi_\epsilon * f$, for any $\epsilon > 0$. In fact it is, but f cannot be *stably* reconstructed from these averages.

***Example* 5.1.9.** Suppose, in the previous example, we also knew that $\hat{f}$ vanishes if $|\xi| > \pi(2\epsilon)^{-1}$. In this case the function $\operatorname{sinc}(\epsilon\xi)$ has a positive lower bound on the support of $\hat{f}$ and therefore it is allowable to use Fourier inversion to recover f from $\psi * f$,

$$f(x) = \int_{-\frac{\pi}{2\epsilon}}^{\frac{\pi}{2\epsilon}} \frac{\hat{f}(\xi)e^{ix\xi}d\xi}{2\pi \operatorname{sinc}(\epsilon\xi)}.$$

Exercises

***Exercise* 5.1.14.** For a positive real number, a, define the function

$$t_a(x) = \frac{a - |x|}{a^2}\chi_{[-a,a]}(x),$$

and for two positive real numbers, a, b, define

$$t_{a,b}(x) = \frac{1}{2}\left[t_a(x) + t_b(x)\right].$$

Graph $t_{a,b}(x)$ for several different choices of (a, b). Show that, for appropriate choices of a and b, the Fourier transform $\hat{t}_{a,b}(\xi)$ does not vanish for any value of ξ.

***Exercise* 5.1.15.** Define a function

$$f(x) = \chi_{[-1,1]}(x)(1 - |x|)^2.$$

Compute the Fourier transform of this function and show that it does not vanish anywhere. Let $f_j = f *_j f$ (the j-fold convolution of f with itself). Show that the Fourier transforms, $\hat{f}_j$, are also nonvanishing.

***Exercise* 5.1.16.** Explain the following statement: If $\hat{\psi}$ vanishes or tends to zero as ξ tends to infinity, then it is not possible to *stably* recover f from the averages $\psi * f$.

***Exercise* 5.1.17.** Show that an even, real-valued function φ, with bounded support, can be expressed as $\varphi = \psi * \psi$ if and only if $\hat{\varphi}$ is nonnegative.

5.2 Convolution and Regularity

Generally speaking, the averages of a function are smoother than the function itself. If f is a locally integrable function and φ is continuous, with bounded support, then $f * \varphi$ is continuous. Let $\boldsymbol{\tau}$ be a vector in $\mathbb{R}^n$. Then

$$\lim_{\boldsymbol{\tau}\to\mathbf{0}}[f * \varphi(\boldsymbol{x} + \boldsymbol{\tau}) - f * \varphi(\boldsymbol{x})] = \lim_{\boldsymbol{\tau}\to\mathbf{0}} \int\limits_{\mathbb{R}^n} f(\boldsymbol{y})[\varphi(\boldsymbol{x} + \boldsymbol{\tau} - \boldsymbol{y}) - \varphi(\boldsymbol{x} - \boldsymbol{y})]\, d\boldsymbol{y}.$$

Because φ has bounded support, it follows that the limit on the right can be taken inside the integral, showing that

$$\lim_{\boldsymbol{\tau}\to\mathbf{0}} f * \varphi(\boldsymbol{x} + \boldsymbol{\tau}) = f * \varphi(\boldsymbol{x}).$$

This argument can be repeated with difference quotients to prove the following result.

Proposition 5.2.1. *Suppose that f is locally integrable, φ has bounded support, and k continuous derivatives. Then $f * \varphi$ also has k continuous derivatives. For any multi-index $\boldsymbol{\alpha}$ with $|\boldsymbol{\alpha}| \leq k$, we have*

$$\partial_{\boldsymbol{x}}^{\boldsymbol{\alpha}}(f * \varphi) = f * (\partial_{\boldsymbol{x}}^{\boldsymbol{\alpha}}\varphi). \tag{5.17}$$

***Remark* 5.2.1.** This result is also reasonable from the point of view of the Fourier transform. Suppose that φ has k integrable derivatives. Then Proposition 4.5.3 shows that

$$|\hat{\varphi}(\boldsymbol{\xi})| \leq \frac{C}{(1 + \|\boldsymbol{\xi}\|)^k}.$$

If f is either integrable or square integrable, then the Fourier transform of $f * \varphi$ satisfies an estimate of the form

$$|\mathscr{F}(f * \varphi)(\boldsymbol{\xi})| \leq \frac{C|\hat{f}(\boldsymbol{\xi})|}{(1 + \|\boldsymbol{\xi}\|)^k}.$$

This shows that the Fourier transform of $f * \varphi$ has a definite improvement in its rate of decay over that of f and therefore $f * \varphi$ is commensurately smoother.

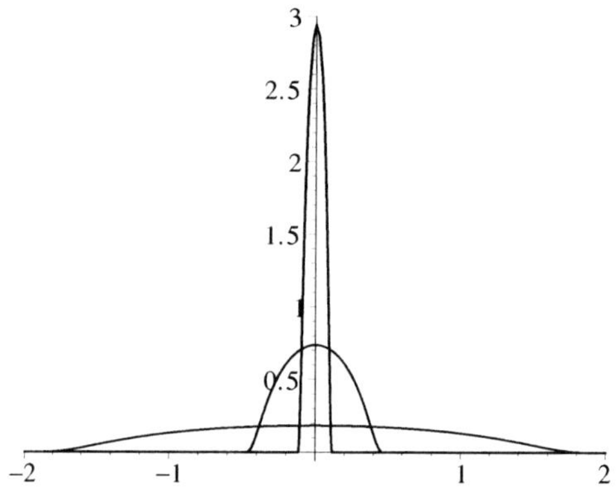

Figure 5.3. Graphs of φ_ϵ, with $\epsilon = .5, 2, 8$.

5.2.1 Approximation by Smooth Functions

Convolution provides a general method for approximating integrable (or locally integrable) functions by smooth functions. Beyond that, it gives a way to define approximate derivatives of functions that are not differentiable. We begin with a definition:

***Definition* 5.2.1.** For a function φ, defined on $\mathbb{R}^n$, and ϵ, a positive real number, define the *scaled function* φ_ϵ by

$$\varphi_\epsilon(\boldsymbol{x}) = \epsilon^{-n}\varphi(\frac{\boldsymbol{x}}{\epsilon}). \tag{5.18}$$

While this notation is quite similar to that used, in Definition 5.1.2, for the translation of a function, the meaning should be clear from the context. A one-dimensional example is shown in Figure 5.3.

Let φ be an infinitely differentiable function with total integral one:

$$\int_{\mathbb{R}^n} \varphi(\boldsymbol{x})\,d\boldsymbol{x} = 1.$$

If φ is supported in the ball of radius 1, then φ_ϵ is supported in the ball of radius ϵ and also has total integral one: Using the change of variables $\epsilon\,\boldsymbol{y} = \boldsymbol{x}$ gives

$$\int_{\mathbb{R}^n} \varphi_\epsilon(\boldsymbol{x})\,d\boldsymbol{x} = \int_{\mathbb{R}^n} \frac{1}{\epsilon^n}\varphi(\frac{\boldsymbol{x}}{\epsilon})\,d\boldsymbol{x} = \int_{\mathbb{R}^n} \varphi(\boldsymbol{y})\,d\boldsymbol{y} = 1. \tag{5.19}$$

For small ϵ the convolution $\varphi_\epsilon * f$ is an approximation to f. Using (5.19), the difference between f and $\varphi_\epsilon * f$ is expressed in a convenient form:

$$\varphi_\epsilon * f(\boldsymbol{x}) - f(\boldsymbol{x}) = \int_{B_\epsilon(\boldsymbol{x})} [f(\boldsymbol{y}) - f(\boldsymbol{x})]\varphi_\epsilon(\boldsymbol{x} - \boldsymbol{y})\,d\boldsymbol{y}. \tag{5.20}$$

The integral is over the ball of radius ϵ, centered at $\boldsymbol{x}$. It is therefore reasonable to expect that, as ϵ goes to zero, $\varphi_\epsilon * f$ converges, in some sense, to f.

The fact that φ has total integral one implies that $\hat{\varphi}(0) = 1$. This gives another route for understanding the relationship between $\varphi_\epsilon * f$ and f as ϵ tends to 0. It follows from Theorem 5.1.1 that

$$\begin{aligned} \mathcal{F}(\varphi_\epsilon * f)(\boldsymbol{\xi}) &= \hat{\varphi}_\epsilon(\boldsymbol{\xi})\hat{f}(\zeta) \\ &= \hat{\varphi}(\epsilon\boldsymbol{\xi})\hat{f}(\boldsymbol{\xi}). \end{aligned} \tag{5.21}$$

For each fixed $\boldsymbol{\xi}$, the limit of $\mathcal{F}(\varphi_\epsilon * f)(\boldsymbol{\xi})$, as ϵ goes to zero, is $\hat{f}(\boldsymbol{\xi})$. Because both noise *and* fine detail are carried by the high-frequency components, convolution with φ_ϵ tends to average out the noise while, at the same time, blurring the fine detail. The size of ϵ determines the degree of blurring. This can be seen quantitatively in the Fourier representation. Since $\hat{\varphi}(0) = 1$ and $\hat{\varphi}$ is a smooth function, we see that

$$\hat{\varphi}(\epsilon\boldsymbol{\xi}) \approx 1 \text{ for } \|\boldsymbol{\xi}\| << \epsilon^{-1}.$$

Hence, for low frequencies—that is $\|\boldsymbol{\xi}\| << \epsilon^{-1}$,—the Fourier transform $\mathcal{F}(\varphi_\epsilon * f)(\boldsymbol{\xi})$ closely approximates $\hat{f}(\boldsymbol{\xi})$. On the other hand, $\hat{\varphi}(\boldsymbol{\xi})$ tends rapidly to zero as $\|\boldsymbol{\xi}\| \to \infty$, and therefore the high-frequency content of f is suppressed in $\varphi_\epsilon * f$. Using convolution to suppress noise inevitably destroys fine detail.

***Example* 5.2.1.** Consider the function of two variables f shown (as a density plot) in Figure 5.4(a). The convolution of f with a smooth function is shown in Figures 5.4(b–c). Near points where f is slowly varying, $\varphi_\epsilon * f$ is quite similar to f. Near points where f is rapidly varying, this is not the case.

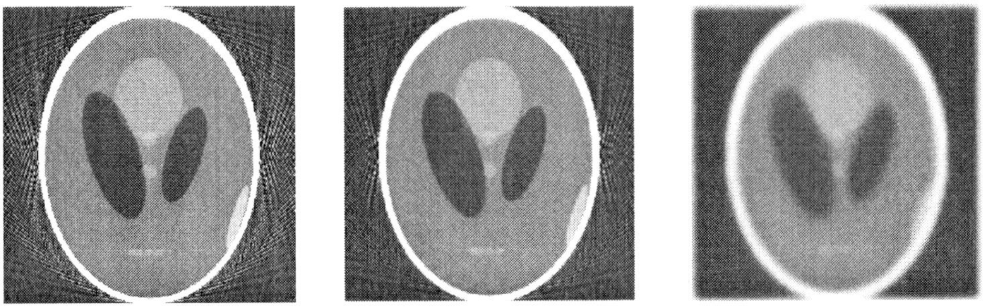

(a) A reconstruction of the Shepp-Logan phantom; see Figure 3.7.

(b) The function in (a) convolved with φ_ϵ with a small ϵ.

(c) The function in (a) convolved with φ_ϵ with a large ϵ.

Figure 5.4. Convolving f reduces the noise but blurs the detail.

***Remark* 5.2.2.** In practice, infinitely differentiable functions can be difficult to work with. To simplify computations, a finitely differentiable function may be preferred. For example, given $k \in \mathbb{N}$, define the function

$$\psi_k(x) = \begin{cases} c_k(1-x^2)^k & \text{if } |x| \leq 1, \\ 0 & \text{if } |x| > 1. \end{cases} \tag{5.22}$$

The constant, c_k, is selected so that ψ_k has total integral one. The function ψ_k has $k-1$ continuous derivatives. If

$$\psi_{k,\epsilon}(x) = \epsilon^{-1}\psi_k(\frac{x}{\epsilon})$$

and f is locally integrable, then $< \psi_{k,\epsilon} * f >$ is a family of $(k-1)$ times differentiable functions, which converge, in an appropriate sense to f.

Exercises

***Exercise* 5.2.1.** Let f be an integrable function with support in the interval $[a, b]$ and g an integrable function with support in $[-\epsilon, \epsilon]$. Show that the support of $f * g$ is contained in $[a-\epsilon, b+\epsilon]$.

***Exercise* 5.2.2.** For the functions ψ_k, defined in (5.22), find the constants c_k so that

$$\int_{-1}^{1} \psi_k(x)\, dx = 1.$$

5.2.2 Some Convergence Results*

We now prove some precise results describing different ways in which $\varphi_\epsilon * f$ converges to f. For most of these results it is only necessary to assume that φ is an L^1-function with total integral one. The sense in which $\varphi_\epsilon * f$ converges depends on regularity and decay properties of f. The square-integrable case is the simplest.

Proposition 5.2.2. *Suppose that φ is an L^1-function with*

$$\int_{\mathbb{R}^n} \varphi(\mathbf{x})\, d\mathbf{x} = 1.$$

If $f \in L^2(\mathbb{R}^n)$, then $\varphi_\epsilon * f$ converges to f in $L^2(\mathbb{R}^n)$.

Proof. The Plancherel formula implies that

$$\|\varphi_\epsilon * f - f\|_{L^2} = \frac{1}{[2\pi]^{\frac{n}{2}}} \|\widehat{\varphi_\epsilon * f} - \hat{f}\|_{L^2}.$$

The Fourier transform of φ_ϵ at $\boldsymbol{\xi}$, computed using (4.37), is

$$\mathcal{F}(\varphi_\epsilon)(\boldsymbol{\xi}) = \hat{\varphi}(\epsilon\boldsymbol{\xi}). \tag{5.23}$$

From Theorem 5.1.1 we obtain

$$\|\widehat{\varphi_\epsilon * f} - \hat{f}\|_{L^2} = \|\hat{f}(\hat{\varphi}_\epsilon - 1)\|_{L^2}.$$

The Lebesgue dominated convergence theorem, (5.23), and the fact that $\hat{\varphi}(0) = 1$ imply that

$$\lim_{\epsilon\to\infty} \|\hat{f}(\hat{\varphi}_\epsilon - 1)\|_{L^2} = 0.$$

□

A similar result holds in the L^1-case.

Proposition 5.2.3. *Suppose that φ is an L^1-function with*

$$\int_{\mathbb{R}^n} \varphi(x)\,dx = 1.$$

If f belongs to $L^1(\mathbb{R}^n)$, then $\varphi_\epsilon * f$ converges to f in the L^1-norm.

Proof. The proof of this result is quite different from the L^2-case. Proposition 5.2.3 is proved with the following lemma:

Lemma 5.2.1. *If f belongs to $L^1(\mathbb{R}^n)$, then*

$$\lim_{\boldsymbol{\tau}\to\mathbf{0}} \|f_{\boldsymbol{\tau}} - f\|_{L^1} = 0.$$

In other words, the translation operator, $(\boldsymbol{\tau}, f) \mapsto f_{\boldsymbol{\tau}}$ is a continuous map of $\mathbb{R}^n \times L^1(\mathbb{R}^n)$ to $L^1(\mathbb{R}^n)$. The proof of this statement is left to the Exercises. The triangle inequality shows that

$$\begin{aligned}
\|\varphi_\epsilon * f - f\|_{L^1} &= \int_{\mathbb{R}^n} \left| \int_{\mathbb{R}^n} [f(\mathbf{x} - \epsilon\mathbf{t}) - f(\mathbf{x})]\varphi(\mathbf{t})\,d\mathbf{t} \right| d\mathbf{x} \\
&\leq \int_{\mathbb{R}^n} |\varphi(\mathbf{t})| \left[\int_{\mathbb{R}^n} |f(\mathbf{x} - \epsilon\mathbf{t}) - f(\mathbf{x})|\,d\mathbf{x} \right] d\mathbf{t} \\
&= \int_{\mathbb{R}^n} |\varphi(\mathbf{t})| \|f_{\epsilon\mathbf{t}} - f\|_{L^1}\,d\mathbf{t}.
\end{aligned} \tag{5.24}$$

The last integrand is bounded by $2\|f\|_{L^1}|\varphi(\mathbf{t})|$, and therefore the limit, as ϵ goes to zero, can be brought inside the integral. The conclusion of the proposition follows from the lemma. □

Finally, it is useful to examine $\varphi_\epsilon * f(\mathbf{x})$ at points where f is smooth. Here we use a slightly different assumption on φ.

Proposition 5.2.4. *Let f be a locally integrable function and φ an integrable function with bounded support and total integral one. If f is continuous at $\mathbf{x}$, then*

$$\lim_{\epsilon\downarrow 0} \varphi_\epsilon * f(\mathbf{x}) = f(\mathbf{x}).$$

Proof. As f is continuous at $\mathbf{x}$, given $\eta > 0$, there is a $\delta > 0$ so that

$$\|\mathbf{x} - \mathbf{y}\| < \delta \Rightarrow |f(\mathbf{x}) - f(\mathbf{y})| < \eta. \tag{5.25}$$

This implies that $|f(\mathbf{y})|$ is bounded for $\mathbf{y}$ in $B_\delta(\mathbf{x})$. If ϵ is sufficiently small, say less than ϵ_0, then the support of φ_ϵ is contained in the ball of radius δ and therefore $\varphi_\epsilon * f(\mathbf{x})$ is defined by an absolutely convergent integral. Since the total integral of φ is 1 we have, for an $\epsilon < \epsilon_0$, that

$$
\begin{aligned}
|\varphi_\epsilon * f(\boldsymbol{x}) - f(\boldsymbol{x})| &= \left| \int\limits_{B_\delta} \varphi_\epsilon(\boldsymbol{y})(f(\boldsymbol{x}-\boldsymbol{y}) - f(\boldsymbol{x}))\, d\boldsymbol{y} \right| \\
&\leq \int\limits_{B_\delta} |\varphi_\epsilon(\boldsymbol{y})||f(\boldsymbol{x}-\boldsymbol{y}) - f(\boldsymbol{x})|\, d\boldsymbol{y} \\
&\leq \int\limits_{B_\delta} |\varphi_\epsilon(\boldsymbol{y})|\eta\, d\boldsymbol{y} \\
&\leq \|\varphi\|_{L^1}\eta.
\end{aligned}
\tag{5.26}
$$

In the third line we use the estimate (5.25). Since $\eta > 0$ is arbitrary, this completes the proof of the proposition. □

***Remark* 5.2.3.** There are many variants of these results. The main point of the proofs is that φ is absolutely integrable. Many similar-*looking* results appear in analysis, though with much more complicated proofs. In most of these cases φ is *not* absolutely integrable. For example, the Fourier inversion formula in one dimension amounts to the statement that $\varphi_\epsilon * f$ converges to f, where $\varphi(x) = \pi^{-1}\operatorname{sinc}(x)$. As we have noted several times before, $\operatorname{sinc}(x)$ is not absolutely integrable.

We close this section by applying the approximation results to complete the proof of the Fourier inversion formula. Thus far, Theorems 4.2.1 and 4.5.1 were proved with the additional assumption that f is continuous.

Proof of the Fourier inversion formula, completed. Suppose that f and $\hat{f}$ are absolutely integrable and φ_ϵ is as previously. Note that $\hat{f}$ is a continuous function. For each $\epsilon > 0$ the function $\varphi_\epsilon * f$ is absolutely integrable and continuous. Its Fourier transform, $\hat{\varphi}(\epsilon\boldsymbol{\xi})\hat{f}(\boldsymbol{\xi})$, is absolutely integrable. As ϵ goes to zero, it converges locally uniformly to $\hat{f}(\xi)$. Since these functions are continuous, we can apply the Fourier inversion formula to conclude that

$$\varphi_\epsilon * f(\boldsymbol{x}) = \frac{1}{2\pi}\int\limits_{\mathbb{R}^n} \hat{\varphi}(\epsilon\boldsymbol{\xi})\hat{f}(\boldsymbol{\xi})^{i\langle \boldsymbol{x},\boldsymbol{\xi}\rangle} d\boldsymbol{\xi}.$$

This is a locally uniformly convergent family of continuous functions and therefore has a continuous limit. The right-hand side converges pointwise to

$$F(\boldsymbol{x}) = \int\limits_{\mathbb{R}^n} \hat{f}(\boldsymbol{\xi})^{i\langle \boldsymbol{x},\boldsymbol{\xi}\rangle} d\boldsymbol{\xi}.$$

Proposition 5.2.3 implies that $\|\varphi_\epsilon * f - f\|_{L^1}$ also goes to zero as ϵ tends to 0 and therefore $F = f$. (To be precise we should say that after modification on a set of measure 0, $F = f$.) This completes the proof of the Fourier inversion formula. □

Exercises

***Exercise* 5.2.3.** Use Corollary A.5.1 to prove Lemma 5.2.1.

***Exercise* 5.2.4.** Give the details of the argument, using Lemma 5.2.1, to show that if f is an L^1-function, then

$$\lim_{\epsilon\to 0}\int_{\mathbb{R}^n} \varphi(\boldsymbol{t})\|f_{\epsilon\boldsymbol{t}} - f\|_{L^1}\, d\boldsymbol{t} = 0.$$

***Exercise* 5.2.5.** Use the method used to prove Proposition 5.2.4 to show that if $f \in L^p(\mathbb{R})$ for a $1 \leq p < \infty$, then $\varphi_\epsilon * f$ converges to f in the L^p-norm. Give an example to show that if f is a bounded, though discontinuous function, then $\|\varphi_\epsilon * f - f\|_\infty$ may fail to tend to zero.

***Exercise* 5.2.6.** Let $\psi_\epsilon(x) = [2\epsilon]^{-1}\chi_{[-\epsilon,\epsilon]}(x)$. Show by direct computation that if $f \in L^2(\mathbb{R})$, then $\psi_\epsilon * f$ converges to f in $L^2(\mathbb{R})$.

5.2.3 Approximating Derivatives and Regularized Derivatives

If either f or φ is a differentiable function then $\varphi * f$ is as well. In this section we assume that φ is a bounded function with support in B_1 and total integral one. If f has k continuous derivatives in $B_\delta(\boldsymbol{x})$, then, for $\epsilon < \delta$ the convolution $\varphi_\epsilon * f$ is k-times differentiable. For each $\boldsymbol{\alpha}$ with $|\boldsymbol{\alpha}| \leq k$, Proposition 5.2.1 implies that

$$\partial_{\boldsymbol{x}}^{\alpha}(\varphi_\epsilon * f)(\boldsymbol{x}) = \varphi_\epsilon * \partial_{\boldsymbol{x}}^{\boldsymbol{\alpha}} f(\boldsymbol{x}).$$

Proposition 5.2.4 can be applied to conclude that

$$\lim_{\epsilon\to 0}\partial_{\boldsymbol{x}}^{\alpha}(\varphi_\epsilon * f)(\boldsymbol{x}) = \partial_{\boldsymbol{x}}^{\boldsymbol{\alpha}} f(\boldsymbol{x}).$$

On the other hand, if f is not a differentiable function but φ is, then

$$\partial_{x_j}(\varphi_\epsilon * f) = (\partial_{x_j}\varphi_\epsilon) * f$$

can be used to *define* regularized approximations to the partial derivatives of f. This can be useful if f is the result of a noisy measurement of a smooth function that, for one reason or another, must be differentiated. Precisely this situation arises in the reconstruction process used in x-ray CT-imaging. We illustrate this idea with an example.

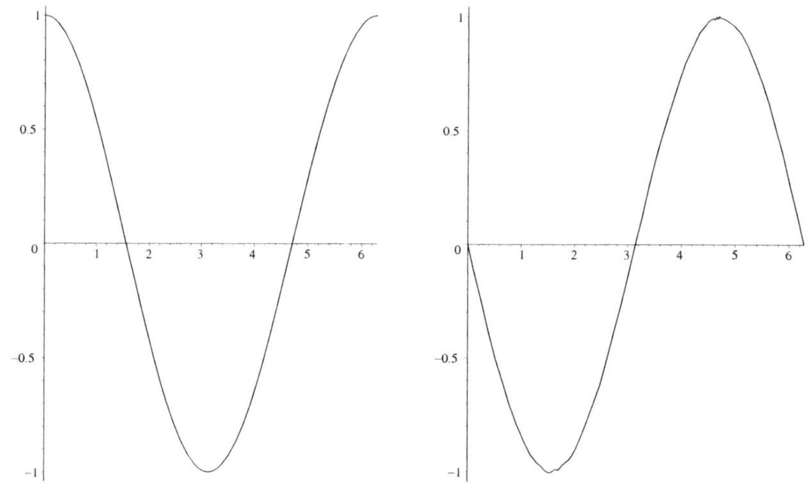

(a) A noisy function smoothed by convolution with a triangle function, $\epsilon = .1$.

(b) The regularized derivative of the noisy function, $\epsilon = .1$.

Figure 5.5. Using convolution to find a regularized approximate derivative.

***Example* 5.2.2.** Let f be the noise corrupted version of $\cos(x)$ depicted in Figure 5.1(c). To smooth f, we use the "triangle function"

$$t_\epsilon(x) = \begin{cases} \frac{\epsilon - |x|}{\epsilon^2} & \text{if } |x| \le \epsilon, \\ 0 & \text{if } |x| > \epsilon. \end{cases}$$

The derivative of $f * t_\epsilon$ is computed using the weak derivative of t_ϵ. The result of computing $f * t_{.1}$ is shown in Figure 5.5(a), while $f * t'_{.1}$ is shown in Figure 5.5(b). The difference $|-\sin(x) - f * t'_{.1}(x)|$ is quite small, provided x is not too close to a zero of the cosine.

Exercise

***Exercise* 5.2.7.** * For k a positive integer, suppose that f and $\xi^k \hat{f}(\xi)$ belong to $L^2(\mathbb{R})$. By approximating f by smooth functions of the form $\varphi_\epsilon * f$, show that f has k L^2-derivatives.

5.2.4 The Support of $f * g$

Suppose that f and g have bounded support. For applications to medical imaging it is important to understand how the support of $f * g$ is related to the supports of f and g. To that end we define the *algebraic sum* of two subsets of $\mathbb{R}^n$.

***Definition* 5.2.2.** Suppose A and B are subsets of $\mathbb{R}^n$. The algebraic sum of A and B is defined as the set

$$A + B = \{\boldsymbol{a} + \boldsymbol{b} \in \mathbb{R}^n \ : \ \boldsymbol{a} \in A, \text{ and } \boldsymbol{b} \in B\}.$$

Using this concept, we can give a quantitative result describing the way in which convolution "smears" out the support of a function.

Proposition 5.2.5. *The support of* $f * g$ *is contained in* $\operatorname{supp} f + \operatorname{supp} g$.

Proof. Suppose that $\boldsymbol{x}$ is not in $\operatorname{supp} f + \operatorname{supp} g$. This means that no matter which $\boldsymbol{y}$ is selected, either $f(\boldsymbol{y})$ or $g(\boldsymbol{x}-\boldsymbol{y})$ is zero. Otherwise, $\boldsymbol{x} = \boldsymbol{y} + (\boldsymbol{x}-\boldsymbol{y})$ would belong to $\operatorname{supp} f + \operatorname{supp} g$. This implies that $f(\boldsymbol{y})g(\boldsymbol{x}-\boldsymbol{y})$ is zero for all $\boldsymbol{y} \in \mathbb{R}^n$ and therefore

$$f * g(\boldsymbol{x}) = \int_{\mathbb{R}^n} f(\boldsymbol{y})g(\boldsymbol{x}-\boldsymbol{y})\,d\boldsymbol{y} = 0$$

as well. Because $\operatorname{supp} f + \operatorname{supp} g$ is a closed set, there is an $\eta > 0$ such that $B_\eta(\boldsymbol{x})$ is disjoint from $\operatorname{supp} f + \operatorname{supp} g$. The argument showing that $f * g(\boldsymbol{x})$ equals 0 applies to any point $\boldsymbol{x}'$ in $B_\eta(\boldsymbol{x})$ and therefore proves the proposition. □

If φ is a function supported in the ball of radius one, then φ_ϵ is supported in the ball of radius ϵ. According to Proposition 5.2.5, the support of $\varphi_\epsilon * f$ is contained in the set

$$\{\boldsymbol{x} + \boldsymbol{y} \;:\; \boldsymbol{x} \in \operatorname{supp} f \text{ and } \boldsymbol{y} \in B_\epsilon\}.$$

These are precisely the points that are within distance ϵ of the support of f, giving another sense in which ϵ reflects the resolution available in $\varphi_\epsilon * f$. Figure 5.6 shows a one-dimensional example.

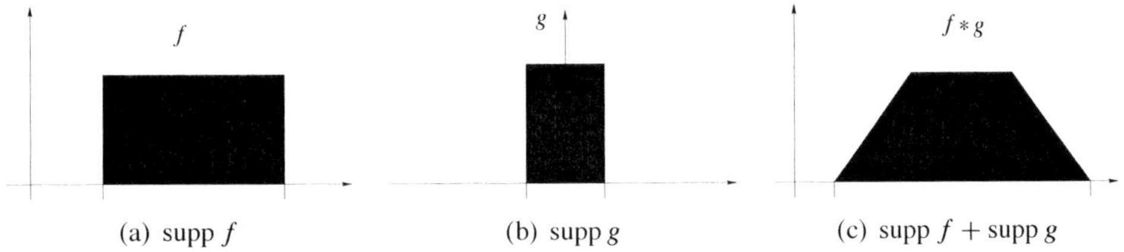
f
g
f * g
(a) supp f
(b) supp g
(c) supp f + supp g

Figure 5.6. The support of $f * g$ is contained in $\operatorname{supp} f + \operatorname{supp} g$.

***Example* 5.2.3.** Suppose that ψ is a nonnegative function that vanishes outside the interval $[-\epsilon, \epsilon]$ and has total integral 1,

$$\int_{-\infty}^{\infty} \psi(x)\,dx = 1.$$

If f is a locally integrable function, then $f * \psi(x)$ is the weighted average of the values of f over the interval $[x-\epsilon, x+\epsilon]$. Note that $\psi * \psi$ also has total integral 1

$$\begin{aligned}\int_{-\infty}^{\infty} \psi * \psi(x)\,dx &= \int_{-\infty}^{\infty}\int_{-\infty}^{\infty} \psi(y)\psi(x-y)\,dy\,dx \\ &= \int_{-\infty}^{\infty}\int_{-\infty}^{\infty} \psi(y)\psi(t)\,dt\,dy \\ &= 1 \cdot 1 = 1.\end{aligned} \tag{5.27}$$

In the second to last line we reversed the order of the integrations and set $t = x - y$.

Thus $f * (\psi * \psi)$ is again an average of f. Note that $\psi * \psi(x)$ is generally nonzero for $x \in [-2\epsilon, 2\epsilon]$, so convolution with $\psi * \psi$ produces more blurring than convolution with ψ alone. Indeed we know from the associativity of the convolution product that

$$f * (\psi * \psi) = (f * \psi) * \psi,$$

so we are averaging the averages, $f * \psi$. This can be repeated as many times as we like. The j-fold convolution $\psi *_j \psi$ has total integral 1 and vanishes outside the interval $[-j\epsilon, j\epsilon]$. Of course, the Fourier transform of $\psi *_j \psi$ is $[\hat{\psi}(\xi)]^j$, which therefore decays j times as fast as $\hat{\psi}(\xi)$.

We could also use the *scaled* j-fold convolution $\delta^{-1}\psi *_j \psi(\delta^{-1}x)$ to average our data. This function vanishes outside the interval $[-j\delta\epsilon, j\delta\epsilon]$ and has Fourier transform $[\hat{\psi}(\delta\xi)]^j$. If we choose $\delta = j^{-1}$, then convolution with this function will not blur details any more than convolution with ψ itself but better suppresses high-frequency noise. By choosing j and δ, we can control, to some extent, the trade-off between blurring and noise suppression.

5.3 The δ-Function*

See: A.4.5.

The convolution product defines a multiplication on $L^1(\mathbb{R}^n)$ with all the usual properties of a product except that there is no unit. If i were a unit, then $i * f = f$ for every function in $L^1(\mathbb{R}^n)$. Taking the Fourier transform, this would imply that, for every $\boldsymbol{\xi}$,

$$\hat{i}(\boldsymbol{\xi})\hat{f}(\boldsymbol{\xi}) = \hat{f}(\boldsymbol{\xi}).$$

This shows that $\hat{i}(\boldsymbol{\xi}) \equiv 1$ and therefore i cannot be an L^1-function. Having a multiplicative unit is so useful that engineers, physicists, and mathematicians have all found it necessary to define one. It is called the *δ-function* and is defined by the property that for any continuous function f

$$f(0) = \int_{\mathbb{R}^n} \delta(\boldsymbol{y}) f(\boldsymbol{y})\, d\boldsymbol{y}. \tag{5.28}$$

Proceeding formally, we see that

$$\begin{aligned} \delta * f(\boldsymbol{x}) &= \int_{\mathbb{R}^n} \delta(\boldsymbol{y}) f(\boldsymbol{x} - \boldsymbol{y})\, d\boldsymbol{y} \\ &= f(\boldsymbol{x} - \mathbf{0}) = f(\boldsymbol{x}). \end{aligned} \tag{5.29}$$

So at least for continuous functions, $\delta * f = f$.

It is important to remember that *the δ-function is not a function.* In the mathematics literature the δ-function is an example of a *distribution* or *generalized function.* The basic properties of generalized functions are introduced in Appendix A.4.5. In the engineering and physics literature, the δ-function is sometimes called a *unit impulse.* In Section 4.4.4 the Fourier transform is extended to generalized functions (at least in the one-dimensional case). The Fourier transform of δ is, as expected, identically equal to 1:

$$\mathcal{F}(\delta) \equiv 1.$$

While (5.28) only makes sense for functions continuous at 0, the convolution of δ with an arbitrary locally integrable function is well defined and satisfies $\delta * f = f$. This is not too different from the observation that if f and g are L^1-functions, then $f * g(x)$ may not be defined at every point; nonetheless, $f * g$ is a well-defined element of $L^1(\mathbb{R}^n)$.

5.3.1 Approximating the δ-Function in One-Dimension

In both mathematics and engineering it is useful to have approximations for the δ-function. There are two complementary approaches to this problem: One is to use functions like φ_ϵ, defined in (5.18), to approximate δ in $\boldsymbol{x}$-space. The other is to approximate $\hat{\delta}$ in $\boldsymbol{\xi}$-space. To close this chapter we consider some practical aspects of approximating the δ-function in one dimension and formalize the concept of resolution.

Suppose that φ is an even function with bounded support and total integral one. The Fourier transform of φ_ϵ is $\hat{\varphi}(\epsilon\xi)$. Because φ_ϵ vanishes outside a finite interval, its Fourier transform is a smooth function and $\hat{\varphi}(0) = 1$. As φ is a nonnegative, even function, its Fourier transform is real valued and assumes its maximum at zero. In applications it is important that the difference $1 - \hat{\varphi}(\epsilon\xi)$ remain small over a specified interval $[-B, B]$. It is also important that $\hat{\varphi}(\epsilon\xi)$ tend to zero rapidly outside a somewhat larger interval. As φ is non-negative, $\partial_\xi\hat{\varphi}(0) = 0$; this means that the behavior of $\hat{\varphi}(\xi)$ for ξ near to zero is largely governed by the "second moment" of φ :

$$\partial_\xi^2\hat{\varphi}(0) = -\int_{-\infty}^{\infty} x^2\varphi(x)\,dx.$$

We would like this number to be small. This is accomplished by putting more of the mass of φ near to $x = 0$. On the other hand, the rate at which $\hat{\varphi}$ decays as $|\xi| \to \infty$ is determined by the smoothness of φ. If $\varphi = \frac{1}{2}\chi_{[-1,1]}$, then $\hat{\varphi}$ decays like $|\xi|^{-1}$. Better decay is obtained by using a smoother function. For the stability of numerical algorithms it is often important that $\hat{\varphi}$ is absolutely integrable. In one dimension this is the case if φ is continuous and piecewise differentiable.

The other approach to constructing approximations to the δ-function is to approximate its Fourier transform. We use a sequence of functions that are approximately 1 in an interval $[-B, B]$ and vanish outside a larger interval. Again a simple choice is $\chi_{[-B,B]}(\xi)$. The inverse Fourier transform of this function is $\psi_B(x) = \pi^{-1}B\operatorname{sinc}(Bx)$. In this context it is called a *sinc pulse.* Note that ψ_B assumes both positive and negative values. A sinc pulse is not absolutely integrable; the fact that the improper integral of ψ_B over the whole real line equals 1 relies on subtle cancellations between the positive and negative parts of the integral. Because ψ_B is not absolutely integrable, it is often a poor choice for approximating the δ-function. Approximating $\hat{\delta}$ by $(2B)^{-1}\chi_{[-B,B]} * \chi_{[-B,B]}(\xi)$ gives a sinc^2 pulse, $(2B)^{-1}\psi_B^2(x)$, as an approximation to δ. This function has better properties: It does not assume negative values, is more sharply peaked at 0, and is absolutely integrable. These functions are graphed in Figure 5.7.

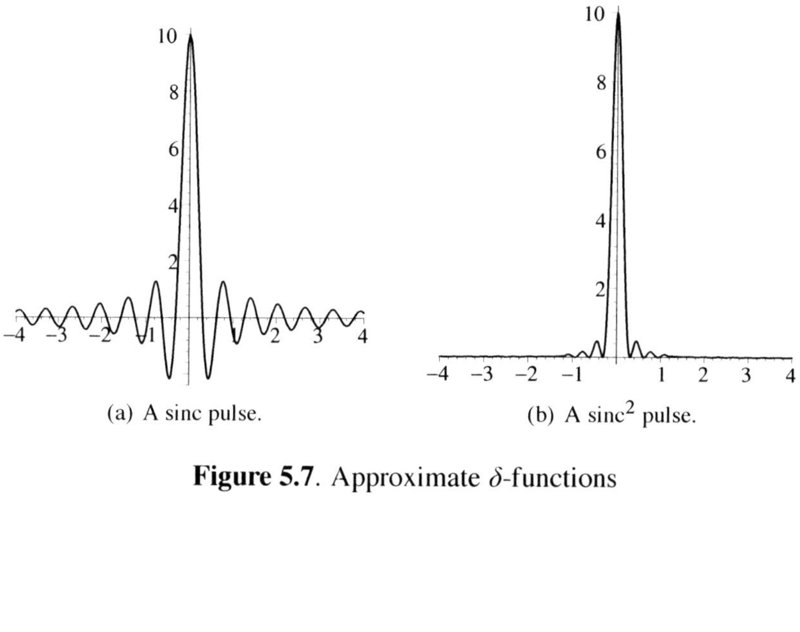

(a) A sinc pulse.

(b) A sinc^2 pulse.

Figure 5.7. Approximate δ-functions

Neither the sinc nor sinc^2 has bounded support; both functions have oscillatory "tails" extending to infinity. In the engineering literature these are called *side lobes*. Side lobes result from the fact that the Fourier transforms of sinc and sinc^2 have bounded support; see Section 4.4.3. The convolutions of these functions with $\chi_{[-1,1]}$ are shown in Figure 5.8. In Figure 5.8(a) notice that the side lobes produce large oscillations near the jump. This is an example of the Gibbs phenomenon. It results from using a discontinuous cutoff function in the Fourier domain. This effect is analyzed in detail, for the case of Fourier series, in Section 7.5.

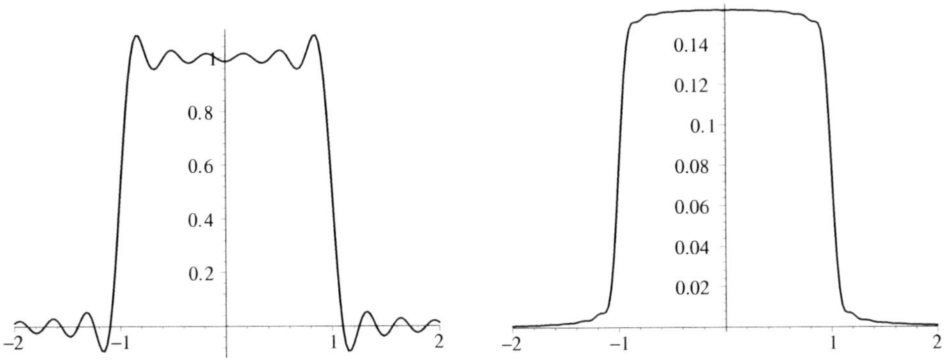

(a) $\operatorname{sinc} * \chi_{[-1,1]}$. (b) $\operatorname{sinc}^2 * \chi_{[-1,1]}$.

Figure 5.8. Approximate δ-functions convolved with $\chi_{[-1,1]}$.

Exercises

***Exercise* 5.3.1.** Suppose that f is a continuous L^1-function and φ is absolutely integrable with $\int_{\mathbb{R}} \varphi = 1$. Show that $< \varphi_\epsilon * f >$ converges pointwise to f.

***Exercise* 5.3.2.** Suppose that φ is an integrable function on the real line with total integral 1 and f is an integrable function such that, for a $k > 1$,

$$|\hat{f}(\xi)| \leq \frac{C}{(1+|\xi|)^k}.$$

Use the Fourier inversion formula to estimate the error $\|\varphi_\epsilon * f(x) - f(x)|$.

***Exercise* 5.3.3.** Suppose that φ is a smooth, non-negative function with bounded support, and total integral 1 defined on $\mathbb{R}$. Let $\delta > 0$ be fixed. Show that, for any fixed $\epsilon > 0$, there exists a non-zero function f in $L^2(\mathbb{R})$ so that

$$\|\varphi_\epsilon * f - f\|_{L^2} > \delta \|f\|_{L^2}.$$

Why does this observation not contradict Proposition 5.2.2?

5.3.2 Resolution and the Full-Width Half-Maximum

We now give a standard definition for the resolution present in a measurement of the form $\psi * f$. Resolution is a subtle and, in some senses, subjective concept. Crudely speaking the resolution available in $\psi * f$ is determined by how well ψ approximates the δ-function. A quantitative measure of resolution is mostly useful for purposes of comparison. The definition presented here is just one of many possible definitions. We return to the problem of quantifying resolution in Chapter 9.

Suppose that ψ is a nonnegative function with a single hump similar to those shown in Figure 5.3. The important features of this function are as follows:

$$\begin{aligned}&\text{1. It is nonnegative}\\&\text{2. It has a single maximum value, which it attains at 0}\\&\text{3. It is monotone increasing to the left of the maximum}\\&\quad\text{and monotone decreasing to the right.}\end{aligned} \tag{5.30}$$

***Definition* 5.3.1.** Let ψ satisfy the conditions in (5.30) and let M be the maximum value it attains. Suppose that $x_1 < 0 < x_2$ are, respectively, the smallest and largest numbers so that

$$\psi(x_1) = \psi(x_2) = \frac{M}{2}.$$

The difference $x_2 - x_1$ is called the *full-width half-maximum* of the function ψ. It is denoted FWHM(ψ).

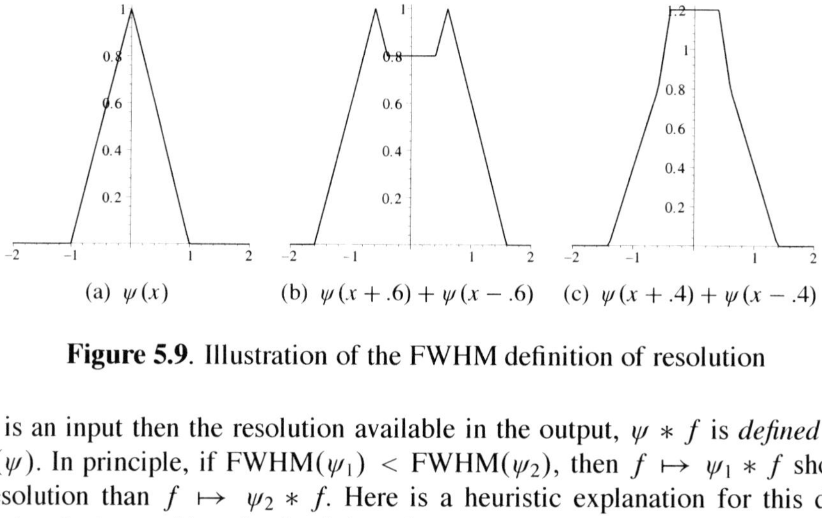

(a) $\psi(x)$ (b) $\psi(x+.6)+\psi(x-.6)$ (c) $\psi(x+.4)+\psi(x-.4)$

Figure 5.9. Illustration of the FWHM definition of resolution

If f is an input then the resolution available in the output, $\psi * f$ is *defined* to be the FWHM(ψ). In principle, if FWHM(ψ_1) $<$ FWHM(ψ_2), then $f \mapsto \psi_1 * f$ should have better resolution than $f \mapsto \psi_2 * f$. Here is a heuristic explanation for this definition. Suppose that the input f is pair of unit impulses separated by distance d,

$$f(x) = \delta(x) + \delta(x-d).$$

Convolving ψ with f produces two copies of ψ,

$$\psi * f(x) = \psi(x) + \psi(x-d).$$

If $d >$ FWHM(ψ), then $\psi * f$ has two distinct maxima separated by a valley. If $d \leq$ FWHM(ψ), then the distinct maxima disappear. If the distance between the impulses is greater than the FWHM(ψ), then we can "resolve" them in the filtered output. More generally, the FWHM(ψ) is considered to be the smallest distance between distinct "features" in f which can be "seen" in $\psi * f$. In Figure 5.9 we use a triangle function for ψ. The

FWHM of this function is 1; the graphs show ψ and the results of convolving ψ with a pair of unit impulses separated, respectively, by $1.2 > 1$ and $.8 < 1$.

The FWHM definition of resolution is often applied to filters defined by functions that do not satisfy all the conditions in (5.30) but are qualitatively similar. For example, the characteristic function of an interval $\chi_{[-B,B]}$ has a unique maximum value and is monotone to the right and left of the maximum. The FWHM$(\chi_{[-B,B]})$ is therefore $2B$. Another important example is the sinc function. It has a unique maximum and looks correct near to it. This function also has large side lobes that considerably complicate the behavior of the map $f \mapsto \operatorname{sinc} * f$. The FWHM(sinc) is taken to be the full-width half-maximum of its central peak; it is given by

$$\text{FWHM(sinc)} \approx 3.790988534.$$

Exercises

***Exercise* 5.3.4.** Numerically compute the FWHM$(\operatorname{sinc}^2(x))$. How does it compare to the FWHM$(\operatorname{sinc}(x))$?

***Exercise* 5.3.5.** Suppose that

$$h_j(x) = \left[\frac{\sin(x)}{x}\right]^j.$$

Using the Taylor expansion for the sine function show that, as j gets large,

$$\text{FWHM}(h_j) \simeq 2\sqrt{\frac{6\log 2}{j}}.$$

***Exercise* 5.3.6.** Show that there is a constant C so that, for $B > 0$,

$$\text{FWHM}(\operatorname{sinc}(Bx)) = \frac{C}{B}.$$

Compute C to 4 decimal places.

***Exercise* 5.3.7.** For $a > 0$, let $g_a(x) = e^{-\frac{x^2}{a^2}}$. Compute FWHM$(g_a)$ and FWHM$(g_a * g_b)$.

5.4 Conclusion

The convolution product provides a flexible model for a wide range of measurement and filtering processes with a very simple representation in terms of the Fourier transform. As these models are expressed in terms of integrals over $\mathbb{R}$ or $\mathbb{R}^n$, we call them *continuum models*. Because we have the tools of calculus at our disposal, continuum models are the easiest to analyze. The very real problem of how to approximate such operations on discretely sampled data remains to be considered. Convolution provides a uniform approach to approximating non-smooth functions by smooth functions and to studying the problem of convergence. Any process reducing the "noise" in a function also smears out the fine detail. We introduced the full-width half-maximum definition of resolution as a means for quantifying the amount of detail remaining in the smoothed approximants.

In the next chapter we derive the inversion formula for the Radon transform and find a description for it as the composition of a shift invariant filter and the back-projection operation. In Chapters 7–10 we introduce the techniques needed to make the transition from continuum models to finite algorithms. This is done in two steps: We first introduce Fourier series and sampling theory as tools for analyzing models with *infinitely* many discrete samples. In Chapter 10 we take the second step from models with infinitely many samples to algorithmic implementations of shift invariant filters, which, of necessity, involve only finite amounts of data.

More general linear operations can be expressed as families of convolutions; for example,

$$\mathcal{A}f(\boldsymbol{x}) = \int_{\mathbb{R}^n} a(\boldsymbol{x}, \boldsymbol{x} - \boldsymbol{y}) f(\boldsymbol{y})\, d\boldsymbol{y}.$$

This larger class of linear transforms can be used to model operations that are not shift invariant. Such models require considerably more computation to approximately implement and are therefore replaced, whenever possible, by ordinary convolutions. Nonetheless, many classes of operators of this sort are mathematically well understood. These go by the names of pseudodifferential operators and Fourier integral operators. The interested reader is referred to [123], [124], or [60].

Chapter 6

The Radon Transform

The Radon transform is at the center of the mathematical model for the measurements made in x-ray CT. In this chapter we prove the central slice theorem, establishing a connection between the Fourier and Radon transforms. Using it, we derive the *filtered back-projection formula*, which provides an exact inverse for the Radon transform. This formula is the basis of essentially all reconstruction algorithms used in x-ray CT today.

As its name suggests, the filtered back-projection formula is the composition of a one-dimensional filter with the back-projection operation, defined in Section 3.4.2. The filtration step is built out of two simpler operations: the Hilbert transform and differentiation. While differentiation is conceptually simple, it leads to difficulties when attempting to implement the filtered back-projection formula on noisy, measured data. The Hilbert transform is a nonlocal, bounded operation that is harder to understand intuitively but rather simple to implement using the Fourier representation. We analyze it in detail.

This chapter begins by reviewing the definition of the Radon transform and establishing properties of the Radon transform that are analogues of those proved for the Fourier transform in the two preceding chapters. After analyzing the exact inversion formula, we consider methods for approximately inverting the Radon transform that are relevant in medical imaging. The chapter concludes with a short introduction to the higher-dimensional Radon transform.

6.1 The Radon Transform

In Section 1.2.1 we identified $\mathbb{R} \times S^1$ with the space of oriented lines in $\mathbb{R}^2$. The pair (t, ω) corresponds to the line

$$l_{t,\omega} = \{x \,:\, \langle \omega, x \rangle = t\} = \{t\omega + s\hat{\omega} \,:\, s \in \mathbb{R}\}.$$

Here $\hat{\omega}$ is the unit vector perpendicular to ω with the orientation determined by

$$\det(\omega\hat{\omega}) > 0.$$

The variable t is called the *affine parameter*; it is the oriented distance of the line $l_{t,\omega}$ to the origin.

Representing the point $\boldsymbol{\omega} \in S^1$ as

$$\boldsymbol{\omega}(\theta) = (\cos\theta, \sin\theta)$$

allows an identification of $\mathbb{R} \times S^1$ with $\mathbb{R} \times [0, 2\pi)$. With this identification $d\theta$ can be used as a line element in the S^1-direction. This is often denoted by $d\boldsymbol{\omega}$ in the sequel. The integral of a function h over $S^1 \times \mathbb{R}$ is given by

$$\int_0^{2\pi}\int_{-\infty}^{\infty} h(t, \boldsymbol{\omega}(\theta))\, dt\, d\theta,$$

which is often denoted

$$\int_0^{2\pi}\int_{-\infty}^{\infty} h(t, \boldsymbol{\omega})\, dt\, d\boldsymbol{\omega}.$$

***Definition* 6.1.1.** The set $L^2(\mathbb{R} \times S^1)$ consists of locally integrable, real-valued functions for which the square integral,

$$\|h\|^2_{L^2(\mathbb{R}\times S^1)} = \int_0^{2\pi}\int_{-\infty}^{\infty} |h(t, \boldsymbol{\omega}(\theta))|^2\, dt\, d\theta, \tag{6.1}$$

is finite.

A function h on $\mathbb{R} \times S^1$ is continuous if $h(t, \theta) \stackrel{d}{=} h(t, \boldsymbol{\omega}(\theta))$ is 2π-periodic in θ and continuous as a function on $\mathbb{R} \times [0, 2\pi]$. Similarly, h is differentiable if it is 2π-periodic and differentiable on $\mathbb{R} \times [0, 2\pi]$ and $\partial_\theta h$ is also 2π-periodic. Higher orders of differentiability have similar definitions.

Recall that the Radon transform of f at $(t, \boldsymbol{\omega})$ is defined by the integral

$$\mathscr{R}f(t, \boldsymbol{\omega}) = \int_{-\infty}^{\infty} f(t\boldsymbol{\omega} + s\hat{\boldsymbol{\omega}})\, ds.$$

For the moment we restrict our attention to piecewise continuous functions with bounded support. Because $l_{t,\boldsymbol{\omega}}$ and $l_{-t,-\boldsymbol{\omega}}$ are the same line, the Radon transform is an even function

$$\mathscr{R}f(-t, -\boldsymbol{\omega}) = \mathscr{R}f(t, \boldsymbol{\omega}). \tag{6.2}$$

The Radon transform has several properties analogous to those established for the Fourier transform in Chapter 4. Suppose that f and g are functions with bounded supported. There is a simple formula relating $\mathscr{R}(f * g)$ to $\mathscr{R}f$ and $\mathscr{R}g$.

Proposition 6.1.1. *Let f and g be piecewise continuous functions with bounded support. Then*

$$\mathscr{R}[f * g](t, \boldsymbol{\omega}) = \int_{-\infty}^{\infty} \mathscr{R}f(s, \boldsymbol{\omega})\, \mathscr{R}g(t - s, \boldsymbol{\omega})\, ds. \tag{6.3}$$

***Remark* 6.1.1.** Colloquially, we say that the Radon transform converts convolution in the plane to convolution in the affine parameter.

Proof. The proof is a calculation. Fix a direction ω. Coordinates (s, t) for the plane are defined by the assignment

$$(s, t) \mapsto s\hat{\omega} + t\omega.$$

This is an orthogonal change of variables, so the area element on $\mathbb{R}^2$ is given by $ds\,dt$. In these variables the convolution of f and g becomes

$$f * g(s\hat{\omega} + t\omega) = \int_{-\infty}^{\infty}\int_{-\infty}^{\infty} f(a\hat{\omega} + b\omega)g((s-a)\hat{\omega} + (t-b)\omega)\,da\,db.$$

The Radon transform of $f * g$ is computing by switching the order of the integrations:

$$\begin{aligned}\mathscr{R}(f * g)(\tau, \omega) &= \int_{-\infty}^{\infty} f * g(\tau\omega + s\hat{\omega})\,ds \\ &= \int_{-\infty}^{\infty}\int_{-\infty}^{\infty}\int_{-\infty}^{\infty} f(a\hat{\omega} + b\omega)g((s-a)\hat{\omega} + (\tau-b)\omega)\,da\,db\,ds \\ &= \int_{-\infty}^{\infty}\int_{-\infty}^{\infty}\int_{-\infty}^{\infty} f(a\hat{\omega} + b\omega)g((s-a)\hat{\omega} + (\tau-b)\omega)\,ds\,da\,db \\ &= \int_{-\infty}^{\infty} \mathscr{R}f(b, \omega)\,\mathscr{R}g(\tau - b, \omega)\,db.\end{aligned} \tag{6.4}$$

In the second to last line we interchanged the s-integration with the a and b integrations. □

***Remark* 6.1.2.** The smoothness of a function with bounded support is reflected in the decay properties of its Fourier transform. From Proposition 6.1.1 it follows that the smoothness of a function of bounded support is also reflected in the smoothness of its Radon transform in the affine parameter. To see this suppose that f is a continuous function of bounded support and φ is a radially symmetric function, with bounded support and k-continuous derivatives. The convolution $f * \varphi$ has bounded support and k-continuous derivatives. The Radon transform of φ is only a function of t; the Radon transform of the convolution,

$$\mathscr{R}(f * \varphi)(t, \omega) = \int_{-\infty}^{\infty} \mathscr{R}f(\tau, \omega)\,\mathscr{R}\varphi(t - \tau)\,d\tau,$$

is at least as smooth in t as $\mathscr{R}\varphi$. Regularity of f is also reflected in smoothness of $\mathscr{R}f$ in the angular variable, though it is more difficult to see explicitly; see Exercise 6.1.7.

For $\boldsymbol{v}$ a vector in $\mathbb{R}^2$ the translate of f by $\boldsymbol{v}$ is the function $f_{\boldsymbol{v}}(\boldsymbol{x}) = f(\boldsymbol{x} - \boldsymbol{v})$. There is a simple relation between the Radon transform of f and that of $f_{\boldsymbol{v}}$.

Proposition 6.1.2. *Let f be a piecewise continuous function with bounded support. Then*

$$\mathcal{R} f_{\boldsymbol{v}}(t, \boldsymbol{\omega}) = \mathcal{R} f(t - \langle \boldsymbol{\omega}, \boldsymbol{v} \rangle, \boldsymbol{\omega}). \tag{6.5}$$

Using this formula, we can relate the Radon transform of f to that of its partial derivatives.

Lemma 6.1.1. *If f is a function with bounded support and continuous first partial derivatives, then $\mathcal{R} f(t, \boldsymbol{\omega})$ is differentiable in t and*

$$\mathcal{R} \partial_x f(t, \boldsymbol{\omega}) = \omega_1 \partial_t \mathcal{R} f(t, \boldsymbol{\omega}), \quad \mathcal{R} \partial_y f(t, \boldsymbol{\omega}) = \omega_2 \partial_t \mathcal{R} f(t, \boldsymbol{\omega}). \tag{6.6}$$

Proof. We consider only the x-derivative; the proof for the y-derivative is identical. Let $\boldsymbol{e}_1 = (1, 0)$. The x-partial derivative of f is defined by

$$\partial_x f(\boldsymbol{x}) = \lim_{h \to 0} \frac{f_{h\boldsymbol{e}_1}(\boldsymbol{x}) - f(\boldsymbol{x})}{-h}.$$

From (6.5) and the linearity of the Radon transform, we conclude that

$$\mathcal{R}\left[\frac{f_{h\boldsymbol{e}_1} - f}{-h}\right](t, \boldsymbol{\omega}) = \frac{\mathcal{R} f(t - h\omega_1, \boldsymbol{\omega}) - \mathcal{R} f(t, \boldsymbol{\omega})}{-h}.$$

The lemma follows by allowing h to tend to zero. □

This result extends, by induction to higher partial derivatives.

Proposition 6.1.3. *Suppose that f has bounded support and continuous partial derivatives of order k. Then $\mathcal{R} f(t, \boldsymbol{\omega})$ is k-times differentiable in t and, for nonnegative integers i, j with $i + j \leq k$, we have the formula*

$$\mathcal{R}\left[\partial_x^i \partial_y^j f\right](t, \boldsymbol{\omega}) = \omega_1^i \omega_2^j \partial_t^{i+j} \mathcal{R} f(t, \boldsymbol{\omega}). \tag{6.7}$$

Let $A : \mathbb{R}^2 \to \mathbb{R}^2$ be a rigid rotation of the plane; that is, A is a linear map such that

$$\langle A\boldsymbol{v}, A\boldsymbol{w} \rangle = \langle \boldsymbol{v}, \boldsymbol{w} \rangle \qquad \text{for all } \boldsymbol{v}, \boldsymbol{w} \in \mathbb{R}^2.$$

If f is a piecewise continuous function with bounded support, then

$$f_A(\boldsymbol{x}) = f(A\boldsymbol{x})$$

is as well. The Radon transform of f_A is related to that of f in a simple way.

Proposition 6.1.4. *Let A be an rigid rotation of $\mathbb{R}^2$ and f a piecewise continuous function with bounded support. Then*

$$\mathcal{R} f_A(t, \boldsymbol{\omega}) = \mathcal{R} f(t, A\boldsymbol{\omega}). \tag{6.8}$$

Proof. The result follows from the fact that $\langle A\omega, A\hat{\omega}\rangle = \langle \omega, \hat{\omega}\rangle = 0$ and therefore

$$\begin{aligned} \mathcal{R}f_A(t,\omega) &= \int_{-\infty}^{\infty} f(tA\omega + sA\hat{\omega})\,ds \\ &= \mathcal{R}f(t, A\omega). \end{aligned} \tag{6.9}$$

□

The results in this section are stated for piecewise continuous functions with bounded supported. As discussed in Chapter 3, the Radon transform extends to sufficiently regular functions with enough decay at infinity. A function belongs to the *natural domain* of the Radon transform if the restriction of f to every line $l_{t,\omega}$ is an absolutely integrable function. If, for example, f is a piecewise continuous function, satisfying an estimate of the form

$$|f(\boldsymbol{x})| \leq \frac{M}{(1 + \|\boldsymbol{x}\|)^{1+\epsilon}},$$

for an $\epsilon > 0$, then f belongs to the natural domain of the Radon transform. The results in this section extend to functions in the natural domain of $\mathcal{R}$. The proofs in this case are left to the reader. Using functional analytic methods, the domain of the Radon transform can be further extended, allowing functions with both less regularity and slower decay. An example of such an extension was presented in Section 3.4.3. We return to this in Section 6.6.

Exercises

***Exercise* 6.1.1.** Prove formula (6.5). The argument is similar to that used in the proof of (6.3).

***Exercise* 6.1.2.** Suppose that φ is a k-times differentiable function with bounded support defined on $\mathbb{R}^2$. Show that $\mathcal{R}\varphi$ is a k-times differentiable of t.

***Exercise* 6.1.3.** Give the details of the argument in the proof of Lemma 6.1.1, showing that $\mathcal{R}f(t,\omega)$ is differentiable in the t-variable.

***Exercise* 6.1.4.** Show how to derive formula (6.7) from (6.6).

***Exercise* 6.1.5.** The Laplace operator Δ is defined by $\Delta f = -(\partial_x^2 f + \partial_y^2 f)$. Find a formula for $\mathcal{R}[\Delta f]$ in terms of $\mathcal{R}f$.

***Exercise* 6.1.6.** Suppose that $A : \mathbb{R}^2 \to \mathbb{R}^2$ is an arbitrary invertible linear transformation. How is $\mathcal{R}f_A$ related to $\mathcal{R}f$?

***Exercise* 6.1.7.** Let A_θ denote the rotation through the angle θ. Setting $\omega(\theta) = (\cos\theta, \sin\theta)$, let $\mathcal{R}f(t,\theta) = \mathcal{R}f(t,\omega(\theta))$ so that

$$\mathcal{R}f_{A_\phi}(t,\theta) = \mathcal{R}f(t,\theta+\phi).$$

Using these formulæ show that

$$\mathcal{R}\big[(y\partial_x - x\partial_y)f\big](t,\theta) = (\partial_\theta\mathcal{R})f(t,\theta).$$

6.2 Inversion of the Radon Transform

Now we are ready to use the Fourier transform to invert the Radon transform.

6.2.1 The Central Slice Theorem*

The Fourier transform and Radon transform are connected in a very simple way. In medical imaging this relationship is called the *central slice theorem.*

Theorem 6.2.1 (Central slice theorem). *Let f be an absolutely integrable function in the natural domain of $\mathcal{R}$. For any real number r and unit vector $\boldsymbol{\omega}$, we have the identity*

$$\int_{-\infty}^{\infty} \mathcal{R}f(t, \boldsymbol{\omega})e^{-itr}\,dt = \hat{f}(r\boldsymbol{\omega}). \tag{6.10}$$

Proof. Using the definition of the Radon transform, we compute the integral on the left:

$$\int_{-\infty}^{\infty} \mathcal{R}f(t, \boldsymbol{\omega})e^{-itr}\,dt = \int_{-\infty}^{\infty}\int_{-\infty}^{\infty} f(t\boldsymbol{\omega} + s\hat{\boldsymbol{\omega}})e^{-itr}\,ds\,dt. \tag{6.11}$$

This integral is absolutely convergent, and therefore we may make the change of variables, $\boldsymbol{x} = t\boldsymbol{\omega} + s\hat{\boldsymbol{\omega}}$. Checking that the Jacobian determinant is 1 and noting that

$$t = \langle \boldsymbol{x}, \boldsymbol{\omega} \rangle,$$

the preceding integral therefore becomes

$$\begin{aligned}\int_{-\infty}^{\infty}\int_{-\infty}^{\infty} f(t\boldsymbol{\omega} + s\hat{\boldsymbol{\omega}})e^{-itr}\,ds\,dt &= \int_{\mathbb{R}^2} f(\boldsymbol{x})e^{-i\langle \boldsymbol{x}, \boldsymbol{\omega}\rangle r}\,d\boldsymbol{x} \\ &= \hat{f}(r\boldsymbol{\omega}).\end{aligned} \tag{6.12}$$

This completes the proof of the central slice theorem. □

For a given vector $\boldsymbol{\xi} = (\xi_1, \xi_2)$ the inner product, $\langle \boldsymbol{x}, \boldsymbol{\xi} \rangle$ is constant along any line perpendicular to the direction of $\boldsymbol{\xi}$. The central slice theorem interprets the computation of the Fourier transform at $\boldsymbol{\xi}$ as a two-step process:

1. First we integrate the function along lines perpendicular to $\boldsymbol{\xi}$; this gives us a function of the affine parameter alone.

2. Compute the *one-dimensional* Fourier transform of this function of the affine parameter.

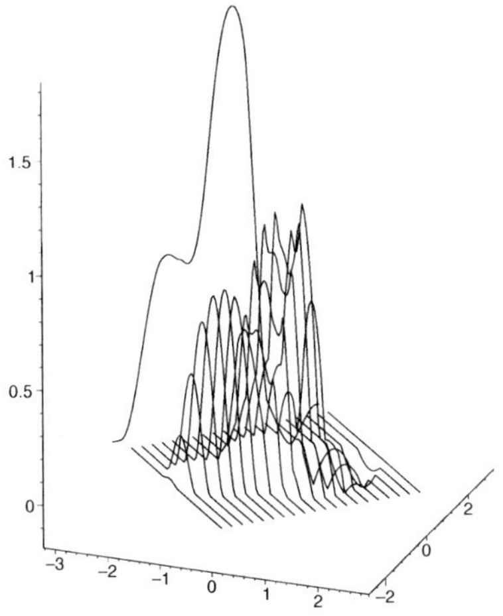

(a) The slices of a function, f, along the family of lines $\langle \boldsymbol{x}, \boldsymbol{\omega} \rangle = t$ and $\mathcal{R}f(t, \boldsymbol{\omega})$.

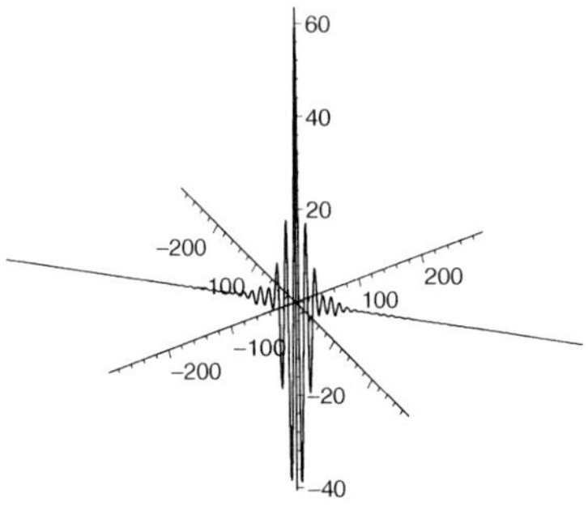

(b) The real part of $\hat{f}$ in the direction $\boldsymbol{\omega}$.

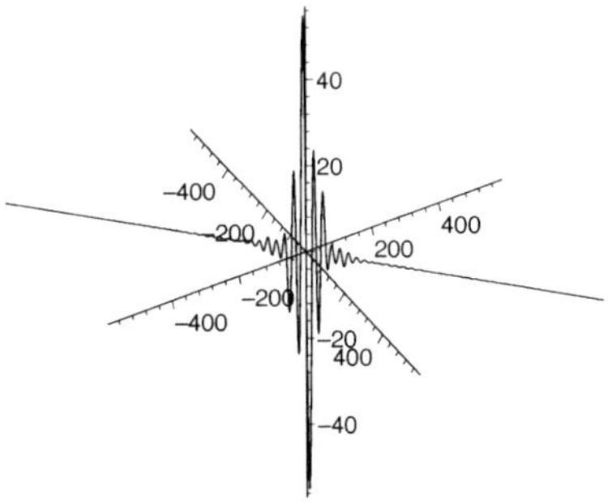

(c) The imaginary part of $\hat{f}$ in the direction $\boldsymbol{\omega}$.

Figure 6.1. According to the central slice theorem the 2D Fourier transform, $\hat{f}(r\boldsymbol{\omega})$, is the 1D Fourier transform of $\mathcal{R}f(t, \boldsymbol{\omega})$.

To understand this better, we consider an example. Let $\boldsymbol{e}_1 = (1, 0)$ and $\boldsymbol{e}_2 = (0, 1)$ and $(t, \boldsymbol{\omega}) = (x, \boldsymbol{e}_1)$. Since $\hat{\boldsymbol{e}}_1 = \boldsymbol{e}_2$, the Radon transform at $(x, \boldsymbol{e}_1)$ is given by

$$\begin{aligned}\mathcal{R}f(x, \boldsymbol{e}_1) &= \int_{-\infty}^{\infty} f(x\boldsymbol{e}_1 + y\boldsymbol{e}_2)\, dy \\ &= \int_{-\infty}^{\infty} f(x, y)\, dy.\end{aligned}$$

The Fourier transform of $\mathcal{R}f(x, \boldsymbol{e}_1)$ is

$$\int_{-\infty}^{\infty} \mathcal{R}f(x, \boldsymbol{e}_1)e^{-irx}\,dx = \int_{-\infty}^{\infty}\int_{-\infty}^{\infty} f(x, y)e^{-irx}\,dy\,dx.$$

As $\langle r\boldsymbol{e}_1, (x, y)\rangle = rx$, this is the definition of $\hat{f}(r\boldsymbol{e}_1)$.

The operations in the central slice theorem are depicted in Figure 6.1. On the left we have a function, f, of two variables depicted as slices along lines in a family, $\{\langle \boldsymbol{x}, \boldsymbol{\omega}\rangle = t\}$. Beyond the graph of f, the integrals of these functions of a single variable are plotted. This, of course, is just the Radon transform $\mathcal{R}f(t, \boldsymbol{\omega})$. To the right and below are the real and imaginary parts of the Fourier transform, in t, of $\mathcal{R}f(t, \boldsymbol{\omega})$. According to the central slice theorem, this gives $\hat{f}$ along the line, through $\mathbf{0}$, in the direction $\boldsymbol{\omega}$.

To simplify the formulæ that follow, we introduce notation for the one-dimensional Fourier transform, in the affine parameter, of a function $h(t, \boldsymbol{\omega})$ defined on $\mathbb{R} \times S^1$:

$$\widetilde{h}(r, \boldsymbol{\omega}) \overset{d}{=} \int_{-\infty}^{\infty} h(t, \boldsymbol{\omega})e^{-itr}\,dt. \tag{6.13}$$

If $h(t, \boldsymbol{\omega})$ belongs to $L^2(\mathbb{R})$ for a fixed $\boldsymbol{\omega}$, then the one-dimensional Parseval formula implies that

$$\int_{-\infty}^{\infty} |h(t, \boldsymbol{\omega})|^2\,dt = \frac{1}{2\pi}\int_{-\infty}^{\infty} |\widetilde{h}(r, \boldsymbol{\omega})|^2\,dr. \tag{6.14}$$

The Parseval formula for the 2D-Fourier transform and the central slice theorem give a Parseval formula for the Radon transform.

Theorem 6.2.2 (Parseval formula for the Radon transform). *Suppose that f is in the natural domain of the Radon transform and is square integrable. Then*

$$\int_{\mathbb{R}^2} |f(\boldsymbol{x})|^2\,d\boldsymbol{x} = \frac{1}{[2\pi]^2}\int_0^{\pi}\int_{-\infty}^{\infty} |\widetilde{\mathcal{R}f}(r, \boldsymbol{\omega})|^2|r|\,dr\,d\boldsymbol{\omega}. \tag{6.15}$$

Proof. We begin by assuming that f is also absolutely integrable. The central slice theorem applies to show that

$$\begin{aligned}\int_{\mathbb{R}^2} |f(\boldsymbol{x})|^2\,d\boldsymbol{x} &= \frac{1}{[2\pi]^2}\int_0^{2\pi}\int_0^{\infty} |\hat{f}(r\boldsymbol{\omega})|^2 r\,dr\,d\boldsymbol{\omega} \\ &= \frac{1}{[2\pi]^2}\int_0^{\pi}\int_{-\infty}^{\infty} |\widetilde{\mathcal{R}f}(r, \boldsymbol{\omega})|^2|r|\,dr\,d\boldsymbol{\omega}.\end{aligned} \tag{6.16}$$

In the last line we use the fact that the evenness of $\mathcal{R}f$ implies that

$$\widetilde{\mathcal{R}f}(r,\omega) = \widetilde{\mathcal{R}f}(-r,-\omega). \tag{6.17}$$

This proves (6.15) with the additional assumption. To remove this assumption we need to approximate f by absolutely integrable functions. Let φ be a nonnegative, infinitely differentiable, radial function with support in the disk of radius 1 with total integral one. As usual, for $\epsilon > 0$ set $\varphi_\epsilon(x) = \epsilon^{-2}\varphi(\epsilon^{-1}x)$. A smooth function with bounded support approximating f is given by

$$f_\epsilon = \left[\chi_{[0,\epsilon^{-1}]}(r) f\right] * \varphi_\epsilon. \tag{6.18}$$

For $\epsilon > 0$ these functions satisfy the hypotheses of both Theorem 6.2.2 and the central slice theorem; the preceding argument therefore applies to f_ϵ. The proof is completed by showing that

$$\frac{1}{[2\pi]^2}\int_0^\pi\int_{-\infty}^\infty |\widetilde{\mathcal{R}f}(r,\omega)|^2|r|\,dr\,d\omega = \lim_{\epsilon\downarrow 0}\frac{1}{[2\pi]^2}\int_0^\pi\int_{-\infty}^\infty |\widetilde{\mathcal{R}f_\epsilon}(r,\omega)|^2|r|\,dr\,d\omega, \tag{6.19}$$

and that, as ϵ goes to 0, f_ϵ converges in $L^2(\mathbb{R}^2)$ to f. These claims are left as exercises for the reader. □

Remark* 6.2.1. Formula (6.15) has two interesting consequences for the map $f \mapsto \mathcal{R}f$ as a map between L^2-spaces. It shows that $\mathcal{R}$ does *not* have an extension as a continuous mapping from $L^2(\mathbb{R}^2)$ to $L^2(\mathbb{R}\times S^1)$ and that $\mathcal{R}^{-1}$ also cannot be a continuous map from $L^2(\mathbb{R}\times S^1)$ to $L^2(\mathbb{R}^2)$. These assertions follow from the observation that

$$\|h\|^2_{L^2(\mathbb{R}\times S^1)} = \frac{1}{2\pi}\int_0^{2\pi}\int_{-\infty}^\infty |\tilde{h}(r,\omega)|^2\,dr\,d\omega.$$

Because $|r|$ varies between zero and infinity in (6.15), we see that there cannot exist constants M or M' so that either estimate,

$$\|\mathcal{R}f\|_{L^2(\mathbb{R}\times S^1)} \le M\|f\|_{L^2(\mathbb{R}^2)} \quad\text{or}\quad \|\mathcal{R}f\|_{L^2(\mathbb{R}\times S^1)} \ge M'\|f\|_{L^2(\mathbb{R}^2)}$$

is valid for f in a dense subset of $L^2(\mathbb{R}^2)$.

To express the Parseval formula as an integral over the space of oriented lines, we define a half-derivative operator

$$D_{\frac{1}{2}}\mathcal{R}f(t,\omega) = \frac{1}{2\pi}\int_{-\infty}^\infty \widetilde{\mathcal{R}f}(r,\omega)|r|^{\frac{1}{2}}e^{irt}\,dr.$$

The Parseval formula can then be rewritten as

$$\int_{\mathbb{R}^2}|f|^2\,dx\,dy = \frac{1}{2\pi}\int_0^\pi\int_{-\infty}^\infty |D_{\frac{1}{2}}\mathcal{R}f(t,\omega)|^2\,dt\,d\omega. \tag{6.20}$$

This implies that in order for a function on the space of lines to be the Radon transform of a square-integrable function, it must have a half-derivative in the affine parameter. Unlike the Fourier transform, the Radon transform is not defined on all of $L^2(\mathbb{R}^2)$.

Exercises

***Exercise* 6.2.1.** If $f \in L^2(\mathbb{R}^2)$ and f_ϵ is defined in (6.18), show that $\operatorname{LIM}_{\epsilon \downarrow 0} f_\epsilon = f$. *Hint:* Use Proposition 5.2.2 to handle the convolution. Do not forget the $\chi_{[0,\epsilon^{-1}]}$-term!

***Exercise* 6.2.2.** If f is in the natural domain of $\mathscr{R}$ and f_ϵ is defined in (6.18) prove (6.19). *Hint:* Use Proposition 6.1.1 and the Plancherel formula.

6.2.2 The Radon Inversion Formula*

The central slice theorem and the inversion formula for the Fourier transform, (4.82) give an inversion formula for the Radon transform.

Theorem 6.2.3 (Radon inversion formula). *If f is an absolutely integrable function in the natural domain of the Radon transform and $\hat{f}$ is absolutely integrable, then*

$$f(x) = \frac{1}{[2\pi]^2} \int_0^\pi \int_{-\infty}^\infty e^{ir\langle x,\omega\rangle} \widetilde{\mathscr{R}f}(r,\omega)|r|\,dr\,d\omega \tag{6.21}$$

Proof. Because $\mathscr{R}f$ is an even function, it follows that its Fourier transform satisfies

$$\widetilde{\mathscr{R}f}(r,\omega) = \widetilde{\mathscr{R}f}(-r,-\omega). \tag{6.22}$$

As f and $\hat{f}$ are absolutely integrable, Theorem 4.5.1 applies to show that

$$f(x) = \frac{1}{[2\pi]^2} \int_{\mathbb{R}^2} \hat{f}(\xi) e^{i\langle x,\xi\rangle} d\xi.$$

Reexpressing the Fourier inversion formula using polar coordinates gives

$$\begin{aligned} f(x) &= \frac{1}{[2\pi]^2} \int_{\mathbb{R}^2} e^{i\langle x,\xi\rangle} \hat{f}(\xi)\,d\xi \\ &= \frac{1}{[2\pi]^2} \int_0^{2\pi} \int_0^\infty e^{ir\langle x,\omega\rangle} \hat{f}(r\omega) r\,dr\,d\omega \\ &= \frac{1}{[2\pi]^2} \int_0^{2\pi} \int_0^\infty e^{ir\langle x,\omega\rangle} \widetilde{\mathscr{R}f}(r,\omega) r\,dr\,d\omega \end{aligned}$$

The central slice theorem is used in the last line. Using the relation (6.22) we can rewrite this to obtain (6.21). □

***Remark* 6.2.2.** As was the case with the Fourier transform, the inversion formula for the Radon transform holds under weaker hypotheses than those stated in Theorem 6.2.3. Under these hypotheses all the integrals involved are absolutely convergent and therefore do not require any further interpretation. In imaging applications the data are usually piecewise continuous, vanishing outside a bounded set. As we know from our study of the Fourier transform, this does not imply that $\hat{f}$ is absolutely integrable, and so the Fourier inversion formula requires a careful interpretation in this case. Such data are square integrable, and therefore it follows from the results in Section 4.5.3 that

$$f = \operatorname*{LIM}_{\rho\to\infty} \frac{1}{[2\pi]^2} \int_0^{\pi} \int_{-\rho}^{\rho} e^{ir\langle x,\omega\rangle} \widetilde{\mathscr{R}f}(r,\omega)|r|\,dr\,d\omega. \tag{6.23}$$

In most cases of interest, at a point x, where f is continuous, the integral

$$\frac{1}{[2\pi]^2} \int_0^{\pi} \int_{-\infty}^{\infty} e^{ir\langle x,\omega\rangle} \widetilde{\mathscr{R}f}(r,\omega)|r|\,dr\,d\omega$$

exists as an improper Riemann integral and equals $f(x)$. Additional care is required in manipulating these expressions.

***Remark* 6.2.3.** Formula (6.21) allows the determination of f from its Radon transform. This formula completes a highly idealized, mathematical model for x-ray CT-imaging:

- We consider a two-dimensional slice of a three-dimensional object, the physical parameter of interest is the attenuation coefficient f of the two-dimensional slice. According to Beer's law, the intensity $I_{(t,\omega)}$ of x-rays (of a given energy) traveling along a line, $l_{t,\omega}$, is attenuated according the differential equation:

$$\frac{dI_{(t,\omega)}}{ds} = -fI_{(t,\omega)}.$$

  Here s is arclength along the line.

- By comparing the intensity of an incident beam of x-rays to that emitted, we measure the Radon transform of f :

$$\mathscr{R}f(t,\omega) = -\log\left[\frac{I_{o,(t,\omega)}}{I_{i,(t,\omega)}}\right].$$

- Using formula (6.21), the attenuation coefficient f is reconstructed from the measurements $\mathscr{R}f$.

The most obvious flaw in this model is that, in practice, $\mathscr{R}f(t,\omega)$ can only be measured for a finite set of pairs (t,ω). Nonetheless, formula (6.21) provides a good starting point for the development of more practical algorithms. Figure 6.2(a) shows the Shepp-Logan phantom; Figure 6.2(b) is its Radon transform. Figure 6.2(c) is the reconstruction of (a) using (an approximation to) formula (6.21). Within the white ellipse, the reconstructed image is nearly identical to the original. In Chapter 11 we present a detailed analysis of the approximate algorithm used to make this image.

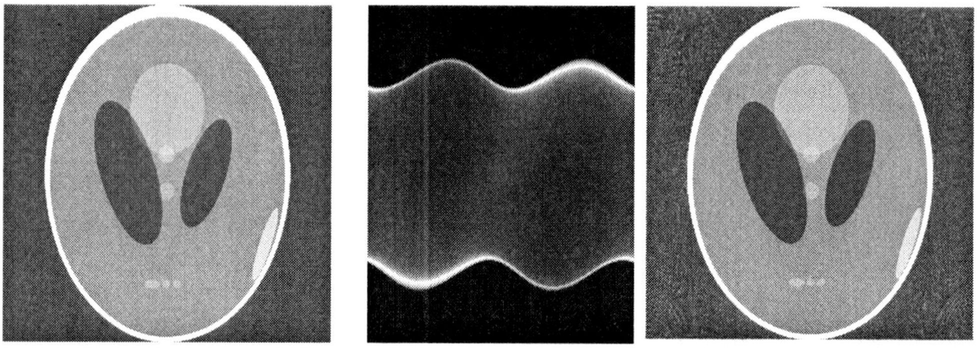

(a) A density plot of the Shepp-Logan phantom.

(b) The line integrals for the function shown in (a).

(c) The (approximate) inverse Radon transform applied to the data in (b).

Figure 6.2. Graphical depiction of the Radon transform.

6.2.3 Filtered Back-Projection*

We now undertake a detailed analysis of the Radon inversion formula. The inversion formula can be understood as a two-step process:

1. The radial integral is interpreted as a *filter* applied to the Radon transform. The filter acts only in the affine parameter; the output of the filter is denoted by

$$\mathcal{G}\mathcal{R}f(t,\omega) = \frac{1}{2\pi}\int_{-\infty}^{\infty} \widetilde{\mathcal{R}f}(r,\omega)e^{irt}|r|\,dr. \tag{6.24}$$

2. The angular integral is then interpreted as the back-projection of the *filtered* Radon transform. The function f is expressed as

$$f(x) = \frac{1}{2\pi}\int_{0}^{\pi} (\mathcal{G}\mathcal{R})f(\langle x,\omega\rangle,\omega)\,d\omega. \tag{6.25}$$

For this reason the Radon inversion formula is often called the *filtered back-projection formula.*

Back-projection is both conceptually and computationally simple, whereas the filtering step requires a more careful analysis. If we were to omit the $|r|$ factor, then it would follow from the one-dimensional Fourier inversion formula applied to $\widetilde{\mathcal{R}f}$ that f would be given by

$$f(x) = \frac{1}{[2\pi]^2} \int_0^{\pi} \int_{-\infty}^{\infty} e^{ir\langle x,\omega\rangle} \hat{f}(r\omega)\, dr\, d\omega$$
$$= \frac{1}{2\pi} \int_0^{\pi} \mathcal{R}f(\langle x, \omega\rangle, \omega)\, d\omega.$$

Note that the line in the family $\{l_{t,\omega} \mid t \in (-\infty, \infty)\}$ passing through the point x is the one with affine parameter $t = \langle x, \omega\rangle$. The value at x obtained this way is half the average of the Radon transform of f over all lines passing through this point. This is the back-projection formula introduced in Section 3.4.2. By comparison with the true inversion formula (6.21), it is now clear why the back-projection formula cannot be correct. In the true formula the low-frequency components are suppressed by $|r|$ whereas the high-frequency components are amplified.

The actual filter is comprised of two operations. Recall that the Fourier transform of the derivative of a function g is equal to the Fourier transform of g multiplied by $i\xi$: $\widehat{\partial_t g}(\xi) = (i\xi)\hat{g}(\xi)$. If, in the inversion formula (6.21), we had r instead of $|r|$, then the formula would give

$$\frac{1}{2\pi i} \int_0^{\pi} \partial_t \mathcal{R}f(\langle x, \omega\rangle, \omega)\, d\omega.$$

This is the back-projection of the t-derivative of $\mathcal{R}f$. If f is real valued, then this function is purely imaginary! Because differentiation is a *local operation,* this is a relatively easy formula to understand. The subtlety in (6.21) therefore stems from the fact that $|r|$ appears and not r itself.

To account for the difference between r and $|r|$, we define another operation on functions of a single variable, which is called the *Hilbert transform.* The *signum* function is defined by

$$\operatorname{sgn}(r) = \begin{cases} 1 & \text{if } r > 0, \\ 0 & \text{if } r = 0, \\ -1 & \text{if } r < 0. \end{cases}$$

***Definition* 6.2.1.** Suppose that g is an L^2-function defined on $\mathbb{R}$. The Hilbert transform of g is defined by

$$\mathcal{H}g = \mathcal{F}^{-1}(\operatorname{sgn} \hat{g}).$$

If $\hat{g}$ is also absolutely integrable, then

$$\mathcal{H}g(t) = \frac{1}{2\pi} \int_{-\infty}^{\infty} \hat{g}(r) \operatorname{sgn}(r) e^{itr}\, dr. \tag{6.26}$$

The Hilbert transform of g is the function whose Fourier transform is $\operatorname{sgn}\hat{g}$. For any given point t_0, the computation of $\mathcal{H}g(t_0)$ requires a knowledge of $g(t)$ for *all* values of t. Unlike differentiation, the Hilbert transform is not a *local* operation. Conceptually, the Hilbert transform is the most difficult part of the Radon inversion formula. On the other hand, because the Hilbert transform has a very simple expression in terms of the Fourier transform, it is easy to implement efficiently.

We compute a couple of examples of Hilbert transforms.

***Example* 6.2.1.** Let

$$f(x) = \frac{\sin(x)}{\pi x}.$$

Its Fourier transform is

$$\hat{f}(\xi) = \chi_{[-1,1]}(\xi) = \begin{cases} 1 \text{ if } & |\xi| \leq 1, \\ 0 \text{ if } & |\xi| > 1. \end{cases}$$

The Hilbert transform of f is expressed as a Fourier integral by

$$\begin{aligned} \mathcal{H}\left(\frac{\sin(x)}{\pi x}\right) &= \frac{1}{2\pi}\left[\int_0^1 e^{ix\xi}dx - \int_{-1}^0 e^{ix\xi}dx\right] \\ &= i\frac{1-\cos(x)}{\pi x}. \end{aligned} \tag{6.27}$$

This pair of functions is graphed in Figure 6.3(a).

***Example* 6.2.2.** The next example is of interest in medical imaging. It is difficult to do this example by a direct calculation. A method to do this calculation, using the theory of functions of a complex variable is explained in the final section of this chapter. Let

$$f(x) = \begin{cases} \sqrt{1-x^2} & \text{for } |x| < 1, \\ 0 & \text{for } |x| \geq 1. \end{cases}$$

The Hilbert transform of f is given by

$$\mathcal{H}(f) = \begin{cases} ix & \text{for } |x| < 1, \\ i(x + \sqrt{x^2-1}) & \text{for } x < -1, \\ i(x - \sqrt{x^2-1}) & \text{for } x > 1. \end{cases} \tag{6.28}$$

Notice the very different character of $\mathcal{H}f(x)$ for $|x| < 1$ and $|x| > 1$. For $|x| < 1$, $\mathcal{H}f(x)$ is a smooth function with a bounded derivative. Approaching ± 1 from the set $|x| > 1$, the derivative of $\mathcal{H}f(x)$ blows up. This pair of functions is graphed in Figure 6.3(b).

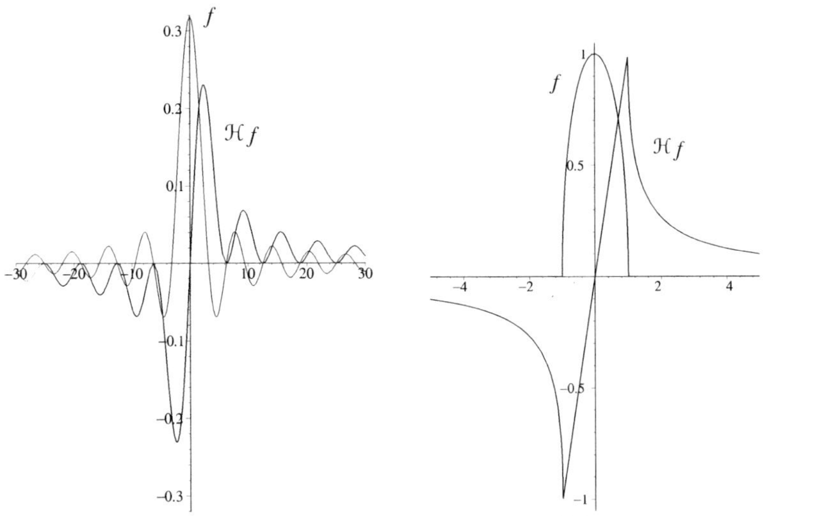

(a) $\frac{\sin(x)}{\pi x}$ and its Hilbert transform.

(b) $\chi_{[-1,1]}\sqrt{1-x^2}$ and its Hilbert transform.

Figure 6.3. Hilbert transform pairs.

From the differentiation formula for the Fourier transform, we conclude that

$$\widetilde{\partial_t \mathcal{R} f}(r) = ir\widetilde{\mathcal{R} f}(r).$$

The Hilbert transform of $\partial_t \mathcal{R} f$ is given by

$$\begin{aligned}
\mathcal{H}(\partial_t \mathcal{R} f)(t, \omega) &= \frac{1}{2\pi} \int_{-\infty}^{\infty} \widetilde{\partial_t \mathcal{R} f}(r, \omega) \operatorname{sgn}(r) e^{itr} \, dr \\
&= \frac{1}{2\pi} \int_{-\infty}^{\infty} i|r|\widetilde{\mathcal{R} f}(r, \omega) e^{irt} \, dr.
\end{aligned}$$

Since $\operatorname{sgn}(r)r = |r|$, we can identify the filtration step in (6.25):

$$\mathcal{G}\mathcal{R} f(t, \omega) = \frac{1}{i}\mathcal{H}(\partial_t \mathcal{R} f)(t, \omega); \tag{6.29}$$

putting this into (6.25), we obtain

$$f(x) = \frac{1}{2\pi i} \int_0^{\pi} \mathcal{H}(\partial_t \mathcal{R} f)(\langle x, \omega \rangle, \omega) \, d\omega. \tag{6.30}$$

The function f is reconstructed by back-projecting the Hilbert transform of $\frac{1}{i}\partial_t \mathcal{R} f$.

***Remark* 6.2.4.** The Fourier transform of the function

$$F = \frac{1}{2}(f + \mathcal{H}f)$$

vanishes for $\xi < 0$, and therefore F has an analytic extension to the upper half-plane; see Theorem 4.4.4. This explains why the Hilbert transform is intimately connected to the theory of analytic functions. Using the Fourier representation, it is easy to see that $\hat{F}(\xi) = \chi_{[0,\infty)}(\xi)\hat{f}(\xi)$, and therefore, if $y > 0$, then

$$F(x + iy) = \frac{1}{2\pi}\int_0^\infty \hat{f}(\xi)e^{-y\xi}e^{ix\xi}d\xi$$

is an absolutely convergent integral. The function $F(x)$ is the boundary value of a analytic function. A basic theorem in analytic function theory states that such a function cannot vanish on an open interval; see [97]. This shows that if f has bounded support, then $\mathcal{H}f$ cannot.

This observation has important implications in image reconstruction. Formula (6.30) expresses f as the back-projection of $-i\mathcal{H}\partial_t\mathcal{R}f$. If f has bounded support, then so does $\partial_t\mathcal{R}f$ and therefore $-i\mathcal{H}\partial_t\mathcal{R}f$ *does not*. If $\boldsymbol{x}$ lies outside the support of f, then this means that the integrand in (6.30) is, generally speaking, not zero. The integral vanishes due to subtle cancellations between the positive and negative parts of $-i\mathcal{H}(\partial_t\mathcal{R}f)(\langle\boldsymbol{x},\boldsymbol{\omega}\rangle,\boldsymbol{\omega})$. We return to this question in Section 12.3.2. For more on the connection between the Hilbert transform and analytic function theory, see Section 6.8.

Exercises

***Exercise* 6.2.3.** Suppose that f is a differentiable function with bounded supported. Show that $\mathcal{H}(\partial_t f) = \partial_t(\mathcal{H}f)$.

Exercise* 6.2.4. Use the previous exercise and a limiting argument to show that $\mathcal{H}(\partial_t f) = \partial_t(\mathcal{H}f)$ functions in $L^2(\mathbb{R})$ that have an L^2-derivative.

***Exercise* 6.2.5.** Suppose that f is an L^2-function with bounded support such that $\hat{f}(0) \neq 0$. Show that $\mathcal{H}f$ does *not* have bounded support. *Hint:* Use Proposition 4.2.3. Can you give an estimate for the rate at which $\mathcal{H}f$ decays?

Exercise* 6.2.6. Show that in the previous problem it is not necessary to assume that $\hat{f}(0) \neq 0$.

Exercise* 6.2.7. Use the Schwarz reflection principle to prove the statement that if $F(x+iy)$ is an analytic function in $y > 0$ such that, for $a < x < b$,

$$\lim_{y\downarrow 0} F(x + iy) = 0,$$

then $F \equiv 0$.

6.2.4 Inverting the Radon Transform, Two Examples

Before continuing our analysis of $\mathcal{R}^{-1}$, we compute the inverse of the Radon transform in two examples.

***Example* 6.2.3.** For our first example, we set $f = \chi_{B_1}$, the characteristic function of the unit disk. Using the rotational symmetry, we check that

$$\mathcal{R}f(t,\omega) = \begin{cases} 2\sqrt{1-t^2} & |t| \leq 1 \\ 0 & |t| > 1. \end{cases} \tag{6.31}$$

Note that $\mathcal{R}f$ satisfies

$$\sup_{h,t} \left| \frac{\mathcal{R}f(t+h) - \mathcal{R}f(t)}{\sqrt{|h|}} \right| < \infty.$$

In other words, $\mathcal{R}f$ is a Hölder-$\frac{1}{2}$ function of t.

To apply the filtered back-projection formula, we need to compute either $\partial_t \mathcal{H}\mathcal{R}f$ or $\mathcal{H}\partial_t \mathcal{R}f$. It is instructive to do both. In Section 6.8 it is shown that

$$\frac{1}{i}\mathcal{H}\mathcal{R}f(t,\omega) = \begin{cases} 2t & \text{for } |t| < 1, \\ 2(t + \sqrt{t^2-1}) & \text{for } t < -1, \\ 2(t - \sqrt{t^2-1}) & \text{for } t > 1. \end{cases} \tag{6.32}$$

Even though this function is not differentiable at $t = \pm 1$, it does have an absolutely integrable, weak derivative given by

$$\frac{1}{i}\partial_t \mathcal{H}\mathcal{R}f(t,\omega) = \begin{cases} 2 - \frac{2|t|}{\sqrt{t^2-1}} & \text{for } |t| \geq 1 \\ 2 & \text{for } |t| < 1. \end{cases} \tag{6.33}$$

On the other hand, we could first compute the weak derivative of $\mathcal{R}f$:

$$\partial_t \mathcal{R}f(t,\omega) = \begin{cases} \frac{-2t}{\sqrt{1-t^2}} & |t| < 1 \\ 0 & |t| > 1. \end{cases}$$

Unfortunately, this function does not belong to $L^2(\mathbb{R})$. Thus far we have only defined the Hilbert transform for L^2-functions. It is also possible to define the Hilbert transform of a function in $L^p(\mathbb{R})$ for any $1 < p \leq 2$; see [79]. As

$$\int |\partial_t \mathcal{R}f|^p < \infty \qquad \text{for } p < 2$$

the Hilbert transform of $\partial_t \mathcal{R}f$ is still defined and can be computed using the complex variable method described in Section 6.8. It is given by the formula

$$\frac{1}{i}\mathcal{H}(\partial_t \mathcal{R}f)(t) = \begin{cases} 2 - \frac{2|t|}{\sqrt{t^2-1}} & \text{for } |t| \geq 1 \\ 2 & \text{for } |t| < 1. \end{cases} \tag{6.34}$$

Now we do the back-projection step. If $\boldsymbol{x}$ is inside the unit disk, then

$$|\langle \boldsymbol{x}, \boldsymbol{\omega} \rangle| \leq 1.$$

At such points, the inverse of the Radon transform is quite easy to compute:

$$\frac{1}{2\pi i} \int_0^{\pi} \mathcal{H}(\partial_t \mathcal{R} f)(\langle \boldsymbol{x}, \boldsymbol{\omega} \rangle, \boldsymbol{\omega})\, d\omega = \frac{1}{2\pi} \int_0^{\pi} 2\, d\omega = 1.$$

This is precisely the value of f for $\|\boldsymbol{x}\| \leq 1$. On the other hand, if $\|\boldsymbol{x}\| > 1$, then the needed calculation is more complicated. Since f is radially symmetric, it suffices to consider $f(x, 0)$. If $x > 1$, then there is an angle $0 < \theta_x < \frac{\pi}{2}$ so that $x \cos \theta_x = 1$; the inversion formula can be written

$$f(x, 0) = \frac{1}{2\pi} \left[4 \int_0^{\theta_x} \left(1 - \frac{|x \cos \theta|}{\sqrt{x^2 \cos^2 \theta - 1}} \right) d\theta - 2 \int_{\theta_x}^{\pi - \theta_x} d\theta \right].$$

This is a much more complicated formula. From the point of view of computation, it is notable that the Radon inversion formula now involves an *unbounded* integrand. It is, of course, absolutely integrable, but difficult to numerically integrate.

The important lesson of this example is the qualitative difference in the filtered back-projection formula between points inside and outside the unit disk. This fact has significant consequences in medical imaging; see Section 12.3.2.

***Example* 6.2.4.** Our next example is a bit smoother than the characteristic function of the disk. Let $r = \sqrt{x^2 + y^2}$ and define g by

$$g(x, y) = \begin{cases} 1 - r^2 & |r| < 1, \\ 0 & |r| \geq 1. \end{cases}$$

Again using the rotational symmetry, we obtain

$$\mathcal{R} g(t, \boldsymbol{\omega}) = \begin{cases} \frac{4}{3}(1 - t^2)^{3/2} & |t| \leq 1, \\ 0 & |t| > 1. \end{cases}$$

As a function of t, $\mathcal{R} g$ is classically differentiable, which is a "half" a derivative smoother than g itself. It is a general fact that the Radon transform has better regularity in the affine parameter than the original function by half a derivative. The derivative of $\mathcal{R} g$ is

$$\partial_t \mathcal{R} g(t, \boldsymbol{\omega}) = \begin{cases} -4t(1 - t^2)^{1/2} & |t| \leq 1, \\ 0 & |t| > 1. \end{cases}$$

It satisfies

$$\sup_{h,t} \left| \frac{\partial_t \mathcal{R} g(t + h) - \partial_t \mathcal{R} g(t)}{|h|^{\frac{1}{2}}} \right| < \infty$$

and therefore $\partial_t \mathcal{R}g$ is a Hölder-$\frac{1}{2}$ function. The Hilbert transform of $\partial_t \mathcal{R}g$ is

$$\frac{1}{i}\mathcal{H}(\partial_t \mathcal{R}g)(t) = \begin{cases} 2 - 4t^2 & |t| \leq 1, \\ 4[\, 4|t|(t^2-1)^{1/2} - (2t^2-1)\,] & |t| > 1. \end{cases}$$

Once again we see that the back-projection formula for points inside the unit disk is, numerically, a bit simpler than for points outside. While $\sqrt{t^2-1}$ is continuous, it is not differentiable at $t = \pm 1$. This makes the numerical integration in the back-projection step more difficult for points outside the disk.

Exercises

***Exercise* 6.2.8.** Prove that (6.33) gives the weak derivative of $\mathcal{H}\mathcal{R}f$ defined in (6.32).

***Exercise* 6.2.9.** Use Simpson's rule to numerically integrate $\sqrt{1-t^2}$ from 0 to 1. Determine how the accuracy of the result depends on the mesh size and compare it to the accuracy when instead $1 - t^2$ is integrated.

Exercise* 6.2.10. Give an algorithm to numerically integrate the function $\frac{1}{\sqrt{1-t^2}}$ from -1 to 1. Provide an estimate for the accuracy of your method.

Exercise* 6.2.11. Generalize the method in the previous exercise to functions of the form $\frac{f}{\sqrt{1-t^2}}$, where f is differentiable on an interval containing $[-1, 1]$.

6.2.5 Back-Projection*

See: A.2.5.

The operation of back-projection has a nice mathematical interpretation. If $(X, \langle\cdot,\cdot\rangle_X)$ and $(Y, \langle\cdot,\cdot\rangle_Y)$ are inner product spaces and $A : X \to Y$ is a linear map, recall that the *adjoint* of A, $A^* : Y \to X$ is defined by the relations

$$\langle A\boldsymbol{x}, \boldsymbol{y}\rangle_Y = \langle \boldsymbol{x}, A^*\boldsymbol{y}\rangle_Y \qquad \text{for all } \boldsymbol{x} \in X \text{ and } \boldsymbol{y} \in Y.$$

If we use the L^2-inner product for functions on $\mathbb{R}^2$ and the inner product for functions on $\mathbb{R} \times S^1$ compatible with the L^2-norm defined in (6.1),

$$\langle h, k\rangle_{\mathbb{R}\times S^1} = \int_0^{2\pi}\int_{-\infty}^{\infty} h(t,\omega)k(t,\omega)\,dt\,d\omega,$$

then back-projection is $[4\pi]^{-1}$ times the formal adjoint of the Radon transform. It is only a formal adjoint because, as noted previously, the Radon transform does not extend to define a continuous map from $L^2(\mathbb{R}^2)$ to $L^2(\mathbb{R} \times S^1)$. The proof is a calculation; for the sake of simplicity, assume that f is a function of bounded support on $\mathbb{R}^2$ and h is a function of bounded support on $\mathbb{R} \times S^1$:

$$\begin{aligned}\langle \mathcal{R}f, h\rangle_{\mathbb{R}\times S^1} &= \int_0^{2\pi}\int_{-\infty}^{\infty} \mathcal{R}f(t,\omega)h(t,\omega)\,dt\,d\omega \\ &= \int_0^{2\pi}\int_{-\infty}^{\infty}\int_{-\infty}^{\infty} f(t\omega + s\hat{\omega})h(t,\omega)\,ds\,dt\,d\omega.\end{aligned} \tag{6.35}$$

Let $x = t\omega + s\hat{\omega}$ so that

$$t = \langle x, \omega\rangle.$$

Interchanging the ω- and the x-integrals, we obtain

$$\begin{aligned}\langle \mathcal{R}f, h\rangle_{\mathbb{R}\times S^1} &= \int_{\mathbb{R}^2}\int_0^{2\pi} f(x)h(\langle x,\omega\rangle,\omega)\,d\omega\,dx \\ &= \langle f, \mathcal{R}^* h\rangle_{\mathbb{R}^2}.\end{aligned} \tag{6.36}$$

This verifies the assertion that back-projection is $[4\pi]^{-1}$ times the formal adjoint of the Radon transform. The fact that $\mathcal{R}^* \neq \mathcal{R}^{-1}$ is a reflection of the fact that $\mathcal{R}$ is not a unitary transformation from $L^2(\mathbb{R}^2)$ to $L^2(\mathbb{R} \times S^1)$.

Using the identification of back-projection with the adjoint, along with the Parseval formula, (4.5.2), we can derive an interesting relationship between $\widehat{\mathcal{R}^*\mathcal{R}f}$ and $\hat{f}$.

Proposition 6.2.1. *Suppose that f is an absolutely integrable and square-integrable function in the natural domain of the Radon transform. Then*

$$\frac{r}{4\pi}\widehat{\mathcal{R}^*\mathcal{R}f}(r\omega) = \hat{f}(r\omega). \tag{6.37}$$

Proof. The proof of this proposition uses the basic principle that, in an inner product space X, an element x is zero if and only if $\langle x, y\rangle_X = 0$ for all y belonging to a dense subset of X. Let f and g be two functions satisfying the hypotheses of the proposition. From the definition of the adjoint, it follows that

$$\langle \mathcal{R}f, \mathcal{R}g\rangle_{\mathbb{R}\times S^1} = \langle f, \mathcal{R}^*\mathcal{R}g\rangle_{\mathbb{R}^2}. \tag{6.38}$$

Using the Parseval formula, we get the relations

$$\begin{aligned}\langle f, \mathcal{R}^*\mathcal{R}g\rangle_{\mathbb{R}^2} &= \frac{1}{[2\pi]^2}\langle \hat{f}, \widehat{\mathcal{R}^*\mathcal{R}g}\rangle_{\mathbb{R}^2} \\ &= \frac{1}{[2\pi]^2}\int_0^{2\pi}\int_0^{\infty} \hat{f}(r\omega)\overline{\widehat{\mathcal{R}^*\mathcal{R}g}(r\omega)}r\,dr\,d\omega,\end{aligned} \tag{6.39}$$

and

$$
\begin{aligned}
\langle \mathcal{R} f, \mathcal{R} g\rangle_{\mathbb{R}\times S^1} &= \frac{1}{2\pi}\int_0^{2\pi}\int_{-\infty}^{\infty} \widetilde{\mathcal{R} f}(r,\omega)\overline{\widetilde{\mathcal{R} g}(r,\omega)}\,dr\,d\omega \\
&= \frac{1}{\pi}\int_0^{2\pi}\int_0^{\infty} \hat{f}(r\omega)\overline{\hat{g}(r\omega)}\,dr\,d\omega.
\end{aligned} \tag{6.40}
$$

In the last line we use the central slice theorem and the evenness of the Radon transform. Since these formulæ hold for all f and g with bounded support, a dense subset of L^2, it follows that

$$
\frac{r}{4\pi}\widehat{\mathcal{R}^* \mathcal{R} g}(r\omega) = \hat{g}(r\omega). \tag{6.41}
$$

□

Proposition 6.2.1 leads to an alternate formula for $\mathcal{R}^{-1}$. In this approach, the back-projection is done first. Then a filter, acting on functions defined on $\mathbb{R}^2$, is applied to $\mathcal{R}^* \mathcal{R} f$. If f is a piecewise continuous function of bounded support, then Proposition 6.2.1 states that

$$
\hat{f}(r\omega) = \frac{r}{4\pi}\widehat{\mathcal{R}^* \mathcal{R} f}(r\omega).
$$

If $\hat{f}$ is absolutely integrable, then the Fourier inversion formula therefore implies that

$$
\begin{aligned}
f(x) &= \frac{1}{[2\pi]^2}\int_0^{2\pi}\int_0^{\infty} \frac{r}{4\pi}\widehat{\mathcal{R}^* \mathcal{R} f}(r\omega)e^{i\langle r\omega, x\rangle} r\,dr\,d\omega \\
&= \frac{1}{[2\pi]^2}\int_{\mathbb{R}^2} \frac{\|\xi\|}{4\pi}\widehat{\mathcal{R}^* \mathcal{R} f}(\xi)e^{i\langle \xi, x\rangle}\,d\xi.
\end{aligned} \tag{6.42}
$$

The Laplace operator on $\mathbb{R}^2$ is defined as the second-order differential operator

$$
\Delta f = (\partial_x^2 f + \partial_y^2 f).
$$

As a constant coefficient differential operator, it can be expressed in terms of the Fourier transform by

$$
-\Delta f(x) = \frac{1}{[2\pi]^2}\int_{\mathbb{R}^2} \|\xi\|^2 \hat{f}(\xi)e^{i\langle \xi, x\rangle}\,d\xi.
$$

This formula motivates a definition for the nonnegative powers of the Laplace operator. For $s \geq 0$ and f, a smooth function with bounded support, define

$$
[-\Delta]^s f(x) = \frac{1}{[2\pi]^2}\int_{\mathbb{R}^2} \|\xi\|^{2s} \hat{f}(\xi)e^{i\langle \xi, x\rangle}\,d\xi. \tag{6.43}
$$

Using the Parseval formula, this operation can be extended to all functions in $L^2(\mathbb{R}^2)$ such that $\|\boldsymbol{\xi}\|^s \hat{f}(\boldsymbol{\xi})$ is square integrable. With this definition for $[-\Delta]^s$, we can rewrite (6.42) as

$$4\pi f(\boldsymbol{x}) = ([-\Delta]^{\frac{1}{2}} \mathcal{R}^*(\mathcal{R} f))(\boldsymbol{x}). \tag{6.44}$$

***Remark* 6.2.5.** Note that $-\Delta\, \mathcal{R}^*(\mathcal{R} f) = 4\pi [-\Delta]^{\frac{1}{2}} f$. This gives an expression for $[-\Delta]^{\frac{1}{2}} f$ that, given $\mathcal{R} f$, can be computed using entirely elementary operations; that is, back-projection and differentiation. The functions f and $[-\Delta]^{\frac{1}{2}} f$ have the same singularities. As edges are discontinuities, this formula gives a straightforward way to find the edges in an image described by a density function f. I thank Gunther Uhlmann for this observation.

***Remark* 6.2.6.** Thus far we have produced a left inverse for the Radon transform. If f is a function in the plane satisfying appropriate regularity and decay hypotheses then, for example,

$$f = ([-\Delta]^{\frac{1}{2}} \mathcal{R}^*) \mathcal{R} f.$$

We have *not* said that if h is an even function on $\mathbb{R} \times S^1$, then

$$h = \mathcal{R}([-\Delta]^{\frac{1}{2}} \mathcal{R}^*) h.$$

That is, we have not shown that $([-\Delta]^{\frac{1}{2}} \mathcal{R}^*)$ is also a right inverse for $\mathcal{R}$. Under some mild hypotheses on h, this is in fact true. The proof of this statement involves characterizing the range of the Radon transform and is beyond the scope of this book. Treatments of this problem can found in [90], [83], and [95].

Exercises

***Exercise* 6.2.12.** Let g be a continuous function with bounded support on $\mathbb{R} \times S^1$. Show that there is a constant C so that

$$|\mathcal{R}^* g(\boldsymbol{x})| \leq \frac{C}{1 + \|\boldsymbol{x}\|}.$$

Show that if g is a nonnegative function that is not identically zero, then there is also a constant $C' > 0$ so that

$$|\mathcal{R}^* g(\boldsymbol{x})| \geq \frac{C'}{1 + \|\boldsymbol{x}\|}.$$

***Exercise* 6.2.13.** Explain how we arrived at the limits of integration in the second line of (6.40).

***Exercise* 6.2.14.** Using the definition, (6.43), show that

1. If s is a positive integer, then the two definitions of $[-\Delta]^s$ agree.

2. For s and t nonnegative numbers,

$$[-\Delta]^s [-\Delta]^t = [-\Delta]^{s+t}. \tag{6.45}$$

3. Conclude from the previous part that

$$-\Delta\, \mathcal{R}^* \mathcal{R} f = [-\Delta]^{\frac{1}{2}} f.$$

6.3 The Hilbert Transform

See: A.4.5, B.7.

The filter $\mathcal{G}$ defined in (6.29) acts on functions of single variable and has a simple Fourier representation:

$$\mathcal{G}f(t) = \frac{1}{2\pi}\int_{-\infty}^{\infty} \hat{f}(r)|r|e^{irt}\,dr.$$

As noted in the previous section, this filter can be expressed as a composition of two simpler operations: differentiation and the Hilbert transform. For implementation on real measured data, each operation presents its own difficulties. In one sense differentiation is a simple operation because it is *local;* to compute $\partial_t f(t)$ we only need to know the values of f in a small neighborhood of t. On the other hand, real data are corrupted with noise, and this requires the data to be smoothed before they are differentiated. The Hilbert transform is a *nonlocal* operation; to determine $\mathcal{H}f(t)$ we need to know the values of f for all t. The difficulty here is that we can only compute integrals over *finite* intervals.

The Hilbert transform is defined by

$$\mathcal{H}f = \mathcal{F}^{-1}(\hat{f}(\xi)\operatorname{sgn}(\xi)), \text{ which implies that } \widehat{\mathcal{H}f}(\xi) = \operatorname{sgn}(\xi)\hat{f}(\xi).$$

In general, the inverse Fourier transform of a product is a convolution; that is,

$$\mathcal{F}^{-1}(\hat{f}\hat{g}) = f * g.$$

Hence, if there existed a nice function h such that $\hat{h}(\xi) = \operatorname{sgn}(\xi)$, then the Hilbert transform would be just $h * f$. Unfortunately, the signum function is not the Fourier transform of a nice function because it does not go to zero, in any sense, as $|\xi| \to \infty$. In this section we further analyze the Hilbert transform. We describe in what sense it is a convolution and give different approaches to approximating it.

It is easy to approximate the Hilbert transform in the Fourier representation and the L^2-norm.

Theorem 6.3.1. *Suppose that $< \phi_\epsilon >$ is a uniformly bounded family of locally integrable functions that converges pointwise to $\operatorname{sgn}(\xi)$ as $\epsilon \to 0$. If f is square integrable, then the Hilbert transform of f is given by the limit in the mean*

$$\mathcal{H}f(t) = \underset{\epsilon\downarrow 0}{\text{LIM}}\,\mathcal{F}^{-1}(\phi_\epsilon\hat{f}).$$

Proof. The Parseval formula shows that

$$\|\mathcal{H}f - \mathcal{F}^{-1}(\phi_\epsilon f)\|_{L^2} = \frac{1}{2\pi}\int_{-\infty}^{\infty} |(\operatorname{sgn}(\xi) - \phi_\epsilon(\xi))\hat{f}(\xi)|^2\,d\xi.$$

As ϕ_ϵ is uniformly bounded and converges to sgn, the conclusion follows from the Lebesgue dominated convergence theorem. $\square$

***Remark* 6.3.1.** If f is sufficiently smooth, so that $\hat{f}$ decays, then $\mathcal{H}f(t)$ is given by the pointwise limit

$$\mathcal{H}f(t) = \lim_{\epsilon \downarrow 0} \int_{-\infty}^{\infty} \phi_\epsilon(\xi)\hat{f}(\xi)e^{it\xi}\frac{d\xi}{2\pi}.$$

By using a particular family of functions $< \phi_\epsilon >$, we can find a representation for $\mathcal{H}$ as a convolution. Modify the signum function by setting

$$\phi_\epsilon(\xi) = \hat{h}_\epsilon(\xi) \stackrel{d}{=} \operatorname{sgn}(\xi)e^{-\epsilon|\xi|} \qquad \text{for } \epsilon > 0.$$

This family satisfies the hypotheses of the theorem. The inverse Fourier transform of $\hat{h}_\epsilon$ is

$$h_\epsilon = \frac{i}{\pi}\frac{t}{t^2 + \epsilon^2}.$$

This function behaves like $1/t$ as t goes to infinity, which is not fast enough for integrability, but at least it goes to zero and has no singularities. For each $\epsilon > 0$ we can therefore define an approximate Hilbert transform:

$$\mathcal{H}_\epsilon f = \mathcal{F}^{-1}(\hat{f}\hat{h}_\epsilon) = f * h_\epsilon. \tag{6.46}$$

6.3.1 The Hilbert Transform as a Convolution

Theorem 6.3.1 implies that if f is an L^2-function, then $h_\epsilon * f$ converges, in $L^2(\mathbb{R})$, to $\mathcal{H}f$. Letting ϵ go to 0, we see that h_ϵ converges pointwise to $i[t\pi]^{-1}$. Formally, this seems to imply that, if f is sufficiently smooth, then

$$\mathcal{H}f(t) = \frac{i}{\pi}\int_{-\infty}^{\infty} \frac{f(s)\,ds}{t-s}. \tag{6.47}$$

Because $1/|t|$ is not integrable in any neighborhood of 0, this expression is not an absolutely convergent integral and therefore requires interpretation. It turns out that in the present instance, the correct interpretation for this formula is as a Cauchy principal value:

$$\mathcal{H}f(t) = \frac{i}{\pi}\,\text{P.V.}(f * \frac{1}{s}) = \frac{i}{\pi}\lim_{\epsilon \to 0}\left[\int_{-\infty}^{-\epsilon} + \int_{\epsilon}^{\infty} \frac{f(t-s)}{s}ds\right]. \tag{6.48}$$

The Cauchy principal value is finite, at least if f has bounded support and is smooth enough. First note that if t lies outside the support of f, then the integral in (6.47) is absolutely convergent. For t in the support of f, a more careful analysis is required. Since the function $1/s$ is odd, we have

$$\left(\int_{-R}^{-\epsilon} + \int_{\epsilon}^{R}\right)\frac{ds}{s} = 0.$$

We can multiply this by $f(t)$ and still get zero:

$$\left(\int_{-R}^{-\epsilon} + \int_{\epsilon}^{R}\right) f(t)\frac{ds}{s} = 0.$$

Assuming that the support of f is contained in $[-\frac{R}{2}, \frac{R}{2}]$ and that t belongs to this interval, then

$$\left(\int_{-\infty}^{-\epsilon} + \int_{\epsilon}^{\infty}\right) \frac{f(t-s)}{s} ds = \left(\int_{-R}^{-\epsilon} + \int_{\epsilon}^{R}\right) \left[\frac{f(t-s) - f(t)}{s}\right] ds. \tag{6.49}$$

If f is once differentiable, then the integrand in (6.49) remains bounded as ϵ goes to 0 and therefore

$$\text{P.V.}(f * \frac{1}{s})(t) = \int_{-R}^{R} \frac{f(t-s) - f(t)}{s} ds \qquad \text{for } t \in [-\frac{R}{2}, \frac{R}{2}]. \tag{6.50}$$

Indeed if, for some $\alpha > 0$, f satisfies the α-Hölder condition,

$$\frac{|f(t) - f(s)|}{|t-s|^{\alpha}} \le M, \tag{6.51}$$

then the integral in (6.49) is absolutely convergent.

Theorem 6.3.2. *If f is a function with bounded support, satisfying* (6.51) *for an $\alpha > 0$, then*

$$\mathcal{H}(f) = \frac{i}{\pi} \text{P.V.}(f * \frac{1}{s}).$$

Proof. Assume that f is supported in $[-\frac{R}{2}, \frac{R}{2}]$. If t is outside this interval, then the limit of $f * h_\epsilon(t)$, as ϵ goes to zero, is the absolutely convergent integral

$$\frac{i}{\pi} \int_{-\infty}^{\infty} \frac{f(t-s)}{s} ds.$$

To complete the proof of the theorem, we need to show that for $t \in [-\frac{R}{2}, \frac{R}{2}]$ we have

$$\frac{i}{\pi} \text{P.V.}(f * \frac{1}{s})(t) = \lim_{\epsilon \downarrow 0} h_\epsilon * f(t).$$

As h_ϵ is also an odd function, this difference is given by

$$\begin{aligned} \frac{i}{\pi} \text{P.V.}(f * \frac{1}{s})(t) - \lim_{\epsilon \downarrow 0} h_\epsilon * f(t) = \frac{i}{\pi} \lim_{\epsilon \downarrow 0} \Bigg[\left(\int_{-R}^{-\epsilon} + \int_{\epsilon}^{R}\right) \left[\frac{\epsilon^2 (f(t-s) - f(t))}{s(s^2 + \epsilon^2)}\right] ds - \\ \int_{-\epsilon}^{\epsilon} \frac{s(f(t-s) - f(s))}{s^2 + \epsilon^2} ds \Bigg]. \end{aligned} \tag{6.52}$$

It is left as an exercise to show that the second integral on the right-hand side of (6.52) tends to zero as ϵ goes to zero. To handle the first term, we let $\epsilon\sigma = s$, obtaining

$$\frac{i}{\pi}\lim_{\epsilon\downarrow 0}\left[\left(\int\limits_{-R\epsilon^{-1}}^{-1} + \int\limits_{1}^{R\epsilon^{-1}}\right)\frac{(f(t-\epsilon\sigma)) - f(t))}{\sigma(\sigma^2+1)}d\sigma\right].$$

Because this integrand decays like σ^{-3} at infinity and is uniformly (in ϵ) integrable for σ near zero, the limit as ϵ goes to zero can be taken inside the integral. As

$$\lim_{\epsilon\downarrow 0}(f(t-\epsilon\sigma) - f(t)) = 0$$

this completes the proof of the theorem. □

As a corollary we obtain a representation formula for $\mathcal{H}$ as an absolutely convergent integral.

Corollary 6.3.1. *If f is an α-Hölder continuous function with support in $[-\frac{R}{2}, \frac{R}{2}]$, then for t in this interval*

$$\mathcal{H}f(t) = \frac{i}{\pi}\int\limits_{-R}^{R}\frac{f(t-s)-f(t)}{s}ds. \tag{6.53}$$

The cancellation due to the symmetric interval used in the definition of the principal value is critical to obtain this result. There are other ways to regularize convolution with $1/t$. For example, we could add an imaginary number to the denominator to make it non-vanishing,

$$\lim_{\epsilon\downarrow 0}\frac{i}{\pi}\int\limits_{-R}^{R}\frac{f(t-s)}{s\pm i\epsilon}ds.$$

A computation shows that

$$\frac{i}{\pi}\frac{1}{2}\left(\frac{1}{s+i\epsilon}+\frac{1}{s-i\epsilon}\right) = \frac{i}{\pi}\frac{s}{s^2+\epsilon^2} = h_\epsilon(s).$$

This shows that the average of the two regularizations, $(s\pm i\epsilon)^{-1}$, results in the same approximation as before. The difference of these two regularizations is

$$\frac{i}{\pi}\cdot\frac{1}{2}\left(\frac{1}{s+i\epsilon}-\frac{1}{s-i\epsilon}\right) = \frac{1}{\pi}\frac{\epsilon}{s^2+\epsilon^2},$$

which does not tend to zero as ϵ tends to zero. As an example, we "test" the characteristic function of the interval $\chi_{[-1,1]}$ by evaluating the limit at $t=0$,

$$\lim_{\epsilon\downarrow 0}\int\limits_{-\infty}^{\infty}\chi_{[-1,1]}(-s)\frac{\epsilon}{s^2+\epsilon^2}ds = \lim_{\epsilon\downarrow 0}\int\limits_{-1}^{1}\frac{\epsilon}{s^2+\epsilon^2}ds = \lim_{\epsilon\downarrow 0}\int\limits_{-1/\epsilon}^{1/\epsilon}\frac{dt}{t^2+1} = \pi.$$

So we see that, in general,

$$\lim_{\epsilon\downarrow 0}\frac{i}{\pi}\int_{-\infty}^{\infty}\frac{f(t-s)\,ds}{s\pm i\epsilon}\neq \mathcal{H}f(t).$$

The lesson is that care must be exercised in choosing a regularization for convolution with $1/t$. Different regularizations lead to different results.

***Remark* 6.3.2.** This discussion shows that there are at least three different approaches to approximating the Hilbert transform and therefore the Radon inversion formula. On the one hand, we can use the convolution formula for $\mathcal{H}$ and directly approximate P.V.$(f*\frac{1}{s})$. On the other hand, we can use the Fourier integral representation and instead approximate $\operatorname{sgn}(\xi)$ as described in Theorem 6.3.1. For sufficiently smooth functions with bounded support, we could use (6.53). Mathematically these approaches are equivalent; computationally they can lead to vastly different results. In most real applications the Fourier representation is used because it is more efficient and does not involve regularizing a divergent expression. Instead an integral over $\mathbb{R}$ must be approximated by an integral over a finite interval.

Exercises

***Exercise* 6.3.1.** Suppose that f and g are continuous functions with bounded support. Show that

$$\mathcal{H}(f*g)=(\mathcal{H}f)*g=f*(\mathcal{H}g).$$

***Exercise* 6.3.2.** Suppose that f is an L^2-function that satisfies an α-Hölder condition for an $0<\alpha\leq 1$. Show that

$$\lim_{\epsilon\downarrow 0}h_\epsilon*f=\frac{i}{\pi}\,\text{P.V.}(f*\frac{1}{s}).$$

***Exercise* 6.3.3.** For f an α-Hölder continuous function, show that

$$\lim_{\epsilon\downarrow 0}\int_{-\epsilon}^{\epsilon}\frac{s(f(t-s)-f(s))}{s^2+\epsilon^2}=0.$$

***Exercise* 6.3.4.** Compute that Fourier transforms of $f_{\pm}(x)=(x\pm i\epsilon)^{-1}$ for $\epsilon>0$.

***Exercise* 6.3.5.** The following are linear operators defined in terms of the Fourier transform. Reexpress these operators in terms of differentiations and the Hilbert transform. For example, if Af is defined by

$$Af(x)=\frac{1}{2\pi}\int_{-\infty}^{\infty}\xi\hat{f}(\xi)e^{ix\xi}d\xi,$$

then the answer to this question is

$$Af(x)=-i\partial_x f(x).$$

Do not worry about convergence.

1.

$$A_1 f(x) = \frac{1}{2\pi} \int_{-\infty}^{\infty} |\xi|^3 \hat{f}(\xi) e^{ix\xi} d\xi$$

2.

$$A_2 f(x) = \frac{1}{2\pi} \int_{-\infty}^{\infty} (\xi^4 + |\xi| + 1) \hat{f}(\xi) e^{ix\xi} d\xi$$

3. In this exercise take note of the lower limit of integration.

$$A_3 f(x) = \frac{1}{2\pi} \int_{0}^{\infty} \hat{f}(\xi) e^{ix\xi} d\xi$$

***Exercise* 6.3.6.** If $f \in L^2(\mathbb{R})$ then show that

$$\int_{-\infty}^{\infty} |f(x)|^2 dx = \int_{-\infty}^{\infty} |\mathcal{H} f(x)|^2 dx.$$

***Exercise* 6.3.7.** This exercise addresses the spectral theory of the Hilbert transform.

1. Which real numbers are eigenvalues of the Hilbert transform? That is, for which real numbers λ does there exist a function f_λ in $L^2(\mathbb{R})$ so that

$$\mathcal{H} f = \lambda f?$$

Hint: Use the Fourier transform.

2. Can you describe the eigenspaces? That is, if λ is an eigenvalue of $\mathcal{H}$, describe the set of all functions in $L^2(\mathbb{R})$ that satisfy

$$\mathcal{H} f = \lambda f.$$

3. Show that $\mathcal{H} \circ \mathcal{H} f = \mathcal{H}(\mathcal{H}(f)) = f$ for any $f \in L^2(\mathbb{R})$.

6.3.2 Mapping Properties of the Hilbert Transform*

See: A.4.1.

The Hilbert transform has very good mapping properties with respect to most function spaces. Using the Parseval formula, we easily establish the L^2-result.

Proposition 6.3.1. *If $f \in L^2(\mathbb{R})$, then $\mathcal{H} f \in L^2(\mathbb{R})$ and in fact*

$$\|f\|_{L^2} = \|\mathcal{H} f\|_{L^2}.$$

The Hilbert transform also has good mapping properties on other L^p-spaces as well as Hölder spaces, though the proofs of these results requires more advanced techniques.

Proposition 6.3.2. *For each $1 < p < \infty$, the Hilbert transform extends to define a bounded map $\mathcal{H} : L^p(\mathbb{R}) \to L^p(\mathbb{R})$.*

Proposition 6.3.3. *Suppose that f is α-Hölder continuous for an $\alpha \in (0, 1)$ and vanishes outside a bounded interval. Then $\mathcal{H} f$ is also α-Hölder continuous.*

Notice that the case of $\alpha = 1$ is excluded in this proposition. The result is false in this case. There exist differentiable functions f such that $\mathcal{H} f$ is not 1-Hölder continuous. Proofs of these propositions can be found in [110].

Exercise

***Exercise* 6.3.8.** By using formula (6.53), which is valid for a Hölder continuous function vanishing outside a bounded interval, prove Proposition 6.3.3.

6.4 Approximate Inverses for the Radon Transform

To exactly invert the Radon transform, we need to compute the Hilbert transform of a derivative. The measured data is a function, g_m, on the space of lines. Measured data are rarely differentiable, and the exact Radon inverse entails the computation of $\partial_t g_m$. Indeed, the Parseval formula, (6.15), implies that unless g_m has a half an L^2-derivative, then it is not the Radon transform of an L^2-function. Thus it is important to investigate how to approximate the inverse of the Radon transform in a way that is usable with realistic data. Each approximation of the Hilbert transform leads to an approximation of the Radon inverse. Because the approximate inverses involve some sort of smoothing, they are often called *regularized inverses*.

Recall that a convolution has the following useful properties with respect to derivatives:

$$\partial_x (f * g) = \partial_x f * g = f * \partial_x g.$$

Using formula (6.46), we get an approximate inverse for the Radon transform:

$$\begin{aligned} f(\boldsymbol{x}) &\approx \frac{1}{2\pi i} \int_0^\pi \mathcal{H}_\epsilon(\partial_t \mathcal{R} f)(\langle \boldsymbol{x}, \boldsymbol{\omega} \rangle, \boldsymbol{\omega})\, d\boldsymbol{\omega} \\ &= \frac{1}{2\pi i} \int_0^\pi h_\epsilon * (\partial_t \mathcal{R} f)(\langle \boldsymbol{x}, \boldsymbol{\omega} \rangle, \boldsymbol{\omega})\, d\boldsymbol{\omega}. \end{aligned} \tag{6.54}$$

Using the formula for h_ϵ and the fact that $f * \partial_t g = \partial_t f * g$, we get

$$\begin{aligned} f(\mathbf{x}) &\approx \frac{1}{2\pi i} \int_0^\pi \int_{-\infty}^\infty \mathcal{R}f(s, \boldsymbol{\omega}) \partial_t h_\epsilon(\langle \mathbf{x}, \boldsymbol{\omega} \rangle - s)\, ds \\ &= \frac{1}{2\pi^2} \int_0^\pi \int_{-\infty}^\infty \left[\mathcal{R}f(s, \boldsymbol{\omega}) \frac{\epsilon^2 - (t-s)^2}{(\epsilon^2 + (t-s)^2)^2} ds \bigg|_{t=\langle \mathbf{x}, \boldsymbol{\omega} \rangle} \right] d\boldsymbol{\omega}. \end{aligned} \tag{6.55}$$

The expression in (6.55) has an important practical advantage: We have moved the t-derivative from the potentially noisy measurement $\mathcal{R}f$ over to the smooth, exactly known function h_ϵ. This means that we do not have to approximate the derivatives of $\mathcal{R}f$.

In most applications, convolution operators, such as derivatives and the Hilbert transform, are computed using the Fourier representation. Theorem 6.3.1 suggests approximating the filtering step, (6.24), in the exact inversion formula by cutting off the high-frequency components. Let $\hat{\psi}(r)$ be a bounded, even function, satisfying the conditions

$$\begin{aligned} \hat{\psi}(0) &= 1, \\ \hat{\psi}(r) &= 0 \qquad \text{for } |r| > W. \end{aligned} \tag{6.56}$$

For l a function on $\mathbb{R} \times S^1$ define

$$\mathcal{G}_\psi(l)(t, \boldsymbol{\omega}) = \frac{1}{2\pi} \int_{-\infty}^\infty \tilde{l}(r, \boldsymbol{\omega}) e^{irt} \hat{\psi}(r) |r|\, dr, \tag{6.57}$$

and

$$\mathcal{R}_\psi^{-1} l(\mathbf{x}) = \frac{1}{2\pi} \int_0^\pi \mathcal{G}_\psi(l)(\langle \mathbf{x}, \boldsymbol{\omega} \rangle, \boldsymbol{\omega})\, d\boldsymbol{\omega}. \tag{6.58}$$

For notational convenience let

$$f_\psi = \mathcal{R}_\psi^{-1} \circ \mathcal{R}f.$$

How is $\mathcal{R}_\psi^{-1} f$ related to f? The answer to this question is surprisingly simple. The starting point for our analysis is Proposition 6.1.1, which says that if f and g are functions on $\mathbb{R}^2$, then

$$\mathcal{R}(f * g)(t, \boldsymbol{\omega}) = \int_{-\infty}^\infty \mathcal{R}f(t - \tau, \boldsymbol{\omega}) \mathcal{R}g(\tau, \boldsymbol{\omega})\, d\tau.$$

Using the convolution theorem for the Fourier transform, we see that

$$\widetilde{\mathcal{R}f * g}(r, \boldsymbol{\omega}) = \widetilde{\mathcal{R}f}(r, \boldsymbol{\omega}) \widetilde{\mathcal{R}g}(r, \boldsymbol{\omega}).$$

Suppose now that g is a radial function so that $\mathcal{R}g$ is independent of $\boldsymbol{\omega}$. The filtered back-projection formula for $f * g$ reads

$$f * g(\boldsymbol{x}) = \frac{1}{4\pi^2} \int_0^{\pi} \int_{-\infty}^{\infty} \widetilde{\mathcal{R}f}(r, \boldsymbol{\omega}) \widetilde{\mathcal{R}g}(r) e^{ir\langle \boldsymbol{x}, \boldsymbol{\omega}\rangle} |r| \, dr \, d\boldsymbol{\omega}. \tag{6.59}$$

Comparing (6.59) with the definition of f_ψ, we see that, if we can find a radial function k_ψ, defined on $\mathbb{R}^2$, so that

$$\mathcal{R}(k_\psi)(t, \boldsymbol{\omega}) = \psi(t),$$

then

$$f_\psi(\boldsymbol{x}) = k_\psi * f(\boldsymbol{x}). \tag{6.60}$$

The existence of such a function is a consequence of the results in Section 3.5. Because $\hat{\psi}$ has bounded support, ψ is an infinitely differentiable function, with all derivatives bounded. To apply Proposition 3.5.1, we need to know that ψ and ψ' are absolutely integrable. This translates into a requirement that $\hat{\psi}$ is sufficiently continuous. In this case, the function k_ψ is given by the formula

$$k_\psi(\rho) = -\frac{1}{\pi} \int_{\rho}^{\infty} \frac{\psi'(t)\, dt}{\sqrt{t^2 - \rho^2}}. \tag{6.61}$$

This completes the proof of the following proposition.

Proposition 6.4.1. *Suppose that $\hat{\psi}$ satisfies the conditions in* (6.56) *and ψ is absolutely integrable. Then*

$$f_\psi(\boldsymbol{x}) = k_\psi * f(\boldsymbol{x}),$$

where k_ψ is given by (6.61).

***Remark* 6.4.1.** Replacing f by f_ψ produces a somewhat blurred image. Increasing the support of $\hat{\psi}$ leads, in general, to a more sharply peaked ψ and therefore a more sharply peaked k_ψ. This reduces the blurring but also reduces the suppression of noise in the data. This discussion is adapted from [113].

6.4.1 Addendum*

See: A.3.3.

The analysis in the previous section is unsatisfactory in one particular: We explicitly exclude the possibility that $\hat{\psi}_W(r) = \chi_{[-W,W]}(r)$. The problem is that $\psi_W(t) = \sin(Wt)/(\pi t)$ is not absolutely integrable and so the general inversion result for radial functions does not apply. In this special case the integral defining k_ψ is a convergent, improper integral, which can be computed exactly.

We use the formula

$$\int_1^\infty \frac{\sin(xt)\,dt}{\sqrt{t^2-1}} = \frac{\pi}{2} J_0(x)$$

for the J_0-Bessel function; see [93]. Putting this into the inversion formula and using the fact that $J_0' = -J_1$, we obtain

$$k_W(x) = \frac{W}{2\pi x} J_1(Wx).$$

The power series for $J_1(x)$ about $x = 0$ is

$$J_1(x) = \frac{x}{2} \sum_{k=0}^{\infty} \frac{(-1)^k x^{2k}}{2^{2k} k!(k+1)!},$$

from which it follows easily that $k_W(x)$ is a smooth function of x^2. The standard asymptotic expansion for J_1 as $|x|$ tends to infinity implies that

$$|k_W(x)| \leq \frac{C}{(1+|x|)^{\frac{3}{2}}}$$

and therefore the integrals defining $\mathcal{R}k_W$ converge absolutely. As the Radon transform is linear, we can extend the result of the previous section to allow functions of the form

$$\hat{\psi}(r) = \chi_{[-W,W]}(r) + \hat{\psi}_c(r),$$

where $\psi_c = \mathcal{F}^{-1}\hat{\psi}_c$ satisfies the hypotheses of Proposition 3.5.1. In this case,

$$f_\psi = (k_W + k_{\psi_c}) * f. \tag{6.62}$$

Exercise

***Exercise* 6.4.1.** Justify the computations for the function $\hat{\psi} = \chi_{[-W,W]}$ leading up to formula (6.62).

6.5 The Radon Transform on Data with Bounded Support

In medical imaging the data under consideration usually have bounded support. The Radon transform of a function with bounded support satisfies an infinite set of *moment* conditions. From the point of view of measurements, these can be viewed as consistency conditions. Mathematically, this is a part of the problem of characterizing the *range* of the Radon transform on data with bounded support. The general problem of describing the range of the Radon transform is well beyond the scope of this text. The interested reader is referred to [83], [90], or [95].

Suppose that f is a function that vanishes outside the disk of radius R. As observed previously, this implies that $\mathcal{R}f(t,\boldsymbol{\omega}) = 0$ if $|t| > R$. For a nonnegative integer, n consider the integral,

$$M_n(f)(\boldsymbol{\omega}) = \int_{\mathbb{R}^2} f(\boldsymbol{x})[\langle \boldsymbol{x}, \boldsymbol{\omega}\rangle]^n \, d\boldsymbol{x}. \tag{6.63}$$

If f has bounded support, then these integrals are well defined for any $n \in \mathbb{N} \cup \{0\}$. If f does not vanish outside a disk of finite radius, then, for sufficiently large n, these integral may not make sense.

Changing coordinates with $\boldsymbol{x} = t\boldsymbol{\omega} + s\hat{\boldsymbol{\omega}}$, we can rewrite this integral in terms of $\mathcal{R}f$,

$$\begin{aligned} M_n(f)(\boldsymbol{\omega}) &= \int_{\mathbb{R}^2} f(t\boldsymbol{\omega} + s\hat{\boldsymbol{\omega}})t^n \, ds \, dt \\ &= \int_{-\infty}^{\infty} \mathcal{R}f(t,\boldsymbol{\omega})t^n \, dt. \end{aligned} \tag{6.64}$$

The function $M_n(f)(\boldsymbol{\omega})$ is called the *nth moment* of the Radon transform of f. If $\mathcal{R}f(t,\boldsymbol{\omega})$ vanishes for $|t| > R$, then this integral is well defined for all n. In Example 3.4.7 we showed that there are functions, which do *not* have bounded support, for which the Radon transform is defined and vanishes for large enough values of t. If f itself has bounded support, then $M_n(f)(\boldsymbol{\omega})$ depends on $\boldsymbol{\omega}$ in a very special way.

It is useful to express $\boldsymbol{\omega}$ as a function of the angle θ,

$$\boldsymbol{\omega}(\theta) = (\cos(\theta), \sin(\theta)).$$

Using the binomial theorem, we obtain

$$\begin{aligned} \langle \boldsymbol{x}, \boldsymbol{\omega}(\theta)\rangle^n &= (x\cos\theta + y\sin\theta)^n \\ &= \sum_{j=0}^{n} \binom{n}{j} (x\cos\theta)^j (y\sin\theta)^{n-j} \\ &= \sum_{j=0}^{n} \binom{n}{j} \cos^j\theta \sin^{n-j}\theta x^j y^{n-j}. \end{aligned}$$

Putting the sum into formula (6.63), we see that this integral defines a trigonometric polynomial of degree n.

$$\begin{aligned} M_n(f)(\theta) &= \sum_{j=0}^{n} \binom{n}{j} \cos^j\theta \sin^{n-j}\theta \iint_{\mathbb{R}^2} f(x,y)x^j y^{n-j} \, dx \, dy \\ &= \sum_{j=0}^{n} a_{nj} \sin^j\theta \cos^{n-j}\theta \end{aligned} \tag{6.65}$$

where

$$a_{nj} = \binom{n}{j} \iint\limits_{\mathbb{R}^2} f(x, y) x^j y^{n-j}\, dx\, dy.$$

If f has bounded support, then $M_n(f)(\theta)$ is a trigonometric polynomial of degree n. We summarize these computations in a proposition.

Proposition 6.5.1. *Suppose that f is a function with bounded support. Then*

1. *$\mathcal{R} f(t, \omega)$ has bounded support.*

2. *For all nonnegative integers, n, there exist constants $\{a_{n0}, \dots, a_{nn}\}$ such that*

$$\int_{-\infty}^{\infty} \mathcal{R} f(t, \omega(\theta)) t^n dt = \sum_{j=0}^{n} a_{nj} \sin^j \theta \cos^{n-j} \theta.$$

The proposition suggests the following question: Suppose that $h(t, \omega)$ is a function on $\mathbb{R} \times S^1$ such that

1. $h(t, \omega) = h(-t, -\omega)$,

2. $h(t, \omega) = 0$ if $|t| > R$,

3. For each nonnegative integer n

$$m_n(h)(\theta) = \int_{-\infty}^{\infty} h(t, \omega(\theta)) t^n\, dt$$

   is a trigonometric polynomial of degree n,

4. $h(t, \omega)$ is a sufficiently smooth function of (t, ω).

Does there exist a function f in the domain of the Radon transform, vanishing outside of the disk of radius R such that

$$h = Rf?$$

In other words, does h belong to the range of the Radon transform, acting on smooth functions with bounded support? According to a theorem of Helgason and Ludwig, the answer to this question turns out to be yes: however, the proof of this result requires techniques beyond the scope of this text. Similar results were obtained earlier by Gelfand, Graev and Vilenkin. Alan Cormack, inventor of the x-ray CT scanner, also had a version of this result. For a detailed discussion of this question, and its connections to x-ray tomography, the reader is referred to [95]. More material can be found in [51], [83], [90], [44], [24], or [36].

We model the data measured in CT imaging as the Radon transform of a piecewise continuous function with bounded support. If we could make measurements for all (t, ω), then it probably would not be the exact Radon transform of such a function. This is because all measurements are corrupted by errors and noise. In particular, the patient's movements, both internal (breathing, heart beat, blood circulation, etc.) and external, affect the measurements. The measured data would therefore be inconsistent and may fail to satisfy the aforementioned moment conditions.

6.6 Continuity of the Radon Transform and Its Inverse*

In order for the measurement process in x-ray tomography to be stable, the map $f \mapsto \mathcal{R}f$ should be continuous in a reasonable sense. Estimates for the continuity of this map quantify the sensitivity of the output, $\mathcal{R}f$, of a CT scanner to changes in the input. The *less* continuous the map, the *more* sensitive the measurements are to changes in the input. Estimates for the continuity of inverse, $h \mapsto \mathcal{R}^{-1}h$, quantify the effect of errors in the measured data on the quality of the reconstructed image. Because we actually *measure* the Radon transform, estimates for the continuity of $\mathcal{R}^{-1}$ are more important for the problem of image reconstruction. To discuss the continuity properties of either transform, we need to select norms for functions in the domain and range. Using the L^2-norms on both, the Parseval formula, (6.15), provides a starting point for this discussion.

The Parseval formula says that if $f \in L^2(\mathbb{R}^2)$, then $D_{\frac{1}{2}}\mathcal{R}f \in L^2(\mathbb{R}\times S^1)$. This estimate has somewhat limited utility; as $|r|$ vanishes at $r = 0$, we cannot conclude that $\mathcal{R}f$ is actually in $L^2(\mathbb{R} \times S^1)$. In medical applications the data have bounded support, and in this case additional estimates are available. For the inverse transform, the Parseval formula says that in order to control the L^2-norm of the reconstructed image, we need to have control on the half-order L^2-derivative of the measured data. Due to noise this is, practically speaking, not possible. After discussing the continuity properties of the forward transform for data with bounded support, we consider the continuity properties of the *approximate inverse* described in Section 6.4.

6.6.1 Data With Bounded Support

Functions with bounded support satisfy better L^2-estimates.

Proposition 6.6.1. *Let $f \in L^2(\mathbb{R}^2)$ and suppose that f vanishes outside the disk of radius L. Then, for each ω, we have the estimate*

$$\int_{-\infty}^{\infty} |\mathcal{R}f(t,\omega)|^2\,dt \leq 2L\|f\|_{L^2}^2. \tag{6.66}$$

Proof. The proof of the proposition is a simple application of the Cauchy-Schwarz inequality. Because f vanishes outside the disk of radius L, we can express $\mathcal{R}f$ as

$$\mathcal{R}f(t,\omega) = \int_{-L}^{L} f(t\omega + s\hat{\omega})\,ds.$$

Computing the L^2-norm of $\mathscr{R}f$ in the t-variable, we obtain

$$\begin{aligned}\int_{-\infty}^{\infty} |\mathscr{R}f(t,\boldsymbol{\omega})|^2\,dt &= \int_{-L}^{L}\left|\int_{-L}^{L} f(t\boldsymbol{\omega}+s\hat{\boldsymbol{\omega}})\,ds\right|^2 dt \\ &\leq 2L\int_{-L}^{L}\int_{-L}^{L} |f(t\boldsymbol{\omega}+s\hat{\boldsymbol{\omega}})|^2\,ds\,dt.\end{aligned} \tag{6.67}$$

In the second line we used the Cauchy-Schwarz inequality. □

The proposition shows that, if f vanishes outside a bounded set, then we control not only the overall L^2-norm of $\mathscr{R}f$ but the L^2-norm in each direction, $\boldsymbol{\omega}$ separately. Using the support properties of f more carefully gives a weighted estimate on the L^2-norm of $\mathscr{R}f$.

Proposition 6.6.2. *Let $f \in L^2(\mathbb{R}^2)$ and suppose that f vanishes outside the disk of radius L. Then, for each $\boldsymbol{\omega}$, we have the estimate*

$$\int_{-\infty}^{\infty} \frac{|\mathscr{R}f(t,\boldsymbol{\omega})|^2\,dt}{\sqrt{L^2-t^2}} \leq 2\|f\|^2_{L^2}. \tag{6.68}$$

Proof. To prove this estimate, observe that

$$f(x,y) = \chi_{[0,L^2]}(x^2+y^2)f(x,y).$$

The Cauchy-Schwarz inequality therefore implies that, for $|t| \leq L$, we have the estimate

$$\begin{aligned}|\mathscr{R}f(t,\boldsymbol{\omega})|^2 &= \left|\int_{-L}^{L} f(t\boldsymbol{\omega}+s\hat{\boldsymbol{\omega}})\chi_{[0,L^2]}(s^2+t^2)\,ds\right|^2 \\ &\leq 2\int_{-L}^{L} |f(t\boldsymbol{\omega}+s\hat{\boldsymbol{\omega}})|^2\,ds \int_{0}^{\sqrt{L^2-t^2}} ds \\ &= 2\sqrt{L^2-t^2}\int_{-L}^{L} |f(t\boldsymbol{\omega}+s\hat{\boldsymbol{\omega}})|^2\,ds.\end{aligned} \tag{6.69}$$

Thus

$$\begin{aligned}\int_{-L}^{L} \frac{|\mathscr{R}f(t,\boldsymbol{\omega})|^2\,dt}{\sqrt{L^2-t^2}} &\leq \int_{-L}^{L} \frac{2\sqrt{L^2-t^2}}{\sqrt{L^2-t^2}}\int_{-L}^{L} |f(t\boldsymbol{\omega}+s\hat{\boldsymbol{\omega}})|^2\,ds\,dt \\ &= 2\|f\|^2_{L^2}.\end{aligned} \tag{6.70}$$

□

A function in $f \in L^2(\mathbb{R}^2)$ with support in the disk of radius L can be approximated, in the L^2-norm, by a sequence of smooth functions $< f_n >$. This sequence can also be taken to have support in the disk of radius L. The Radon transforms of these functions satisfy the estimates

$$\int_{-\infty}^{\infty} |\mathcal{R}f_n(t,\omega)|^2\,dt \le 2L\|f_n\|_{L^2}^2$$

and

$$\frac{1}{[2\pi]^2}\int_0^{\pi}\int_{-\infty}^{\infty} |\widetilde{\mathcal{R}f_n}(r,\omega)|^2|r|\,dr\,d\omega = \|f_n\|_{L^2(\mathbb{R}^2)}^2.$$

In a manner analogous to that used to extend the Fourier transform to L^2-functions, we can now extend the Radon transform to L^2-functions with support in a fixed bounded set.

For bounded functions on $\mathbb{R}\times S^1$ vanishing for $|t| > L$, a norm is defined by

$$\|h\|_{2,L}^2 = \sup_{\omega\in S^1}\int_{-L}^{L} |h(t,\omega)|^2\,dt + \frac{1}{[2\pi]^2}\int_0^{\pi}\int_{-\infty}^{\infty} |\tilde{h}(r,\omega)|^2|r|\,dr\,d\omega.$$

The closure of $\mathcal{C}^0([-L,L]\times S^1)$ in this norm is a Hilbert space that can be identified with a subspace of $L^2([-L,L]\times S^1)$. For f, an L^2-function with support in B_L, $\mathcal{R}f$ is defined as the limit of $\mathcal{R}f_n$ in this norm. Evidently, the estimates, (6.66), (6.68), hold for $\mathcal{R}f$. On the other hand, the elementary formula for $\mathcal{R}f(t,\omega)$ may not be meaningful as f may not be absolutely integrable over $l_{t,\omega}$.

While it is well beyond the scope of this text, it is nonetheless true that a function on $\mathbb{R}\times S^1$ with support in the set $|t| \le L$ and finite $\|\cdot\|_{2,L}$-norm that satisfies the moment conditions is the generalized Radon transform of function in $L^2(\mathbb{R}^2)$ with support in the disk of radius L. A proof can be found in [51] or [95].

Exercises

***Exercise* 6.6.1.** Suppose that $f \in L^2(\mathbb{R}^2)$ and that f vanishes outside the disk of radius L. Show that $\|\mathcal{R}f(\cdot,\omega_1)-\mathcal{R}f(\cdot,\omega_2)\|_{L^2(\mathbb{R})}$ tends to zero as ω_1 approaches ω_2. In other words, the map $\omega \mapsto \mathcal{R}f(\cdot,\omega)$ is a continuous map from the circle into $L^2(\mathbb{R})$. This shows that, if we measure errors in the L^2-norm, then the Radon transform is not excessively sensitive to small changes in the measurement environment.

***Exercise* 6.6.2.** Suppose that $< f_n >$ is a sequence of smooth functions with support in a fixed disk converging to f in $L^2(\mathbb{R}^2)$. For the terms in the approximating sequence, $< \mathcal{R}f_n >$, the moments $\{m_k(\mathcal{R}f_n)\}$ satisfy the conditions in Proposition 6.5.1. Show that for the limiting function, the moments $\{m_k(\mathcal{R}f)\}$ are well defined and also satisfy these conditions.

6.6.2 Estimates for the Inverse Transform

The question of more immediate interest is the continuity properties of the *inverse* transform. This is the more important question because we actually measure an approximation, Rf_m to $\mathcal{R}f$. It would appear that to estimate the error in the reconstructed image, we would need to estimate

$$\mathcal{R}^{-1}Rf_m - f = \mathcal{R}^{-1}(Rf_m - \mathcal{R}f). \tag{6.71}$$

There are several problems that arise immediately. The most obvious problem is that Rf_m may not be in the range of the Radon transform. If $Rf_m(t, \omega)$ does not have an L^2-half-derivative in the t-direction, that is,

$$\int_0^{2\pi}\int_{-\infty}^{\infty} |\widetilde{Rf_m}(r, \omega)|^2 |r|\, dr\, d\omega = \infty,$$

then according to the Parseval formula, (6.15) Rf_m is *not* the Radon transform of a function in $L^2(\mathbb{R}^2)$. In order to control the L^2-error,

$$\| \mathcal{R}^{-1}(Rf_m - \mathcal{R}f)\|_{L^2(\mathbb{R}^2)},$$

it is necessary that measurements have a half-derivative and the difference

$$\|D_{\frac{1}{2}}(Rf_m - \mathcal{R}f)\|_{L^2(\mathbb{R}\times S^1)}$$

is small. This means that we need to control the high-frequency content of Rf_m; in practice this is not possible. While the mathematical problem of estimating the Radon inverse is quite interesting and important, it has little bearing on the problem of practical image reconstruction. A very nice treatment of the mathematical question is given in [95]. We now turn our attention to understanding the continuity of the *approximate* inverses defined in Section 6.4.

An approximate inverse is denoted by $\mathcal{R}_\psi^{-1}$, where ψ is a regularizing function. This is an even function whose Fourier transform satisfies the conditions

$$\begin{aligned} \hat{\psi}(0) &= 1, \\ \hat{\psi}(r) &= 0 \qquad \text{for } |r| > W. \end{aligned} \tag{6.72}$$

It is also assumed that the radial function k_ψ defined in (6.61) is in the domain of the Radon transform and

$$\mathcal{R}k_\psi = \psi.$$

In this case,

$$\mathcal{R}_\psi^{-1}\mathcal{R}f = k_\psi * f. \tag{6.73}$$

***Example* 6.6.1.** Let $\hat{\psi}$ be the piecewise linear function

$$\hat{\psi}(r) = \begin{cases} 1 & \text{for } |r| < W - C, \\ \frac{W-|r|}{C} & \text{for } W - C \le |r| \le W, \\ 0 & \text{for } |r| > W. \end{cases}$$

Radial graphs of ψ and k_ψ are shown in Figure 6.4.

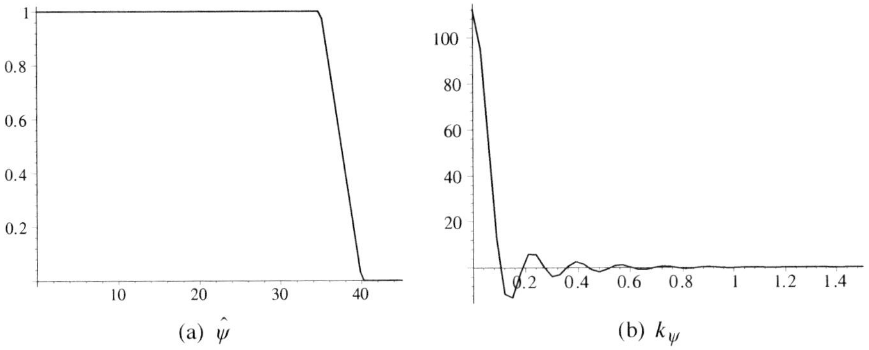

Figure 6.4. Graphs of $\hat{\psi}$ and k_ψ, with $W = 40$, $C = 5$.

The reconstructed image is

$$f_\psi = \mathcal{R}_\psi^{-1} R f_m;$$

therefore, we need to estimate the difference $f - f_\psi$. As $k_\psi * f = \mathcal{R}_\psi^{-1} \mathcal{R} f$, we can rewrite this difference as

$$f - f_\psi = (f - k_\psi * f) + \mathcal{R}_\psi^{-1}(\mathcal{R} f - R f_m). \tag{6.74}$$

The first term on the right-hand side is the error caused by using an approximate inverse. It is present even if we have perfect data. Bounds for this term depend in an essential way on the character of the data. If f is assumed to be a continuous function of bounded support, then, by taking W very large, the pointwise error,

$$\|f - k_\psi * f\|_\infty = \sup_{x \in \mathbb{R}^2} |f(x) - k_\psi * f(x)|$$

can be made as small as desired. It is more realistic to model f as a piecewise continuous function. In this case the difference, $|f(x) - k_\psi * f(x)|$, can be made small at points where f is continuous. Near points where f has a jump, the approximate reconstruction may display an oscillatory artifact. Figure 6.5 is a radial graph of the reconstruction of $\chi_{B_1}(x)$ using the regularizing function graphed in Figure 6.4.

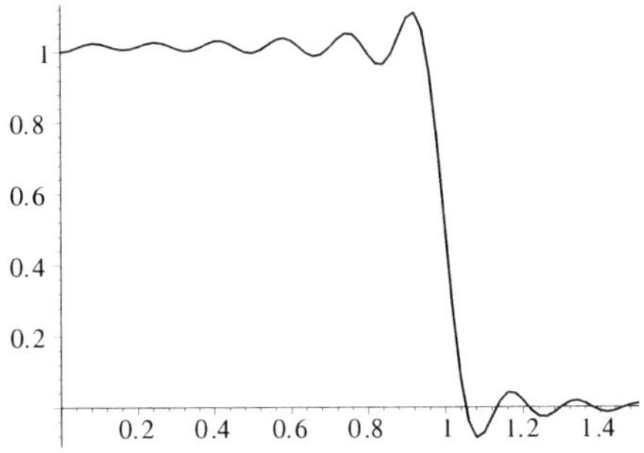

Figure 6.5. Radial graph of $k_\psi * \chi_{B_1}$, with $W = 40$, $C = 5$.

Robust estimates for the second term are less dependent on the precise nature of f. For h a function on $\mathbb{R} \times S^1$ with bounded support, the approximate inverse is given by

$$
\begin{aligned}
(\mathcal{R}_\psi^{-1} h)(\boldsymbol{x}) &= \frac{1}{4\pi^2} \int_0^\pi \int_{-\infty}^\infty \widetilde{h}(r, \boldsymbol{\omega}) e^{ir\langle \boldsymbol{x}, \boldsymbol{\omega}\rangle} \hat{\psi}(r)|r|\, dr \\
&= \frac{1}{2\pi} \int_0^\pi (g_\psi *_t h)(\langle \boldsymbol{x}, \boldsymbol{\omega}\rangle, \boldsymbol{\omega})\, d\boldsymbol{\omega}.
\end{aligned}
\tag{6.75}
$$

Here $g_\psi = \mathcal{F}^{-1}(\hat{\psi}(r)|r|)$ and $*_t$ indicates convolution in the t-variable.

A simple estimate for the sup norm of $\mathcal{R}_\psi^{-1} h$ follows from the sup-norm estimate for a convolution:

$$\|l * k\|_{L^\infty} \leq \|l\|_{L^\infty} \|k\|_{L^1}.$$

Applying this estimate gives

$$\|\mathcal{R}_\psi^{-1} h\|_{L^\infty} \leq \frac{\|g_\psi\|_{L^\infty}}{2\pi} \int_0^\pi \int_{-\infty}^\infty |h(t, \boldsymbol{\omega})|\, dt\, d\boldsymbol{\omega} \tag{6.76}$$

If $\hat{\psi}$ is nonnegative, then

$$|g_\psi(t)| \leq |g_\psi(0)| = \int_{-\infty}^\infty |r| \hat{\psi}(r)\, dr.$$

Assuming that $0 \leq \hat{\psi}(t) \leq M$ and that it vanishes outside the interval $[-W, W]$ leads to the estimate

$$\|g_\psi\|_{L^\infty} \leq MW^2.$$

Combining this with (6.76) gives

$$\|\mathcal{R}_\psi^{-1} h\|_{L^\infty} \leq \frac{MW^2}{2\pi} \|h\|_{L^1(\mathbb{R}\times S^1)}. \tag{6.77}$$

This estimate shows that the sup norm of the error in the approximate reconstructed image, $\mathcal{R}_\psi^{-1}(\mathcal{R}f - Rf_m)$, can be controlled if the measurement errors can be controlled in the L^1-norm. It also shows that the error increases as W increases.

To summarize, the error in the approximate reconstruction is bounded by

$$|f - f_\psi| \leq |f - k_\psi * f| + \frac{\|g_\psi\|_{L^\infty}}{2\pi} \|\mathcal{R}f - Rf_m\|_{L^1(\mathbb{R}\times S^1)}. \tag{6.78}$$

Recall that

$$\mathcal{F}(k_\psi) = \hat{\psi} \text{ and } \mathcal{F}(g_\psi) = |r|\hat{\psi}.$$

The function k_ψ is rapidly decreasing and sharply peaked if $\hat{\psi}$ is smooth and W is taken large. On the other hand, g_ψ cannot decay faster than $O(t^{-2})$. This is a consequence of the fact that $|r|\hat{\psi}(r)$ is singular at $r = 0$.

Exercises

***Exercise* 6.6.3.** Prove that $\|l * k\|_{L^\infty} \leq \|l\|_{L^\infty}\|k\|_{L^1}$.

***Exercise* 6.6.4.** Suppose that $\hat{\psi}$ is a smooth function with bounded support such that $\hat{\psi}(0) \neq 0$ and let

$$g_\psi(t) = \frac{1}{2\pi}\int_{-\infty}^{\infty} \hat{\psi}(\xi)|\xi|e^{it\xi}d\xi.$$

Show that there is a constant $C > 0$ so that the following *lower bound* holds for large enough t:

$$|g_\psi(t)| \geq \frac{C}{1+t^2}. \tag{6.79}$$

***Exercise* 6.6.5.** Use the central slice theorem to give a formula for k_ψ as a Bessel transform of $\hat{\psi}(r)$.

***Exercise* 6.6.6.** Use Hölder's inequality to show that

$$\|l * k\|_{L^\infty} \leq \|l\|_{L^2}\|k\|_{L^2}.$$

Use this estimate to prove that

$$\|\mathcal{R}_\psi^{-1}h\|_{L^\infty} \leq \frac{\|g_\psi\|_{L^2(\mathbb{R})}}{\sqrt{4\pi}}\|h\|_{L^2(\mathbb{R}\times S^1)}.$$

Under the assumptions used previously to estimate $\|g_\psi\|_{L^\infty}$, show that

$$\|g_\psi\|_{L^2} \leq \sqrt{\frac{2}{3}}MW^{\frac{3}{2}}.$$

6.7 The Higher-Dimensional Radon Transform*

See: A.2.1.

For the sake of completeness, we briefly present the theory of the Radon transform in higher dimensions. The parameterization of the affine hyperplanes in $\mathbb{R}^n$ is quite similar to that used for lines in $\mathbb{R}^2$. Let $\boldsymbol{\omega}$ be a unit vector in $\mathbb{R}^n$ (i.e., a point on S^{n-1}) and let $t \in \mathbb{R}$. Each affine hyperplane has a representation in the form

$$l_{t,\boldsymbol{\omega}} = \{\boldsymbol{x} \in \mathbb{R}^n : \langle \boldsymbol{x}, \boldsymbol{\omega}\rangle = t\}.$$

As in the two-dimensional case, $l_{t,\boldsymbol{\omega}} = l_{-t,-\boldsymbol{\omega}}$ and the choice of vector $\boldsymbol{\omega}$ defines an orientation on the hyperplane.

In order to define the Radon transform, it is useful to choose vectors $\{\boldsymbol{e}_1, \ldots, \boldsymbol{e}_{n-1}\}$ so that

$$\langle \boldsymbol{\omega}, \boldsymbol{e}_j \rangle = 0 \text{ and } \langle \boldsymbol{e}_i, \boldsymbol{e}_j \rangle = \delta_{ij} \qquad \text{for } i, j = 1, \ldots, n-1.$$

The n-vectors $\langle \boldsymbol{\omega}, \boldsymbol{e}_1, \ldots, \boldsymbol{e}_{n-1} \rangle$ are an orthonormal basis for $\mathbb{R}^n$. Define new orthogonal coordinates, $(t, s_1, \ldots, s_{n-1})$, on $\mathbb{R}^n$ by setting

$$\boldsymbol{x} = t\boldsymbol{\omega} + \sum_{j=1}^{n-1} s_j \boldsymbol{e}_j.$$

The n-dimensional Radon transform is defined by

$$\mathscr{R}f(t, \boldsymbol{\omega}) = \int_{l_{t,\omega}} f \, d\sigma_{n-1} = \int_{\mathbb{R}^{n-1}} f(t\boldsymbol{\omega} + \sum s_j \boldsymbol{e}_j) \, ds_1 \cdots ds_{n-1}.$$

As before, the Radon transform is an even function

$$\mathscr{R}f(t, \boldsymbol{\omega}) = \mathscr{R}f(-t, -\boldsymbol{\omega}).$$

With this definition, the n-dimensional analogue of the Central slice theorem is as follows:

Theorem 6.7.1 (Central slice theorem). *If f is an absolutely integrable function on $\mathbb{R}^n$ then*

$$\widetilde{\mathscr{R}f}(r, \boldsymbol{\omega}) = \int_{-\infty}^{\infty} \mathscr{R}f(t, \boldsymbol{\omega}) e^{-irt} dt = \hat{f}(r\boldsymbol{\omega}). \tag{6.80}$$

The central slice theorem and the Fourier inversion formula give the Radon inversion formula.

Theorem 6.7.2 (The Radon inversion formula). *Suppose that f is a smooth function with bounded support on $\mathbb{R}^n$. Then*

$$f(\boldsymbol{x}) = \frac{1}{2(2\pi)^n} \int_{S^{n-1}} \int_{-\infty}^{\infty} \widetilde{\mathscr{R}f}(r, \boldsymbol{\omega}) r^{n-1} e^{ir\langle \boldsymbol{\omega}, \boldsymbol{x} \rangle} \, dr \, d\boldsymbol{\omega}. \tag{6.81}$$

***Remark* 6.7.1.** This formula holds in much greater generality. Under the hypotheses in the theorem, all the integrals converge absolutely and the simplest form of the Fourier inversion formula applies.

This formula takes a very simple form if the dimension is odd: Set $n = 2k + 1$. In this case the r-integral in (6.81) can be computed explicitly:

$$\frac{1}{2\pi}\int_{-\infty}^{\infty} \widetilde{\mathscr{R}f}(r,\boldsymbol{\omega})r^{n-1}e^{ir\langle\boldsymbol{\omega},\boldsymbol{x}\rangle}\,dr = (-1)^k\partial_t^{2k}\mathscr{R}f(t,\langle\boldsymbol{\omega},\boldsymbol{x}\rangle). \tag{6.82}$$

Using this expression in (6.81), we obtain

$$f(\boldsymbol{x}) = \frac{(-1)^k}{2(2\pi)^{2k}}\int_{S^{n-1}} (\partial_t^{2k}\mathscr{R}f)(\langle\boldsymbol{\omega},\boldsymbol{x}\rangle,\boldsymbol{\omega})\,d\boldsymbol{\omega}.$$

Thus in odd dimensions the inverse of the Radon transform is differentiation in t followed by back-projection.

The Laplace operator on $\mathbb{R}^n$ is defined by

$$\Delta_{\mathbb{R}^n} f = \sum_{j=1}^{n}\partial_{x_j}^2 f.$$

This differential operator is invariant under rotations so it follows that, for the coordinates $(t, s_1, \ldots, s_{n-1})$ introduced previously, we also have the formula

$$\Delta_{\mathbb{R}^n} f = \partial_t^2 f + \sum_{j=1}^{n-1}\partial_{s_j}^2 f. \tag{6.83}$$

This formula allows us to establish a connection between $\mathscr{R}(\Delta_{\mathbb{R}^n} f)$ and $\mathscr{R}f$.

Proposition 6.7.1. *Suppose that f is a twice differentiable function of bounded support on $\mathbb{R}^n$, then*

$$\mathscr{R}(\Delta_{\mathbb{R}^n} f) = \partial_t^2\mathscr{R}f. \tag{6.84}$$

We close our discussion by explaining how the Radon transform can be applied to solve the wave equation. Let τ denote the time variable and c the speed of sound. The wave equation for a function $u(\boldsymbol{x};\tau)$ defined on $\mathbb{R}^n \times \mathbb{R}$ is

$$\partial_\tau^2 u = c^2\Delta_{\mathbb{R}^n} u.$$

If u satisfies this equation, then it follows from the proposition that, for each $\boldsymbol{\omega} \in S^{n-1}$, $\mathscr{R}u(t,\boldsymbol{\omega};\tau)$ satisfies the equation

$$\partial_\tau^2\mathscr{R}u = c^2\partial_t^2\mathscr{R}u.$$

Here $\mathscr{R}u(t,\boldsymbol{\omega};\tau)$ is the Radon transform of $u(\boldsymbol{x};\tau)$ in the $\boldsymbol{x}$-variables with τ the time parameter. In other words, the Radon transform translates the problem of solving the wave equation in n dimensions into the problem of solving a family of wave equations in one dimension.

The one-dimensional wave equation is solved by any function of the form

$$v(t;\tau) = g(ct+\tau) + h(ct-\tau).$$

The initial data are usually $v(t;0)$ and $v_\tau(t;0)$; they are related to g and h by

$$\begin{aligned} g(ct) &= \frac{1}{2}\left[v(t;0) + c\int_{-\infty}^{t} v_\tau(s;0)\,ds \right], \\ h(ct) &= \frac{1}{2}\left[v(t;0) - c\int_{-\infty}^{t} v_\tau(s;0)\,ds \right]. \end{aligned} \tag{6.85}$$

If $u(\boldsymbol{x};0) = u_0(\boldsymbol{x})$ and $u_\tau(\boldsymbol{x};0) = u_1(\boldsymbol{x})$, then we see that

$$\mathcal{R}u(t,\boldsymbol{\omega};\tau) = g(ct+\tau;\boldsymbol{\omega}) + h(ct-\tau;\boldsymbol{\omega}),$$

where

$$\begin{aligned} g(ct;\boldsymbol{\omega}) &= \frac{1}{2}\left[\mathcal{R}u_0(t;\boldsymbol{\omega}) + c\int_{-\infty}^{t} \mathcal{R}u_1(s;\boldsymbol{\omega})\,ds \right], \\ h(ct;\boldsymbol{\omega}) &= \frac{1}{2}\left[\mathcal{R}u_0(t;\boldsymbol{\omega}) - c\int_{-\infty}^{t} \mathcal{R}u_1(s;\boldsymbol{\omega})\,ds \right]. \end{aligned} \tag{6.86}$$

Using these formulæ along with (6.81), we can obtain an explicit formula for the solution of the wave equation.

Exercises

***Exercise* 6.7.1.** Prove the central slice theorem.

***Exercise* 6.7.2.** Let $n = 2k+1$ and suppose that f is a function for which

$$\mathcal{R}f(t,\boldsymbol{\omega}) = 0 \qquad \text{if } |t| < R.$$

Prove that $f(\boldsymbol{x}) = 0$ if $\|\boldsymbol{x}\| < R$. Is this true in even dimensions?

***Exercise* 6.7.3.** Prove formula (6.83) and formula 6.84.

***Exercise* 6.7.4.** Prove Proposition (6.7.1). *Hint:* Integrate by parts.

***Exercise* 6.7.5.** Use the simplified version of the Radon inversion formula available for $n = 3$ to derive an explicit formula for the solution of the wave equation in three space dimensions in terms of the initial data $u_0(x)$ and $u_1(x)$.

6.8 The Hilbert Transform and Complex Analysis*

Earlier in this chapter, we used several explicit Hilbert transforms. Here we explain how these computations are done. We restrict our discussion to the case of square-integrable functions. If $f \in L^2(\mathbb{R})$ with Fourier transform $\hat{f}$, then, as a limit-in-the-mean,

$$f(x) = \frac{1}{2\pi} \int_{-\infty}^{\infty} e^{ix\xi} \hat{f}(\xi)\, d\xi.$$

Define two L^2-functions:

$$\begin{aligned} f_+(x) &= \frac{1}{2\pi} \int_{0}^{\infty} e^{ix\xi} \hat{f}(\xi)\, d\xi, \\ f_-(x) &= \frac{1}{2\pi} \int_{-\infty}^{0} e^{ix\xi} \hat{f}(\xi)\, d\xi. \end{aligned} \tag{6.87}$$

Obviously, we have $f = f_+ + f_-$ and $\mathcal{H} f = f_+ - f_-$. This decomposition is useful because the function $f_+(x)$ has an extension as an analytic function in the upper half-plane, $H_+ = \{x + iy \; : \; y > 0\}$:

$$f_+(x + iy) = \frac{1}{2\pi} \int_{0}^{\infty} e^{i(x+iy)\xi} \hat{f}(\xi)\, d\xi.$$

The Fourier transform of $f_+(x + iy)$ in the x-variable is just $\hat{f}(\xi)\chi[0, \infty)(\xi)e^{-y\xi}$. Since $y\xi > 0$, we see that $f_+(x + iy)$ is in $L^2(\mathbb{R})$ for each $y \geq 0$. A similar analysis shows that f_- has an analytic extension to the lower half-plane, $H_- = \{x + iy \; : \; y < 0\}$, such that $f_-(x + iy) \in L^2(\mathbb{R})$ for each $y \leq 0$. Indeed, it is not hard to show that this decomposition is unique. The precise statement is the following.

Proposition 6.8.1. *Suppose that $F(x + iy)$ is an analytic function in H_+ such that for $y \geq 0$,*

1.

$$\int_{-\infty}^{\infty} |F(x + iy)|^2 dx < M,$$

2.

$$\lim_{y \downarrow 0} \int_{-\infty}^{\infty} |F(x + iy)|^2 dx = 0,$$

then $F \equiv 0$.

Proof. By Theorem 4.4.4, a function satisfying the L^2-boundedness condition has the following property:

$$\hat{F}(\cdot + iy) = \hat{f}(\xi)e^{-y\xi},$$

where $\hat{f}(\xi)$ is the Fourier transform $F(x)$. Moreover, $\hat{f}(\xi) = 0$ if $\xi < 0$. By the Parseval formula,

$$\int_{-\infty}^{\infty} |F(x+iy)|^2\, dx = \int_{0}^{\infty} |\hat{f}(\xi)|^2 e^{-2y\xi}\, d\xi.$$

The second condition implies that $\hat{f}(\xi) = 0$ and therefore $F \equiv 0$. □

If the functions $f_{\pm}$ can be explicitly determined, then $\mathcal{H}f$ can also be computed. If f is a "piece" of an analytic function, then this determination is often possible. The following example is typical.

***Example* 6.8.1.** Let

$$f(x) = \begin{cases} \sqrt{1-x^2} & \text{for } |x| < 1, \\ 0 & \text{for } |x| \geq 1. \end{cases}$$

The analytic function, $\sqrt{1-z^2}$, has a single valued determination in the complex plane minus the subset of $\mathbb{R}$, $\{x : |x| \geq 1\}$. Denote this function by $F(z)$. Of course, $F(x) = f(x)$ for $x \in (-1, 1)$, and the restrictions of F to the upper and lower half-planes, $F_{\pm}$ are analytic. Moreover, for $|x| > 1$ we easily compute that

$$\lim_{\epsilon \downarrow 0}]F_{+}(x + i\epsilon) + F_{-}(x - i\epsilon)] = 0.$$

This would solve our problem but for the fact that $F(x+iy)$ is not in L^2 for any $y \neq 0$. To fix this problem we need to add a correction term that reflects the asymptotic behavior of $F(z)$ for large z. Indeed, if we set

$$f_{\pm}(z) = \frac{1}{2}[F_{\pm}(z) \pm iz],$$

then a simple calculation shows that

$$f_{+}(x) + f_{-}(x) = f(x) \qquad \text{for all real } x$$

and that

$$f_{\pm}(x \pm iy) \simeq \frac{1}{x} \qquad \text{for large } x$$

and therefore $f_{\pm}(x \pm iy) \in L^2(\mathbb{R})$ for all $y > 0$. This allows us to compute the Hilbert transform of f:

$$\mathcal{H}f(x) = f_{+}(x) - f_{-}(x) = \begin{cases} ix & \text{for } |x| < 1, \\ i(x + \sqrt{x^2 - 1}) & \text{for } x < -1, \\ i(x - \sqrt{x^2 - 1}) & \text{for } x > 1. \end{cases} \tag{6.88}$$

Exercise

***Exercise* 6.8.1.** Compute the Hilbert transform of $\chi_{[-1,1]}(x)$. A good place to start is with the formula $\mathcal{H}f = \lim_{\epsilon \downarrow 0} h_\epsilon * f$; see formula (6.48) in Section 6.3.

6.9 Conclusion

We have now completed our analysis of the Radon transform and its inverse. As we shall see in Chapter 11, the filtered back-projection formula is basis for most practical algorithms used in medical image reconstruction. It is built out of three basic operations: the Hilbert transform, differentiation, and back-projection. The Hilbert transform is intimately tied to the theory of analytic functions of a complex variable. Using this connection, we were able to compute several examples and deduce certain properties of the Hilbert transform. Of particular importance for medical imaging is the fact that the Hilbert transform of a function with bounded support never has bounded support. As we saw in several examples, this implies that the reconstruction process for piecewise continuous functions with bounded support entails subtle cancellation between positive and negative parts of the filtered Radon transform.

In addition to the exact inversion formula, we also considered several approaches to finding approximations that could be used on realistic data. Using the convolution theorem for the Radon transform, the approximate reconstruction formulæ were shown to produce smeared-out versions of the original function. An important feature of both the exact and approximate formulæ is that they can be meaningfully applied to a large class of inputs and have good continuity properties.

At this point we have developed a complete though idealized model for the measurement and reconstruction process used in x-ray CT. It is a continuum model that assumes that the Radon transform of the attenuation coefficient can be measured for all (t, ω). In the next chapter we introduce Fourier series and take the first serious step toward a discrete model. Fourier series is an analogue of the Fourier transform for functions defined in bounded intervals or bounded rectangles. Of particular note, for our latter applications, are the facts that the Fourier coefficients are a sequence of numbers and the inversion formula for the Fourier series is expressed as a sum rather than an integral.

Chapter 7

Introduction to Fourier Series

In applications data are never collected along the whole real line or from the entire plane. Real data can only be collected from a bounded interval or planar domain. In order to use the Fourier transform to analyze and filter this type of data, we can either:

- extend the data by cutting it off to equal zero outside of the set over which the data was collected, or
- extend the data periodically.

If the data are extended "by zero" then the Fourier transform is available. If the data are extended periodically then the data do not vanish at infinity, and hence their Fourier transform is not a function.

Fourier series provide a tool for the analysis of functions defined on finite intervals in $\mathbb{R}$ or products of intervals in $\mathbb{R}^n$. The goal of Fourier series is to express an "arbitrary" periodic function as a infinite *sum* of complex exponentials. Fourier series serve as a bridge between continuum models, like that presented in the previous chapter, and *finite* models, which can be implemented on a computer. The theory of Fourier series runs parallel to that of the Fourier transform presented in Chapter 4. After running through the basic properties of Fourier series in the one-dimensional case, we consider the problem of approximating a function by finite partial sums of its Fourier series. This leads to a discussion of the Gibbs phenomenon, which describes the failure of Fourier series to represent functions near to jump discontinuities. The chapter concludes with a brief introduction to Fourier series in $\mathbb{R}^n$.

7.1 Fourier Series in One Dimension*

See: A.5.1, B.2.

To simplify the exposition, we begin with functions defined on the interval [0, 1]. This does not limit the generality of our analysis, for if g is a function defined on an interval

$[a, b]$, then setting $f(x) = g(a+(b-a)x)$ gives a function defined on $[0, 1]$ that evidently contains the same information as g.

***Definition* 7.1.1.** Let f be a complex-valued, *absolutely integrable* function defined on $[0, 1]$. The L^1-norm of f is defined to be

$$\|f\|_{L^1} = \int_0^1 |f(x)|\, dx < \infty.$$

The set of such functions, $L^1([0, 1])$, is a complete normed, linear space with norm defined by $\|\cdot\|_{L^1}$.

***Definition* 7.1.2.** The *Fourier coefficients* of a function f in $L^1([0, 1])$ is the bi-infinite sequence of numbers, $< \hat{f}(n) >$, defined by

$$\hat{f}(n) = \int_0^1 f(x)e^{-2\pi inx}\, dx \quad \text{for} \quad n \in \mathbb{Z}. \tag{7.1}$$

***Example* 7.1.1.** If $f(x) = \cos(2\pi mx)$, then, using the formula $\cos(y) = 2^{-1}(e^{iy} + e^{-iy})$, we easily compute that

$$\hat{f}(n) = \begin{cases} \frac{1}{2} & \text{if } n = \pm m, \\ 0 & \text{if } n \neq \pm m. \end{cases}$$

***Example* 7.1.2.** Let $0 \leq a < b < 1$. Then the Fourier coefficients of $\chi_{[a,b]}(x)$ are

$$\begin{aligned} \widehat{\chi}_{[a,b]}(n) &= \int_a^b e^{-2\pi inx}\, dx \\ &= \begin{cases} (b-a) & \text{if } n = 0, \\ \frac{e^{-2\pi ina} - e^{-2\pi inb}}{2\pi in} & \text{if } n \neq 0. \end{cases} \end{aligned} \tag{7.2}$$

***Example* 7.1.3.** Let $f(x) = \sin(\pi x)$, again using the expression for the sine in terms of exponentials, we compute

$$\hat{f}(n) = \frac{-2}{\pi}\left[\frac{1}{4n^2 - 1}\right]. \tag{7.3}$$

A function f defined on the interval $[0, 1]$ is *even* if

$$f(x) = f(1-x)$$

and *odd* if

$$f(x) = -f(1-x).$$

The symmetry properties of a function are reflected in its Fourier coefficients. These results are summarized in a proposition.

Proposition 7.1.1. *Let f be an L^1-function on* $[0, 1]$.

1. *If f is real valued, then*
$$\hat{f}(-n) = \overline{\hat{f}(n)}.$$
2. *If f is real valued and even, then its Fourier coefficients are real.*
3. *If f is real valued and odd, then its Fourier coefficients are purely imaginary.*

The proof of the proposition is left as an exercise.

To help study the reconstruction of function from its Fourier coefficients, we introduce the *partial sum operator.*

***Definition* 7.1.3.** Let f be an absolutely integrable function on $[0, 1]$. For each positive integer N, define the *Nth-partial sum* of the Fourier series of f to be

$$S_N(f) = \sum_{n=-N}^{N} \hat{f}(n) e^{2\pi i n x}. \tag{7.4}$$

The evaluation of $S_N(f)$ at x is denoted $S_N(f; x)$. For each N the partial sum is a linear operation

$$S_N(f + g) = S_N(f) + S_N(g) \text{ and } S_N(af) = aS_N(f) \text{ for } a \in \mathbb{C}.$$

In applications we work with a fixed partial sum, so it is important to understand in what sense $S_N(f)$ is an approximation to f. The best we might hope for is that

$$\lim_{N\to\infty} S_N(f; x) = f(x)$$

at every point x. At discontinuities of f, such a statement is unlikely to be true. In fact, it can even fail at points where f is continuous. The pointwise convergence of Fourier series is a subtle problem. For the simplest result we make a strong hypothesis about the rate of decay of the Fourier coefficients.

Proposition 7.1.2 (Fourier inversion formula). *If f is a continuous function defined on* $[0, 1]$ *such that the Fourier coefficients of f satisfy*

$$\sum_{n=-\infty}^{\infty} |\hat{f}(n)| < \infty, \tag{7.5}$$

then f is represented, at every point, by its uniformly convergent Fourier series:

$$f(x) = \sum_{n=-\infty}^{\infty} \hat{f}(n) e^{2\pi i n x}, \qquad \text{for all } x \in [0, 1]. \tag{7.6}$$

Proof. The hypothesis (7.5) and the comparison test for infinite sums, Theorem B.2.2, implies that the infinite sum in (7.6) is uniformly convergent for all real numbers. We need to show that it converges to f. For $0 < r < 1$, define the absolutely convergent series

$$P_r(x) = \sum_{n=-\infty}^{\infty} r^{|n|} e^{2\pi i n x} = 1 + 2\,\mathrm{Re}\left[\sum_{n=1}^{\infty} r^n e^{2\pi i n x}\right]. \tag{7.7}$$

Using the second expression and the formula for the sum of a geometric series, we see that

$$P_r(x) = \frac{1 - r^2}{1 - 2r\cos(2\pi x) + r^2}.$$

For $0 \le r < 1$ this formula implies that $P_r(x) > 0$. For each such r define

$$f_r(x) = \int_0^1 P_r(x - y) f(y)\, dy.$$

From the representation of P_r as an infinite sum, we deduce that

$$f_r(x) = \sum_{j=-\infty}^{\infty} \hat{f}(n) r^{|n|} e^{2\pi i n x}.$$

In light of (7.5), the comparison test for infinite sums implies that

$$\lim_{r\uparrow 1} f_r(x) = \sum_{j=-\infty}^{\infty} \hat{f}(n) e^{2\pi i n x},$$

with uniform convergence for $x \in \mathbb{R}$. To complete the proof, we now show that $\lim_{r\uparrow 1} f_r(x) = f(x)$.

For each r

$$\int_0^1 P_r(x)\, dx = 1.$$

This fact and the positivity of P_r imply that

$$\begin{aligned} |f_r(x) - f(x)| &= \left| \int_0^1 P_r(x - y)(f(y) - f(x))\, dy \right| \\ &\le \int_0^1 P_r(x - y)|f(y) - f(x)|\, dy. \end{aligned} \tag{7.8}$$

If $x \neq 0$, then

$$\lim_{r\uparrow 1} P_r(x) = 0.$$

In fact, if $\epsilon > 0$ is fixed, then there is a $0 < \delta$ so that

$$P_r(x) < \epsilon \text{ if } r > 1 - \delta \text{ and } \epsilon < x < 1 - \epsilon. \tag{7.9}$$

The concentration, as r tends to 1, of $P_r(x)$ near $x = 0$ is evident in figure 7.1.

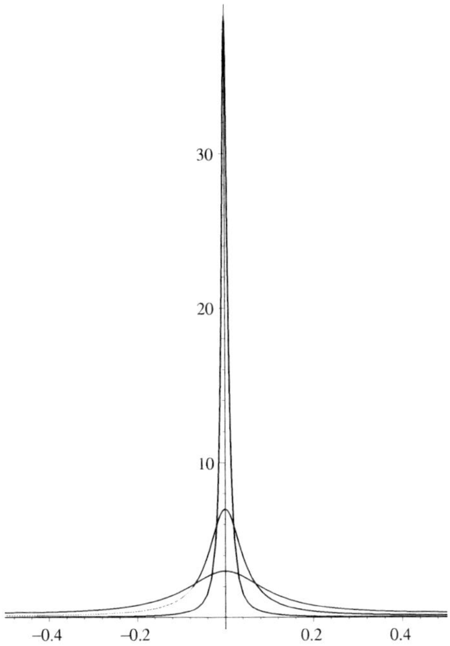

Figure 7.1. Graph of the function P_r for $r = .5, .75$ and $.95$.

To show that the difference $f(x) - f_r(x)$ becomes small, we break the integral into two pieces. One piece is small because f is continuous at x; the other is small because f is bounded and $P_r(t)$ is small, if t is far enough from 0. Since f is continuous, given $\eta > 0$ there is an $\epsilon' > 0$ so that

$$|x - y| < \epsilon' \text{ implies that } |f(x) - f(y)| < \eta.$$

There is also an M so that $|f(y)| \leq M$ for all y. Let $\epsilon = \min\{\epsilon', \eta\}$. From (7.9) it follows that there is a $\delta > 0$ so that $r > 1 - \delta$ implies that

$$\begin{aligned} |f_r(x) - f(x)| &\leq \int_0^1 P_r(x-y)|f(y) - f(x)|\,dy \\ &= \int_{|x-y|<\epsilon} P_r(x-y)|f(y) - f(x)|\,dy + \int_{|x-y|>\epsilon} P_r(x-y)|f(y) - f(x)|\,dy \\ &\leq \eta \int_{|x-y|<\epsilon} P_r(x-y)\,dy + 2M\epsilon \\ &\leq (1 + 2M)\eta. \end{aligned} \tag{7.10}$$

This estimate shows that $\lim_{r \uparrow 1} f_r(x) = f(x)$ and thereby completes the proof of the Proposition. □

***Remark* 7.1.1.** The argument in the second part of the proof does not require f to be everywhere continuous. It shows that, if f is bounded, then $< f_r(x) >$ converges to $f(x)$ at any point of continuity of f. This does not mean that the Fourier series of f also converges to f at such a point. The hypothesis (7.5), on the Fourier coefficients of f, implies that f is a continuous function. Exercise 7.3.16 outlines a proof of the Fourier inversion formula, without the assumption that f is continuous.

As was the case with the Fourier integral, the Fourier coefficients of an absolutely integrable function f may fail to satisfy (7.5). For example, let

$$f(x) = \begin{cases} 1 & \text{if } x \in [0, \frac{1}{2}], \\ 0 & \text{if } x \in (\frac{1}{2}, 1]. \end{cases}$$

This function is piecewise smooth with jump discontinuities at 0 and $\frac{1}{2}$. A calculation shows that

$$\hat{f}(n) = \begin{cases} 0 & \text{if } n \text{ is even and } n \neq 0, \\ \frac{1}{2} & \text{for } n = 0, \\ \frac{1}{\pi i n} & \text{if } n \text{ is odd.} \end{cases}$$

If the Fourier coefficients satisfy (7.5), then the partial sums, $< S_N(f) >$, converge uniformly to f. Since each partial sum is a continuous function, it follows from an elementary result in analysis that the limit is also a continuous function. Because discontinuous data are common in imaging applications, the difficulties of representing such functions in terms of Fourier series is an important topic. It is considered, in detail, in Section 7.5.

Exercises

***Exercise* 7.1.1.** Compute the Fourier coefficients of $\sin(2\pi m x)$.

***Exercise* 7.1.2.** Find a more explicit formula for $\widehat{\chi}_{[\frac{1}{4},\frac{3}{4}]}$.

***Exercise* 7.1.3.** Compute the Fourier coefficients of $\cos(\frac{\pi x}{2})$.

***Exercise* 7.1.4.** Prove Proposition 7.1.1.

***Exercise* 7.1.5.** Show that P_r, defined in (7.7), is nonnegative and has total integral 1 for any $0 \leq r < 1$.

***Exercise* 7.1.6.** Show that if $\epsilon > 0$ is fixed, then there is a $0 < \delta$ so that

$$P_r(x) < \epsilon \text{ if } r > 1 - \delta \text{ and } \epsilon < x < 1 - \epsilon. \tag{7.11}$$

***Exercise* 7.1.7.** Suppose that f is an L^1-function defined on $\mathbb{R}$. For each positive integer N define

$$f_N(x) = f(Nx - (1 - x)N) \text{ for } x \in [0, 1].$$

Show that if $< m_N >$ is a sequence such that $\lim_{N\to\infty} \frac{\pi m_N}{N} = \xi$ then

$$\lim_{N\to\infty} 2N e^{\frac{\pi i m_N}{N}} \hat{f}_N(m_N) = \hat{f}(\xi).$$

***Exercise* 7.1.8.** Suppose that f is a function defined on $[a, b]$. Give definitions for such a function to be even or odd.

***Exercise* 7.1.9.** Use the concept of a set of measure zero to explain why it is unreasonable to expect that $S_N(f; x)$ will converge to $f(x)$ at a point where f is discontinuous.

7.2 The Decay of Fourier Coefficients

See: A.4.1, B.2.

As with the Fourier transform, the rate of decay of the Fourier coefficients is determined by the smoothness of the function. The only general result on the decay of Fourier coefficients for absolutely integrable functions is the Riemann-Lebesgue lemma.

Theorem 7.2.1 (Riemann-Lebesgue lemma). *If f belongs to $L^1([0, 1])$, then $\hat{f}(n) \to 0$ as $|n| \to \infty$.*

Proof. Extending a function f by zero defines a function F in $L^1(\mathbb{R})$:

$$F(x) = \begin{cases} f(x) & \text{for } x \in [0, 1], \\ 0 & \text{for } x \notin [0, 1]. \end{cases}$$

The Fourier transform of F is given by

$$\hat{F}(\xi) = \int_0^1 f(x) e^{-ix\xi} \, dx,$$

and therefore the Fourier coefficients of f can be expressed in terms of $\hat{F}$ by

$$\hat{f}(n) = \hat{F}(2\pi n).$$

Using this relation and Theorem 4.2.2, applied to $\hat{F}$, we conclude that

$$\lim_{|n| \to \infty} \hat{f}(n) = 0.$$

□

The Fourier coefficients satisfy a simple estimate, which we record as a proposition.

Proposition 7.2.1. *If f belongs to $L^1([0, 1])$, then*

$$|\hat{f}(n)| \leq \|f\|_{L^1} \qquad \text{for all } n \in \mathbb{N}. \tag{7.12}$$

The Riemann-Lebesgue lemma does not say that $< \hat{f}(n) >$ goes to zero at some particular rate, say faster than $< n^{-1/3} >$. In fact, there is a theorem saying that for any bi-infinite sequence, $<a_n>$, with

$$\lim_{|n|\to\infty} a_n = 0,$$

there exists an integrable function, f, whose Fourier coefficients $< \hat{f}(n) >$ satisfy $|\hat{f}(n)| \geq |a_n|$ for all n. This shows that Fourier coefficients can go to zero arbitrarily slowly; see [79, section I.4]. To obtain more precise results, we need to make assumptions about the regularity of f.

Exercises

***Exercise* 7.2.1.** Explain why (7.12) is a "continuity" result for the map $f \mapsto < \hat{f}(n) >$.

***Exercise* 7.2.2.** Use summation by parts *twice* to show that

$$f(x) = \sum_{n=2}^{\infty} \frac{\cos(2\pi nx)}{\log n}$$

represents a nonnegative, integrable function. In light of this, it is a remarkable fact that

$$\sum_{n=2}^{\infty} \frac{\sin(2\pi nx)}{\log n}$$

does *not* represent an absolutely integrable function!

7.2.1 Periodic Extension

The definition of the Fourier coefficients, (7.1), only requires that f be defined on the interval $[0, 1]$. A function *represented* by a Fourier series is automatically defined for all real numbers as a periodic function. This is clear if the sum in (7.6) converges uniformly. The fact that

$$e^{2\pi in(x+1)} = e^{2\pi inx} \qquad \text{for all } n \text{ and } x \in \mathbb{R}$$

implies that the infinite sum defines a function on $\mathbb{R}$ that is continuous and 1-periodic; that is,

$$f(x+1) = f(x) \qquad \text{for all } x \in \mathbb{R}.$$

This shows that, *when discussing Fourier series,* we should only consider a function on $[0, 1]$ to be continuous if both

$$\lim_{y\to x} f(y) = f(x) \qquad \text{for } x \in (0, 1), \text{ and}$$
$$\lim_{x\to 0^+} f(x) = f(0) = f(1) = \lim_{x\to 1^-} f(x).$$

That is, we think of f as a 1-periodic function *restricted to* the interval $[0, 1]$.

On the other hand, a function defined on $[0, 1)$ can be *extended* to define a 1-periodic function. If x is in $[1, 2)$, then we define

$$f(x) \stackrel{d}{=} f(x-1);$$

now for x in $[-2, 0)$ we can define

$$f(x) \stackrel{d}{=} f(x+2), \text{ and so on.}$$

It is clear that the values of f in $[0, 1)$ uniquely specify a 1-periodic function on the whole real line.

***Definition* 7.2.1.** If a function, f, is defined on $[0, 1)$, then its 1-periodic extension to $\mathbb{R}$ is the function $\tilde{f}$ such that $\tilde{f}(x) = f(x)$ for x in $[0, 1)$ and

$$\tilde{f}(x+n) = \tilde{f}(x) \qquad \text{for all } n \in \mathbb{Z}. \tag{7.13}$$

The 1-periodic extension of f is also denoted by f.

If f is a continuous function on $[0, 1]$ in the usual sense, then the condition $f(0) = f(1)$ is equivalent to the condition that the 1-periodic extension of f to $\mathbb{R}$ is continuous.

***Example* 7.2.1.** The function $f(x) = x$ is a continuous function on $[0, 1]$ but its 1-periodic extension, shown in Figure 7.2(a), is not.

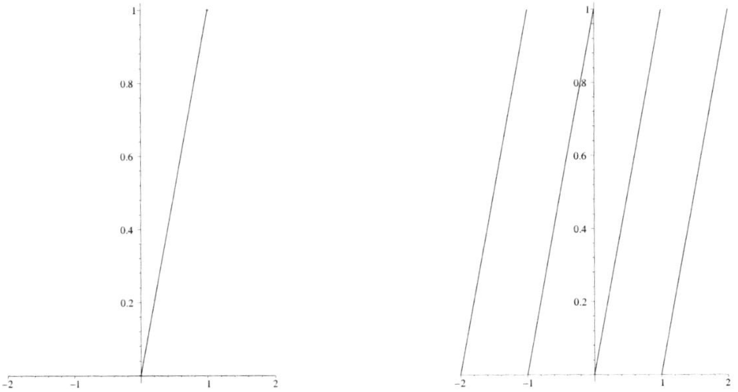

(a) One period of x. (b) The 1-periodic extension of x.

Figure 7.2. Periodic extension may turn a continuous function into discontinuous function.

These considerations easily extend to the derivatives of f. Suppose that f is differentiable on $[0, 1]$. This means that f has continuous derivative in $(0, 1)$ and the following limits exist:

$$\begin{aligned} f'_{+}(0) &= \lim_{x\to 0^{+}} \frac{f(x)-f(0)}{x}, \text{ where } \lim_{x\to 0^{+}} f'(x) = f'_{+}(0) \text{ and} \\ f'_{-}(1) &= \lim_{x\to 1^{-}} \frac{f(1)-f(x)}{1-x}, \text{ where } \lim_{x\to 1^{-}} f'(x) = f'_{-}(1); \end{aligned} \tag{7.14}$$

$f'_+(0)$ and $f'_-(1)$ are right and left derivatives, respectively. The 1-periodic extension of f has both left and right derivatives at each integer point. Using the definition, we see that the left derivative at 1 is $f'_-(1)$ while the right derivative is $f'_+(0)$. The 1-periodic extension has a continuous derivative provided that $f'_+(0) = f'_-(1)$. Higher-order left and right derivatives are defined recursively. In order for a k-times differentiable function f defined on $[0, 1]$ to be k-times differentiable *as a 1-periodic function*, it is necessary that

$$f_+^{[j]}(0) = f_-^{[j]}(1) \qquad \text{for } j = 0, 1, \dots, k. \tag{7.15}$$

***Example* 7.2.2.** The 1-periodic extension, shown in Figure 7.3(b), of the function $\sin(\pi x)$ is continuous but *not* differentiable.

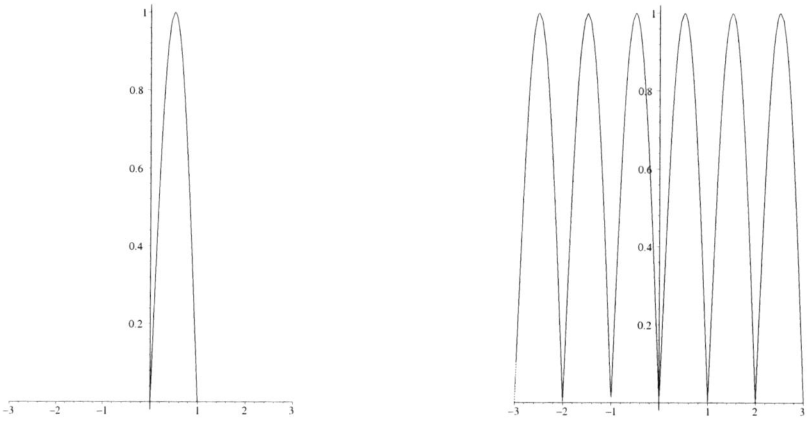

(a) One period of $\sin(\pi x)$.

(b) The 1-periodic extension of $\sin(\pi x)$

Figure 7.3. A continuous periodic extension may not be differentiable.

Exercises

***Exercise* 7.2.3.** Let f be defined as the function $f(x) = x$ for x in the interval $[-\frac{1}{2}, \frac{1}{2})$. Plot its 1-periodic extension.

***Exercise* 7.2.4.** Give precise definitions for the higher-order left and right derivatives of a function.

***Exercise* 7.2.5.** Show that if f is a k-times differentiable function defined on $[0, 1]$ that satisfies (7.15), then its 1-periodic extension is a $\mathscr{C}^k$-function on $\mathbb{R}$.

7.2.2 Fourier Coefficients of Differentiable Functions

Suppose that f is continuously differentiable on $[0, 1]$; for the moment we do *not* assume that $f(0) = f(1)$. Integrating by parts gives

$$\begin{aligned}\hat{f}(n) = \int_0^1 f(x)e^{-2\pi inx}\,dx &= \frac{1}{-2\pi in} f(x)e^{-2\pi inx}\Big|_0^1 + \frac{1}{2\pi in}\int_0^1 f'(x)e^{-2\pi inx}\,dx \\ &= \frac{f(1) - f(0)}{-2\pi in} + \frac{1}{2\pi in}\int_0^1 f'(x)e^{-2\pi inx}\,dx.\end{aligned} \tag{7.16}$$

In other words,

$$\hat{f}(n) = \frac{1}{2\pi in}[f(0) - f(1)] + \frac{\widehat{f'}(n)}{2\pi in}.$$

Since f is continuously differentiable in $[0, 1]$, its first derivative is an integrable function. By the Riemann-Lebesgue lemma, the Fourier coefficients of f' go to zero as n goes to infinity. This shows that if $f(0) \neq f(1)$, then $\hat{f}(n)$ decays exactly as $1/n$. The $1/n$ rate of decay is characteristic of a function with a simple jump discontinuity. If $f(0) = f(1)$, then the Fourier coefficients of f are given by

$$\hat{f}(n) = \frac{\widehat{f'}(n)}{2\pi in},$$

which therefore go to zero faster than $1/n$.

If f has $(k - 1)$-derivatives, as a 1-periodic function, then the integration by parts in (7.16) can be repeated to obtain the following:

Theorem 7.2.2. *If f belongs to $\mathcal{C}^k([0, 1])$ and*

$$f_+(0) = f_-(1),\ f'_+(0) = f'_-(1), \ldots, f_+^{(k-1)}(0) = f_-^{(k-1)}(1),$$

then

$$\hat{f}(n) = \frac{\widehat{f^{(k)}}(n)}{(2\pi in)^k} \quad \text{for } n \neq 0. \tag{7.17}$$

The Riemann-Lebesgue lemma then implies that if f is k-times continuously differentiable on $[0, 1]$, periodic in the appropriate sense, then $< \hat{f}(n) >$ decays *faster* than $< n^{-k} >$. This result has a partial converse.

Theorem 7.2.3. *If f is in $L^1([0, 1])$ and there is a constant C and an $\epsilon > 0$, so that*

$$|\hat{f}(n)| \leq \frac{C}{(1 + |n|)^{k+\epsilon}},$$

then f is in $\mathcal{C}^{(k-1)}([0, 1])$ with $f(0) = f(1)$, $f'_+(0) = f'_-(1), \ldots, f_+^{(k-1)}(0) = f_-^{(k-1)}(1)$.

Proof. Using the comparison test, Theorem B.2.2, the estimates on $<\hat{f}(n)>$ imply that the series $\sum \hat{f}(n)e^{2\pi inx}$ and its jth derivatives, for $0 \leq j \leq k-1$, converge absolutely and uniformly. Theorem A.4.1 implies that

$$\sum_{n=-\infty}^{\infty} \hat{f}(n)e^{2\pi ix}$$

represents a $(k-1)$-times, continuously differentiable function and that we can differentiate, term by term, to obtain

$$f^{[j]}(x) = \sum_{n=-\infty}^{\infty} (2\pi in)^j \hat{f}(n)e^{2\pi inx}.$$

See [111]. □

As before, formula (7.17) can be viewed as a formula for the Fourier coefficients of $f^{[j]}$ in terms of those of f.

Corollary 7.2.1. *If f has k integrable, derivatives on $[0,1]$ with $f_+^{[j]}(0) = f_-^{[j]}(1)$ for $0 \leq j \leq k-1$, then, for $j \leq k$,*

$$\widehat{f^{[j]}}(n) = [2\pi in]^j \hat{f}(n). \tag{7.18}$$

For the case of the Fourier series, it is important that the derivatives are also periodic functions.

***Example* 7.2.3.** Let $f(x) = x(1-x)$ for $x \in [0,1]$. Then $f(0) = f(1)$ but $f'(x) = 1-2x$, is not continuous as a 1-periodic function. The Fourier coefficients of f are

$$\hat{f}(n) = \begin{cases} \frac{1}{6}, & \text{for } n = 0, \\ \frac{-1}{2\pi^2 n^2}, & \text{for } n \neq 0. \end{cases} \tag{7.19}$$

The Fourier coefficients of f' are

$$\widehat{f'}(n) = \begin{cases} 0, & \text{for } n = 0, \\ \frac{1}{\pi in}, & \text{for } n \neq 0, \end{cases} \tag{7.20}$$

showing that $\widehat{f'}(n) = (2\pi in)\hat{f}(n)$. Note the relationship between the smoothness of f, as a 1-periodic function, and the decay of its Fourier coefficients.

***Example* 7.2.4.** If we set

$$f(x) = x - n \text{ for } x \in (n, n+1],$$

then f does not have *any* periodic derivatives. The Fourier coefficients of f are given by

$$\hat{f}(n) = \begin{cases} \frac{1}{2} \text{ for} & n = 0, \\ \frac{i}{2\pi n} \text{ for} & n \neq 0. \end{cases}$$

They display the $\frac{1}{n}$-rate of decay, which is characteristic of functions with simple jump discontinuities. On the interval $(0,1)$, $f'(x)$ equals 1; note, however, that $\widehat{f'}(n) \neq (2\pi in)\hat{f}(n)$, for *any* n.

7.3 L^2-Theory

Theorem 7.2.3 is not the exact converse to Theorem 7.2.2, though it is closely analogous to the results for the Fourier transform and reflects the subtlety of pointwise convergence for Fourier series. Simpler statements are obtained by using the L^2-norm. Recall that

$$L^2([0,1]) = \{f : \int_0^1 |f(x)|^2\,dx < \infty\};$$

with the norm defined on $L^2([0,1])$ by

$$\|f\|_2 = \left[\int_0^1 |f(x)|^2 dx\right]^{1/2}.$$

With this norm $L^2([0,1])$ is a complete normed linear space. An element of $L^2([0,1])$ is called a square-integrable or square-summable function.

7.3.1 Geometry in $L^2([0,1])$

See: A.2.5.

The norm on $L^2([0,1])$ is defined by the inner product,

$$\langle f, g\rangle_{L^2} = \int f(x)\overline{g(x)}\,dx \text{ by } \|f\|_2^2 = \langle f, f\rangle_{L^2}.$$

The Cauchy-Schwarz inequality holds for functions in $L^2([0,1])$.

Theorem 7.3.1 (Cauchy-Schwarz inequality). *If f, g are two functions in $L^2([0,1])$, then*

$$|\langle f, g\rangle_{L^2}| \leq \|f\|_2\|g\|_2. \tag{7.21}$$

The proof is exactly as for the $\mathbb{C}^n$-case given in Section 2.3.2. A consequence of the Cauchy-Schwarz inequality is that $L^2([0,1])$ is a subspace of $L^1([0,1])$. If f belongs to $L^2([0,1])$, then

$$\begin{aligned}\|f\|_{L^1} &= \int_0^1 |f(x)|\,dx \\ &\leq \left[\int_0^1 |f(x)|^2\,dx\right]^{\frac{1}{2}}\left[\int_0^1 1\cdot dx\right]^{\frac{1}{2}}.\end{aligned} \tag{7.22}$$

Because the L^2-norm is defined by an inner product, it is possible to do many finite-dimensional Euclidean geometric constructions in L^2. As earlier, the Cauchy-Schwarz inequality implies that

$$\frac{|\langle f, g\rangle_{L^2}|}{\|f\|_2\|g\|_2} \leq 1.$$

We can therefore define an angle θ between f and g in $L^2([0, 1])$ by setting

$$\cos\theta = \frac{|\langle f, g\rangle_{L^2}|}{\|f\|_2\|g\|_2}.$$

This angle is usually assumed to take values in $[0, \frac{\pi}{2}]$. Two functions $f, g \in L^2([0, 1])$ are orthogonal if the angle between them is $\frac{\pi}{2}$; that is,

$$\langle f, g\rangle_{L^2} = 0.$$

For example,

$$\langle e^{2\pi inx}, e^{2\pi imx}\rangle_{L^2} = \int_0^1 e^{2\pi i(n-m)x}\,dx = \begin{cases} 1 & \text{if } n = m, \\ 0 & \text{if } n \neq m. \end{cases} \tag{7.23}$$

In other words, the functions $\{e^{2\pi inx} \mid n \in \mathbb{Z}\}$ are pairwise orthogonal and each has length 1.

If V is a subspace of $\mathbb{C}^n$, then the set of vectors orthogonal to V is also a subspace. It is called the *orthogonal complement* of V and is usually denoted by $V^\perp$:

$$V^\perp = \{z \in \mathbb{C}^n : \langle z, v\rangle = 0 \quad \text{for every } v \in V\}.$$

Every vector in $\mathbb{C}^n$ has a unique representation as

$$z = z_0 + z_1, \quad \text{where } z_0 \in V \text{ and } z_1 \in V^\perp. \tag{7.24}$$

Define a map from $\mathbb{C}^n$ to itself by setting

$$P_V(z) = z_0.$$

This is a linear map with range equal to V. It has several properties:

1. $P_V^2 = P_V$.

2. If $z \in V$, then $P_V(z) = z$.

3. If $z \in V^\perp$, then $P_V(z) = 0$.

Any linear map with the first property is called a projection. A map with all these properties is called the orthogonal projection onto V. The vector $P_V(\boldsymbol{z})$ has a variational characterization: $P_V(\boldsymbol{z})$ is the vector in V that minimizes the distance, $\|\boldsymbol{z}-\boldsymbol{v}'\|$, among all $\boldsymbol{v}'$ in V. This construction generalizes to L^2.

If $S \subset L^2([0, 1])$ is a subspace then we say that f is orthogonal to S if

$$\langle f, g\rangle_{L^2} = 0 \qquad \text{for every } g \in S.$$

***Definition* 7.3.1.** Let S be a subspace of $L^2([0, 1])$. The *orthogonal complement* of S, denoted $S^\perp$, is the subspace of $L^2([0, 1])$ consisting of all functions orthogonal to S,

$$S^\perp = \{g \in L^2([0, 1]) \, : \, \langle f, g\rangle = 0 \qquad \text{for all } f \in S\}.$$

As in finite dimensions, a subspace S defines a linear map P_S, satisfying the properties enumerated previously. That is, $P_S^2 = P_S$; for $f \in S$, $P_S(f) = f$; and for $f \in S^\perp$, $P_S(f) = 0$. In finite-dimensional linear algebra the existence of orthogonal projections is an algebraic fact; it requires no analysis. Using the variational characterization of $P_S(f)$, it can be shown, using estimates, that the orthogonal projection onto a closed subspace of $L^2([0, 1])$ always exists; see [42].

If f is an L^2-function on $[0, 1]$, then it is also absolutely integrable; hence its Fourier coefficients are defined. In this case the Fourier coefficients of f go to zero sufficiently fast to make $\sum_{-\infty}^{\infty} |\hat{f}(n)|^2$ converge. Once again, we have a Parseval formula.

Theorem 7.3.2 (Parseval formula). *If f belongs to $L^2([0, 1])$, then*

$$\int_0^1 |f(x)|^2 dx = \sum_{n=-\infty}^{\infty} |\hat{f}(n)|^2.$$

Proof. Again the theorem is simple to prove for a special class of L^2-functions; in this instance, the trigonometric polynomials,

$$\mathcal{T} = \{\sum_{j=-N}^{N} c_j e^{2\pi i j x} \, : \, c_j \in \mathbb{C}, \quad N \in \mathbb{N}\}.$$

If f belongs to $\mathcal{T}$, then multiplying out the finite sum defining $f\bar{f}$ gives

$$|f(x)|^2 = \sum_{j,k=-N}^{N} c_j \bar{c}_k e^{2\pi i (j-k)x}. \tag{7.25}$$

Integrating both sides of (7.25), using (7.23), gives

$$\int_0^1 |f(x)|^2 dx = \sum_{j=-N}^{N} |c_j|^2.$$

This is the Parseval formula for f in $\mathcal{T}$.

To complete the proof we need two additional facts. The first is that an arbitrary L^2-function is well approximated (in the L^2-norm) by trigonometric polynomials.

Lemma 7.3.1. *If f belongs to $L^2([0,1])$ and $\epsilon > 0$ is given, then there is a trigonometric polynomial, g so that $\|f - g\|_{L^2} < \epsilon$.*

The second is Bessel's inequality. It states that among functions of the form

$$g_N = \sum_{n=-N}^{N} c_n e^{2\pi i n x}$$

the Nth-partial sum of the Fourier series of f minimizes the error,

$$\|f - g_N\|_{L^2}.$$

The lemma is proved in Section 7.5.3 and Bessel's inequality is proved in Section 7.3.3.

The definition of the Fourier coefficients implies that

$$0 \leq \|f - S_N(f)\|_{L^2}^2 = \|f\|_{L^2}^2 - \|S_N(f)\|_{L^2}^2$$

and therefore

$$\|S_N(f)\|_{L^2}^2 \leq \|f\|_{L^2}^2.$$

In particular, using the result for trigonometric polynomials and letting N tend to infinity, we deduce that

$$\sum_{n=-\infty}^{\infty} |\hat{f}(n)|^2 \leq \int_0^1 |f(x)|^2\, dx. \tag{7.26}$$

On the other hand, the triangle inequality gives the estimate

$$\|f\|_{L^2} \leq \|f - S_N(f)\|_{L^2} + \|S_N(f)\|_{L^2}.$$

Bessel's inequality, Lemma 7.3.1, and the result for trigonometric polynomials now show that, for any $\epsilon > 0$,

$$\int_0^1 |f(x)|^2\, dx \leq \sum_{n=-\infty}^{\infty} |\hat{f}(n)|^2 + \epsilon.$$

Together these inequalities complete the proof of Parseval's formula. □

A bi-infinite sequence $< a_n >$ is *square summable* if

$$\| < a_n > \|_{l_2} = \sqrt{\sum_{n=-\infty}^{\infty} |a_n|^2}$$

is finite. The set of square summable sequences is denoted l^2. With the norm $\|\cdot\|_{l_2}$ it is a complete normed linear space. The Parseval formula states that the map from $L^2([0,1])$ to l^2 defined by $f \mapsto < \hat{f}(n) >$ is an isometry. It can be viewed as a criterion for a sequence to be the Fourier coefficients of a square-summable function. For example, $< |n|^{-1/2} >$ cannot be the Fourier coefficients of an L^2 function because

$$\sum_{n=1}^{\infty} \frac{1}{n} = \infty.$$

The Parseval formula has a simple geometric interpretation.

Theorem 7.3.3. *The set of exponentials* $\{e^{2\pi inx} : n = -\infty, \dots, \infty\}$ *is an orthonormal basis for* $L^2([0,1])$.

***Remark* 7.3.1.** The Parseval formula should therefore be regarded as an infinite-dimensional version of Pythagoras' theorem.

Exercises

***Exercise* 7.3.1.** Let $V \subset \mathbb{C}^n$ be a subspace. Prove that $V^\perp$ is a subspace and formula (7.24). Find an explicit formula for P_V. *Hint:* Introduce orthonormal bases.

***Exercise* 7.3.2.** Let $V \subset \mathbb{C}^n$ be a subspace. Prove that $P_V(z)$ is the vector in V closest to z.

***Exercise* 7.3.3.** Let $V, W \subset \mathbb{C}^n$ be subspaces; suppose that $V \cap W = \{0\}$ and $\dim V + \dim W = n$. Show that every vector $z \in \mathbb{C}^n$ has a unique representation $z = \boldsymbol{v} + \boldsymbol{w}$, where $\boldsymbol{v} \in V$ and $\boldsymbol{w} \in W$. Show that the linear map $M_V(z) = \boldsymbol{v}$ is a projection. When is M_V the orthogonal projection onto V?

***Exercise* 7.3.4.** Let V be a one-dimensional subspace of $\mathbb{C}^2$. Show that there exist subspaces, W, satisfying the hypotheses of Exercise 7.3.3 so that the ratio $\|M_V z\| \|z\|^{-1}$ can be made as large as you like.

***Exercise* 7.3.5.** Let $S \subset L^2([0,1])$ be a subspace. Show that $S^\perp$ is also a subspace.

***Exercise* 7.3.6.** Let $S \subset L^2([0,1])$ be a subspace. Show that $\mathrm{Id} - P_S$ is the orthogonal projection onto $S^\perp$

***Exercise* 7.3.7.** Let $S \subset L^2([0,1])$ be a subspace. Show that for any f, g in $L^2([0,1])$, the following identity holds:

$$\langle P_S f, g\rangle_{L^2} = \langle f, P_S g\rangle_{L^2}.$$

In other words, the orthogonal projection onto S, P_S, is a self-adjoint operator. Show that $\|P_S f\|_{L^2} \leq \|f\|$, for every $f \in L^2([0,1])$.

***Exercise* 7.3.8.** Suppose that $S \subset L^2([0,1])$ is the subspace defined as the scalar multiples of a single function f. Find an explicit formula for P_S.

***Exercise* 7.3.9.** Prove Hölder's inequality for l^2: If $< a_n >$ and $< b_n >$ are square summable sequences, then

$$\left|\sum_{n=-\infty}^{\infty} a_n \bar{b}_n\right| \leq \sqrt{\sum_{n=-\infty}^{\infty} |a_n|^2 \sum_{n=-\infty}^{\infty} |b_n|^2}.$$

***Exercise* 7.3.10.** Using the Parseval formula and the function in Example 7.2.4, prove that

$$\sum_{n=1}^{\infty} \frac{1}{n^2} = \frac{\pi^2}{6}. \tag{7.27}$$

***Exercise* 7.3.11.** Let $< a_n >$ belong to l^2. Show that there is a function $f \in L^2[0,1]$ with $\hat{f}(n) = a_n$, for all n.

7.3.2 The L^2-Inversion formula

For any f in $L^2([0, 1])$, each partial sum, $S_N(f)$, is a very nice function; it is infinitely differentiable and all of its derivatives are periodic. For a general function, f, in $L^2([0, 1])$ and point x in $[0, 1]$, we therefore do *not* expect that

$$f(x) = \lim_{N\to\infty} S_N(f; x).$$

In fact, the partial sums may not converge pointwise to a limit at all.

***Example* 7.3.1.** Define a sequence of coefficients by setting $\hat{f}(n) = n^{-\frac{3}{4}}$ for $n > 0$ and zero otherwise. Because

$$\sum_{n=1}^{\infty} [n^{-\frac{3}{4}}]^2 < \infty,$$

these are the Fourier coefficients of an L^2-function. However,

$$\sum_{n=1}^{\infty} n^{-\frac{3}{4}} = \infty.$$

The Fourier coefficients of an L^2-function $< \hat{f}(n) >$ do not generally go to zero fast enough to make the series

$$\sum_{n=-\infty}^{\infty} \hat{f}(n)e^{2\pi inx}$$

converge pointwise.

This means that we need to find a different way to understand the convergence of the Fourier series for L^2-functions. Parseval's formula implies that the Fourier series of an L^2-function converges to the function in the L^2-norm.

Proposition 7.3.1 (L^2-inversion formula). *If f belongs to $L^2([0, 1])$, then*

$$\lim_{M,N\to\infty} \|f - \sum_{j=-M}^{N} \hat{f}(j)e^{2\pi ijx}\|_{L^2} = 0. \tag{7.28}$$

***Remark* 7.3.2.** As before, it is said that the Fourier series of f converges to f *in the mean*; this is denoted

$$\operatorname*{LIM}_{M,N\to\infty} \sum_{j=-M}^{N} \hat{f}(j)e^{2\pi ijx} = f(x).$$

Proof. Given the Parseval formula, the proof is a simple computation:

$$\|f - \sum_{j=-M}^{N} \hat{f}(j)e^{2\pi ijx}\|_{L^2}^2 = \|f\|_{L^2}^2 - \sum_{j=-M}^{N} |\hat{f}(j)|^2. \tag{7.29}$$

From the Parseval formula it follows that

$$\|f\|_{L^2}^2 - \sum_{j=M}^{N} |\hat{f}(j)|^2 = \sum_{j=-\infty}^{-(M+1)} |\hat{f}(j)|^2 + \sum_{j=N+1}^{\infty} |\hat{f}(j)|^2. \tag{7.30}$$

As the sum $\sum_{-\infty}^{\infty} |\hat{f}(j)|^2$ is finite, the right-hand side in (7.30) tends to zero as M and N tend to infinity. □

If we measure the distance between two functions in the L^2 sense, then it is shown in Proposition 7.3.1 that the distance between f and the partial sums of its Fourier series goes to zero as N tends to ∞. This means that, in some average sense, $< S_N(f) >$ converges to f. While not as simple as pointwise convergence, this concept is well adapted to problems in which measurement is a serious consideration. Given a function f and a point x, the difference

$$|f(x) - S_N(f;x)|$$

cannot be exactly measured. A reasonable mathematical model for what actually can be measured is an average of such differences. For example, we let

$$g_\epsilon(y) = \begin{cases} 0 & \text{if } |x-y| > \epsilon, \\ \frac{1}{2\epsilon} & \text{if } |x-y| \leq \epsilon. \end{cases}$$

Note that $g_\epsilon(y) \geq 0$ for all y and

$$\int_0^1 g_\epsilon(y)\,dy = 1;$$

the positive number ϵ reflects the resolution of the measuring apparatus. A reasonable measure for the size of the error $|S_N(f;x) - f(x)|$ is given by the average

$$\int_0^1 |f(y) - S_N(f;y)| g_\epsilon(y)\,dy \leq \|f - S_N(f)\|_2 \|g_\epsilon\|_2 = \frac{1}{\sqrt{2\epsilon}} \|f - S_N(f)\|_2.$$

This estimate for the error is an application of (7.21). For a *fixed resolution*, we see that, as $N \to \infty$, the *measured* difference between f and $S_N(f)$ goes to zero. Any function g in $L^2([0,1])$ defines a measurement,

$$m_g(f) = \langle f, g \rangle_{L^2}.$$

In a precise sense any continuous, scalar-valued, linear measurement is of this form. For any fixed function g,

$$\lim_{N\to\infty} m_g(S_N(f)) = m_g(f).$$

Exercises

***Exercise* 7.3.12.** Prove (7.29).

***Exercise* 7.3.13.** Given any function $g \in L^2([0, 1])$, define a *measurement* by setting $m_g(f) = \langle f, g\rangle_{L^2}$. Show that for any $f \in L^2([0, 1])$,

$$\lim_{N\to\infty} m_g(S_N(f)) = m_g(f).$$

7.3.3 Bessel's Inequality

See: A.5.1.

A fundamental issue, both in proving theorems and applying mathematical techniques to real-world problems, is that of approximation. In general, a function in $L^2([0, 1])$ has an infinite number of nonzero Fourier coefficients. This means that most functions cannot be *exactly* represented by a finite sum of exponentials. As we can only handle a finite amount of data, we often need to find the "best" way to approximate a function by a finite sum of exponential functions. What is meant by the "best approximation" is determined by how the error in the approximation is measured.

For each N we define the space of exponential polynomials of degree N to be

$$\mathcal{T}_N = \{\sum_{n=-N}^{N} a_n e^{2\pi i n x} \ : \ a_n \in \mathbb{C}\}.$$

Let l denote a norm on a space of functions defined on $[0, 1]$. The norm defines a distance by setting $d_l(f, g) = l(f - g)$. Each choice of a norm l gives an approximation problem:

> Given a function f with $l(f) < \infty$, find the function $g_f \in \mathcal{T}_N$ such that
>
> $$d_l(f, g_f) = \min\{d_l(f, g) \ : \ g \in \mathcal{T}_N\}.$$
>
> That is, find the point in $\mathcal{T}_N$ whose d_l-distance to f is as small as possible. The minimum value $d_l(f, g_f)$ is called the *error* in the approximation.

The ease with which such a problem is solved depends largely on the choice of l. For most choices of norm this problem is very difficult to solve; indeed is not solvable in practice. We usually have to settle for finding a sequence of approximants $< g_N >$ for which the errors $< d_l(f, g_N) >$ go to zero at essentially the same *rate* as the optimal error. A notable exception is L^2. The following theorem gives the answer if the error is measured in the L^2-norm.

Theorem 7.3.4 (Bessel's inequality). *Given a function f in $L^2([0,1])$ and complex numbers $\{a_{-N},\ldots,a_N\}$, the following inequality holds:*

$$\|f - \sum_{n=-N}^{N} \hat{f}(n)e^{2\pi inx}\|_2 \le \|f - \sum_{n=-N}^{N} a_n e^{2\pi inx}\|_2$$

with equality if and only if $a_n = \hat{f}(n)$ for all $n \in \{-N,\ldots,N\}$.

Proof. Using the relation

$$\|f+g\|_2^2 = \langle f+g, f+g\rangle_{L^2} = \|f\|_2^2 + 2\,\mathrm{Re}\langle f,g\rangle_{L^2} + \|g\|_2^2,$$

we have

$$\|f - \sum_{n=-N}^{N} a_n e^{2\pi inx}\|_2^2 - \|f - \sum_{n=-N}^{N} \hat{f}(n)e^{2\pi inx}\|_2^2 = \|\sum_{n=-N}^{N} (a_n - \hat{f}(n))e^{2\pi inx}\|_2^2 \ge 0.$$

The equality holds if and only if $a_n = \hat{f}(n)$ for $-N \le n \le N$. □

Another way to say this is that for every N, the partial sum $S_N(f)$ gives the best L^2-approximation to f among functions in $\mathcal{T}_N$. A consequence of the proof of Bessel's inequality is that

$$\langle f - S_N(f), g\rangle_{L^2} = 0 \qquad \text{for any } g \in \mathcal{T}_N.$$

That is, the error $f - S_N(f)$ is orthogonal to the subspace $\mathcal{T}_N$. This gives another description of $S_N(f)$ as the L^2-orthogonal projection of f onto the subspace $\mathcal{T}_N$.

Proposition 7.3.2. *The map $f \mapsto S_N(f)$ is the L^2-orthogonal projection onto $\mathcal{T}_N$.*

Exercises

***Exercise* 7.3.14.** Prove Proposition 7.3.2.

***Exercise* 7.3.15.** For each of the norms

$$\|f\|_p = \left[\int_0^1 |f(x)|^p dx\right]^{\frac{1}{p}}, \qquad 1 < p < \infty$$

find the variational condition characterizing the function $g_N \in \mathcal{T}_N$ that minimizes the error $\|f - g_N\|_p$. Explain why these problems are very difficult to solve if $p \ne 2$.

7.3.4 L^2-Derivatives*

See: A.4.3, A.4.5.

In Section 4.3 we introduced notions of weak and L^2-derivatives for functions defined on $\mathbb{R}$. Here we consider what this means for functions defined on a finite interval. We could once again use integration by parts to define a weak derivative, but, on a finite interval, this is complicated by the presence of boundary terms. There are, in fact, several different notions of weak differentiability for functions defined on a bounded interval. As we are mostly interested in the relationship between L^2-differentiability and the behavior of the Fourier coefficients, we handle the boundary terms by regarding f as a 1-periodic function. That is, we use the boundary condition $f(0) = f(1)$.

***Definition* 7.3.2.** An L^2-function f is said to have a derivative in $L^2([0, 1])$ if there is a function $g \in L^2([0, 1])$ such that

$$f(x) = f(0) + \int_0^x g(s)\,ds \qquad \text{for every } x \in [0, 1]$$

and $f(0) = f(1)$.

The definition can be applied recursively to define the class of functions with k L^2-derivatives.

***Definition* 7.3.3.** A periodic function $f \in L^2([0, 1])$ has k L^2-derivatives if there are functions $f_j \in L^2([0, 1])$ for $j = 1, \dots k$ such that

- $$f(0) = f(1) \text{ and } f_j(0) = f_j(1) \qquad \text{for } j = 1, \dots, k-1,$$
- $$f(x) = f(0) + \int_0^x f_1(s)\,ds \qquad \text{for every } x \in [0, 1], \tag{7.31}$$
- $$f_{j-1}(x) = f_{j-1}(0) + \int_0^x f_j(s)\,ds \qquad \text{for every } x \in [0, 1] \text{ and } j = 2, \dots, k. \tag{7.32}$$

The function f_j is the jth L^2-derivative of f.

As before, the standard notations, $f^{[j]}$, $\partial_x^j f$, and so on, are used for L^2-derivatives. There is a close connection between having L^2-derivatives and the behavior of the Fourier coefficients.

Theorem 7.3.5. *A function f in $L^2([0,1])$ has k L^2-derivatives if and only if*

$$\sum_{n=-\infty}^{\infty} (1+|n|)^{2k}|\hat{f}(n)|^2 < \infty. \tag{7.33}$$

In this case, we have

$$\begin{aligned} \int_0^1 |f^{[j]}(x)|^2\,dx &= \sum_{n=-\infty}^{\infty} |2\pi n|^{2j}|\hat{f}(n)|^2 < \infty \text{ and} \\ \widehat{f^{[j]}}(n) &= [2\pi i n]^j \hat{f}(n) \text{ for } j = 1,\dots,k. \end{aligned} \tag{7.34}$$

Sketch of proof. If (7.33) holds, then Parseval's formula implies that the sequences

$$< [2\pi i n]^j \hat{f}(n) >, \quad j = 0,\dots,k$$

are the Fourier coefficients of the functions $f_0, f_1, \dots, f_k$ in $L^2([0,1])$. Integrating, *formally* it is not difficult to show that these functions are the L^2-derivatives of f. On the other hand, if f has k L^2-derivatives, then, using the alternate definition given in Exercise 7.3.17 and test functions defined by trigonometric polynomials, we deduce that

$$\widehat{f^{[j]}}(n) = [2\pi i n]^j \hat{f}(n) \qquad \text{for } j = 1,\dots,k. \tag{7.35}$$

The estimate (7.33) is a consequence of these formulæ and Parseval's formula. □

As before, there is a relationship between the classical notion of differentiability and having an L^2-derivative. For clarity, let g denote the L^2-derivative of f. It follows from the definition that, for any pair of numbers $0 \le x < y \le 1$, we have

$$f(y) - f(x) = \int_x^y g(s)\,ds.$$

The Cauchy-Schwarz inequality applies to show that

$$\begin{aligned} \left|\int_x^y g(s)\,ds\right| &\le \sqrt{\int_x^y 1\cdot ds}\sqrt{\int_x^y |g(s)|^2\,ds} \\ &\le \sqrt{|x-y|}\|g\|_{L^2}. \end{aligned} \tag{7.36}$$

If we put together this estimate with the previous equation, we have that

$$\frac{|f(x)-f(y)|}{\sqrt{|x-y|}} \le \|g\|_{L^2}.$$

In other words, if a function of one variable has an L^2-derivative, then it is Hölder-$\frac{1}{2}$.

In (7.4) we defined the partial sums, $S_N(f)$, of the Fourier series of a function f. If f has an L^2-derivative, then it follows from Theorem 7.3.5 and the Cauchy-Schwarz inequality that

$$\sum_{n\neq 0}^{\infty}|\hat{f}(n)| \leq \sqrt{\sum_{n=-\infty}^{\infty} n^2|\hat{f}(n)|^2}\sqrt{\sum_{n\neq 0}\frac{1}{n^2}} < \infty. \tag{7.37}$$

As we already know that a function with an L^2-derivative is continuous, we can apply Proposition 7.1.2 to conclude that the Fourier series of such a function converges pointwise to the function. Indeed it is not difficult to estimate the pointwise error $|f(x) - S_N(f;x)|$,

$$\begin{aligned}|f(x) - S_N(f;x)| &= \left|\sum_{|n|>N}\hat{f}(n)e^{2\pi i n x}\right| \\ &\leq \sum_{|n|>N}|\hat{f}(n)| \\ &\leq \sqrt{\sum_{|n|>N}\|n\hat{f}(n)|^2}\sqrt{\sum_{|n|>N}\frac{1}{n^2}} \\ &\leq \|f'\|_{L^2}\sqrt{\frac{2}{N}}.\end{aligned} \tag{7.38}$$

In the first line we used the inversion formula and in the last line the Parseval formula.

Exercises

***Exercise* 7.3.16.** Show that the hypothesis, in Proposition 7.1.2, that f is continuous is unnecessary by using the observation that $< \hat{f}(n) >$ is a square-summable sequence and therefore $f \in L^2([0,1])$. The conclusion needs to be modified to say that f can be modified on a set of measure zero so that (7.6) holds.

***Exercise* 7.3.17.** A different definition for the L^2-derivative is as follows: A function f defined on $[0,1]$ has an L^2-derivative provided that there is a function $f_1 \in L^2([0,1])$ so that

$$\int_0^1 f(x)\varphi'(x)\,dx = -\int_0^1 f_1(x)\varphi(x)\,dx \tag{7.39}$$

for every 1-periodic, once differentiable function φ. Show that this definition is equivalent to Definition 7.3.2.

***Exercise* 7.3.18.** Suppose we use the condition in the previous exercise to define L^2-derivatives but without requiring the test functions φ be 1-periodic. That is, (7.39) should hold for *every* once differentiable function φ, without assuming 1-periodicity. Show that we do not get the same class of functions. What boundary condition must a function satisfy to be differentiable in this sense?

***Exercise* 7.3.19.** Provide the details for the derivations of the formulæ in (7.35).

7.4 General Periodic Functions

Up to this point we have only considered functions of period 1. Everything can easily be generalized to functions with arbitrary periods. A function, defined on the real line, is periodic of period P, or P-periodic, if

$$f(x+P) = f(x).$$

A P-periodic function is determined by its values on any interval of length P. Conversely, a function defined on an interval of length P has a well-defined P-periodic extension to $\mathbb{R}$. For an L^1-function of period P, define the Fourier coefficients by

$$\hat{f}(n) = \int_0^P f(x) e^{-\frac{2\pi i n x}{P}} \, dx.$$

The various results proved previously have obvious analogues in this case:

1. Inversion formula: If f is continuous and $\sum_{n=-\infty}^{\infty} |\hat{f}(n)| < \infty$, then

$$f(x) = \frac{1}{P} \sum_{n=-\infty}^{\infty} \hat{f}(n) e^{\frac{2\pi i n x}{P}}.$$

2. Parseval formula: If f is in $L^2([0, P])$, then

$$\int_0^P |f(x)|^2 \, dx = \frac{1}{P} \sum_{n=-\infty}^{\infty} |\hat{f}(n)|^2.$$

3. Convergence in the mean: If f is in $L^2([0, P])$, then

$$\lim_{M,N\to\infty} \| f(x) - \frac{1}{P} \sum_{j=-M}^{N} \hat{f}(j) e^{\frac{2\pi i j x}{P}} \|_{L^2}^2 = 0.$$

Exercise

***Exercise* 7.4.1.** Derive these P-periodic formulæ from the case $P = 1$.

7.4.1 Convolution and Partial Sums*

The notion of the convolution product can be adapted to periodic functions. If f and g are P-periodic, bounded functions, then their convolution is the P-periodic function defined by

$$f * g(x) = \int_0^P f(y)g(x-y)\,dy.$$

Evaluating these integrals requires a knowledge of $g(s)$ for $s \in [-P, P]$; this is where the P-periodicity of g is used. As in the case of the real line, convolution satisfies the estimate

$$\|f * g\|_{L^1} \leq \|f\|_{L^1}\|g\|_{L^1}; \tag{7.40}$$

here $\|\cdot\|_{L^1}$ is the L^1-norm over $[0, P]$. Using this estimate, the convolution product is extended to define a continuous map from $L^1([0, P]) \times L^1([0, P])$ to $L^1([0, P])$. Its basic properties are summarized in the following proposition:

Proposition 7.4.1. *If f, g, and h are P-periodic, L^1-functions, then $f * g$ is also P-periodic. The periodic convolution has the usual properties of a multiplication:*

$$f * g = g * f, \quad f * (g * h) = (f * g) * h, \quad f * (g + h) = f * g + f * h.$$

Periodic convolution and Fourier series are connected in the same way as convolution and the Fourier transform.

Theorem 7.4.1. *If f and g are P-periodic, L^1-functions, then the Fourier coefficients of $f * g$ are given by*

$$\widehat{f * g}(n) = \hat{f}(n)\hat{g}(n). \tag{7.41}$$

The proof is an elementary computation.

There is also a definition of convolution for sequences. It is motivated by the result of multiplying trigonometric polynomials. If $f = \sum_{j=-N}^{N} a_j e^{2\pi i j x}$ and $g = \sum_{j=-N}^{N} b_j e^{2\pi i j x}$, then

$$f \cdot g = \sum_{l=-2N}^{2N} \left[\sum_{\max\{-N-l,-N\} \leq j \leq \min\{N, N+l\}} a_j b_{l-j} \right] e^{2\pi i l x}. \tag{7.42}$$

***Definition* 7.4.1.** Let $A =< a_n >$ and $B =< b_n >$ be square-summable, bi-infinite sequences. The convolution of A with B is the sequence defined by

$$[A \star B]_n = \sum_{j=-\infty}^{\infty} a_j b_{n-j}.$$

Hölder's inequality for l^2 implies that $A \star B$ is a bounded sequence; see Exercise 7.3.9.

If f and g are L^2-functions, then fg is a L^1-function. Using the notion of convolution of sequences, we get a formula for the Fourier coefficients of the pointwise product fg.

Proposition 7.4.2. *If f, g are in $L^2([0, P])$, then the Fourier coefficients of fg are given by*

$$\widehat{fg}(n) = \frac{1}{P}\hat{f} \star \hat{g}(n) = \frac{1}{P}\sum_{j=-\infty}^{\infty} \hat{f}(j)\hat{g}(n-j). \tag{7.43}$$

Proof. For finite sums this result is (7.42). Without worrying about the limits of summation, we obtain

$$\begin{aligned} f(x)g(x) &= \frac{1}{P^2}\sum_{j=-\infty}^{\infty} \hat{f}(j)e^{\frac{2\pi ijx}{P}} \sum_{k=-\infty}^{\infty} \hat{g}(k)e^{\frac{2\pi ikx}{P}} \\ &= \frac{1}{P^2}\sum_{j=-\infty}^{\infty}\sum_{k=-\infty}^{\infty} \hat{f}(j)\hat{g}(k)e^{\frac{2\pi i(k+j)x}{P}} \\ &= \frac{1}{P}\sum_{l=-\infty}^{\infty}\left[\frac{1}{P}\sum_{j=-\infty}^{\infty} \hat{f}(j)\hat{g}(l-j)\right]e^{\frac{2\pi ilx}{P}}. \end{aligned} \tag{7.44}$$

To get to the last line, we set $l = j + k$.

To complete the proof, we need to show that if f, g belong to L^2, then

$$\lim_{N\to\infty} \widehat{S_N(f)S_N(g)}(n) = \widehat{fg}(n). \tag{7.45}$$

Briefly, if f, g belong to L^2, then the Cauchy-Schwarz inequality implies that fg is in L^1. Moreover,

$$fg - S_N(f)S_N(g) = (f - S_N(f))g + S_N(f)(g - S_N(g)),$$

and therefore the triangle inequality and another application of the Cauchy-Schwarz inequality give

$$\begin{aligned} \|(f - S_N(f))g + S_N(f)(g-S_N(g))\|_{L^1} & \\ &\leq \|(f - S_N(f))g\|_{L^1} + \|S_N(f)(g - S_N(g))\|_{L^1} \\ &\leq \|(f - S_N(f))\|_{L^2}\|g\|_{L^2} + \|S_N(f)\|_{L^2}\|(g - S_N(g))\|_{L^2}. \end{aligned} \tag{7.46}$$

This shows that $S_N(f)S_N(g)$ converges to fg in L^1. The proof is completed by using Proposition 7.1.1 to verify (7.45). □

Exercises

***Exercise* 7.4.2.** Prove the estimate (7.40) and explain how to use it to extend the definition of convolution to pairs of L^1-functions.

***Exercise* 7.4.3.** Prove Proposition 7.4.1.

***Exercise* 7.4.4.** Prove Theorem 7.4.1.

***Exercise* 7.4.5.** Show that for square-summable sequences A and B the convolution $A \star B$ is a bounded sequence. *Hint:* Use the Cauchy-Schwarz inequality for l^2.

***Exercise* 7.4.6.** Use Proposition 7.4.2 to show that if A and B are in l^2, then

$$\lim_{n\to\pm\infty} [A \star B]_n = 0.$$

***Exercise* 7.4.7.** Prove formula (7.42).

***Exercise* 7.4.8.** Give a complete proof of (7.43) by showing that

$$\lim_{N\to\infty} S_N\widehat{(f)S_N}(g)(n) = \widehat{fg}(n).$$

***Exercise* 7.4.9.** A discrete analogue of the Hilbert transform is defined by

$$\widehat{\mathcal{H}f}(n) = \operatorname{sgn}(n)\hat{f}(n).$$

Where $\operatorname{sgn}(n) = 1$ if $n > 0$, $\operatorname{sgn}(0) = 0$ and $\operatorname{sgn}(n) = -1$ if $n < 0$. As before, it can expressed as a principal value integral.

$$\mathcal{H}f(x) = \text{P.V.}\int_0^{2\pi} h(x-y)f(y)dy.$$

1. By considering the approximate Hilbert transforms defined by

$$\widehat{\mathcal{H}_\epsilon f}(n) = e^{-\epsilon|n|}\operatorname{sgn}(n)\hat{f}(n),$$

find the function, h, appearing in the above formula.

2. Show that if

$$f(x) = \sum_{n=1}^{\infty} a_n \sin(nx),$$

then

$$\mathcal{H}f(x) = -i\sum_{n=1}^{\infty} a_n \cos(nx).$$

3. Let $f \in L^2([0, P])$, and show that

$$\lim_{y\to 0}\int_0^P |f(x-y) - f(x)|^2 dx = 0. \tag{7.47}$$

7.4.2 Dirichlet Kernel

The partial sums of the Fourier series can be expressed as a periodic convolution. Let

$$\hat{d}_N(n) = \begin{cases} 1 & \text{if } |n| \le N, \\ 0 & \text{if } |n| > N. \end{cases}$$

The Nth-partial sum of the Fourier series of f is just the inverse Fourier transform of the sequence $< \hat{f}\hat{d}_N >$.

***Definition* 7.4.2.** For each N, define the *Dirichlet* kernel

$$D_N = \frac{1}{P}\sum_{n=-N}^{N} e^{\frac{2\pi i n x}{P}} = \frac{\sin(\frac{2\pi(N+\frac{1}{2})x}{P})}{P\sin(\frac{\pi x}{P})}. \tag{7.48}$$

Figure 7.4 shows a graph of D_3.

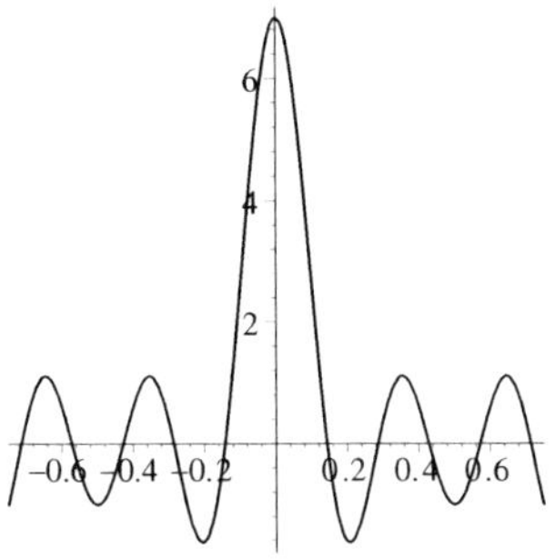

Figure 7.4. Graph of the Dirichlet kernel, D_3

It is clear from the definition that $\widehat{D_N} = \hat{d}_N$. Theorem 7.4.1 shows that for f in $L^1([0, P])$,

$$S_N(f) = f * D_N. \tag{7.49}$$

The zeroth Fourier coefficient of D_N is 1; that is,

$$\int_0^P D_N(x)\,dx = 1.$$

The Dirichlet kernel is oscillatory and assumes both positive and negative values. It is not difficult to show that

$$\lim_{N\to\infty}\int_0^P |D_N(x)|\,dx = \infty. \tag{7.50}$$

This fact underlies the difficulties in analyzing the pointwise convergence of the partial sums of the Fourier series. Even if f is a continuous function, it is *not* always true that

$$\lim_{N\to\infty} S_N(f;x) = f(x).$$

In the next several sections we explore these issues in detail. First we consider what happens to the partial sums of the Fourier series near a jump discontinuity. Then we find a replacement for the partial sums that has better pointwise convergence properties.

Exercises

***Exercise* 7.4.10.** Prove formula (7.48) by using the formula for the sum of a geometric series.

***Exercise* 7.4.11.** Use the explicit formula for $D_N(x)$ to prove (7.50). *Hint:* Compare

$$\int_0^P |D_N(x)|\,dx$$

to the harmonic series.

7.5 The Gibbs Phenomenon

See: B.2.

Let f be a function with a jump discontinuity at x_0; that is, the left- and right-hand limits

$$\lim_{x \to x_0^+} f(x) \text{ and } \lim_{x \to x_0^-} f(x)$$

both exist but are not equal. Since the partial sums $< S_N(f) >$ are continuous functions, it is a foregone conclusion that they cannot provide a good pointwise approximation to f near x_0. In fact, they do an especially poor job. What is notable is that the way in which they fail does not depend very much on f and can be completely analyzed. We begin by considering an example.

***Example* 7.5.1.** Consider the 2π-periodic function

$$g(x) = \begin{cases} \frac{\pi - x}{2} & 0 \le x \le \pi, \\ -\frac{\pi + x}{2} & -\pi \le x < 0 \end{cases} \tag{7.51}$$

whose graph is shown in figure 7.5 along with two partial sums of its Fourier series.

From the graphs it is apparent that, even as N increases, the pointwise approximation does not improve near the jump. In fact, the graphs of the partial sums, $S_N(g)$, "overshoot" the graph of g near the discontinuity. The partial sums, $S_N(g)$, are highly oscillatory near the jump. In the graphs it appears that the amount of overshoot does not decrease as N increases. This collection of bad behaviors is called the *Gibbs phenomenon*. In the engineering literature it is also called *overshoot*.

7.5.1 An Example of the Gibbs Phenomenon

To analyze the Gibbs phenomenon, we first consider, in detail, the partial sums of the function g defined in (7.51). This function has a jump discontinuity of size π at $x = 0$, and its Fourier series is given by

$$g(x) = \sum_{k=-\infty}^{\infty} \hat{g}(k)e^{ikx} = \sum_{k=1}^{\infty} \frac{\sin kx}{k}.$$

If x is not a multiple of 2π, then $S_N(g; x)$ converges to $g(x)$; see Section 7.6. At $x = 0$ the series converges to 0, which is the average of $\lim_{x\to 0^+} g(x)$ and $\lim_{x\to 0^-} g(x)$.

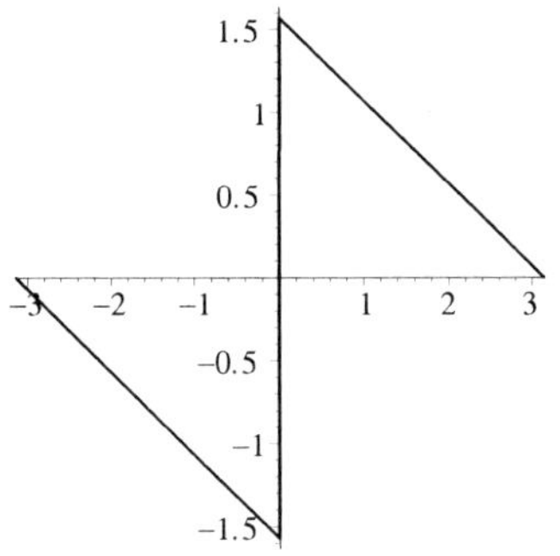

(a) A periodic function with a jump discontinuity.

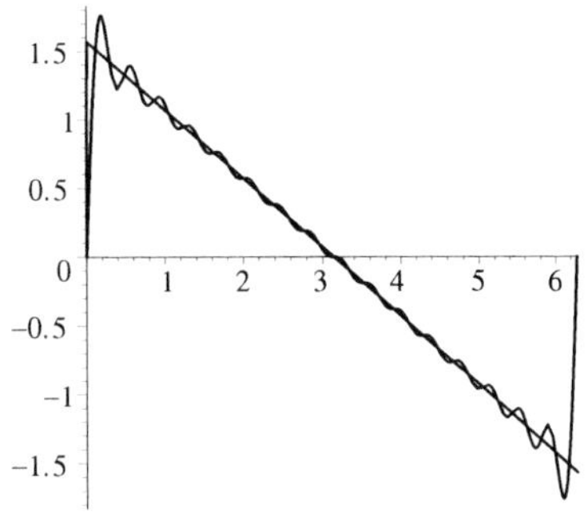

(b) Fourier reconstruction with $N = 16$.

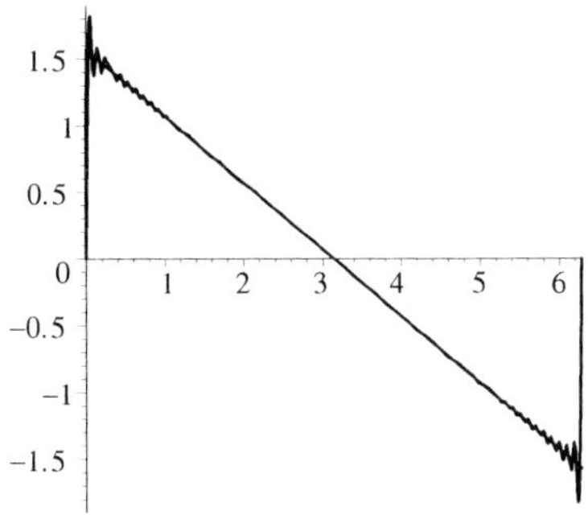

(c) Fourier reconstruction with $N = 64$.

Figure 7.5. An example of the Gibbs phenomenon.

The partial sum, $S_N(g)$, can be reexpressed as

$$S_N(g) = \sum_{k=1}^{N} \frac{\sin kx}{k} = \int_0^x \sum_{k=1}^{N} \cos ky \, dy = \int_0^x \frac{1}{2}\left[\frac{\sin(N+\frac{1}{2})y}{\sin \frac{1}{2}y} - 1\right] dy, \tag{7.52}$$

since

$$\frac{\sin kx}{k} = \int_0^x \cos ky\, dy.$$

We are looking for the maximum of the difference $S_N(g; x) - g(x)$. From elementary calculus we know that at a point where the maximum occurs the derivative vanishes:

$$\frac{d}{dx}[S_N(g; x) - g(x)] = 0. \tag{7.53}$$

Away from its jumps, g is a linear function of slope $-\frac{1}{2}$; hence we are looking for points where

$$S_N'(g; x) = -\frac{1}{2}. \tag{7.54}$$

Let x_N denote the smallest, positive x where this holds. This is a reasonable place to look for the worst behavior as it is the local maximum error closest to the jump. Evaluating $S_N(g; x_N) - g(x_N)$ then gives a lower bound for the maximum difference.

From equation (7.52), we have

$$S_N'(g; x) = \frac{1}{2}\left[\frac{\sin(N + \frac{1}{2})x}{\sin \frac{1}{2}x} - 1\right].$$

Equation (7.54) holds if

$$\sin(N + \frac{1}{2})x = 0.$$

The number

$$x_N = \frac{\pi}{N + \frac{1}{2}}$$

is the smallest positive solution of this equation. The partial sum at x_N is given by

$$\begin{aligned} S_N(g; \frac{\pi}{N + \frac{1}{2}}) &= \frac{1}{2}\int_0^{\frac{\pi}{N+\frac{1}{2}}} \left[\frac{\sin(N + \frac{1}{2})y}{\sin \frac{1}{2}y} - 1\right] dy \\ &= \frac{1}{2}\int_0^{\frac{\pi}{N+\frac{1}{2}}} \frac{\sin(N + \frac{1}{2})y}{\sin \frac{1}{2}y} dy - \frac{1}{2}\frac{\pi}{N + \frac{1}{2}} \\ &= \frac{1}{2}\int_0^1 \frac{\sin(t\pi)}{\sin\left(\frac{1}{2}\frac{t\pi}{N+\frac{1}{2}}\right)} \frac{\pi\, dt}{N + \frac{1}{2}} - \frac{1}{2}\frac{\pi}{N + \frac{1}{2}}. \end{aligned} \tag{7.55}$$

In the last line we used the change of variable, $y = \frac{t\pi}{N+\frac{1}{2}}$. Using the Taylor expansion for $\sin x$ gives

$$(N + \frac{1}{2}) \sin \frac{t\pi}{2(N+\frac{1}{2})} = (N + \frac{1}{2})\left[\frac{t\pi}{2(N+\frac{1}{2})} - \frac{1}{6}\left(\frac{t\pi}{2(N+\frac{1}{2})}\right)^3 + \cdots\right].$$

Hence, as $N \to \infty$, the denominator of the integrand converges to $\frac{t\pi}{2}$ and the numerator converges to $\sin t\pi$. Therefore, we have

$$\lim_{N\to\infty} S_N(g; \frac{\pi}{N+\frac{1}{2}}) = \frac{1}{2}\int_0^1 \frac{\sin t\pi}{\frac{t\pi}{2}} \pi\, dt = \int_0^1 \frac{\sin t\pi}{t} dt. \tag{7.56}$$

From the definition of g,

$$\lim_{N\to\infty} g(\frac{\pi}{N+\frac{1}{2}}) = \frac{\pi}{2}.$$

Evaluating the preceding integral numerically, we obtain that

$$\lim_{N\to\infty} S_N(g; \frac{\pi}{N+\frac{1}{2}}) = \left(\frac{\pi}{2}\right) 1.178979744\cdots. \tag{7.57}$$

This implies that

$$\lim_{N\to\infty} [S_N(g; \frac{\pi}{N+\frac{1}{2}}) - g(\frac{\pi}{N+\frac{1}{2}})] = \frac{\pi}{2} 0.178979744\cdots. \tag{7.58}$$

From the graphs in Figure 7.5 it is clear that, as N increases, the oscillations in the partial sums become more and more concentrated near the jump. The preceding discussion shows that the local maxima and minima of $S_N(g; x) - g(x)$ occur at the set of points $\{x_{N,k}\}$ that satisfy

$$\sin(N + \frac{1}{2})x_{N,k} = 0.$$

These are simply $x_{N,k} = (k\pi)(N + \frac{1}{2})^{-1}$ for $k \in \mathbb{Z}$. The number of oscillations of a given size is essentially independent of N, but the region in which they occur scales with N. The oscillations in $S_N(g)$ are concentrated in a region of size N^{-1} around the jump. The graphs in Figure 7.6 show the original function, its partial sums and its "Fejer means." These are the less oscillatory curves lying below the graphs of g and are explained in Section 7.5.3. In these graphs we have rescaled the x-axis to illustrate that the Gibbs oscillations near the discontinuity remain of a constant size.

In the next section we show that the phenomenon exhibited here is universal for the partial sums of the Fourier series of piecewise differentiable but discontinuous functions. Suppose that f is a piecewise differentiable function, with a jump discontinuity at the point x_0. This means that the left and right limits both exist,

$$\lim_{x\to x_0^-} f(x) = L \text{ and } \lim_{x\to x_0^+} f(x) = R,$$

but $L \neq R$. Suppose that $L < R$. Fixing any sufficiently small $\epsilon > 0$, we show that

$$\lim_{N\to\infty} \max_{0<x-x_0<\epsilon} (S_N(f;x) - f(x)) = (G-1)\frac{R-L}{2}.$$

The coefficient G is a universal constant defined by

$$G = \frac{2}{\pi}\int_0^1 \frac{\sin \pi t}{t} dt = 1.178979744\cdots. \tag{7.59}$$

In fact, there is a sequence of points $< x_N >$ that converge to x_0 so that the error $|S_N(f;x_N) - f(x_N)|$ is about 9% of the size of the jump.

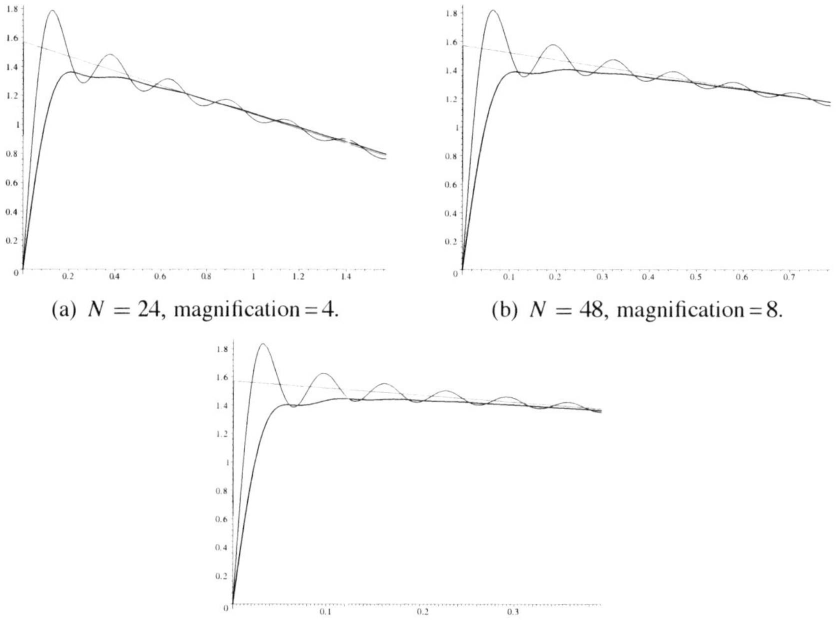

(a) $N = 24$, magnification = 4.

(b) $N = 48$, magnification = 8.

(c) $N = 96$, magnification = 16.

Figure 7.6. Detail showing equi-oscillation property in Gibbs phenomenon.

7.5.2 The General Gibbs Phenomenon*

We prove the theorem describing the Gibbs phenomenon for general piecewise differentiable functions.

Theorem 7.5.1. *Let f be a piecewise $\mathscr{C}^1$-function with jump discontinuities at $\{x_1, \dots, x_k\}$ of sizes $\{h_1, \dots, h_k\}$, that is,*

$$h_j = \lim_{x\to x_j^+} f(x) - \lim_{x\to x_j^-} f(x),$$

and set

$$\eta_j = \operatorname{sgn} h_j.$$

For any sufficiently small $\epsilon > 0$, *we have that*

$$\lim_{N\to\infty} \max_{0<\eta_j(x-x_j)<\epsilon} (S_N(f;x) - f(x)) = (G-1)\frac{|h_j|}{2} \qquad \text{for each } j \in \{1,\ldots,k\}.$$

The constant G *is given by* (7.59).

Proof. First we consider $g_j(x) = \frac{h_j}{\pi} g(x - x_j)$, which is a scaled and translated version of g. From the analysis of g, we have that

$$\lim_{N\to\infty} \max_{0<\eta_j(x-x_j)<\epsilon} (S_N(g_j;x) - g_j(x)) = (G-1)\frac{|h_j|}{2}. \tag{7.60}$$

In Section 7.6 it is shown that for $i \neq j$ we have

$$\lim_{N\to\infty} \max_{0<|x-x_j|<\epsilon} |S_N(g_i;x) - g_i(x)| = 0.$$

Rewrite f as

$$f(x) = f_c(x) + \sum_{j=1}^{k} \frac{h_j}{\pi} g(x - x_j), \text{ where } f_c(x) = f(x) - \sum_{j=1}^{k} \frac{h_j}{\pi} g(x - x_j).$$

Note that f_c is a continuous and piecewise C^1-function. In Exercise 7.5.2 it is shown that f_c has an L^2-derivative, and it therefore follows from (7.37) that the Fourier series of f_c converges uniformly to f_c. The jumps in the function f have been "transferred" to the sum of the g_js:

$$S_N(f;x) - f(x) = S_N(f_c;x) - f_c(x) + \sum_{j=1}^{k} S_N(g_j;x) - \sum_{j=1}^{k} g_j(x).$$

Since $\lim_{N\to\infty} S_N(f_c;x) = f_c(x)$ for every x, it follows from (7.60) that

$$\lim_{N\to\infty} \max_{0<\eta_j(x-x_j)<\epsilon} (S_N(f;x) - f(x)) = \lim_{N\to\infty} \max_{0<\eta_j(x-x_j)<\epsilon} (S_N(g_j;x) - g_j(x)) = (G-1)\frac{|h_j|}{2}.$$

This completes the proof of the theorem. □

***Remark* 7.5.1.** The proof of the theorem shows that, near the jumps, the partial sums $S_N(f)$ look like the partial sums of $S_N(g)$ near its jump. In (7.38) we showed that $|S_N(f_c;x) - f_c(x)|$ behaves like $N^{-\frac{1}{2}}$ for all x. This implies that $S_N(f)$ has the same oscillatory artifacts near the jump points as are present in $S_N(g)$. The Gibbs phenomenon places an inherent limitation on the utility of the Fourier transform when working with discontinuous data. In imaging applications, such function often arise. Taking higher and higher partial sums does not lead, in and of itself, to a better reconstructed image near such points. In the next section we describe a different partial inverse for the Fourier series that eliminates the Gibbs phenomenon.

Exercises

***Exercise* 7.5.1.** In Section 7.6 it is shown that for $x \neq 0$, $\lim_{N\to\infty} S_N(g; x) = g(x)$. Assuming this, use partial summation to show that there is a constant, M, that depends on x so that

$$|S_N(g; x) - g(x)| \leq \frac{M}{N}.$$

Explain why M *must* depend on x.

***Exercise* 7.5.2.** Show that a piecewise differentiable, continuous periodic function has an L^2-derivative.

7.5.3 Fejer Means

The Nth-partial sum is expressible as the convolution of f with the Dirichlet kernel:

$$S_N(f) = D_N * f.$$

What makes the convergence of the partial sums so delicate is the fact that the Dirichlet kernel assumes both positive and negative values. This means that the convergence (or non-convergence) of the partial sums relies on subtle cancellations (or their absence). There is a general technique to obtain a more stable pointwise approximation of functions by finite trigonometric sums that does not sacrifice too much of the favorable L^2-approximation properties of the partial sums.

***Definition* 7.5.1.** The Nth *Fejer mean* of the partial sums is the average of the first $N+1$ partial sums,

$$C_N(f; x) = \frac{S_0(f; x) + \cdots + S_N(f; x)}{N+1}.$$

***Definition* 7.5.2.** The Nth *Fejer kernel* is the average of the first $N+1$ Dirichlet kernels:

$$F_N(x) = \frac{D_0(x) + \cdots + D_N(x)}{N+1}.$$

A calculation shows that for functions periodic of period P, the Nth Fejer kernel is given by

$$F_N(x) = \frac{1}{P(N+1)} \left[\frac{\sin\left(\frac{\pi(N+1)x}{P}\right)}{\sin\left(\frac{\pi x}{P}\right)} \right]^2. \tag{7.61}$$

A graph of F_5 is shown in Figure 7.7.

It follows from (7.49) that

$$C_N(f) = F_N * f.$$

The important difference between the Fejer kernel and the Dirichlet kernel is that the Fejer kernel does not assume negative values. Fejer's theorem is a consequence of this fact.

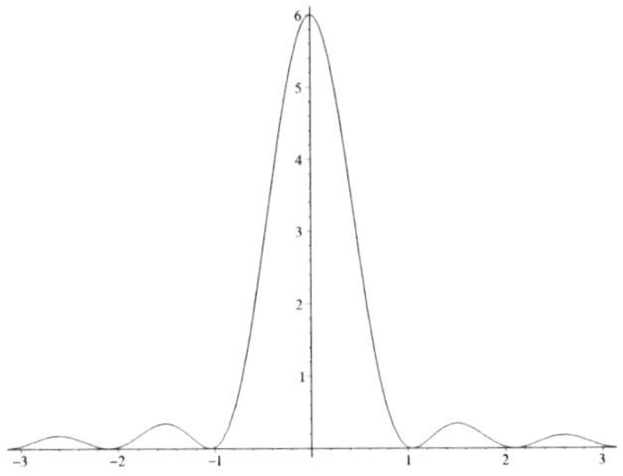

Figure 7.7. Graph of the Fejer kernel, F_5

Theorem 7.5.2 (Fejer's theorem). *If f is an absolutely integrable function that is continuous at x, then*

$$\lim_{N\to\infty} C_N(f;x) = f(x).$$

***Remark* 7.5.2.** As remarked previously, the analogous statement for the partial sums is false.

Proof. The proof is very similar to the proof of the Fourier inversion formula. Note that the Fejer kernel shares three properties with P_r:

1. The Fejer kernel is nonnegative.

2. For every N, $\int_0^1 F_N(x)\,dx = 1$.

3. Given $\epsilon > 0$, there is an M so that if $N > M$, then

$$F_N(x) < \epsilon \qquad \text{for } \epsilon < x < 1-\epsilon. \tag{7.62}$$

The proof of Fejer's theorem follows from these properties.

The first two properties imply that

$$\begin{aligned}|C_N(f;x) - f(x)| &= \left|\int_0^1 F_N(x-y)(f(y)-f(x))\,dy\right| \\ &\le \int_0^1 F_N(x-y)|f(y)-f(x)|\,dy.\end{aligned} \tag{7.63}$$

As f is continuous at x, given $\epsilon > 0$, there is a $\delta > 0$ so that

$$|y-x| < \delta \Rightarrow |f(x) - f(y)| < \epsilon.$$

Using the third property of the Fejer kernel, there is an M so that if $N > M$, then

$$F_N(x) < \epsilon \qquad \text{provided } \delta < x < 1-\delta.$$

We split the integral into two parts:

$$
\begin{aligned}
|C_N(f;x) - f(x)| &\leq \int_{|x-y|<\delta} F_N(x-y)|f(y)-f(x)|\,dy + \int_{|x-y|\geq\delta} F_N(x-y)|f(y)-f(x)|\,dy \\
&\leq \epsilon \int_{|x-y|<\delta} F_N(x-y)\,dy + \int_{|x-y|\geq\delta} \epsilon(|f(y)|+|f(x)|)\,dy \\
&\leq \epsilon(1+|f(x)|+\|f\|_{L^1}).
\end{aligned}
\tag{7.64}
$$

As $\epsilon > 0$ is arbitrary, this completes the proof of the theorem. □

***Remark* 7.5.3.** The proof shows that if f is a continuous, periodic function, then $C_N(f)$ provides a uniformly accurate approximation. That is, given $\epsilon > 0$, there is an N such that $|C_N(f;x) - f(x)| < \epsilon$ for all $x \in [0,1]$. The smoother curves in Figure 7.6 are the Fejer means of g. They are not nearly as oscillatory, near the jump points, as the partial sums. In general, if a function f satisfies

$$m \leq f(x) \leq M \qquad \text{for } x \in [0,1],$$

then for every N

$$m \leq C_N(f;x) \leq M \qquad \text{for } x \in [0,1],$$

as well.

A further computation shows that the Fourier coefficients of $C_N(f)$ are given by

$$
\widehat{C_N}(f;n) = \begin{cases} \left[1 - \frac{|n|}{N+1}\right]\hat{f}(n) & \text{for } |n| \leq N, \\ 0 & \text{for } |n| > N. \end{cases}
\tag{7.65}
$$

From this is not hard to see that $\|C_N(f) - f\|_2$ goes to zero as $N \to \infty$, at worst, half as fast as $\|S_{\frac{N}{2}}(f) - f\|_2$; see Exercise 7.5.6.

***Example* 7.5.2.** We quantitatively compare the L^2-errors in the Fejer means with the L^2-errors in the partial sums for the function g, defined in (7.51). The comparisons, listed in Table 7.1, show that $\|f - S_N(f)\|_2$ is about half the size of $\|f - C_N(f)\|_2$.

N	$\|f - S_N(f)\|_{L^2}$	$\|f - C_N(f)\|_{L^2}$
4	.695306572	.942107038
8	.369174884	.570026118
12	.251193672	.411732444
16	.190341352	.323158176
20	.153218060	.266320768
24	.128210482	.226669996
28	.110220018	.197397242
32	.096656766	.174881126

Table 7.1. Comparison of the mean square errors

Using the Fejer kernel, we can now prove the lemma used in the proof of the Parseval formula.

Lemma 7.5.1. *Let $f \in L^2([0,1])$ and fix an $\epsilon > 0$. There is a trigonometric polynomial g such that*

$$\|f - g\|_{L^2} < \epsilon.$$

Proof. In Corollary A.5.1 it is established that there is a continuous, periodic function g_0 so that

$$\|f - g_0\|_{L^2} < \frac{\epsilon}{2}.$$

Fejer's theorem implies that there is an N so that, for all $x \in [0,1]$,

$$|g - C_N(g; x)| \leq \frac{\epsilon}{2}.$$

With this N, we use the triangle inequality to obtain

$$\|f - C_N(g)\|_{L^2} \leq \|f - g\|_{L^2} + \|C_N(g) - g\|_{L^2} < \frac{\epsilon}{2} + \frac{\epsilon}{2}.$$

This completes the proof of the lemma. □

Exercises

***Exercise* 7.5.3.** Prove that if $m \leq f(x) \leq M$ for $x \in [0,1]$, then $m \leq C_N(f; x) \leq M$ for $x \in [0,1]$ is as well.

***Exercise* 7.5.4.** Prove the closed form expression for the Fejer kernel (7.61).

***Exercise* 7.5.5.** Prove the formulæ (7.65) for the Fourier coefficients of F_N.

***Exercise* 7.5.6.** If $f \in L^2([0,1])$, show that

$$\|C_{2N}(f) - f\|_{L^2} \leq 2\|S_N(f) - f\|_{L^2}.$$

***Exercise* 7.5.7.** Show that the Fourier transform also suffers from the Gibbs phenomenon by analyzing $S_R(\chi_{[-1,1]})$ as $R \to \infty$. The partial inverse S_R is defined in (5.6).

***Exercise* 7.5.8.** Explain why the kernel $f_B = \mathcal{F}^{-1}([2B]^{-1}\chi_{[B,B]} * \chi_{[B,B]})$ is the Fourier transform analogue of the Fejer kernel. *Hint:* Express $f_B * g$ in terms of the Fourier transform of g.

7.5.4 Resolution in the Partial Sums of the Fourier Series

The Fejer means produce visually more appealing images near to jump points. However, this is at the expense of reducing the overall resolution. From (7.65) it is apparent that convolution with the Fejer kernel progressively attenuates the high frequencies, eventually cutting them off entirely at $n = \pm(N+1)$. Employing Fejer means to reduce the Gibbs effect near to jumps inevitably results in an overall decrease in the available resolution. A calculation using Taylor's formula shows that, as N tends to infinity,

$$\text{FWHM}(D_N) \simeq \frac{.86}{\pi N} \tag{7.66}$$

whereas

$$\text{FWHM}(F_N) \simeq \frac{1.33}{\pi N}. \tag{7.67}$$

Here we are measuring the FWHM of the central peaks. Using the same number of Fourier coefficients, the partial sum has about $\frac{3}{2}$ times as much FWHM-resolution as the Fejer mean.

***Example* 7.5.3.** In the graphs in Figure 7.8 we have an interesting function that we reconstruct first using a partial sum of its Fourier series and then with Fejer means. It is clear that the oscillatory artifact has disappeared in the Fejer means reconstruction. The resolution of the latter reconstruction is also evidently lower.

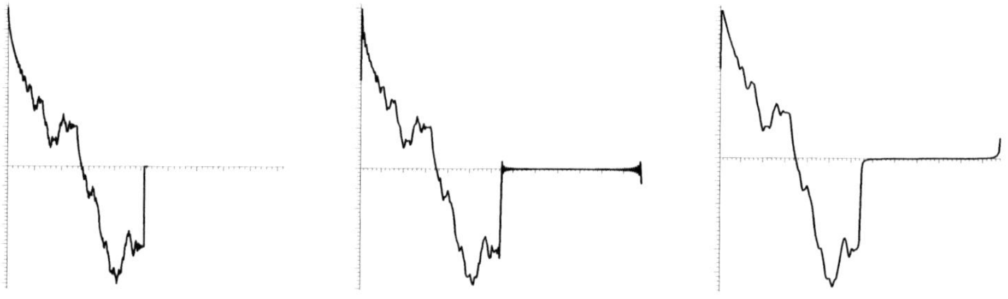

(a) A function (b) Partial sum (c) Fejer mean

Figure 7.8. Graphs comparing the partial sums and Fejer means.

The graphs in Figure 7.9 are expanded views of these functions between .1 and .3. Here the loss of resolution in the Fejer means, at points away from the jump, is quite evident.

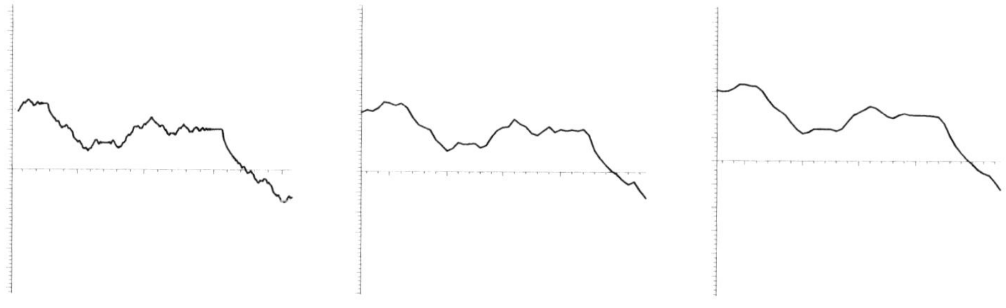

(a) A function (b) Partial sum (c) Fejer mean

Figure 7.9. Expanded view showing the loss of resolution in the Fejer means.

Finally, in Figure 7.8, we compare the behavior of these approximations near the jump discontinuity. Both the Gibbs oscillations and higher resolution in the partial sums are again evident.

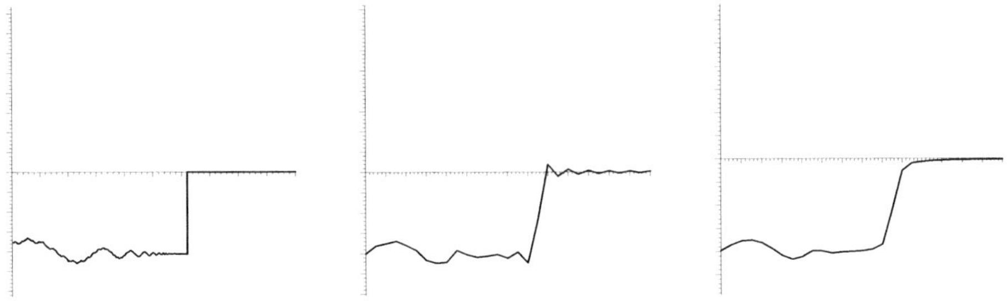

(a) A function (b) Partial sum (c) Fejer mean

Figure 7.10. Expanded view, near the jump, showing the Gibbs phenomenon in the partial sums.

Exercise

***Exercise* 7.5.9.** Derive (7.66) and (7.67).

7.6 The Localization Principle

Each Fourier coefficient is defined by an integral of f over its whole domain of definition and therefore depends on the value of the function everywhere. Like the one-dimensional Fourier transform, the Fourier series is very sensitive to the local behavior of a function. This is the content of the following theorem.

Theorem 7.6.1 (The localization principle). *Let f and g be L^1-functions on $[0, P]$. Suppose that for some x, $S_N(g; x)$ converges to $g(x)$ as $N \to \infty$. If there exists a positive ϵ so that $f(t)$ equals $g(t)$ for t in an interval $[x - \epsilon, x + \epsilon]$, then*

$$\lim_{N\to\infty} S_N(f; x) = f(x).$$

Proof. In the following computations, recall that x is a fixed point. Let $s_N = D_N * f(x)$, and

$t_N = D_N * g(x)$. By linearity, $(s_N - t_N) = D_N * (f - g)(x)$, and we compute

$$\begin{aligned} D_N * (f-g)(x) &= \int_0^P \sin\left(\frac{\pi(2N+1)(x-y)}{P}\right)\left[\frac{f(y)-g(y)}{P\sin\left(\frac{\pi(x-y)}{P}\right)}\right]dy \\ &= \operatorname{Im}\left\{\int_0^P e^{\frac{2\pi i N(x-y)}{P}} e^{\frac{i\pi(x-y)}{P}}\left[\frac{f(y)-g(y)}{P\sin\left(\frac{\pi(x-y)}{P}\right)}\right]dy\right\} \\ &= \operatorname{Im}\left\{e^{\frac{\pi i(2N+1)x}{P}}\int_0^P e^{-\frac{2\pi i N y}{P}}\left[\frac{f(y)-g(y)}{P\sin\left(\frac{\pi(x-y)}{P}\right)}\right]e^{-\frac{i\pi y}{P}}dy\right\}. \end{aligned} \tag{7.68}$$

Since $f(y)$ equals $g(y)$ for y in $[x-\epsilon, x+\epsilon]$, the expression in brackets in the last line is absolutely integrable:

$$\int \left|\frac{f(y)-g(y)}{\sin\left(\frac{2\pi(x-y)}{P}\right)} e^{-\frac{i\pi y}{P}}\right| dy < \infty.$$

Therefore, the last integral in (7.68) is the Nth Fourier coefficient of an integrable function. By the Riemann-Lebesgue lemma, it goes to zero as $N \to \infty$. By the hypothesis, we know that $\lim_{N\to\infty} t_N = g(x) = f(x)$. Rewrite this as

$$\lim_{N\to\infty} s_N = \lim_{N\to\infty}(s_N - t_N) + \lim_{N\to\infty} t_N.$$

Since $s_N - t_N$ goes to zero as $N \to \infty$, we are done. □

This result shows that if an absolutely integrable function f is well behaved in the neighborhood of a point x_0, then $S_N(f; x_0)$ converges to $f(x_0)$. Suppose that f is continuously differentiable in an interval $(x_0 - \epsilon, x_0 + \epsilon)$. Let φ be an infinitely differentiable function that satisfies

$$\varphi(x) = \begin{cases} 1 & \text{for } |x - x_0| < \frac{\epsilon}{2}, \\ 0 & \text{for } |x - x_0| > \frac{3\epsilon}{4}. \end{cases}$$

If we set g equal to $f\varphi$, then g is a continuously differentiable, periodic function and

$$f(x) = g(x) \qquad \text{for } x \in (x_0 - \frac{\epsilon}{2}, x_0 + \frac{\epsilon}{2}).$$

Because g is continuously differentiable, it follows easily that

$$\sum_{n=-\infty}^{\infty} |\hat{g}(n)| < \infty.$$

Therefore, we can apply (7.6) to conclude that

$$g(x) = \sum_{n=-\infty}^{\infty} \hat{g}(n)e^{2\pi i n x}, \qquad \text{for any } x.$$

No matter how wildly f behaves outside the interval $(x_0 - \epsilon, x_0 + \epsilon)$, the localization principle states that

$$\lim_{N\to\infty} \sum_{n=-N}^{N} \hat{f}(n)e^{2\pi i n x_0} = f(x_0).$$

Notice that the asymptotic behavior of the sequence $< \hat{f}(n) >$ is, in general, completely different from that of $< \hat{g}(n) >$. This just makes the localization principle all the more remarkable!

Exercise

***Exercise* 7.6.1.** In the proof of Theorem 7.6, show that it is not necessary to assume that $f(t)$ equals $g(t)$ in an interval containing x by explaining why it suffices to assume that

$$h(y) = \frac{f(y) - g(y)}{x - y}$$

is an integrable function.

7.7 Higher-Dimensional Fourier Series

The theory of Fourier series extends without difficulty to functions defined on the unit cube in $\mathbb{R}^n$. For completeness we include statements of the basic results. Proofs can be found in [42] or [120]. The Fourier coefficients are now labeled by vectors $\boldsymbol{k} \in \mathbb{Z}^n$; that is, vectors of the form

$$\boldsymbol{k} = (k_1, \dots, k_n), \quad \text{where } k_j \in \mathbb{Z} \text{ for } j = 1, \dots, n.$$

If f is an absolutely integrable function defined on

$$[0,1]^n = \underbrace{[0,1] \times \cdots \times [0,1]}_{n-\text{times}}$$

then its Fourier coefficients are defined by

$$\hat{f}(\boldsymbol{k}) = \int\limits_{[0,1]^n} f(\boldsymbol{x})e^{-2\pi i \langle \boldsymbol{k},\boldsymbol{x}\rangle} d\boldsymbol{x}.$$

Many aspects of the theory are quite similar in higher dimensions; however, the theory of pointwise convergence is much more involved and has not yet been completely worked out.

If the Fourier coefficients of f tend to zero rapidly enough, then we have an inversion formula:

Proposition 7.7.1. *Suppose that f is an absolutely integrable function on $[0,1]^n$ such that*

$$\sum_{\boldsymbol{k}\in\mathbb{Z}^n} |\hat{f}(\boldsymbol{k})| < \infty. \tag{7.69}$$

Then

$$f(x) = \sum_{\boldsymbol{k} \in \mathbb{Z}^n} \hat{f}(\boldsymbol{k}) e^{2\pi i \langle \boldsymbol{k}, \boldsymbol{x} \rangle}.$$

In general, the Fourier coefficients of an absolutely integrable function may not satisfy (7.69). Indeed, as the dimension increases, this is a more and more restrictive condition. In order for the infinite sum

$$\sum_{\boldsymbol{k} \in \mathbb{Z}^n} \frac{1}{(1 + \|\boldsymbol{k}\|)^\alpha}$$

to converge, it is necessary to take $\alpha > n$. The Riemann-Lebesgue lemma generalizes to n dimensions.

Proposition 7.7.2 (Riemann-Lebesgue lemma). *If f is an absolutely integrable function on $[0, 1]^n$, then*

$$\lim_{\|\boldsymbol{k}\| \to \infty} \hat{f}(\boldsymbol{k}) = 0.$$

Once again the proof is effected by approximating L^1-functions by an n-dimensional analogue of step functions. In this generality there is, as before, no estimate on the rate at which the Fourier coefficients go to zero.

As in the one-dimensional case, when working with Fourier series, we need to consider f as a periodic function of period 1 in each variable. That is we extend f to all of $\mathbb{R}^n$ by using the condition

$$f(\boldsymbol{x}) = f(\boldsymbol{x} + \boldsymbol{k}) \qquad \text{for every } \boldsymbol{k} \in \mathbb{Z}^n.$$

The inversion formula defines a function on all of $\mathbb{R}^n$ with this property. As before, in the context of Fourier series, a function is considered continuous if its periodic extension to $\mathbb{R}^n$ is continuous and differentiable if its periodic extension to $\mathbb{R}^n$ is differentiable, etc.

If the Fourier coefficients do not satisfy (7.69), then the problem of summing the Fourier series can be quite subtle. The question of the pointwise convergence for the partial sums is considerably more complicated in higher dimensions than in one dimension. In the one-dimensional case there is, in essence, only one reasonable way to define partial sums. In n dimensions there are many different possible choices. The simplest way is to define the Nth partial sum to be

$$S_N(f) = \sum_{k_1=-N}^{N} \cdots \sum_{k_n=-N}^{N} \hat{f}(\boldsymbol{k}) e^{2\pi i \langle \boldsymbol{k}, \boldsymbol{x} \rangle}.$$

Because there is a very fast algorithm to do this calculation (at least for N a power of 2), this is the usual meaning of "partial sums" of the Fourier series in applications. However, it is by no means the only way to partially invert the higher-dimensional Fourier series. We could equally well consider the sum over all vectors $\boldsymbol{k}$ such that $\|\boldsymbol{k}\| \leq R$. Let

$$\Sigma_R(f) = \sum_{\{\boldsymbol{k} : \|\boldsymbol{k}\| < R\}} \hat{f}(\boldsymbol{k}) e^{2\pi i \langle \boldsymbol{k}, \boldsymbol{x} \rangle}$$

denote this sum. While not as useful in applications, this partial inverse is easier to analyze. From this analysis, it is known that the localization principle fails in higher dimensions. The convergence of the Fourier series at $\boldsymbol{x}$ *is* sensitive to the behavior of f at points distant from $\boldsymbol{x}$. The relationship between $S_N(f)$ and $\Sigma_R(f)$ has, so far, not been completely elucidated. An analysis of $\Sigma_R(f)$ is given in [103].

The Gibbs phenomenon also persists in higher dimensions but is, as expected, more complicated to analyze. If a piecewise smooth function, f, has a simple jump along a smooth hypersurface, S, then the behavior of the partial sums near $\boldsymbol{x} \in S$ is determined in part by the size of the jump at $\boldsymbol{x}$, as well as the *curvature* of S at $\boldsymbol{x}$. Asymptotic formulæ for $\Sigma_R(f; \boldsymbol{x})$ are given in [103]. If S itself is not smooth, then even more complicated phenomena arise. As the techniques involved are far beyond the scope of this text, we content ourselves with giving an example of an S_N-type partial sum for the Fourier series of $\chi_{[-1,1]}(x)\chi_{[-1,1]}(y)$. In Figure 7.11 note the Gibbs oscillations parallel to the edges of the square and the "Gibbs shadow" near the corner. The absence of oscillations in the shadow region is an extreme example of the effect of the curvature of the jump curve on the Gibbs phenomenon.

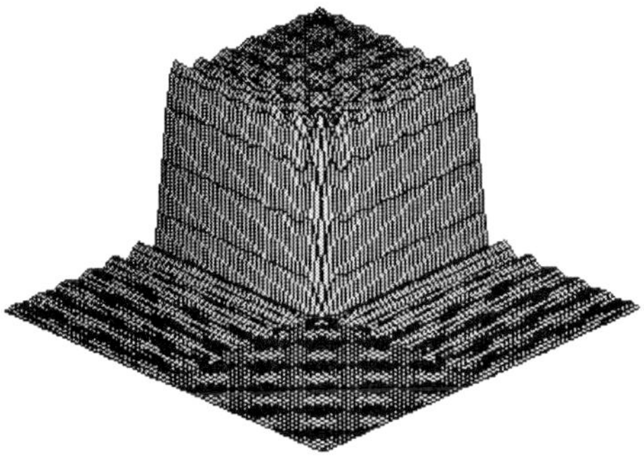

Figure 7.11. Illustration of the 2D Gibbs phenomenon.

There is an obvious generalization of the notion convolution for periodic functions on $\mathbb{R}^n$ given by

$$f * g(\boldsymbol{x}) = \int_{[0,1]^n} f(\boldsymbol{x} - \boldsymbol{y}) g(\boldsymbol{y})\, d\boldsymbol{y}.$$

Periodic convolution in n dimensions is connected to the Fourier series just as in one dimension:

$$\widehat{f * g}(\boldsymbol{k}) = \hat{f}(\boldsymbol{k})\hat{g}(\boldsymbol{k}) \qquad \text{for all } \boldsymbol{k} \in \mathbb{Z}^n. \tag{7.70}$$

We can also define the convolution of two sequences $A = < a_k >$, $B = < b_k >$ indexed by $\mathbb{Z}^n$ by setting

$$(A \star B)_k = \sum_{j \in \mathbb{Z}^n} a_{k-j} b_j.$$

The Fourier series of a pointwise product is then given by

$$\widehat{fg}(\boldsymbol{k}) = \hat{f} \star \hat{g}(\boldsymbol{k}). \tag{7.71}$$

Exercises

***Exercise* 7.7.1.** Let $r_1, \ldots, r_n$ be numbers between 0 and 1. Compute the Fourier coefficients of

$$f(x_1, \ldots, x_n) = \chi_{[-r_1, r_1]}(x_1) \cdots \chi_{[-r_n, r_n]}(x_n).$$

***Exercise* 7.7.2.** Show that

$$\sum_{\boldsymbol{k} \in \mathbb{Z}^n} \frac{1}{(1 + \|\boldsymbol{k}\|)^{\alpha}}$$

converges if $\alpha > n$ and diverges if $\alpha \leq n$. *Hint:* Compare this sum to an integral.

***Exercise* 7.7.3.** Find the periodic function, $D_{N,n}$, on $\mathbb{R}^n$ for which

$$S_N(f) = D_{N,n} * f.$$

***Exercise* 7.7.4.** Find the periodic function, $B_{R,n}$, on $\mathbb{R}^n$ for which

$$\Sigma_R(f) = B_{R,n} * f.$$

7.7.1 L^2-Theory

See: A.4.6.

As in the one-dimensional case, a much more complete theory is available for square-integrable functions. The basic result is the Parseval formula.

Proposition 7.7.3 (Parseval formula). *An L^1-function f belongs to $L^2([0,1]^n)$ if and only if its Fourier coefficients are square summable. In this case we have*

$$\int_{[0,1]^n} |f(\boldsymbol{x})|^2 \, d\boldsymbol{x} = \sum_{\boldsymbol{k} \in \mathbb{Z}^n} |\hat{f}(\boldsymbol{k})|^2.$$

The sense in which the Fourier series converges can also be made precise in this case.

Proposition 7.7.4 (L^2-Inversion formula). *If f belongs to $L^2([0,1]^n)$, then the partial sums of the Fourier series of f converge to f in the L^2-norm. That is,*

$$\lim_{N\to\infty} \|f - S_N(f)\|_{L^2} = 0. \tag{7.72}$$

As in one dimension, the partial sums may fail to converge pointwise.

By defining L^2-partial derivatives, we generalize the theory of L^2-derivatives to higher dimensions. In this case we use the integration by parts formulation.

***Definition* 7.7.1.** A function f in $L^2([0,1]^n)$ has an L^2-partial derivative in the x_j-direction if there is a function f_j in $L^2([0,1]^n)$ such that, for every *periodic*, once differentiable function φ, we have

$$\int_{[0,1]^n} f(\boldsymbol{x})\partial_{x_j}\varphi(\boldsymbol{x})\,d\boldsymbol{x} = -\int_{[0,1]^n} f_j(\boldsymbol{x})\varphi(\boldsymbol{x})\,d\boldsymbol{x}.$$

The restriction to *periodic* test functions is important. More generally, we can define higher L^2-derivatives.

***Definition* 7.7.2.** A function $f \in L^2([0,1]^n)$ has m L^2-derivatives if for each multi-index $\boldsymbol{\alpha}$ with $|\boldsymbol{\alpha}| \leq m$ there is a function $f_{\boldsymbol{\alpha}} \in L^2([0,1]^n)$ such that for every m-times differentiable *periodic* test function φ we have

$$\int_{[0,1]^n} f(\boldsymbol{x})\partial_{\boldsymbol{x}}^{\boldsymbol{\alpha}}\varphi(\boldsymbol{x})\,d\boldsymbol{x} = (-1)^{|\boldsymbol{\alpha}|}\int_{[0,1]^n} f_{\boldsymbol{\alpha}}(\boldsymbol{x})\varphi(\boldsymbol{x})\,d\boldsymbol{x}.$$

As before, the standard notations are used to denote the L^2-derivatives (i.e., $\partial_{\boldsymbol{x}}^{\boldsymbol{\alpha}} f$).

The existence of L^2-derivatives is intimately tied to the rate of decay of the Fourier coefficients.

Proposition 7.7.5. *A function f in $L^2([0,1]^n)$ has m L^2-derivatives if and only if*

$$\sum_{\boldsymbol{k}\in\mathbb{Z}^n} \|\boldsymbol{k}\|^{2m}|\hat{f}(\boldsymbol{k})|^2 < \infty.$$

In this case

$$\int_{\mathbb{R}^n} |\partial_{\boldsymbol{x}}^{\boldsymbol{\alpha}} f(\boldsymbol{x})|^2\,d\boldsymbol{x} = \sum_{\boldsymbol{k}\in\mathbb{Z}^n} |\boldsymbol{k}^{\alpha}\hat{f}(\boldsymbol{k})|^2$$

and

$$\widehat{\partial_{\boldsymbol{x}}^{\alpha} f}(\boldsymbol{k}) = (i\boldsymbol{k})^{\boldsymbol{\alpha}}\hat{f}(\boldsymbol{k}),$$

for every multi-index $\boldsymbol{\alpha}$ with $|\boldsymbol{\alpha}| \leq m$.

As noted previously, in higher dimensions, a faster rate of decay is needed to conclude that the Fourier coefficients are absolutely summable. In one dimension we showed that a function with one L^2-derivative is continuous. In dimension n, slightly more that $\frac{n}{2}$ L^2-derivatives are required for this conclusion.

Functions that are defined on products of intervals $[a_1, b_1] \times \cdots \times [a_n, b_n]$ can be rescaled to be defined on $[0, 1]^n$ and can therefore be expanded in Fourier series as well. We leave the details of this discussion to the interested reader. While intervals are the only connected subsets of the real line, higher-dimensional spaces have a rich array of such subsets. The Fourier series in higher dimensions is only defined for functions that are defined on products of intervals. For example, we cannot directly apply Fourier series to study functions defined in the unit disk. The analysis of functions defined in other sorts of regions requires more sophisticated mathematical techniques. The interested reader is referred to [43].

Exercises

***Exercise* 7.7.5.** Prove (7.70) and (7.71).

***Exercise* 7.7.6.** Let $[n/2]$ be the largest integer smaller than $n/2$. Show that a periodic function with $[n/2] + 1$ L^2-derivatives is a continuous function. *Hint:* Use the Cauchy-Schwarz inequality.

***Exercise* 7.7.7.** If f is in $L^2([0, 1]^n)$, show that

$$\lim_{N\to\infty} \|f - \Sigma_N(f)\|_{L^2} = 0.$$

7.8 Conclusion

Fourier series is a tool for analyzing functions defined on finite intervals in $\mathbb{R}$ or finite rectangles in $\mathbb{R}^n$. As with the Fourier transform, the rate of decay of the Fourier coefficients is related to the smoothness of the function. In the Fourier series case it is the smoothness of the *periodic extension* that determines the rate of decay of the Fourier coefficients. The Fourier coefficients of a function of a single variable with simple jump discontinuities display a characteristic $\frac{1}{n}$ rate of decay.

The partial sum operator S_N provides an approximation to a function f as a exponential polynomial,

$$S_N(f; x) = \sum_{j=-N}^{N} \hat{f}(j)e^{2\pi ijx}.$$

If the error is measured in the L^2-norm, then $S_N(f)$ is the best approximation to f among exponential polynomials of degree N. The Parseval formula implies that $S_N(f)$ converges to f in the L^2-norm. The story of pointwise convergence is more complicated. Even if f is continuous, the sequence $< S_N(f; x) >$ may not converge to $f(x)$. If f has a jump discontinuity, then the behavior of the partial sums is governed by the Gibbs phenomenon. Near to a jump discontinuity the partial sums are oscillatory, overshooting the graph of f

by about 9% of the height of the jump. The size of the overshoot does not diminish as N tends to infinity, but does concentrate in an interval around the jump of size N^{-1}. The Fejer means, $C_N(f)$ were introduced as a way to reduce the Gibbs phenomenon. Moreover, if f is continuous at x, then $C_N(f; x)$ converges to $f(x)$. For a fixed N, the Fejer mean $C_N(f)$ has about $\frac{2}{3}$ the FWHM-resolution of the partial sum $S_N(f)$.

In a realistic situation a function, f, of a continuous variable is measured at a discrete sequence of points, $< x_j >$. In the simplest model for such a measurement the function is evaluated at these points, so that the measurements are simply the values $< f(x_j) >$. In the next chapter we consider what these measurements say about the function at points not in the sample set. The answer we find is neatly expressed in the language of the Fourier transform and Fourier series.

Chapter 8

Poisson Summation, Sampling and Nyquist's Theorem

See: A.6.1, A.5.2.

In Chapters 4 through 7, we developed the mathematical tools needed to describe functions of continuous variables and methods to analyze and reconstruct them. This chapter continues the transition from the world of pure mathematics to its application to problems in image reconstruction. In the first sections of this chapter, we imagine that our "image" is a function, f, of a single real variable, x. In a purely mathematical context, x is a real number that can assume any value along a continuum of numbers. The function also takes values in a continuum, either in $\mathbb{R}$, $\mathbb{C}$, or perhaps $\mathbb{R}^n$. In practical applications we can only evaluate f at a finite set of points $\{x_j\}$. This is called *sampling*. As most of the processing takes place in digital computers, both the points $\{x_j\}$ and the measured values $\{f(x_j)\}$ are forced to lie in the preassigned, finite set of numbers known to the computer. This is called *quantization*. The reader is urged to review Section A.1, where these ideas are discussed in some detail.

Except for a brief discussion of quantization, this chapter is about the consequences of sampling. We examine the fundamental question: How much information about a function is contained in a finite or infinite set of samples? Central to this analysis is the *Poisson summation formula*. This formula is a bridge between the Fourier transform and the Fourier series. While the Fourier transform is well suited to an abstract analysis of image or signal processing, it is the Fourier series that is actually used to do the work. The reason is quite simple: The Fourier transform and its inverse require integrals over the entire real line, whereas the Fourier series is phrased in terms of infinite sums and integrals *over finite intervals*, both of which are eventually approximated by finite sums. Indeed, we also introduce the *finite Fourier transform*, an analogue of the Fourier transform for finite sequences. This chapter covers the next step from the abstract world of the infinite and the infinitesimal to the real world of the finite and discrete.

8.1 Sampling and Nyquist's Theorem*

See: A.5.2, B.1.

Recall that our basic model for a measurement is the evaluation of a function at a point. A set of points $\{x_j\}$ contained in an interval (a, b) is *discrete* if no subsequence converges to a point in (a, b). Evaluating a function on a discrete set of points is called *sampling*. Practically speaking, a function can only be evaluated at points where it is continuous. From the perspective of measurement, the value of a function at points of *dis*continuity is not well defined.

***Definition* 8.1.1.** Suppose that f is a function defined in an interval (a, b) and $\{x_j\}$ is a discrete set of points in (a, b). The points $\{x_j\}$ are called the *sample points*. The values $\{f(x_j)\}$ are called the *samples* of f at the points $\{x_j\}$.

In most applications the discrete set is of the form $\{x_0 + jl \mid j \in \mathbb{Z}\}$, where l is a fixed positive number. These are called *equally spaced samples*; the number l is called the *sample spacing*. The reciprocal l^{-1} of l is called the *sampling rate*. Sampling theory studies the problem of reconstructing functions of a continuous variable from a set of samples and the relationship between these reconstructions and the idealized data.

8.1.1 Bandlimited Functions and Nyquist's Theorem

A model for measurement, more realistic than pointwise evaluation, is the evaluation of a convolution $f * \varphi$. Here φ is an L^1 weight function that models the measuring apparatus. For most reasonable choices of weight function, $f * \varphi$ is a continuous function, so its value is well defined (from the point of view of measurement) at all x. As the Fourier transform of $f * \varphi$ is $\hat{f}\hat{\varphi}$, the Riemann-Lebesgue lemma, Theorem 4.2.2, implies that $\hat{\varphi}$ tends to zero as $|\xi|$ tends to infinity. This means that the measuring apparatus attenuates the high-frequency information in f. In applications we often make the assumption that there is "no high-frequency information" or that is has been filtered out.

***Definition* 8.1.2.** Let f be a function defined on $\mathbb{R}$. If its Fourier transform, $\hat{f}$, is supported in a finite interval, then f is a *bandlimited function*. If $\hat{f}$ is supported in $[-L, L]$, then f is an *L-bandlimited* function.

***Definition* 8.1.3.** Let f be a function defined on $\mathbb{R}$. If its Fourier transform, $\hat{f}$, is supported in a finite interval, then f is said to have *finite bandwidth*. If $\hat{f}$ is supported in an interval of length W, then f is said to have *bandwidth W*.

A bandlimited function is always infinitely differentiable. If f is either L^1 or L^2, then $\hat{f}$ is in L^1 and the Fourier inversion formula states that

$$f(x) = \frac{1}{2\pi} \int_{-L}^{L} \hat{f}(\xi) e^{ix\xi} d\xi. \tag{8.1}$$

As the integrand is supported in a finite interval, the integral in (8.1) can be differentiated as many times as we like. This, in turn, shows that f is infinitely differentiable. If f is in L^2, then so are all of its derivatives.

Nyquist's theorem states that a bandlimited function is determined by a set of uniformly spaced samples, provided that the sample spacing is sufficiently small.

Theorem 8.1.1 (Nyquist's theorem). *If f is a square integrable function and*

$$\hat{f}(\xi) = 0 \text{ for } |\xi| > L,$$

then f is determined by the samples $\{f(\frac{\pi n}{L}) \,:\, n \in \mathbb{Z}\}$.

***Remark* 8.1.1.** As the proof shows, f is determined by any collection of uniformly spaced samples with sample spacing $\frac{\pi n}{L}$, i.e.$\{f(x_0 + \frac{\pi n}{L})\}$ for any $x_0 \in \mathbb{R}$.

Proof. If we think of $\hat{f}$ as a function *defined* on the interval $[-L, L]$, then it follows from (8.1) that the numbers $\{2\pi f(\frac{\pi n}{L})\}$ are the Fourier *coefficients* of $\hat{f}$. The inversion formula for Fourier series then applies to give

$$\hat{f}(\xi) = \left(\frac{\pi}{L}\right) \operatorname*{LIM}_{N\to\infty} \left[\sum_{n=-N}^{N} f(\frac{n\pi}{L})e^{-\frac{n\pi i\xi}{L}}\right] \quad \text{if } |\xi| < L. \tag{8.2}$$

For the remainder of the proof we use the notation

$$\left(\frac{\pi}{L}\right) \sum_{n=-\infty}^{\infty} f(\frac{n\pi}{L})e^{-\frac{n\pi i\xi}{L}}$$

to denote this LIM. The function defined by this infinite sum is periodic of period $2L$; we can use it to express $\hat{f}(\xi)$ in the form

$$\hat{f}(\xi) = \left(\frac{\pi}{L}\right) \left[\sum_{n=-\infty}^{\infty} f(\frac{n\pi}{L})e^{-\frac{n\pi i\xi}{L}}\right] \chi_{[-L,L]}(\xi). \tag{8.3}$$

This proves Nyquist's theorem, for a function in $L^2(\mathbb{R})$ is determined by its Fourier transform. □

The exponentials $e^{\pm iLx}$ have period $\frac{2\pi}{L}$ and frequency $\frac{L}{2\pi}$. If a function is L-bandlimited, then $\frac{L}{2\pi}$ is the highest frequency appearing in its Fourier representation. Nyquist's theorem states that we must sample such a function at the rate $\frac{L}{\pi}$; that is, at twice its highest frequency. As we shall see, sampling at a lower rate does not provide enough information to completely determine f.

***Definition* 8.1.4.** The optimal sampling rate for an L-bandlimited function, $\frac{L}{\pi}$, is called the *Nyquist rate*. Sampling at a lower rate is called *undersampling*, and sampling at a higher rate is called *oversampling*.

If this were as far as we could go, Nyquist's theorem would be an interesting result of little practical use. However, the original function f can be explicitly reconstructed using (8.3) in the Fourier inversion formula. To justify our manipulations, we assume that f tends to zero rapidly enough so that

$$\sum_{n-\infty}^{\infty} |f(\frac{n\pi}{L})| < \infty. \tag{8.4}$$

With this understood, we use (8.3) in (8.1) to obtain

$$\begin{aligned} f(x) &= \frac{1}{2\pi}\int_{-\infty}^{\infty} \chi_{[-L,L]}(\xi)\hat{f}(\xi)e^{i\xi x}d\xi \\ &= \frac{1}{2\pi}\frac{\pi}{L}\int_{-L}^{L}\sum_{n=-\infty}^{\infty} f(\frac{n\pi}{L})e^{ix\xi-\frac{n\pi i\xi}{L}}d\xi \\ &= \frac{1}{2L}\sum_{n=-\infty}^{\infty} f(\frac{n\pi}{L})\int_{-L}^{L} e^{ix\xi-\frac{n\pi i\xi}{L}}d\xi \\ &= \sum_{n=-\infty}^{\infty} f(\frac{n\pi}{L})\operatorname{sinc}(Lx-n\pi). \end{aligned}$$

This formula expresses the value of f, for every $x \in \mathbb{R}$, in terms of the samples $\{f(\frac{n\pi}{L}) : n \in \mathbb{Z}\}$:

$$f(x) = \sum_{n=-\infty}^{\infty} f(\frac{n\pi}{L})\operatorname{sinc}(Lx-n\pi). \tag{8.5}$$

***Remark* 8.1.2.** If f belongs to $L^2(\mathbb{R})$, then $\hat{f}$ is also square integrable. This, in turn, implies that $\sum_{-\infty}^{\infty} |f(L^{-1}n\pi)|^2$ is finite. Using the Cauchy-Schwarz inequality for l^2, we easily show that the sum on the right-hand side of (8.5) converges locally uniformly. The Fourier transform of a bandlimited function is absolutely integrable. The Riemann-Lebesgue lemma therefore implies that a bandlimited function is bounded and tends to zero as $|x|$ goes to infinity. Indeed, a bounded, integrable function is also square integrable. Note, however, that a square-integrable, bandlimited function need not be absolutely integrable.

Exercises

***Exercise* 8.1.1.** Suppose that f is an L-bandlimited function. Show that it is determined by the set of samples $\{f(x_0 + \frac{\pi n}{L}) : n \in \mathbb{Z}\}$, for any $x_0 \in \mathbb{R}$.

***Exercise* 8.1.2.** Explain why, after possible modification on a set of measure zero, a bandlimited L^2-function is given by (8.1).

***Exercise* 8.1.3.** Show that a bounded integrable function defined on $\mathbb{R}$ is also square integrable.

***Exercise* 8.1.4.** Give an example of a bandlimited function in $L^2(\mathbb{R})$ that is not absolutely integrable.

***Exercise* 8.1.5.** Suppose that f is a bandlimited function in $L^2(\mathbb{R})$. Show that the infinite sum in (8.5) converges locally uniformly.

8.1.2 Shannon-Whittaker Interpolation

See: A.5.2.

The explicit interpolation formula, (8.5), for f in terms of its samples at $\{\frac{n\pi}{L} \mid n \in \mathbb{Z}\}$ is sometimes called the Shannon-Whittaker interpolation formula. In Section A.5.2 we consider other methods for interpolating a function from sampled values. These formulæ involve finite sums and only give exact reconstructions for a finite-dimensional family of functions. The Shannon-Whittaker formula gives an exact reconstruction for all L-bandlimited functions. Since it requires an infinite sum, it is mostly of theoretical significance. In practical applications only a finite part of this sum can be used. That is, we would set

$$f(x) \approx \sum_{n=-N}^{N} f(\frac{n\pi}{L}) \operatorname{sinc}(Lx - n\pi).$$

Because

$$\operatorname{sinc}(Lx - n\pi) \simeq \frac{1}{n}$$

and $\sum n^{-1} = \infty$, the partial sums of the series in (8.5) may converge to $f(x)$ very slowly. In order to get a good approximation to $f(x)$, we would therefore need to take N very large. This difficulty can be often be avoided by oversampling.

Formula (8.5) is only one of an infinite family of similar interpolation formulæ . Suppose that f is an $(L-\eta)$-bandlimited function for an $\eta > 0$. Then it is also an L-bandlimited function. This makes it possible to use oversampling to obtain more rapidly convergent interpolation formulæ . To find such formulæ select a function φ such that

1. $\hat{\varphi}(\xi) = 1$ for $|\xi| \leq L - \eta$
2. $\hat{\varphi}(\xi) = 0$ for $|\xi| > L$

A function of this sort is often called a *window function.*

From (8.2) it follows that

$$\hat{f}(\xi) = \left(\frac{\pi}{L}\right) \sum_{n=-\infty}^{\infty} f(\frac{\pi n}{L}) e^{-\frac{n\pi i\xi}{L}} \text{ for } |\xi| < L. \tag{8.6}$$

Since $\hat{f}$ is supported in $[\eta - L, L - \eta]$ and $\hat{\varphi}$ satisfies condition (1), it follows that

$$\hat{f}(\xi) = \hat{f}(\xi)\hat{\varphi}(\xi).$$

Using this observation and (8.2) in the Fourier inversion formula gives

$$
\begin{aligned}
f(x) &= \frac{1}{2\pi}\int_{-\infty}^{\infty} \hat{f}(\xi)\hat{\varphi}(\xi)e^{i\xi x}d\xi \\
&= \frac{1}{2\pi}\frac{\pi}{L}\sum_{n=-\infty}^{\infty} f(\frac{n\pi}{L})\int_{-L}^{L}\hat{\varphi}(\xi)e^{ix\xi-\frac{n\pi i\xi}{L}}d\xi \\
&= \frac{1}{2L}\sum_{n=-\infty}^{\infty} f(\frac{n\pi}{L})\varphi(x-\frac{n\pi}{L}).
\end{aligned}
\tag{8.7}
$$

This is a different interpolation formula for f; the sinc-function is replaced by $[2L]^{-1}\varphi$. The Shannon-Whittaker formula, (8.5), corresponds to the choice $\hat{\varphi} = \chi_{[-L,L]}$.

Recall that more *smoothness* in the Fourier transform of a function is reflected in faster decay of the function itself. Using a smoother function for $\hat{\varphi}$ therefore leads to a more rapidly convergent interpolation formula for f. There is a small price to pay for using a different choice of $\hat{\varphi}$. The first issue is that φ, given by

$$
\varphi(x) = \frac{1}{2\pi}\int_{-\infty}^{\infty}\hat{\varphi}(\xi)e^{ix\xi}d\xi,
$$

may be more difficult to accurately compute than the sinc function. The second is that we need to sample f above the Nyquist rate. In this calculation, f is an $(L-\eta)$-bandlimited function, but we need to use a sample spacing

$$
\frac{\pi}{L} < \frac{\pi}{L-\eta}.
$$

On the other hand, a little oversampling and additional computational overhead often leads to superior results.

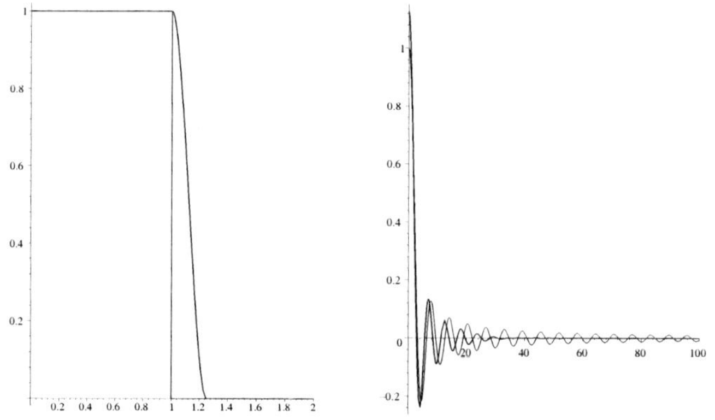

(a) Window functions in Fourier space.

(b) The corresponding interpolation functions.

Figure 8.1. Window functions in Fourier space and their inverse Fourier transforms.

***Example* 8.1.1.** We compare the Shannon-Whittaker interpolant to the interpolant obtained using a second order smoothed window, with 25% oversampling. The second-order window is defined by

$$s_2(\xi) = \begin{cases} 1 & \text{if } |\xi| \le 1, \\ 128|\xi|^3 - 432|\xi|^2 + 480|\xi| - 175 & \text{if } 1 < |\xi| < \frac{5}{4}, \\ 0 & \text{if } |\xi| \ge \frac{5}{4}. \end{cases} \tag{8.8}$$

It has one continuous derivative and a weak second derivative, its inverse Fourier transform is

$$\mathcal{F}^{-1}(s_2)(x) = 768\left[\frac{\cos(x) - \cos(5x/4)}{\pi x^4}\right] - 96\left[\frac{\sin(5x/4) + \sin(x)}{\pi x^3}\right].$$

Figure 8.1(a) shows the two window functions, and Figure 8.1(b) shows their inverse Fourier transforms. Observe that $\mathcal{F}^{-1}(s_2)$ converges to zero as x tends infinity much faster than the sinc function.

The graphs in Figure 8.2 compare the results of reconstructing *non*-bandlimited functions using the Shannon-Whittaker interpolant and that obtained using (8.7) with $\hat{\varphi}$ a suitably scaled version of s_2. The sample spacing for the Shannon-Whittaker interpolation is .1, for the "oversampled" case we use .08. The original function is shown in the first column; the second column shows both interpolants; and the third column is a detailed view near to ± 1. This is a point where the original function or one of its derivatives is discontinuous. Two things are evident in these graphs: The error in the oversampled interpolant tends to zero much faster as x tends to infinity, and the smoother the function, the easier it is to interpolate.

Exercises

***Exercise* 8.1.6.** Use the Shannon-Whittaker formula to reconstruct the function

$$f(x) = \frac{\sin(Lx)}{\pi x}$$

from the samples $\{f(\frac{n\pi}{L})\}$.

***Exercise* 8.1.7.** How should the Shannon-Whittaker formula be modified if, instead of $\{f(\frac{\pi n}{L}) : n \in \mathbb{Z}\}$, the samples $\{f(x_0 + \frac{\pi n}{L}) : n \in \mathbb{Z}\}$ are collected?

***Exercise* 8.1.8.** Show that for each $n \in \mathbb{N}$, function, $\operatorname{sinc}(Lx - n\pi)$ is L-bandlimited. The Shannon-Whittaker formula therefore expresses an L-bandlimited function as a sum of such functions.

***Exercise* 8.1.9.** The Fourier transform of

$$f(x) = \frac{1 - \cos(x)}{2\pi x}$$

is $\hat{f}(\xi) = i \operatorname{sgn} \xi \chi_{[-1,1]}(\xi)$. Use the Shannon-Whittaker formula to reconstruct f from the samples $\{f(n\pi)\}$.

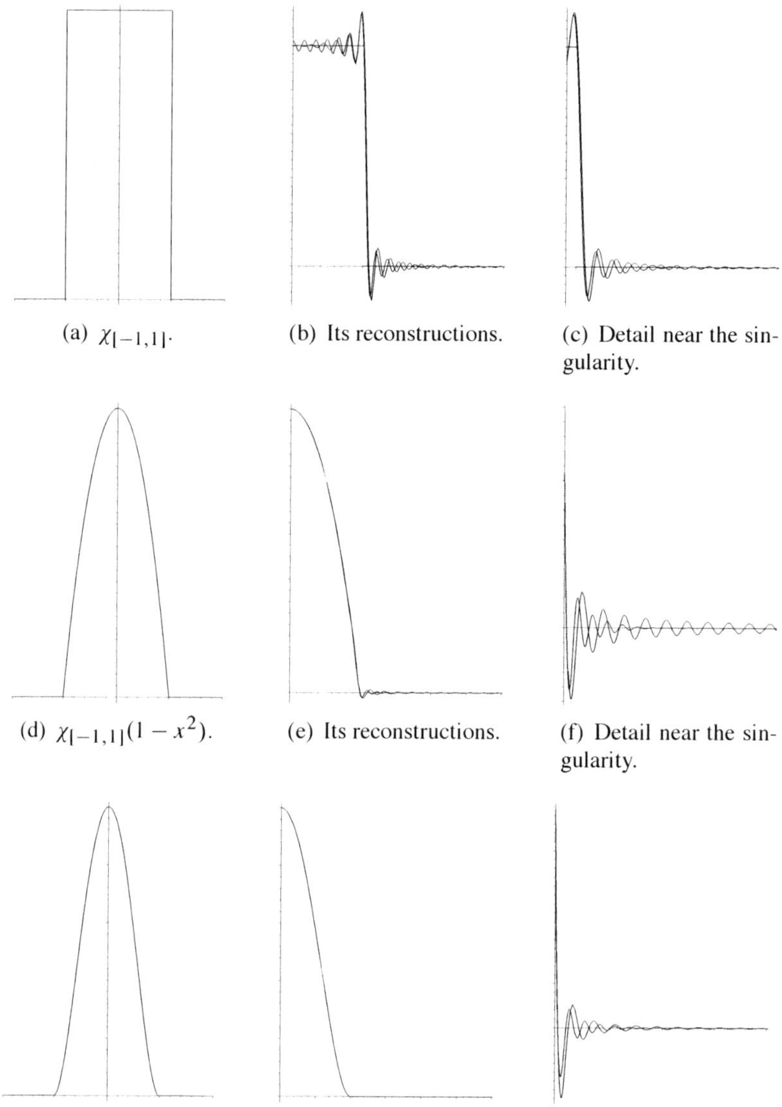

(a) $\chi_{[-1,1]}$.
(b) Its reconstructions.
(c) Detail near the singularity.
(d) $\chi_{[-1,1]}(1 - x^2)$.
(e) Its reconstructions.
(f) Detail near the singularity.
(g) $\chi_{[-1,1]}(1 - x^2)^2$.
(h) Its reconstructions.
(i) Detail near the singularity.

Figure 8.2. Shannon-Whittaker and generalized Shannon-Whittaker interpolation for several functions

8.2 The Poisson Summation Formula

What happens if we do not have enough samples to satisfy the hypotheses of Nyquist's theorem? For example, what if our signal is not bandlimited? Functions that describe images in medical applications generally have bounded support, so they cannot be bandlimited and therefore we are always undersampling (see Chapter 4, Proposition 4.4.1). To analyze the effects of undersampling, we introduce the Poisson summation formula. It gives a relationship between the Fourier transform and the Fourier series.

8.2.1 The Poisson Summation Formula

Assume that f is a continuous function that decays reasonably fast as $|x| \to \infty$. We construct a periodic function out of f by summing the values of f at its integer translates. Define f_p by

$$f_p(x) = \sum_{n=-\infty}^{\infty} f(x+n). \tag{8.9}$$

This is a periodic function of period 1, $f_p(x+1) = f_p(x)$. If f is absolutely integrable on $\mathbb{R}$, then it follows from Fubini's theorem that f_p is absolutely integrable on $[0, 1]$.

The Fourier *coefficients* of f_p are closely related to the Fourier *transform* of f:

$$\begin{aligned}
\hat{f}_p(m) &= \int_0^1 f_p(x) e^{-2\pi imx} dx \\
&= \int_0^1 \sum_{n=-\infty}^{\infty} f(x+n) e^{-2\pi imx} dx = \sum_{n=-\infty}^{\infty} \int_n^{n+1} f(x) e^{-2\pi imx} dx \\
&= \int_{-\infty}^{\infty} f(x) e^{-2\pi imx} dx = \hat{f}(2\pi m).
\end{aligned}$$

The interchange of the integral and summation is easily justified if f is absolutely integrable on $\mathbb{R}$.

Proceeding formally, we use the Fourier inversion formula for periodic functions, Theorem 7.1.2, to find Fourier series representation for f_p,

$$f_p(x) = \sum_{n=-\infty}^{\infty} \hat{f}_p(n) e^{2\pi inx} = \sum_{n=-\infty}^{\infty} \hat{f}(2\pi n) e^{2\pi inx}.$$

Note that $\{\hat{f}_p(n)\}$ are the Fourier *coefficients* of the 1-periodic function f_p whereas $\hat{f}$ is the Fourier *transform* of the absolutely integrable function f defined on all of $\mathbb{R}$. To justify these computations, it is necessary to assume that the coefficients $\{\hat{f}(2\pi n)\}$ go to zero sufficiently rapidly. If f is smooth enough, then this will be true. The *Poisson summation formula* is a precise formulation of these observations.

Theorem 8.2.1 (Poisson summation formula). *If f is an absolutely integrable function such that*

$$\sum_{n=-\infty}^{\infty} |\hat{f}(2\pi n)| < \infty$$

then, at points of continuity of f_p, we have

$$\sum_{n=-\infty}^{\infty} f(x+n) = \sum_{n=-\infty}^{\infty} \hat{f}(2\pi n)e^{2\pi inx}. \tag{8.10}$$

***Remark* 8.2.1.** The hypotheses in the theorem are not quite optimal. Some hypotheses are required, as there are examples of absolutely integrable functions f such that both sums,

$$\sum_{n=-\infty}^{\infty} |f(x+n)| \text{ and } \sum_{n=-\infty}^{\infty} |\hat{f}(2\pi n)|$$

converge but (8.10) does *not* hold. A more detailed discussion can be found in [79].

Using the preceding argument and rescaling, we easily find the Poisson summation formula for $2L$-periodic functions:

$$\sum_{n=-\infty}^{\infty} f(x+2nL) = \frac{1}{2L} \sum_{n=-\infty}^{\infty} \hat{f}(\frac{\pi n}{L})e^{\frac{\pi inx}{L}}. \tag{8.11}$$

Suitably scaled, the hypotheses are the same as those in Theorem 8.2.1.

As an application of (8.10) we can prove an x-space version of Nyquist's theorem. Suppose that f equals 0 outside the interval $[-L, L]$ (i.e., f is a *space-limited function*). For each $x \in [-L, L]$, only the $n = 0$ term on the left-hand side of (8.11) is nonzero. The Poisson summation formula states that

$$f(x) = \frac{1}{2L} \sum_{n=-\infty}^{\infty} \hat{f}(\frac{\pi n}{L})e^{\frac{\pi inx}{L}} \qquad \text{for } x \in [-L, L].$$

Therefore, if f is supported in $[-L, L]$, then it can be reconstructed from the samples of its Fourier transform

$$\{\hat{f}(\frac{\pi n}{L}) \mid n \in \mathbb{Z}\}.$$

This situation arises in magnetic resonance imaging (MRI). In this modality, we directly measure samples of the Fourier transform of the image function. That is, we measure $\{\hat{f}(n\Delta\xi)\}$. On the other hand, the function is known, a priori, to be supported in a fixed, bounded set $[-L, L]$. In order to reconstruct f without introducing errors, we need to take

$$\Delta\xi \leq \frac{\pi}{L}. \tag{8.12}$$

Thus, if we measure samples of the Fourier transform of a space-limited function, then Nyquist's theorem places a constraint on the sample spacing in the *Fourier domain*. An extensive discussion of sampling in MRI can be found in [50]. Figure 8.3 shows the result, in MRI, of undersampling the Fourier transform. Note that portions of the original image have been "folded over" at the top and right-hand side of the reconstructed image.

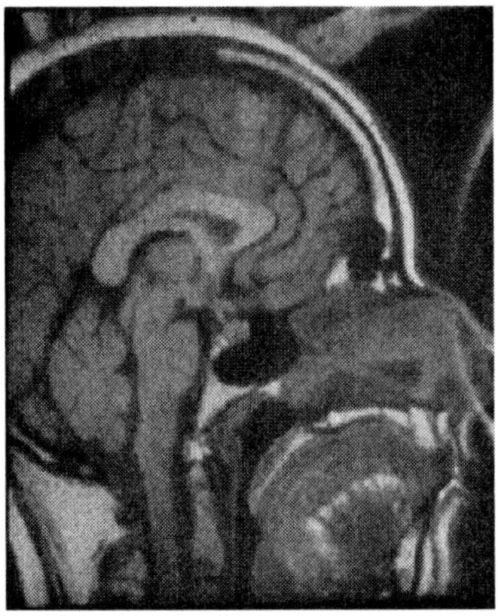

Figure 8.3. Aliasing artifacts produced by undersampling in magnetic resonance imaging. (Image courtesy of Dr. Felix Wehrli.)

It most applications we sample a function rather than its Fourier transform. The analysis of undersampling in this situation requires the *dual Poisson summation formula*. Let f be a function such that the sum

$$\sum_{-\infty}^{\infty} \hat{f}(\xi + 2nL)$$

converges. Considering

$$\hat{f}_p(\xi) = \sum_{-\infty}^{\infty} \hat{f}(\xi + 2nL)$$

and its Fourier coefficients in the same manner as previously we obtain the following:

Theorem 8.2.2 (The dual Poisson summation formula). *If f is a function such that $\hat{f}$ is absolutely integrable and*

$$\sum_{n=-\infty}^{\infty} |f(\frac{\pi n}{L})| < \infty$$

then, at a point of continuity of $\hat{f}_p$,

$$\sum_{n=-\infty}^{\infty} \hat{f}(\xi + 2nL) = \left(\frac{\pi}{L}\right) \sum_{n=-\infty}^{\infty} f(\frac{\pi n}{L}) e^{-\frac{n\pi i\xi}{L}}. \tag{8.13}$$

Exercises

***Exercise* 8.2.1.** Give the details for the proof of Theorem 8.2.2. You may assume that f is smooth and rapidly decreasing.

***Exercise* 8.2.2.** Explain formula (8.12).

Exercise* 8.2.3. This exercise requires a knowledge of the Fourier transform for generalized functions (see Section 4.4.4). Suppose that f is a periodic function of period 1. The generalized function l_f has a Fourier transform that is a generalized function. Using the dual Poisson summation formula, show that

$$\widehat{l_f} = 2\pi \sum_{n=-\infty}^{\infty} \hat{f}(n)\delta(2\pi n - \xi); \tag{8.14}$$

here $\{\hat{f}(n)\}$ are the Fourier coefficients defined in (7.1).

Exercise* 8.2.4. What is the analogue of formula (8.14) for a $2L$-periodic function?

8.2.2 Undersampling and Aliasing*

Using the Poisson summation formula, we analyze the errors introduced by undersampling. Whether or not f is an L-bandlimited function, the Shannon-Whittaker formula defines an L-bandlimited function:

$$F_L(x) = \sum_{n=-\infty}^{\infty} f(\frac{n\pi}{L}) \operatorname{sinc}(Lx - n\pi).$$

As $\operatorname{sinc}(0) = 1$ and $\operatorname{sinc}(n\pi) = 0$, for nonzero integers it follows that F_L interpolates f at the sample points,

$$F_L(\frac{n\pi}{L}) = f(\frac{n\pi}{L}) \qquad \text{for } n \in \mathbb{Z}.$$

Reversing the steps in the derivation of the Shannon-Whittaker formula, and applying formula (8.13) we see that Fourier transform of F_L is given by

$$\widehat{F_L}(\xi) = \sum_{n=-\infty}^{\infty} \hat{f}(\xi + 2nL)\chi_{[-L,L]}(\xi). \tag{8.15}$$

If f is L-bandlimited then for all ξ, we have

$$\hat{f}(\xi) = \widehat{F_L}(\xi).$$

On the other hand, if f is *not* L-bandlimited, then

$$\hat{f}(\xi) - \widehat{F_L}(\xi) = \begin{cases} \hat{f}(\xi) & \text{if } |\xi| > L, \\ -\sum_{n\neq 0} \hat{f}(\xi + 2nL) & \text{if } |\xi| \leq L. \end{cases} \tag{8.16}$$

The function F_L or its Fourier transform $\hat{F}_L$ encodes all the information present in the sequence of samples. Formula (8.16) shows that there are two distinct sources of error in F_L. The first is *truncation error*; as F_L is L-bandlimited, the high-frequency information in f is no longer available in F_L. The second source of error arises from the fact that the high-frequency information in f *reappears* at low frequencies in the function F_L. This latter type of distortion is called *aliasing*. The high-frequency information in the original signal is not simply "lost" but resurfaces, corrupting the low frequencies. Hence F_L faithfully reproduces neither the high-frequency nor the low-frequency information in f.

Aliasing is familiar in everyday life: If we observe the rotation of the wheels of a fast-moving car in a movie, it appears that the wheels rotate very slowly. A movie image is actually a sequence of samples (24 frames/second). This sampling rate is below the Nyquist rate needed to accurately reproduce the motion of the rotating wheel.

***Example* 8.2.1.** If a car is moving at 60 mph and the tires are 3 ft in diameter, then the angular velocity of the wheels is

$$\omega = 58\frac{1}{3}\frac{\text{rotations}}{\text{second}}.$$

We can model the motion of a point on the wheel as $(r\cos((58\frac{1}{3})2\pi t), r\sin((58\frac{1}{3})2\pi t)$. The Nyquist rate is therefore

$$2\cdot 58\frac{1}{3}\frac{\text{frames}}{\text{second}} \simeq 117\frac{\text{frames}}{\text{second}}.$$

Sampling only 24 times per second leads to aliasing. As $58\frac{1}{3} = 10\frac{1}{3} + 2 * 24$, the aliased frequencies are $\pm(10\frac{1}{3})$.

The previous example is useful to conceptualize the phenomenon of aliasing but has little direct bearing on imaging. To better understand the role of aliasing in imaging we rewrite F_L in terms of its Fourier transform,

$$F_L(x) = \frac{1}{2\pi}\int_{-L}^{L} \hat{f}(\xi)e^{ix\xi}d\xi + \frac{1}{2\pi}\int_{-L}^{L}\sum_{n\neq 0}\hat{f}(\xi + 2nL)e^{ix\xi}d\xi.$$

The first term is the partial Fourier inverse of f. For a function with jump discontinuities this term produces Gibbs oscillations. The second term is the aliasing error itself. In many examples of interest in imaging, the Gibbs artifacts and the aliasing error are about the same size. What makes either term a problem is slow decay of the Fourier transform of f.

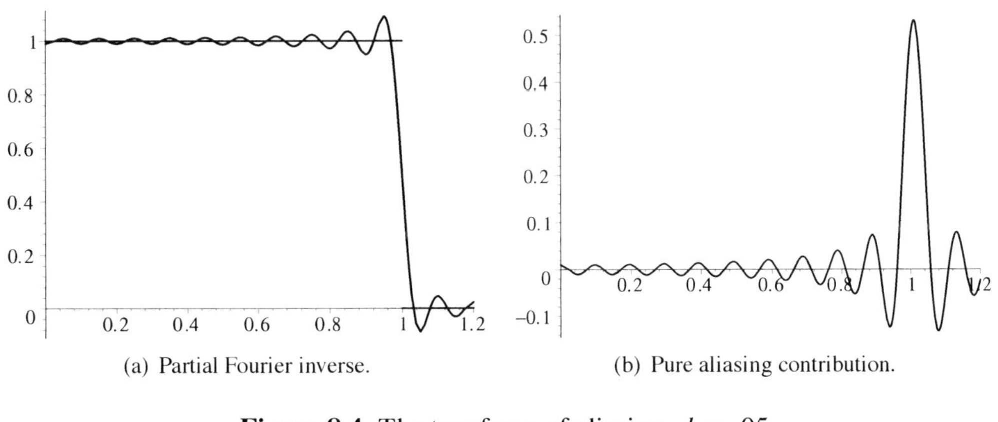

(a) Partial Fourier inverse. (b) Pure aliasing contribution.

Figure 8.4. The two faces of aliasing, $d = .05$.

***Example* 8.2.2.** In Figure 8.4 the two contributions to $f - F_L$ are shown separately, for the rectangle function $f = \chi_{[-1,1]}$. Figure 8.4(a) shows the Gibbs contribution; Figure 8.4(b) shows the "pure aliasing" part. Figure 8.5 shows the original function, its partial Fourier inverse, $\mathcal{F}^{-1}(\hat{f}\chi_{[-L,L]})$, and its Shannon-Whittaker interpolant. The partial Fourier inverse is the medium weight line. The curve slightly to the right of this line is the Shannon-Whittaker interpolant. In this example, the contributions of the Gibbs artifact and the pure aliasing error are of about the same size and have same general character. It is evident that the Shannon-Whittaker interpolant is more distorted than the partial inverse of the Fourier transform, though visually they are quite similar.

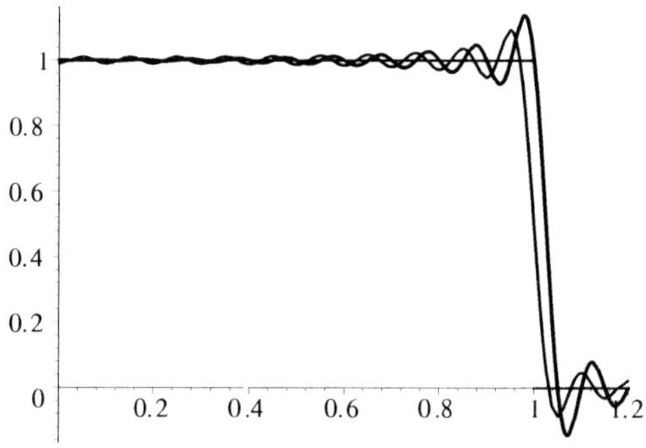

Figure 8.5. Partial Fourier inverse and Shannon-Whittaker interpolant.

***Example* 8.2.3.** For comparison, consider the continuous function $g(x) = \chi_{[-1,1]}(x)(1 - x^2)$ and its reconstruction using the sample spacing $d = .1$. In Figure 8.6(a) it is just barely possible to distinguish the original function from its approximate reconstruction. The worst errors occur near the points ± 1, where g is finitely differentiable. Figure 8.6(b) shows the graph of the difference, $g - G_L$ (note the scale along the y-axis).

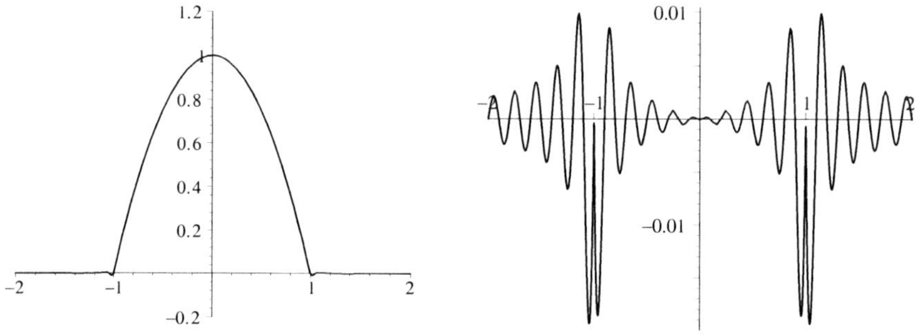

(a) The Shannon-Whittaker interpolation.

(b) The difference.

Figure 8.6. What aliasing looks like for a continuous, piecewise differentiable function, $d = 0.1$.

***Example* 8.2.4.** As a final example we consider the effect of sampling on a "fuzzy function." Here we use a function of the sort introduced in Example4.2.5. These are continuous functions with "sparse," but slowly decaying, Fourier transforms. Figure 8.7(a) is the graph of such a function, and Figure 8.7(b) shows the Shannon-Whittaker interpolants with $d = .1, .05$, and $.025$. For a function of this sort, Shannon-Whittaker interpolation appears to produce *smoothing*.

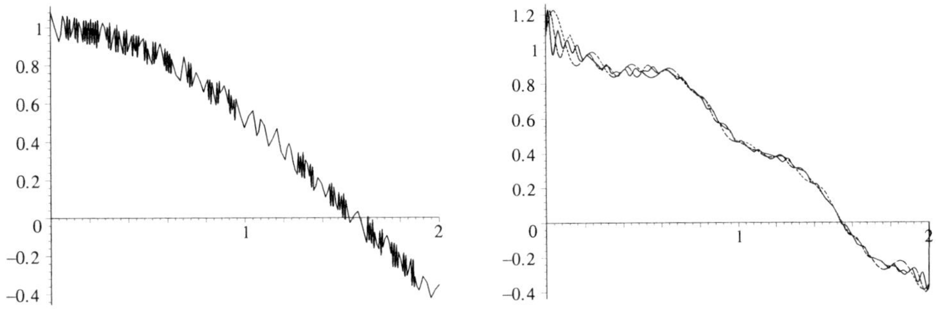

(a) A fuzzy function. (b) Shannon-Whittaker interpolants.

Figure 8.7. What aliasing looks like for a fuzzy function, $d = .1, .05, .025$.

Remark* 8.2.2. The functions encountered in imaging applications are usually spatially limited and therefore cannot be bandlimited. However, if a function f is smooth enough, then its Fourier transform decays rapidly and therefore, by choosing L sufficiently large, the difference, $\hat{f} - \hat{F}_L$ can be made "small." If this is so, then the *effective support* of $\hat{f}$ is contained in $[-L, L]$, and f is an *effectively bandlimited function.* Though effective support is an important concept in imaging, it does not have a precise definition. In most applications it does not suffice to have $\hat{f}$ itself small outside of $[-L, L]$. Usually the sampling rate must be large enough so that the aliasing error,

$$\sum_{n\neq 0} \hat{f}(\xi + 2nL),$$

is also small. This is a somewhat heuristic principle because the meaning of *small* is dictated by the application.

Examples 8.2.2 and 8.2.3 illustrate what is meant by effective bandlimiting. Neither function is actually bandlimited. No matter how large L is taken, a Shannon-Whittaker interpolant for f, in Example 8.2.2, displays large oscillatory artifacts. In most applications this function would not be considered effectively L-bandlimited, for any L. However, it should be noted that, away from the jumps, the Shannon-Whittaker interpolant does a good job reconstructing f. For most purposes, the function g would be considered effectively bandlimited, though the precise effective bandwidth would depend on the application.

To diminish the effects of aliasing, an analogue signal may be passed through a "lowpass filter" *before* it is sampled. In general terms, a lowpass filter is an operation that attenuates the high-frequency content of a signal without introducing too much distortion into the low-frequency content. In this way the sampled data accurately represent the low-frequency information present in the original signal without corruption from the high frequencies. An *ideal* lowpass filter removes all the high-frequency content in a signal outside of given band, leaving the data within the *passband* unchanged. An ideal lowpass filter replaces f with the signal f_L, defined by the following properties:

$$\begin{aligned}\hat{f}_L(\xi) &= \hat{f}(\xi) \text{ if } |\xi| \leq L,\\ \hat{f}_L(\xi) &= 0 \text{ if } |\xi| \geq L.\end{aligned} \tag{8.17}$$

The samples $\{f_L(\frac{n\pi}{L})\}$ contain all the low frequency-information in f *without* the aliasing errors. Using the Shannon-Whittaker formula to reconstruct a function, with these samples, gives f_L for all x. This function is just the partial Fourier inverse of f,

$$f_L(x) = \frac{1}{2\pi}\int_{-L}^{L} \hat{f}(\xi)e^{ix\xi}\,d\xi,$$

and is still subject to unpleasant artifacts like the Gibbs phenomenon.

A realistic measurement consists of samples of a convolution $\varphi * f$, with φ a function with total integral one. If it has support in $[-\eta, \eta]$, then the measurement at $\frac{n\pi}{L}$ is the average,

$$\varphi * f(\frac{n\pi}{L}) = \int_{-\eta}^{\eta} f(\frac{n\pi}{L} - x)\varphi(x)\,dx.$$

Hence "measuring" the function f at $x = \frac{n\pi}{L}$ is the same thing as sampling the convolution $\varphi * f$ at $\frac{n\pi}{L}$. The Fourier transform of φ goes to zero as the frequency goes to infinity; the smoother φ is, the faster this occurs. As the Fourier transform of $\varphi * f$ is

$$\widehat{\varphi * f}(\xi) = \hat{\varphi}(\xi)\hat{f}(\xi),$$

the measurement process itself attenuates the high-frequency content of f. On the other hand,

$$\hat{\varphi}(0) = \int_{-\infty}^{\infty} \varphi(x)\,dx = 1$$

and therefore $\widehat{\varphi * f}$ resembles $\hat{f}$ for sufficiently low frequencies. Most measurement processes provide a form of lowpass filtering.

The more sharply peaked φ is, the larger the interval over which the "measurement error,"

$$\widehat{\varphi * f}(\xi) - \hat{f}(\xi) = (1 - \hat{\varphi}(\xi))\hat{f}(\xi)$$

can be controlled. The aliasing error in the measured samples is

$$\sum_{n\neq 0} \hat{\varphi}(\xi + 2nL)\hat{f}(\xi + 2nL).$$

By choosing φ to be smooth, this can be made as small as we like. If φ is selected so that $\hat{\varphi}(nL) = 0$ for $n \in \mathbb{Z} \setminus \{0\}$, then the Gibbs-like artifacts that result from truncating the Fourier transform to the interval $[-L, L]$ can also be eliminated.

Remark **8.2.3.** A detailed introduction to *wavelets* that includes interesting generalizations of the Poisson formula and the Shannon-Whittaker formula can be found in [58].

Exercises

Exercise **8.2.5.** Derive formula (8.15) for $\widehat{F}_L$.

Exercise **8.2.6.** Compute the Fourier transform of $g(x) = \chi_{[-1,1]}(x)(1 - x^2)$.

Exercise **8.2.7.** What forward velocity of the car in Example8.2.1 corresponds to the apparent rotational velocity of the wheels? What if the car is going 40 mph?

Exercise **8.2.8.** Sometimes in a motion picture or television image the wheels of a car appear to be going clockwise, even though the car is moving forward. Explain this by giving an example.

Exercise **8.2.9.** Explain why the artifact produced by aliasing looks like the Gibbs phenomenon. For the function $\chi_{[-1,1]}$, explain why the size of the pointwise error, in the Shannon-Whittaker interpolant, does not go to zero as the sample spacing goes to zero.

Exercise **8.2.10.** Experiment with the family of functions

$$f_\alpha(x) = \chi_{[-1,1]}(x)(1 - x^2)^\alpha$$

to understand effective bandlimiting. For a collection of $\alpha \in [0, 2]$, see whether there is a Gibbs-like artifact in the Shannon-Whittaker interpolants and, if not, at what sample spacing is the Shannon-Whittaker interpolant visually indistinguishable from the original function (over $[-2, 2]$).

Exercise **8.2.11.** The ideal lowpass filtered function, f_L, can be expressed as a convolution

$$f_L(x) = f * k_L(x).$$

Find the function k_L. If the variable x is "time," explain the difficulty in implementing an ideal lowpass filter.

Exercise **8.2.12.** Suppose that φ is a non-negative, even, real valued function such that $\hat{\varphi}(0) = 1$; explain why the interval over which $|\hat{f}(\xi) - \widehat{\varphi * f}(\xi)|$ is small is controlled by

$$\int_{-\infty}^{\infty} x^2 \varphi(x)\, dx.$$

Exercise **8.2.13.** Show that if

$$\psi(x) = \varphi * \chi_{[-\frac{\pi}{L}, \frac{\pi}{L}]}(x),$$

then $\hat{\psi}(nL) = 0$ for all $n \in \mathbb{Z} \setminus \{0\}$.

8.2.3 Subsampling

Subsampling is a way to take advantage of aliasing to "demodulate" a bandlimited signal whose Fourier transform is supported in a set of the form $[-\omega - B, -\omega + B]$ or $[\omega - B, \omega + B]$. In this context ω is called the *carrier frequency* and $2B$ the bandwidth of the signal. This situation arises in FM radio as well as in MR imaging. For simplicity, suppose that there is a positive integer N so that

$$\omega = NB.$$

Let f be a function whose Fourier transform is supported in $[\omega - B, \omega + B]$. If we sample this function at the points $\{\frac{n\pi}{B}, : n \in \mathbb{Z}\}$ and use formula (8.5) to obtain F_L, then (8.13) implies that

$$\widehat{F_L}(\xi) = \hat{f}(\omega + \xi). \tag{8.18}$$

The function F_L is called the demodulated version of f; the two signals are very simply related:

$$f(x) = e^{-i\omega x} F_L(x).$$

From the formula relating f and F_L, it is clear that if F_L is real valued, then, in general, the measured signal, f, is not. A similar analysis applies to a signal with Fourier transform supported in $[-\omega - B, -\omega + B]$.

Exercises

***Exercise* 8.2.14.** Suppose that ω is not an integer multiple of B and that f is a signal whose Fourier transform is supported in $[\omega - B, \omega + B]$. If F_L is constructed as previously from the samples $\{f(\frac{n\pi}{B})\}$, then determine $\widehat{F_L}$. What should be done to get a faithfully demodulated signal by sampling? Keep in mind that normally $\omega >> B$.

***Exercise* 8.2.15.** Suppose that f is a real-valued function whose Fourier transform is supported in $[-\omega - B, -\omega + B] \cup [\omega - B, \omega + B]$. Assuming that $\omega = NB$ and f is sampled at $\{\frac{n\pi}{B}\}$, how is $\widehat{F_L}$ related to $\hat{f}$?

8.3 Periodic Functions and the Finite Fourier Transform*

See: 2.3.2.

We now adapt the discussion of sampling to the context of periodic functions. In this case a natural *finite* analogue of the Fourier series plays a role analogous to that played, in the study of sampling for functions defined on $\mathbb{R}$, by the Fourier series itself. We begin with the definition and basic properties of the *finite Fourier transform.*

***Definition* 8.3.1.** Suppose that $< x_0, \ldots, x_{m-1} >$ is a sequence of complex numbers. The *finite Fourier transform* of this sequence is the sequence $< \hat{x}_0, \ldots, \hat{x}_{m-1} >$ of complex numbers defined by

$$\hat{x}_k = \frac{1}{m} \sum_{j=0}^{m-1} x_j e^{-\frac{2\pi ijk}{m}}. \tag{8.19}$$

Sometimes it is denoted by

$$\mathcal{F}_m(< x_0, \ldots, x_{m-1} >) = (< \hat{x}_0, \ldots, \hat{x}_{m-1} >).$$

Using the formula for the sum of a geometric series and the periodicity of the exponential function, we easily obtain the formulæ

$$\sum_{j=0}^{m-1} \exp\left(\frac{2\pi ij}{m}(k-l)\right) = \begin{cases} m & k = l, \\ 0 & k \neq l. \end{cases} \tag{8.20}$$

These formulæ have a nice geometric interpretation: The set of vectors

$$\{(1, e^{\frac{2\pi ik}{m}}, e^{\frac{4\pi ik}{m}}, \ldots, e^{\frac{2(m-1)\pi ik}{m}}) : k = 0, \ldots, m-1\}$$

is an orthogonal basis for $\mathbb{C}^m$. These vectors are obtained by sampling the functions $\{e^{2\pi ikx} : \quad k = 0, \ldots, m-1\}$ at the points $\{\frac{j}{m} : j = 0, \ldots, m-1\}$.

The computations in (8.20) show that the inverse of the finite Fourier transform is given by

$$x_j = \sum_{k=0}^{m-1} \hat{x}_k e^{\frac{2\pi ijk}{m}}. \tag{8.21}$$

The formulæ (8.19) and (8.21) defining the sequences $< \hat{x}_k >$ and $< x_j >$, respectively, make sense with k or j any integer. In much the same way, as it is often useful to think of a function defined on $[0, L]$ as an L-periodic functions, it is useful to think of $< x_j >$ and $< \hat{x}_k >$ as bi-infinite sequences satisfying

$$x_{j+m} = x_j \text{ and } \hat{x}_{k+m} = \hat{x}_k. \tag{8.22}$$

Such a sequence is called an *m-periodic sequence.*

The summation in (8.21) is quite similar to that in (8.19); the exponential multipliers have been replaced by their complex conjugates. This means that a fast algorithm for computing $\mathcal{F}_m$ automatically provides a fast algorithm for computing $\mathcal{F}_m^{-1}$. Indeed, if m is a power of 2, then there is a fast algorithm for computing both $\mathcal{F}_m$ and $\mathcal{F}_m^{-1}$. Either transformation requires about $3m \log_2 m$ computations, which should be compared to the $O(m^2)$ computation generally required to multiply an m-vector by an $m \times m$-matrix. This algorithm, called the *fast Fourier transform* or *FFT*, is outlined in Section 10.5.

We now return to our discussion of sampling for periodic functions. Let f be an L-periodic function with Fourier coefficients $< \hat{f}(n) >$.

***Definition* 8.3.2.** A periodic function f is called N-bandlimited if $\hat{f}(n) = 0$ for all n with $|n| \geq N$.

In this case, the inversion formula for Fourier series implies that

$$f(x) = \frac{1}{L} \sum_{n=1-N}^{N-1} \hat{f}(n) e^{\frac{2\pi i n x}{L}}.$$

This is a little simpler than the continuum case since f already lies in the finite-dimensional space of functions spanned by

$$\{e^{\frac{2\pi i n x}{L}} \, : \, 1 - N \leq n \leq N - 1\}.$$

Suppose that f is sampled at $\{\frac{jL}{2N-1} \, : \, j = 0, \ldots, 2N - 2\}$. Substituting the Fourier representation of f into the sum defining the finite Fourier transform gives

$$\begin{aligned}
\sum_{j=0}^{2N-2} f(\frac{jL}{2N-1}) e^{-\frac{2\pi i k j}{2N-1}} &= \sum_{j=0}^{2N-2} \frac{1}{L} \sum_{n=1-N}^{N-1} \hat{f}(n) \exp\left(\frac{2\pi i n j L}{(2N-1)L} - \frac{2\pi i k j}{2N-1}\right) \\
&= \frac{1}{L} \sum_{n=1-N}^{N-1} \hat{f}(n) \sum_{j=0}^{2N-2} \exp\left(\frac{2\pi i j}{2N-1}(n-k)\right) \\
&= \frac{2N-1}{L} \hat{f}(k), \quad k \in \{1-N, 2-N, \ldots, N-2, N-1\}.
\end{aligned} \tag{8.23}$$

The relations in (8.20), with m replaced by $2N - 1$, are used to go from the second to the third line. This computation shows that if f is an N-bandlimited function, then, but for an overall multiplicative factor, the *finite* Fourier transform of the sequence of samples $< f(0), \ldots, f(\frac{(2N-2)L}{2N-1}) >$ computes the nonzero Fourier coefficients of f itself. From the periodicity of $\mathcal{F}_{2N-1}(< f(0), \ldots, f(\frac{(2N-2)L}{2N-1}) >)$ it follows that

$$\begin{aligned}
\mathcal{F}_{2N-1}(< f(0), \ldots, f(\frac{(2N-2)L}{2N-1}) >) &= \\
< \hat{f}(0), \hat{f}(1), \ldots, \hat{f}(N-1), \hat{f}(1-N), \ldots, \hat{f}(-2), \hat{f}(-1) > .
\end{aligned} \tag{8.24}$$

The inversion formula for the finite Fourier transform implies the periodic analogue of Nyquist's theorem.

Theorem 8.3.1 (Nyquist's theorem for periodic functions). *If f is an L-periodic function and $\hat{f}(n) = 0$ for $|n| \geq N$, then f can be reconstructed from the equally spaced samples $\{f(\frac{jL}{2N-1}) \, : \, j = 0, 1, \ldots, (2N-2)\}$.*

From equation (8.23) and the Fourier inversion formula, we derive an interpolation formula analogous to (8.5):

$$\begin{aligned}
f(x) &= \frac{1}{L}\sum_{n=1-N}^{N-1} \hat{f}(n) e^{\frac{2\pi i n x}{L}} \\
&= \frac{1}{L}\sum_{n=1-N}^{N-1} \frac{L}{2N-1}\sum_{j=0}^{2N-2} f\left(\frac{jL}{2N-1}\right) e^{-\frac{2\pi i n j}{2N-1}} e^{\frac{2\pi i n x}{L}} \\
&= \frac{1}{2N-1}\sum_{j=0}^{2N-2} f\left(\frac{jL}{2N-1}\right) \sum_{n=1-N}^{N-1} e^{-\frac{2\pi i n j}{2N-1}} e^{\frac{2\pi i n x}{L}} \\
&= \frac{1}{2N-1}\sum_{j=0}^{2N-2} f\left(\frac{jL}{2N-1}\right) \frac{\sin \pi (2N-1)\left(\frac{x}{L} - \frac{j}{2N-1}\right)}{\sin \pi \left(\frac{x}{L} - \frac{j}{2N-1}\right)}
\end{aligned} \tag{8.25}$$

Even if f is not bandlimited, the last line in (8.25) defines an N-bandlimited function,

$$F_N(x) = \frac{1}{2N-1}\sum_{j=0}^{2N-2} f\left(\frac{jL}{2N-1}\right) \frac{\sin \pi (2N-1)\left(\frac{x}{L} - \frac{j}{2N-1}\right)}{\sin \pi \left(\frac{x}{L} - \frac{j}{2N-1}\right)}.$$

As before, this function interpolates f at the sample points

$$F_N\left(\frac{jL}{2N-1}\right) = f\left(\frac{jL}{2N-1}\right), \quad j = 0, 1, \ldots, (2N-2).$$

The Fourier coefficients of F_N are related to those of f by

$$\hat{F}_N(k) = \sum_{n=-\infty}^{\infty} \hat{f}(k + n(2N-1)) = \hat{f}(k) + \sum_{n \neq 0} \hat{f}(k + n(2N-1)) \quad 1 - N \leq k \leq N - 1.$$

If f is not N-bandlimited, then F_N has aliasing distortion: High-frequency data in f distort the low frequencies in F_N. Of course, if f is discontinuous, then F_N also displays Gibbs oscillations.

Exercises

***Exercise* 8.3.1.** Prove (8.20). Remember to use the Hermitian inner product!

***Exercise* 8.3.2.** Explain formula (8.24). What happens if f is not N-bandlimited?

Exercise* 8.3.3. As an m-periodic sequence, $< x_0, \ldots, x_{m-1} >$ is even if

$$x_j = x_{m-j} \text{ for } j = 1, \ldots, m$$

and odd if

$$x_j = -x_{m-j} \text{ for } j = 1, \ldots, m.$$

Show that the finite Fourier transform of a real-valued, even sequence is real valued and the finite Fourier transform of a real-valued, odd sequence is imaginary valued.

***Exercise* 8.3.4.** Suppose that f is an N-bandlimited, L-periodic function. For a subset $\{x_1, \ldots, x_{2N-1}\}$ of $[0, L)$ such that

$$x_j \neq x_k \text{ if } j \neq k,$$

show that f can be reconstructed from the samples

$$\{f(x_j) \ : \ j = 1, \ldots, 2N-1\}.$$

From the point of view of computation, explain why equally spaced samples are preferable.

***Exercise* 8.3.5.** Prove that F_N interpolates f at the sample points:

$$F_N(\frac{jL}{2N-1}) = f(\frac{jL}{2N-1}), \ j = 0, 1, \ldots, 2N-2.$$

***Exercise* 8.3.6.** Find analogues of the generalized Shannon-Whittaker formula in the periodic case.

8.4 Quantization Errors

See: A.1.

In the foregoing sections it is implicitly assumed that we have a continuum of numbers at our disposal to make measurements and do computations. As digital computers are used to implement the various filters, this is not the case. In Section A.1.2 we briefly discuss how numbers are actually stored and represented in a computer. For simplicity we consider a base 2, fixed-point representation of numbers. Suppose that we have $(n+1)$ bits and let the binary sequence $(b_0, b_1, \ldots, b_n)$ correspond to the number

$$(b_0, b_1, \ldots, b_n) \leftrightarrow (-1)^{b_0} \frac{\sum_{j=0}^{n-1} b_{j+1} 2^j}{2^n}. \tag{8.26}$$

This allows us to represent numbers between -1 and $+1$ with a maximum error of 2^{-n}. There are several ways to map the continuum onto $(n+1)$-bit binary sequences. Such a correspondence is called a *quantization* map. In essentially any approach, numbers greater than or equal to 1 are mapped to $(0, 1, \ldots, 1)$, and those less than or equal to -1 are mapped to $(1, 1, \ldots, 1)$. This is called clipping and is very undesirable in applications. To avoid clipping, the data are usually scaled before they are quantized.

The two principal quantization schemes are called *rounding* and *truncation*. For a number x between -1 and $+1$, its rounding is defined to be the number of the form in (8.26) closest to x. If we denote this by $Q_r(x)$, then clearly

$$|Q_r(x) - x| \leq \frac{1}{2^{n+1}}.$$

There exist finitely many numbers that are equally close to two such numbers; for these values a choice simply has to be made. If

$$x = (-1)^{b_0} \frac{\sum_{j=-\infty}^{n-1} b_{j+1} 2^j}{2^n}, \quad \text{where } b_j \in \{0, 1\},$$

then its $(n+1)$-bit truncation corresponds to the binary sequence $(b_0, b_1, \ldots, b_n)$. If we denote this quantization map by $Q_t(x)$, then

$$0 \leq x - Q_t(x) \leq \frac{1}{2^n}.$$

We use the notation $Q(x)$ to refer to either quantization scheme.

Not only are measurements quantized, but arithmetic operations are as well. The usual arithmetic operations must be followed by a quantization step in order for the result of an addition or multiplication to fit into the same number of bits. The machine uses

$$Q(Q(x) + Q(y)) \text{ and } Q(Q(x) \cdot Q(y))$$

for addition and multiplication, respectively. The details of these operations depend on both the quantization scheme and the representation of numbers (i.e., fixed point or floating point). We consider only the fixed point representation. If $Q(x)$, $Q(y)$ and $Q(x) + Q(y)$ all lie between -1 and $+1$, then no further truncation is needed to compute the sum. If $Q(x) + Q(y)$ is greater than $+1$, we have an *overflow* and if the sum is less than -1 an *underflow*. In either case the value of the sum is clipped. On the other hand, if

$$Q(x) = (-1)^{b_0} \frac{\sum_{j=0}^{n-1} b_{j+1} 2^j}{2^n} \text{ and } Q(y) = (-1)^{c_0} \frac{\sum_{j=0}^{n-1} c_{j+1} 2^j}{2^n},$$

then

$$Q(x)Q(y) = (-1)^{b_0+c_0} \frac{\sum_{j,k=1}^{n-1} b_{j+1} c_{k+1} 2^{j+k}}{2^{2n}}.$$

This is essentially a $(2n+1)$-bit binary representation and therefore must be re-quantized to obtain an $(n+1)$-bit representation. Because all numbers lie between $+1$ and -1, overflows and underflows cannot occur in fixed-point multiplication.

It is not difficult to find numbers x and y between -1 and 1 so that $x + y$ is also between -1 and 1 but

$$Q(x + y) \neq Q(x) + Q(y).$$

This means that quantization is not a linear map!

***Example* 8.4.1.** Using truncation as the quantization method and three binary bits, we see that $Q(\frac{3}{16}) = 0$ but $Q(\frac{3}{16} + \frac{3}{16}) = \frac{1}{4}$.

Because it is *nonlinear,* quantization is difficult to analyze. An *exact* analysis requires entirely new techniques. Another approach is to regard the error $e(x) = x - Q(x)$ as *quantization noise*. If $\{x_j\}$ is a sequence of samples, then $\{e_j = x_j - Q(x_j)\}$ is the quantization noise sequence. For this approach to be useful, we need to assume that the sequence $\{e_j\}$ has good statistical properties (e.g., it is of mean zero and the successive values are not highly correlated). If the original signal is sufficiently complex, then this is a good approximation. However, if the original signal is too slowly varying, then these assumptions may not hold. This approach is useful because it allows an analysis of the effect on the signal-to-noise ratio of the number of bits used in the quantization scheme. It is beyond the scope of this text to consider these problems in detail; a thorough treatment and references to the literature can be found in Chapter 9 of [100].

8.5 Higher-Dimensional Sampling

In imaging applications, we usually work with functions of two or three variables. Let f be a function defined on $\mathbb{R}^n$ and let $\{\boldsymbol{x}_k\}$ be a discrete set of points in $\mathbb{R}^n$. As before, the values $\{f(\boldsymbol{x}_k)\}$ are the *samples* of f at the *sample points* $\{\boldsymbol{x}_k\}$. Parts of the theory of sampling in higher dimensions exactly parallel the one-dimensional theory, though the problems of sampling and reconstruction are considerably more complicated.

As in the one-dimensional case, samples are usually collected on a uniform grid. In this case it is more convenient to label the sample points using vectors with integer coordinates. As usual, boldface letters are used to denote such vectors, that is,

$$\boldsymbol{j} = (j_1, \ldots, j_n), \quad \text{where } j_i \in \mathbb{Z}, \quad i = 1, \ldots, n.$$

***Definition* 8.5.1.** The *sample spacing* for a set of uniformly spaced samples in $\mathbb{R}^n$ is a vector $\boldsymbol{h} = (h_1, \ldots, h_n)$ with positive entries. The index $\boldsymbol{j}$ corresponds to the sample point

$$\boldsymbol{x}_{\boldsymbol{j}} = (j_1 h_1, \ldots, j_n h_n).$$

A set of values, $\{f(\boldsymbol{x}_{\boldsymbol{j}})\}$, at these points is a uniform sample set.

A somewhat more general definition of uniform sampling is sometimes useful: Fix n orthogonal vectors $\{\boldsymbol{v}_1, \ldots, \boldsymbol{v}_n\}$. For each $\boldsymbol{j} = (j_1, \ldots, j_n) \in \mathbb{Z}^n$, define the point

$$\boldsymbol{x}_{\boldsymbol{j}} = j_1 \boldsymbol{v}_1 + \cdots j_n \boldsymbol{v}_n. \tag{8.27}$$

The set of points $\{\boldsymbol{x}_{\boldsymbol{j}} : \boldsymbol{j} \in \mathbb{Z}^n\}$ defines a uniform sample set. This sample set is the result of applying a rotation to a uniform sample set with sample spacing $(\|\boldsymbol{v}_1\|, \ldots, \|\boldsymbol{v}_n\|)$. As in the one-dimensional case, the definitions of sample spacing and uniform sampling depend on the choice of coordinate system. A complication in several variables is that there are many different coordinate systems that naturally arise.

***Example* 8.5.1.** Let $(h_1, \ldots, h_n)$ be a vector with positive coordinates. The set of points

$$\{(j_1 h_1, \ldots, j_n h_n) : (j_1, \ldots, j_n \in \mathbb{Z}^n\}$$

is a uniform sample set.

***Example* 8.5.2.** Let (r, θ) denote polar coordinates for $\mathbb{R}^2$; they are related to rectangular coordinates by

$$x = r\cos\theta, \quad y = r\sin\theta.$$

In CT imaging we often encounter functions that are uniformly sampled on a *polar grid*. Let $f(r, \theta)$ be a function on $\mathbb{R}^2$ in terms of polar coordinates and let $\rho > 0$ and $M \in \mathbb{N}$ be fixed. The set of values

$$\{f(j\rho, \frac{2k\pi}{M}) \ : \ j \in \mathbb{Z},\ k = 1, \dots M\}$$

consists of uniform samples of f, in polar coordinates; however, the points

$$\{(j\rho\cos\left(\frac{2k\pi}{M}\right), \quad j\rho\sin\left(\frac{2k\pi}{M}\right))\}$$

are *not* a uniform sample set as defined previously.

In more than one dimension there are many different, reasonable notions of finite bandwidth, or bandlimited data. If D is any convex subset in $\mathbb{R}^n$ containing 0, then a function is D-bandlimited if $\hat{f}$ is supported in D. The simplest such regions are boxes and balls.

***Definition* 8.5.2.** Let $\boldsymbol{B} = (B_1, \dots, B_n)$ be an n-tuple of positive numbers. A function f defined in $\mathbb{R}^n$ is $\boldsymbol{B}$-bandlimited if

$$\hat{f}(\xi_1, \dots, \xi_n) = 0 \text{ if } |\xi_j| > B_j \qquad \text{for } j = 1, \dots, n. \tag{8.28}$$

***Definition* 8.5.3.** A function f defined in $\mathbb{R}^n$ is R-bandlimited if

$$\hat{f}(\xi_1, \dots, \xi_n) = 0 \qquad \text{if } \|\xi\| > R \tag{8.29}$$

The Nyquist theorem and Shannon-Whittaker interpolation formula carry over easily to $\boldsymbol{B}$-bandlimited functions. However, these generalizations are often inadequate to handle problems that arise in practice. The generalization of Nyquist's theorem is as follows:

Theorem 8.5.1 (Higher-dimensional Nyquist theorem). *Let $\boldsymbol{B} = (B_1, \dots, B_n)$ be an n-tuple of positive numbers. If f is a square-integrable function that is $\boldsymbol{B}$-bandlimited, then f can be reconstructed from the samples*

$$\{f(\frac{j_1\pi}{B_1}, \dots, \frac{j_n\pi}{B_n}) \ : \ (j_1, \dots, j_n) \in \mathbb{Z}^n\}.$$

This result is "optimal."

In order to apply this result to an R-bandlimited function, we would need to collect the samples:

$$\{f(\frac{j_1\pi}{R}, \dots, \frac{j_n\pi}{R}) \ : \ (j_1, \dots, j_m) \in \mathbb{Z}^n\}.$$

As $\hat{f}$ is known to vanish in a large part of $[-R, R]^n$, this would appear to be some sort of oversampling.

Neither Theorem 8.1.1 nor Theorem 8.5.1 say anything about nonuniform sampling. It is less of an issue in one dimension. If the vectors, $\{\boldsymbol{v}_1, \dots, \boldsymbol{v}_n\}$ are linearly independent but not orthogonal, then formula (8.27) defines a set of sample points $\{\boldsymbol{x}_j\}$. Unfortunately, Nyquist's theorem is not directly applicable to decide whether or not the set of samples $\{f(\boldsymbol{x}_j) : \boldsymbol{j} \in \mathbb{Z}^n\}$ suffices to determine f. There are many results in the mathematics literature that state that a function whose Fourier transform has certain support properties is determined by its samples on appropriate subsets, though few results give an explicit interpolation formula like (8.5). The interested reader is referred to [66] and [102].

The Poisson summation formula also has higher-dimensional generalizations. If f is a rapidly decreasing function, then

$$f_p(\boldsymbol{x}) = \sum_{\boldsymbol{j} \in \mathbb{Z}^n} f(\boldsymbol{x} + \boldsymbol{j})$$

is a periodic function. The Fourier coefficients of f_p are related to the Fourier transform of f in much the same way as in one dimension:

$$\begin{aligned} \widehat{f_p}(\boldsymbol{k}) &= \int\limits_{[0,1]^n} f_p(\boldsymbol{x}) e^{-2\pi i \langle \boldsymbol{x}, \boldsymbol{k} \rangle} d\boldsymbol{x} \\ &= \int\limits_{\mathbb{R}^n} f(\boldsymbol{x}) e^{-2\pi i \langle \boldsymbol{x}, \boldsymbol{k} \rangle} d\boldsymbol{x} \\ &= \hat{f}(2\pi i \boldsymbol{k}). \end{aligned} \tag{8.30}$$

Applying the Fourier series inversion formula with a function that is smooth enough and decays rapidly enough shows that

$$\sum_{\boldsymbol{j} \in \mathbb{Z}^n} f(\boldsymbol{x} + \boldsymbol{j}) = \sum_{\boldsymbol{k} \in \mathbb{Z}^n} \hat{f}(2\pi i \boldsymbol{k}) e^{2\pi i \langle \boldsymbol{x}, \boldsymbol{k} \rangle}. \tag{8.31}$$

This is the n-dimensional Poisson summation formula.

The set of sample points is sometimes determined by the physical apparatus used to make the measurements. As such, we often have samples $\{f(\boldsymbol{y}_k)\}$, of a function, on a nonuniform grid $\{\boldsymbol{y}_k\}$. To use computationally efficient methods, it is often important to have samples on a uniform grid $\{\boldsymbol{x}_j\}$. To that end, *approximate* values for f, at these points, are obtained by interpolation. Most interpolation schemes involve averaging the known values at nearby points. For example, suppose that $\{\boldsymbol{y}_{k_1}, \dots, \boldsymbol{y}_{k_l}\}$ are the points in the nonuniform grid closest to $\boldsymbol{x}_j$ and there are numbers $\{\lambda_i\}$, all between 0 and 1, so that

$$\boldsymbol{x}_j = \sum_{i=1}^{l} \lambda_i \boldsymbol{y}_{k_i}.$$

A reasonable way to assign a value to f at $\boldsymbol{x}_j$ is to set

$$f(\boldsymbol{x}_j) \stackrel{d}{=} \sum_{i=1}^{l} \lambda_i f(\boldsymbol{y}_{k_i}).$$

This sort of averaging is not the result of convolution with an L^1-function and does not produce smoothing. The success of such methods depends critically on the smoothness of f. A somewhat more robust and efficient method for multi-variable interpolation is discussed in Section 11.8. Another approach to nonuniform sampling is to find a computational scheme adapted to the nonuniform grid. An example of this is presented in Section 11.5.

Exercises

***Exercise* 8.5.1.** Prove Theorem 8.5.1.

***Exercise* 8.5.2.** Find an n-dimensional generalization of the Shannon-Whittaker interpolation formula (8.5).

***Exercise* 8.5.3.** Give a definition of oversampling and a generalization of formula (8.7) for the n-dimensional case.

***Exercise* 8.5.4.** For a set of linearly independent vectors $\{\boldsymbol{v}_1, \ldots, \boldsymbol{v}_n\}$, find a notion of $\boldsymbol{V}$-bandlimited so that a $\boldsymbol{V}$-bandlimited function is determined by the samples $\{f(\boldsymbol{x}_j) : \boldsymbol{j} \in \mathbb{Z}^n\}$ where $\boldsymbol{x}_j = j_1\boldsymbol{v}_1 + \cdots + j_n\boldsymbol{v}_n$. Show that your result is optimal.

***Exercise* 8.5.5.** Using the results proved earlier about Fourier series, give hypotheses on the smoothness and decay of f that are sufficient for (8.31) to be true.

8.6 Conclusion

Data acquired in medical imaging are usually described as samples of a function of continuous variables. Nyquist's theorem and the Poisson summation formula provide a precise and quantitative description of the errors, known as aliasing errors, introduced by sampling. The Shannon-Whittaker formula and its variants give practical methods for approximating (or reconstructing) a function from a discrete set of samples. The finite Fourier transform, introduced in Section 8.3, is the form in which Fourier analysis is finally employed in applications. In the next chapter we reinterpret (and rename) many of the results from earlier chapters in the context and language of filtering theory. In Chapter 10, we analyze how the finite Fourier transform provides an approximation to the both the Fourier transform and Fourier series and use this analysis to approximately implement (continuum) shift invariant filters on finitely sampled data. This constitutes the final step in the transition from abstract continuum models to finite algorithms.

Chapter 9

Filters

Building on the brief introduction to shift invariant filters in Section 5.1.2, this chapter discusses basic concepts in filtering theory. As noted before, a *filter* is the engineering term for any process that maps an *input* or collection of inputs to an *output* or collection of outputs. As inputs and outputs are generally functions of a variable (or variables), in mathematical terms, a filter is nothing more or less than *a map from one space of functions to another space of functions*. A large part of our discussion is devoted to linear filters, recasting our treatment of the Fourier transform in the language of linear filtering theory.

In the first part of the chapter we introduce engineering vocabulary for concepts already presented from a mathematical standpoint. In imaging applications the functions of interest usually depend on two or three *spatial* variables. The measurements themselves are, of necessity, also functions of time, though this dependence is often suppressed or ignored. In most of this chapter we consider filters acting on inputs that are functions of a single variable. There are three reasons for doing this: (1) The discussion is simpler; (2) this development reflects the origins of filtering theory in radio and communications; and (3) filters acting on functions of several variables are usually implemented "one variable at a time." Most linear filters are expressed, at least formally, as integrals. The fact that higher-dimensional filters are implemented one variable at a time reflects the fact that higher-dimensional integrals are actually computed as iterated, one-dimensional integrals, that is,

$$\int\limits_{[a_1,b_1]\times[a_2,b_2]} f(x,y)\,dx\,dy = \int\limits_{a_1}^{b_1}\left[\int\limits_{a_2}^{b_2} f(x,y)\,dy\right]dx.$$

In Section 9.4 we present some basic concepts of image processing and a filtering theoretic analysis of some of the hardware used in x-ray tomography.

9.1 Basic Definitions

See: A.3, A.4.1, A.4.4.

Most of the tools and concepts of filtering theory have already been discussed in Chapters 4–8. In this section we present some standard engineering terminology and revisit mathematical concepts in light of their applications in filtering theory. Much of the terminology in filtering theory is connected to the intended application. A single mathematical concept, when connected to a filter, has one name if the filter is used to process radio signals and a different name if the filter is used in imaging. In this section the input is usually denoted by x. It is often a function of a single variable denoted by t. Functional notation, similar to that used for linear transformations, is often used to denote the action of a filter; that is, a filter $\mathscr{A}$ takes the input, x, to the output, $\mathscr{A}x$.

9.1.1 Examples of Filters

In applications we rarely consider "arbitrary" filters. Before beginning an analysis of filters, we consider some typical examples acting on functions of a single variable.

***Example* 9.1.1.** The operation of scaling defines a filter $\mathscr{A}$ given by

$$\mathscr{A}x(t) = ax(t);$$

here a is a positive number often called the amplification factor.

***Example* 9.1.2.** Shifting an input in time defines a filter. Let τ denote a constant and define

$$\mathscr{A}_\tau x(t) = x(t-\tau).$$

***Example* 9.1.3.** Multiplying a input by a function defines a filter. Let ψ denote a function and define

$$\mathscr{M}_\psi x(t) = \psi(t)x(t).$$

***Example* 9.1.4.** Convolution with a function defines a filter. Let φ be a function and define

$$\mathscr{C}_\varphi x(t) = \int_{-\infty}^{\infty} \varphi(t-s)x(s)\,ds.$$

***Example* 9.1.5.** If x is an input depending on a single variable, then differentiation defines a filter

$$\mathscr{D}x(t) = \frac{dx}{dt}(t).$$

***Example* 9.1.6.** The Fourier transform, $\mathscr{F} : x \to \hat{x}$, is a filter, as is its inverse.

From the last three examples it is clear that some filters can only be applied to certain kinds of inputs (e.g., inputs that are sufficiently regular for $\mathscr{D}$ and inputs that decay sufficiently rapidly for $\mathscr{C}_\varphi$ and $\mathscr{F}$). An important difference between the mathematical and engineering approaches to filtering lies in the treatment of the spaces of inputs and outputs. Before a mathematician starts to discuss a map from one space of functions to another, he or she likes to have well-defined domain and target spaces, often equipped with norms. By contrast, engineers often describe a process or write down a formula without stipulating the exact nature of the inputs or the expected properties of the outputs. Of course, the implementation of the filter requires that real inputs produce meaningful outputs. For this to be so, the filter must be continuous in an appropriate sense, though it is unusual for this to be made explicit.

All the examples considered so far are linear filters; in mathematical language these are linear transformations or linear operators. But for the vagueness about the domain and range, the definition of a linear filter is identical to the definition of a linear transformation:

***Definition* 9.1.1.** A *linear filter* $\mathscr{A}$ is an operation mapping inputs, taking values in a vector space, to outputs, taking values in a vector space, that satisfies the following conditions:

1. If x_1 and x_2 are a pair of inputs, then
$$\mathscr{A}(x_1 + x_2) = \mathscr{A}(x_1) + \mathscr{A}(x_2).$$
2. If x is an input and α is a scalar, then
$$\mathscr{A}(\alpha x) = \alpha\, \mathscr{A}(x).$$

In order to be clear about this distinction, we consider some examples of *nonlinear* filters.

***Example* 9.1.7.** The squaring operation, $\mathscr{S}x(t) = (x(t))^2$, is a nonlinear filter. This filter is the basis of AM radio demodulation.

***Example* 9.1.8.** Suppose that the input is a pair of functions, (x_1, x_2), and the output is their product
$$\mathscr{P}(x_1, x_2)(t) = x_1(t)x_2(t).$$
A simple calculation shows that this is a nonlinear filter:
$$\begin{aligned}\mathscr{P}(x_1 + y_1, x_2 + y_2)(t) = (x_1(t) + y_1(t))(x_2(t) + y_2(t)) \neq \\ x_1(t)y_1(t) + x_2(t)y_2(t) = \mathscr{P}(x_1, x_2)(t) + \mathscr{P}(y_1, y_2)(t).\end{aligned} \tag{9.1}$$

***Example* 9.1.9.** An electrical diode is a circuit element that only passes current moving in one direction. Its action is modeled by the formula
$$\mathscr{R}x(t) = \chi_{[0,\infty)}(x(t))x(t).$$
In electrical engineering this filter is called a *rectifier*.

***Example* 9.1.10.** The process of quantization is defined by a nonlinear filter. Suppose that for each t the binary representation of $x(t)$ is given by

$$x(t) = [\operatorname{sgn} x(t)] \sum_{j=-\infty}^{\infty} b_j(x(t))2^j.$$

One scheme used for quantization is called truncation. We let

$$\mathcal{Q}_{N,M}x(t) = [\operatorname{sgn} x(t)] \sum_{j=-M}^{N} b_j(x(t))2^j.$$

***Example* 9.1.11.** The function x might represent an input that we would like to measure, and $\mathcal{A}x$ is the result of our measurement. The measurement process itself defines the filter $\mathcal{A}$, which can either be linear or nonlinear. A simple linear model for measurement is evaluation of a weighted average,

$$\mathcal{A}_l x(t) = \int_{-\infty}^{\infty} \psi(t-s)x(s)\,ds.$$

On the other hand, many "detectors" become saturated when the signal is too strong. A model for such a device might be

$$\mathcal{A}_{nl} x(t) = \int_{-\infty}^{\infty} \psi(t-s)G[x(s)]\,ds.$$

Here G is a nonlinear function that models the saturation of the detector; for example,

$$G(x) = \begin{cases} x & \text{if } |x| < T, \\ T & \text{if } x \geq T, \\ -T & \text{if } x \leq -T. \end{cases}$$

A slightly different model is given by $\mathcal{A}'_{nl}x(t) = G[(\mathcal{A}_l x)(t)]$.

We close this section with a few examples of filters of a rather different character in that the input and output are functions on different spaces.

***Example* 9.1.12.** Suppose that f is a function defined on $\mathbb{R}^2$ in the natural domain of the Radon transform. The map $f \mapsto \mathcal{R}f$ defines a filter where the input is a function on $\mathbb{R}^2$ and the output is a function on $\mathbb{R} \times S^1$.

***Example* 9.1.13.** Let τ be a fixed positive number. For a function x defined on $\mathbb{R}$ we define the filter $\mathcal{S}_\tau$ that associates to x the sequence of samples of x at the points $\{n\tau : n \in \mathbb{Z}\}$,

$$\mathcal{S}_\tau x = < x(n\tau) : n \in \mathbb{Z} > .$$

This filter maps functions defined on $\mathbb{R}$ to functions defined on $\mathbb{Z}$.

***Example* 9.1.14.** The Shannon-Whittaker interpolation formula defines a filter that goes in the opposite direction. To a bi-infinite sequence $< a_n >$ we associate the function of the real variable t given by

$$\mathcal{A}_\tau(< a_n >)(t) = \sum_{n=-\infty}^{\infty} a_n \operatorname{sinc}(t - n\tau).$$

Exercises

***Exercise* 9.1.1.** Show that the filters in Examples 9.1.1– 9.1.6 are linear.

***Exercise* 9.1.2.** Show that Examples 9.1.7– 9.1.10 are nonlinear.

9.1.2 Linear filters

See: A.4.5, A.4.6.

As a filter is a map from a space of functions to a space of functions, it should come as no surprise that *linear* filters are, in general, much easier to analyze than *nonlinear* filters. This analysis often begins by expressing the action of the filter as an integral:

$$\mathcal{A}\, x(t) = \int_{-\infty}^{\infty} a(t, s)x(s)\, ds. \tag{9.2}$$

The function $a(t, s)$ is called the *kernel function;* it describes completely the action of the filter on an "arbitrary" input.

***Example* 9.1.15.** Let $\mathcal{I}$ denote the linear filter that maps a function defined on $\mathbb{R}$ to its anti-derivative. The kernel function for $\mathcal{I}$ is

$$a(t, s) = \begin{cases} 1 & \text{if } 0 \leq s \leq t, \\ 0 & \text{if } s > t, \end{cases}$$

so that

$$\mathcal{I}(x)(t) = \int_{-\infty}^{\infty} a(t, s)x(s)\, ds = \int_{-\infty}^{t} x(s)\, ds.$$

Sometimes the action of a filter is expressed as an integral even though the integral does not, strictly speaking, make sense. For example, the Hilbert transform is often "defined" by the convolution integral

$$\mathcal{H}x(t) = \frac{i}{\pi} \int_{-\infty}^{\infty} \frac{x(s)}{t - s} ds.$$

The kernel function for the Hilbert transform would appear to be $i/[\pi(t-s)]$. Because this function is not locally integrable, this integral is not unambiguously defined and therefore requires interpretation. As discussed in Section 6.3, the Hilbert transform is actually the Cauchy principal value (P.V.) of this integral. In this case the kernel of $\mathcal{H}$ is really the generalized function on $\mathbb{R}^2$ given by P.V.$[i/\pi(t-s)]$.

When a filter is described by a formula, we need to be careful that the formula makes sense for the inputs one has in mind. For example, if an input x does not have bounded support, then an integral of the form

$$\mathcal{A}x = \int_{-\infty}^{\infty} a(t,s)x(s)\,ds$$

only makes unambiguous sense if $a(t,s)x(s)$ is an absolutely integrable function of s. A meaning can often be assigned to such expressions even when the integrals involved are not absolutely convergent. The Fourier transform provides a good illustration. The operation $x \mapsto \hat{x}$ is initially defined for absolutely integrable functions by the formula

$$\hat{x}(t) = \int_{-\infty}^{\infty} x(s)e^{-ist}\,ds;$$

the kernel function is $a(t,s) = e^{-ist}$. The Parseval formula allows an extension of the Fourier transform to L^2-functions. This operation is well defined as a map from $L^2(\mathbb{R})$ to itself even though the integral may not exist. Using the Parseval formula and duality, the Fourier transform can be further extended to generalized functions. For these extensions of the Fourier transform the preceding integral is a *formal* expression, which means it is *not* defined as a limit of Riemann sums.

Perhaps the simplest general class of filters are the *multiplication* filters. If ψ is a function, then the operation

$$\mathcal{M}_\psi : x(t) \mapsto \psi(t)x(t)$$

defines a filter. This operation makes sense for very general types of inputs. In applications the multiplier ψ is sometimes taken to equal one for t in an interval and zero for t outside a larger interval. In this case M_ψ *windows* the input. If ψ_1 and ψ_2 are two functions, then we have the relations

$$\mathcal{M}_{\psi_1}(\mathcal{M}_{\psi_2}(x)) = \mathcal{M}_{\psi_1\psi_2}(x) = \mathcal{M}_{\psi_2}(\mathcal{M}_{\psi_1}(x)).$$

In other words, the order in which multiplication filters are applied does not affect the outcome. Mathematically we say that multiplication filters *commute*. The next section treats another class of filters that have this property.

9.1.3 Shift Invariant Filters and the Impulse Response

Shift invariant filters were defined in Section 5.1.2. Recall that a linear filter $\mathcal{A}$ is *shift invariant* if for all real numbers τ we have the identity

$$\mathcal{A}(x_\tau) = (\mathcal{A}x)_\tau, \qquad \text{where } x_\tau(t) = x(t-\tau).$$

In Section 5.1.2 it is shown that a filter defined by convolution, as in Example 9.1.4, is linear and shift invariant. We now give a formal argument to show that any linear, shift invariant filter can be expressed as a convolution. Suppose that $\mathscr{A}$ is such a filter with kernel function $a(t,s)$. Changing variables gives the following equalities:

$$\mathscr{A}(x_\tau)(t) = \int_{-\infty}^{\infty} a(t,s)x_\tau(s)\,ds = \int_{-\infty}^{\infty} a(t,s)x(s-\tau)\,ds = \int_{-\infty}^{\infty} a(t,\sigma+\tau)x(\sigma)\,d\sigma.$$

On the other hand,

$$(\mathscr{A}x)_\tau(t) = \mathscr{A}(x)(t-\tau) = \int_{-\infty}^{\infty} a(t-\tau,\sigma)x(\sigma)\,d\sigma.$$

Since $\mathscr{A}$ is shift invariant, $\mathscr{A}x_\tau$ must equal $(\mathscr{A}x)_\tau$ for every τ *and* every input; that is,

$$\int_{-\infty}^{\infty} a(t,\sigma+\tau)x(\sigma)\,d\sigma = \int_{-\infty}^{\infty} a(t-\tau,\sigma)x(\sigma)\,d\sigma \qquad \text{for all } \tau \in \mathbb{R}, \tag{9.3}$$

and all inputs x. The only way this can be true is if

$$a(t,\sigma+\tau) = a(t-\tau,\sigma) \text{ for all } t,\sigma,\tau \in \mathbb{R}.$$

Setting $\sigma = 0$ gives

$$a(t,\tau) = a(t-\tau,0).$$

In other words, the kernel function $a(t,s)$, which describes the action of $\mathscr{A}$, only depends on the difference $t-s$. Setting $k(t) = a(t,0)$, we can express $\mathscr{A}x$ as a convolution:

$$\mathscr{A}x(t) = \int_{-\infty}^{\infty} k(t-s)x(s)\,ds = k * x(t). \tag{9.4}$$

Combining this argument with Proposition 5.1.2 gives the following:

Proposition 9.1.1. *A linear filter is shift invariant if and only if it can be represented as a convolution.*

Often the δ-function is thought of as the input to a linear system. In electrical engineering the δ-function is called the *unit impulse*. If $\mathscr{A}$ is shift invariant with kernel function $k(t-s)$, then

$$\mathscr{A}\delta(t) = k * \delta(t) = k(t).$$

***Definition* 9.1.2.** The response of a linear, shift invariant filter to the unit impulse is called the *impulse response* of the filter.

The terms *unit impulse* and *impulse response* are common in electrical engineering. In imaging applications the δ-function is called a "point source." The output $\mathcal{A}\delta$ describes how the filter spreads out a point source and therefore k is called the *point spread function*, or *PSF*.

We stress the fact that

$$\boxed{\text{The } \delta\text{-"function" is } \textit{not} \text{ a function.}} \tag{9.5}$$

We need to exercise care when working with δ; for example, its square δ^2 has no well-defined meaning.

We consider some simple examples of filters and their impulses responses:

***Example* 9.1.16.** Let $\mathcal{A}_\eta$ be the moving average

$$\mathcal{A}_\eta(x)(t) = \frac{1}{2\eta} \int\limits_{x-\eta}^{x+\eta} x(s)\, ds.$$

The impulse response is $[2\eta]^{-1}\chi_{[-\eta,\eta]}$.

***Example* 9.1.17.** Define a linear shift invariant filter by setting

$$\mathcal{A}_\varphi(x) = \partial_t^2 \int\limits_{-\infty}^{\infty} \varphi(t-s)x(s)\, ds.$$

The impulse response is φ''.

***Example* 9.1.18.** Define a linear shift invariant filter by setting

$$\mathcal{A}_d(x) = \frac{x(t) - x(t-d)}{2}.$$

The impulse response is $\frac{1}{2}[\delta(t) - \delta(t-d)]$.

The kernel function, which completely determines the action of $\mathcal{A}$, is itself determined by the output of $\mathcal{A}$ when the input is a unit impulse. This fact can be applied in the following way: Suppose that we are confronted by an *unknown* physical system that is, however, known to be linear and shift invariant. If we can accurately measure the response of this system to a δ-function, then we have a complete model for the system. In actual applications the δ-function is approximated by a highly localized, energetic input. For example, in an acoustical measurement a unit impulse might be approximated by a very short duration spark or explosion. In optical imaging a point source might be a bright, tiny light source, while in x-ray imaging the δ-function might be approximated by a small lead sphere. If the input is an *approximate* δ-function, then the measured output provides an approximation to the impulse response.

Mathematically, if φ is a nonnegative function of a single variable that satisfies the conditions

$$\int_{-\infty}^{\infty} \varphi(t)\,dt = 1 \text{ and } \begin{cases} \varphi(0) & = 1, \\ \varphi(t) & = 0, \text{ if } |t| > 1, \end{cases} \tag{9.6}$$

then the family of functions

$$\varphi_\epsilon(t) = \frac{1}{\epsilon}\varphi\left(\frac{t}{\epsilon}\right), \quad \epsilon > 0$$

gives an approximation to the δ-function (see Section 5.3). If φ_ϵ can be used as an input to $\mathscr{A}$, then, for small ϵ,

$$k_\epsilon(t) = \mathscr{A}\,\varphi_\epsilon(t)$$

provides an approximation to the impulse response of $\mathscr{A}$. Whether or not φ_ϵ can be used as an input to $\mathscr{A}$ usually depends on its smoothness.

***Example* 9.1.19.** Let $\mathscr{A} f = \partial_t f$; then the function $\chi_{-\frac{1}{2},\frac{1}{2}}(t)$ is not a good input and the corresponding scaled family does not provide a usable approximation for the impulse response. On the other hand, the continuous, piecewise differentiable function

$$\varphi(t) = \begin{cases} 1 - |t| & \text{for } |t| < 1, \\ 0 & \text{for } |t| \geq 1 \end{cases}$$

has a weak derivative and

$$\varphi'_\epsilon = \begin{cases} \frac{1}{\epsilon^2} & \text{for } -\epsilon < t < 0, \\ -\frac{1}{\epsilon^2} & \text{for } 0 \leq t < \epsilon, \\ 0 & \text{otherwise} \end{cases}$$

is often used as an approximation to δ'.

Exercises

***Exercise* 9.1.3.** The *identity filter* is defined as the filter that takes an input to itself,

$$\operatorname{Id} x(t) = x(t).$$

Show that the impulse response of the identity filter is $\delta(t)$.

Exercise* 9.1.4. Closely related to the δ-function is the *Heaviside function*. It is defined by

$$H(x) = \begin{cases} 0 \text{ if } x < 0, \\ 1 \text{ if } 0 \leq x. \end{cases}$$

Formally H is the indefinite integral of the δ-function

$$H(x) = \int_{-\infty}^{x} \delta(y)\,dy.$$

With this interpretation it is clear that the Fourier transform of H should be

$$\mathcal{F}(H)(\xi) = \frac{1}{i\xi}.$$

Because ξ^{-1} is not locally integrable, this requires interpretation. By considering H as the limit

$$H(x) = \lim_{\epsilon \downarrow 0} H(x)e^{-\epsilon x},$$

show that, as a generalized function,

$$\mathcal{F}(H)(\xi) = \lim_{\epsilon \downarrow 0} \frac{1}{i\xi + \epsilon}.$$

Show that the filter defined by

$$\mathscr{A}\, x(t) = H(t)x(t)$$

is represented in the Fourier domain as a convolution by

$$\mathscr{A}\, x(t) = \mathcal{F}^{-1}(\hat{H} * \hat{x}) \stackrel{d}{=} \lim_{\epsilon \downarrow 0} \frac{1}{2\pi} \int_{-\infty}^{\infty} \frac{\hat{x}(\eta)e^{it\eta}d\eta}{\epsilon + i(\xi - \eta)}.$$

9.1.4 Harmonic Components

We often assume that the input x has a Fourier transform, $\hat{x}$. If the Fourier transform is written in terms of polar coordinates in the complex plane,

$$\hat{x}(\xi) = |\hat{x}(\xi)|e^{i\phi(\xi)},$$

then $|\hat{x}(\xi)|$ is called the *amplitude* of the input at frequency $\xi/2\pi$ and the real number $\phi(\xi)$ is called the *phase*. Because the complex exponential is 2π-periodic, that is

$$e^{ia} = e^{i(a+2n\pi)} \qquad \text{for any } n \in \mathbb{Z},$$

there is an ambiguity in the definition of the phase. The rule used to choose the phase depends on the context. Often we fix an interval of length 2π,—for example, $[-\pi, \pi)$ or $[0, 2\pi)$,—and then insists that $\phi(\xi)$ belong to this interval. A different method is to take the phase to be a *continuous* function of ξ. The latter approach may run into difficulties if $\hat{x}$ vanishes.

***Example* 9.1.20.** Suppose that $x(t) = e^{it^2}$. Using the first approach, the phase $\phi_1(t)$ is computed by finding an integer n so that

$$0 \leq t^2 - 2\pi n < 2\pi;$$

the phase is then $\phi_1(t) = t^2 - 2\pi n$. In the second approach the phase is simply $\phi_2(t) = t^2$.

(a) An image described by f_a.

(b) An image described by f_b.

(c) The image obtained as $\mathscr{F}^{-1}(m_b e^{i\phi_a})$.

(d) The image obtained as $\mathscr{F}^{-1}(m_a e^{i\phi_b})$.

Figure 9.1. An experiment that indicates that *among images* the important information resides in the phase of the Fourier transform.

It is reasonable to inquire where the "information" in the Fourier transform lies. Is it more important to get the amplitude or phase correct? The answer is *highly* dependent on the intended application. In many imaging applications it turns out that accurate phase information is more important than the amplitude. In Figure 9.1 we replicate a striking experiment from [68]. Figures 9.1(a) and (b) are two, very different, gray scale images, described by the functions f_a and f_b. Let

$$\hat{f}_a = m_a e^{i\phi_a} \text{ and } \hat{f}_b = m_b e^{i\phi_b}.$$

Figure 9.1(c) shows the image obtained as the inverse Fourier transform of $m_b e^{i\phi_a}$, while Figure 9.1(d) is obtained by Fourier inverting $m_a e^{i\phi_b}$.

(a) The image defined by $\mathscr{F}^{-1}(m_1 e^{i\phi_a})$.

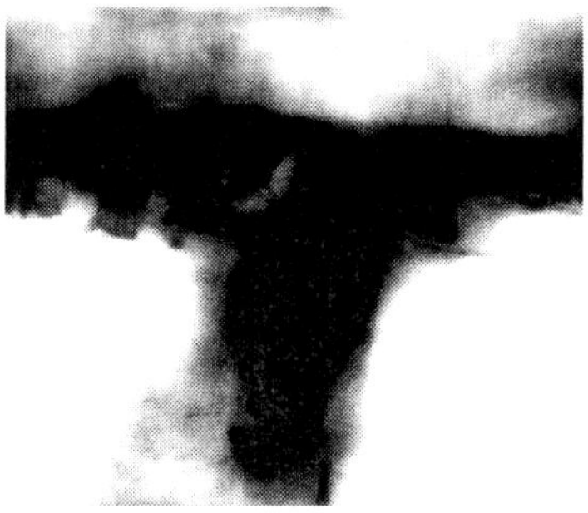

(b) The image defined by $\mathscr{F}^{-1}(m_2 e^{i\phi_a})$.

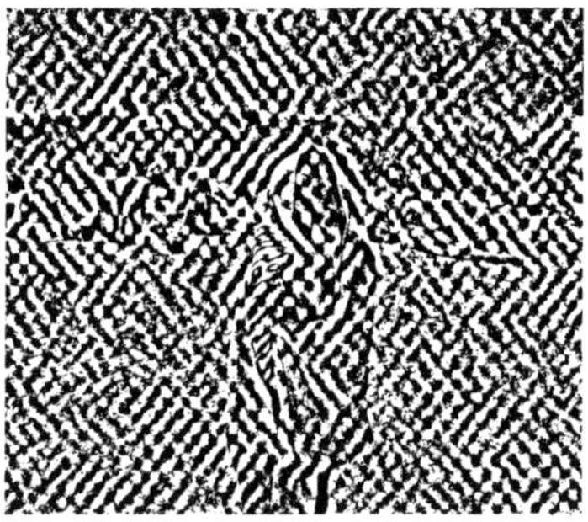

(c) The image obtained as $\mathscr{F}^{-1}(m_3 e^{i\phi_a})$.

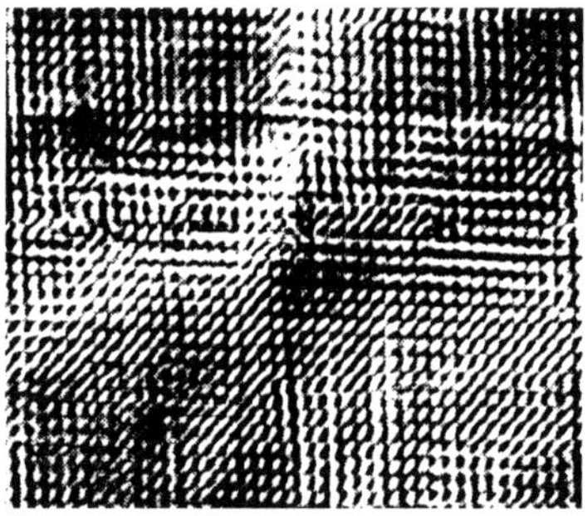

(d) The image obtained as $\mathscr{F}^{-1}(m_4 e^{i\phi_a})$.

Figure 9.2. Reconstructing an image with a variety of magnitudes for the Fourier transform.

How are we to understand these results? The correct interpretation seems to be that the general character of the magnitude of the Fourier transform of a function *representing a natural scene* is largely independent of the actual scene. This is a statement about the Fourier transforms of a very specific set of functions and *not* a general fact about the Fourier transforms of functions of two variables.

Roughly speaking, the magnitude of the Fourier transform of an image is an even function assuming its maximum at 0 and decaying like $\|\xi\|^{-1}$. To demonstrate this we reconstruct images using the phase of $\hat{f}_a$ and several choices for the magnitude. In Figure 9.2(a) the magnitude is $m_1 = (1 + |\xi_1| + |\xi_2|)^{-\frac{1}{2}}$; in Figure 9.2(b) the magnitude is $m_2 = (1 + |\xi_1| + |\xi_2|)^{-2}$; in Figure 9.2(c) we use $m_3 = (1 + ||\xi_1| - 20| + ||\xi_2| - 30|)^{-1}$; and finally in Figure 9.2(d) we use

$$m_4(\xi) = \begin{cases} \xi_1 > 0,\ \xi_2 > 0 (1 + |\xi_1 - 30| + |\xi_2 - 40|)^{-1}, \\ \xi_1 > 0,\ \xi_2 < 0 (1 + |\xi_1 - 30| + |\xi_2|)^{-1}, \\ \xi_1 < 0,\ \xi_2 > 0 (1 + |\xi_1| + |\xi_2 - 40|)^{-1}, \\ \xi_1 < 0,\ \xi_2 < 0 (1 + |\xi_1| + |\xi_2|)^{-1}. \end{cases}$$

Clearly, the reconstructions in (c) and (d) are the worst. The magnitude used in (c) does not assume its maximum at 0, whereas that used in (d) does not have the symmetries of the Fourier transform of a real-valued function (i.e., $|f(\xi)| = |f(-\xi)|$). None of these images is as good as Figure 9.1(c), indicating that $|\hat{f}_b|$ is a better approximation to $|\hat{f}_a|$ than any of the functions m_1 through m_4.

Physically we think of the Fourier transform as giving a decomposition of an input into its *harmonic components*. The actual operation of the Fourier transform is somewhat at variance with an intuitive understanding of this concept. Returning to the one-dimensional case, the determination of $\hat{x}(\xi)$, for one value of ξ, requires a knowledge of the input x, for *all* times. The number $|\hat{x}(\xi)|$ measures the "amount" of the input, x at frequency $\frac{\xi}{2\pi}$, while $\arg \hat{x}(\xi)$ gives the phase of this component. Intuitively, we think of an input as having an "instantaneous frequency." For example, if x equals $\cos(\omega t)$ for t in an interval $[t_1, t_2]$, then we would probably say that the frequency of x, in that time interval, is $\frac{\omega}{2\pi}$. The idealized signal, x_i, equals $\cos(\omega t)$ for *all* t, its Fourier transform is the generalized function

$$\hat{x}_i(\xi) = \pi[\delta(\xi - \omega) + \delta(\xi + \omega)].$$

This formula can be heuristically justified by putting $\hat{x}_i$ into the Fourier inversion formula. A more realistic model for a signal "at frequency $\frac{\omega}{2\pi}$" is

$$x_r(t) = \psi(t)\cos(\omega t),$$

where ψ equals 1 over an interval $[t_1, t_2]$ and is zero outside a larger interval. The Fourier transform of the real signal is

$$\hat{x}_r(\xi) = \frac{1}{2}[\hat{\psi}(\xi - \omega) + \hat{\psi}(\xi + \omega)].$$

If $\psi = 1$ over a long interval and vanishes smoothly outside it, then $|\hat{\psi}|$ is sharply peaked at zero and decreases rapidly as $|\xi| \to \infty$. If ψ is shifted by an amount t_0, then its Fourier transform is multiplied by $\exp(it_0\xi)$. Hence the phase of $\hat{x}_r$ indicates where the signal is nonzero. If ψ is an even function, supported on a long interval, then the Fourier transform of x_r is a good approximation to that of x_i.

Thus far we have used the Fourier transform and Fourier series as tools. The Fourier transform of a function has been viewed as an indicator of the qualitative features of the original function, without much significance of its own or, as a convenient way to represent convolution operators. In many applications it is not the function itself but its Fourier transform that contains the information of interest. This is the case if x describes the state of a system that is composed of collection of resonant modes. In nuclear magnetic resonance (NMR) spectroscopy a complicated molecule is caused to vibrate and emit a radio frequency signal x. This signal is composed of a collection of exponentially damped vibrations. Figure 9.3(a) shows a typical measurement of x as a function of time. The useful information in x is extracted by taking the magnitude of the Fourier transform, as shown in Figure 9.3(b). The locations of the peaks determine the frequencies of the different vibrational modes and their widths give a measure of the damping. This information can, in turn, be used to deduce the structure of the molecule.

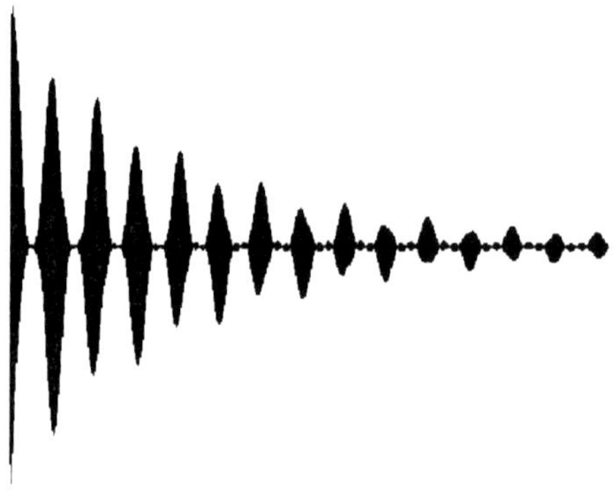

(a) The signal x as a function of time.

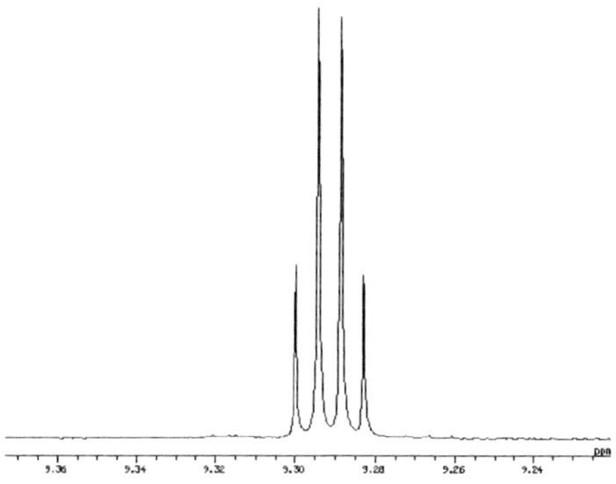

(b) The magnitude of $\hat{x}$.

Figure 9.3. Using magnetic resonance to determine the vibrational modes of a molecule. (Images courtesy of Dr. Walter Bauer, Universität Erlangen-Nürnberg, Institut für Organische Chemie.)

Spectral analysis of this sort is used throughout science and engineering; it provides a different perspective on the meaning of the Fourier transform. In magnetic resonance spectroscopy the signal decays exponentially, so little error results from cutting it off after a finite time and computing the Fourier transform of the time limited signal. In other applications, the input does not decay in any reasonable sense and is regarded instead as a periodic input. Such inputs are analyzed using Fourier series. In the world of real measurements and computation, where everything is done with finite data sets, the practical distinctions between the Fourier transform and Fourier series disappear, though these distinctions remain important in the design of algorithms and the interpretation of results.

Exercises

***Exercise* 9.1.5.** Find an example of a continuous, complex-valued function $\hat{x}(\xi)$ such that the phase cannot be defined as a continuous function of ξ.

***Exercise* 9.1.6.** Suppose that ψ equals 1 over a large interval $[t_1, t_2]$ and vanishes smoothly outside a slightly larger interval. How is the phase of $\hat{\psi}$ related to t_1 and t_2?

***Exercise* 9.1.7.** By considering the functions

$$x_{\omega,\alpha} = e^{-\alpha t} e^{i\omega t} \chi_{[0,\infty)}(t),$$

explain why the location of the peak in $|\hat{x}_{\omega,\alpha}|$ locates the resonance and the width of the peak determines the rate of damping. Here ω is a real number and α is a positive real number.

***Exercise* 9.1.8.** For positive real numbers $\{\alpha_j\}$ and real numbers $\{\omega_j\}$ let

$$x = \chi_{[0,\infty)}(t) \sum_{j=1}^{N} a_j e^{-\alpha_j t} e^{i\omega_j t}.$$

Compute $\hat{x}$ and explain the physical significance of $|\hat{x}|$. *Hint:* Consider what happens as the α_j tend to zero.

***Exercise* 9.1.9.** Why might it be preferable to use a window function that goes smoothly from 1 to 0?

***Exercise* 9.1.10.** Is the Fourier transform a shift invariant filter?

***Exercise* 9.1.11.** For an input measured over a finite interval $[t_1, t_2]$, the Fourier series can be used to decompose the signal into harmonic components. This would appear to solve the problems entailed using the Fourier transform; however, it introduces a different problem. Explain what it is.

***Exercise* 9.1.12.** For an input represented by

$$x(t) = r(t)e^{i\phi(t)}$$

the instantaneous frequency is often defined to be $\phi'(t)$. When is this a reasonable definition? What are some of its limitations?

9.1.5 The Transfer Function

See: A.2.3.

The action of a shift invariant filter is expressed as a convolution in equation (9.4). Suppose that both the impulse response k and the input x have Fourier transforms. The Fourier transform of the output of such a filter is then the product these Fourier transforms:

$$\widehat{k * x}(\xi) = \hat{k}(\xi)\hat{x}(\xi).$$

***Definition* 9.1.3.** If $\mathscr{A}$ is the shift invariant filter defined by convolution with k, then $\hat{k}$, the Fourier transform of k, is called the *transfer function* of the filter. In imaging applications $\hat{k}$ is called the *modulation transfer function,* or *MTF.*

If the input to $\mathscr{A}$ is the exponential $x_\xi = e^{it\xi}$, then (at least formally)

$$\mathscr{A}x_\xi(t) = \int_{-\infty}^{\infty} e^{is\xi}k(t-s)\,ds = \hat{k}(\xi)e^{it\xi}.$$

The transfer function describes the action of $\mathscr{A}$ on the harmonic component with frequency $\frac{\xi}{2\pi}$. If we write the transfer function in polar coordinates,

$$\hat{k}(\xi) = A(\xi)e^{i\theta(\xi)},$$

where $A(\xi)$ is a nonnegative real number and $\theta(\xi)$ is a real number, then $A(\xi)$ is called the *amplitude* of the response and $e^{i\theta(\xi)}$ is called the *phase shift,*

$$\mathscr{F}(k * x) = A(\xi)|\hat{x}(\xi)|e^{i(\theta(\xi)+\phi(\xi))}.$$

If $A(\xi) > 1$, then the filter amplifies the component of the input at frequency $\frac{\xi}{2\pi}$, and if $A(\xi) < 1$, it attenuates the component of the input at this frequency. The phase of the output at frequency $\frac{\xi}{2\pi}$ is shifted by $\theta(\xi)$,

$$\phi(\xi) \mapsto \phi(\xi) + \theta(\xi).$$

The description of a shift invariant filter in terms of its transfer function is called the *frequency space description.*

In terms of its transfer function and the inverse Fourier transform the output, $\mathcal{A}x$ is given by

$$\mathcal{A}x(t) = \mathcal{F}^{-1}(\hat{k}\hat{x})(t) = \frac{1}{2\pi}\int_{-\infty}^{\infty} \hat{k}(\xi)\hat{x}(\xi)e^{ix\xi}d\xi. \tag{9.7}$$

This representation is one reason that linear, shift invariant filters play such a large role in applications. The Fourier transform and its inverse have very efficient *approximate* implementations that lead to very efficient *approximate* implementations of linear, shift invariant filters. The Fourier transform converts a shift invariant filter into a multiplication filter.

The impulse response of a filter can be a generalized function. Even when this is so, the transfer function may be an ordinary function. In this case, the frequency space description of the filter is much simpler to use than its spatial domain description. We consider some examples.

***Example* 9.1.21.** The Hilbert transform, $\mathcal{H}$, defines a shift invariant filter. As noted previously, the impulse response of $\mathcal{H}$ is the generalized function P.V. $\left[\frac{i}{\pi t}\right]$. The transfer function is the ordinary function $\operatorname{sgn}\xi$. The frequency space representation of $\mathcal{H}$ is

$$\mathcal{H}x(t) = \frac{1}{2\pi}\int_{-\infty}^{\infty} \operatorname{sgn}(\xi)\hat{x}(\xi)e^{ix\xi}d\xi.$$

***Example* 9.1.22.** The filter defined by $\mathcal{A}_\tau x(t) = x(t-\tau)$ has impulse response δ_τ and transfer function $e^{-i\tau\xi}$. Its frequency space representation is

$$\mathcal{A}_\tau x(t) = \frac{1}{2\pi}\int_{-\infty}^{\infty} e^{-i\tau\xi}\hat{x}(\xi)e^{it\xi}d\xi.$$

***Example* 9.1.23.** The filter defined by the first derivative $\mathcal{D}x(t) = x'(t)$ has impulse response δ' and transfer function $i\xi$. Its frequency space representation is

$$\mathcal{D}x(t) = \frac{1}{2\pi}\int_{-\infty}^{\infty} i\xi\hat{x}(\xi)e^{it\xi}d\xi.$$

***Example* 9.1.24.** If $a(t)$ is a non-constant function, then the filter defined by $\mathcal{D}_a x(t) = a(t)x'(t)$ is *not* shift invariant. The kernel function of this is filter is $a(t)\delta'(t-s)$.

***Example* 9.1.25.** Suppose that $\mathscr{A}$ is a filter that maps an input x to an output y such that x and y are related by a constant coefficient, linear differential equation

$$\sum_{j=0}^{m} a_j \frac{d^j x}{dt^j} = \sum_{j=0}^{n} b_j \frac{d^j y}{dt^j}. \tag{9.8}$$

Assuming that both x and y have Fourier transforms, we can take the Fourier transform of this relation to obtain

$$\left[\sum_{j=0}^{m} a_j (i\xi)^j\right] \hat{x}(\xi) = \left[\sum_{j=0}^{n} b_j (i\xi)^j\right] \hat{y}(\xi). \tag{9.9}$$

Still proceeding formally, we see that

$$\widehat{\mathscr{A}x}(\xi) = \frac{\left[\sum_{j=0}^{m} a_j (i\xi)^j\right]}{\left[\sum_{j=0}^{n} b_j (i\xi)^j\right]} \hat{x}(\xi). \tag{9.10}$$

This relation might require interpretation if the denominator has real zeros. At least formally, the transfer function of $\mathscr{A}$ is the rational function on the right-hand side of (9.10). The filter may also fail to be shift invariant if $y = \mathscr{A}x$ is defined as a solution to (9.9) with boundary conditions specified at a finite point. If boundary conditions are specified at $\pm\infty$, then $\mathscr{A}$ is again shift invariant.

For the moment, let us take for granted that the Fourier transform can be efficiently approximated and consider the relationship between the computations required to implement a shift invariant filter and a general linear filter. A general linear filter is represented as an integral:

$$\mathscr{A}x(t) = \int_{-\infty}^{\infty} a(t, s) x(s)\, ds.$$

As discussed in Chapter 8, a input is usually sampled at a finite set of points $\{s_1, \dots, s_N\}$; that is, we measure $\{x(s_j)\}$. Suppose that we would like to approximate $\mathscr{A}x(t_i)$ for N points $\{t_1, \dots, t_N\}$. A reasonable way to do this is to approximate the integral defining $\mathscr{A}$ by a Riemann sum

$$\mathscr{A}x(t_i) \approx \sum_{j=1}^{N} a(t_i, s_j) x(s_j)(s_j - s_{j-1}).$$

Examining this formula we see that the action of the filter has been approximated by the product of the $N \times N$ matrix, $a_{ij} = a(t_i, s_j)$, and the N vector, $(x(s_1), \dots, x(s_N))$. For a general $N \times N$ matrix, this computation requires $O(N^2)$ arithmetic operations to perform.

The analogous computation for a shift invariant filter, *in the frequency space description*, is the approximate determination of the pointwise product, $\hat{k}(\xi)\hat{x}(\xi)$. This is done by evaluating $\hat{k}(\xi_j)$ and $\hat{x}(\xi_j)$ for N values of ξ and then computing the products $\{\hat{k}(\xi_j)\hat{x}(\xi_j)\}$. The matrix analogue is multiplying the vector $(\hat{x}(\xi_1), \dots, \hat{x}(\xi_N))$ by the *diagonal* matrix

$$k_{ij} = \hat{k}(\xi_i)\delta_{ij}.$$

This requires $O(N)$ arithmetic operations. As we shall see, the approximate computation of the Fourier transform requires $O(N \log_2 N)$ operations, provided that N is a power of 2. In applications $N = 2^{10}$ is not unusual; in this case

$$N^2 = 2^{20}, \qquad N \log_2(N) \approx 2^{13}, \qquad \frac{N^2}{N \log_2 N} \approx 102.$$

Exercises

***Exercise* 9.1.13.** A multiplication filter has a frequency space description as a convolution. If ψ is a function such that $\hat{\psi}$ is absolutely integrable, show that

$$\mathcal{M}_\psi x(t) = \mathcal{F}^{-1}(\hat{\psi} * \hat{x})(t).$$

***Exercise* 9.1.14.** Let ψ be a smooth function and let $\mathcal{D}$ denote the shift invariant filter $\mathcal{D}x = \partial_t x$. Compute the difference

$$\mathcal{D}\mathcal{M}_\psi - \mathcal{M}_\psi\mathcal{D}.$$

***Exercise* 9.1.15.** Suppose that a, b, c are real numbers such that $-a\xi^2 + ib\xi + c \neq 0$ for any real value of ξ. We define the following operation on continuous functions with bounded support: Let f be such a function then define $\mathcal{A}f$ to be the function g that solves the second-order ordinary differential equation

$$ag'' + bg' + cg = f$$

such that

$$\lim_{|t|\to\infty} g(t) = 0. \tag{9.11}$$

Show that $f \mapsto \mathcal{A}f$ defines a linear, shift invariant filter. What is the transfer function of this filter? Suppose that instead of (9.11) we use the conditions

$$g(0) = 0, \quad g'(0) = 0$$

to uniquely determine a solution to the preceding O.D.E. Let $\mathcal{A}_0 f$ denote this solution. Show that $\mathcal{A}_0$ is not a shift invariant filter.

9.1.6 Cascades of Filters

See: A.2.3, A.2.6.

Suppose that $\{\mathscr{A}_1, \dots, \mathscr{A}_k\}$ are linear filters. The output of one can be used as the input to another, this is the filtering analogue of the composition of linear maps,

$$x \to \mathscr{A}_1 x \to \mathscr{A}_2(\mathscr{A}_1 x) \to \cdots \to \mathscr{A}_k(\mathscr{A}_{k-1}(\dots(\mathscr{A}_1 x)\dots)) = \mathscr{A}_k \circ \cdots \circ \mathscr{A}_1 x.$$

In this way, complex filtering operations are built up out of simpler pieces. A filter built in this way is called a *cascade of filters*. For general linear filters the order in which the operations are performed is quite important. In Section 9.1.5 we saw that the action of a linear filter is analogous to multiplying a vector by a matrix. Cascading filters is then analogous to matrix multiplication. It is a familiar fact from linear algebra that if A and B are non-diagonal matrices, then generally

$$AB \neq BA.$$

For general linear filters $\mathscr{A}_1$ and $\mathscr{A}_2$, it is also the case that

$$\mathscr{A}_1 \circ \mathscr{A}_2 \neq \mathscr{A}_2 \circ \mathscr{A}_1.$$

The shift invariant case, which is analogous to multiplication of a vector by a *diagonal matrix*, is much simpler. If the filters $\{\mathscr{A}_j\}$ have impulses responses $\{h_j\}$, then the cascade $\mathscr{A}_k \circ \cdots \circ \mathscr{A}_1$ is given by

$$x \mapsto (h_k * \cdots * h_1) * x.$$

This is again a shift invariant filter with impulse response $h_k * \cdots * h_1$. From the commutativity of the convolution product, $f * g = g * f$, it follows that the result of applying a cascade of shift invariant, linear filters is, *in principle*, independent of the order in which the component filters are applied. In actual practice, where the actions of the filters can only be approximated, this is often not the case. Different orders of processes can produce substantially different outputs.

***Example* 9.1.26.** Suppose that

$$\mathscr{A}_1 x(t) = \partial_t x(t)$$

and

$$\mathscr{A}_2 x(t) = \int_{-\infty}^{\infty} \varphi(t-s) x(s)\, ds,$$

where φ is a differentiable function, vanishing outside a bounded interval. The two possible compositions are

$$\mathscr{A}_1 \circ \mathscr{A}_2 x(t) = \int_{-\infty}^{\infty} \partial_t \varphi(t-s) x(s)\, ds,$$

and

$$\mathcal{A}_2 \circ \mathcal{A}_1 x(t) = \int_{-\infty}^{\infty} \varphi(t-s)\partial_s x(s)\, ds.$$

To implement the first case we need to approximate the convolution $\varphi_t * x$, whereas in the second case we first need to approximate $\partial_t x$ and then the convolution $\varphi * x_t$. Because of the difficulties in approximating the derivative of an input, the composition $\mathcal{A}_1 \circ \mathcal{A}_2$ is easier to implement than $\mathcal{A}_2 \circ \mathcal{A}_1$.

***Example* 9.1.27.** If $\mathcal{A}_1$ is shift invariant and $\mathcal{A}_2$ is not, then generally $\mathcal{A}_1 \circ \mathcal{A}_2 \neq \mathcal{A}_2 \circ \mathcal{A}_1$. As an example, let $\mathcal{A}_1 x = \partial_t x$ and

$$\mathcal{A}_2 x(t) = \int_{-\infty}^{\infty} (s+t)x(s)\, ds.$$

A direct computation shows that

$$\mathcal{A}_1 \circ \mathcal{A}_2 x(t) = \int_{-\infty}^{\infty} x(s)\, ds$$

whereas integrating by parts gives

$$\mathcal{A}_2 \circ \mathcal{A}_1 x(t) = -\int_{-\infty}^{\infty} x(s)\, ds.$$

The transfer function for the cascade of filters defined by the impulse response $h = h_k * \cdots * h_1$ is the product of the transfer functions

$$\hat{h}(\xi) = \hat{h}_k(\xi) \cdots \hat{h}_1(\xi).$$

In the implementation of a cascade, using the Fourier representation, it is important to account for the limitations of finite precision arithmetic when selecting the order in which to multiply the terms in the transfer function. By grouping terms carefully, we can take advantage of cancellations between large and small factors, thereby avoiding overflows or underflows.

Exercise

***Exercise* 9.1.16.** Assuming that x and x_t are absolutely integrable, prove the formulæ in Example 9.1.27.

9.1.7 Causal Filters

In the context of time-dependent inputs, there is a special subclass of filters called *causal filters*.

***Definition* 9.1.4.** A filter is causal if the output, at a given time, depends only on the behavior of the input at earlier times. For a linear filter this means that

$$\mathscr{A}x(t) = \int_{-\infty}^{t} a(t,s)x(s)\,ds.$$

A linear, shift invariant filter is causal if and only if its impulse response k vanishes for $t < 0$.

This condition is important when working with time-dependent inputs if a filter must be implemented in "real time." In the context of image processing this distinction is often less important because an image is represented as a function of spatial variables. To avoid aliasing in the data acquisition step, it is useful to attenuate the high-frequency components *before* the input is sampled. This lowpass filtering must often be realized by a causal filter. The transfer function of a causal filter has an important analytic property.

Proposition 9.1.2. *If the filter $\mathscr{A}$ defined by $\mathscr{A}x = k * x$ is causal, then the $\hat{k}$ has a complex analytic extension to the lower half-plane.*

Proof. The hypothesis that $\mathscr{A}$ is causal implies that $k(t) = 0$ for $t < 0$ and therefore

$$\hat{k}(\xi) = \int_{0}^{\infty} k(t)e^{-it\xi}\,dt.$$

If we replace ξ by $z = \xi + i\sigma$, with $\sigma < 0$, then

$$\operatorname{Re}[-it(\xi + i\sigma)] = t\sigma < 0$$

in the domain of the integration. Thus, if $\sigma < 0$, then

$$\begin{aligned}\hat{k}(z) &= \int_{0}^{\infty} k(t)e^{-itz}\,dt \\ &= \int_{0}^{\infty} k(t)e^{\sigma t}e^{-it\xi}\,dt;\end{aligned} \tag{9.12}$$

the real exponential in the integrand is decaying. Differentiating under the integral sign shows that $\partial_{\bar{z}}\hat{k} = 0$ in the lower half-plane. □

Exercises

***Exercise* 9.1.17.** Prove that a linear, shift invariant filter is causal if and only if its impulse response k vanishes for $t < 0$.

***Exercise* 9.1.18.** Among the Examples 9.1.1–9.1.10 which are causal and which are not?

9.1.8 Bandpass Filters

Filtering theory frequently employs certain idealized filters. A basic example is the *bandpass filter*. It is defined in the Fourier representation by

$$\mathcal{B}_{[\alpha,\beta]}x = \mathcal{F}^{-1}[\chi_{[\alpha,\beta]}(|\xi|)\hat{x}(\xi)],$$

with $0 \leq \alpha < \beta$. The filtered signal contains the part of the input with absolute frequencies in the band $[[2\pi]^{-1}\alpha, [2\pi]^{-1}\beta]$. This is called the *passband* of the filter. Computing the inverse Fourier transform, we see that $\mathcal{B}_{[\alpha,\beta]}$ is represented as a convolution with

$$b_{[\alpha,\beta]}(t) = 2\,\mathrm{Re}\left[e^{-i\frac{t(\alpha+\beta)}{2}} \frac{\sin\frac{t(\beta-\alpha)}{2}}{\pi t}\right].$$

Because the sine function is nonzero for positive and negative arguments, an ideal bandpass filter is *never* causal! If the passband is of the form $[0, \beta]$, then the filter is called an *ideal lowpass filter*. An *ideal highpass filter* has a transfer function of the form $1 - \chi_{[0,\beta]}(|\xi|)$.

***Example* 9.1.28.** Let x denote a time signal and $\hat{x}$ its Fourier transform. The action of an ideal lowpass filter $\mathcal{B}_{[0,\beta]}$ with passband $[0, \beta]$ is given by

$$\mathcal{B}_{[0,\beta]}x(t) = \frac{1}{2\pi}\int_{-\beta}^{\beta} \hat{x}(\xi)e^{it\xi}d\xi.$$

This is a "partial inverse" for the Fourier transform; its point spread function is given by

$$h_\beta(t) = \frac{\beta\,\mathrm{sinc}(\beta t)}{\pi}.$$

In Section 7.5, we described the Gibbs phenomenon. This phenomenon also occurs for the partial inverse of the Fourier integral. If the function x has a jump discontinuity at t_0 and is otherwise smooth, then the lowpass filtered functions $\{\mathcal{B}_{[0,\beta]}x(t)\}$ oscillate for t near to t_0 and the size of these oscillations does not decrease as β tends to infinity. On the other hand, the oscillations are concentrated in a region of size β^{-1} around the jump in x. The underlying cause of these oscillations is the fact that the point spread functions $\{h_\beta(t)\}$ are integrable but not absolutely integrable. As in the case of the Fourier series, these oscillations can be damped by using a smooth approximation to $\chi_{[-\beta,\beta]}$ to define the transfer function of an *approximate* lowpass filter. In imaging applications, a filter that attenuates the high frequencies and passes low frequencies with little change is called an *apodizing filter*. Its transfer function is called an *apodizing function*. We consider two examples of such filters.

***Example* 9.1.29.** A simple example is an analogue of the Fejer mean. Instead of $\chi_{[-\beta,\beta]}$ we use the "tent" function

$$\hat{t}_\beta(\xi) = \frac{1}{\beta}\chi_{[-\frac{\beta}{2},\frac{\beta}{2}]} * \chi_{[-\frac{\beta}{2},\frac{\beta}{2}]}(\xi).$$

Figure 9.4 shows a graph of $\hat{t}_4$. This function is continuous, piecewise linear and satisfies

$$\hat{t}_\beta(0) = 1, \quad \hat{t}_\beta(\xi) = 0 \qquad \text{if } |\xi| \geq \beta.$$

Its point spread function is easily computed using the convolution theorem for the Fourier transform:

$$t_\beta(t) = \frac{1}{\beta}\left[\frac{\sin\left(\frac{\beta t}{2}\right)}{\pi t}\right]^2.$$

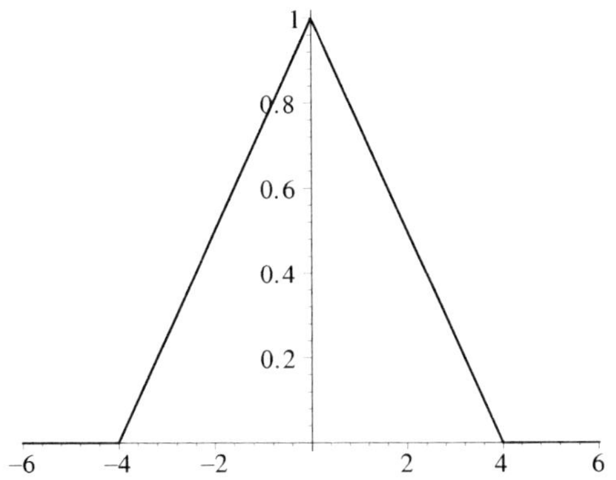

Figure 9.4. Transfer function for a tent filter.

***Example* 9.1.30.** A second example is called the Hanning window. Its transfer function is

$$\hat{h}_\beta(\xi) = \begin{cases} \cos^2\left(\frac{\pi\xi}{2\beta}\right) & \text{if } |\xi| < \beta, \\ 0 & \text{if } |\xi| > \beta. \end{cases}$$

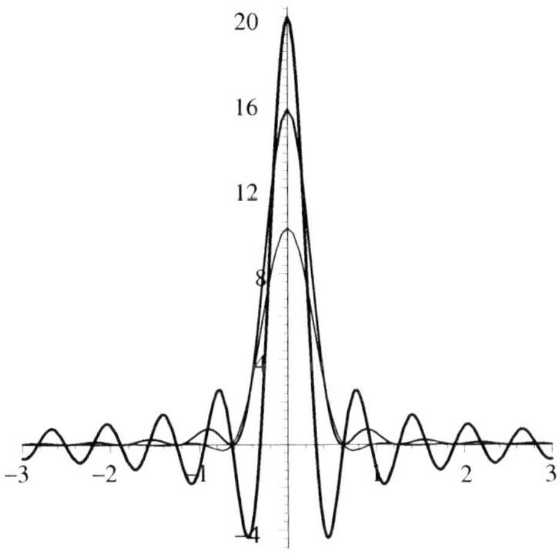

Figure 9.5. Point spread functions for lowpass filters.

This function is even smoother than the tent function, having a continuous first derivative. To compute its point spread function, we use the identity

$$\cos^2(x) = \frac{\cos(2x)+1}{2},$$

obtaining

$$h_\beta(t) = \left[\frac{\pi}{2\beta^2}\right] \frac{\sin(t\beta)}{t\left[\left(\frac{\pi}{\beta}\right)^2 - t^2\right]}.$$

This function decays like t^{-3} as t tends to infinity and is therefore absolutely integrable. The point spread functions for the three lowpass filters are shown in Figure 9.5. The tallest peak corresponds to the ideal lowpass filter, the middle is the tent function, and the shortest comes from the Hanning window.

Exercises

***Exercise* 9.1.19.** Show that a highpass filter cannot be causal.

***Exercise* 9.1.20.** The function $\chi_{[0,\infty)}(\xi)$ defines the transfer function of a filter $\mathcal{P}_+$ that removes all negative frequency components. Show that

$$\mathcal{P}_+ = \frac{1}{2}(\mathrm{Id} + \mathcal{H});$$

here $\mathcal{H}$ is the Hilbert transform.

***Exercise* 9.1.21.** If E is a subset of the real numbers, define the generalized bandpass filter with passband E by setting

$$\mathcal{B}_E x(t) = \frac{1}{2\pi} \int_{-\infty}^{\infty} \chi_E(\xi)\hat{x}(\xi)\, d\xi.$$

Prove the following statement about generalized bandpass filters: If E is a subset of $\mathbb{R}$ and the convolution $h * x$ is defined, then

$$\mathcal{B}_E(h * x) = (\mathcal{B}_E h) * x = h * (\mathcal{B}_E x) = (\mathcal{B}_E h) * (\mathcal{B}_E x). \tag{9.13}$$

9.1.9 The Inverse Filter

Let k denote the impulse response of a shift invariant filter $\mathcal{A}$, $\mathcal{A}x = k * x$. Suppose that x is an input that we would like to determine and $k * x$ is the output of a measurement device. How can x be reconstructed given only the available measurements? What is required is a filter that *undoes* the action of $\mathcal{A}$. Such a filter is called an *inverse filter*. For some types of filters it is very clear that it is not possible to recover the input from the filtered output. For example, if we apply the bandpass filter $B_{[\alpha,\beta]}$ to x, then all the information about the input at frequencies outside the passband is irrevocably lost. In other cases the Fourier transform suggests a way to try to recover x from the knowledge of the filtered output $k * x$. In the Fourier representation, the *inverse filter* should be given by

$$\mathscr{A}^{-1} : k * x \longrightarrow \mathscr{F}^{-1}\left[\frac{\widehat{k * x}}{\hat{k}}\right]. \tag{9.14}$$

From the examples we have studied, it is clear that this formula often does not define a useful operation.

If k is function that goes to zero as $|t| \to \infty$ in a reasonable way, for example, k is absolutely integrable, then $\hat{k}$ goes to 0 as $|\xi| \to \infty$. This means that the process of dividing by $\hat{k}$ takes the measured data and increasingly amplifies the high-frequency components. If the measured data behaved like the convolution, $k * x$, then this would not be a problem: The high frequencies in the input will have been attenuated. In a real situation there is noise; the measurement is then modeled as $k * x + n$, where n is the noise. The noise part is *not* the result of sending an input through the measurement device. In this case

$$\frac{\mathscr{F}(k * x + n)}{\hat{k}}(\xi) = \hat{x}(\xi) + \frac{\hat{n}(\xi)}{\hat{k}(\xi)}.$$

The high-frequency content in the noise is amplified by this attempt to reverse the measurement process. One way to try to avoid this problem is to cut off the transfer function of the inverse filter outside a bounded interval. If $\hat{k}(\xi) \neq 0$ for $\xi \in [-a, a]$, then an approximate inverse filter is given by

$$\mathscr{F}^{-1}[\frac{\mathrm{rect}_{[-a,a]}(\xi)}{\hat{k}(\xi)}\hat{x}]. \tag{9.15}$$

This gives a perfect reconstruction for data whose Fourier transform vanishes outside of $[-a, a]$ and otherwise suppresses the amplification of high-frequency noise.

Though real inputs are rarely bandlimited, they are usually considered to be *effectively bandlimited.* This means that all the "useful information" in the input is contained in a finite frequency band $[-a, a]$; that is, the effective support of the Fourier transform of the input is $[-a, a]$. The *effective bandwidth* in the measurement is then $2a$. Frequency components in the measurement from outside this band are regarded as coming from noise. By having estimates for the spectral properties of the data, the measuring apparatus and the noise, we can formulate quantitative criteria for the effective bandwidth of the data (see Remark 8.2.2).

Another problem is that $\hat{k}$ might vanish at finite frequencies.

***Example* 9.1.31.** For each integer l define

$$\mathrm{rect}_l(x) = l\chi_{[-\frac{1}{2l},\frac{1}{2l}]}(x);$$

the filter defined by rect_l averages an input over an interval of length l^{-1}. The Fourier transform of rect_l is

$$\widehat{\text{rect}_l}(\xi) = \int\limits_{-1/(2l)}^{1/(2l)} l e^{-i\xi x} dx = -\frac{l}{i\xi}[e^{-\frac{i\xi}{2l}} - e^{\frac{i\xi}{2l}}] = \text{sinc}\left(\frac{\xi}{2l}\right).$$

This function vanishes at the points $\{4\pi l m\}$, where $m \in \mathbb{Z} \setminus \{0\}$. If there were no noise, then the Fourier transform of $\text{rect}_l * x$ would also vanish at the zeros of $\widehat{\text{rect}_l}$, and dividing $\widehat{\text{rect}_l * x}$ by $\widehat{\text{rect}_l}$ would reproduce the input. In practice, this is not a good idea, as division by $\widehat{\text{rect}_l}$ infinitely amplifies anything supported on its zero set. One approach would be to simply cut off $[\text{rect}_l(\xi)]^{-1}$ outside of an interval $[-a, a]$ contained in $(-4\pi l, 4\pi l)$. If the effective support of the data is larger than $[-4\pi l, 4\pi l]$, then a less drastic approach would be to modify $[\mathcal{F}(\text{rect}_l)(\xi)]^{-1}$ in intervals containing the zeros of $\widehat{\text{rect}_l}$; for example, we could let

$$\mathcal{F}(\text{rect}_l)_\epsilon(\xi) = \begin{cases} \mathcal{F}(\text{rect}_l)(\xi) & \text{if } |\mathcal{F}(\text{rect}_l)(\xi)| > \epsilon, \\ \epsilon & \text{if } 0 < \mathcal{F}(\text{rect}_l)(\xi) \leq \epsilon, \\ -\epsilon & \text{if } -\epsilon \leq \mathcal{F}(\text{rect}_l)(\xi) \leq 0. \end{cases}$$

An approximate inverse filter is then given by

$$\mathcal{F}^{-1}\left[\frac{\hat{x}}{\mathcal{F}(\text{rect}_l)_\epsilon}\right].$$

An even better idea is to combine the two approaches, repairing the transfer function near its zeros and cutting it off entirely outside the effective support of the data.

Designing an inverse filter is largely an engineering problem. A formula such as (9.14) provides a starting point. As the model is never exact, and the measured data always contain noise, this is generally not a bounded operation and *must* therefore be approximated. The fine detail in an input is contained in the high-frequency components of its Fourier transform. A measurement process usually involves averaging, which suppresses this information. This is reflected in the various definitions of the resolution available in the output of a filter considered in Section 9.1.10. The implementation of an inverse filter is constrained, on the one hand, by a desire to retain as much of this high-frequency information as possible and, on the other hand, by the presence of noise. A characteristic feature of noise is its irregularity, which is reflected in the slow decay of its Fourier transform (see Example 4.2.5). Finding an "optimal" approximation for the inverse filter begins by modeling the noise in the input and the measurement process but ultimately requires empirical adjustment of the parameters.

***Example* 9.1.32.** Suppose that x is an L-bandlimited input. Nyquist's theorem says that in order to perfectly reconstruct the input, we must sample x at equally spaced points that are no further apart than $\frac{\pi}{L}$. Of course, a real measurement is an average and not a point evaluation. Let $\epsilon > 0$, and define the averaging function

$$h_\epsilon(t) = \frac{1}{2\epsilon}\chi_{[-1,1]}(\frac{t}{\epsilon}).$$

Observe that if x is L-bandlimited, then so is $h_\epsilon * x$. To see this we compute the Fourier transform:

$$\widehat{h_\epsilon * x} = \hat{h}_\epsilon(\xi)\hat{x}(\xi). \tag{9.16}$$

If we sample the filtered function at the points $\{\frac{n\pi}{L} \mid n \in \mathbb{Z}\}$, then we can reconstruct $\widehat{h_\epsilon * x}$ using (8.2):

$$\widehat{h_\epsilon * x}(\xi) = \chi_{[-L,L]}(\xi) \sum_{n=-\infty}^{\infty} h_\epsilon * x(\frac{n\pi}{L})e^{-\frac{n\pi i\xi}{L}}. \tag{9.17}$$

On the other hand,

$$\hat{h}_\epsilon(\xi) = \operatorname{sinc}\left(\frac{\epsilon\xi}{2}\right)$$

has its first zero at $\xi_0 = 2\pi/\epsilon$. If ϵ is chosen so that

$$\frac{2\pi}{\epsilon} \geq L,$$

then it follows from (9.16) that $\widehat{h_\epsilon * x}$ can be divided by $\hat{h}_\epsilon$, leading to an exact reconstruction of $\hat{x}$. The estimate for ϵ can be rewritten

$$\frac{\epsilon}{2} < \frac{\pi}{L}.$$

For an exactly L-bandlimited function, $\epsilon = \frac{2\pi}{L}$ works but does not give a stable method for reconstructing the original input from the measurements. This is because the function $\hat{h}_\epsilon$ vanishes at $\xi = \pm L$. Notice also that the consecutive intervals

$$[\frac{n\pi}{L} - \frac{\epsilon}{2}, \frac{n\pi}{L} + \frac{\epsilon}{2}] \text{ and } [\frac{(n+1)\pi}{L} - \frac{\epsilon}{2}, \frac{(n+1)\pi}{L} + \frac{\epsilon}{2}]$$

overlap. A more stable algorithm results if we take $\epsilon = \frac{\pi}{L}$. In this case the consecutive intervals do not overlap, and therefore the values of x that are averaged to determine the consecutive measurements $h_\epsilon * x(\frac{n\pi}{L})$ and $h_\epsilon * x(\frac{(n+1)\pi}{L})$ do not overlap. The smallest value that $\hat{h}_{\frac{\pi}{L}}$ attains on $[-L, L]$ is

$$\hat{h}_{\frac{\pi}{L}}(L) = \frac{2}{\pi} \simeq 0.63661977\ldots.$$

This example shows that a bandlimited function can be exactly and stably reconstructed from "realistic" measurements, provided the resolution of the measuring device is sufficiently high.

***Example* 9.1.33.** The Hilbert transform, $\mathcal{H}$, is a filter that has a well-behaved inverse. In fact, $\mathcal{H}$ is its own inverse. In the Fourier representation,

$$\mathcal{H}x = \frac{1}{2\pi}\int_{-\infty}^{\infty} \operatorname{sgn}\xi\,\hat{x}(\xi)e^{ix\xi}d\xi.$$

The assertion that $\mathcal{H} = \mathcal{H}^{-1}$ follows from the equation $\operatorname{sgn}\xi \cdot \operatorname{sgn}\xi = 1$.

Exercise

***Exercise* 9.1.22.** Keeping in mind the Gibbs phenomenon, explain why (9.15) might be a poor choice for an approximate inverse filter. Suggest a modification likely to produce better results.

9.1.10 Resolution

In applications it is important to have a notion of resolution. This is not a purely mathematical concept and may be defined in a variety of ways. Let $\mathscr{A}$ denote a filter. The resolution in the output of $\mathscr{A}$ is given as a length, $R_{\mathscr{A}}$. This length has a variety of interpretations:

1. The size of the smallest feature that is discernible in the output,

2. The minimum separation between just discernible features,

3. The extent to which a pointlike object is spread out by the filter.

Whichever interpretation is used, the resolution *increases* as $R_{\mathscr{A}}$ *decreases*. In Section 5.3.2 we discussed the full-width half-maximum definition of resolution. In this section we consider several other definitions applicable to linear, shift invariant filters.

Suppose that x is an input and $\mathscr{A}$ is a linear, shift invariant filter with point spread function k and transfer function $\hat{k}$. If $|\hat{k}|$ is nonvanishing and stays uniformly away from zero as $|\xi| \to \infty$, then the input can be reconstructed by performing a bounded operation:

$$x = \mathscr{F}^{-1}\left[\frac{\mathscr{F}(\mathscr{A}x)}{\hat{k}}\right].$$

In this case the output $\mathscr{A}x = k * x$ has the same resolution as the input. For the remainder of this section we therefore suppose that both k and $\hat{k}$ are ordinary functions that tend to zero as their arguments go to infinity. We now give several definitions for the resolution in the output of a linear shift invariant filter.

FULL-WIDTH κ-MAXIMUM: This is a family of definitions that apply to filters whose point spread functions assume their maxima at zero and decay as $|t| \to \infty$. The case $\kappa = \frac{1}{2}$ is the full-width half-maximum definition considered previously. Let $M = k(0)$ denote the maximum value attained by k. For a number $0 < \kappa < 1$, let $t_-(\kappa) < 0 < t_+(\kappa)$ denote the largest negative and smallest positive values (respectively), where $k(t) = \kappa M$. The *full-width* κ*-maximum* of the filter $\mathscr{A}$ is defined to be

$$\Delta_{\mathscr{A},\kappa} = t_+(\kappa) - t_-(\kappa).$$

The numbers $\Delta_{\mathscr{A},\kappa}$ increase as κ decreases; hence this definition of resolution is more stringent for smaller values of κ. In imaging applications it is very common to see resolution quoted in units of "full width tenth maximum," (i.e., the number $\Delta\mathscr{A}, .1$).

***Example* 9.1.34.** An extreme case is the rectangle function $(2d)^{-1}\chi_{[-d,d]}$. The corresponding filters $\mathscr{A}_d$ average the input over an interval of length of $2d$. In these cases

$$\Delta_{\mathscr{A}_d,\kappa} = 2d$$

for all values of κ.

***Example* 9.1.35.** A less extreme case is provided by the tent functions

$$t_d(t) = \begin{cases} 0 & \text{if } |t| \geq d, \\ t + d & \text{if } -d < t < 0, \\ d - t & \text{if } 0 \geq t < d. \end{cases}$$

Letting T_d denote the corresponding filters, we see that

$$\Delta_{T_d,\kappa} = 2(1 - \kappa)d.$$

FIRST ZERO: If the point spread function of a filter vanishes, then the locations of the first positive and negative zeros can be used to give another definition of resolution. This is just the previous definition with $\kappa = 0$. Suppose that k is the point spread function of a filter $\mathscr{A}$ and it vanishes at positive and negative values. Let $t_- < 0$ be the largest negative zero and $t_+ > 0$ be the smallest positive zero. Define

$$\Delta_{\mathscr{A},0} = t_+ - t_-.$$

***Example* 9.1.36.** Let F_β denote the Fejer mean filter with transfer function $\hat{t}_\beta$ and point spread function

$$t_\beta(t) = \frac{1}{\beta}\left[\frac{\sin\left(\frac{\beta t}{2}\right)}{\pi t}\right]^2;$$

see Example 9.1.29. The ideal lowpass filter with passband $[0, \beta]$ has point spread function

$$d_\beta(t) = \frac{\beta \operatorname{sinc}(\beta t)}{\pi}.$$

We see that

$$\Delta_{F_\beta,0} = \frac{2\pi}{\beta} \text{ and } \Delta_{D_\beta,0} = \frac{\pi}{\beta}.$$

By this measure, the output of D_β has twice the resolution of the output of F_β. This should be compared with the computations of the full-width half-maxima given in Section 5.3.2.

EQUIVALENT WIDTH: Suppose that $\mathscr{A}$ is a filter with point spread function k that assumes its maximum value at 0 and is an integrable function. We then define the *equivalent width* resolution of $\mathscr{A}$ to be

$$\Delta_{\mathscr{A},\text{ew}} = \frac{\int_{-\infty}^{\infty} k(t)\,dt}{k(0)} = \frac{\hat{k}(0)}{k(0)}.$$

This is a measure of how much the filter smears out a point source. The number $\Delta_{\mathscr{A},\text{ew}}$ is the width of the rectangle function enclosing the same *signed area* as the graph of $k/k(0)$. The fact that we use signed area has a significant effect on the value and interpretation of this number. It is not difficult to construct an example of a function k so that $k(0) = 1$ and

$$\int_{-\infty}^{\infty} k(t)\,dt = 0.$$

For the filter $\mathscr{A}$ defined by this example, $\Delta_{\mathscr{A},\text{ew}} = 0$; in other words, by this measure there is no loss in resolution.

More pertinent examples are provided by D_β and F_β; for these filters we have

$$\Delta_{F_\beta,\text{ew}} = \frac{4\pi^2}{\beta} \text{ and } \Delta_{D_\beta,\text{ew}} = \frac{\pi}{\beta}.$$

Using the previous definition, D_β has twice the resolution of F_β, whereas, using the equivalent width definition, D_β has $4\pi \approx 12$ times the resolution of F_β. This is a reflection of the fact that D_β assumes both positive and negative values while F_β has a nonnegative point spread function. The equivalent width definition rewards filters with negative sidelobes. In light of this, it might be more meaningful to modify this definition by instead using the integral of $|k|$. With the latter definition, the equivalent width of D_β would be infinite! Equivalent absolute width only gives a useful measure of resolution for filters with absolutely integrable point spread functions. For point spread functions with sidelobes, this definition is sometimes applied to the central peak.

ϵ-NYQUIST WIDTH: Suppose that the transfer function, $\hat{k}$, of a filter $\mathscr{A}$ is nonvanishing in an interval $(-\beta, \beta)$ but vanishes at $+\beta$ or $-\beta$. The Nyquist criterion gives a heuristic for measuring the resolution in the output of such a filter. To reconstruct perfectly a β-bandlimited input from uniformly spaced samples requires that the sample spacing d satisfy $d \leq \pi\beta^{-1}$. On the other hand, if an input is uniformly sampled at points, separated by a distance d, then it is quite clear that the resolution in the sampled data is at most d. With this is mind, we define the *Nyquist width* of the filter $\mathscr{A}$ to be

$$\Delta_{\mathscr{A},\text{ny}} = \frac{\pi}{\beta}.$$

With this definition, D_β and F_β satisfy

$$\Delta_{D_\beta,\text{ny}} = \frac{\pi}{\beta} = \Delta_{F_\beta,\text{ny}}.$$

This definition evidently leaves something to be desired. If we suppose that k is real and that $|\hat{k}|$ assumes its maximum at $\xi = 0$, then we can get a more satisfactory definition by adding a parameter $0 < \epsilon < 1$. Let ξ_- be the largest negative value of ξ so that $|\hat{k}(\xi)| < \epsilon|\hat{k}(0)|$ and let ξ_+ be the smallest positive value of ξ so that $|\hat{k}(\xi)| \leq \epsilon|\hat{k}(0)|$. Define the *$\epsilon$-Nyquist width* to be

$$\Delta_{\mathscr{A},\mathrm{ny},\epsilon} = \frac{\pi}{\min\{|\xi_+|, |\xi_-|\}}.$$

With this definition, we see that

$$\Delta_{D_\beta,\mathrm{ny},\frac{1}{2}} = \frac{\pi}{\beta} \text{ while } \Delta_{F_\beta,\mathrm{ny},\frac{1}{2}} = \frac{2\pi}{\beta},$$

which is in better agreement with our intuitive notion of resolution.

Exercises

***Exercise* 9.1.23.** Determine the resolution, according to each definition, for the lowpass filters defined in Section 9.1.8.

***Exercise* 9.1.24.** Suppose that k is the point spread function of a filter. For $0 < \beta$, define $k_\beta(t) = \beta^{-1}k(\beta^{-1}t)$. For each definition of resolution, how is the resolution of the filter defined by k_β related to that defined by k?

9.1.11 The Resolution of a Cascade of Filters

In Section 9.1.10 we discussed a variety of definitions for the resolution available in the output of a linear, shift invariant filter. If $\mathscr{A}_1$ and $\mathscr{A}_2$ are such filters, it is reasonable to inquire how the resolution of $\mathscr{A}_1 \circ \mathscr{A}_2$ is related to the resolution of the components. The answers depend on the type of filter and the definition of resolution.

FULL-WIDTH κ-MAXIMUM: For values of κ between 0 and 1 and general filters, it is difficult to relate $\Delta_{\mathscr{A}_1 \circ \mathscr{A}_2,\kappa}$ to $\Delta_{\mathscr{A}_1,\kappa}$ and $\Delta_{\mathscr{A}_2,\kappa}$. For the special case of filters with Gaussian point spread functions, there is a simple relation. For each $a > 0$, set

$$g_a(t) = e^{-at^2}$$

and let $\mathscr{G}_a$ denote the filter with this point spread function. A simple calculation shows that

$$\Delta_{\mathscr{G}_a,\kappa} = \sqrt{\frac{-\log\kappa}{a}}.$$

Using the fact that

$$\mathscr{F}(g_a)(\xi) = \sqrt{\frac{\pi}{a}}e^{-\frac{\xi^2}{4a}},$$

we obtain the relation

$$\Delta_{\mathscr{G}_a \circ \mathscr{G}_b,\kappa} = \sqrt{\Delta^2_{\mathscr{G}_a,\kappa} + \Delta^2_{\mathscr{G}_b,\kappa}}. \tag{9.18}$$

FIRST ZERO: If k_1 and k_2, the point spread functions of filters $\mathscr{A}_1$ and $\mathscr{A}_2$, have the additional properties that

- k_1 and k_2 are even functions, and
- each function is positive in an interval $(-t_j, t_j)$ and vanishes outside this interval,

then

$$\Delta_{\mathscr{A}_1 \circ \mathscr{A}_2,0} = \Delta_{\mathscr{A}_1,0} + \Delta_{\mathscr{A}_2,0}. \tag{9.19}$$

This follows from the fact that the point spread function of $\mathscr{A}_1 \circ \mathscr{A}_2$ is $k_1 * k_2$ and the support properties of convolutions given in Proposition 5.2.5.

EQUIVALENT WIDTH: If k_1 and k_2, the point spread functions of filters $\mathscr{A}_1$ and $\mathscr{A}_2$, are nonnegative and assume their maximum values at 0, then

$$\Delta_{\mathscr{A}_1 \circ \mathscr{A}_2,\mathrm{ew}} \geq \sqrt{\Delta_{\mathscr{A}_1,\mathrm{ew}} \Delta_{\mathscr{A}_2,\mathrm{ew}}}.$$

The proof of this estimate uses the mean value theorem for integrals, Theorem B.6.4. We can suppose that $k_1(0) = k_2(0) = 1$ as this does not affect the equivalent width. Because both functions are nonnegative, the MVT for integrals applies to give numbers t_1 and t_2 such that

$$k_1 * k_2(0) = \int_{-\infty}^{\infty} k_1(t)k_2(-t)\,dt = k_1(t_1) \int_{-\infty}^{\infty} k_2(-t)\,dt = k_2(t_2) \int_{-\infty}^{\infty} k_1(t)\,dt.$$

On the other hand,

$$\int_{-\infty}^{\infty} k_1 * k_2(t)\,dt = \int_{-\infty}^{\infty} k_1(t)\,dt \int_{-\infty}^{\infty} k_2(t)\,dt.$$

Thus

$$\begin{aligned}
\Delta_{\mathscr{A}_1 \circ \mathscr{A}_2,\mathrm{ew}} &= \frac{\int_{-\infty}^{\infty} k_1 * k_2(t)\,dt}{k_1 * k_2(0)} \\
&= \frac{\int_{-\infty}^{\infty} k_1(t)\,dt \int_{-\infty}^{\infty} k_2(t)\,dt}{\sqrt{k_1(t_1)k_2(t_2) \int k_1 dt \int k_2\,dt}} \\
&= \sqrt{\frac{\Delta_{\mathscr{A}_1,\mathrm{ew}} \Delta_{\mathscr{A}_2,\mathrm{ew}}}{k_1(t_1)k_2(t_2)}}.
\end{aligned} \tag{9.20}$$

The proof is completed by observing that $k_1(t_1)k_2(t_2) \leq 1$.

ϵ-NYQUIST WIDTH: The simplest and most general relation holds for the 0-Nyquist width. If $\hat{k}_1$ and $\hat{k}_2$ are the transfer functions for filters $\mathscr{A}_1$ and $\mathscr{A}_2$, then the transfer function of $\mathscr{A}_1 \circ \mathscr{A}_2$ is $\hat{k}_1\hat{k}_2$. Since $\hat{k}_1(\xi)\hat{k}_2(\xi) \neq 0$ if and only if each factor is nonvanishing, it follows that

$$\Delta_{\mathscr{A}_1 \circ \mathscr{A}_2,\text{ny},0} = \max\{\Delta_{\mathscr{A}_1,\text{ny},0}, \Delta_{\mathscr{A}_2,\text{ny},0}\}. \tag{9.21}$$

For $\epsilon > 0$ there is a somewhat less precise result:

$$\max\{\Delta_{\mathscr{A}_1,\text{ny},\epsilon^2}, \Delta_{\mathscr{A}_2,\text{ny},\epsilon^2}\} \leq \Delta_{\mathscr{A}_1 \circ \mathscr{A}_2,\text{ny},\epsilon^2} \leq \max\{\Delta_{\mathscr{A}_1,\text{ny},\epsilon}, \Delta_{\mathscr{A}_2,\text{ny},\epsilon}\}. \tag{9.22}$$

Exercises

***Exercise* 9.1.25.** Prove (9.18).

***Exercise* 9.1.26.** Provide the details of the proof (9.19).

***Exercise* 9.1.27.** Suppose that for $j = 1, 2$ the point spread functions k_j are positive in an interval $(-t_{j-}, t_{j+})$ where $t_{j-} < 0 < t_{j+}$ and vanish otherwise. Show that

$$\Delta_{\mathscr{A}_1 \circ \mathscr{A}_2,0} \geq \Delta_{\mathscr{A}_1,0} + \Delta_{\mathscr{A}_2,0}.$$

***Exercise* 9.1.28.** Prove (9.21) and (9.22).

9.2 Filtering Periodic Inputs

In the previous section we considered the fundamentals of filtering for inputs that are represented as functions of a real variable. To use the Fourier representation we need to assume that the inputs under consideration have Fourier transforms. For example, the inputs might be bounded and supported in a bounded interval, or perhaps in $L^2(\mathbb{R})$. For many applications these hypotheses are not appropriate. A case of particular interest is that of periodic inputs. An input x, defined for $t \in \mathbb{R}$, is L-periodic if $x(t + L) = x(t)$. If, on the other hand, x is defined only in the interval $[0, L)$, then it can be extended to $\mathbb{R}$ as an L-periodic function.

In this section we briefly consider the modifications needed to analyze linear, shift invariant filters acting on periodic functions. A filter $\mathscr{A}$ is "L-periodic" if it carries L-periodic functions to L periodic functions. For $\tau \in \mathbb{R}$, the shift operation

$$x \mapsto x(t - \tau) = x_\tau(t)$$

is evidently an L-periodic filter. An L-periodic filter, $\mathscr{A}$, is shift invariant if $\mathscr{A}(x_\tau) = (\mathscr{A}x)_\tau$.

Recall that if f and g are L-periodic, then their periodic convolution is also L-periodic. The "L-periodic" unit impulse, δ_L, is given formally as the sum

$$\delta_L = \sum_{j=-\infty}^{\infty} \delta(t + jL).$$

If $\mathscr{A}$ is an L-periodic, shift invariant, linear filter, then it is determined by its impulse response, k, which is given by

$$k(t) = \mathscr{A}(\delta_L).$$

The impulse response can be either an ordinary function or a generalized function. As before, $\mathscr{A}$ has a representation as periodic convolution with its impulse response,

$$\mathscr{A} f(t) = \int_0^L k(t-s) f(s)\, ds.$$

Instead of the Fourier transform, the Fourier series now provides a spectral representation for a shift invariant filter. In this case the "transfer function" is the bi-infinite sequence $< \hat{k}(n) >$, defined for $n \in \mathbb{Z}$ by applying the filter directly to the complex exponentials

$$\mathscr{A}(e^{\frac{2\pi int}{L}}) = \hat{k}(n) e^{\frac{2\pi int}{L}}.$$

If k is an ordinary function, then this agrees with the usual definition for its Fourier coefficients. If k is a *periodic* generalized function, then this formula still makes sense. Formula (7.41) implies that the Fourier coefficients of $\mathscr{A}x = k * x$ are given by $< \hat{k}(n)\hat{x}(n) >$, and therefore the Fourier space representation of $\mathscr{A}$ is

$$\mathscr{A} x(t) = \frac{1}{L} \sum_{n=-\infty}^{\infty} \hat{k}(n)\hat{x}(n) e^{\frac{2\pi int}{L}}. \tag{9.23}$$

As in the case of the real line, a composition of linear, shift invariant, L-periodic filters is again a filter of the same type. It is also true that linear, shift invariant, L-periodic filters commute. As before, we need to exercise caution when implementing cascades and choose an ordering that avoids numerical difficulties.

***Example* 9.2.1.** A periodic function of period L has an expansion as a Fourier series

$$f(t) = \frac{1}{L} \sum_{j=-\infty}^{\infty} \hat{f}(j) e^{\frac{2\pi ijt}{L}}.$$

Recall that the Nth-partial sum operator is defined by

$$S_N(f;t) = \frac{1}{L} \sum_{j=-N}^{N} \hat{f}(j) e^{\frac{2\pi ijt}{L}}.$$

The Nth-partial sum operator is a shift invariant filter given by convolution with

$$d_N(t) = \frac{\sin\left(\frac{2\pi(2N+1)t}{L}\right)}{L \sin\left(\frac{\pi t}{L}\right)}.$$

The impulse response of S_N is $d_N(t)$.

***Example* 9.2.2.** An L-periodic analogue of the Hilbert transform is defined by

$$\begin{aligned}\mathcal{H}_L x(t) &= \lim_{\epsilon \downarrow 0} \left[\int_{-\frac{L}{2}}^{-\epsilon} + \int_{\epsilon}^{\frac{L}{2}} \right] \frac{\cos\left(\frac{\pi s}{L}\right)}{2\sin\left(\frac{\pi s}{L}\right)} f(t-s)\,ds \\ &= \text{P.V.} \int_{-\frac{L}{2}}^{\frac{L}{2}} \frac{\cos\left(\frac{\pi s}{L}\right)}{2\sin\left(\frac{\pi s}{L}\right)} f(t-s)\,ds.\end{aligned} \tag{9.24}$$

***Example* 9.2.3.** The transfer function for the Hilbert transform is given by

$$\hat{h}(n) = \begin{cases} 1 & n > 0, \\ 0 & n = 0, \\ -1 & n < 0. \end{cases}$$

By convention, $\hat{h}(0) = 0$. The Hilbert transform, in the Fourier representation, is given by

$$\mathcal{H}_L x(t) = \frac{1}{L}\left[\sum_{n=1}^{\infty} \hat{x}(n) e^{\frac{2\pi i n t}{L}} - \sum_{n=-\infty}^{-1} \hat{x}(n) e^{\frac{2\pi i n t}{L}} \right].$$

***Example* 9.2.4.** The bandpass filter with passband $[a, b]$ is defined in the Fourier representation by the transfer function

$$\hat{k}(n) = \begin{cases} 1 & |n| \in [a, b], \\ 0 & k \notin [a, b]. \end{cases}$$

***Example* 9.2.5.** As noted previously, for each N the Nth-partial sum of the Fourier series $S_N(f)$ defines a shift invariant, linear filter. It is a bandpass filter with passband $[0, N]$. Its transfer function is therefore

$$\hat{d}_N(n) - \begin{cases} 1 & \text{for } |n| \le N, \\ 0 & \text{for } |n| \ge N. \end{cases}$$

Exercises

***Exercise* 9.2.1.** Show that the shift $\mathcal{A}_\tau x(t) = x(t-\tau)$ defines an L-periodic shift invariant filter. What is its transfer function?

***Exercise* 9.2.2.** Suppose that k_1 and k_2 are the impulse responses of a pair of L-periodic, shift invariant filters. Show that the impulse response of the composition is $k_1 * k_2$. What is the transfer function?

***Exercise* 9.2.3.** Show that the first derivative $x \mapsto \partial_t x$ defines an L-periodic shift invariant filter. What is its transfer function?

***Exercise* 9.2.4.** The time reversal filter is defined by $\mathscr{A}x(t) = x(-t)$. It certainly carries L-periodic functions to L-periodic functions. Is it shift invariant?

***Exercise* 9.2.5.** The "periodizing" map defined by

$$x \mapsto \sum_{j=-\infty}^{\infty} x(t + jL)$$

defines a filter that maps functions on $\mathbb{R}$ to L-periodic functions. Show that, in an appropriate sense, this is a shift invariant filter. What are its impulse response and transfer function?

***Exercise* 9.2.6.** Show that the filters defined in Examples 9.2.2 and 9.2.3 agree.

***Exercise* 9.2.7.** Show that the Fejer means $C_N(f)$ (see Definition 7.5.1) are periodic, linear shift invariant filters. For each N, what are the impulse response and transfer function?

Exercise* 9.2.8. Give a definition for a periodic, generalized function.

9.2.1 Resolution of Periodic Filters

Let $\mathscr{A}$ denote a shift invariant, linear L-periodic filter with point spread function k. Some of the definitions of resolution given in Section 9.1.10 can be adapted to the context of periodic filters and inputs. The *full-width* κ*-maximum* definitions carry over in an obvious way, at least for point spread functions with a well-defined maximum at zero. We can also use the *first zero* definition for point spread functions that vanish. The *equivalent width* definition can be adapted if we use the integral of k over a single period:

$$\Delta_{\mathscr{A},\mathrm{ew}} = \frac{\int_0^L k(t)\,dt}{k(0)} = \frac{\hat{k}(0)}{k(0)}.$$

Applying the Nyquist criterion for periodic functions, we can also define the ϵ-Nyquist width for an L-periodic filter. Let $< \hat{k}(n) >$ denote the Fourier coefficients of k and suppose that $0 < \epsilon < 1$. If

$$|\hat{k}(j)| \geq \epsilon |\hat{k}(0)| \qquad \text{for } |j| \leq N$$

but either $|\hat{k}(N+1)| < \epsilon|\hat{k}(0)|$ or $|\hat{k}(-(N+1))| < \epsilon|\hat{k}(0)|$, then the ϵ-Nyquist width of $\mathscr{A}$ is defined to be

$$\Delta_{\mathscr{A},\mathrm{ny},\epsilon} = \frac{L}{2N+1}.$$

9.2.2 The Comb Filter and Poisson Summation*

See: A.4.5.

In Chapter 8 we saw that the result of sampling an input defined on $\mathbb{R}$ is to pass from the realm of the Fourier transform to the realm of the Fourier series. Sampling is not, in any reasonable sense, a shift invariant operation. It is rather a generalized multiplication filter that can be analyzed using a formalism quite similar (actually dual) to that used to analyze shift invariant filters. To that end, recall the properties of the δ-function:

- $x * \delta(L) = \int x(t)\delta(L-t)\,dt = x(L)$
- $\hat{\delta}(\xi) = 1$
- $\hat{\delta}_L(\xi) = \int \delta(t-L)e^{i\xi t}dt = e^{i\xi L}$

Multiplying the delta function by a continuous function x gives

$$x\delta(t-L) = x(L)\delta(t-L).$$

Repeating this for a sum of shifted delta functions gives

$$x\sum_{n=-\infty}^{\infty}\delta(t-nL) = \sum_{n=-\infty}^{\infty}x(nL)\delta(t-nL).$$

This gives a model for a sequence of samples as a train of impulses located at the sample points. Sampling is therefore defined as multiplication by the generalized function

$$C_L(t) = \sum_{n=-\infty}^{\infty}\delta(t-nL).$$

This generalized function is sometimes called a *Comb filter*. The fact that sampling is modeled as multiplication by a generalized function is a reflection of the non-physical nature of pointwise evaluation of a function. Real sampling processes always involve averaging a function over an interval of positive length.

Integrating the output of the comb filter gives

$$\int_{n=-\infty}^{\infty} x(t)\sum_{n=-\infty}^{\infty}\delta(t-nL) = \sum_{n=-\infty}^{\infty}x(nL).$$

Parseval's formula, for L^2 functions f and g, is

$$\int f(t)\overline{g(t)}\,dt = \frac{1}{2\pi}\int \hat{f}(\xi)\overline{\hat{g}(\xi)}\,d\xi. \tag{9.25}$$

On the other hand, the Poisson summation formula states that

$$\sum_{n=-\infty}^{\infty} f(nL) = \frac{1}{L} \sum_{n=-\infty}^{\infty} \hat{f}(\frac{2\pi n}{L}). \tag{9.26}$$

The δ-function is not a function, but arguing by analogy and comparing (9.25) to (9.26) gives

$$\int_{-\infty}^{\infty} x(t) C_L(t)\, dt = \frac{1}{2\pi} \int_{-\infty}^{\infty} \hat{x}(\xi) \overline{\hat{C}_L(\xi)}\, d\xi.$$

Thus the Fourier transform of the comb filter is also a generalized function:

$$\mathcal{F}[\sum_{n=-\infty}^{\infty} \delta(t - nL)] = \frac{1}{L} \sum_{n=-\infty}^{\infty} \delta(\xi - \frac{2\pi n}{L}).$$

As the comb filter is defined as the product of x and C_L, it has a Fourier representation as a convolution:

$$\mathcal{F}(x \cdot C_L)(\xi) = \hat{x} * \hat{C}_L(\xi) = \frac{1}{L} \sum_{n=-\infty}^{\infty} \hat{x}(\xi - \frac{2\pi n}{L}).$$

This formula is another way to write the Poisson summation formula.

The operation of sampling, which is done in the time domain, is frequently followed by windowing in the frequency domain and then reconstruction or interpolation, done in the time domain. As this is a composition of a multiplication filter and a shift invariant filter, it is not itself shift invariant; nonetheless there is a very simple formula for the kernel function of the composite operation. Let L denote the time domain sample spacing and $\hat{\varphi}$ the frequency domain windowing function. The sampling step takes

$$\mathcal{S}_l : x \longrightarrow \sum_{n=-\infty}^{\infty} x(nl)\delta(t - nL);$$

the output of the windowing and reconstruction steps is

$$\mathcal{WR} : \mathcal{S}_l(x) \longrightarrow \mathcal{F}^{-1}\left[\hat{\varphi}(\xi)\mathcal{F}\left(\sum_{n=-\infty}^{\infty} x(nl)\delta(t - nL)\right)\right].$$

Using the formula for the inverse Fourier transform of a product gives

$$\mathcal{WR} \circ \mathcal{S}_l(x) = \varphi * \left[\sum_{n=-\infty}^{\infty} x(nl)\delta(t - nL)\right] = \sum_{n=-\infty}^{\infty} x(nl)\varphi(t - nL). \tag{9.27}$$

This is the generalized Shannon-Whittaker formula, (8.7). Formula (9.27) shows that the kernel function for $\mathcal{WR} \circ \mathcal{S}_l$ is the generalized function

$$a_{\mathcal{WRS}_l}(t, s) = \sum_{n=-\infty}^{\infty} \varphi(t - nL)\delta(nL - s).$$

In a more realistic model for measurement, the samples $\{x(nL)\}$ are replaced by the samples of an average $\{\psi * x(nL)\}$. Here ψ models the sampling device. The output of the composite filter becomes

$$\sum_{n=-\infty}^{\infty} \varphi(t - nL)\psi * x(nL).$$

This modification is easily incorporated into the kernel function for $\mathcal{WR} \circ \mathcal{S}_l$. Letting $\mathcal{C}_\psi$ denote convolution with ψ, the kernel function for $\mathcal{WR} \circ \mathcal{S}_l \circ \mathcal{C}_\psi$ is

$$a_{\mathcal{WRS}_l\psi} = \sum_{n=-\infty}^{\infty} \varphi(t - nL)\psi(nL - s). \tag{9.28}$$

That the kernel function for $\mathcal{WR} \circ \mathcal{S} \circ \mathcal{C}_\psi$ is an ordinary function reflects the more realistic models for sampling and interpolation embodied in this filter.

Exercises

***Exercise* 9.2.9.** Show that

$$\int_{-\infty}^{\infty} a_{\mathcal{WRS}_l}(t, s)x(s)\,ds = (\mathcal{WR} \circ \mathcal{S}_l f)(x).$$

***Exercise* 9.2.10.** While the filter $\mathcal{WR} \circ \mathcal{S}_l \circ \mathcal{C}_\psi$ is fairly realistic, it still involves an infinite sum. A further refinement is to collect only a finite number of samples. Let χ denote a function with bounded support.

1. Find the kernel function for the filter

$$x \mapsto \mathcal{WR} \circ \mathcal{S}_l \circ \mathcal{C}_\psi(\chi x).$$

2. Find the kernel function for the filter

$$x \mapsto \mathcal{WR} \circ \mathcal{S}_l \circ (\chi \mathcal{C}_\psi x).$$

9.3 Higher-Dimensional Filters

See: A.4.6.

In imaging applications the data are not usually a function of a time variable but rather a function of several spatial variables. Typically they are functions defined on $\mathbb{R}^2$ and $\mathbb{R}^3$. The theory of filtering functions of several variables is formally quite similar to that for functions of a single variable, though a concept like causality has no useful analogue.

The material in this section and the following is not required for our subsequent study of algorithms in x-ray CT. These algorithms are expressed, from the outset, as compositions of one-dimensional filtering operations.

A linear filter acting on functions of n variables is usually represented as an integral

$$\mathscr{A} f(\boldsymbol{x}) = \int_{\mathbb{R}^n} a(\boldsymbol{x}, \boldsymbol{y}) f(\boldsymbol{y})\, d\boldsymbol{y}. \tag{9.29}$$

As before, $a(\boldsymbol{x}, \boldsymbol{y})$ is called the *kernel function* defining the filter $\mathscr{A}$; it may be an ordinary function or a generalized function.

***Example* 9.3.1.** The *identity filter* is the filter that carries a function to itself. It is usually denoted by Id, so that $(\text{Id}\, f)(\boldsymbol{x}) = f(\boldsymbol{x})$. The kernel function for the identity acting on functions of n variables is $\delta(\boldsymbol{x} - \boldsymbol{y})$, where δ is the n-dimensional delta function; see (5.28).

***Example* 9.3.2.** Suppose that $f(\boldsymbol{x})$ is a function of n variables and $\boldsymbol{\tau} \in \mathbb{R}^n$ is a fixed vector. The "shift by $\boldsymbol{\tau}$" is the filter defined by

$$\mathscr{A}_{\boldsymbol{\tau}} f(\boldsymbol{x}) = f(\boldsymbol{x} - \boldsymbol{\tau}) = f_{\boldsymbol{\tau}}(\boldsymbol{x}).$$

The kernel function for $\mathscr{A}_{\boldsymbol{\tau}}$ is $\delta(\boldsymbol{x} - \boldsymbol{y} - \boldsymbol{\tau})$.

Recall that a filter $\mathscr{A}$ acting on functions of n variables is shift invariant if

$$\mathscr{A} f_{\boldsymbol{\tau}} = (\mathscr{A} f)_{\boldsymbol{\tau}}$$

for all inputs f and vectors $\boldsymbol{\tau} \in \mathbb{R}^n$. A shift invariant filter is expressible as convolution.

Proposition 9.3.1. *Let $\mathscr{A}$ be a shift invariant filter acting on functions defined on $\mathbb{R}^n$. If $k(\boldsymbol{x}) = \mathscr{A}\delta(\boldsymbol{x})$ is the impulse response and f is an "arbitrary" input, then*

$$\mathscr{A} f(\boldsymbol{x}) = \int_{\mathbb{R}^n} k(\boldsymbol{x} - \boldsymbol{y}) f(\boldsymbol{y})\, d\boldsymbol{y}.$$

As in the one-dimensional case, the impulse response may be a generalized function, in which case this formula requires careful interpretation. The Fourier transform of the impulse response, $\hat{k}$, is called the transfer function (in electrical engineering). In imaging it is called the modulation transfer function. It provides a frequency space description for the action of a shift invariant filter:

$$\mathscr{A} f(\boldsymbol{x}) = \frac{1}{[2\pi]^n} \int_{\mathbb{R}^n} \hat{k}(\boldsymbol{\xi}) \hat{f}(\boldsymbol{\xi}) e^{i\langle \boldsymbol{x}, \boldsymbol{\xi}\rangle} d\boldsymbol{\xi}.$$

Suppose that f is a function of n variables with Fourier transform $\hat{f}$. The are many ways to define the bandwidth for a function of several variables. The simplest definition would be to say that f is R*-bandlimited* is $\hat{f}(\boldsymbol{\xi}) = 0$ for $\boldsymbol{\xi}$ with $\|\boldsymbol{\xi}\| > R$. If the "important" information in $\hat{f}$ is contained in the ball of radius R, then we would say that f is *effectively*

R-bandlimited. Though we usually use this definition, another possible definition would be to say that f is R-bandlimited if $\hat{f}(\xi_1, \dots, \xi_n) = 0$ whenever $\max\{|\xi_j| : j = 1, \dots, n\} > R$. Indeed, any choice of norm for $\mathbb{R}^n$ leads to a definition of bandwidth.

Let $\boldsymbol{x} = (x_1, \dots, x_n)$ be coordinates for $\mathbb{R}^n$. The simplest functions on $\mathbb{R}^n$ are functions that can be expressed as products of functions on $\mathbb{R}^1$. In filtering theory such a function is called *separable.* If

$$k(\boldsymbol{x}) = k_1(x_1) \cdots k_n(x_n)$$

is the impulse response of a filter $\mathscr{A}$, then $\mathscr{A}$ is said to be a *separable filter.* The transfer function of a separable filter is also a product:

$$\hat{k}(\boldsymbol{\xi}) = \hat{k}_1(\xi_1) \cdots \hat{k}_n(\xi_n).$$

***Example* 9.3.3.** Because the frequency, in n dimensions is a vector, a "lowpass" filter in n dimensions can be defined in many different ways. The transfer function $s_R(\boldsymbol{\xi}) = \chi_{[0,R]}(\|\boldsymbol{\xi}\|)$ defines a filter that removes all harmonic components whose frequencies have *length* greater than $R/2\pi$. Another possibility is to use

$$m_R(\boldsymbol{\xi}) = \prod_{j=1}^{n} \chi_{[0,R]}(|\xi_j|).$$

This filter removes harmonic components whose frequency *in any coordinate direction* exceeds $R/2\pi$. These functions define shift invariant filters

$$S_R(f) = \mathscr{F}^{-1}(s_R(\boldsymbol{\xi})\hat{f}(\boldsymbol{\xi})), \quad M_R(f) = \mathscr{F}^{-1}(m_R(\boldsymbol{\xi})\hat{f}(\boldsymbol{\xi})).$$

Their impulse responses (in two dimensions) are shown in Figure 9.6. The filter M_R is separable, whereas the filter S_R is not.

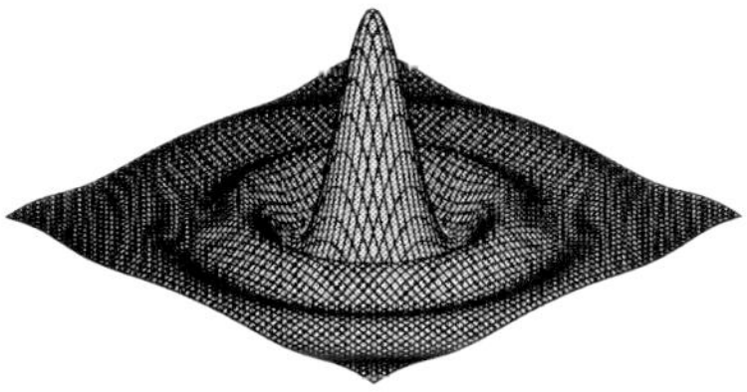

(a) Impulse response for S_R.

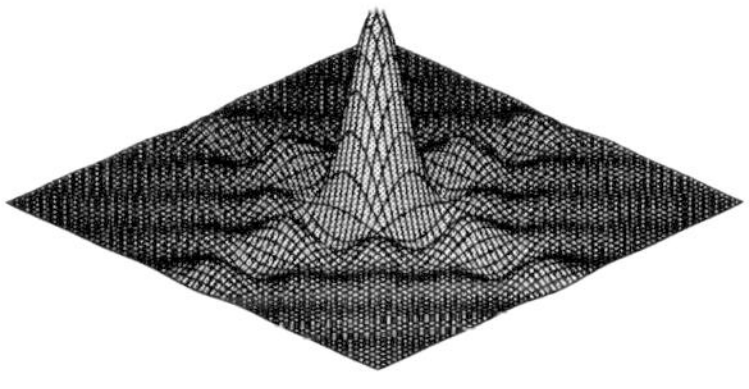

(b) Impulse response for M_R.

Figure 9.6. Impulse responses for two-dimensional lowpass filters.

Each of the filters in Example 9.3.3 defines a partial inverse to the Fourier transform, in the sense that either $S_R f$ or $M_R f$ converges to f as R tends to infinity. Figure 9.7 shows the results of applying these filters to the characteristic function of a square,

$$\chi_{[-1,1]^2}(\boldsymbol{x}) = \chi_{[-1,1]}(x_1)\chi_{[-1,1]}(x_2).$$

As the data have jump discontinuities, the filtered image exhibits Gibbs artifacts. Note the ringing artifact parallel to the edges of the square and also its absence in the "Gibbs shadow" formed by the vertex. This is an indication of the fact that the detailed analysis of the Gibbs phenomenon is more complicated in higher dimensions than it is in one dimension. Note also that the Gibbs artifact is a little more pronounced in (a).

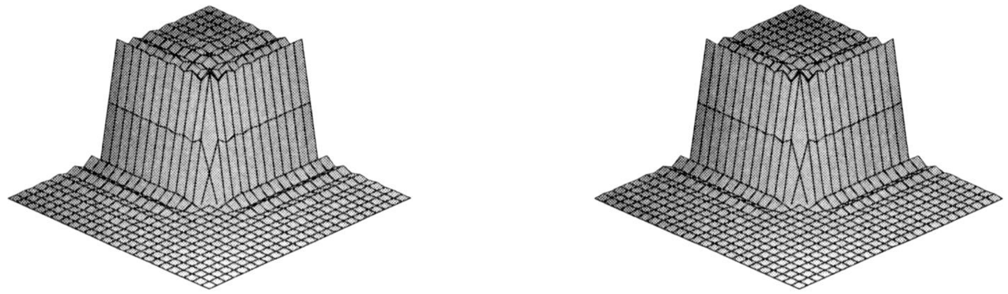

(a) S_R applied to $\chi_{[-1,1]^2}$. (b) M_R applied to $\chi_{[-1,1]^2}$.

Figure 9.7. Lowpass filters in two dimensions.

***Example* 9.3.4.** A filter can also act selectively in different directions. For example, a lowpass filter in the x_1-direction is defined by the transfer function $\chi_{[0,R]}(|\xi_1|)$. Given a unit vector $\boldsymbol{\omega} \in S^{n-1}$, the transfer function $\chi_{[0,R]}(|\langle\boldsymbol{\omega}, \boldsymbol{\xi}\rangle|)$ removes all harmonic components whose frequency in the $\boldsymbol{\omega}$-direction exceeds $R/2\pi$.

Complex, higher-dimensional filtering operations are often assembled out of simpler pieces. If $\mathscr{A}_1$ and $\mathscr{A}_2$ are linear filters then their composite, $\mathscr{A}_1 \circ \mathscr{A}_2$ is as well. While, in general, $\mathscr{A}_1 \circ A_2 \neq \mathscr{A}_2 \circ \mathscr{A}_1$, shift invariant filters do commute. If k_i, $i = 1, 2$ are the impulse responses of shift invariant filters $\mathscr{A}_i$, $i = 1, 2$, then the impulse response of the cascade $\mathscr{A}_1 \circ \mathscr{A}_2$ is

$$k_1 * k_2 = k_2 * k_1.$$

The transfer function is the product $\hat{k}_1(\boldsymbol{\xi})\hat{k}_2(\boldsymbol{\xi})$.

Exercises

***Exercise* 9.3.1.** Prove Proposition 9.3.1, assuming that f is a smooth function with bounded support.

***Exercise* 9.3.2.** Prove that if $\mathscr{A}_1$ and $\mathscr{A}_2$ are shift invariant filters then $\mathscr{A}_1 \circ \mathscr{A}_2 = \mathscr{A}_2 \circ \mathscr{A}_1$.

9.3.1 Isotropic Filters

A linear transformation $U : \mathbb{R}^n \to \mathbb{R}^n$ is a rigid rotation if

$$\|U\boldsymbol{x}\| = \|\boldsymbol{x}\| \tag{9.30}$$

for all vectors $\boldsymbol{x} \in \mathbb{R}^n$. Fixing an orthogonal coordinate system for $\mathbb{R}^n$, the rigid rotations are defined by matrices U, which satisfy the matrix equation

$$U^t U = \mathrm{Id} = U U^t.$$

Here U^t denotes the transpose of the matrix U. This collection of matrices is denoted $O(n)$. The rigid rotations of $\mathbb{R}^2$ are given by the matrices

$$O(2) = \left\{ \begin{pmatrix} \cos\theta & \sin\theta \\ -\sin\theta & \cos\theta \end{pmatrix}, \begin{pmatrix} -\sin\theta & \cos\theta \\ \cos\theta & \sin\theta \end{pmatrix} \text{ for } \theta \in [0, 2\pi) \right\}.$$

A linear transformation U defines an operation on functions by setting

$$f_U(\boldsymbol{x}) = f(U\boldsymbol{x}).$$

***Definition* 9.3.1.** A filter $\mathcal{A}$ acting on functions on $\mathbb{R}^n$ is *isotropic* if it commutes with all rotations; that is,

$$\mathcal{A} f_U(\boldsymbol{x}) = (\mathcal{A} f)_U(\boldsymbol{x}) = \mathcal{A} f(U\boldsymbol{x}).$$

An isotropic filter can be linear or nonlinear. For example, the filter that takes a function f to its absolute value, $|f|$, is a nonlinear isotropic filter. Linear, shift invariant isotropic filters have a very simple characterization.

Proposition 9.3.2. *A linear, shift invariant filter $\mathcal{A}$ is isotropic if and only if its impulse response (or transfer function) is a radial function.*

Proof. Let $k = \mathcal{A}\delta$ denote the impulse response of $\mathcal{A}$. From the results in Section 4.5.4 it follows that k is a radial function if and only if $\hat{k}$ is a radial function. If k is a radial function, then there is a function κ such that

$$k(\boldsymbol{x}) = \kappa(\|\boldsymbol{x}\|).$$

Let U be a rigid rotation. Then

$$\begin{aligned} \mathcal{A} f_U(\boldsymbol{x}) &= \int\limits_{\mathbb{R}^n} \kappa(\|\boldsymbol{x} - \boldsymbol{y}\|) f(U\boldsymbol{y})\, d\boldsymbol{y} \\ &= \int\limits_{\mathbb{R}^n} \kappa(\|U(\boldsymbol{x} - \boldsymbol{y})\|) f(U\boldsymbol{y})\, d\boldsymbol{y} \\ &= \mathcal{A} f(U(\boldsymbol{x})). \end{aligned} \tag{9.31}$$

The substitution $\boldsymbol{y}' = U\boldsymbol{y}$ is used to go from the second line to the last line. This shows that a radial impulse response defines an isotropic filter.

The converse statement is even easier because $\delta_U = \delta$, as follows by formally changing variables in the integral defining δ_U:

$$\int_{\mathbb{R}^n} \delta(U\boldsymbol{x}) f(\boldsymbol{x})\, d\boldsymbol{x} = \int_{\mathbb{R}^n} \delta(\boldsymbol{x}) f(U^{-1}\boldsymbol{x})\, d\boldsymbol{x} = f(0).$$

The definition of an isotropic filter implies that

$$k(\boldsymbol{x}) = \mathscr{A}\,\delta(\boldsymbol{x}) = \mathscr{A}\,\delta_U(\boldsymbol{x}) = \mathscr{A}\,\delta(U\boldsymbol{x}) = k_U(\boldsymbol{x}) \qquad \text{for all } U \in O(n).$$

This shows that $k(\boldsymbol{x})$ only depends on $\|\boldsymbol{x}\|$. □

***Example* 9.3.5.** If ψ is a smooth, nonnegative function with support contained in the ball of radius ϵ and total integral 1, then the formula

$$\mathscr{A}_\psi f(\boldsymbol{x}) = \int_{\mathbb{R}^n} \psi(\boldsymbol{x} - \boldsymbol{y}) f(\boldsymbol{y})\, d\boldsymbol{y}$$

defines a smoothing filter. Its transfer function is $\hat{\psi}(\boldsymbol{\xi})$. As

$$\int_{\mathbb{R}^n} \psi(\boldsymbol{x})\, d\boldsymbol{x} = 1,$$

it follows that $\hat{\psi}(0) = 1$. As ψ is smooth and vanishes outside a bounded set, its Fourier transform tends to zero as $\|\boldsymbol{\xi}\| \to \infty$. Thus $\mathscr{A}_\psi$ is an approximate lowpass filter. As in one dimension, filters like $\mathscr{A}_\psi$ provide models for measuring devices. If ψ is a radial function, then the filter $\mathscr{A}_\psi$ is isotropic.

Exercises

***Exercise* 9.3.3.** Suppose that $\mathscr{A}_1$ and $\mathscr{A}_2$ are isotropic filters. Show that $\mathscr{A}_1 \circ \mathscr{A}_2$ is as well.

***Exercise* 9.3.4.** Show that the squared length of the gradient

$$\mathscr{A} f = \sum_{j=1}^{n} \left(\frac{\partial f}{\partial x_j} \right)^2$$

is an isotropic filter.

***Exercise* 9.3.5.** Show that the following filters are *not* isotropic:

$$\begin{aligned} \mathscr{A}_\infty f &= \max\left\{ \left| \frac{\partial f}{\partial x_j} \right| : j = 1, \dots, n \right\}, \\ \mathscr{A}_1 f &= \sum_{j=1}^{n} \left| \frac{\partial f}{\partial x_j} \right|. \end{aligned} \tag{9.32}$$

9.3.2 Resolution

The discussion of resolution for one-dimensional filters in Section 9.1.10 can be repeated almost verbatim for higher-dimensional, isotropic filters. If the filter is not isotropic, then the situation is more complicated. For example, let $\mathscr{A}$ be a filter acting on functions of n variables with impulse response, a. Suppose that a achieves its maximum at 0 and decays to zero as $\|\boldsymbol{x}\|$ tends to infinity. The set of points

$$\text{FWHM}(a) = \{\boldsymbol{x} : a(\boldsymbol{x}) = \frac{1}{2}|a(0)|\},$$

where a assumes half its maximum value, is a hypersurface, the *half-maximum hypersurface*. If a is a radial function, then this hypersurface is a sphere centered at zero and the FWHM resolution in the output of $\mathscr{A}$ can be defined as the radius of this sphere. If a is not a radial function, then there are many possible ways to assign a single number that quantifies the resolution in the output of $\mathscr{A}$. A conservative assessment of the resolution in a non-isotropic filter is to use the largest distance between two points on the half maximum hypersurface. This is called the *diameter* of this hypersurface.

***Example* 9.3.6.** Figure 9.8 shows level curves for the impulse responses of the filters S_R and M_R.

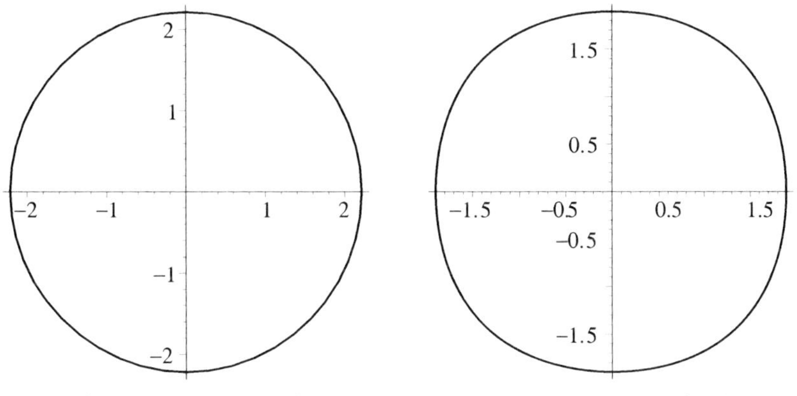

(a) Half-maximum curve for the isotropic filter S_R.

(b) Half-maximum curve for the non-isotropic filter M_R.

Figure 9.8. Half-maximum curves for two-dimensional lowpass filters.

***Definition* 9.3.2.** Let $\mathscr{A}$ be a linear, shift invariant filter with impulse response a. Suppose that $|a|$ assumes it maximum at 0 and tends to zero as $\|\boldsymbol{x}\|$ tends to infinity. The hypersurface FWHM(a) may have several components. Let FWHM$(a)_0$ be the component bounding a region that contains 0. The *full-width half-maximum* of $\mathscr{A}$ is defined to be the diameter of FWHM$(a)_0$.

Generalizing the other definitions is left as an exercise for the curious reader. It should be kept in mind that the output of a non-isotropic filter generally has different resolution in different directions.

Exercises

***Exercise* 9.3.6.** Generalize the definitions of resolution given for 1-dimensional filters in Section 9.1.10 to shift invariant filters acting on functions defined on $\mathbb{R}^n$.

***Exercise* 9.3.7.** Compute the FWHM resolution of the filters S_R and M_R.

9.4 Some Applications of Filtering Theory

In this section we apply filtering concepts to the study two subjects of direct relevance to medical imaging. The first subject we consider is image processing. This is a wide-ranging and growing field that, among other things, encompasses image reconstruction, image compression, and various sorts of enhancement or noise reduction. We give an overview of elementary methods for enhancing images, finding edges and reducing noise, seen as a subdiscipline of higher-dimensional filtering theory. In the second part of this section, we apply filtering concepts to analyze several pieces of hardware used in x-ray imaging. The material in this section is not required for the later chapters in the book.

9.4.1 Image Processing

(a) An "image."

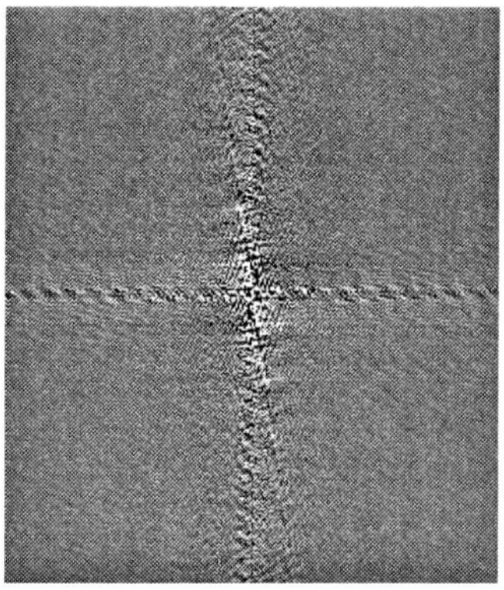

(b) The real part of its Fourier transform.

Figure 9.9. The Fourier transform of an image is not usually an image.

We consider methods for *postprocessing* images to improve their sharpness, reduce noise, and correct for other sorts of distortion. In this context "images" are mathematical representations of pictures of things in the natural world. The goal of much of this sort of processing is to make the information in pictures more accessible to the human visual system. Another related aim is to help machines to "see." In the sense that we are now using the word *image*, the Fourier transform of an image is usually *not* an image, as it is not generally recognizable as a "picture" of anything. This distinction is illustrated in Figure 9.9. Image processing operations and the language used to describe them closely mirror these origins and intents. While the filters described in this section can be applied to any sufficiently regular function of the correct number of variables, the interpretation of the output is closely tied to *a priori* assumptions about the inputs.

The sort of image processing discussed in this section is an important component of medical imaging, though these operations are more important in the "postprocessing" phase rather than in the measurement and image formation processes, which are our main topics. We present this material both because of its importance to medical imaging per se and because of its rich mathematical content. It shows how particular aims shape the design and implementation of filters. Our presentation is adapted from the very extensive treatment of this subject in [68].

Basic Concepts and Operations

We first introduce basic image processing operations and examples of filters that implement them. Many image processing operations are nonlinear. For simplicity we consider planar images that are represented by scalar functions of two variables. The value $f(x_1, x_2)$ represents the "gray level" or density of the image at (x_1, x_2). With this interpretation it is reasonable to assume that f takes only nonnegative values. Similar considerations apply to color images, which are usually represented by a triple of functions $[r(x_1, x_2), g(x_1, x_2), b(x_1, x_2)]$. These functions represent the intensities of three "independent" colors at (x_1, x_2).

A visual output device, such as a monitor or printer, is required to pass from a functional description of an image (i.e., f) to a picture, in the ordinary sense. Such a device has a fixed size and dynamic range. Mappings must be specified between the coordinates used in the parameterization of f and coordinates in the output device as well as between the values f assumes and gray levels (or colors) available in the output. In the first part of this section we consider filtering operations defined on functions of continuous variables; at the end we briefly discuss the problems of sampling continuous images and implementation of the filters.

The basic operations we consider fall into several classes:

Coordinate transformations: Images are sometimes distorted because of systematic modeling (or measurement) errors. In this connection it is useful to make a distinction between the "image plane" and the "measurement plane." In this paragraph the image plane is the plane in which the image itself lies; let (y_1, y_2) denote orthogonal coordinates in this plane. This means that if $f(y_1, y_2)$ is displayed as a gray level image in the y_1y_2-plane, then the image appears undistorted. The measurement plane refers to the coordinates defined by the apparatus used to *measure* the image; let (x_1, x_2) denote coordinates in this plane. Suppose that $f(x_1, x_2)$, for (x_1, x_2) lying in a subset D of $\mathbb{R}^2$, are measurements of an image. The parameters (x_1, x_2) in the measurement plane may differ from the coordinates in the image plane. Displaying $f(x_1, x_2)$ in the x_1x_2-plane would then result in a distorted image. This is called *geometric distortion*. Figure 9.10 shows an image displayed in image plane rectangular coordinates and image plane polar coordinates.

(a) f displayed in the x_1x_2-plane.

(b) $\mathscr{A}_{pr} f$ displayed in the image plane.

Figure 9.10. Removing geometric distortion.

Suppose that the measuring device is *calibrated* in polar coordinates with (x_1, x_2) corresponding to the point in the image plane with Cartesian coordinates

$$y_1 = x_1 \cos x_2, \quad y_2 = x_1 \sin x_2. \tag{9.33}$$

Displaying f in the x_1x_2-plane results in a distorted image [see Figure 9.10(a)]. Define a filter $\mathscr{A}_{pr}$ with

$$(\mathscr{A}_{pr} f)(y_1, y_2) = f\left(\sqrt{y_1^2 + y_2^2}, \tan^{-1}\left[\frac{y_2}{y_1}\right]\right).$$

The filtered image is shown in Figure 9.10(b). This is a linear filter whose output is the image parameterized by Cartesian coordinates in the image plane. The actual calibration of the measuring equipment determines the choice of branch of $\tan^{-1}$. The value of $(\mathscr{A}_{pr} f)(y_1, y_2)$ is not defined if $(y_1, y_2) = (0, 0)$; notwithstanding that, the inverse transformation, (9.33), is defined in the whole "polar coordinate plane." It is, however, not one to one.

The example illustrates most of the features of the general case. Let (x_1, x_2) be coordinates in the measurement plane, and suppose that these differ from those in the image plane. Denote the latter coordinates by (y_1, y_2). The two coordinate systems are functionally related with

$$x_1 = g(y_1, y_2), \quad x_2 = h(y_1, y_2),$$

defined for (y_1, y_2) belonging to a subset, D_i, in the image plane. This defines a mapping Φ from D_i to $D_m = \Phi(D_i)$, a subset of the measurement plane. Let $f(x_1, x_2)$ for $(x_1, x_2) \in D$ denote the measurements of an image; then

$$(\mathscr{A}_\Phi f)(y_1, y_2) = f \circ \Phi(y_1, y_2) = f(g(y_1, y_2), h(y_1, y_2)) \tag{9.34}$$

defines a filter that maps the portion of the measured image lying over $D_m \cap D$ to an image defined in the corresponding part of D_i. As with the example of polar coordinates, this transformation may not be defined in the the entire image plane and there may be choices involved in the definition of the map $\Phi : (y_1, y_2) \mapsto (g(y_1, y_2), h(y_1, y_2))$. This operation defines a linear filter that is usually not translation invariant. The kernel function for $\mathscr{A}_\Phi$ is the generalized function

$$a_\Phi(y_1, y_2; x_1, x_2) = \delta(x_1 - g(y_1, y_2), x_2 - h(y_1, y_2)),$$

here δ is the two-dimensional delta function. Let $g(y_1, y_2), h(y_1, y_2)$ be a pair of functions defined in a subset $D \subset \mathbb{R}^2$. Setting

$$\Phi(y_1, y_2) = (g(y_1, y_2), h(y_1, y_2))$$

defines a map of D into a subset $D' = \Phi(D)$ of $\mathbb{R}^2$. Whether or not Φ is a change of coordinates, formula (9.34) defines a linear filter carrying functions defined on D' to functions defined on D.

Exercises

***Exercise* 9.4.1.** Find conditions on (g, h) that imply that $\mathscr{A}_\Phi$ is translation invariant.

***Exercise* 9.4.2.** Suppose that the image is a transparency lying in the y_1y_2-plane; $f(y_1, y_2)$ describes the amount of incident light transmitted through the point (y_1, y_2). Suppose that the measurement is made by projecting the transparency onto a screen that lies in the plane $y_2 = y_3$ using a light source that produces light rays orthogonal to the y_1y_2-plane.

1. Using $(x_1, x_2) = (y_1, \sqrt{2}y_3)$ as coordinates for the measurement plane, find an expression for the amount of light incident at each point of the measurement plane (see Section 3.2).

2. Why are these good coordinates for the measurement plane?

3. Find a filter that removes the geometric distortion resulting from the projection.

***Exercise* 9.4.3.** Let $\Phi = (g, h)$ be a pair of functions defined in $D \subset \mathbb{R}^2$ and let $D' = \Phi(D)$. What are necessary and sufficient conditions on Φ for the filter A_Φ to be invertible as a map from functions defined on D' to functions defined on D?

In the remainder of this section the measurement plane and image plane are assumed to agree.

Noise reduction: Images can be corrupted by noise. There are two main types of noise: *uniform* noise and *binary* noise. In the first case the noise is uniformly distributed and locally of mean zero, whereas binary noise consists of sparsely, but randomly, distributed large errors. It is often caused by sampling or transmission errors. In this section techniques are described for reducing the effects of uniform noise; binary noise is discussed in Section 9.4.1. Because uniform noise is locally of mean zero, replacing f by a weighted average generally has the effect of reducing the noise content.

A shift invariant filter of this type can be defined by convolution with a nonnegative weight function φ, satisfying the usual condition:

$$\int_{\mathbb{R}^2} \varphi \, dx_1 \, dx_2 = 1. \tag{9.35}$$

Let $\mathscr{A}_\varphi f = \varphi * f$ be the convolution filter defined by φ. If φ has support in a small ball, then the output of $\mathscr{A}_\varphi f$ at (x_1, x_2) only depends on values of $f(y_1, y_2)$ for (y_1, y_2), near to (x_1, x_2) and the filter acts in a localized manner. If the noise has a directional dependence, then this can be incorporated into φ. If, for example, the noise is isotropic, then it is reasonable to use a radial function.

The frequency space representation of such a filter is

$$\mathscr{A}_\varphi f = \mathscr{F}^{-1}\left[\hat{\varphi}\hat{f}\right]. \tag{9.36}$$

As φ is integrable, its Fourier transform tends to zero as $\|\xi\|$ tends to infinity. This means that $\mathscr{A}_\varphi$ is an approximate lowpass filter. If $\hat{\varphi}$ is any function satisfying the conditions

$$\begin{aligned} \hat{\varphi}(0, 0) &= 1, \\ \lim_{\xi \to \infty} \hat{\varphi}(\xi) &= 0, \end{aligned} \tag{9.37}$$

then (9.36) defines an approximate lowpass filter. Even if the inverse Fourier transform, φ, assumes negative values, the effect of $\mathscr{A}_\varphi$ is to reduce uniform noise. In this generality there can be problems interpreting the output as representing an image. The filtered output, $\mathscr{A}_\varphi f$, may assume negative values, even if f is pointwise positive; to represent $\mathscr{A}_\varphi f$ as an image, it is necessary to remap its range to the allowable range of densities. This operation is discussed later in the paragraph on contrast enhancement (page 358). If $\hat{\varphi}$ vanishes outside a bounded set, then φ cannot have bounded support (see Proposition 4.4.1). From the representation of $\mathscr{A}_\varphi f$ as a convolution, it follows that the value of $\mathscr{A}_\varphi f$ at a point (x_1, x_2) would depend on values of f at points distant from (x_1, x_2).

As observed previously, reduction of the available resolution is an undesirable side-effect of lowpass filtering. In an image this appears as blurring. Using statistical properties of the noise, an "optimal filter" can be designed that provides a balance between noise reduction and blurring. This is discussed in Chapter 17. The noise might be locally uniform but non-uniform across the image plane. In this case it might be better to use a nonshift invariant filter to reduce the effects of noise. A heuristic, used in image processing to retain detail in a filtered image, is to represent an image as a linear combination of the lowpass filtered output and the original image. That is, instead of using either $\mathscr{A}_\varphi f$ or f alone, we use

$$\mu \mathscr{A}_\varphi f + (1 - \mu) f, \tag{9.38}$$

where $0 \leq \mu \leq 1$. If the noise is nonuniform across the image plane, then μ could be taken to depend on (x_1, x_2).

Sharpening: An image may be blurred during acquisition; it is sometimes possible to filter the image and recapture some of the fine detail. In essence, this is an inverse filtering operation. The measured image is modeled as $\mathscr{A}f$, where f denotes the "original" image and $\mathscr{A}$ models the measurement process. If $\mathscr{A}$ is a shift invariant, linear filter, then the methods discussed in Section 9.1.9 can be applied to try to approximately invert $\mathscr{A}$ and restore details present in f that were "lost" in the measurement process. In general, this leads to an amplification of the high frequencies, which can, in turn, exacerbate problems with noise.

The fine detail in an image is also "high-frequency information." A slightly different approach to the de-blurring problem is to simply remove, or attenuate, the low-frequency information. In the frequency domain representation, such a filter is given by

$$\mathcal{A}_\varphi f = \mathcal{F}^{-1}\left[\hat{\varphi}\hat{f}\right],$$

where, instead of (9.37), $\hat{\varphi}$ satisfies

$$\begin{aligned} \hat{\varphi}(0) &= 0, \\ \lim_{\xi \to \infty} \hat{\varphi}(\boldsymbol{\xi}) &= 1. \end{aligned} \tag{9.39}$$

If the function $\hat{\varphi} - 1$ has an inverse Fourier transform, ψ, then this filter has spatial representation as

$$\mathcal{A}_\varphi f = f - \psi * f.$$

In other words, A_φ is the difference between the identity filter and an approximate lowpass filter; hence it is an approximate highpass filter.

Edge detection: Objects in an image are delimited by their edges. Edge detection is the separation of the boundaries of objects from other, more slowly varying features of an image. The rate at which a smooth function varies in direction ω is measured by its directional derivative

$$\mathcal{D}_\omega f(\boldsymbol{x}) = \frac{d}{dt} f(\boldsymbol{x} + t\omega)\Big|_{t=0} = \langle \nabla f, \omega \rangle.$$

Here

$$\nabla f = (\partial_{x_1} f, \partial_{x_2} f)$$

is the gradient of f. The Euclidean length of ∇f provides an isotropic measure of the variation of f (i.e., it is equally sensitive to variation in all directions). Thus points where $\|\nabla f\|$ is large should correspond to edges. An approach to separating the edges from other features is to set a threshold, t_{edge}, so that points with $\|\nabla f(\boldsymbol{x})\| > t_{\text{edge}}$ are considered to belong to edges. A filtered image, showing only the edges, would then be represented by

$$\mathcal{A}_1 f(\boldsymbol{x}) = \chi_{(t_{\text{edge}},\infty)}(\|\nabla f(\boldsymbol{x})\|). \tag{9.40}$$

While this is a nonlinear filter, it is shift invariant, and the approximate computation of ∇f can be done very efficiently using the fast Fourier transform.

This approach is not robust as the gradient of f is also large in highly textured or noisy regions. If D is a region in the plane with a smooth boundary and $f = \chi_D$, then $\nabla f(\boldsymbol{x}) = 0$ if $\boldsymbol{x} \notin bD$. From the point of view of sampling, it may be quite difficult to detect such a sharply defined edge. These problems can be handled by smoothing f before computing its gradient. Let φ denote a smooth, nonnegative function, with small support satisfying (9.35). In regions with a lot of texture or noise, but no boundaries, the gradient of f varies "randomly" so that cancellation in the integral defining

$$\nabla(\varphi * f) = \varphi * \nabla f$$

should lead to a relatively small result. On the other hand, along an edge, the gradient of f is dominated by a component in the direction orthogonal to the edge. Therefore, the weighted average, $\varphi * \nabla f$, should also have a large component in that direction. Convolution smears the sharp edge in χ_D over a small region, and therefore points where $\|\nabla(\varphi * \chi_D)\|$ is large are more likely to show up in a sample set. This give a second approach to edge detection implemented by the filter

$$\mathcal{A}_2 f(\boldsymbol{x}) = \chi_{(t_{\text{edge}},\infty)}(\|\nabla(\varphi * f)(\boldsymbol{x})\|). \tag{9.41}$$

Once again, $f \mapsto \nabla\varphi * f$ is a linear shift invariant filter that can be computed efficiently using the Fourier transform. In order not to introduce a preference for edges in certain directions, a radial function should be used to do the averaging.

The second filter helps find edges in regions with noise or texture but may miss edges where the adjoining gray levels are very close. In this case it might be useful to compare the size of the gradient to its average over a small region. In this approach the $\|\nabla f\|$ (or perhaps $\|\nabla\varphi * f\|$) is computed and is then convolved with a second averaging function to give $\psi * \|\nabla f\|$ (or $\psi * \|\nabla\varphi * f\|$). A point $\boldsymbol{x}$ then belongs to an edge if the ratio

$$\frac{\|\nabla f(\boldsymbol{x})\|}{(\psi * \|\nabla f\|)(\boldsymbol{x})}$$

exceeds a threshold τ_{edge}. A filter implementing this idea is given by

$$\mathcal{A}_3 f(\boldsymbol{x}) = \chi_{(\tau_{\text{edge}},\infty)}\left(\frac{\|\nabla f(\boldsymbol{x})\|}{\psi * \|\nabla f\|(\boldsymbol{x})}\right). \tag{9.42}$$

Once again, the filter is nonlinear, but its major components can be efficiently implemented using the Fourier transform. The image produced by the output of these filters shows only the edges.

There is a related though somewhat more complicated approach to edge detection that entails the use of (minus) the Laplace operator,

$$-\Delta f = \partial_{x_1}^2 f + \partial_{x_2}^2 f,$$

as a way to measure the local variability of f. (We call this $-\Delta$ for consistency with earlier usages of this symbol.) There are three reasons for this approach: (1) The Laplace operator is rotationally invariant; (2) the singularities of the function Δf are the same as those of f; and (3) there is some evidence that animal optical tracts use a filtering operation of this sort to detect edges. The first statement is easily seen in the Fourier representation:

$$\mathcal{F}(-\Delta f)(\xi_1, \xi_2) = -(\xi_1^2 + \xi_2^2)\hat{f}(\xi_1, \xi_2).$$

The point of (2) is that the sharp edges of objects are discontinuities of the density function and will therefore remain discontinuities of Δf. This property requires more advanced techniques to prove; see [43]. For the reasons discussed previously, the Laplace operator is often combined with a Gaussian smoothing operation,

$$\mathcal{G}_\sigma f(\boldsymbol{x}) = \iint\limits_{\mathbb{R}^2} e^{-\frac{|\boldsymbol{x}-\boldsymbol{y}|^2}{\sigma}} f(\boldsymbol{y})\, d\boldsymbol{y}$$

to get

$$\mathcal{A}_\sigma f = -\Delta \mathcal{G}_\sigma f.$$

The transfer function of $\mathcal{A}_\sigma$ is

$$\hat{a}_\sigma(\xi_1, \xi_2) = -\pi\sigma(\xi_1^2 + \xi_2^2)e^{-\frac{\sigma(\xi_1^2+\xi_2^2)}{4}}$$

and its impulse response response is

$$a_\sigma(x_1, x_2) = \frac{4}{\sigma^2}(x_1^2 + x_2^2 - \sigma)e^{-\frac{x_1^2+x_2^2}{\sigma}}.$$

These are radial functions; the graphs of a radial section are shown in Figure 9.11. Note that the impulse response assumes both positive and negative values.

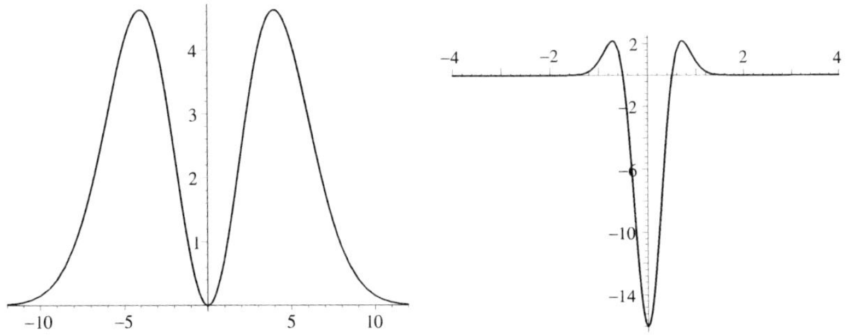

(a) The transfer function. (b) The impulse response.

Figure 9.11. Radial graphs of the transfer function and impulse response for $\mathcal{A}_{.25}$.

It is not immediately obvious how to use the output of $\mathcal{A}_\sigma$ to locate edges. This is clarified by considering the special case of a sharp edge. Applying $\mathcal{A}_\sigma$ to the function $\chi_{[0,\infty)}(x_1)$ gives

$$\mathcal{A}_\sigma \chi_{[0,\infty)}(x_1, x_2) = \frac{2cx_1}{\sqrt{\sigma}}e^{-\frac{x_1^2}{\sigma}};$$

here c is a positive constant. The zero crossing of $\mathcal{A}_\sigma \chi_{[0,\infty)}$ lies on the edge; nearby are two sharp peaks with opposite signs. Figure 9.12 shows a cross section in the output of $A_\sigma \chi_{[0,\infty)}$ orthogonal to the edge. The parameter σ takes the values .01, .1, .25 and 1; smaller values correspond to sharper peaks.

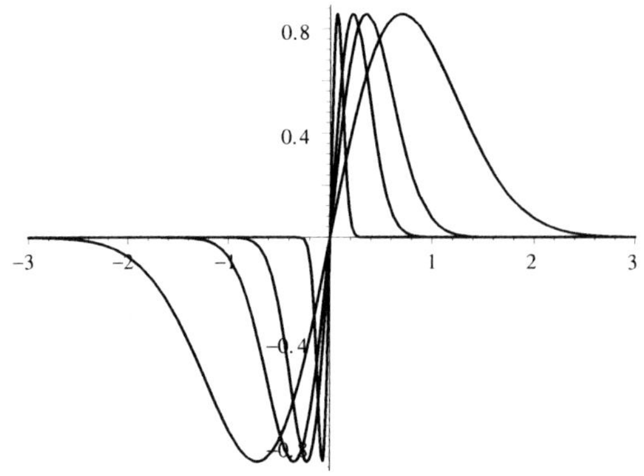

Figure 9.12. Output of Laplacian edge detection filters.

This example suggests that the absolute value of the output of filter $\mathcal{A}_\sigma$ can be used to locate edges. An edge shows up as a black arc (the zero crossings) flanked by two nearly parallel white arcs. A different approach is to define

$$\mathcal{B}_\sigma f(x) = \frac{1}{2}\left[\operatorname{sgn}((\mathcal{A}_\sigma f)(x)) + 1\right].$$

The preceding example indicates that $\mathcal{B}_\sigma f$ should be 0 on one side of an edge and 1 on the other. The image produced by $\mathcal{B}_\sigma f$ is then a "schematic" showing the outlines of the objects with all the fine detail removed. An edge image can be combined with the original image, as in (9.38) to obtain a composite image with fine detail and slowly varying features as well as enhanced contrast between objects.

Exercise

***Exercise* 9.4.4.** Let U be a 2×2 matrix defining a rigid rotation. If f is a function defined in the plane, then set

$$f_U(x_1, x_2) = f(U(x_1, x_2)).$$

Show that

$$\|\nabla f_U(0,0)\| = \|\nabla f(0,0)\|.$$

Explain the statement "∇f provides an isotropic measure of the variation of f."

Contrast enhancement: The equipment used to display an image has a fixed dynamic range. Suppose that the gray values that can be displayed are parameterized by the interval $[d_{\min}, d_{\max}]$. To display an image, a mapping needs to be fixed from the range of f to $[d_{\min}, d_{\max}]$. That is, the values of f need to be scaled to fit the dynamic range of the output device. Suppose that f assumes values in the interval $[m, M]$. Ordinarily, the scaling map is a monotone map $\gamma : [m, M] \to [d_{\min}, d_{\max}]$. This map is usually nonlinear and needs to be adapted to the equipment being used. By choosing γ carefully, different aspects of the image can be emphasized or *enhanced*.

Suppose that there is a region R in the image where f varies over the range $[a, A]$. If $A - a$ is very small compared to $M - m$, then a linear scaling function,

$$\gamma(t) = d_{\max}\frac{t-m}{M-m} + d_{\min}\frac{M-t}{M-m},$$

would compress the information in R into a very small part of $[d_{\min}, d_{\max}]$. In the output this region would have very low contrast and so the detail present there would not be visible. The contrast in R can be enhanced by changing γ to emphasize values of f lying in $[a, A]$, though necessarily at the expense of values outside this interval. For example, a piecewise linear scaling function

$$\gamma_{aA}(t) = \begin{cases} d_{\min} & \text{if } t < a, \\ d_{\max}\frac{t-a}{A-a} + d_{\min}\frac{A-t}{A-a} & \text{if } a \leq t \leq A, \\ d_{\max} & \text{if } t > A \end{cases}$$

would make the detail in R quite apparent. On the other hand, detail with gray values outside of $[a, A]$ is entirely lost.

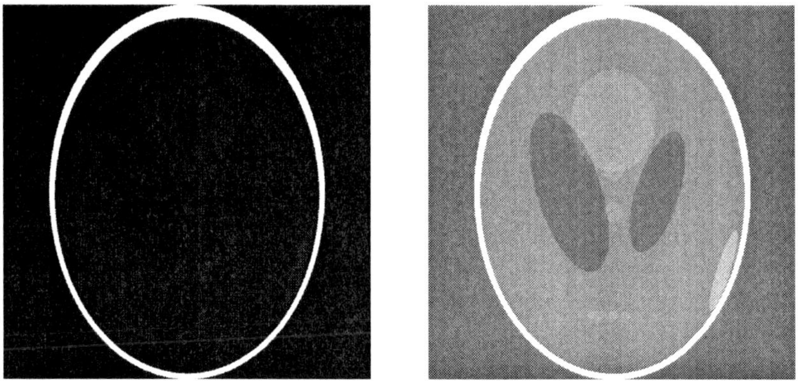

(a) Low-contrast display. (b) High-contrast display.

Figure 9.13. The effect of rescaling gray values in the Shepp-Logan phantom.

In Figure 9.13 an example of a CT-phantom is shown with two different choices of γ. In medical imaging this is called windowing or thresholding. Another method used to make information present in an image more apparent in a visual display is to map the gray values to colors. In either case, no more information is actually "there," but, due to the physiology of human vision, it becomes more apparent.

Discretized Images

So far, an "image" has been represented by a real-valued function of a pair of continuous variables. In image processing, functions describing images are usually sampled on a discrete set of points and quantized to take values in a finite set. In this section we briefly discuss sampling of images and implementation of filters on the sampled data. As before, we restrict our attention to two-dimensional images. The sample set in imaging is usually a uniform rectangular grid,

$$\{(jh_1, kh_2) \; : \; j, k \in \mathbb{Z}\},$$

where h_1 and h_2 are positive numbers. The individual rectangles with corners

$$\{[(jh_1, kh_2), ((j+1)h_1, kh_2), (jh_1, (k+1)h_2), ((j+1)h_1, (k+1)h_2)] \; : \; j, k \in \mathbb{Z}\}$$

are called "picture elements" or *pixels*. To simplify the discussion, we do not consider questions connected with quantization and work instead with the full range of real numbers.

Suppose that the function f describing the image is supported in the unit square, $[0, 1] \times [0, 1]$. If the image has bounded support, then this can always be arranged by scaling. The actual measurements of the image consist of a finite collection of samples, collected on a uniform rectangular grid. Let M and N denote the number of samples in the x_1- and x_2-directions, respectively. The simplest model for measurements of an image is the set of samples

$$f_{jk} = f(\frac{j}{M}, \frac{k}{N}) \qquad \text{for } 0 \le j \le M-1, \quad 0 \le k \le N-1.$$

The discussion of higher-dimensional sampling presented in Section 8.5 is directly applicable in this situation. In particular, Nyquist's theorem provides lower bounds on the sampling rates in terms of the bandwidth, needed to avoid aliasing. Because the bandwidth (or effective bandwidth) can vary with direction, aliasing can also have a directional component. The so-called *moiré effect* is the result of undersampling a directional periodic structure. The figure shown in 9.14(a) is sampled in the vertical direction by selecting every third pixel leading to aliasing, which appears as the moiré effect in Figure 9.14(b). Figure 9.14(c) shows the result of selecting every third pixel in both directions.

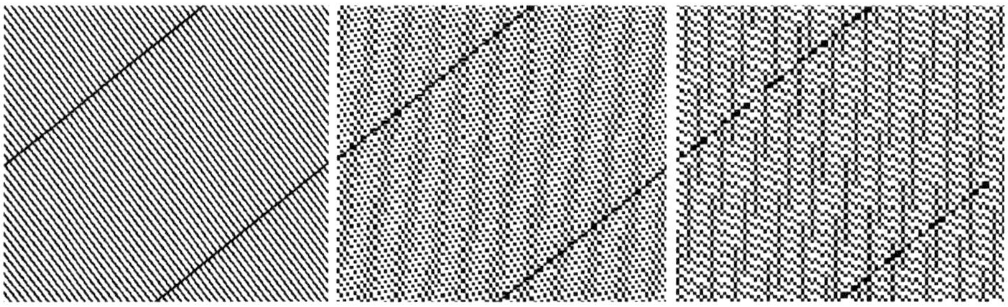

(a) The original image. (b) The result of sampling in the y-direction. (c) The result of sampling in the both directions.

Figure 9.14. The moiré effect is directional aliasing.

As was the case with one-dimensional inputs, lowpass filtering, prior to sampling, reduces aliasing artifacts due to undersampling. Most practical measurements involve some sort of averaging and therefore the measurement process itself incorporates a lowpass filter.

A shift invariant measurement process is modeled as samples of a convolution $\varphi * f$. In either case the samples are the result of evaluating *some* function. For the remainder of this section we consider the sample set to be the values of a function at the sample points.

Suppose that $\{f_{jk}\}$ are samples of an image. To use these samples to reconstruct an image, the image plane is divided into rectangles. With the samples above the jkth pixel, denoted p_{jk}, is the rectangle

$$p_{jk} = \{(x_1, x_2) \, : \, \frac{j}{M} \leq x_1 < \frac{j+1}{M}, \quad \frac{k}{N} \leq x_2 < \frac{k+1}{N}\}.$$

A simple way to use the samples to define an image is to set

$$\tilde{f}(x_1, x_2) = \sum_{j=0}^{M-1} \sum_{k=0}^{N-1} f_{jk} \chi_{p_{jk}}(x_1, x_2). \tag{9.43}$$

In this image, each pixel has a constant gray level. More sophisticated methods, using interpolation, can be used to get a smoother-looking result.

Coordinate transformations: Filters that employ changes of coordinate are difficult to implement on sampled data. Suppose that $\Phi = (g(x_1, x_2), h(x_1, x_2))$ is a map defined on a domain D' whose image is contained in the unit square. As previously, let $\{(\frac{j}{M}, \frac{k}{N})\}$ denote the sample points in $[0, 1]^2$ and $\{(\frac{l}{M'}, \frac{m}{N'})\}$ uniformly spaced sample points in D'. If f is defined everywhere in $[0, 1]^2$, then there is no difficulty sampling the values of $\mathscr{A}_\Phi f$, obtaining the samples

$$(\mathscr{A}_\Phi f)_{lm} = \{f(\Phi(\frac{l}{M'}, \frac{m}{N'})\}.$$

If f has already been sampled, then a problem immediately presents itself: The images of sample points in D', $\{\Phi(\frac{l}{M'}, \frac{m}{N'})\}$ are unlikely to belong to the set $\{(\frac{j}{M}, \frac{k}{N})\}$. Several strategies can be used to approximate $\mathscr{A}_\Phi$ on sampled data. The easiest approach is to use the piecewise constant image function, $\tilde{f}$, defined in (9.43), setting

$$(\mathscr{A}_\Phi f)_{lm} \overset{d}{=} \tilde{f}(\Phi(\frac{l}{M'}, \frac{m}{N'})).$$

Another, more exact approach is to use some sort of "nearest neighbor" interpolation. This is a three-step process:

- For each $(\frac{l}{M'}, \frac{m}{N'})$ in D', find the sample points in $[0, 1]^2$ "nearest" to $\Phi(\frac{l}{M'}, \frac{m}{N'})$. Denote these points $\{\boldsymbol{x}_1^{lm}, \dots, \boldsymbol{x}_{j_{lm}}^{lm}\}$.

- Assign weights $\{w_q^{lm}\}$ to these point according to some measure of their distance to $\Phi(\frac{l}{M'}, \frac{m}{N'})$. The weights are usually nonnegative and add up to 1.

- Define $(\mathscr{A}_\Phi f)_{lm}$ as a weighted average:

$$(\mathscr{A}_\Phi f)_{lm} = \sum_{q=1}^{j_{lm}} w_q^{lm} f(\boldsymbol{x}_q^{lm}).$$

A definition of "nearest neighbors" and a scheme for assigning weights is needed to implement this procedure.

Exercise

***Exercise* 9.4.5.** Explain how the first procedure used to define $(\mathscr{A}_\Phi f)_{lm}$ can be interpreted as an interpolation scheme.

Linear, shift invariant filters: Suppose that a is the impulse response of a linear shift invariant, two-dimensional filter $\mathscr{A}$. Its action in the continuous domain is given by

$$\mathscr{A} f(x_1, x_2) = \iint_{\mathbb{R}^2} a(x_1 - y_1, x_2 - y_1) f(y_1, y_2)\, dy_1\, dy_2.$$

A Riemann sum approximation for this filter, at sample points, is

$$\mathscr{A} f(\frac{j}{M}, \frac{k}{N}) \approx \mathscr{A}_s f_{jk} \stackrel{d}{=} \sum_{l=0}^{M-1} \sum_{m=0}^{N-1} a_{(j-l)(k-m)} f_{lm} \frac{1}{MN}, \tag{9.44}$$

where $a_{jk} = a(\frac{j}{M}, \frac{k}{N})$. The finite Fourier transform gives an efficient way to implement the discrete convolution on the right-hand side of (9.44).

In image processing, many operations on sampled data are naturally defined by sums like those on the right-hand side of (9.44) with only a small number of nonzero coefficients. Highly localized operations of this sort are directly implemented using this sum. A filter with this property is called a *local operation*. In this context the array of coefficients (a_{jk}) is called the *filter mask*. A local operation has a *small* filter mask. The mask can be represented as a matrix showing the nonzero elements, with a specification of the index corresponding to the "center" of the matrix.

***Example* 9.4.1.** A standard method to reduce uniform noise in a uniformly sampled image is to average the nearest neighbors. A simple example of such a filter is

$$(\mathscr{S}^a f)_{jk} = \frac{1}{1+4a}[f_{jk} + a(f_{(j-1)k} + f_{(j+1)k} + f_{j(k-1)} + f_{j(k+1)})].$$

Its mask is nicely expressed as a matrix, showing only the nonzero elements, with the center of the matrix corresponding to (0, 0):

$$s^a_{jk} = \frac{1}{1+4a}\begin{pmatrix} 0 & a & 0 \\ a & 1 & a \\ 0 & a & 0 \end{pmatrix}.$$

***Example* 9.4.2.** Partial derivatives can be approximated by finite differences. The x_1-partial derivative has three different, finite difference approximations:

FORWARD DIFFERENCE: $\partial_{x_1} f(\frac{j}{M}, \frac{k}{N}) \approx M[f(\frac{j+1}{M}, \frac{k}{N}) - f(\frac{j}{M}, \frac{k}{N})] \stackrel{d}{=} \mathfrak{D}^f_1 f(j, k)$

BACKWARD DIFFERENCE: $\partial_{x_1} f(\frac{j}{M}, \frac{k}{N}) \approx M[f(\frac{j}{M}, \frac{k}{N}) - f(\frac{j-1}{M}, \frac{k}{N})] \stackrel{d}{=} \mathfrak{D}_1^b f(j,k)$

SYMMETRIC DIFFERENCE: $\partial_{x_1} f(\frac{j}{M}, \frac{k}{N}) \approx 2M[f(\frac{j+1}{M}, \frac{k}{N}) - f(\frac{j-1}{M}, \frac{k}{N})] \stackrel{d}{=} \mathfrak{D}_1^s f(j,k)$

Let $d^f_{jk}, d^b_{jk}, d^s_{jk}$ denote the corresponding filter masks. Each has only two nonzero entries in the $k=0$ row; as 1×3 matrices they are

$$\begin{aligned} d^f_{j0} &=(M, -M, 0), \\ d^b_{j0} &=(0, M, -M), \\ d^s_{j0} &=(-\frac{M}{2}, 0, \frac{-M}{2}). \end{aligned} \tag{9.45}$$

In each, the center entry corresponds to $(0,0)$. Using any of these approximations requires $O(MN)$ operations to approximate $\partial_{x_1} f$ as compared to $O(MN \log_2 M)$ operations, if M is a power of 2 and the FFT is used. For large M, the finite difference approximation is more efficient.

***Example* 9.4.3.** Suppose that $M = N$. A standard finite difference approximation for the Laplace operator is $-\Delta \approx \mathfrak{D}_1^b \circ \mathfrak{D}_1^f + \mathfrak{D}_2^b \circ \mathfrak{D}_2^f$. Let Δ_s denote this finite difference operator. Its filter mask is the 3×3 matrix

$$[\Delta_s]_{jk} = N^2 \begin{pmatrix} 0 & 1 & 0 \\ 1 & -4 & 1 \\ 0 & 1 & 0 \end{pmatrix}.$$

The center of the matrix corresponds to $(0,0)$.

An important theoretical feature of the smoothing and edge detection filters defined in the previous section is that they are isotropic. An-isotropy is an unavoidable consequence of sampling, even if the sample spacings are equal in all directions. The coordinates axes become preferred directions in a uniformly sampled image. The direct implementation of an isotropic, shift invariant filter then leads to an an-isotropic filter.

***Example* 9.4.4.** The Laplace operator provides an example of this phenomenon. Let Δ_s denote the finite difference approximation to $-\Delta$ defined in Example 9.4.3. The direct implementation of this filter on an $N \times N$-grid of data points require about $10N^2$ operation. The an-isotropy of this operator is quite apparent in the Fourier representation, wherein

$$(\Delta^s f)_{jk} = (\mathscr{F}_2^{-1} < \hat{d}_{mn} \cdot \hat{f}_{mn} >)_{jk}.$$

Here $\mathscr{F}_2^{-1}$ is the inverse of the two-dimensional finite Fourier transform and $< \hat{d}_{mn} >$ are the Fourier coefficients of Δ_s thought of as an $(N+2) \times (N+2)$-periodic matrix. A simple calculation shows that

$$\begin{aligned} \hat{d}_{mn} &= (2N^2)[\cos\left(\frac{2\pi l}{N+2}\right) + \cos\left(\frac{2\pi m}{N+2}\right) - 2] \\ &= (-4N^2)[\frac{\pi^2(l^2+m^2)}{(N+2)^2} - \frac{2\pi^4 l^4}{3(N+2)^4} - \frac{2\pi^4 m^4}{3(N+2)^4} + O(l^6+m^6)]. \end{aligned} \tag{9.46}$$

The fourth-order terms are *not* radial. In Proposition 9.3.2 it is shown that a linear, shift invariant filter is isotropic if and only if its transfer function is radial (that is, can be expressed as a function of $\xi_1^2 + \xi_2^2$). An implementation of the Laplace operator in the Fourier representation, obtained by directly sampling $\xi_1^2 + \xi_2^2$, the transfer function for Δ, is isotropic. For large N this implementation requires considerably more computation than the non-isotropic, finite difference approximation.

Binary noise: Suppose that f_{jk} with $0 \le j, k \le N - 1$ represents a discretized image. Due to transmission or discretization errors, for a sparse, "random" subset of indices, the values $\{f_{j_1 k_1}, \ldots, f_{j_m k_m}\}$ are dramatically wrong. This kind of noise is called *binary noise*. As it is not, in any sense, of "mean zero," it is not attenuated by lowpass filtering. A different approach is needed to correct such errors. *Rank value* filtering is such a method. These are filters that compare the values that f_{jk} assumes at nearby pixels.

A simple example is called the *median filter*. Fix a pair of indices jk. A pixel p_{lm} is a *neighbor* of p_{jk} if $\max\{|l - j|, |m - k|\} \le 1$. With this definition of neighbor, each pixel has nine neighbors, including itself. The values of f_{lm} for the neighboring pixels are listed in increasing order. The fifth number in this list is called the median value; denote it by m_{jk}. The median filter, $\mathcal{M}f$, replaces f_{jk} by m_{jk}. This removes wild oscillations from the image and otherwise produces little change in the local variability of f_{jk}. Other schemes replace f_{jk} by the maximum, minimum, or average value. Any of these operations can be modified to leave the value of f_{jk} unchanged unless it differs dramatically, in the context of the neighboring pixels, from the median.

This concludes our very brief introduction to image processing. Complete treatments of this subject can be found in [81] or [68].

Exercises

***Exercise* 9.4.6.** Show that the median filter is shift invariant.

***Exercise* 9.4.7.** Show that the median filter is nonlinear.

***Exercise* 9.4.8.** Suppose that the image $\{f_{jk}\}$ contains a sharp edge so $f_{jk} = 1$ for values on one side and 0 for values on the other side. A point is on the edge if it has neighbor whose value differs by ± 1. Considering the different possible orientations for an edge, determine the effect of the median filter at points on an edge.

***Exercise* 9.4.9.** Let $\{a_0(t), \ldots, a_n(t)\}$ be non-constant functions on $\mathbb{R}$. The differential operator

$$\mathcal{D}f = \sum_{j=0}^{n} a_j(t) \frac{d^j f}{dt^j}$$

is a non–shift invariant filter. Compare implementations of this filter on sampled data obtained using the discrete Fourier transform and finite difference operators to approximate the derivatives. Which is more efficient?

9.4.2 Linear Filter Analysis of Imaging Hardware

A very important question in the design of an imaging device is its resolution. There are two rather different limitations on the resolution. The first derives from the physical limitations of real measuring devices, and the second arises from the sampling and processing done to reconstruct an image. From the point of view of signal processing, a precise result is Nyquist's theorem, which relates the sample spacing to the bandwidth of the sampled data. In imaging applications the data are spatially limited, so they cannot be bandlimited. We therefore introduced the concept of effective support as the frequency band where "most" of the energy in the data lies. In this section we use linear filtering theory to analyze the distortion of the data that results from using real, physical measuring devices. These are limitations that are present in the measurements before any attempt is made to process the data and reconstruct the image. In this section we use geometric optics to model x-rays as a diverging flux of particles in much the same spirit as in Section 3.2. In this discussion we have borrowed heavily from [6], where the reader can find, *inter alia,* a careful discussion of γ-ray detectors, x-ray sources, and collimators.

The Transfer Function of the Scanner

In this section we consider a three-dimensional situation, beginning with the simple setup in Figure 9.15. A source of radiation is lying in the *source plane* with a distribution $f(\boldsymbol{r})$ [i.e., $f(\boldsymbol{r})\,d\boldsymbol{r}$ is the number of photons per unit time emitted by an area on the source of size $d\boldsymbol{r}$ located at position $\boldsymbol{r}$]. The source output is assumed to be independent of time.

In a *parallel* plane, at distance s_1 from the source plane, is an object that is described by a transmittance function $g(\boldsymbol{r}')$. The fraction of the incident photons transmitted by an area $d\boldsymbol{r}'$ located at the point $\boldsymbol{r}'$ in the object plane is given by $g(\boldsymbol{r}')d\boldsymbol{r}'$. Usually $g(\boldsymbol{r}')$ takes values between 0 and 1. It is sometimes useful to think of g as the probability that a photon incident at $\boldsymbol{r}'$ will be transmitted. The object plane is sometimes thought of as a thin slice of a three-dimensional object. If the width of the slice is ϵ, then the transmittance is related to the attenuation coefficient in Beer's law by

$$g(\boldsymbol{r}') = 1 - \exp[-\int_0^{\epsilon} \mu\, ds].$$

Here μ is an attenuation coefficient and the integral is along the line perpendicular to object plane through the point $\boldsymbol{r}'$. Finally, a detector lies in a second *parallel* plane, at distance s_2 from the object plane. For the moment assume that the detector is perfect (i.e., everything incident on the detector is measured). Later, a more realistic detector will be incorporated into the model. To analyze the source-object-detector geometry, first assume that the object is transparent; that is, $g(\boldsymbol{r}') = 1$.

It is convenient to use a different systems of coordinates in each plane, $\boldsymbol{r}, \boldsymbol{r}', \boldsymbol{r}''$. As previously, $d\boldsymbol{r}, d\boldsymbol{r}'$, and $d\boldsymbol{r}''$ denote the corresponding area elements. A point source is isotropic if the flux through an area A on the sphere of radius ρ, centered on the source, is proportional to the ratio of A to the area of the whole sphere $4\pi\rho^2$. The constant of proportionality is the intensity of the source. The *solid angle* Ω subtended by a region D,

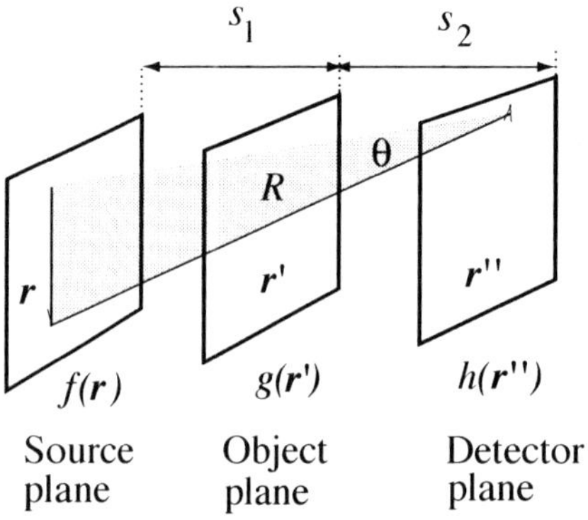

Figure 9.15. Arrangement of an imaging device with a source, object, and detector.

relative to a point p, is defined by projecting the region onto the sphere of radius 1 centered at p, along lines through the center of the sphere. If D' is the projected image of D, then the solid angle it subtends is defined to be

$$\Omega = \text{area of the region } D'.$$

From the definition it is clear that the solid angle assumes values between 0 and 4π.

To find the flux through a region, D, due to a planar distribution of isotropic sources with density $f(\boldsymbol{r})$, we need to find the contribution of each infinitesimal area element $d\boldsymbol{r}$. If $\Omega(D, \boldsymbol{r})$ is the solid angle subtended at $\boldsymbol{r}$ by D, then the contribution of the area element centered at $\boldsymbol{r}$ to the flux through D is $f(\boldsymbol{r})\, d\boldsymbol{r}\Omega(D, \boldsymbol{r})/4\pi$. In our apparatus we are measuring the area element on a plane parallel to the source plane. As shown in Figure 9.16, the solid angle subtended at a point $\boldsymbol{r}$ in the source plane by an infinitesimal area $d\boldsymbol{r}''$ at $\boldsymbol{r}''$ in the detector plane is

$$\frac{\cos\theta\, d\boldsymbol{r}''}{R^2}, \quad \text{where } R = \frac{s_1 + s_2}{\cos\theta}.$$

Therefore, the infinitesimal solid angle is

$$d\Omega = \frac{\cos^3\theta}{(s_1 + s_2)^2} d\boldsymbol{r}''.$$

If no absorbing material is present, a detector at $\boldsymbol{r}''$ of planar area $d\boldsymbol{r}''$ absorbs $d\Omega/4\pi$ of the emitted radiation. If the intensity of the source at $\boldsymbol{r}$ is $f(\boldsymbol{r})\, d\boldsymbol{r}$, then we get

$$f(\boldsymbol{r})\, d\boldsymbol{r}\frac{d\Omega}{4\pi} = f(\boldsymbol{r})\frac{\cos^3\theta}{4\pi(s_1 + s_2)^2} d\boldsymbol{r}\, d\boldsymbol{r}''$$

as the measured flux. Notice that θ is a function of $\boldsymbol{r}$ and $\boldsymbol{r}''$.

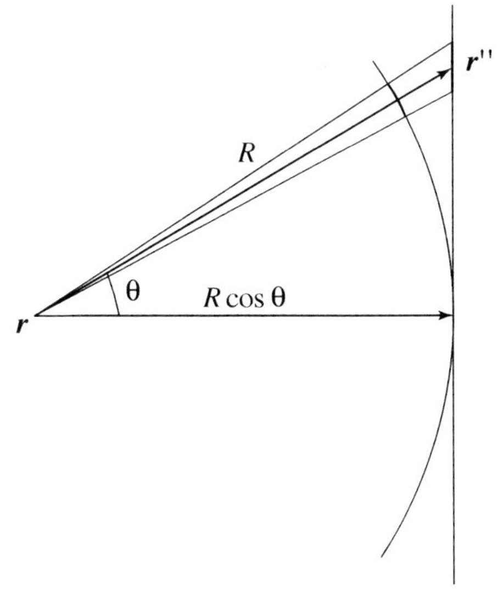

Figure 9.16. Computing the solid angle.

Now we include the effect of an absorbing material. The measured flux at $\boldsymbol{r}''$ is an integral along the source plane given by

$$h(\boldsymbol{r}'') = \frac{1}{4\pi(s_1+s_2)^2}\int_{\text{source}} \cos^3\theta(\boldsymbol{r},\boldsymbol{r}'')f(\boldsymbol{r})g(\boldsymbol{r}'(\boldsymbol{r},\boldsymbol{r}''))\,d\boldsymbol{r}. \tag{9.47}$$

Here $h(\boldsymbol{r}'')$ is the measured photon flux density at the detector point $\boldsymbol{r}''$. We see that $\boldsymbol{r}'$ is a function of $\boldsymbol{r}$ and $\boldsymbol{r}''$. To obtain this relation, we can think of the two-dimensional plane containing the three vectors $\boldsymbol{r}$, $\boldsymbol{r}'$, and $\boldsymbol{r}''$, as in Figure 9.17. Since we have similar triangles, we see that

$$\frac{\boldsymbol{r}'-\boldsymbol{r}}{s_1} = \frac{\boldsymbol{r}''-\boldsymbol{r}'}{s_2},$$

or

$$\boldsymbol{r}' = \frac{s_2}{s_1+s_2}\boldsymbol{r} + \frac{s_1}{s_1+s_2}\boldsymbol{r}'' = a\boldsymbol{r}'' + b\boldsymbol{r},$$

where

$$a = \frac{s_1}{s_1+s_2}, \qquad b = \frac{s_2}{s_1+s_2} = 1-a.$$

Now equation (9.47) reads

$$h(\boldsymbol{r}'') = \frac{1}{4\pi(s_1+s_2)^2}\int \cos^3\theta f(\boldsymbol{r})g(a\boldsymbol{r}''+b\boldsymbol{r})\,d\boldsymbol{r}. \tag{9.48}$$

In a transmission imaging problem, the source function $f(\boldsymbol{r})$ is assumed to be known. Relation (9.48) states that the output along the detector plane is a linear filter applied to f. It is more or less a convolution of the known source function $f(\boldsymbol{r})$ with the unknown transmittance function $g(\boldsymbol{r}')$. In a transmission imaging problem we are trying to determine

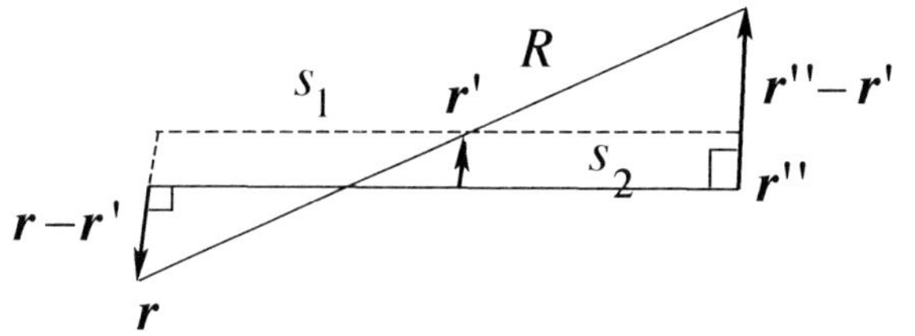

Figure 9.17. The similar triangle calculation.

$g(\boldsymbol{r}')$. As it stands, formula (9.48) is not quite in the form of a convolution for two reasons: (1) $\theta(\boldsymbol{r}, \boldsymbol{r}'')$ is not a function of $\boldsymbol{r} - \boldsymbol{r}''$, and (2) there are two coordinates systems, one in the source and one in the detector plane. Letting

$$\boldsymbol{r}_0'' = -\frac{b\boldsymbol{r}}{a},$$

we express everything in the coordinates of the detector plane. Let $\tilde{f}$ be the scaled source function and $\tilde{g}$ be the scaled transmission function given by

$$\hat{f}(\boldsymbol{r}_0'') = f(-a\boldsymbol{r}_0''/b), \quad \tilde{g}(\boldsymbol{r}_0'') = g(a\boldsymbol{r}_0'')$$

Now the measurement is expressed as

$$h(\boldsymbol{r}'') = \left(\frac{a}{b}\right)^2 \frac{1}{4\pi(s_1 + s_2)^2} \int \tilde{f}(\boldsymbol{r}_0'')\tilde{g}(\boldsymbol{r}'' - \boldsymbol{r}_0'') \cos^3\theta d\boldsymbol{r}_0''.$$

But for the $\cos^3\theta$-term, this would be a convolution. In many applications, the angle θ is close to 0 and therefore the cosine term can be approximated by 1. With this approximation, a shift invariant linear system relates the source and transmittance to the measured output,

$$h(\boldsymbol{r}'') = \left(\frac{a}{b}\right)^2 \frac{1}{4\pi(s_1 + s_2)^2} \int \tilde{f}(\boldsymbol{r}_0'')\tilde{g}(\boldsymbol{r}'' - \boldsymbol{r}_0'')\, d\boldsymbol{r}_0''. \tag{9.49}$$

This formula is not only useful for the analysis of transmission imaging but is also useful in nuclear medicine. Instead of a known x-ray source, we imagine that $f(\boldsymbol{r})$ describes an unknown distribution of radioactive sources. In this context we could use a pinhole camera to form an image of the source distribution. Mathematically, this means that g is taken to equal 1 in a tiny disk and zero everywhere else. Let us consider this situation geometrically: Only lines drawn from source points to detector points that pass through the support of g contribute to the image formed in the detector plane. If the support of g is a single point, then the image formed would be a copy of f scaled by the ratio s_1/s_2. This is because each point $\boldsymbol{r}''$ in the detector plane is joined to a unique point $\boldsymbol{r}(\boldsymbol{r}'')$ in the source plane by a line passing through the support of g. Note however that the intensity of the image at a point $\boldsymbol{r}''$ is proportional to $|\boldsymbol{r}'' - \boldsymbol{r}(\boldsymbol{r}'')|^{-2}$ (see Figure 9.18).

Now suppose that the support of g is the disk $B_0(d)$ for a very small d. The image formed at a point $\boldsymbol{r}''$ in the detector plane is the result of averaging the source intensities

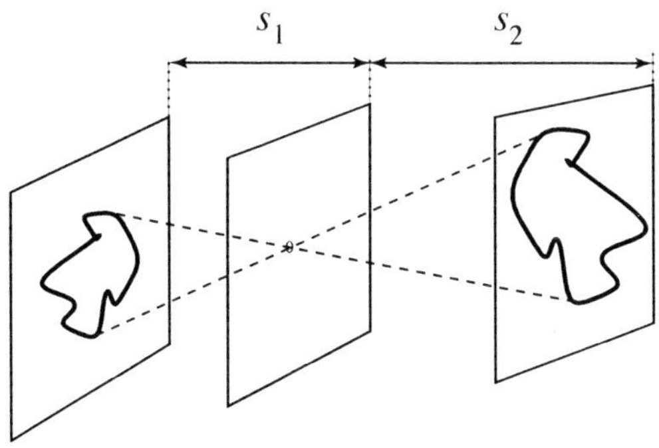

Figure 9.18. A pinhole camera.

over the disk in the source plane visible from the point $\boldsymbol{r}''$ through the pinhole, $B_0(d)$. Of course the actual image is a weighted average. The result of a positive diameter pinhole is a blurred image.

Exercises

***Exercise* 9.4.10.** Find a more accurate model for the "output" of a pinhole camera by letting $g(\boldsymbol{r}') = \delta(\boldsymbol{r}')$ in (9.48).

***Exercise* 9.4.11.** Find a formula for the output of a pinhole camera with

$$g(\boldsymbol{r}') = \chi_D(\boldsymbol{r}')$$

using the simplified formula (9.49).

The Resolution of an Imaging System

We now estimate the resolution of the transmission imaging system described in Section 9.4.2 using the FWHM criterion. We consider the problem of resolving two spots in the object plane. The source is assumed to have constant intensity over a disk of diameter d_{fs}; in this case f is the characteristic function of a disk $f(\boldsymbol{r}) = \chi_{B_1}(2|\boldsymbol{r}|/d_{\text{fs}})$. In this section B_1 denotes the disk of radius 1 centered at $(0, 0)$. The source is projected onto the detector plane, passing it through an object composed of two identical opaque spots separated by some distance. Mathematically, it is equivalent to think of the object as being entirely opaque but for two disks of perfect transmittance separated by the same distance. It is clear that the outputs of the two configurations differ by a constant. For the calculation the object is modeled as the sum of two δ-functions,

$$g(\boldsymbol{r}') = \delta(\boldsymbol{r}' - \boldsymbol{r}_1') + \delta(\boldsymbol{r}' - \boldsymbol{r}_2').$$

Of course this is not a function taking values between 0 and 1. This situation is approximated by a transmittance given by

$$g_\epsilon = \chi_{B_1}\left(\frac{|\boldsymbol{r}' - \boldsymbol{r}_1'|}{\epsilon}\right) + \chi_{B_1}\left(\frac{|\boldsymbol{r}' - \boldsymbol{r}_2'|}{\epsilon}\right),$$

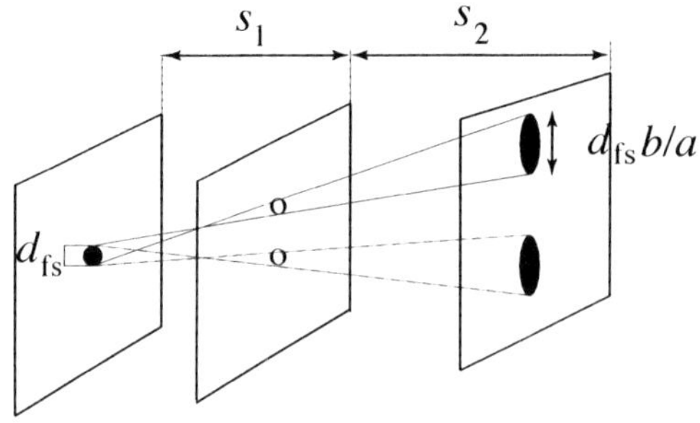

Figure 9.19. The image of two dots.

letting ϵ tend to zero *and* rescaling the output by ϵ^{-2} so that the total measured intensity is constant. The limiting case is mathematically equivalent to using the sum of δ-functions as the transmittance.

The image of a single tiny spot on the object plane located at $\boldsymbol{r}_1'$ is given by formula (9.49):

$$\begin{aligned} h(\boldsymbol{r}'', \boldsymbol{r}_1') &= \frac{1}{4\pi(s_1+s_2)^2}\int \chi_{B_1}\left(\frac{2r}{d_{fs}}\right)\delta(a\boldsymbol{r}''+b\boldsymbol{r}-\boldsymbol{r}_1')\,d\boldsymbol{r} \\ &= \frac{1}{4\pi(s_1+s_2)^2}\frac{1}{b^2}\chi_{B_1}\left(\frac{2|a\boldsymbol{r}''-\boldsymbol{r}_1'|}{bd_{fs}}\right) \\ &= \frac{1}{4\pi(s_1+s_2)^2}\frac{1}{b^2}\chi_{B_1}\left(\frac{2|\boldsymbol{r}''-(\boldsymbol{r}_1'/a)|}{bd_{fs}/a}\right) \end{aligned}$$

The image is a disk centered at $\boldsymbol{r}_1'/a$ with radius bd_{fs}/a. This is the impulse response of the source-detector pair. The FWHM for the point spread function of this system is just the diameter of the image disk,

$$d_{fs}'' = \frac{b}{a}d_{fs}.$$

According to this definition, two points sources in the object plane are resolvable if their images are completely non-overlapping, which might appear to be overly stringent.

We would actually like to know the minimum separation in the *object plane* for two point sources to be resolvable. This distance is found by projecting to the object plane,

$$\delta_{fs}' = bd_{fs} = \frac{s_2}{s_1+s_2}d_{fs}.$$

If $s_2 = 0$ then $\delta_{fs}' = 0$. If the object is sitting on the detector, then no matter how close the two point sources are, they can be distinguished. As $s_1 \to 0$, $\delta_{fs}' \to d_{fs}$, so the resolution is better for objects closer to the detector.

This simple model is completed by including a detector response function. Suppose that $k(\boldsymbol{r}'')$ is the impulse response of the detector. The complete imaging system with source f, transmittance g, and detector response k is given by

$$h(\boldsymbol{r}'') = \left(\frac{a}{b}\right)^2 \frac{1}{4\pi(s_1+s_2)^2} \int\int \tilde{f}(\boldsymbol{r}_0'')\tilde{g}(\boldsymbol{r}_1'' - \boldsymbol{r}_0'')k(\boldsymbol{r}'' - \boldsymbol{r}_1'')\cos^3\theta d\boldsymbol{r}_0'' d\boldsymbol{r}_1''.$$

This models what is called an "imaging detector," such as photographic film or an array of scintillation counters and photo-multiplier tubes. Setting $\boldsymbol{r}'' = 0$ gives a model for a "counting detector." Such a detector only records the number of incident photons, making no attempt to record the location on the detector where the photon arrives. In this case $k(\boldsymbol{r}'')$ is the probability that the detector responds to a photon arriving at $\boldsymbol{r}''$.

We now consider the sensitivity of the detector. The bigger the area on the detector subtended by the image, the more photons we count. By taking $s_2 = 0$, the object is in contact with the detector. This gives the best resolution but, for a pair of infinitesimal objects, no measurable data. There is a trade-off between the *resolution* of the image and the *sensitivity* of the detector. A larger detector captures more photons but gives less information about where they came from. In general, it is a very difficult problem to say which configuration is optimal. Probablisitic models for the source and detector are frequently employed in the analysis of this sort of problem (see Sections 16.1.1 and 17.4.3).

We consider the question of optimizing the placement of the object when both the source and detector are Gaussian. This means that

$$\hat{f} \text{ is proportional to } \exp\left[-\pi\left(\frac{|\xi|}{\rho_f}\right)^2\right],$$

$$\hat{k} \text{ is proportional to } \exp\left[-\pi\left(\frac{|\xi|}{\rho_d''}\right)^2\right].$$

Such a source is said to have a *Gaussian focal spot*. When referred back to the object plane, we get the following modulation transfer function:

$$\text{MTF} = \exp\left[-\pi|\xi|^2\left(\frac{1}{(\lambda\rho_d'')^2} + \frac{(\lambda-1)^2}{\lambda^2\rho_f^2}\right)\right]. \tag{9.50}$$

The parameter $\lambda = \frac{s_1+s_2}{s_1}$ is the magnification factor. In this context the optimal configuration is the one that distorts the object the least; this is the one for which the Gaussian in (9.50) decays as slowly as possible. For this family of functions this corresponds to

$$\frac{1}{(\lambda\rho_d'')^2} + \frac{(\lambda-1)^2}{\lambda^2\rho_f^2}$$

assuming its minimum value. Differentiating and setting the derivative equal to zero, we find that

$$\lambda^{opt} = 1 + \left(\rho_f/\rho_d''\right)^2.$$

Note that a large spot corresponds to $\rho_f \to 0$ and a poor detector corresponds to $\rho_d'' \to 0$.

Collimators

In the previous section, we discussed a simple geometry for a source, object, and detector. The object was simplified to be two dimensional. In medical imaging applications, we can think of the object as being made of many slices. The analysis presented in the previous section indicates that the spreading of the x-ray beam causes distortion in the projected image, which depends on how far the object is from the source. As illustrated in Figure 9.20, the images of points in the plane P1 are spread out more than the images of points in P3. A diverging beam magnifies objects more in the slices closer to the source than those in slices farther away. We could try reducing this effect by making the distance from the source to the detector much larger than the size of object. This has the undesirable effect of greatly attenuating the x-ray beam incident on the detector.

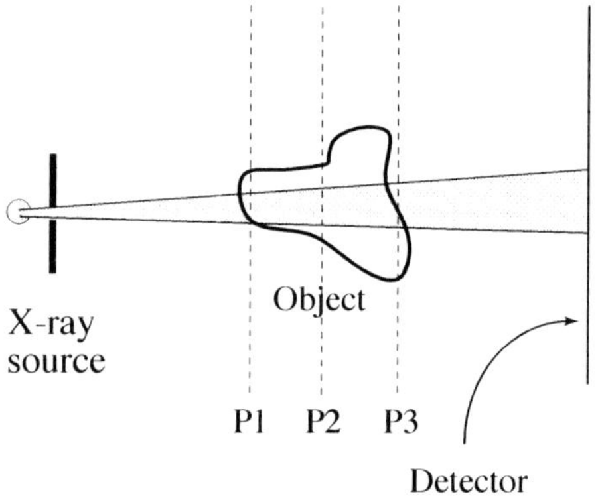

Figure 9.20. Beam spreading.

To control beam spreading distortion, we can reject x-rays arriving from certain directions by using a *collimator*. Collimators are an essential part of most x-ray imaging systems. In simple terms, a collimator is a cylindrical (or conical) hole bored through x-ray absorbing material. There are two physical parameters describing a collimator:

$$D_b\text{: the diameter of the hole,} \quad L_b\text{: the height of sides,}$$

called the *bore diameter* and *bore length* (see Figure 9.21). Often collimators are used in arrays. In this case, each collimator is better modeled as narrow pipe made of x-ray absorbing material. In our discussion it is assumed that only photons that pass through the collimator bore reach the detector.

This is also a three-dimensional analysis. The collimator is circularly symmetric about its axis, which implies that the response to a point source depends only on the distance, z, from the point source to the front of the collimator and the distance, r_s, from the source to the axis of the collimator.

The impulse response of a collimator can be deduced from the the analysis in Section 9.4.2. The effect of the collimator on a x-ray beam is identical to the effect of two concentric pinholes of diameter D_b, cut into perfectly absorbing plates, lying in parallel

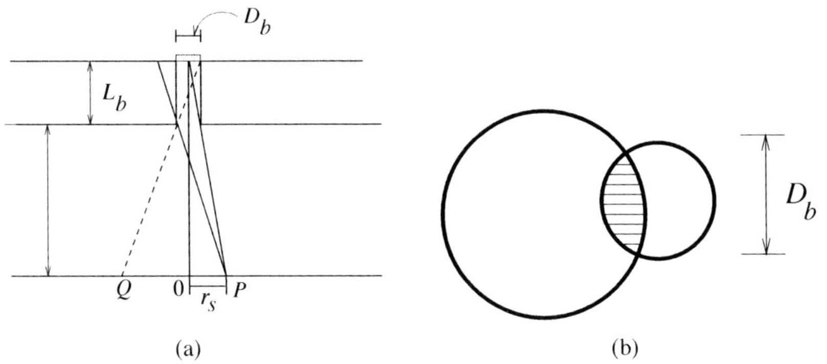

Figure 9.21. The geometry of a collimator.

planes a distance L_b apart. Let $f(\boldsymbol{r})$ model the x-ray source, $g_1(\boldsymbol{r}')$ the lower pinhole, and $g_2(\boldsymbol{r}'')$ the upper pinhole. From Section 9.4.2, we have that the photon flux incident on the upper plate is

$$h(\boldsymbol{r}'') = \frac{1}{4\pi(L_b+z)^2}\int f(\boldsymbol{r})g_1(a\boldsymbol{r}''+b\boldsymbol{r})\,d\boldsymbol{r};$$

where

$$a = \frac{z}{L_d+z}, \quad b = \frac{L_b}{L_b+z}.$$

The flux incident on the detector (through the upper pinhole) is therefore

$$\int h(\boldsymbol{r}'')g_2(\boldsymbol{r}'')\,d\boldsymbol{r}'' = \frac{1}{4\pi(L_b+z)^2}\int\int f(\boldsymbol{r})g_1(a\boldsymbol{r}''+b\boldsymbol{r})g_2(\boldsymbol{r}'')\,d\boldsymbol{r}''d\boldsymbol{r}.$$

It is assumed that the angle θ the beam makes with the detector plane is approximately 0 and therefore $\cos\theta \approx 1$.

Now suppose that the source is a point source, P, located a distance r_s from the axis of the collimator [i.e., $f(\boldsymbol{r}) = \delta(\boldsymbol{r}-\boldsymbol{r}_s)$] and g_1 is the characteristic function of a disk:

$$g_1(\boldsymbol{r}') = \chi_{B_1}\left(\frac{2|\boldsymbol{r}'|}{D_b}\right).$$

For the sort of collimator considered here, g_2 is the same:

$$g_2(\boldsymbol{r}'') = \chi_{B_1}\left(\frac{2|\boldsymbol{r}''|}{D_b}\right).$$

The output of our detector is

$$p(\boldsymbol{r}_s; z) = \frac{1}{4\pi(L_b+z)^2}\int \chi_{B_1}\left(\frac{2|\boldsymbol{r}''|}{D_b}\right)\chi_{B_1}\left(\frac{2a|\boldsymbol{r}''+b\boldsymbol{r}_s/a|}{D_b}\right)d\boldsymbol{r}''.$$

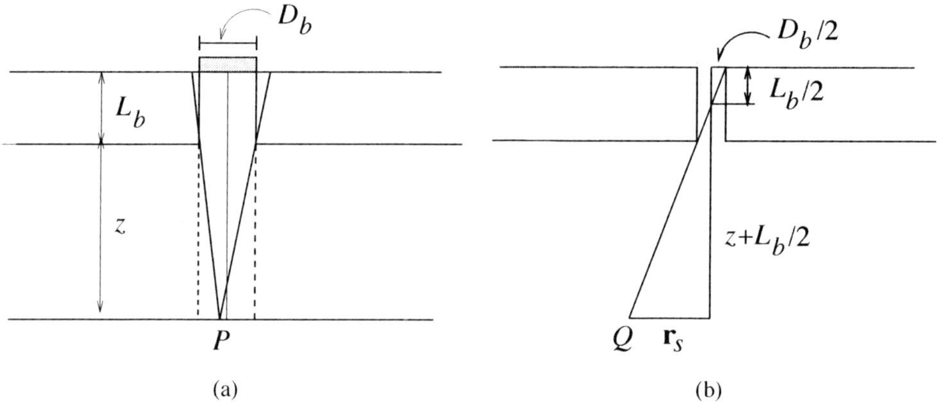

Figure 9.22. Evaluating the point spread function of a collimator.

This integral is exactly the area of the intersection of the two disks, as shown in Figure 9.21(b). The larger circle is the projection of the lower circle from the source point onto the upper plate. The $1/z^2$ scaling accounts for beam spreading.

In two special regions, $p(\boldsymbol{r}_s; z)$ is easy to determine. First, if $|\boldsymbol{r}_s| < \frac{1}{2}D_b$, as in Figure 9.22(a), then the projected circle covers the upper circle; hence $p(\boldsymbol{r}_s; z)$ is the area of the smaller disk; that is,

$$p(\boldsymbol{r}_s; z) = \pi \left(\frac{D_b}{2}\right)^2.$$

On the other hand, if P lies far enough from the origin, then the two circles do not meet. The location of the first such point, Q, is found using similar triangles:

$$\frac{L_b}{2} : \frac{D_b}{2} = \frac{L_b}{2} + z : \boldsymbol{r}_s \Rightarrow \boldsymbol{r}_s = \frac{D_b}{2}\left(\frac{L_b + 2z}{L_b}\right);$$

see Figure 9.22(b). Hence if $|\boldsymbol{r}_s| > \frac{D_b}{2}\left(\frac{L_b+2z}{L_b}\right)$, the area of the intersection is zero and $p(\boldsymbol{r}_s; z) = 0$. Figure (9.23) show the graph of $p(|\boldsymbol{r}_s|; z)$ for a fixed value of z.

To approximate the FWHM, we approximate p by a piecewise linear function. Let $\delta(z)$ be the resolution, defined to be

$$\delta(z) = \text{FWHM}(p(\cdot; z)) = D_b\frac{(L_b + 2z)}{L_b}.$$

As z increases, the resolution gets worse, as observed in Section 9.4.2. If the object is sitting right on the collimator face (i.e., $z = 0$), then $\delta(0) = D_b$, which is just the diameter of the collimator bore.

The other parameter of interest for a detector is its sensitivity. The sensitivity of a detector is important as it determines how intense a source is needed to make a usable image. To analyze the sensitivity, we use a uniform planar source instead of the point source used in the resolution analysis. Imagine that a uniform radioactive source is spread

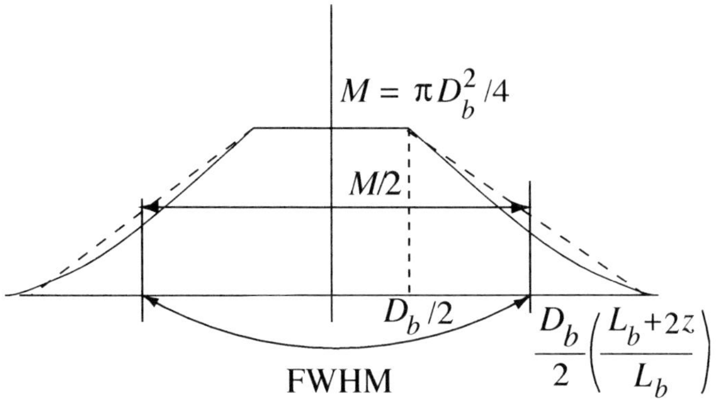

Figure 9.23. The graph of $p(\boldsymbol{r}_s; z)$, for a fixed z.

on a plane at distance z from the face of the collimator. The intensity of the "standard source" is $1\mu C/\text{cm}^2$. For a typical radioactive material, this translates into a photon flux of $3.7 \times 10^4/\text{cm}^2 \cdot \text{s}$. The sensitivity of the measuring device is given by the number of photons arriving at the detector. For the collimator described previously, we obtain

$$\begin{aligned} S &= \frac{3.7 \times 10^4}{4\pi (L_b + z)^2} \iint \chi_{B_1}\left(\frac{2\boldsymbol{r}''}{D_b}\right) \chi_{B_1}\left(\frac{2|a\boldsymbol{r}'' + b\boldsymbol{r}|}{D_b}\right) d\boldsymbol{r}\, d\boldsymbol{r}'' \\ &= \frac{3.7 \times 10^4}{4\pi (L_b + z)^2} \frac{\pi D_b^2}{4b^2} \frac{\pi D_b^2}{4} \\ &= \frac{3.7 \times 10^4}{64} \frac{\pi D_b^4}{L_b^2}. \end{aligned}$$

The sensitivity does not depend on z; this is the consequence of using a uniform source of *infinite* extent.

With L_b and z fixed, the resolution, δ is proportional to D_b, while the sensitivity is proportional to D_b^4. Thus we see that

$$S \text{ is proportional to } \delta^4.$$

To form a usable image, the required dosage is roughly proportional to $1/S$. This shows that

$$\textit{dosage} \text{ is proportional to } \delta^{-4}.$$

To increase the resolution by a factor of 2 (i.e., $\delta \to \frac{1}{2}\delta$), the dosage must be increased by a factor of 16.

Exercises

***Exercise* 9.4.12.** Find the transfer function for a collimator with a cylindrical bore. This can be modeled as two concentric, transparent disks of radii D_{b1} and D_{b2} lying in parallel planes a distance L apart.

***Exercise* 9.4.13.** What is the effect of the collimator in the previous exercise if $D_{b1} < D_{b2}$? How about if $D_{b1} > D_{b2}$?

9.5 Conclusion

Most of this chapter is devoted to the analysis of shift invariant, linear filters. Many actual filtering operations are not shift invariant, but, for computational reasons, are approximated by such filters. The theory of shift invariant filters is very simple because, in the Fourier representation, a shift invariant filter becomes a multiplication filter. This has a variety of theoretical and practical consequences. Primary among them are the facts that (1) a pair of shift invariant filters commute, (2) an approximate inverse filter (if it exists) is easily constructed, and (3) because the (approximate) Fourier transform and its inverse have efficient implementations, so does any shift invariant filter. This describes in a nutshell why shift invariant filters are so special and so important in applications.

Two important functions are defined by a shift invariant linear filter: its impulse response (or point spread function) and its transfer function (or modulation transfer function). The first is the output of the filter when the input is a unit impulse, and the second is its Fourier transform. These functions completely describe the action of the filter. By using a short-duration, intense pulse as an input, an approximation to the impulse response can be *measured*. Using these functions, we defined various notions of the resolution available in the output. The definitions of resolution are fairly simple for filters acting on functions of a single variable and become more complicated in higher dimensions. In all but the simplest cases, quantitative measures of resolution are only meaningful for purposes of comparison.

In the next chapter we take the final step in the long path from continuum models to finite algorithms that approximately implement them. The main point is to understand the relationship between the finite Fourier transform of the samples of a function and that function's Fourier transform. After analyzing this problem, we detail the implementation of an "arbitrary," linear shift invariant filter.

Chapter 10

Implementing Shift Invariant Filters

See: A.5.2, A.6.1.

In practical applications we cannot measure an input continuously; rather it is sampled at a discrete sequence of points. The models for physical systems presented thus far are in the language of functions of continuous variables, so the question then arises as to how such models can be utilized in a real situation. In this section we analyze the transition from continuous parameter, shift invariant linear filters to their *approximate* realizations on finitely sampled data. For brevity we refer to this as "implementing" a shift invariant filter, with the understanding that such implementations are, of necessity, approximate. The main tool is the finite Fourier transform, and the main thing to understand is the sense in which it provides an approximation to the continuum Fourier transform. The one-dimensional case is treated in detail; higher-dimensional filters are considered briefly.

In this section x denotes an input and $\mathcal{H}$ a shift invariant linear filter with impulse response h and transfer function $\hat{h}$. We assume that x is sampled at the discrete set of equally spaced points

$$\{t_0 + j\tau \mid j \in \mathbb{Z}\}.$$

The positive number τ is the sample spacing. In certain situations the physical measuring apparatus makes it difficult to collect equally spaced samples. In such situations the data are often interpolated to create an equally spaced set of samples for further analysis. This is done, even though it introduces a new source of error, because algorithms for equally spaced samples are so much simpler than those for unequal sample spacing.

In terms of the continuous parameters t and ξ, there are two different representations for $\mathcal{H}$: the spatial domain representation as a convolution

$$\mathcal{H}x(t) = h * x(t) = \int_{-\infty}^{\infty} h(t-s)x(s)\,ds, \tag{10.1}$$

and the frequency domain representation as a Fourier integral

$$\mathcal{H}x(t) = \frac{1}{2\pi}\int_{-\infty}^{\infty} \hat{h}(\xi)\hat{x}(\xi)e^{it\xi}d\xi. \tag{10.2}$$

Each representation leads to a different discretization scheme. For several reasons, the frequency domain representation is usually employed: Certain filters, like the Hilbert transform or differentiation, are difficult to represent as convolutions, since their impulse responses are generalized functions. For most shift invariant filters, usage of the fast Fourier transform algorithm (FFT) makes the frequency domain computation vastly more efficient than the spatial domain computation. The structure of the Cooley-Tukey FFT algorithm is outlined in an appendix to this chapter.

10.1 Sampled Data

Sampling a function entails evaluating it at points. From the point of view of measurements, a function only has a well-defined value at points of continuity, and therefore it is assumed throughout most of this analysis that both the signal x and the impulse response of the filter h are continuous functions. This is not a serious restriction for, as discussed in Chapter 8, real measurements are modeled as the evaluation of an average, which is usually continuous. With τ the sample spacing in the spatial domain, let

$$h_s(j) = h(j\tau) \qquad \text{for } j \in \mathbb{Z},$$

denote samples of the impulse response and

$$x_s(j) = x(j\tau) \qquad \text{for } j \in \mathbb{Z},$$

samples of the input as a functions of an integer variable. The spatial domain representation of the filter can be approximated by using a Riemann sum

$$\begin{aligned}\mathcal{H}x(j\tau) &= h * x(j\tau) \\ &\approx \sum_{k=-\infty}^{\infty} h_s(j-k)x_s(k)\tau \\ &= \tau h_s \star x_s(j).\end{aligned} \tag{10.3}$$

In the last line of (10.3), $h_s \star x_s$ denotes the discrete convolution operation, defined in Section 7.4. A Riemann sum is just one possible approximation for the convolution; if the integrand is smooth, a higher-order, numerical integration method might give superior results (see Section A.6.1).

For the first step of the analysis it is assumed that x is absolutely integrable and that an infinite sequence of samples is collected. In imaging applications the data are usually supported in a bounded interval, so this is not an unreasonable assumption. The sequence of samples has a *sample Fourier transform*

$$\hat{x}_s(\xi) = \tau \sum_{j=-\infty}^{\infty} x_s(j) e^{-ij\tau\xi}, \tag{10.4}$$

which is a $\frac{2\pi}{\tau}$-periodic function.

The sample Fourier transform is connected to discrete convolution in a simple way:

$$\begin{aligned} \tau \widehat{h_s \star x_s}(\xi) &= \tau^2 \sum_{j=-\infty}^{\infty} h_s \star x_s(j\tau) e^{-ij\tau\xi} \\ &= \tau^2 \sum_{j=-\infty}^{\infty} \sum_{k=-\infty}^{\infty} h((j-k)\tau) e^{-i(j-k)\tau\xi} x(k\tau) e^{-ik\tau\xi} \\ &= \hat{h}_s(\xi)\hat{x}_s(\xi). \end{aligned} \tag{10.5}$$

In other words, the sample Fourier transform of the sequence $\tau h_s \star x_s$ is simply $\hat{h}_s \hat{x}_s$.

The dual Poisson summation formula, (8.13), relates $\hat{x}_s$ to $\hat{x}$:

$$\hat{x}_s(\xi) = \sum_{j=-\infty}^{\infty} \hat{x}(\xi + \frac{2\pi j}{\tau}). \tag{10.6}$$

On the other hand, formula (10.4) is a Riemann sum for the integral defining $\hat{x}$. Comparing (10.6) to (10.4), it is apparent that careful consideration is needed to understand in what sense $\hat{x}_s$ is an approximation to $\hat{x}$. When the input is $\frac{\pi}{\tau}$-bandlimited, Nyquist's theorem implies that

$$\hat{x}(\xi) = \hat{x}_s(\xi) \chi_{[-\frac{\pi}{\tau}, \frac{\pi}{\tau}]}(\xi).$$

If the transfer function for the filter $\mathcal{H}$ is also supported in this interval, then

$$\mathcal{H}x(t) = \frac{1}{2\pi} \int_{-\frac{\pi}{\tau}}^{\frac{\pi}{\tau}} \hat{x}_s(\xi)\hat{h}_s(\xi) e^{it\xi} d\xi.$$

This indicates that, for this application, the windowed function

$$\hat{x}_s(\xi) \chi_{[-\frac{\pi}{\tau}, \frac{\pi}{\tau}]}(\xi)$$

should *always* be regarded as an approximation for $\hat{x}$. The accuracy of this approximation depends on the high-frequency behavior of $\hat{x}$ and the sample spacing.

From the Fourier inversion formula it follows that

$$h_s \star x_s(l)\tau = \frac{1}{2\pi} \int_{-\frac{\pi}{\tau}}^{\frac{\pi}{\tau}} \hat{h}_s(\xi)\hat{x}_s(\xi) e^{il\tau\xi} d\xi. \tag{10.7}$$

Thus if x and h both are $\frac{\pi}{\tau}$-bandlimited, then

$$h * x(l\tau) = \tau h_s \star x_s(l) \qquad \text{for all } l \in \mathbb{Z},$$

and the discrete convolution of the sampled sequences consists of samples of the continuous convolution. In any case, the integral on the right-hand side of (10.7) gives another approximation to $h * x(j\tau)$:

$$h * x(l\tau) \approx \frac{1}{2\pi} \int_{-\frac{\pi}{\tau}}^{\frac{\pi}{\tau}} \hat{h}_s(\xi)\hat{x}_s(\xi)e^{il\tau\xi}d\xi. \tag{10.8}$$

An important variant on (10.8) is to use the exact transfer function $\hat{h}$ for the filter (which is often known) instead of the sample Fourier transform, $\hat{h}_s$; this gives a different approximation

$$h * x(l\tau) \approx \frac{1}{2\pi} \int_{-\frac{\pi}{\tau}}^{\frac{\pi}{\tau}} \hat{h}(\xi)\hat{x}_s(\xi)e^{il\tau\xi}d\xi. \tag{10.9}$$

Provided $\hat{h}$ is an ordinary function, this approximation is valid even if h is a generalized function.

Equations (10.3), (10.8), (10.9), and the relation

$$\hat{x}(\xi) \approx \hat{x}_s(\xi)\chi_{[-\frac{\pi}{\tau},\frac{\pi}{\tau}]}(\xi) \tag{10.10}$$

form the foundation for the implementation of linear shift invariant filters on sampled data.

***Remark* 10.1.1 (Important remark).** In imaging applications the input is usually zero outside a finite interval, which implies that it cannot be bandlimited (see Section 4.4.3). The sample spacing is chosen to attain a certain degree of resolution in the final result. From (10.6) we see that in order for $\hat{x}_s$ to provide a good approximation to $\hat{x}$ over the interval $[-\frac{\pi}{\tau}, \frac{\pi}{\tau}]$, it is necessary for x to be effectively $\frac{\pi}{\tau}$-bandlimited. In the present application this means that the aliasing error,

$$\sum_{j\neq 0} \hat{x}(\xi - \frac{2\pi j}{\tau}) \qquad \text{for } \xi \in [-\frac{\pi}{\tau}, \frac{\pi}{\tau}],$$

must be uniformly small. This is a stronger condition than simply requiring $|\hat{x}(\xi)|$ to be small if $|\xi|$ is greater than $\frac{\pi}{\tau}$. To ensure that the aliasing error is small, an input is usually passed through a lowpass filter *before it is sampled*. In any case, once the input is sampled, the sample spacing fixes the *usable bandwidth* in all subsequent *computations* to be $\frac{2\pi}{\tau}$. It should be noted that this does *not* mean that the data itself had an effective bandwidth of $\frac{2\pi}{\tau}$.

10.2 Implementing Periodic Convolutions

See: 2.3.2, 8.3.

The finite Fourier transform was introduced in Section 8.3. In this section we establish some further properties of this operation and explain how it is used to implement shift invariant *periodic* filters. This material is quite elementary but essential for a correct understanding of the implementation of (10.1) with finitely sampled data.

10.2.1 Further Properties of the Finite Fourier Transform

Just as a function defined on the finite interval $[0, L]$ defines an L-periodic function, recall that a finite sequence can be extended to define a periodic, bi-infinite sequence. The N-periodic extension of $< x_0, \ldots, x_{N-1} >$ is defined by

$$x_{j+lN} = x_j \qquad \text{for } 0 \leq j \leq N-1, l \in \mathbb{Z}.$$

Periodic sequences can be convolved.

***Definition* 10.2.1.** Given two N-periodic sequences, $< x >$, $< y >$, their *periodic convolution* is defined by

$$(x \star y)_k = \sum_{j=0}^{N-1} x_j y_{k-j} \qquad \text{for } k = 0, \ldots, N-1.$$

The periodic convolution of sequences is related to the finite Fourier transform in precisely the same way the periodic convolution of functions is related to the Fourier coefficients.

Proposition 10.2.1. *Let $< x >$ and $< y >$ be sequences of length N. Then*

$$\mathscr{F}_N(x \star y) = N < \hat{x}_0 \hat{y}_0, \ldots, \hat{x}_{N-1} \hat{y}_{N-1} > . \tag{10.11}$$

Proof. The proof follows by changing the order of summations:

$$\begin{aligned} \mathscr{F}_N(x \star y)_k &= \frac{1}{N} \sum_{j=0}^{N-1} (x \star y)_j e^{-\frac{2\pi ijk}{N}} \\ &= \frac{1}{N} \sum_{l=0}^{N-1} (\sum_{j=0}^{N-1} x_l y_{j-l}) e^{-\frac{2\pi ijk}{N}}. \end{aligned} \tag{10.12}$$

Letting $j - l = m$ in the last summation and using the periodicity of both $< y >$ and the exponential function gives

$$\mathscr{F}_N(x \star y)_k = \frac{1}{N} \sum_{l=0}^{N-1} x_l e^{-\frac{2\pi ilk}{N}} \sum_{m=0}^{N-1} y_m e^{-\frac{2\pi imk}{N}} = N \hat{x}_k \hat{y}_k. \tag{10.13}$$

□

A periodic finite convolution can therefore be computed using the finite Fourier transform and its inverse.

Corollary 10.2.1. *Let $< x >$ and $< y >$ be a sequences of length N. Their periodic convolution, $x \star y$ is given by*

$$x \star y = N\mathcal{F}_N^{-1}(< \hat{x}_0\hat{y}_0, \dots, \hat{x}_{N-1}\hat{y}_{N-1} >). \tag{10.14}$$

If $N = 2^k$, then this is the most efficient way to do such calculations. The computation of the FFT and its inverse require about $6N \log_2 N$ operations; multiplying the two FFTs requires N operations. Altogether it requires $6N \log_2 N + N$ operations to compute $x \star y$ using the FFT. This should be compared to the $N(2N - 1)$ operations needed to do this calculation directly.

10.2.2 The Approximation of Fourier Coefficients

We now consider in what sense the finite Fourier transform of a sequence of samples approximates the Fourier coefficients of the sampled function. Let x be a function defined on $[0, 1]$. In this section we let $N = 2M + 1$. Set

$$x_j = x\left(\frac{j}{2M+1}\right), \; j = 0, \dots, 2M,$$

and let $< \hat{x}_j >$ denote the finite Fourier transform of this sequence. The sequence $< \hat{x}_j >$ is defined by a Riemann sum for an integral:

$$\hat{x}_k = \frac{1}{2M+1}\sum_{j=0}^{2M} x_j e^{-\frac{2\pi ijk}{2M+1}} \approx \int_0^1 x(t)e^{-2\pi ikt}dt = \hat{x}(k). \tag{10.15}$$

It is therefore reasonable to expect that $\hat{x}_k$ provides some sort of approximation for the kth Fourier coefficient of x. Ones needs to be careful, because, for any integer l, this sum is also a Riemann sum for the integral defining $\hat{x}(k + l(2M + 1))$. In fact, the approximation in (10.15) is accurate for values of k between $-M$ and M, but its accuracy rapidly deteriorates as $|k|$ exceeds M. Recall that formula (8.19) defines $\hat{x}_k$ for all values of k as an $2M + 1$-periodic sequence. In particular, we have

$$\hat{x}_{-k} = \hat{x}_{2M+1-k} \qquad \text{for } k = M + 1, M + 2, \dots, 2M,$$

so what we mean is that for $M < k \leq 2M + 1$, the finite Fourier coefficient $\hat{x}_{2M+1-k}$ is a good approximation for the Fourier coefficient $\hat{x}(-k)$. This is sometimes expressed by saying that the output of the finite Fourier transform, *when indexed by frequency*, should be interpreted to be

$$< \hat{x}_0, \hat{x}_1, \dots, \hat{x}_M, \hat{x}_{-M}, \hat{x}_{1-M}, \dots, \hat{x}_{-1} > .$$

If $|k| \leq M$, then the sum in (10.15) is actually a quadrature formula for the integral. In fact, for a given number of samples, this finite sum provides an essentially optimal approximation to the integral. If $x(t)$ is a periodic function with n continuous periodic derivatives, then, for $|k| \leq M$, we have an estimate of the form:

$$|x_k - \hat{x}(k)| \leq C_{f,M} M^{-n}. \tag{10.16}$$

Here $C_{f,M}$ is a constant depending on f and M, which tends to zero as M tends to infinity. A complete analysis of this question can be found in [35]

On the other hand, even for a very smooth function, if $|k| > M$, then $\hat{x}_k$ is likely to be a poor approximation to $\hat{x}(k)$. Given our remarks above, an approximation to the partial sum of the Fourier series of x should be given by

$$S_M(x; t) \approx \sum_{j=0}^{M} \hat{x}_j e^{2\pi ijt} + \sum_{j=M+1}^{2M} \hat{x}_j e^{2\pi i(j-(2M+1))t}. \tag{10.17}$$

Because the samples are uniformly spaced, additional cancellations take place in this finite sum. The finite sum in (10.17) is almost as accurate an approximation to $x(t)$ as the partial sum, $S_M(x; t)$, of the Fourier series with this number of terms. If x is sufficiently smooth, and M is large enough for $S_M(x; t)$ to be a good approximation to x, then formula (10.17) can be used to accurately interpolate values of x between the sample points.

Using a naive interpretation of $\hat{x}_j$ as an approximation to $\hat{x}(j)$, for all j, leads to another exponential polynomial that interpolates x at the sample points:

$$p_x(t) = \sum_{j=0}^{2M} \hat{x}_j e^{2\pi ijt}. \tag{10.18}$$

Unlike (10.17), formula (10.18) does not provide a good approximation to x away from the sample points. This is easily seen in an example.

***Example* 10.2.1.** The function $\sin(3t)$ is sampled at 11 points, and the finite Fourier transform of this sequence is used for the coefficients in formula (10.18). Figure 10.1 shows the real part of this function along with $\sin(3t)$. As expected, the two functions agree at the sample points, but away from the sample points, the function defined in (10.18) does a very poor job of approximating $\sin(3t)$. Using formula (10.17) with 11 sample points gives an exact reconstruction of $\sin(3t)$, further justifying the interpretation for the finite Fourier transform given previously.

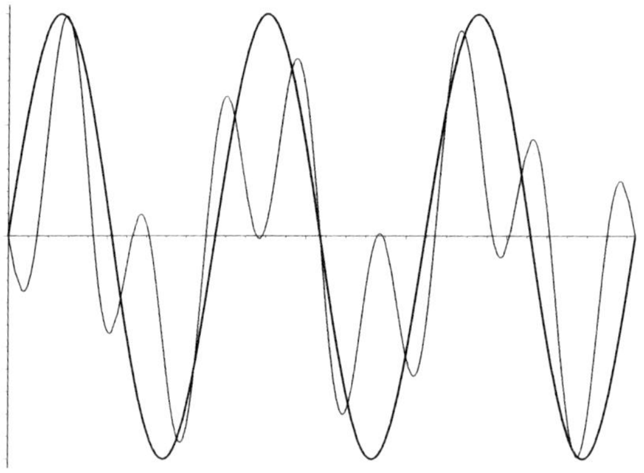

Figure 10.1. Bad interpolation using formula (10.18); p_x is the more oscillatory function.

Using the same reasoning, it follows that if x is a function defined on $[0, L]$ and we collect $2M+1$-equally spaced samples,

$$x_j = x\left(\frac{jL}{2M+1}\right) \qquad \text{for } j = 0, \dots, 2M,$$

then

$$\begin{aligned}\hat{x}_k &= \frac{1}{2M+1}\sum_{j=0}^{2M} x\left(\frac{jL}{2M+1}\right)e^{-\frac{2\pi ijk}{2M+1}} \\ &= \frac{1}{L}\sum_{j=0}^{2M} x\left(\frac{jL}{2M}\right)e^{-\frac{2\pi ijLk}{(2M+1)L}}\frac{L}{2M+1}\end{aligned} \tag{10.19}$$

and therefore

$$\hat{x}_k \approx \begin{cases} \frac{1}{L}\hat{x}(k) & \text{if } 0 \le k \le M, \\ \frac{1}{L}\hat{x}(k-(2M+1)) & \text{if } M+1 \le k \le 2M. \end{cases} \tag{10.20}$$

Exercise

***Exercise* 10.2.1.** Why does the polynomials p_x defined in (10.18) satisfy

$$p_x\left(\frac{j}{2M+1}\right) = x\left(\frac{j}{2M+1}\right) \text{ for } j = 0, \dots, 2M?$$

10.2.3 Approximating Periodic Convolutions

The action of a shift invariant filter $\mathcal{H}$ on a 1-periodic function is defined in terms of its impulse response h (another 1-periodic function) by the periodic convolution

$$\mathcal{H}x(t) = \int_0^1 x(s)h(t-s)\,ds.$$

Let

$$\hat{h}(k) = \int_0^1 h(t)e^{-2\pi ikt}\,dt$$

denote the Fourier coefficients of the impulse response. The Fourier representation of this filter is

$$\mathcal{H}(x)(t) = \sum_{k=-\infty}^{\infty} \hat{x}(k)\hat{h}(k)e^{2\pi ikt}. \tag{10.21}$$

If $< x_j >$ are the samples $< x\left(\frac{j}{2M+1}\right) >$, then (10.20) and (10.21) imply that a finite sum approximation to $\mathcal{H}x$ at the sample point $\frac{j}{2M+1}$ is given by

$$\mathcal{H}x(\frac{j}{2M+1}) \approx \sum_{k=0}^{M} \hat{x}_k \hat{h}(k) e^{\frac{2\pi ijk}{2M+1}} + \sum_{k=M+1}^{2M} \hat{x}_k \hat{h}(k-(2M+1)) e^{\frac{2\pi ijk}{2M+1}}.$$

In this approximation we have used samples of the exact transfer function of $\mathcal{H}$. A different approximation can be obtained by using the finite Fourier transform of the sample sequence $< h(0), h(\frac{1}{2M+1}), \dots, h(\frac{2M}{2M+1}) >$.

***Example* 10.2.2.** The Hilbert transform is defined in the Fourier representation by the transfer function

$$\hat{h}_k = \begin{cases} 1 & k > 0, \\ 0 & k = 0, \\ -1 & k < 0. \end{cases}$$

A finite frequency approximation to the Hilbert transform is given by

$$\mathcal{H}x(\frac{j}{2M+1}) \approx \sum_{k=1}^{M} \hat{x}_k e^{\frac{2\pi ijk}{2M+1}} - \sum_{k=M+1}^{2M} \hat{x}_k e^{\frac{2\pi ijk}{2M+1}}.$$

Exercises

***Exercise* 10.2.2.** Suppose that x is supported on $[a, b]$ and we collect the samples $x_j = x(a + j\frac{b-a}{N})$ for $j = 0, \dots, N-1$. How is the finite Fourier transform of the sequence $< x_0, \dots, x_{N-1} >$ related to samples of the Fourier transform of x?

***Exercise* 10.2.3.** The direct computation of the sums defining $\mathcal{F}_N$ requires $O(N^2)$ arithmetic operations. An algorithm for computing $\mathcal{F}_N$ is "fast" if it uses $O(N^\alpha)$ operations for an $\alpha < 2$. Explain why a fast algorithm for computing $\mathcal{F}_N$ also gives a fast algorithm for computing $\mathcal{F}_N^{-1}$.

***Exercise* 10.2.4.** What is the smallest $N = 2^k$ so that the FFT computation of $x \star y$ is more efficient than the direct computation?

***Exercise* 10.2.5.** Explain why (10.20) agrees with Nyquist's theorem for periodic functions.

***Exercise* 10.2.6.** For x a once differentiable, periodic function, compare the approximations to $\hat{x}(k)$ obtained by sampling the integrands to approximate the following integrals:

$$\begin{aligned}\hat{x}(k) &= \int_0^1 x(t)e^{-2\pi ikt}dt,\\ \hat{x}(k) &= \frac{1}{2\pi ik}\int_0^1 x'(t)e^{-2\pi ikt}dt.\end{aligned} \tag{10.22}$$

10.3 Implementing Filters on Finitely Sampled Data

In this section $\mathcal{H}$ denotes a shift invariant filter with impulse response h acting on functions defined on $\mathbb{R}$. The input x is assumed to have bounded support, and this means that it can be represented in terms of either the Fourier transform or the Fourier series. Indeed both representations are used in this section. Since it is important to distinguish which is intended, we introduce the following *provisional* notation for Fourier coefficients: For x *thought of* as an L-periodic function its kth Fourier coefficient is denoted

$$\hat{x}_{(L)}(k) = \int_0^L x(t)e^{-\frac{2\pi ikt}{L}}dt.$$

For a function x vanishing outside $[0, L]$, we have the relation

$$\hat{x}_{(L)}(k) = \hat{x}(\frac{2\pi k}{L}).$$

This section treats the details of using the finite Fourier transform to implement a shift invariant filter on finitely sampled data.

Let x be a function defined on $\mathbb{R}$ with support in $[0, 1]$ and h the impulse response of a filter $\mathcal{H}$, so that

$$\mathcal{H}x = h * x.$$

In general, h is not assumed to have support in $[0, 1]$. Suppose that x is sampled at the points $\{\frac{j}{N} : j = 0, \ldots, N-1\}$; it is implicit that the remaining "uncollected" samples are zero. A Riemann sum gives an approximate value for $h * x$ at the sample points:

$$h * x(\frac{k}{N}) \approx \sum_{j=0}^{N-1} h(\frac{k-j}{N})x(\frac{j}{N})\frac{1}{N}. \tag{10.23}$$

To do this calculation for $k \in \{0, \ldots, N-1\}$ requires a knowledge of $2N - 1$ values of h,

$$\{h(\frac{1-N}{N}), h(\frac{2-N}{N}), \ldots, h(\frac{0}{N}), \ldots, h(\frac{N-1}{N})\}.$$

In order to use the finite Fourier transform to compute the discrete convolution on the right-hand side of (10.23), this sum must be interpreted as a *periodic convolution* of two sequences of length $2N-1$. This means that the sequence of samples of x must be augmented to get a sequence of length $2N-1$. This is done by adding $(N-1)$-zeros to the end in a process called *zero padding*. The zero padded sequence is

$$x_s = < x(0), x(\frac{1}{N}), \cdots, x(\frac{N-1}{N}), \underbrace{0, 0, \cdots, 0}_{N-1} > .$$

This is consistent with the interpretation of x as a function defined on $\mathbb{R}$ with support contained in $[0, 1]$. Formula (10.20) relates $\hat{x}_s$, the finite Fourier transform of x_s, to the Fourier coefficients of $x_{(L_N)}$,

$$\hat{x}_s(k) \approx \begin{cases} \frac{N}{2N-1}\hat{x}_{(L_N)}(k) & \text{for } 0 \le k \le N-1 \\ \frac{N}{2N-1}\hat{x}_{(L_N)}(k-2N+1) & \text{for } N \le k \le 2N-2. \end{cases} \tag{10.24}$$

Here $L_N = \frac{2N-1}{N}$ and $x_{(L_N)}$ is x restricted to the interval $[0, L_N]$.

If we define

$$h_s = < h(\frac{0}{N}), \ldots, h(\frac{N-1}{N}), h(\frac{1-N}{N}), \ldots, h(\frac{-1}{N}) >,$$

then

$$\sum_{j=0}^{N-1} h(\frac{k-j}{N})x(\frac{j}{N})\frac{1}{N} = \frac{1}{N} h_s \star x_s(k). \tag{10.25}$$

Of course, h_s and x_s are regarded as $(2N-1)$-periodic sequences so that

$$h_s(-1) = h_s(2N-2), \quad h_s(-2) = h_s(2N-3), \text{ and so on.}$$

Using (10.14) and (10.25), this can be rewritten in terms of the finite Fourier transform as

$$\begin{aligned} h * x(\frac{k}{N}) &\approx \sum_{j=0}^{N-1} h(\frac{k-j}{N})x(\frac{j}{N})\frac{1}{N} \\ &= \frac{2N-1}{N}\mathcal{F}_{2N-1}^{-1}(< \hat{h}_s(0)\hat{x}_s(0), \ldots, \hat{h}_s(2N-2)\hat{x}_s(2N-2) >)(k). \end{aligned} \tag{10.26}$$

A different derivation of (10.26) follows by thinking of x and h as being defined on $\mathbb{R}$ with x effectively bandlimited to $[-N\pi, N\pi]$. We use a Riemann sum approximation for the partial Fourier inverse to obtain

$$
\begin{aligned}
h * x\left(\frac{k}{N}\right) &\approx \frac{1}{2\pi} \int_{-N\pi}^{N\pi} \hat{h}(\xi)\hat{x}(\xi) e^{i\frac{k}{N}\xi} d\xi \\
&\approx \frac{1}{2\pi} \sum_{j=1-N}^{N-1} \hat{x}\left(\frac{2\pi j N}{2N-1}\right) \hat{h}\left(\frac{2\pi j N}{2N-1}\right) \exp\left(\frac{2\pi i j N}{2N-1} \cdot \frac{k}{N}\right) \left(\frac{2\pi N}{2N-1}\right) \\
&\approx \frac{2N-1}{N} \mathcal{F}_{2N-1}^{-1}(< \hat{h}_s(0)\hat{x}_s(0), \dots, \hat{h}_s(2N-2)\hat{x}_s(2N-2) >)(k).
\end{aligned}
\tag{10.27}
$$

The last line is a consequence of (10.24) and the fact that

$$
\hat{x}_{(L_N)}(k) = \hat{x}\left(\frac{2\pi k N}{2N-1}\right).
$$

For latter applications it is useful to have formulæ for the approximate Fourier implementation of a linear, shift invariant filter $\mathcal{H}$ without scaling the data to be defined on $[0, 1]$. We sample x at a sequence of equally spaced points and add $N-1$ zeros to get

$$
x_s =< x(t_0), x(t_0+\tau), \dots, x(t_0+(N-1)\tau), 0, \dots, 0 > .
$$

Let $< \hat{x}_s >$ be the finite Fourier transform of x_s and $< \hat{h}_s >$ the finite Fourier transform of

$$
h_s =< h(0), \dots, h((N-1)\tau), h((1-N)\tau), \dots, h(-\tau) > .
$$

The approximate value of $\mathcal{H}x(t_0+k\tau)$, given by a Riemann sum and computed using the finite Fourier transform, is

$$
\begin{aligned}
h * x(t_0+k\tau) &\approx \sum_{j=0}^{N-1} x_s(j) h_s(k-j)\tau \\
&= \tau(2N-1)\mathcal{F}_{2N-1}^{-1}(< \hat{x}_s(0)\hat{h}_s(0), \dots, \hat{x}_s(2N-2)\hat{h}_s(2N-2) >)(k).
\end{aligned}
\tag{10.28}
$$

In many applications the sequence $< \hat{h}_s(0), \dots, \hat{h}_s(2N-2) >$ appearing in (10.28) is *not* computed as the finite Fourier transform of the sample sequence h_s. Instead, it is obtained by directly sampling the transfer function $\hat{h}$. This is because the transfer function may be an ordinary function that can be computed exactly, even when the impulse response is a generalized function. The finite Fourier transform of the sequence h_s is given by

$$
\begin{aligned}
\hat{h}_s(k) &= \frac{1}{2N-1} \sum_{j=0}^{2N-2} h((1-N)\tau + j\tau) e^{-\frac{2\pi i j k}{2N-1}} \\
&= \frac{1}{\tau(2N-1)} \sum_{j=0}^{2N-2} h((1-N)\tau + j\tau) e^{-ij\tau\frac{2\pi k}{\tau(2N-1)}} \tau.
\end{aligned}
\tag{10.29}
$$

The discussion is Section 10.2 shows that $\hat{h}_s$ should be interpreted as

$$\hat{h}_s(k) \approx \begin{cases} \frac{1}{\tau(2N-1)}\hat{h}\left(\frac{2\pi k}{\tau(2N-1)}\right) & \text{if } 0 \leq k \leq N-1, \\ \frac{1}{\tau(2N-1)}\hat{h}\left(\frac{2\pi(k-2N+1)}{\tau(2N-1)}\right) & \text{if } N \leq k \leq 2N-2. \end{cases} \tag{10.30}$$

So that there should be no confusion, note that here $\hat{h}$ denotes the Fourier *transform* of h, as a function (or generalized function) defined on $\mathbb{R}$. This shows that the correct samples of $\hat{h}$ to use in (10.28) are

$$\hat{h}'_k = \frac{1}{\tau(2N-1)}\hat{h}\left(\frac{2\pi k}{\tau(2N-1)}\right), \quad \text{for } k \in \{-(N-1), \dots, (N-1)\}, \tag{10.31}$$

regarded as a $(2N-1)$-periodic sequence. This agrees with the second line in (10.27). With the sequence $< \hat{h}'_k >$ defined as these samples of $\hat{h}$, formula (10.28) provides another approximation to $h * x(t_0 + j\tau)$:

$$h * x(t_0 + k\tau) \approx \tau(2N-1)\mathcal{F}_{2N-1}^{-1}(< \hat{x}_s(0)\hat{h}'_0, \dots, \hat{x}_s(2N-2)\hat{h}'_{2N-2} >)(k). \tag{10.32}$$

The two formulæ, equation (10.28) and equation (10.32), generally give *different* results. Which result is preferable is often an empirical question.

The formulæ obtained previously always involve sequences of odd length. To use the fast Fourier transform these sequences must be augmented to get even-length sequences. The sample sequence of the input is padded with zeros until we reach a power of 2. The same can be done for the impulse response sample sequence, or this sequence can be augmented with further samples of the impulse response itself. If the latter approach is used, then care must be exercised with the indexing so that the two sample sequences are compatible (see Exercise 10.3.3). If the transfer function is sampled in the Fourier domain, then samples of its Fourier transform are added, symmetrically about zero frequency, until we reach a power of 2. To accomplish this we need to have either one more sample at a positive frequency than at negative frequency or vice versa. We simply need to make a choice.

We have given a prescription for approximating $h * x(x_0 + j\tau)$ for $j \in \{0 \dots, N\}$. There is no reason, in principle, why $h * x(x_0 + j\tau)$ cannot be approximated by the same method for $j \in \{0, \dots, M\}$ for an M larger than N. One simply needs to pad the samples of x with $2M - N - 1$ zeros and use $2M - 1$ samples of h (or $\hat{h}$).

Exercises

***Exercise* 10.3.1.** From formula (10.27) it is clear that $\mathcal{F}_{2N-1}^{-1}$ provides an approximation to the inverse Fourier transform. Formula (10.28) is a Riemann sum approximation to the convolution. By replacing the sample sequence $\{x_s(j)\}$ by $\{x_{ws}(j) = w(j)x_s(j)\}$, one can approximate other integration schemes such as the trapezoidal rule or Simpson's rule. How should the weights $\{w(j)\}$ be selected so that

$$\frac{2N-1}{N}\mathcal{F}_{2N-1}^{-1}(< \hat{h}_s(0)\hat{x}_{ws}(0), \dots, \hat{h}_s(2N-2)\hat{x}_{ws}(2N-2) >)(k)$$

provides a trapezoidal rule or Simpson's rule approximation to the integral

$$\int_{-\infty}^{\infty} h\left(\frac{k}{N} - t\right) x(t)dt?$$

Implement these algorithms using MATLAB, Maple, or another similar program. Do the higher order integration schemes produce better or worse results? Can you explain this?

***Exercise* 10.3.2.** If x is a real-valued, even function with support in $[-1, 1]$, then its Fourier transform is real valued. On the other hand, the finite Fourier transform of the zero padded sequence

$$< x(-1), x(-1+\tau), \dots, x(1-\tau), x(1), 0, \dots, 0 >$$

is generally not real valued. Explain why not and give a different way to zero pad the sequence $< x(-1), x(-1+\tau), \dots, x(1-\tau), x(1) >$ so that its finite Fourier transform is real valued. *Hint:* See Exercise 8.3.3.

***Exercise* 10.3.3.** Suppose that x is a function supported in $[0, 1]$ and N is an integer with $2^{k-1} < 2N - 1 < 2^k$. Let x_s be the sequence of length 2^k:

$$x_s =< x(0), x(\frac{1}{N}), \dots, x(\frac{N-1}{N}), \underbrace{0, 0, \cdots, 0}_{2^k - N} > .$$

Let h denote the impulse response of a filter $\mathcal{H}$ that we would like to approximate using the finite Fourier transform

$$\mathcal{H}x(\frac{j}{N}) \approx \frac{2^k}{N} \mathcal{F}_{2^k}^{-1}(< \hat{h}_s \hat{x}_s >)(j),$$

where $< \hat{h}_s >$ is defined as the finite Fourier transform of a sequence of 2^k samples of h. Which samples of h should we use? Suppose instead of $\hat{h}_s$ we would like to use samples of $\hat{h}$; which samples should we use?

***Exercise* 10.3.4.** What is the connection between $\hat{x}_s$ in (10.24) and in (10.4). For $0 \leq k \leq N - 1$ use (10.6) to obtain a formula for the error in (10.24):

$$|\hat{x}_s(k) - \frac{N}{2N-1} \hat{x}_{(L_N)}(k)|.$$

10.3.1 Zero Padding Reconsidered

Padding the sequence of samples of x with $N - 1$ zeros is a purely mechanical requirement for using the finite Fourier transform to evaluate a discrete convolution: The finite sum in (10.25) must be seen as a periodic convolution of two sequences of *equal* length. There is also an analytic interpretation for zero padding. For each positive integer m define the function

$$x_{(m)}(t) = \begin{cases} x(t) & \text{for } t \in [0, 1], \\ 0 & \text{for } t \in (1, m]. \end{cases}$$

Let

$$\hat{x}(\xi) = \int_{-\infty}^{\infty} x(t)e^{-it\xi}\,dt$$

be the Fourier transform of x thought of as a function defined on the whole real line, but supported in $[0, 1]$.

Fix a sample spacing $\tau > 0$ in the spatial domain and collect $N_m = \frac{m}{\tau} + 1$ samples of $x_{(m)}$,

$$x_{(m,s)} = < x(0), x(\tau), \ldots, x((N_m - 1)\tau) > .$$

Of course the samples $x_{(m,s)}(k)$ are zero for $k > \tau^{-1}$. The finite Fourier transform of this sequence is a sequence $\hat{x}_{(m,s)}$ of length N_m. Formula (10.20) implies that the correct interpretation of this sequence, as an approximation to $\hat{x}$, is

$$\begin{aligned}\hat{x}_{(m,s)}(k) &\approx \frac{1}{m}\hat{x}_{(m)}(k) \text{ for } k \leq \frac{N_m}{2}\\ \hat{x}_{(m,s)}(k) &\approx \frac{1}{m}\hat{x}_{(m)}(k - N_m) \text{ for } k > \frac{N_m}{2}.\end{aligned} \tag{10.33}$$

On the other hand

$$\hat{x}_{(m)}(k) = \int_0^m x_{(m)}(t)e^{-\frac{2\pi ikt}{m}}\,dt = \hat{x}(\frac{2\pi k}{m})$$

and therefore

$$\begin{aligned}\hat{x}_{(m,s)}(k) &\approx \frac{1}{m}\hat{x}(\frac{2\pi k}{m}) \text{ for } k \leq \frac{N_m}{2},\\ \hat{x}_{(m,s)}(k) &\approx \frac{1}{m}\hat{x}(\frac{2\pi(k - N_m)}{m}) \text{ for } k > \frac{N_m}{2}.\end{aligned} \tag{10.34}$$

As k varies from 0 to $\frac{m}{\tau}$, the sequence $< \hat{x}_{m,s}(k) >$ consists of approximate samples of $\hat{x}$ for $\xi \in [-\frac{\pi}{\tau}, \frac{\pi}{\tau}]$. Once again, showing that the usable bandwidth in the sampled data is determined by the sample spacing. Note, however, that the sample spacing in the *Fourier domain* is $\frac{2\pi}{m}$. This shows that the effect of adding additional zeros to a sequence of samples of a function with bounded support is to decrease the effective mesh size in the Fourier domain. Practically speaking, the additional values in the finite Fourier transform (approximately) interpolate additional values of $\hat{x}$.

***Example* 10.3.1.** A graph of the function $f = (1 - t^2)\sin(10\pi t)\chi_{[-1,1]}(t)$ is shown in Figure 10.2(a), and the imaginary part of its Fourier transform in Figure 10.2(b). The imaginary part of the finite Fourier transform of f using 256 sample points in the interval $[-1, 1]$ is shown in Figure 10.3(a), while the imaginary part of the finite Fourier transform of f after padding with zeros to get a sequence of length 1024 is shown in Figure 10.3(b). This shows that padding a sequence with zeros leads to a finer mesh size in the Fourier domain. The additional values of the finite Fourier transform appearing in Figure 10.3(b) closely approximate the exact Fourier transform of f.

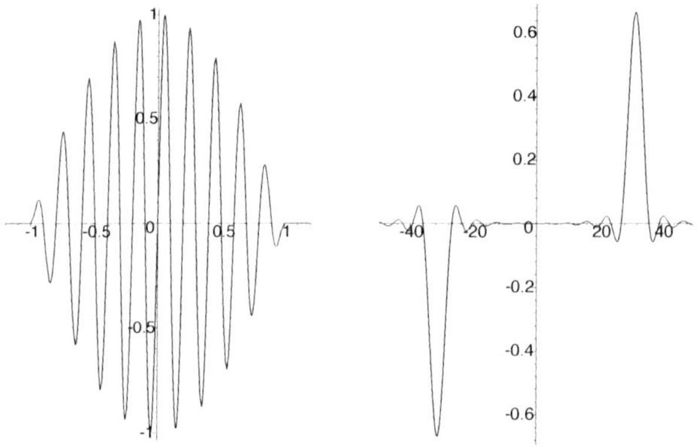

(a) The graph of f. (b) The graph of $\hat{f}$.

Figure 10.2. A function $f = (1 - t^2)\sin(10\pi t)\chi_{[-1,1]}(t)$ and its Fourier transform.

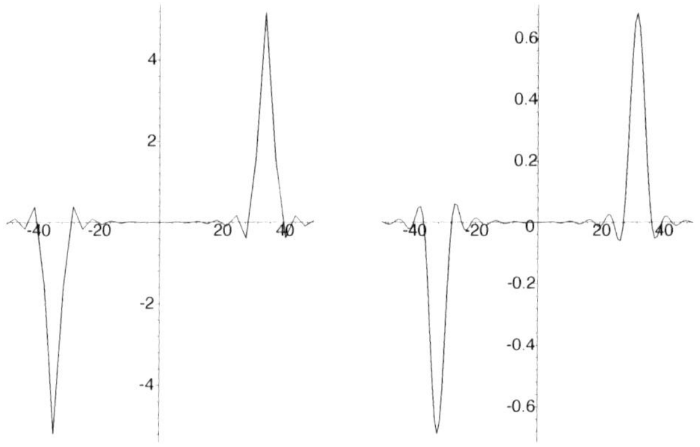

(a) The FFT of f using 256 samples. (b) The FFT of f with zero padding.

Figure 10.3. The effect of zero padding on the approximation to $\hat{f}$ by the finite Fourier transform.

10.4 Higher-Dimensional Filters

Similar considerations apply to implement shift invariant, linear filters acting on inputs that depend on more than one variable. As in the one-dimensional case, convolutions are usually computed using a Fourier representation. As there is no essential difference between the two-dimensional and n-dimensional cases, we consider thc general case. Suppose that x is an input, with bounded support, depending continuously on n real variables. A uniform sample set in $\mathbb{R}^n$ is specified by a vector of positive numbers $\boldsymbol{h} = (h_1, \dots, h_n)$, whose coordinates are the sample spacings in the corresponding coordinate directions. In one dimension the sample points and samples are labeled by an integer; in n dimensions it is more convenient to use n-tuples of integers. The integer vector $\boldsymbol{j} = (j_1, \dots, j_n) \in \mathbb{Z}^n$ labels the sample point

$$\boldsymbol{x}_j = (j_1 h_1, \dots, j_n h_n)$$

and the sample

$$f_j = f(\boldsymbol{x}_j).$$

The subsequent development in this book is largely independent of the material in this section.

10.4.1 Riemann Sum Approximations

Let a denote the impulse response of a shift invariant filter, $\mathscr{A}$, acting on a function of n variables,

$$\mathscr{A}\, x(\boldsymbol{t}) = \int\limits_{\mathbb{R}^n} x(\boldsymbol{s}) a(\boldsymbol{t} - \boldsymbol{s})\, d\boldsymbol{s}.$$

An n-dimensional integral is computed by rewriting it as iterated, one-dimensional integrals,

$$\mathscr{A}\, x(\boldsymbol{t}) = \int\limits_{-\infty}^{\infty} \cdots \int\limits_{-\infty}^{\infty} x(s_1, \dots, s_n) a(t_1 - s_1, \dots, t_n - s_n) ds_1 \cdots ds_n.$$

The iterated integrals can, in turn, be approximated by iterated Riemann sums. Using the uniformly spaced samples defined by $\boldsymbol{h}$ to define a partition, the integral is approximated by

$$\mathscr{A}\, x(\boldsymbol{t}) \approx \sum_{j_1=-\infty}^{\infty} \cdots \sum_{j_n=-\infty}^{\infty} x(j_1 h_1, \dots, j_n h_n) a(t_1 - j_1 h_1, \dots, t_n - j_n h_n) h_1 \cdots h_n.$$

At sample points this can be rewritten using the more economical notation, introduced previously, as

$$\mathscr{A}\, x(\boldsymbol{t_k}) \approx h_1 \cdots h_n \sum_{\boldsymbol{j} \in \mathbb{Z}^n} x_{\boldsymbol{j}} a_{\boldsymbol{k}-\boldsymbol{j}}. \tag{10.35}$$

In the sequel a sample of the output, $\mathscr{A}\, x(\boldsymbol{t_j})$ is denoted by $(\mathscr{A}\, x)_{\boldsymbol{j}}$.

Because the input is assumed to have bounded support, these sums can be replaced by finite sums. By translating the coordinates, it can be assumed that $< x_{\boldsymbol{j}} >$ is only nonzero for $\boldsymbol{j}$ belonging to the set

$$\mathscr{J}_{\boldsymbol{M}} = \{\boldsymbol{j} \,:\, 1 \le j_i \le M_i, \quad i = 1, \dots, n\}, \tag{10.36}$$

here $\boldsymbol{M} = (M_1, \dots, M_n)$. Altogether there are $M_1 \cdots M_n$ potentially nonzero samples. The Riemann sum for $(\mathscr{A}\, x)_{\boldsymbol{k}}$ becomes

$$(\mathscr{A}\, x)_{\boldsymbol{k}} \approx [h_1 \cdots h_n] \sum_{j_1=1}^{M_1} \cdots \sum_{j_n=1}^{M_n} x_{\boldsymbol{j}} a_{\boldsymbol{k}-\boldsymbol{j}}. \tag{10.37}$$

If $< a_{\boldsymbol{k}} >$ is nonzero for most values of $\boldsymbol{k}$ and the numbers $\{M_j\}$ are powers of 2, then this is computed most efficiently using the Fourier representation. Computing the sum in (10.37), for all indices that satisfy (10.36), requires a knowledge of $a_{\boldsymbol{k}}$ for all indices in the set

$$\mathcal{J}_{2\boldsymbol{M}} = \{\boldsymbol{j} \,:\, 1 - M_i \leq k_i \leq M_i, \quad i = 1, \dots, n\}. \tag{10.38}$$

To use the Fourier transform to compute (10.37), the set of samples $\{x_{\boldsymbol{j}}\}$, defined for $\boldsymbol{j} \in \mathcal{J}_{\boldsymbol{M}}$, must be augmented so that $x_{\boldsymbol{j}}$ is defined for all indices in $\mathcal{J}_{2\boldsymbol{M}}$. As x is assumed to vanish outside the sample set, this is done by adding the samples

$$\{x_{\boldsymbol{j}} = 0 \qquad \text{for } \boldsymbol{j} \in \mathcal{J}_{2\boldsymbol{M}} \setminus \mathcal{J}_{\boldsymbol{M}}\}.$$

As in one dimension, this is called *zero padding*. In n dimensions this amounts to adding about $(2^n - 1)M_1 \cdots M_n$ zero samples.

10.4.2 The Finite Fourier Transform in n Dimensions

The n-dimensional finite Fourier transform is defined by iterating the one-dimensional transform. Suppose that $< x_{\boldsymbol{j}} >$ is a collection of numbers parameterized by the set of indices

$$\mathcal{J}_{\boldsymbol{N}} = \{\boldsymbol{j} \,:\, 0 \leq j_i \leq N_i - 1, \quad i = 1, \dots, n\}. \tag{10.39}$$

The finite Fourier transform of $< x_{\boldsymbol{j}} >$ is the indexed set of numbers $< \hat{x}_{\boldsymbol{k}} >$ defined by

$$\hat{x}_{\boldsymbol{k}} = \frac{1}{N_1 \cdots N_n} \sum_{j_1=0}^{N_1-1} \cdots \sum_{j_n=0}^{N_n-1} x_{\boldsymbol{j}} e^{-\frac{2\pi i j_1 k_1}{N_1}} \cdots e^{-\frac{2\pi i j_n k_n}{N_n}}. \tag{10.40}$$

The formula defines $\hat{x}_{\boldsymbol{k}}$ for all $\boldsymbol{k} \in \mathbb{Z}^n$ as a periodic sequence, periodic of period N_j in the jth index. Thus $< \hat{x}_{\boldsymbol{k}} >$ is also naturally parameterized by $\mathcal{J}_{\boldsymbol{N}}$. When it is important to emphasize the index set, this transform is denoted by $\mathcal{F}_{\boldsymbol{N}}(< x_{\boldsymbol{j}} >)$.

For $1 \leq l \leq n$, let ${}^l\mathcal{F}_N$ denote the one-dimensional, N-point Fourier transform acting only in the lth index,

$$({}^l\mathcal{F}_{N_l} < x_{\boldsymbol{j}} >)_{\boldsymbol{k}} = \frac{1}{N_l} \sum_{j=0}^{N_l-1} x_{k_1 \dots k_{l-1} j k_{l+1} \dots k_n} e^{-\frac{2\pi i j k_l}{N_l}}. \tag{10.41}$$

The n-dimensional finite Fourier transform can be computed as an iterated sum. In the notation introduced previously,

$$\mathcal{F}_{\boldsymbol{N}}(< x_{\boldsymbol{j}} >) = {}^n\mathcal{F}_{N_n} \circ \cdots \circ {}^1\mathcal{F}_{N_1}(< x_{\boldsymbol{j}} >). \tag{10.42}$$

Suppose that $< a_{\boldsymbol{j}} >$ and $< b_{\boldsymbol{j}} >$ are sequences parameterized by $\mathcal{J}_{\boldsymbol{N}}$. These sequences can be extended to all of $\mathbb{Z}^n$ be requiring that they be periodic, of period N_j in the jth index, for $j = 1, \dots, n$. A convolution operation is defined for such periodic sequences by setting

$$a \star b_{\boldsymbol{k}} = \sum_{\boldsymbol{j} \in \mathcal{J}_{\boldsymbol{N}}} a_{\boldsymbol{j}} b_{\boldsymbol{k}-\boldsymbol{j}}.$$

As in the one-dimensional case, convolution is intimately connected to the finite Fourier transform:

$$\mathcal{F}_{\boldsymbol{N}}(a \star b)_{\boldsymbol{k}} = N_1 \cdots N_n < \hat{a} \cdot \hat{b} > . \tag{10.43}$$

Here $< \hat{a} \cdot \hat{b} >$ is the sequence whose $\boldsymbol{k}$th-element is the ordinary product $\hat{a}_{\boldsymbol{k}} \hat{b}_{\boldsymbol{k}}$. This relation is the basis for computing convolutions using the finite Fourier transform,

$$a \star b = N_1 \cdots N_n \mathcal{F}_{\boldsymbol{N}}^{-1} < \hat{a}_{\boldsymbol{k}} \cdot \hat{b}_{\boldsymbol{k}} > \tag{10.44}$$

Exercises

***Exercise* 10.4.1.** If $l \neq m$, show that

$${}^l\mathcal{F}_{N_l} \circ {}^m\mathcal{F}_{N_m} = {}^m\mathcal{F}_{N_m} \circ {}^l\mathcal{F}_{N_l}.$$

***Exercise* 10.4.2.** Prove (10.42).

***Exercise* 10.4.3.** If M is a power of 2, then the M-point Fourier transform can be computed using $O(M \log_2 M)$ arithmetic operations. Show that if each M_j, $j = 1, \ldots, n$ is a power of 2, then the "length $(M_1, \ldots, M_n)$," n-dimensional, finite Fourier transform can be computed using $O(M_1 \log_2 M_1 \cdots M_n \log_2 M_n)$ arithmetic operations. Can you give a better estimate?

***Exercise* 10.4.4.** Find a formula for the inverse of the n-dimensional, finite Fourier transform.

***Exercise* 10.4.5.** Prove the identity (10.43).

10.4.3 The Fourier Representation for Shift Invariant Filters

If the sequence $< x_{\boldsymbol{j}} >$ is obtained as uniformly spaced samples of a function x, defined on $\mathbb{R}^n$, then the finite Fourier transform has an interpretation as an approximation to $\hat{x}$, the Fourier transform of x. Let $\boldsymbol{h}$ denote the sample spacing and $\boldsymbol{M} = (M_1, \ldots, M_n)$ the number of samples in each coordinate direction. The sample set and its Fourier transform are parameterized by $\mathcal{J}_{\boldsymbol{M}}$. Because $< \hat{x}_{\boldsymbol{k}} >$ is periodic, it is also possible to parameterize it using the indices

$$\mathcal{J}'_{\boldsymbol{M}} = \{\boldsymbol{k} : -\frac{1 - M_i}{2} \leq k_i \leq \frac{M_i - 1}{2}\},$$

where we assume that the $\{M_i\}$ are odd numbers. For an index $\boldsymbol{k} \in \mathcal{J}'_{\boldsymbol{M}}$ the relationship between $\hat{x}_{\boldsymbol{k}}$ and $\hat{x}$ is easily expressed as

$$\hat{x}_{\boldsymbol{k}} \approx \frac{1}{h_1 \cdots h_n} \hat{x}(\frac{2\pi k_1}{M_1 h_1}, \ldots, \frac{2\pi k_n}{M_n h_n}). \tag{10.45}$$

The usable bandwidth of the sample set, $[-\frac{\pi}{h_1}, \frac{\pi}{h_1}] \times \cdots \times [-\frac{\pi}{h_n}, \frac{\pi}{h_n}]$, depends only on the sample spacing. The number of samples in each direction is then determined by the size of the support of x.

In light of (10.35) and (10.44), the finite Fourier transform can be used to approximate the output of a linear, shift invariant filter. Let $\mathcal{J}_M$ denote the indices satisfying (10.36) and $\mathcal{J}_{2M}$ the augmented index set satisfying (10.38). The samples of x are denoted $\{x_s(\boldsymbol{j}) : \boldsymbol{j} \in \mathcal{J}_M\}$ and the samples of a by $\{a_s(\boldsymbol{j}) : \boldsymbol{j} \in \mathcal{J}_{2M}\}$. In order to compute $a_s \star x_s$, the samples of x have to be zero padded to be defined on $\mathcal{J}_{2M}$. Both sequences should then be considered to be periodic with period $2M_j - 1$ in the jth index. If $\boldsymbol{h}$ is the vector of sample spacings, then

$$\begin{aligned} \mathcal{A}x(j_1h_1, \dots, j_nh_n) &\approx [h_1 \cdots h_n](a_s \star x_s)_j \\ &= [h_1(2M_1 - 1) \cdots h_n(2M_n - 1)]\mathcal{F}_{2M}^{-1}(< \hat{a}_s\hat{x}_s >)_j. \end{aligned} \tag{10.46}$$

As in the one-dimensional case, a slightly different way to approximate shift invariant filters is to bypass the impulse response and sample the transfer function directly. Again this is because the transfer function is often an ordinary function, even when the impulse response is not. Using (10.45), the discussion leading up to (10.31) can be adapted to show that the correct samples of $\hat{a}$ to use in (10.46) are

$$\hat{a}_s(\boldsymbol{j}) = \frac{1}{h_1(2M_1 - 1) \cdots h_n(2M_n - 1)}\hat{a}\left(\frac{2\pi j_1}{h_1(2M_1 - 1)}, \dots, \frac{2\pi j_n}{h_1(2M_n - 1)}\right),$$
$$\text{for } 1 - M_i \leq j_i \leq M_i - 1, \quad i = 1, \dots, n. \tag{10.47}$$

Exercises

***Exercise* 10.4.6.** Show that (10.46) gives a Riemann sum approximation to the Fourier representation of $a * f(j_1h_1, \dots, j_nh_n)$.

***Exercise* 10.4.7.** Give a detailed justification for (10.47).

***Exercise* 10.4.8.** The Laplace operator is defined in two dimensions as

$$\Delta f = \partial_{x_1}^2 f + \partial_{x_2}^2 f.$$

The transfer function for this operator is $-(\xi_1^2 + \xi_2^2)$. The Laplace operator can also be approximated by finite differences; for example,

$$\begin{aligned} \Delta f(x_1, x_2) \approx &\frac{f(x_1 + h, x_2) - 2f(x_1, x_2) + f(x_1 - h, x_2)}{h^2} + \\ &\frac{f(x_1, x_2 + h) - 2f(x_1, x_2) + f(x_1, x_2 - h)}{h^2}. \end{aligned} \tag{10.48}$$

Compare the Fourier domain approximations obtained from sampling the transfer function directly and using the Fourier representation of the finite difference formula.

***Exercise* 10.4.9.** In the previous exercise, what is the impulse response of the finite difference approximation to Δ?

10.5 Appendix: The Fast Fourier Transform

If $N = 2^q$, then there is a very efficient way to compute the finite Fourier transform of a sequence of length N. The fast algorithm for the finite Fourier transform is the Cooley-Tukey or *fast Fourier transform algorithm*, usually referred to as the FFT. Let $\zeta = e^{\frac{2\pi i}{N}}$ be the primitive Nth-root of unity, and let $\zeta_j = \zeta^j$. This makes the notation simpler in what follows. The finite Fourier transform of $< x_0, x_1, \ldots, x_{N-1} >$ is given by

$$\hat{x}_k = \frac{1}{N}\sum_{j=1}^{N-1} x_k e^{-\frac{2\pi ijk}{N}} = \frac{1}{N}\sum_{j=1}^{N-1} x_j \bar{\zeta}_j^k,$$

which can be expressed as a matrix multiplying a vector:

$$\begin{pmatrix} \hat{x}_0 \\ \hat{x}_1 \\ \hat{x}_2 \\ \vdots \\ \hat{x}_{N-1} \end{pmatrix} = \frac{1}{N}\begin{pmatrix} 1 & 1 & \cdots & \cdots & 1 \\ 1 & \bar{\zeta}_1 & \bar{\zeta}_2 & \cdots & \bar{\zeta}_{N-1} \\ 1 & \bar{\zeta}_1^2 & \bar{\zeta}_2^2 & \cdots & \bar{\zeta}_{N-1}^2 \\ 1 & \cdots & & \cdots & \\ 1 & \bar{\zeta}_1^{N-1} & \bar{\zeta}_2^{N-1} & \cdots & \bar{\zeta}_{N-1}^{N-1} \end{pmatrix}\begin{pmatrix} x_0 \\ x_1 \\ x_2 \\ \vdots \\ x_{N-1} \end{pmatrix}.$$

Denote this matrix by C_N. We now show that for $N = 2^q$, the number of calculations involved in multiplying C_N times a vector can be reduced to $3Nq$. For even modest values of q this is a much smaller number than $N(2N-1)$. This reduction comes from several observations about the structure of the matrix C_N.

If A is a square matrix with complex entries a_{ij}, then the adjoint of A, denoted A^*, is the matrix whose ijth-entry is $\bar{a}_{ji}$. The matrix A is *unitary* if

$$A^{-1} = A^*.$$

The matrix

$$\sqrt{N}C_N$$

is a unitary matrix. This is a consequence of formulæ (8.19) and (8.21). In matrix notation we have

$$\sqrt{N}C_N^*\sqrt{N}C_N = I.$$

The inverse of C_N therefore has essentially the same form as C_N^*. A fast algorithm for multiplication by C_N should also give a fast algorithm for multiplication by C_N^{-1}. In other words, if we can compute $\mathcal{F}_N$ efficiently, then we can also compute $\mathcal{F}_N^{-1}$ efficiently.

The following identities among the Nth and $(2N)$th roots of unity lead to the fast Fourier transform algorithm. Let $\mu = e^{\frac{2\pi i}{2N}}$ be the primitive $(2N)$th root of unity; as previously, $\mu_j = \mu^j$. The following identities are elementary:

$$e^{\frac{2\pi i2kj}{2N}} = e^{\frac{2\pi ikj}{N}}, \quad e^{\frac{2\pi ik(j+N)}{N}} = e^{\frac{2\pi ikj}{N}} \text{ and } e^{\frac{2\pi i(2k+1)j}{2N}} = e^{\frac{2\pi ikj}{N}}e^{\frac{2\pi ij}{2N}}.$$

These identities can be rewritten as

$$\mu_j^{2k} = \zeta_j^k, \quad \zeta_j^{k+N} = \zeta_j^k \text{ and } \mu_j^{2k+1} = \mu_j\zeta_j^k. \tag{10.49}$$

From the definition of C_N, the $(2k+1)$st and $(2k+2)$th rows of C_{2N} are given by

$$\begin{aligned}
(2k+1)\text{st}: &\quad (1, \bar{\mu}_1^{2k}, \cdots, \bar{\mu}_{2N-1}^{2k}) \\
(2k+2)\text{th}: &\quad (1, \bar{\mu}_1^{2k+1}, \cdots, \bar{\mu}_{2N-1}^{2k+1}).
\end{aligned}$$

Comparing these with the kth row of C_N and using the relations in (10.49), rows of C_{2N} can be expressed in terms of the rows of C_N as follows:

$$(2k+1)\text{st}: \quad (1, \bar{\mu}_1^{2k}, \cdots, \bar{\mu}_{N-1}^{2k}, \bar{\mu}_N^{2k}, \cdots, \bar{\mu}_{2N-1}^{2k}) = (1, \bar{\zeta}_1^k, \cdots, \bar{\zeta}_{N-1}^k, 1, \bar{\zeta}_1^k, \cdots, \bar{\zeta}_{N-1}^k)$$

and,

$$\begin{aligned}
(2k+2)\text{th}: \quad &(1, \bar{\mu}_1^{2k+1}, \cdots, \bar{\mu}_{N-1}^{2k+1}, \bar{\mu}_N^{2k+1}, \cdots, \bar{\mu}_{2N-1}^{2k+1}) \\
&= (1, \bar{\zeta}_1^k \bar{\mu}_1, \bar{\zeta}_2^k \bar{\mu}_2 \cdots, \bar{\zeta}_{N-1}^k \bar{\mu}_{N-1}, \bar{\mu}_N, \bar{\zeta}_1^k \bar{\mu}_{N+1} \cdots, \bar{\zeta}_{N-1}^k \bar{\mu}_{2N-1}).
\end{aligned}$$

In terms matrices, C_{2N} is essentially obtained by multiplying two copies of C_N by another simple matrix:

$$C_{2N} = C_N^{\#} U_N.$$

Define the $2N \times 2N$ matrix

$$C_N^{\#} = \begin{pmatrix} r_1 & \mathbf{0} \\ \mathbf{0} & r_1 \\ \vdots & \vdots \\ r_N & \mathbf{0} \\ \mathbf{0} & r_N \end{pmatrix},$$

where the $\{r_i\}$ are the rows of C_N and the vector $\mathbf{0} = \underbrace{(0, \ldots, 0)}_{N}$,

$$C_N = \begin{pmatrix} r_1 \\ \vdots \\ r_N \end{pmatrix}.$$

The matrix U_N is defined by

$$U_N = \begin{pmatrix} I & I \\ D_N^1 & D_N^2 \end{pmatrix},$$

where

$$D_N^1 = \begin{pmatrix} 1 & & & & \\ & \bar{\mu}_1 & & 0 & \\ & & \bar{\mu}_2 & & \\ & 0 & & \ddots & \\ & & & & \bar{\mu}_{N-1} \end{pmatrix} \text{ and } D_N^2 = \begin{pmatrix} \bar{\mu}_N & & & & \\ & \bar{\mu}_{N+1} & & 0 & \\ & & \bar{\mu}_{N+2} & & \\ & 0 & & \ddots & \\ & & & & \bar{\mu}_{2N-1} \end{pmatrix}.$$

The important feature of U_N is that it has exactly two nonzero entries per row. If $N = 2^q$ then this argument applies recursively to $C_N^{\#}$ to give a complete factorization.

Theorem 10.5.1. *If $N = 2^q$, then $C_N = E_1 E_2 \cdots E_q$, where each row of the $N \times N$ matrix E_i has two nonzero entries.*

It is not difficult to determine exactly which entries in each row of the matrices E_j are nonzero. For an arbitrary N-vector $\boldsymbol{x} = (x_1, \ldots, x_N)$, the computation of $E_j \boldsymbol{x}$ can be done using exactly $N(2 \text{ multiplications} + 1 \text{ addition})$. Using this factorization and the knowledge of which entries of the E_j are nonzero, we can reduce the number of operations needed to compute the matrix product $\mapsto C_N \boldsymbol{x}$ to $3qN = 3N \log_2 N$. Indeed the combinatorial structures of the matrices $\{E_j\}$ are quite simple, and this has led to very efficient implementations of this algorithm. Each *column* of E_j also has exactly two nonzero entries, and therefore the factorization of C_N gives a factorization of C_N^*:

$$C_N^* = E_q^* E_{q-1}^* \ldots E_1^*.$$

***Example* 10.5.1.** We can factor the matrix $4C_4$ as

$$4C_4 = \begin{pmatrix} 1 & 1 & 0 & 0 \\ 0 & 0 & 1 & 1 \\ 1 & -1 & 0 & 0 \\ 0 & 0 & 1 & -1 \end{pmatrix} \begin{pmatrix} 1 & 0 & 1 & 0 \\ 0 & 1 & 0 & 1 \\ 1 & 0 & -1 & 0 \\ 0 & -i & 0 & i \end{pmatrix}. \tag{10.50}$$

For a more complete discussion of the fast Fourier transform, see [100].

10.6 Conclusion

If x is a function defined on $\mathbb{R}$ with bounded support, then the finite Fourier transform of a set of equally spaced samples of x can be interpreted as approximate samples of the Fourier transform of x. The sample spacing determines, via the Nyquist criterion, the usable bandwidth available in the sampled data. To implement a shift invariant linear filter on a finite sample set consisting of N samples $< x_0, \ldots, x_{N-1} >$, the sample sequence must be augmented by padding it with at least $N - 1$ zeros. If $\mathcal{H}$ is a shift invariant filter and we add $k \geq N - 1$ zeros, then $\mathcal{H}x$, *at sample points,* is approximated by a finite-dimensional linear transformation of the form

$$\mathcal{H}x \approx \tau(N+k)\mathcal{F}_{N+k}^{-1}\Lambda_h \mathcal{F}_{N+k}(< x_0, \ldots, x_{N-1}, 0, \ldots, 0 >).$$

Here $\mathcal{F}_{N+k}$ is the finite Fourier transform on sequences of length $N + k$, τ is the sample spacing and Λ_h is a *diagonal* matrix. The entries of Λ_h are either samples of the transfer function of H or obtained as the finite Fourier transform of $N + k$ samples of the impulse response of $\mathcal{H}$. If $N + k$ is a power of 2, then the linear transformation $\mathcal{F}_{N+k}$ has a very fast implementation.

In the next chapter we discuss how data are actually collected in a variety of x-ray CT machines. Gross features in the design of the a CT machine determine how the Radon transform of the attenuation coefficient is actually sampled. Using the results in this chapter we derive several implementations of approximate filtered back-projection algorithms on finitely sampled data. Since we consider two-dimensional slices of the attenuation coefficient this would appear to be a two-dimensional filtering problem. In fact, the filtration and back-projection steps are usually defined as one-dimensional filters that are applied in succession.

Chapter 11

Reconstruction in X-Ray Tomography

At long last we are returning to the problem of reconstructing images in x-ray tomography. Recall that if f is a function defined on $\mathbb{R}^2$ that is bounded and has bounded support, then its Radon transform, $\mathcal{R}f$, is a function on the space of oriented lines. The oriented lines in $\mathbb{R}^2$ are parameterized by pairs $(t, \boldsymbol{\omega}) \in \mathbb{R} \times S^1$, with

$$l_{t,\omega} \leftrightarrow \{\boldsymbol{x} : \langle \boldsymbol{x}, \boldsymbol{\omega} \rangle = t\}.$$

The positive direction along $l_{t,\omega}$ is defined by the unit vector $\hat{\boldsymbol{\omega}}$ orthogonal to $\boldsymbol{\omega}$ with $\det(\boldsymbol{\omega}\,\hat{\boldsymbol{\omega}}) = +1$. The Radon transform of f is the function defined on $\mathbb{R} \times S^1$ by

$$\mathcal{R}f(t, \boldsymbol{\omega}) = \int_{-\infty}^{\infty} f(t\boldsymbol{\omega} + s\hat{\boldsymbol{\omega}})\, ds.$$

Exact inversion formulæ for the Radon transform are derived in Chapter 6. These formulæ assume that $\mathcal{R}f$ is known for *all* lines. In a real x-ray CT machine, $\mathcal{R}f$ is *approximately* sampled at a finite set of points. The goal is to construct a function defined on a discrete set that is, to the extent possible, samples or averages of the original function f. In this chapter we see how the Radon inversion formulæ lead to methods for approximately reconstructing f from realistic measurements. We call such a method a *reconstruction algorithm*.

In this chapter we derive the reconstruction algorithms used in most modern CT scanners. After a short review of the basic setup in x-ray tomography, we describe the principal designs used in modern CT scanners. This, in turn, determines how the Radon transform of the attenuation coefficient is sampled. We next derive algorithmic implementations of the filtered back-projection formula for parallel beam, fan beam, and spiral CT scanners. A large part of our discussion is devoted to an analysis of the filtration step and its *efficient* implementation.

11.1 Basic Setup in X-Ray Tomography

Before deriving and analyzing the reconstruction algorithms, we review the setup in x-ray CT and fix the notation for the remainder of the chapter. Beer's law is the basic physical principle underlying x-ray tomography. To an object D in $\mathbb{R}^3$ there is associated an *attenuation coefficient* $\mu(\boldsymbol{x})$. This is a nonnegative function that describes the probability that an x-ray photon of a given energy which encounters the object at the point $\boldsymbol{x}$ is absorbed or scattered. Beer's law is phrased as a differential equation describing the change in the intensity of a (one-dimensional) "beam," composed of many photons, traveling along a line ℓ in $\mathbb{R}^3$. If $\boldsymbol{\Omega} \in S^2$ is the direction of ℓ and $\boldsymbol{x}_0$ is a point on ℓ, then the line is given parametrically by

$$\ell = \{s\boldsymbol{\Omega} + \boldsymbol{x}_0 \ : \ s \in \mathbb{R}\}.$$

Let $I(s)$ denote the intensity of the photon beam at the point $s\boldsymbol{\Omega} + \boldsymbol{x}_0$. Beer's law states that

$$\frac{dI}{ds}(s) = -\mu(s\boldsymbol{\Omega} + \boldsymbol{x}_0)I(s).$$

If D is a bounded object, then $\ell \cap D$ is contained in an interval of parameter values: $s \in [a, b]$. In the simplest model for x-ray tomography, the incoming intensity $I(a)$ is known and the outgoing intensity $I(b)$ is measured. Beer's law then relates the ratio $I(a)/I(b)$ to the line integral of μ:

$$\log\left[\frac{I(a)}{I(b)}\right] = \int_a^b \mu(s\boldsymbol{\Omega} + \boldsymbol{x}_0)\, ds. \tag{11.1}$$

Originally *tomography* referred to methods for nondestructively determining the internal structure of two-dimensional slices of a three-dimensional object. Indeed the Greek root τομή means cut or slice [see [6, pp. 8–9]]. Though the term has come to refer to any method for determining the internal structure of a three-dimensional object from external measurements, we largely adhere to the earlier meaning. Suppose that a coordinate system (x_1, x_2, x_3) for $\mathbb{R}^3$ is fixed. The data collected in x-ray CT are approximations to the Radon transforms of the two-dimensional "slices" of the function μ, in the x_3-direction. These are the functions

$$f_c(x_1, x_2) = \mu(x_1, x_2, c),$$

obtained by fixing the last coordinate. In the preceding formulation, this corresponds to taking

$$\begin{aligned} \boldsymbol{x}_0 =&(t\boldsymbol{\omega}, c) \text{ and } \boldsymbol{\Omega} = (\hat{\boldsymbol{\omega}}, 0) \text{ where, } t, c \in \mathbb{R} \text{ and} \\ \boldsymbol{\omega}, \hat{\boldsymbol{\omega}} \in S^1 &= \{(x_1, x_2, 0) \ : \ x_1^2 + x_2^2 = 1\}. \end{aligned} \tag{11.2}$$

For a fixed c, the integrals in (11.1) are nothing but the Radon transform of f_c. With these measurements the function f_c can therefore be reconstructed.

A real x-ray CT machine only measures a finite number of line integrals. In a simple model for the actual measurements there is a finite set of values, $\{c_1, \ldots, c_n\}$ such that the Radon transforms of the functions, $\{f_{c_1}, \ldots, f_{c_n}\}$ are sampled along a finite set of lines

$\{l_{t_j,\omega_j} : j = 1, \ldots, P\}$. The design of the machine determines which line integrals are measured. This chapter considers algorithms for reconstructing a single two-dimensional slice. The problem of using the slices to reassemble a three-dimensional object is not treated in any detail. Note, however, the important fact that the CT machine itself determines a frame of reference, fixing, for example, the "x_3-direction." Once the machine is calibrated, the positions in space that correspond to the different slices and the lines within a slice are known in advance. As we shall see, the design of the machine also singles out a particular coordinate system in the space of oriented lines.

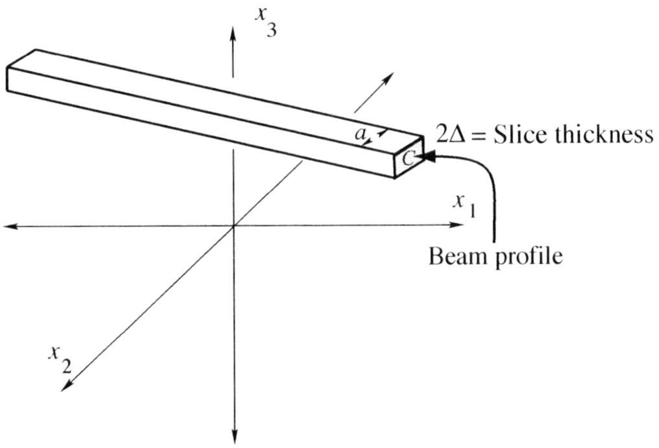

Figure 11.1. A three-dimensional x-ray beam.

A real x-ray beam is not one dimensional but three dimensional. Let C denote the cross section of the beam at right angles to its direction. The cross section is often approximately rectangular:

$$C \approx [-a, a] \times [-\Delta, \Delta],$$

where the second factor lies in the x_3-direction. The width of the second factor 2Δ is called the *slice thickness*; this parameter is usually adjustable at the time the measurements are made. In circa 2002 commercial CT machines, the slice thickness varies between 2 and 10 mm. The beam intensity also varies continuously within C, falling off to zero at the edge. As a practical matter, a larger cross section increases the energy in the beam, which, as we shall see, improves the signal-to-noise ratio in the measurements. On the other hand, poorer spatial resolution is also a consequence of a larger cross section. In our initial discussion of the reconstruction algorithms, we model the measurements as line integrals of a slice, $f_c(x_1, x_2) = \mu(x_1, x_2, c)$. In other words, we assume that the x-ray beam *is* one dimensional. In Sections 12.1 and 12.2 we consider consequences of the finite beam width in the plane of the slice. A linear model for the effect of the *third* dimension is to replace line integrals of f_c by line integrals of a weighted average of these functions in the x_3-variable:

$$\int_{-\Delta}^{\Delta} w_{\text{ssp}}(c-b) f_b \, db. \tag{11.3}$$

The function w_{ssp} is called the *slice selectivity profile* or *SSP*; it encodes information about the resolution of the CT scan in the x_3-direction. Except for our short discussion of spiral scan CT in Section 11.7, the third dimension is rarely mentioned in this and the following chapter.

Let (x, y) denote Cartesian coordinates in the slice $x_3 = c$, which is heretofore fixed. The two-dimensional object we would like to image lies in D_L, the disk of radius L, centered at (0, 0) in this plane. In this chapter the x-ray source is assumed to be monochromatic of energy $\mathscr{E}$ and the object is described by its x-ray attenuation coefficient f at this energy. As the object lies in D_L, f is assumed to vanish outside this set. Our goal is to use samples of $\mathscr{R}f$ to approximately determine the values of f on a uniform *reconstruction* grid,

$$\mathscr{R}_\tau = \{(x_j, y_k) = (j\tau, k\tau) : j, k \in \mathbb{Z}\},$$

in the (x, y)-plane. Here $\tau > 0$ denotes the sample spacing, which is selected to reflect the resolution available in the data. In circa 2002 commercial machines the resolution also varies between 2 and 10 mm.

The reconstruction grid can also be thought of as dividing the plane into a grid of squares of side τ. Each square is called a picture element or *pixel*. As each slice is really a three-dimensional slab, and hence each square is really a cuboid, the elements in the reconstruction grid are often called volume elements or *voxels* (see Figure 11.2). The value reconstructed at a point $(x_j, y_k) \in \mathscr{R}_\tau$ should be thought of as a weighted average of f over the voxel containing this point. Of course, f is only reconstructed at points of $\mathscr{R}_\tau$ lying in $[-L, L] \times [-L, L] \supset D_L$. We assume that f is bounded and regular enough for its Radon transform to be sampled. As the actual measurements involve averaging μ with a continuous function, this is not a restrictive assumption.

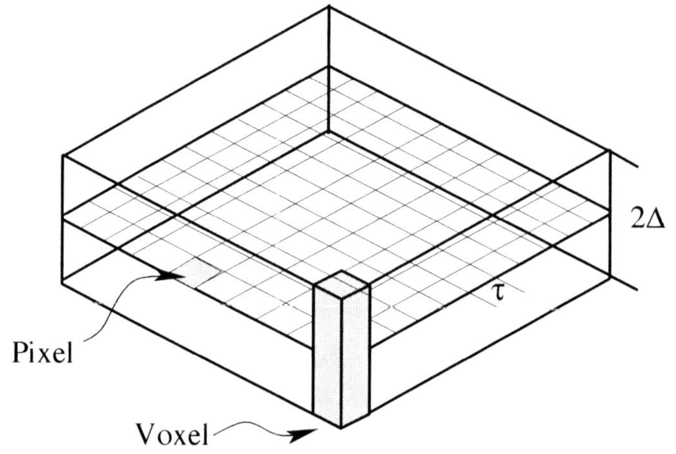

Figure 11.2. The reconstruction grid.

In most of this section it is assumed that measurements are made with infinite precision and that the full set of real numbers is available. Real measurements have errors and are quantized to fit into a finite number of bits. The measurements approximate samples of the Radon transform of a nonnegative function assuming values in a *known* range and supported in a *known* set. A good estimate for the range of measured values is therefore known in advance. Say that the measurements lie in an interval $[0, M]$. If b bits are used in the quantization of the measurements, then the smallest increment available in the measured data is $2^{-b}M$. The number of bits used to quantize the data therefore determines the size of the smallest difference in density that might be discernible in the reconstructed image. In medical imaging this is usually referred to as the *contrast* available in the reconstructed image. This numerical resolution reflects the precision of the measurements themselves and the computations used to process them. It should not be confused with *spatial* resolution, which depends on the sample spacing and the point spread function of the reconstruction algorithm. The accuracy of the measurements is determined by the accuracy and sensitivity of the detectors and the stability of the x-ray source. A detailed discussion of x-ray sources and detectors can be found in [22] or [6].

11.2 The Reconstruction Problem

We now review the inversion formulæ derived earlier and consider how they might be approximated in a real computation. These formulæ are consequences of the Fourier inversion formula and central slice theorem; the later states that

$$\widetilde{\mathcal{R}f}(r,\omega) = \int_{-\infty}^{\infty} \mathcal{R}f(t,\omega)e^{-irt}dt = \hat{f}(r\omega).$$

Writing the Fourier inversion formula in polar coordinates gives

$$\begin{aligned} f(x,y) &= \frac{1}{[2\pi]^2}\int_0^{2\pi}\int_0^{\infty} \hat{f}(r\omega)e^{ir\langle(x,y),\omega\rangle}r\,dr\,d\omega \\ &= \frac{1}{[2\pi]^2}\int_0^{\pi}\int_{-\infty}^{\infty} \hat{f}(r\omega)e^{ir\langle(x,y),\omega\rangle}|r|\,dr\,d\omega. \end{aligned} \tag{11.4}$$

Directly approximating the Fourier inversion formula is one approach to reconstructing f. Note that the polar variable r is allowed to assume negative values with the understanding that if $r < 0$, then

$$r\omega = |r|(-\omega).$$

Using the central slice theorem gives

$$f(x,y) = \frac{1}{[2\pi]^2}\int_0^{\pi}\int_{-\infty}^{\infty} \widetilde{\mathcal{R}f}(r,\omega)|r|e^{ir\langle(x,y),\omega\rangle}dr\,d\omega. \tag{11.5}$$

Interpreting the r integral as a linear, shift invariant filter acting in the t-variable,

$$\begin{aligned}\mathcal{G}(\mathcal{R}f)(t,\omega) &= \frac{1}{2\pi}\int_{-\infty}^{\infty} \widetilde{\mathcal{R}f}(r,\omega)|r|e^{irt}\,dr \\ &= -i\mathcal{H}\partial_t\,\mathcal{R}f(t,\omega),\end{aligned} \tag{11.6}$$

leads to the filtered back-projection formula

$$f(x,y) = \frac{1}{2\pi}\int_{0}^{\pi} \mathcal{G}(\mathcal{R}f)(\langle(x,y),\omega\rangle,\omega)\,d\omega. \tag{11.7}$$

Of course, these formulæ are entirely equivalent from the mathematical point of view. However, approximating mathematically equivalent formulæ can lead to very different algorithms. As previously, τ is the uniform spacing in the reconstruction grid. The object is assumed to lie within the disk of radius L. The number of grid points on the x or y-axis is therefore $2K+1$, where

$$K = \left\lfloor \frac{L}{\tau} \right\rfloor + 1.$$

Here $\lfloor s \rfloor$ denotes the largest integer less than or equal to s. Our immediate goal is to approximately reconstruct the set of values

$$\{f(j\tau,k\tau) : -K \le j,k \le K\}.$$

In order to distinguish the reconstructed values from the actual samples, the reconstructed values are denoted by

$$\{\tilde{f}(j\tau,k\tau) : -K \le j,k \le K\}.$$

While f is known to be zero on grid points lying outside D_L, the reconstructed values at these points are generally non-zero. As remarked previously, real measurements have an interpretation as weighted averages of $\mathcal{R}f$ over the cross section of the x-ray beam. In our initial derivations of the reconstruction algorithms, we overlook this point; the effects of finite beam width are easily incorporated, and we do so in the next chapter.

A good reconstruction algorithm is characterized by *accuracy*, *stability*, and *efficiency*. An accurate algorithm is often found by starting with an exact inversion formula and making judicious approximations. In x-ray CT, stability is the result of lowpass filtering and the continuity properties of the exact inversion formula. Whether an algorithm can be implemented efficiently depends on its overall structure as well as the hardware available to do the work. An x-ray CT-machine is *not* a general-purpose computer, but rather a highly specialized machine designed to do one type of computation quickly and accurately. Such a machine may have many processors, allowing certain parts of the algorithm to be "parallelized." As we shall soon see, the measuring hardware naturally divides the measurements into a collection of *views* or *projections*. The most efficient algorithms allow the data from a view to be processed as soon as it is collected.

In the early days of imaging, machines were calibrated by making measurements of composite objects with known attenuation coefficients. These objects are called *phantoms*. The problem with this approach is that it mixes artifacts caused by physical measurement errors with those caused by algorithmic errors. A very important innovation in medical imaging was introduced by Larry Shepp. In order to isolate the algorithmic errors, he replaced the (physical) phantom with a *mathematical phantom*.

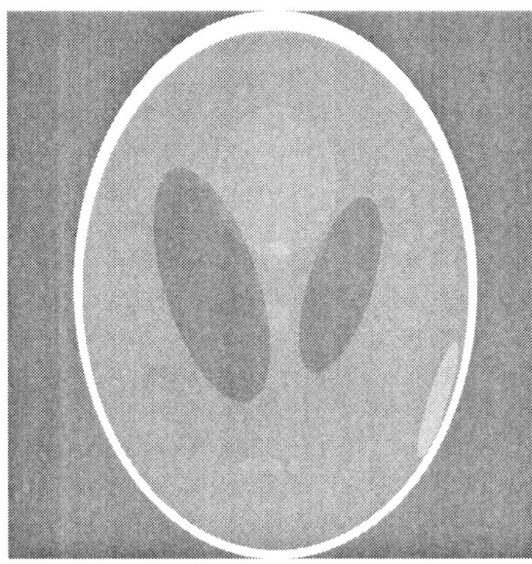

Figure 11.3. A mathematical phantom.

Instead of using real measurements of a known object, Shepp suggested that one give a mathematical description of a phantom to create simulated and *controlled* data. In this way, algorithmic errors could be separated from measurement errors. A mathematical phantom is created as follows: First a simplified model of a slice of the human head (or other object) is described as an arrangement of ellipses and polygons. Each region is then assigned a density or attenuation coefficient (see figure 11.3). Next the continuous model is digitized by superimposing a regular grid and replacing the piecewise continuous densities by their averaged values over the squares that make up the grid. Finally, measurements are simulated by integrating the digitized model over a collection of strips, arranged to model a particular measurement apparatus. The robustness of an algorithm to different sorts of measurement errors (e.g., noise, beam hardening, patient motion, miscalibration, etc.) can be tested by incorporating these errors into the simulated measurements.

By using mathematical phantoms, the consequences of algorithmic errors can be separated from those of measurements errors. A priori, you know exactly what is being measured and can therefore compare the reconstructed image to a known, exact model. Mathematical phantoms are very useful in the study of artifacts caused by sampling errors and noise as well as for comparison of different algorithms. Of course, the fundamental question is how well the algorithms perform on actual measurements. In many of the examples in this and the next chapter, we apply an algorithm to a mathematical phantom.

11.3 Scanner Geometries

The structure of a reconstruction algorithm is dictated by which samples of $\mathcal{R}f$ are available. Before discussing algorithms, we therefore need to consider what kind of measurements are actually made. In broad terms there are two types of x-ray CT-machine, using two dimensional slices: (a) parallel beam scanner [see Figure 11.4(a)], (b) divergent beam scanner [see Figure 11.5]. The earliest scanner was a parallel beam scanner. This case is considered first because the geometry and algorithms are simpler to describe. Because the data can be collected much faster, most modern machines are divergent or *fan beam* scanners. Algorithms for these machines are a bit more involved and are treated later. The most recent machines (*circa* 2002) use a *spiral scan*, so that the actual measurements are not samples of the Radon transforms of two dimensional slices. We briefly discuss this modality in Section 11.7.

In a parallel beam scanner approximate samples of $\mathcal{R}f$ are measured in a finite set of directions, $\{\omega(k\Delta\theta) \quad \text{for } k = 0\ldots, M\}$, where

$$\Delta\theta = \frac{\pi}{M+1} \text{ and } \boldsymbol{\omega}(k\Delta\theta) = (\cos(k\Delta\theta), \sin(k\Delta\theta)).$$

In terms of the angular variable, θ the samples are equally spaced. The measurements made in a given direction are then samples of $\mathcal{R}f$ at a set of equally spaced affine parameters,

$$\{jd + d_0 \;:\; j = -N, \ldots, N\},$$

where d is the sample spacing in the affine parameter and $N = Ld^{-1}$ and d_0 is a fixed offset. Such an offset is sometimes used, though in the foregoing analysis we take $d_0 = 0$.

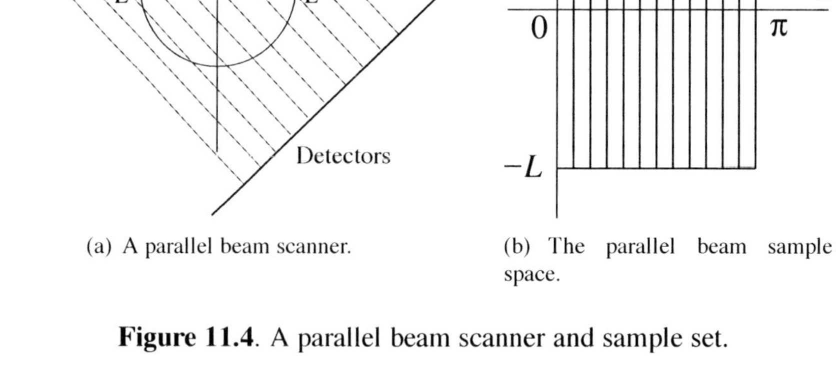

(a) A parallel beam scanner.

(b) The parallel beam sample space.

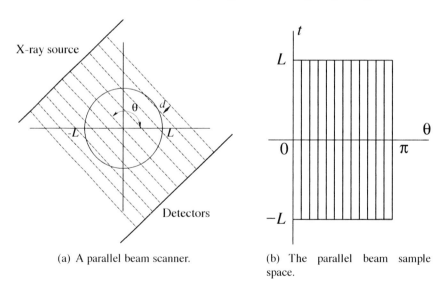

Figure 11.4. A parallel beam scanner and sample set.

Parallel beam data therefore consist of the samples

$$\{\mathcal{R}f(jd, \boldsymbol{\omega}(k\Delta\theta)) : j = -N, \ldots, N, \quad k = 0, \ldots, M\}. \tag{11.8}$$

Because of the symmetry of the Radon transform,

$$\mathcal{R}f(-t, -\boldsymbol{\omega}) = \mathcal{R}f(t, \boldsymbol{\omega}), \tag{11.9}$$

measurements are only required for angles lying in $[0, \pi)$. Sometimes it is useful to make measurements over a full 360°; the extra measurements can then be averaged as a way to reduce the effects of noise and systematic errors.

The individual measurements are called *rays*. A *view*, for a parallel beam machine, consists of the samples of $\mathcal{R}f(t, \boldsymbol{\omega})$ for a fixed $\boldsymbol{\omega}$. These are the integrals of f along the collection of equally spaced parallel lines,

$$\{l_{jd,\boldsymbol{\omega}(k\Delta\theta)} : j = -N, \ldots, N\}.$$

In $(t, \boldsymbol{\omega})$-space, parallel beam data consist of equally spaced samples on the vertical lines shown in Figure 11.4(b).

The other type of scanner in common use is called a *divergent beam* or *fan beam scanner*. A point source of x-rays is moved around a circle centered on the object being measured. The source is pulsed at a discrete sequence of angles, and measurements of $\mathcal{R}f$ are collected for a finite family of lines passing through the source. In a machine of this type, data are collected by detectors that are usually placed on a circular arc. There are two different designs for fan beam machines that, for some purposes, need to be distinguished. In a *third-generation* machine the detectors are placed on a circular arc centered on the source. The detectors and the source are rotated together. In a *fourth-generation* machine the detectors are on a fixed ring, centered on the object. Only the source is rotated, again around a circle centered on the object, within the ring of detectors. These designs are shown schematically in Figure 11.5.

For a fan beam machine it is useful to single out the *central ray*. For a third-generation machine this is the line that passes through the source and the center of rotation. The central ray is well defined no matter where the source is positioned. Since all the rays that are sampled pass through a source position, the natural angular parameter, for this geometry, is the angle, ϕ, between a given ray and the central ray. Suppose that the source is at distance D and the central ray makes an angle ψ with the positive x-axis. The affine parameter for the line passing through the source, making an angle ϕ with the central ray, is given by

$$t = D\sin(\phi) \tag{11.10}$$

[see Figure 11.6(a)]. The angular parameter of this line is

$$\theta = \psi + \phi - \frac{\pi}{2}; \tag{11.11}$$

recall that positive angles are measured counterclockwise.

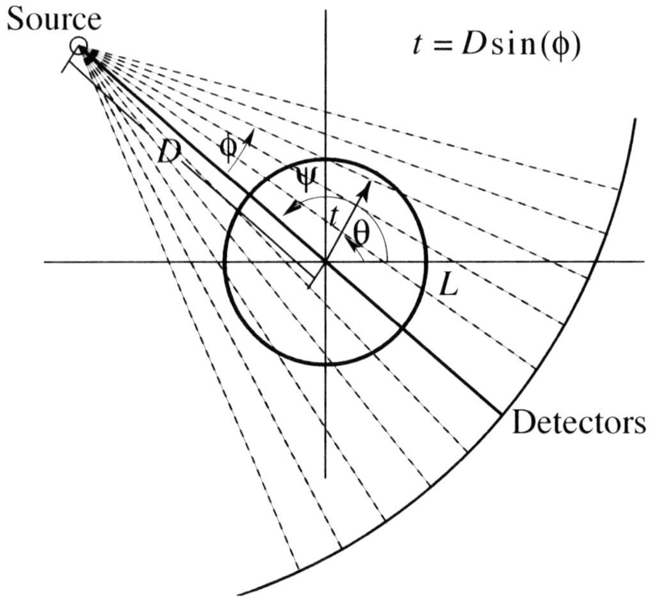

(a) A third-generation scanner.

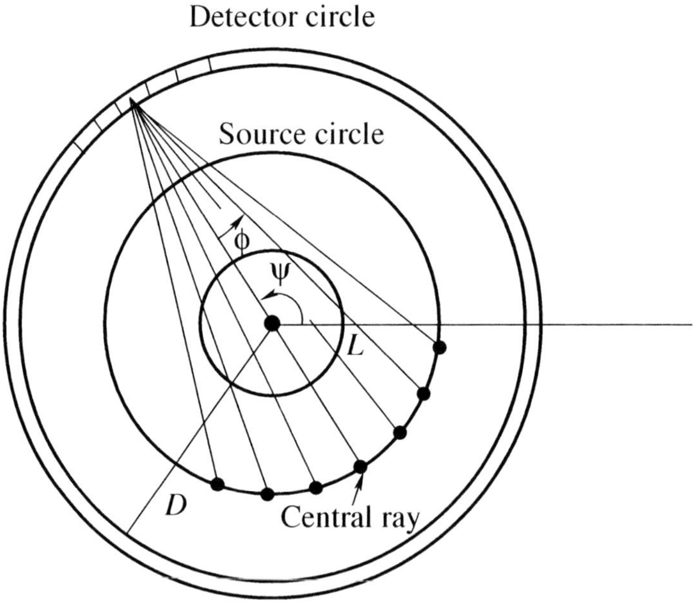

(b) A fourth-generation scanner.

Figure 11.5. The two different divergent beam geometries.

In a third-generation machine the source is placed at a finite number of equally spaced angles,

$$\psi_k \in \{\frac{2\pi k}{M+1} : k = 0, \ldots, M\},$$

and data are collected along a set of lines through the source, equally spaced in the ϕ-parameter,

$$\phi_j \in \{\frac{j\pi}{N} : j = -P, \ldots, P\}.$$

Third-generation, fan beam data are the set of samples

$$\{\mathcal{R}f(D\sin(\frac{j\pi}{N}), \omega\left(\frac{j\pi}{N} + \frac{2\pi k}{M+1} - \frac{\pi}{2}\right)) : j = -P, \ldots, P, \quad k = 0, \ldots, M\}.$$

The maximum and minimum values $\pm\frac{P\pi}{N}$, for the angle ϕ, are determined by the necessity of sampling all lines that meet an object lying in D_L, as shown in Figure 11.5(a). The samples lie on the sine curves shown in Figure 11.6(b). A *view* for a third-generation machine is the set of samples from the rays passing through a given *source* position.

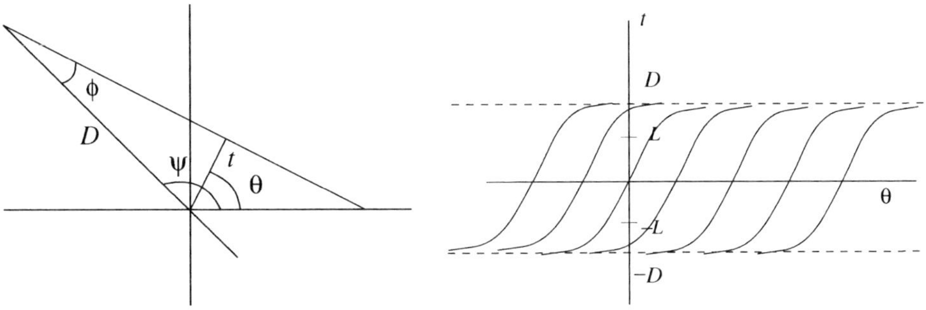

(a) Angular parameters for a fan beam machine.

(b) Sample set for a fan beam machine.

Figure 11.6. Parameters for a fan beam machine.

For a fourth-generation machine the source moves and the detectors remain fixed. Each time the source emits a pulse of x rays, a sample can in principle, be collected by any detector that can "see" the source. In practice, data are collected for rays that pass through the target area, as shown in Figure 11.5(b). In this way the data are grouped into views according to the detector position, *not* the source position.

For many purposes third- and fourth-generation machines can be treated together. Indeed we can still use the coordinates in Figure 11.6(a) to parameterize the data, with the following small changes: The detectors now lie on the circle of radius D and ψ denotes the angle from the positive x-axis to the "central ray," which is now a line joining $(0, 0)$ to a fixed *detector*. A view for a fourth-generation machine consists of the rays passing through a given *detector*. The parameter ϕ measures the angle between lines through a detector and its central ray. With this understood, the sample set and data collected are essentially the same for third- and fourth-generation machines. At a basic level, third- and fourth-generation scanners are quite similar, though they differ in many subtle points. These differences are considered in greater detail in Section 12.4.

The measurements are integrals of a nonnegative function f. In the engineering and medical literature, we sometimes see these raw measurements represented directly in (t, ω) or (ψ, ϕ)-space. Such a diagram is called a *sinogram.* A sinogram is a gray scale plot where the density of image at (t, ω) is a monotone function of the measured value $\mathcal{R}f(t, \omega)$. Sinograms are difficult for a human observer to directly interpret. As they contain *all* the information available in the data set; they may be preferable for machine-based assessments.

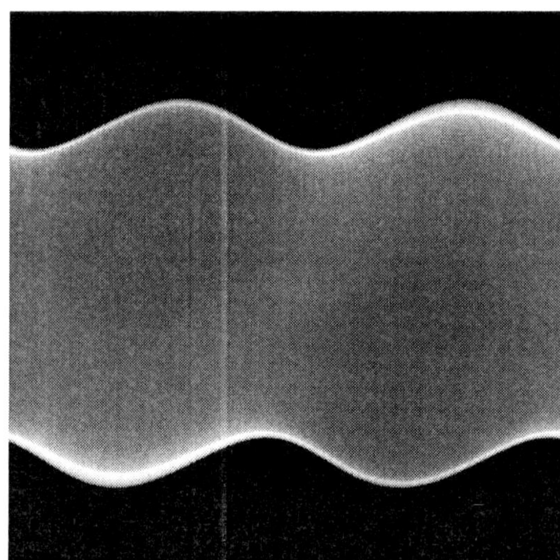

Figure 11.7. An example of a sinogram from a fan beam machine. The affine parameter is along the vertical axis and the angle is along the horizontal axis. The angle ranges from 0 to 2π. Black corresponds to 0 and white to the largest value assumed by $\mathcal{R}f$.

Exercises

***Exercise* 11.3.1.** For a third-generation, fan beam machine, the set of angles $\{\phi_k\}$ could be selected so that the sample spacing in the affine parameter is constant. To get a sample spacing d in the affine parameter, what set of angles $\{\phi_k\}$ should be used?

***Exercise* 11.3.2.** Explain why the sinogram in Figure 11.7 appears to have a reflectional symmetry around its center.

***Exercise* 11.3.3.** The sinogram in Figure 11.7 comes from measurements made on the Shepp-Logan head phantom. Find two features of the sinogram suggesting that this is an image of a head.

11.4 Reconstruction Algorithms for a Parallel Beam Machine

See: A.6.1.

We now consider reconstruction algorithms for a parallel beam machine. A good first step in the derivation of an algorithm is to assume that we can measure *all* the data from a finite set of equally spaced angles. In this case the data would be

$$\{\mathcal{R}f(t, \omega(k\Delta\theta)) : k = 0, \ldots, M, \quad t \in [-L, L]\},$$

where $\Delta\theta = \frac{\pi}{M+1}$. With these data we can apply the central slice theorem to compute angular samples of the two-dimensional Fourier transform of f,

$$\hat{f}(r\omega(k\Delta\theta)) = \int_{-\infty}^{\infty} \mathcal{R}f(t, \omega(k\,\Delta\theta))e^{-irt}dt.$$

***Remark* 11.4.1.** Recall that in this chapter the polar variable r can be either positive or negative; if $r < 0$, then

$$\hat{f}(r\omega(\theta)) = \hat{f}(|r|\omega(\theta + \pi)). \tag{11.12}$$

11.4.1 Direct Fourier Inversion

Formula (11.4) suggests using the two-dimensional Fourier inversion formula directly, to reconstruct f. Using a Riemann sum in the angular direction gives

$$f(x, y) \approx \frac{1}{4\pi(M+1)} \sum_{k=0}^{M} \int_{-\infty}^{\infty} \hat{f}(r\omega(k\,\Delta\theta))e^{ir\langle(x,y),\omega(k\Delta\theta)\rangle}|r|\,dr.$$

Our actual measurements are the samples $\{\mathcal{R}f(jd, \omega(k\Delta\theta))\}$ of $\mathcal{R}f(t, \omega(k\Delta\theta))$. Since the sample spacing in the t-direction is d, the usable bandwidth in the sampled data is $\frac{2\pi}{d}$. In light of the results in Section 10.2, these samples can be used to compute approximations to the following samples of the Fourier transform of f,

$$\hat{f}(r_j\omega(k\,\Delta\theta)) \quad r_j \in \{0, \pm\eta, \pm 2\eta, \ldots, \pm N\eta\}, \text{ where } \eta = \frac{1}{N}\frac{\pi}{d} = \frac{\pi}{L}.$$

This is a set of equally spaced samples of $\hat{f}$ in the *polar* coordinate system. The accuracy of these approximations depends on the effective support of $\hat{f}$.

Approximating the inverse Fourier transform directly in polar coordinates would entail $O((2M+2)(N+1))$ operations to compute the approximate inverse of the Fourier transform for a single point (x, y). As the number of points in the reconstruction grid is $O(K^2)$, the number of computations needed to reconstruct the image at every grid point is

$$O(MNK^2).$$

For realistic data this is a very large number. The size of this computation can be vastly reduced by using the fast Fourier transform algorithm (FFT).

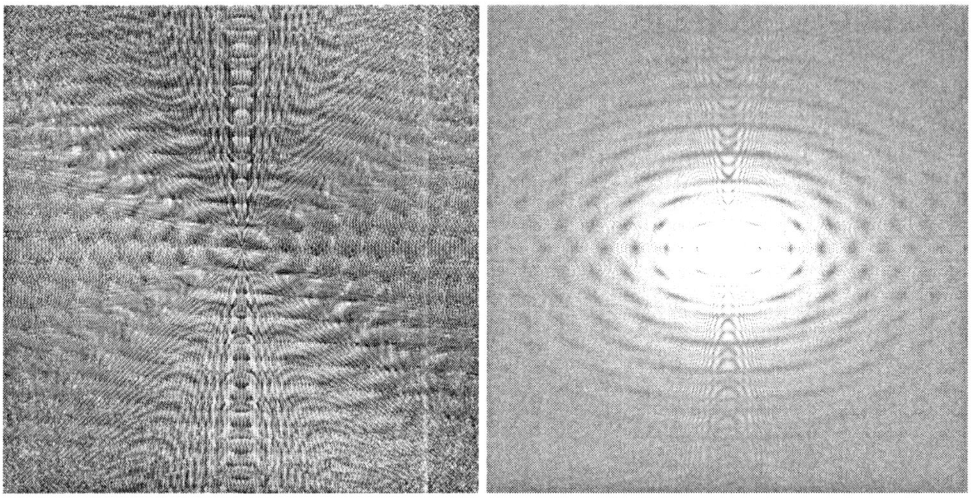

(a) The phase. (b) The magnitude.

Figure 11.8. The phase and magnitude of the Fourier transform of the Shepp-Logan phantom.

The two-dimensional fast Fourier transform is only usable if samples of $\hat{f}$ are known on a uniform grid in a *rectangular* coordinate system. Our data,

$$\{\hat{f}(r_j\omega(k\Delta\theta))\},$$

are samples of $\hat{f}$ on a uniform grid in a *polar* coordinate system. To use the two-dimensional FFT, the data must first be interpolated to get simulated measurements on a uniform, rectangular grid. Using nearest neighbor, linear interpolation, the amount of computation required to do this is a modest $O(K^2)$ calculations. Assuming that K is a power of 2, the FFT leading to $\tilde{f}(x_j, y_k)$ would require $O((K\log_2 K)^2)$ calculations. The full reconstruction of $\tilde{f}$, at the grid points, from the parallel beam data, would therefore require $O(K^2(\log_2 K)^2 + 2MN\log_2 N)$ computations. The $2MN\log_2 N$ term comes from using the one-dimensional FFT to compute $< \hat{f}(r_j\omega(k\Delta\theta)) >$ from $< \mathcal{R}f(jd, \omega(k\Delta\theta)) >$. For realistic values of M and N, this is a much smaller number than what was computed for a direct inversion of the Fourier transform. Indeed, this algorithm is the fastest algorithm for implementation on a general-purpose computer. However, with real data, simple linear interpolation leads to unacceptably large errors in the approximation of $\hat{f}$. This is largely due to the fact that $\hat{f}$ is a complex valued function with a rapidly oscillating phase. Figure 11.8 shows the phase and magnitude of the Fourier transform of the Shepp-Logan phantom. Note that both the phase and magnitude are highly oscillatory, explaining why it is difficult to accurately interpolate $\hat{f}$. A more sophisticated scheme of this sort is presented in Section 11.8.

Exercises

***Exercise* 11.4.1.** The magnitude of the Fourier transform shows a pronounced oscillation in the radial direction. Can you explain this?

***Exercise* 11.4.2.** Explain how to use zero padding to obtain a approximation to $\hat{f}(r\omega)$ on a finer grid in the r-variable. Is this justified in the present instance? Is there any way to reduce the sample spacing in the angular direction?

11.4.2 Filtered Back-Projection

Formula (11.7) organizes the approximate inversion in a different way. The Radon transform is first filtered,

$$\mathcal{G}\mathcal{R}f(t,\omega) = -i\mathcal{H}\partial_t \mathcal{R}f(t,\omega),$$

and then back-projected to find f at (x, y). The operation $\mathcal{R}f \mapsto \mathcal{G}\mathcal{R}f$ is a one-dimensional, linear shift invariant filter. On a parallel beam scanner, the data for a given ω define a single view and are collected with the source-detector array in a fixed position. Once the data from a view have been collected, they can be filtered. In this way, a large part of the processing is done by the time all the data for a slice has been collected. Supposing, as before, that sampling only occurs in the angular variable, then the data set for a parallel beam scanner would be the samples

$$\{\mathcal{R}f(t,\omega(k\Delta\theta)) : k = 0, \ldots, M\}.$$

In a filtered back-projection algorithm, each view, $\mathcal{R}f(t,\omega(k\Delta\theta))$, is filtered immediately after it is measured, giving $\mathcal{G}\mathcal{R}f(t,\omega(k\Delta\theta))$. When all the data have been collected and filtered, the image is approximately reconstructed by using a Riemann sum approximation to the back-projection:

$$\tilde{f}(x, y) = \frac{1}{2(M+1)} \sum_{k=0}^{M} \mathcal{G}\mathcal{R}f(\langle (x, y), \omega(k\Delta\theta)\rangle, \omega(k\,\Delta\theta)); \tag{11.13}$$

here use is made of the symmetry (11.17). Assuming that all the necessary values of $\mathcal{G}\mathcal{R}f$ are known, the back-projection step requires $O(MK^2)$ operations to determine $\tilde{f}$ on the reconstruction grid, $\mathcal{R}_\tau$. This step is also highly parallelizable: With $O(K^2)$ processors the back-projection could be done simultaneously for all points in $\mathcal{R}_\tau$ in $O(M)$ cycles. The serious work of implementing this algorithm is in deciding how to approximate the filter $\mathcal{G}$ on data that are sampled in both t and ω.

In real applications the approximation to the transfer function of $\mathcal{G}$ is chosen to be an approximation to $|r|$. Denote the approximate transfer function by $\hat{\phi}$ and define

$$Q_\phi f(t,\omega) = \frac{1}{2\pi} \int_{-\infty}^{\infty} \widetilde{\mathcal{R}f}(r,\omega)\hat{\phi}(r)e^{irt}\,dr.$$

In order for Q_ϕ to approximate $\mathcal{G}$, its modulation transfer function $\hat{\phi}$ should provide an "approximation" to $|r|$ over the effective support of the data. Exactly what is meant by this statement is a subject of considerable discussion. For example, as $|r|$ is an even, real-valued function $\hat{\phi}$ should also be. In order to have a stable algorithm and suppress noise, it is important for $\hat{\phi}$ to tend to zero as $|r|$ tends to infinity. In the end, whether or not a given choice of ϕ provides a "good" approximation is largely an empirical question.

Once ϕ is chosen, the filtered Radon transform is given by

$$\begin{aligned} Q_\phi f(t,\omega) &= \int_{-\infty}^{\infty} \mathcal{R}f(s,\omega)\phi(t-s)\,ds \\ &= \frac{1}{2\pi}\int_{-\infty}^{\infty} \widetilde{\mathcal{R}f}(r,\omega)\hat{\phi}(r)e^{irt}\,dr. \end{aligned} \tag{11.14}$$

With "complete data" the approximate reconstruction, defined by ϕ, would be

$$f_\phi(x,y) = \frac{1}{2\pi}\int_0^{\pi} Q_\phi f(\langle(x,y),\omega\rangle,\omega)\,d\omega. \tag{11.15}$$

Approximating the integrals in (11.14) on the sampled data, in either the spatial or frequency representation, and using a Riemann sum approximation for the back-projection gives an approximate reconstruction, $\tilde{f}_\phi$, at points $(x_m, y_l) \in \mathcal{R}_\tau$,

$$\tilde{f}_\phi(x_m,y_l) = \frac{d}{2(M+1)}\sum_{k=0}^{M}\sum_{j=-N}^{N} \mathcal{R}f(jd,\omega(k\,\Delta\theta))\phi(\langle(x_m,y_l),\omega(k\,\Delta\theta)\rangle - jd). \tag{11.16}$$

Under the constraints mentioned previously, we try to choose the function ϕ to optimize some aspect of this reconstruction. As always, there are trade-offs among efficiency, resolution, and noise reduction. The function ϕ is often regarded as a *parameter* that can be adjusted to achieve certain aims. In the imaging literature, $\hat{\phi}$ is often expressed as a product

$$\hat{\phi}(r) = A(r)|r|.$$

Here A is a function that tends to zero as $|r| \to \infty$; it is called an *apodizing function.*

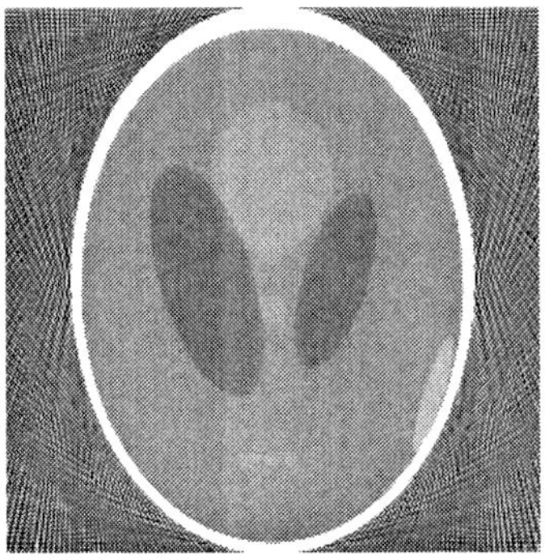

(a) Simple bandlimiting.

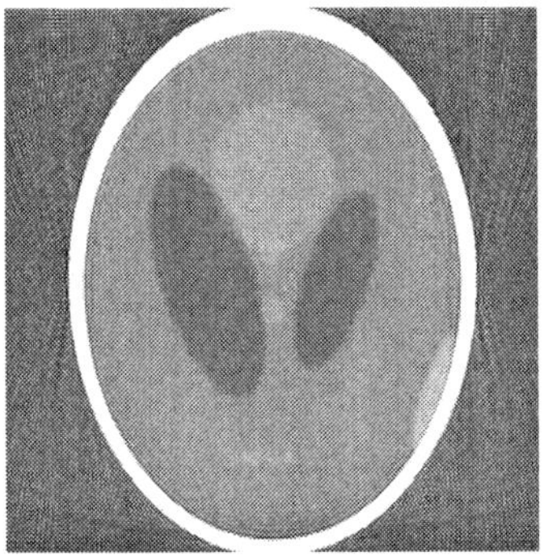

(b) The Hamming filter.

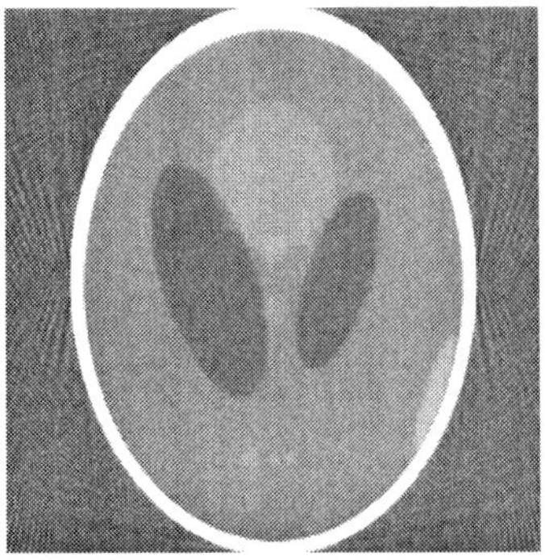

(c) The Hanning filter.

Figure 11.9. Filtered back-projection reconstructions of the Shepp-Logan phantom using different apodizing functions.

***Example* 11.4.1.** In this example we show the results of using different apodizing function in the filtered back-projection algorithm. Figure 11.9 shows the results of applying the filtered back-projection to the Shepp-Logan phantom using three standard filters, while Figure 11.10 shows the differences between these images. In absolute terms the differences are very small, and the gray levels in the difference images have been rescaled to show the fine detail in these images.

In Figure 11.9(a), the apodizing function is $\chi_{[-B,B]}(r)$; this is called simple bandlimiting. In Figure 11.9(b), $A(r) = (.3 + .7\cos\left(\frac{\pi r}{B}\right))\chi_{[-B,B]}(r)$, the "Hamming filter," and in Figure 11.9(c), $A(r) = \cos^2(\frac{\pi r}{2B})\chi_{[-B,B]}(r)$, the "Hanning filter." Notice that the bandlimited image is sharper than either the Hamming or Hanning image. All three images have very pronounced oscillatory artifacts in the exterior of the largest ellipse. From the difference image it is clear that each filter produces different artifacts. The Hamming image also shows a very pronounced Gibbs artifact in the *interior* of the large ellipse that is not present in the other images.

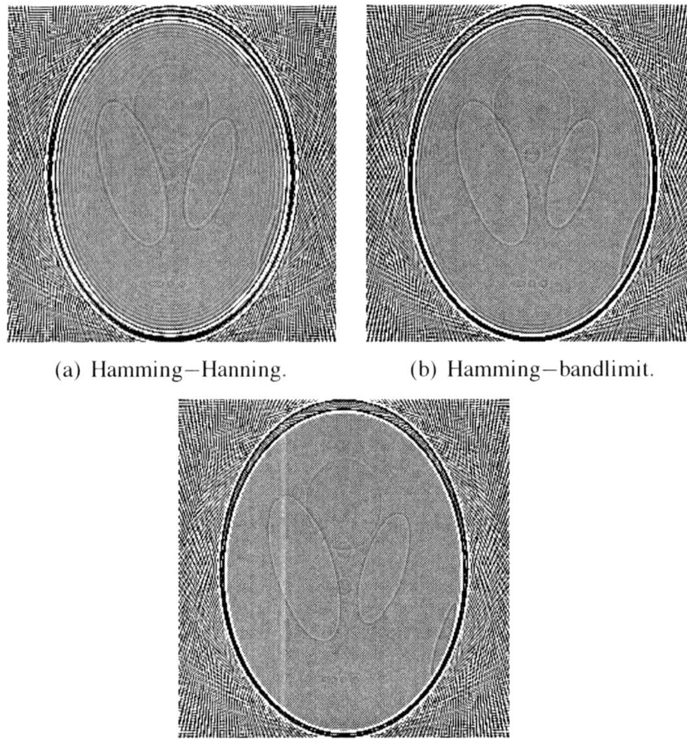

(a) Hamming–Hanning. (b) Hamming–bandlimit.

(c) Hanning–bandlimit.

Figure 11.10. The differences of the images in Figure 11.9 rescaled to show details.

Remark* 11.4.2 *(* Important notational remark*). In the sequel, the notation f_ϕ refers to an approximate reconstruction of f using "complete data" like that in (11.15), whereas $\tilde{f}_\phi$ refers to a discretization of this integral, as in (11.16).

This family of reconstruction formulæ has an important feature that follows from the fact that the quantity

$$\langle (x, y), \boldsymbol{\omega}(k\,\Delta\theta)\rangle - jd$$

is the signed distance from the line $l_{jd,\boldsymbol{\omega}(k\,\Delta\theta)}$ to the point (x, y) The algorithm defined by ϕ can be expressed as

$$\tilde{f}_\phi(x, y) = \frac{d}{2(M+1)} \sum_{k=0}^{M} \sum_{j=-N}^{N} \mathcal{R}f(jd, \boldsymbol{\omega}(k\,\Delta\theta))\phi(\text{dist}[(x, y), l_{jd,\boldsymbol{\omega}(k\,\Delta\theta)}]).$$

This is also a feature of the exact reconstruction formula. We now consider the details of implementing (11.16) on parallel beam data.

Exercise

Exercise **11.4.3.** Show that

$$\mathcal{G}\mathcal{R}f(-t, -\omega) = \mathcal{G}\mathcal{R}f(t, \omega). \tag{11.17}$$

11.4.3 Linearly Interpolated Filters

From formula (11.16) it would appear that the computation of $\tilde{f}_\phi(x_m, y_l)$, for a point $(x_m, y_l) \in \mathcal{R}_\tau$, requires a knowledge of the values

$$\begin{aligned}\{\phi(\langle (x_m, y_l), \omega(k\,\Delta\theta)\rangle - jd) \text{ for}\\ -N \le j \le N,\ -K \le l, m \le K \text{ and } 0 \le k \le M\}.\end{aligned} \tag{11.18}$$

The discrete filtering operation could still be done as soon as all of the data from a view, $\{\mathcal{R}f(jd, \omega(k\,\Delta\theta)) : -N \le j \le N\}$, are collected. But from (11.18), it would appear that a *different* filter function is required for each point in $\mathcal{R}_\tau$. In other words, the filter step would have to be repeated $O(K^2)$ times for each view! Ramachandran and Lakshminarayanan found a computational shortcut so that the filter operation would only have to be done once for each view. Their idea was to fix values of ϕ at the sample points, $\{\phi(jd) : -N \le j \le N\}$ and then *linearly* interpolate ϕ to the intermediate values.

We first consider what this means and why it reduces the computational load in the implementation of the filtered back-projection algorithm so dramatically. The function ϕ is defined to be linear between the sample points hence for any $\alpha \in [0, 1]$,

$$\phi(\alpha(k+1)d + (1-\alpha)(kd)) \stackrel{d}{=} \alpha\phi((k+1)d) + (1-\alpha)\phi(kd).$$

Since

$$\alpha(k+1)d + (1-\alpha)(kd) - jd = \alpha(k+1-j)d + (1-\alpha)(k-j)d,$$

we have the relations

$$\phi(\alpha(k+1)d + (1-\alpha)(kd) - jd) = \alpha\phi((k+1-j)d) + (1-\alpha)\phi((k-j)d). \tag{11.19}$$

For each $(x_m, y_l) \in \mathcal{R}_\tau$ and direction $\omega(k\,\Delta\theta)$ there is an integer $n_{klm} \in [-N, N]$ such that

$$n_{klm}d < \langle (x_m, y_l), \omega(k\,\Delta\theta)\rangle \le (n_{klm}+1)d.$$

Thus there is a number $\alpha_{klm} \in [0, 1]$ so that

$$\langle (x_m, y_l), \omega(k\,\Delta\theta)\rangle = \alpha_{klm}(n_{klm}+1)d + (1-\alpha_{klm})n_{klm}d.$$

The *trivial* but *crucial* observation is that (11.19) implies that, for any integer j,

$$\langle (x_m, y_l), \omega(k\,\Delta\theta)\rangle - jd = \alpha_{klm}[(n_{klm}+1)d - jd] + (1-\alpha_{klm})[n_{klm}d - jd]. \tag{11.20}$$

If ϕ is linearly interpolated between sample points, then

$$\begin{aligned} Q_\phi \tilde{f}(\langle (x_m, y_l), \boldsymbol{\omega}(k\,\Delta\theta)\rangle, \boldsymbol{\omega}(k\,\Delta\theta)) &= \alpha_{klm} Q_\phi \tilde{f}((n_{klm}+1)d, \boldsymbol{\omega}(k\,\Delta\theta)) \\ &\quad + (1-\alpha_{klm}) Q_\phi \tilde{f}(n_{klm}d, \boldsymbol{\omega}(k\,\Delta\theta)). \end{aligned} \tag{11.21}$$

In other words, $Q_\phi \tilde{f}(\langle (x_m, y_l), \boldsymbol{\omega}(k\,\Delta\theta)\rangle, \boldsymbol{\omega}(k\,\Delta\theta))$ is a weighted average of $Q_\phi \tilde{f}$ at *a pair of sample points*. The interpolation can even be done as part of the back-projection step:

$$\tilde{f}_\phi(x_m, y_l) = \frac{1}{2(M+1)} \sum_{k=0}^{M} \left[\alpha_{klm} Q_\phi \tilde{f}((n_{klm}+1)d, \boldsymbol{\omega}(k\,\Delta\theta)) + (1-\alpha_{klm}) Q_\phi \tilde{f}(n_{klm}d, \boldsymbol{\omega}(k\,\Delta\theta))\right]. \tag{11.22}$$

As a practical matter, the sampling angles and reconstruction grid are essentially fixed and therefore the coefficients, $\{\alpha_{klm}, n_{klm}\}$ can be evaluated once and stored in tables.

The filtered back-projection algorithm, using a linearly interpolated filter, is the following sequence of steps:

STEP 1: For each view (fixed k), approximately compute the filtered Radon transform, $Q_\phi \tilde{f}(jd, \boldsymbol{\omega}(k\,\Delta\theta))$ at sample points. The Riemann sum approximation is

$$Q_\phi \tilde{f}(jd, \boldsymbol{\omega}(k\,\Delta\theta)) = d \sum_{n=-N}^{N} \mathcal{R}f(nd, \boldsymbol{\omega}(k\,\Delta\theta))\phi((j-n)d).$$

STEP 2: Back-project using (11.22), with linearly interpolated values for

$$Q_\phi \tilde{f}(\langle (x_m, y_l), \boldsymbol{\omega}(k\,\Delta\theta)\rangle, \boldsymbol{\omega}(k\,\Delta\theta)),$$

to determine the values of $\tilde{f}_\phi(x_m, y_l)$ for $(x_m, y_l) \in \mathcal{R}_\tau$.

Step 1 can be done a view at a time; it requires a knowledge of $\phi(jd)$ for $-(2N-1) \le j \le (2N-1)$. The calculation of $Q_\phi \tilde{f}(jd, \boldsymbol{\omega})$ is a discrete convolution that is usually computed using the FFT.

As noted previously, the coefficients used in the back-projection step can be computed in advance and this step can also be parallelized. The interpolation used in the filtered back-projection is interpolation of a slowly varying, real-valued function and is much less delicate than that needed to do direct Fourier reconstruction. Empirically it leads to an overall blurring of the image but does not introduce complicated oscillatory artifacts or noise. Approximately MK^2 calculations are required once the values $\{Q_\phi \tilde{f}(jd, \boldsymbol{\omega}(k\,\Delta\theta))\}$ are computed. Using an FFT, the computation of Q_ϕ requires about $MN \log_2 N$-steps and using a direct convolution about MN^2-steps. The FFT is clearly faster, but the back-projection step already requires a comparable number of calculations to that needed for a direct convolution.

***Remark* 11.4.3.** In their original work, Ramachandran and Lakshminarayanan used the apodizing function defined by simple bandlimiting so that

$$\phi(jd) = \frac{1}{2\pi} \int_{-B}^{B} |r| e^{irjd} dr.$$

In medical applications this is usually called the Ram-Lak filter.

Exercises

***Exercise* 11.4.4.** How should formula (11.16) be modified if sampling is done around the full circle (i.e., θ varies between 0 and 2π)?

***Exercise* 11.4.5.** Compute $\{\phi(jd)\}$ for the Ram-Lak filter with $B = \frac{\pi}{d}$.

11.4.4 The Shepp-Logan Analysis of the Ram-Lak Filters

See: B.5.

Ramachandran and Lakshminarayanan introduced the linear interpolation method described in the previous section as a way to reduce the computational burden of the filtered back-projection algorithm. Initially it was assumed that using linear interpolation to define ϕ would result in a significant loss in accuracy. However, that turned out not to be the case. Shepp and Logan explained the "surprisingly" high quality of the Ram-Lak reconstructions by analyzing the filter defined by

$$\phi(0) = \frac{4}{\pi d^2}, \quad \phi(kd) = \frac{-4}{\pi d^2(4k^2 - 1)}, \tag{11.23}$$

with ϕ linearly interpolated between sample points. This function has a tall narrow peak at zero and a long *negative* tail. Figure 11.11 shows ϕ with $d = .28$ along with the impulse response for an approximation to $\mathcal{G}$ obtained in Section 6.4.

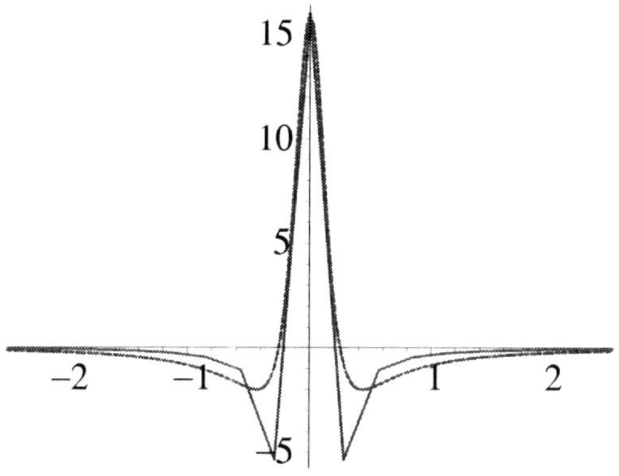

Figure 11.11. The piecewise linear impulse response for the Shepp-Logan filter and a continuous approximation.

The transfer function of the Shepp-Logan filter is given by

$$\hat{\phi}(r) = |r||\operatorname{sinc}\left(\frac{rd}{2}\right)|^3. \tag{11.24}$$

Formula (11.24) is derived in an appendix to this section. From the Taylor series of $\operatorname{sinc} x$,

$$\operatorname{sinc} x = 1 - \frac{x^2}{3!} + \frac{x^4}{5!} + \cdots,$$

we obtain that

$$\begin{aligned}\hat{\phi}(r) &= |r|\left[1 + O(r^2d^2)\right]^3 \\ &= |r|[1 + O(d^2r^2)].\end{aligned}$$

By choosing d small enough, $\hat{\phi}$ can be made close to $|r|$ over any *fixed* interval. Hence the Shepp-Logan filter function is even and resembles $|r|$ for "small" values of r. Here small is in comparison to effective bandwidth of the computation, which equals $\frac{\pi}{d}$. The relative error at $\frac{\pi}{d}$,

$$\frac{|\hat{\phi}(\frac{\pi}{d}) - \frac{\pi}{d}|}{\frac{\pi}{d}} = \left[\frac{2}{\pi}\right]^3 \approx \frac{1}{4},$$

is also not excessive. The transfer function does not have bounded support but decays like $|r|^{-2}$. Nonetheless, $\hat{\phi}$ has the general features described previously for a good filter function.

Earlier approaches to image reconstruction in x-ray CT used $\chi_{[0,\frac{\pi}{d}]}(|r|)$ as the apodizing function and did not use linear interpolation. In this case the point spread function is given by

$$\frac{\pi}{d}\frac{\sin(\frac{\pi t}{d})}{\pi t} - \frac{1 - \cos(\frac{\pi t}{d})}{\pi t^2}.$$

Note that this function decays like t^{-1} as $t \to \infty$. This is because the jump discontinuities in the transfer function at $\pm\frac{\pi}{d}$ determine the asymptotics of the inverse Fourier transform as t tends to infinity.

Formally, the impulse response of the exact filter is a constant times ∂_t P.V. t^{-1}, which decays like t^{-2} at infinity. Another desirable consequence of linear interpolation is that ϕ decays at the same rate. If $\hat{\psi}$, the Fourier transform of a function, behaves like $|r|$ near zero and is otherwise smooth, with absolutely integrable derivatives, then ψ decays at infinity like t^{-2}. The inverse Fourier transform, ψ, is computed by twice integrating by parts:

$$\begin{aligned}
\psi(t) &= \frac{1}{2\pi}\left[\int_{-\infty}^{0} + \int_{0}^{\infty}\right]\hat{\psi}(r)e^{itr}\,dr \\
&= \frac{e^{itr}}{2\pi i t}\hat{\psi}(r)\Big|_{-\infty}^{0} + \frac{e^{itr}}{2\pi i t}\hat{\psi}(r)\Big|_{0}^{\infty} - \frac{1}{2\pi}\left[\int_{-\infty}^{0} + \int_{0}^{\infty}\right]\frac{e^{itr}}{it}\hat{\psi}'(r)\,dr \\
&= \frac{e^{itr}}{2\pi t^2}\hat{\psi}'(r)\Big|_{-\infty}^{0} + \frac{e^{itr}}{2\pi t^2}\hat{\psi}'(r)\Big|_{0}^{\infty} - \frac{1}{2\pi}\left[\int_{-\infty}^{0} + \int_{0}^{\infty}\right]\frac{e^{itr}}{t^2}\hat{\psi}''(r)\,dr \\
&= \frac{\hat{\psi}'(0^-) - \hat{\psi}(0^+)}{2\pi t^2} + o\left(\frac{1}{t^2}\right) = \frac{-1}{\pi t^2} + o\left(\frac{1}{t^2}\right).
\end{aligned}$$

The last line is a consequence of the fact that $\hat{\psi}'(0^-) = -1$ and $\hat{\psi}'(0^+) = 1$.

The transfer function in the Shepp-Logan example is not smooth away from 0 but does have two weak derivatives. This suffices for the argument given previously to apply, and therefore, as t tends to infinity,

$$\phi(t) = \frac{-1}{\pi t^2} + o\left(\frac{1}{t^2}\right).$$

So not only does the transfer function of the Shepp-Logan filter resemble $|r|$ near $r = 0$, but the point spread function also resembles that of $-i\partial_t\mathcal{H}$ near infinity. This explains, in part, why linear interpolation leads to filters that provide a good approximation to the exact filtration step in the filtered back-projection formula. The Shepp-Logan analysis applies to a large class of filters defined by selecting values for ϕ at the sample points and linearly interpolating in between. We close this section with an example showing the output of the Shepp-Logan filter applied to the Radon transform of the characteristic function of a disk.

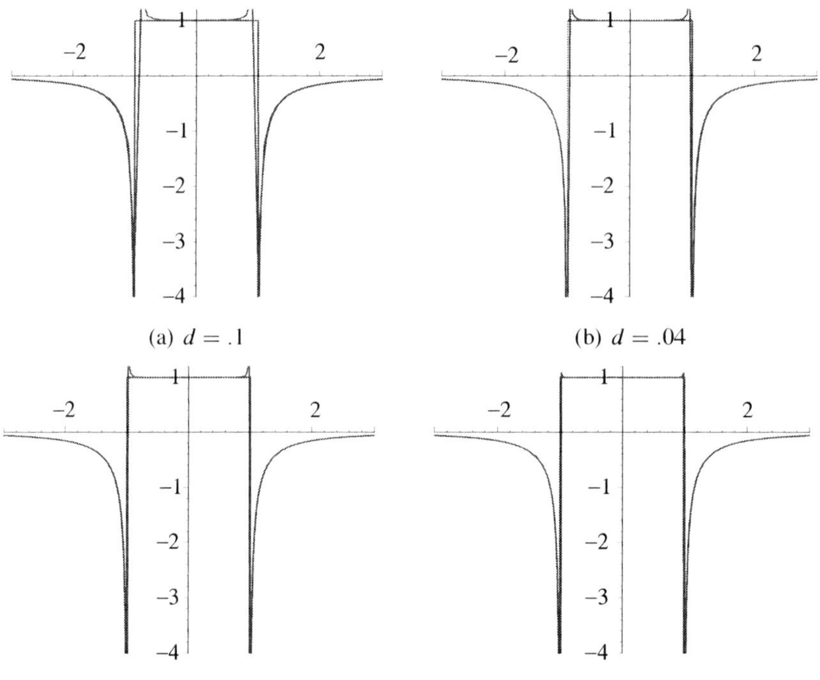

(a) $d = .1$ (b) $d = .04$

(c) $d = .02$ (d) $d = .005$

Figure 11.12. Shepp-Logan filters applied to $\mathcal{R}f$.

***Example* 11.4.2.** Let f be the characteristic function of the disk of radius 1 centered on $(0, 0)$. Its Radon transform is

$$\mathcal{R}f(t, \boldsymbol{\omega}) = 2\sqrt{1 - t^2}\chi_{[-1,1]}(t).$$

Graphs of $Q_\phi f$ are shown in Figure 11.12 with $d = .1, .04, .02, .005$. For comparison, the exact filtered function, $-i\mathcal{H}\partial_t\mathcal{R}f$, computed in (6.33) is also shown. Apart from small intervals around $t = \pm 1$, the two functions are almost indistinguishable.

Exercises

***Exercise* 11.4.6.** Show that $\hat{\phi}$ has a weak second derivative away from $r = 0$.

***Exercise* 11.4.7.** Give an argument showing that $\phi(t) = O(t^{-2})$ as t tend to infinity that does not use the Fourier transform. *Hint:* Use (11.23).

Appendix: The Transfer Function for Linearly Interpolated Filters*.

How is the Fourier transform of the interpolated filter function found? This is a nice application of the theory of generalized functions (see Sections A.4.5 and 4.4.4). We follow the derivation in [74]. Let the sample spacing be d, and suppose that the samples $\phi_n = \phi(nd)$ are specified. The function ϕ is defined at intermediate points by interpolation. To compute $\hat{\phi}$, ϕ is written as a convolution, allowing for different interpolation schemes.

Let P denote the generalized function

$$P(t) = d \sum_{n=-\infty}^{\infty} \phi_n \delta(t - nd)$$

and I a function with support in the interval $[-d, d]$ and total integral 1. The filter function is *defined* at intermediate points by setting

$$\phi(t) = I * P(t).$$

If

$$I(t) = \frac{1}{d}\left(1 - \frac{|t|}{d}\right) \chi_{[0,d]}(|t|),$$

then ϕ is a linearly interpolated filter, while

$$I(t) = \frac{1}{d}\chi_{[-\frac{d}{2},\frac{d}{2}]}(|t|)$$

produces a piecewise constant interpolant. The Fourier transform of P is the $\left(\frac{2\pi}{d}\right)$-periodic function

$$\hat{P}(r) = \sum_{n=-\infty}^{\infty} d\phi_n e^{-indr}$$

(see Exercise 8.2.3).

Using the convolution theorem for the Fourier transform gives

$$\hat{\phi}(r) = \hat{P}(r)\hat{I}(r).$$

For a linearly interpolated filter,

$$\hat{I}(r) = \operatorname{sinc}^2\left[\frac{rd}{2}\right].$$

Piecewise constant interpolation gives

$$\hat{I}(r) = \operatorname{sinc}\left[\frac{rd}{2}\right].$$

Linear interpolation gives a smoother point spread function that is reflected in the faster rate of decay of the corresponding transfer function. In the Shepp-Logan example, the transfer function can be rewritten as

$$\hat{\phi} = \frac{2}{d}|\sin(\frac{rd}{2})| \cdot \operatorname{sinc}^2\left[\frac{rd}{2}\right]$$

with the sinc^2-factor the result of using linear interpolation.

Exercises

***Exercise* 11.4.8.** For the Shepp-Logan example, (11.23), give the details of the computation of $\hat{P}(r)$.

***Exercise* 11.4.9.** Suppose that a filter function is defined at sample points by

$$\phi_n = \frac{1}{2\pi}\int_{-\frac{\pi}{d}}^{\frac{\pi}{d}} |r|e^{irnd}\,dr.$$

What is $\hat{P}(r)$? Does the corresponding point spread function decay like $O(t^{-2})$ as t tends to infinity?

***Exercise* 11.4.10.** Find conditions that the sequence of values $< \phi_n >$ must satisfy in order for the argument in the previous section to apply to show that $\hat{\phi}(r)$ behaves like $|r|$ for small r and $\phi(t)$ behaves like t^{-2} for large t.

11.4.5 Sample Spacing in a Parallel Beam Machine

In a parallel beam scanner with filtered back-projection as the reconstruction algorithm, there are three different sample spacings. We discuss how these parameters are related; our discussion is adapted from [76]. Assume that the attenuation coefficient is supported in the disk of radius L and that the measurements are samples of $\mathcal{R}f$. The finite width of the x-ray beam, which has been ignored thus far, has significant implications for this analysis. We return to this question in Section 12.3.

The sample spacings are as follows:

(1) The reconstruction grid sample spacing, τ: The reconstruction is usually done in a square, say $[-L, L] \times [-L, L]$, which is divided into a $(2K + 1) \times (2K + 1)$ grid, $\tau = LK^{-1}$.

(2) Spacing between consecutive projection angles: $\Delta\theta = \frac{\pi}{M+1}$.

(3) The sample spacing in the affine parameter is $d = LN^{-1}$. There are $2N + 1$ samples of each projection.

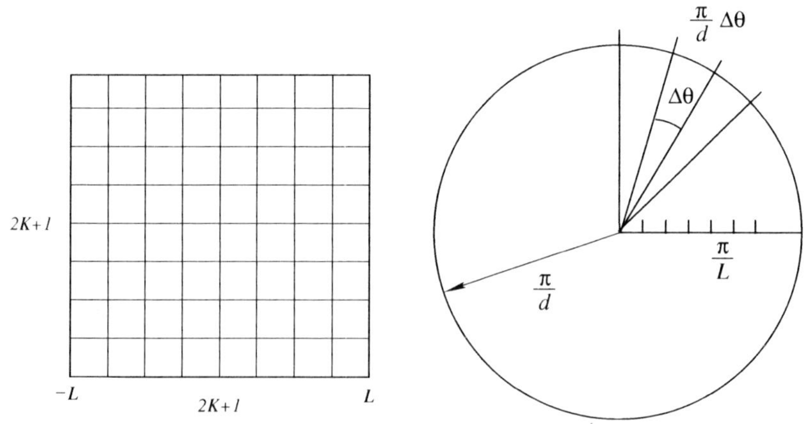

(a) The reconstruction grid. (b) Sampling in frequency space.

Figure 11.13. How to choose sample spacings.

A priori each of these numbers could be chosen independently; in practice once one is fixed, it is possible to determine reasonable values for the others. The data are effectively bandlimited, and the sample spacings are chosen to reflect the essential bandwidth of the data. The sample spacing in the affine parameter is $d = \frac{L}{N}$. From Nyquist's theorem, it follows that, in order to avoid aliasing in this measurement, the effective bandwidth of the object (or at least its Radon transform) should be $\frac{2\pi}{d}$. This implies that $\hat{f}(r\omega)$ should be essentially supported in the disk of radius of $\frac{\pi}{d}$. Along each radial line, $l_{0,\omega(k\,\Delta\theta)}$, the Fourier transform, $\hat{f}$, is sampled at N points and therefore the sample spacing in this direction is about $\frac{\pi}{L}$. The widest sample spacing of $\hat{f}$ in the angular direction, at least within the effective support of $\hat{f}$ is therefore $\Delta\theta\frac{\pi}{d}$ [see Figure 11.13(b)]. A reasonable criterion is to choose parameters so that the worst angular resolution, in the Fourier domain, is equal to the resolution in the radial direction. This means that

$$\Delta\theta\frac{\pi}{d} = \frac{\pi}{L}, \quad \text{implying that } M + 1 = \frac{\pi}{2}(2N + 1).$$

We are sampling a function supported in the square $[-L, L]\times[-L, L]$ and then using a filtered back-projection algorithm to reconstruct the image. To use filtered back-projection, we must zero pad the measured data. As a periodic function, the (zero padded) data are defined by its values in the square $[-2L, 2L] \times [-2L, 2L]$ and then extended periodically. Taking account of the zero padding, we have essentially $4N \times 4N$ samples of the Fourier transform of this function, and therefore we should use an equal number of grid points in the square with side length $4L$. This implies that we should take $K \simeq N$, so that $\tau \approx d$. With $K \approx N$ we are not sacrificing any resolution that is in our measurements, nor are we interpolating by using partial sums of the Fourier series. A slightly different way to view this question is to choose the sample spacing in the reconstruction grid so that the number of measurements equals the number of nontrivial reconstruction points. There are approximately πK^2 grid points in the circle of radius L. About $2MN$ samples of $\mathscr{R}f$ are collected. Using $M \approx \pi N$, from before, implies that K should be approximately $\sqrt{2}N$, so that $\tau \approx \frac{d}{\sqrt{2}}$.

A careful analysis of this question requires a knowledge of the "full" point spread function, which incorporates the measurement process, sampling, and reconstruction algorithm. This is complicated by the observation that the map from the "input" f to the output $\{\tilde{f}(x_m, y_l)\}$ is not, in any reasonable sense, a shift invariant filter and so it does not have a point spread function, per se. We return to this question after we have considered various physical limitations of the measurement process. This problem is sufficiently complicated that a final determination for the various parameters must be done empirically. A circa 2002 commercial fan beam scanner uses 1500 to 2500 views per slice with about 1500 rays/view. This is in reasonably good agreement with the previous computation. The reconstruction is done on a 512×512 or 1024×1024 grid.

Exercise

***Exercise* 11.4.11.** Explain why the effective bandwidth of a function f and the effective bandwidth of its Radon transform $\mathcal{R}f(t, \omega)$, in the t variable, are the same. By effective bandwidth we mean the effective bandwidth of the data itself, not the effective bandwidth of the subsequent computations implied by the sampling rate.

11.5 Filtered Back-Projection in the Fan Beam Case

We now consider reconstruction algorithms for a fan beam scanner. A view for a parallel beam scanner consists of measurements of $\mathcal{R}f$ for a family of parallel lines, and therefore the central slice theorem applies to give an approximation to the Fourier transform of the attenuation coefficient along lines passing through the origin. A view, for either type of fan beam scanner, consists of samples of $\mathcal{R}f$ for a family of lines passing through a point, and so the central slice theorem is not directly applicable. There are two general approaches to reconstructing images from the data collected by a fan beam machine: (1) Re-sort and interpolate to obtain the data needed to apply the parallel beam algorithms discussed earlier, (2) work directly with the fan beam geometry and find algorithms well adapted to it. Herman, Lakshminarayanan, and Naparstek first proposed an algorithm of the second type in [55]. Our derivation of this algorithm closely follows the presentation in [76]. In this section, ξ is used for the Fourier variable so as to avoid confusion with the Euclidean radius function that is denoted by r. This section may be omitted without any loss in continuity, as we use the parallel beam formula in the analyses of imaging artifacts presented in the next chapter.

11.5.1 Fan Beam Geometry

It is convenient to use a different parameterization for the lines in the plane from that used earlier. As before, (t, θ) denotes parameters for the line with oriented normal $\omega(\theta)$ at distance t from the origin. To simplify the notation, we use $\mathcal{R}f(t, \theta)$ to denote $\mathcal{R}f(t, \omega(\theta))$. Consider the geometry shown in Figure 11.14. Here S denotes the intersection point of the lines defining a view. It lies a distance D from the origin. The central ray [through S and $(0, 0)$] makes an angle β with the positive y-axis. The other lines through S are parameterized by the angle, γ, they make with the central ray. These parameters define coordinates on a subset of the space of oriented lines. They are related to the (t, θ) variables by

$$\theta = \gamma + \beta \text{ and } t = D \sin \gamma ; \tag{11.25}$$

we call (β, γ) *fan beam coordinates*.

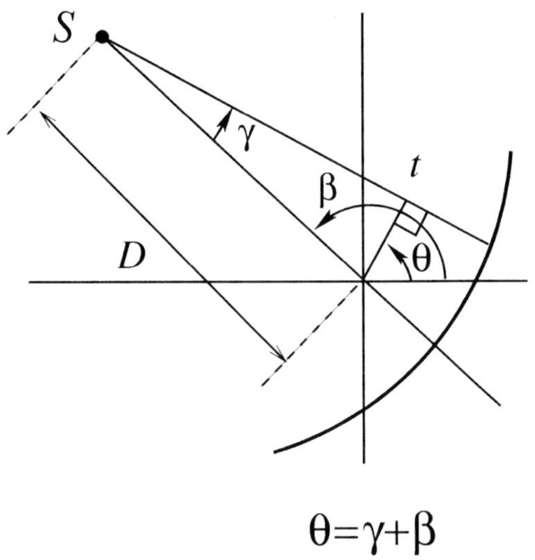

Figure 11.14. Fan beam geometry.

We now derive the continuous form of the approximate reconstruction formula used in fan beam algorithms. This is the analogue of the formula,

$$f_\phi(x, y) = \frac{1}{2\pi} \int_0^\pi \int_{-L}^{L} \mathcal{R}f(t, \theta)\phi(x\cos\theta + y\sin\theta - t)\, dt\, d\theta, \tag{11.26}$$

with ϕ the filter function, used for a parallel beam scanner. In (11.26) the weighting of different lines in the filtering operation depends only on their distance from the point (x, y). This general form is retained for the moment, though by the end of the calculation we end up with approximations that do not satisfy this condition. It is convenient to use polar coordinates, (r, φ) in the reconstruction plane,

$$(x, y) = (r\cos\varphi, r\sin\varphi).$$

So as not to confuse it with the polar angle of the parallel beam filter function, we let κ denote the filter function. This function is assumed to be smooth and decaying at infinity. In polar coordinates (11.26) becomes

$$f_\kappa(r, \varphi) = \frac{1}{2} \int_0^{2\pi} \int_{-L}^{L} \mathcal{R}f(t, \theta)\kappa(r\cos(\theta - \varphi) - t)\, dt\, d\theta. \tag{11.27}$$

Our goal is to reexpress the reconstruction formula in fan beam coordinates, as a filtered back-projection. Because of the geometry of the fan beam coordinates, this is not quite possible. Instead the final formula is a *weighted*, filtered back-projection.

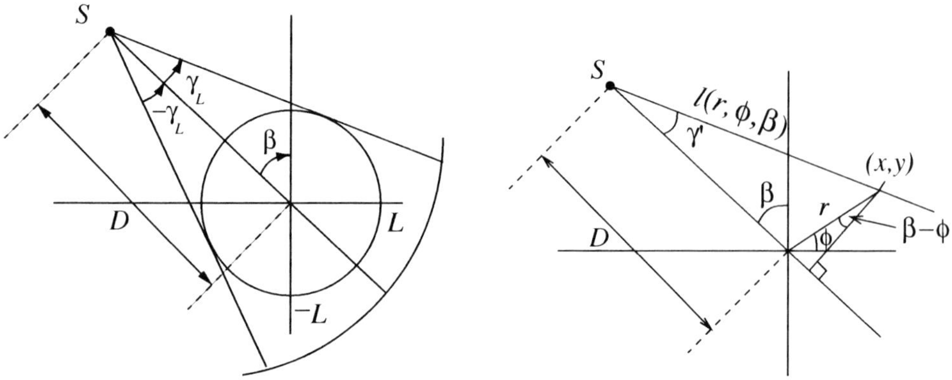

(a) Physical parameters in a fan beam scanner. (b) Variables for the reconstruction formula.

Figure 11.15. Quantities used in the fan beam, filtered back-projection algorithm.

Using the relations in (11.25) to reexpress this integral in fan beam coordinates gives

$$f_\kappa(r,\varphi) = \frac{1}{2}\int_0^{2\pi}\int_{-\gamma_L}^{\gamma_L} \mathscr{R}f(D\sin\gamma,\beta+\gamma)\kappa(r\cos(\beta+\gamma-\varphi)-D\sin\gamma)D\cos\gamma\,d\gamma\,d\beta.$$

The function f is supported in the disk of radius L. The limits of integration in the γ-integral, $\pm\gamma_L$, are chosen so that the lines corresponding to the parameters

$$\{(\beta,\gamma):\ \beta\in[0,2\pi),\quad -\gamma_L\le\gamma\le\gamma_L\}$$

include all those intersecting D_L. The total angle $2\gamma_L$ is called the *fan angle*. The data actually collected with a fan beam machine are an approximation to uniformly spaced samples in the (β,γ)-coordinates. To simplify the notation we introduce

$$Pf(\beta,\gamma)\overset{d}{=}\mathscr{R}f(D\sin\gamma,\beta+\gamma).$$

In terms of these data, the formula reads

$$f_\kappa(r,\varphi) = \frac{1}{2}\int_0^{2\pi}\int_{-\gamma_L}^{\gamma_L} Pf(\beta,\gamma)\kappa(r\cos(\beta+\gamma-\varphi)-D\sin\gamma)D\cos\gamma\,d\gamma\,d\beta.$$

From the trigonometric identity

$$\cos(a+b)=\cos(a)\cos(b)-\sin(a)\sin(b)$$

it follows that

$$r\cos(\beta+\gamma-\varphi)-D\sin\gamma = r\cos(\beta-\varphi)\cos\gamma-[r\sin(\beta-\varphi)+D]\sin\gamma. \tag{11.28}$$

Let $l(r, \varphi, \beta)$ be the distance from S to the point (x, y) and let γ' be the angle between the ray $\overline{SO}$ and the ray $\overline{S(x, y)}$. As S is outside the reconstruction grid, the function $l(r, \varphi, \beta)$ is strictly positive for points of interest, with

$$l \cos \gamma' = D + r \sin(\beta - \varphi), \tag{11.29}$$

$$l \sin \gamma' = r \cos(\beta - \varphi). \tag{11.30}$$

As functions of (r, φ, β), the distance, l and the angle γ' are given by

$$\begin{aligned} l(r, \varphi, \beta) &= \sqrt{[D + r \sin(\beta - \varphi)]^2 + [r \cos(\beta - \varphi)]^2}, \\ \gamma'(r, \varphi, \beta) &= \tan^{-1} \frac{r \cos(\beta - \varphi)}{D + r \sin(\beta - \varphi)}. \end{aligned}$$

Using (11.28) and (11.30), the argument of κ can be rewritten

$$r \cos(\beta + \gamma - \varphi) - D \sin \gamma = l(r, \varphi, \beta) \sin(\gamma'(r, \varphi, \beta) - \gamma). \tag{11.31}$$

With these substitutions, the expression for f_κ becomes

$$f_\kappa(r, \varphi) = \frac{1}{2} \int_0^{2\pi} \int_{-\gamma_L}^{\gamma_L} Pf(\beta, \gamma) \kappa(l(r, \varphi, \beta) \sin(\gamma'(r, \varphi, \beta) - \gamma)) D \cos \gamma \, d\gamma \, d\beta. \tag{11.32}$$

Using a slightly different functional form for the filter function κ leads to a much simpler formula. In the next section we explicitly incorporate the condition that $\hat{\kappa}(\xi) \approx |\xi|$ for small values of $|\xi|$.

Exercise

***Exercise* 11.5.1.** Describe the subset of the space of lines parameterized by (β, γ) in (11.25).

11.5.2 Fan Beam Filtered Back-Projection

Let χ_ϵ be a family of functions of bounded support so that

$$\lim_{\epsilon \to 0} \chi_\epsilon = 1 \qquad \text{for all } \xi.$$

Tracing backward to (11.27) shows that the exact reconstruction of a function supported in D_L is obtained as the limit

$$f(r, \varphi) = \lim_{\epsilon \to 0} \frac{1}{4\pi} \int_0^{2\pi} \int_{-\gamma_L}^{\gamma_L} Pf(\beta, \gamma) D \cos \gamma \left[\int_{-\infty}^{\infty} e^{il \sin(\gamma' - \gamma)\xi} |\xi| \chi_\epsilon(\xi) d\xi \right] d\gamma \, d\beta.$$

The β-integral is essentially a weighted back-projection and is innocuous. For the analysis of the γ and ξ integrals, we let $h(\gamma)$ be a bounded function with bounded support and set

$$H(\gamma') = \lim_{\epsilon \to 0} \frac{1}{4\pi} \int_{-\gamma_L}^{\gamma_L} h(\gamma) D \cos \gamma \left[\int_{-\infty}^{\infty} e^{il \sin(\gamma' - \gamma)\xi} |\xi| \chi_\epsilon(\xi) d\xi \right] d\gamma \, d\beta.$$

We change coordinates in the ξ-integral, letting

$$\eta = \left[\frac{l \sin(\gamma' - \gamma)}{\gamma' - \gamma}\right]\xi,$$

to obtain

$$H(\gamma') = \lim_{\epsilon \to 0} \frac{1}{4\pi} \int_{-\gamma_L}^{\gamma_L} \int_{-\infty}^{\infty} h(\gamma) \left[\frac{\gamma' - \gamma}{l \sin(\gamma' - \gamma)}\right]^2 |\eta| \chi_\epsilon \left(\eta \left[\frac{\gamma' - \gamma}{l \sin(\gamma' - \gamma)}\right]\right) e^{i(\gamma' - \gamma)\eta} \, d\eta \, d\gamma.$$

The function $h(\gamma)$ has bounded support, and therefore the order of integration can be interchanged. As an iterated integral,

$$H(\gamma') = \frac{1}{4\pi} \int_{-\infty}^{\infty} \int_{-\gamma_L}^{\gamma_L} h(\gamma) \left[\frac{\gamma' - \gamma}{l \sin(\gamma' - \gamma)}\right]^2 |\eta| e^{i(\gamma' - \gamma)\eta} d\gamma \, d\eta.$$

This is an exact formula for the filtering step expressed in fan beam coordinates. An approximation to this integral, different from (11.32), is given by

$$H_\epsilon(\gamma') = \frac{1}{2} \int_{-\gamma_L}^{\gamma_L} h(\gamma) \left[\frac{\gamma' - \gamma}{l \sin(\gamma' - \gamma)}\right]^2 \kappa_\epsilon(\gamma' - \gamma) \, d\gamma,$$

where

$$\kappa_\epsilon(\gamma) = \frac{1}{2\pi} \int_{-\infty}^{\infty} |\eta| \chi_\epsilon(\eta) e^{i\eta\gamma} \, d\eta.$$

From this calculation it follows that a reasonable approximation to $f(r, \varphi)$ is given by

$$f_\kappa(r, \varphi) = \frac{1}{2} \int_0^{2\pi} \int_{-\gamma_L}^{\gamma_L} Pf(\beta, \gamma) \left[\frac{\gamma' - \gamma}{l \sin(\gamma' - \gamma)}\right]^2 \kappa(\gamma' - \gamma) D \cos\gamma \, d\gamma \, d\beta,$$

where, as before, κ should be chosen so that $\hat{\kappa}(\xi) \simeq |\xi|$ over the effective bandwidth (in the γ-variable) of $Pf(\beta, \gamma)$. This formula can be rewritten as

$$f_g(r, \varphi) = \int_0^{2\pi} \frac{1}{l^2(r, \varphi, \beta)} \int_{-\gamma_L}^{\gamma_L} Pf(\beta, \gamma) g(\gamma' - \gamma) D \cos\gamma \, d\gamma \, d\beta,$$

where

$$g(\gamma) = \frac{1}{2} \left(\frac{\gamma}{\sin\gamma}\right)^2 \kappa(\gamma).$$

The weight factor, which comes from the geometry of the fan beam variables, is included in the definition of the filter function. To interpret this formula as a weighted back-projection, set

$$
\begin{aligned}
Q_g f(\beta, \gamma') &= \int P' f(\beta, \gamma - \gamma') g(\gamma)\, d\gamma \text{ and} \\
f_g(r, \varphi) &= \int_0^{2\pi} \frac{1}{l^2(r, \varphi, \beta)} Q_g f(\beta, \gamma')\, d\beta.
\end{aligned} \tag{11.33}
$$

Here $P' f(\beta, \gamma) = P f(\beta, \gamma) D \cos \gamma$ and

$$
\gamma'(r, \varphi, \beta) = \tan^{-1}\left[\frac{r \cos(\beta - \varphi)}{D + r \sin(\beta - \varphi)}\right].
$$

11.5.3 Implementing the Fan Beam Algorithm

Using (11.33), we can describe an algorithm for image reconstruction, well adapted to the geometry of a fan beam scanner. The fan beam data are

$$
P f(\beta_j, n\alpha), \text{ where } \beta_j = \frac{2\pi j}{M+1}, \quad j = 0, \dots, M
$$

and n takes integer values. The image is reconstructed in three steps.

STEP 1: Replace the measurements by weighted measurements; that is, multiply by the factor $D \cos n\alpha$ to obtain

$$
P' f(\beta_j, n\alpha) = P f(\beta_j, n\alpha) D \cos n\alpha
$$

STEP 2: Discretely convolve the weighted projection data $P' f(\beta_j, n\alpha)$ with $g(n\alpha)$ to generate the filtered projection at the sample points:

$$
\begin{aligned}
Q_g \tilde{f}(\beta_j, n\alpha) &= \alpha[P' f(\beta_j, \cdot) \star g](n\alpha), \\
\text{where } g(n\alpha) &= \frac{1}{2}\left(\frac{n\alpha}{\sin n\alpha}\right)^2 \kappa(n\alpha).
\end{aligned}
$$

The filter function κ is selected according to the criteria used in the selection of ϕ for the parallel beam case: It should be real, even, and decay at infinity. For ξ in the effective support of the data, $\hat{\kappa}(\xi) \approx |\xi|$.

STEP 3: Perform a weighted back-projection of each filtered projection:

$$
\tilde{f}_g(x_m, y_l) \approx \Delta\beta \sum_{k=0}^{M} \frac{1}{l^2(x_m, y_l, \beta_k)} Q_g f(\beta_k, \gamma'(x_m, y_l, \beta_k)).
$$

As before, the values $\{Q_g \tilde{f}(\beta_k, \gamma'(x_m, y_l, \beta_k))\}$ were *not* computed in the previous step. They are obtained by using interpolation from the values, $\{Q_g f(\beta_k, n\alpha)\}$, which were computed.

If the values of the functions

$$\{l(x_m, y_l, \beta_k), \gamma'(x_m, y_l, \beta_k)\}$$

as well as the interpolation coefficients are precomputed, then the computational load of this algorithm is the same order of magnitude as that of the parallel beam, linearly interpolated algorithm. Note that, as before, steps 1 and 2 can be performed as soon as the data from a given view have been collected and the back-projection step can also be parallelized. For a third-generation machine, a view is defined by the source position, so the filter step can begin almost immediately. For a fourth-generation machine, a view is defined by a detector position; hence the filtering step must wait until all the data for first view have been collected. Once this threshold is reached, the filtering can again be effectively parallelized.

***Example* 11.5.1.** Using the filter function defined by Shepp and Logan for $g(n\alpha)$ gives

$$g_{\mathrm{SL}}(n\alpha) = \begin{cases} \frac{2}{\pi\alpha^2} & \text{if } n = 0, \\ \frac{-2}{\pi \sin^2 n\alpha} \frac{n^2}{4n^2-1} & \text{if } \neq 0. \end{cases} \tag{11.34}$$

The graph of this function with $\alpha = .21$ is shown in Figure 11.16(a).

***Example* 11.5.2.** In their original paper, Herman, Lakshminarayanan, and Naparstek used a slightly different function defined by

$$g_{\mathrm{HLN}}(n\alpha) = \begin{cases} \frac{1}{8\alpha^2} & \text{if } n = 0 \text{ or even}, \\ -\frac{1}{2\pi^2 \sin^2(n\alpha)} & \text{if } n \text{ is odd}. \end{cases} \tag{11.35}$$

The graph of this function for $\alpha = .21$ is shown in Figure 11.16(b).

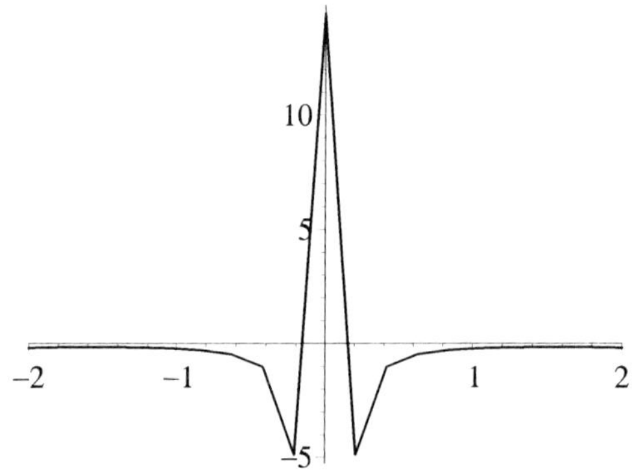

(a) The impulse response for a Shepp-Logan fan beam filter.

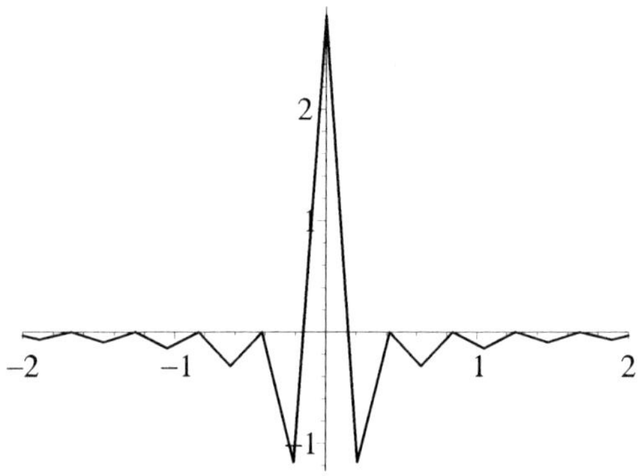

(b) The impulse response for the HLN fan beam filter.

Figure 11.16. The impulse responses for two fan beam filter functions.

11.5.4 Data Collection for a Fan Beam Scanner

In the preceding derivation of the algorithm, it is assumed that data are collected for $\beta \in [0, 2\pi)$. This means that every projection is measured twice, as two pairs of fan beam coordinates (β_1, γ_1) and (β_2, γ_2) define the same line if and only if

$$\begin{aligned} \gamma_1 &= -\gamma_2, \\ \beta_1 - \gamma_1 &= \beta_2 - \gamma_2 + \pi \Rightarrow \beta_1 = \beta_2 + 2\gamma_1 + \pi \end{aligned} \tag{11.36}$$

[see Figure 11.17(a)]. With a parallel beam machine it suffices to collect data for $\theta \in [0, \pi)$, similarly, for a fan beam machine we do not need to have measurements for all $\beta \in [0, 2\pi)$. However, some care is required in collecting and processing smaller data sets. The Radon transform satisfies

$$\mathcal{R}f(t, \theta) = \mathcal{R}f(-t, \pi + \theta).$$

In fan beam coordinates this is equivalent to

$$Pf(\beta, \gamma) = Pf(\beta + 2\gamma + \pi, -\gamma).$$

Sampling Pf on the range

$$\beta \in [0, \pi] \text{ and } -\gamma_L \leq \gamma \leq \gamma_L$$

and using $t = D \sin \gamma$ and $\theta = \beta + \gamma$ gives the diagram, in (t, θ) space, shown in figure 11.17(b). The numbers show how many times a given projection is measured for (β, γ) in this range. Some points are measured once, some twice, and some not at all.

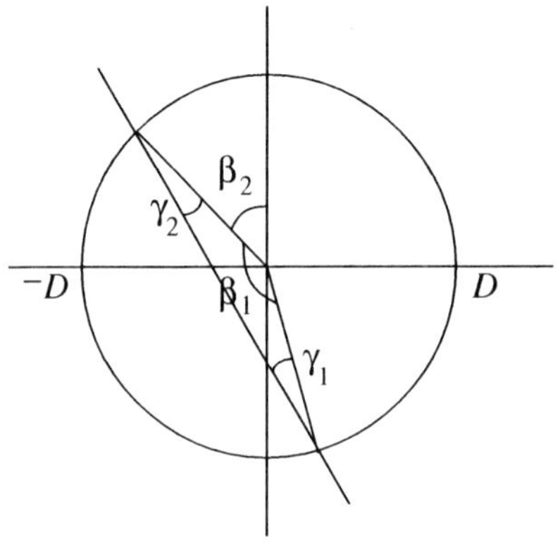

(a) Symmetry in (γ, β)-coordinates.

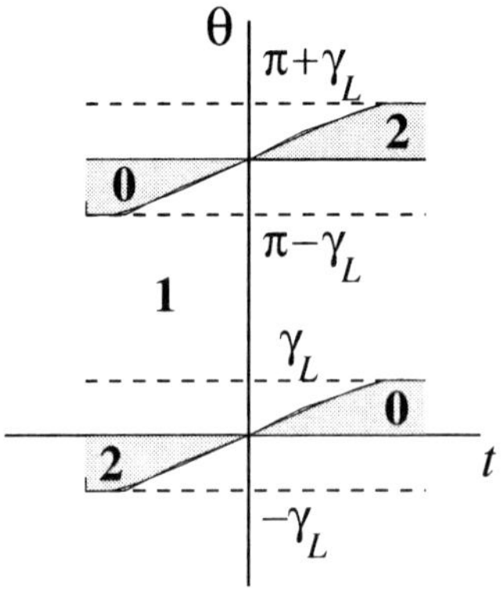

(b) Data collection in (t, θ)-coordinates.

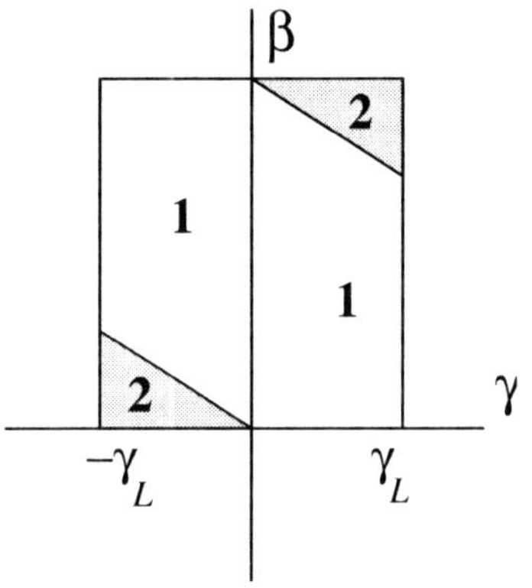

(c) Data collection in (γ, β)-coordinates.

Figure 11.17. Collecting data for fan beam scanners.

In order to gather a complete data set, it is necessary for β to go from 0 to $\pi + 2\gamma_L$. Of course, even more values are now sampled twice. The preceding algorithm can be used with measurements gathered over such a range; however, care must be taken to count each projection exactly once. This can be done by multiplying the projection data by a windowing function. For example, we could use

$$w_\beta(\gamma) = \begin{cases} 0 & 0 \leq \beta \leq 2\gamma_L + 2\gamma, \\ 1 & \text{otherwise.} \end{cases}$$

As usual, a sharp cutoff produces its own artifacts, so it is preferable to use a smoother window. The window should be continuous and satisfy the conditions

(a) $w_{\beta_1}(\gamma_1) + w_{\beta_2}(\gamma_2) = 1$, for pairs (β_1, γ_1), (β_2, γ_2) satisfying (11.36),

(b) $w_\beta(\gamma) \geq 0$.

Given the known regularity of the data, a window function with a bounded, though not necessarily continuous, first derivative usually suffices. An example is given by

$$w_\beta(\gamma) = \begin{cases} \sin^2\left[\frac{\pi\beta}{4(\gamma_L-\gamma)}\right] & 0 \le \beta \le 2\gamma_L - 2\gamma, \\ 1 & 2\gamma_L - 2\gamma \le \beta \le \pi - 2\gamma, \\ \sin^2\left[\frac{\pi}{4}\frac{\pi+2\gamma_L-\beta}{\gamma+\gamma_L}\right] & \pi - 2\gamma \le \beta \le \pi + 2\gamma_L. \end{cases}$$

11.5.5 Rebinning

It is also possible to re-sort the fan beam data into collections of approximately parallel projections, interpolate, and then use a parallel beam algorithm. One such method is called *rebinning*. It requires that $\Delta\beta = \alpha$, from which it follows that

$$Pf(m\alpha, n\alpha) = \mathcal{R}f(D\sin n\alpha, \boldsymbol{\omega}((n+m)\alpha)).$$

If $n + m = q$, then these are samples belonging to a single "parallel beam" view. Since $\sin(n+1)\alpha - \sin n\alpha$ depends on n, these samples are not equally spaced in the t-direction. Equally spaced samples of the parallel projections can be obtained by interpolating in this direction. The interpolated data set could then be processed using a standard parallel beam algorithm to reconstruct the image.

Another possibility is to select angles $< \gamma_{-P}, \dots, \gamma_P >$ so that

$$D(\sin(\gamma_j) - \sin(\gamma_{j-1})) = \Delta t, \text{ for } j = 1 - P, \dots, P.$$

The fan beam data $\{Pf(\beta_j, \gamma_k)\}$ are then equally spaced in the t-direction. Interpolating in the β-parameter gives data that again approximate the data collected by a parallel beam machine. Beyond the errors introduced by interpolations, these algorithms cannot be parallelized effectively. This is because all the data from a slice need to be collected before the interpolation and rebinning can begin.

11.6 Some Mathematical Remarks*

In the foregoing sections we consider algorithms for approximately reconstructing an unknown function f from finitely many samples of its Radon transform $\{\mathcal{R}f(t_j, \boldsymbol{\omega}_k)\}$. It is reasonable to enquire what is the "best" we can do in approximating f from such data and, if g is the "optimal solution," then what does $f - g$ "look like." A considerable amount of work has been done on these two questions. We briefly describe the results of Logan and Shepp; see [89] and [88].

Assume that f is an L^2-function that is supported in the disk of radius 1. It is assumed that the complete projection $\mathcal{R}f(t, \boldsymbol{\omega}_j)$ is known for n-distinct directions. Logan and Shepp examine the following problem: Find the function $g \in L^2(B_1)$ such that

$$\mathcal{R}f(t, \boldsymbol{\omega}_j) = \mathcal{R}g(t, \boldsymbol{\omega}_j), \quad j = 0, \dots, n-1, \tag{11.37}$$

which minimizes

$$\begin{aligned} \operatorname{Var}(g) &= \int_{B_1} (g(x,y) - \bar{g})^2 \, dx \, dy, \text{ where} \\ \bar{g} &= \frac{1}{\pi} \int_{B_1} g(x,y) \, dx \, dy. \end{aligned} \tag{11.38}$$

Briefly, find the function g with the specified projection data and minimum L^2-variation from its mean. In light of the fact that many imaging artifacts are highly oscillatory, the solution to this variational problem is a reasonable candidate for the "optimal reconstruction" from finitely many (complete) projections.

In [89] it is shown that this problem has a unique solution of the form

$$g(x,y) = \sum_{j=0}^{n-1} \alpha_j(\langle (x,y), \boldsymbol{\omega}_j \rangle),$$

where, as indicated, $\{\alpha_0(t), \ldots, \alpha_{n-1}(t)\}$ are functions of one variable. An explicit solution, as Fourier series for the $\{\alpha_j\}$ is obtained in the case that the angles are equally spaced. In the general case, an algorithm is provided to determine these Fourier coefficients. In the process of deriving the formula for the optimal function, necessary and sufficient conditions are found for n-functions $\{r_j(t)\}$ to satisfy

$$r_j(t) = \mathcal{R} f(t, \boldsymbol{\omega}(\frac{j\pi}{n})), \quad j = 0, \ldots, n-1$$

for some function $f \in L^2(B_1)$. These are complicated relations satisfied by the sine-series coefficients of the functions $\{r_j(\cos\tau)\}$.

In [88] the question of how large n must be taken to get a good approximation is considered. The precise answer is rather complicated. Roughly speaking, if the Fourier transform of f is essentially supported in a disk of radius $n - n^{\frac{1}{3}}$, then, at least for n large, n-projections suffice to find a good approximation to f. Moreover the error $e = f - g$ is a "high-frequency" function in the sense that

$$\int_{\|\xi\| < n - n^{\frac{1}{3}}} |\hat{e}(\xi)|^2 \, d\xi_1 \, d\xi_2 \leq \lambda_n \int_{B_1} |f(x,y)|^2 \, dx \, dy.$$

The coefficient λ_n tends to zero as n tends to infinity. These results are precise versions of a heuristic principle in image reconstruction: A nonzero function f for which $\mathcal{R} f(t, \boldsymbol{\omega}) = 0$ for "many" values of $(t, \boldsymbol{\omega})$ is necessarily highly oscillatory. This is a reflection of the general experience that, with a good reconstruction algorithm, the error $f - \tilde{f}_\phi$ is a highly oscillatory function.

11.7 Spiral Scan CT

Second-, third-, and fourth-generation CT machines collect data one slice at a time. For this reason we could simply ignore the x_3-variable and reconstruct each two-dimensional slice separately. In the early 1990s a new design, called *spiral or helical scan CT*, was introduced. To understand this modality we need to consider the earlier modalities in somewhat greater detail. In this section we only describe a continuum model for the measurements and the techniques used to process them. In a real machine these data are sampled and discrete algorithms, similar to those described previously, are used to reconstruct an image. For simplicity we consider only a third-generation machine, which is referred to, in this section, as a "conventional scanner." Our discussion is adapted from [26]. This section can be omitted without any loss in continuity.

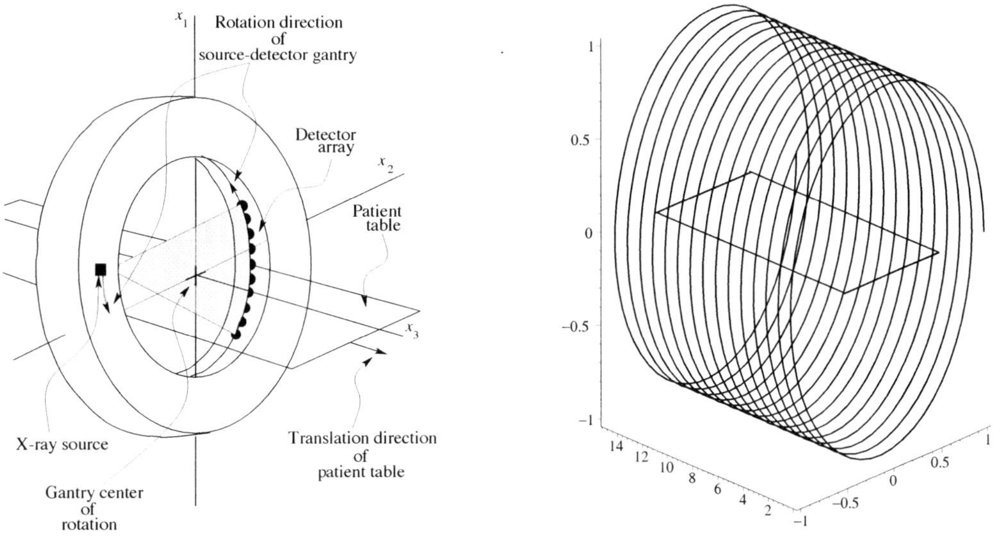

(a) Schematic diagram of a spiral scan CT machine.

(b) The source position as seen by the patient on the moving table.

Figure 11.18. The geometry of spiral scan CT machines.

11.7.1 Interpolation methods

The x-ray source and detectors are mounted on a gantry that rotates in a plane called the *slice plane* P. The patient is placed on a table at right angles to the slice plane [see Figure 11.18(a)]. With a conventional scanner, data for a slice are measured with the table held stationary and the source-detector gantry rotating. Let $P(\beta, \gamma, z')$ denote the (continuum) fan beam projection data for the slice with $x_3 = z'$. After these data are collected, the source-detector gantry stops, the table is translated by an fixed amount Δz perpendicular to P, and $P(\beta, \gamma, z' + \Delta z)$ is measured. This process is repeated as many times as is needed to collect all the required slices, giving the data set

$$\mathcal{D}_{\text{third}} = \{P(\beta, \gamma, z' + m\Delta z) \,:\, \beta \in [0, 2\pi), \quad -\gamma_L \leq \gamma \leq \gamma_L, \quad m = 1, 2, \ldots, M\}.$$

For each value of m, samples of $P(\beta, \gamma, z' + m\Delta z)$ can be used to reconstruct an approximation to the slice $\mu(x, y, z' + m\Delta z)$. Because of the mechanical realities of starting and stopping the gantry's rotation and moving the patient table, this method for collecting data is rather slow. Patient motion can be a significant problem in clinical applications of x-ray CT. To reduce its effect, it is necessary to find faster methods for acquiring the data.

One such method is spiral scan CT. In this modality the source-detector gantry rotates *continuously* while the patient is *continuously* pulled through the source-detector ring. The continuum model for what is measured is the set of projections

$$\mathcal{D}_{\text{spiral}} = \{P(\beta, \gamma, z(\beta)) \,:\, \beta \in [\beta_{\min}, \beta_{\max}], \quad -\gamma_L \leq \gamma \leq \gamma_L\}.$$

Here β is a real number, not restricted to lie in $[0, 2\pi)$; it should be thought of as the total *accumulated* rotation angle of the source. The function $z(\beta)$ is the x_3-coordinate of slice plane when the source has rotated through the angle β. Usually the table moves at a constant speed so that $z(\beta) = c\beta + z'$. From the perspective of the patient on the table, the source traces a spiral in 3-space

$$\{(D\cos\beta, D\sin\beta, z(\beta)) \,:\, \beta \in [-\beta_{\min}, \beta_{\max}]\}$$

[see Figure 11.18(b)]. Mathematically, the data set $\mathcal{D}_{\text{spiral}}$ does not suffice to reconstruct even a single two-dimensional slice of the μ. This is not surprising as it is a function of only two variables. However, using interpolation, we can approximate slice data for a single slice and use the reconstruction algorithms derived in the previous section to obtain an approximation for $\mu(x, y, z)$.

There are many possible ways to define interpolated data, P_{intrp}, from the measurements made in a spiral scan machine. We describe only the simplest; others are considered in [26]. Throughout this section we assume that the table moves at a constant linear speed and the gantry rotates at a constant angular speed such that

$$\partial_\beta z = c.$$

The simplest approach is to simulate data for a single slice in which β ranges through 2π radians by using measurements for β ranging through 4π radians. Suppose we would like to simulate fan beam data for the slice where

$$x_3 = z_0 = z(\beta_0).$$

We use the measurements

$$\{P(\beta, \gamma, z_0 + c(\beta - \beta_0) \,:\, \beta \in [\beta_0 - 2\pi, \beta_0 + 2\pi], \quad -\gamma_L \leq \gamma \leq \gamma_L\}$$

to create an interpolated data set

$$\{P_{\text{intrp}}(\beta, \gamma, z_0) \,:\, \beta \in [\beta_0, \beta_0 + 2\pi], \quad -\gamma_L \leq \gamma \leq \gamma_L\}.$$

This can be accomplished, using linear interpolation, by setting

$$\begin{aligned} P_{\text{intrp}}(\beta, \gamma, z_0) = &\left(\frac{\beta - \beta_0}{2\pi}\right) P(\beta, \gamma, z_0 + c(\beta - \beta_0)) + \\ &\left(\frac{2\pi + \beta_0 - \beta}{2\pi}\right) P(\beta - 2\pi, \gamma, z_0 + c(\beta - 2\pi - \beta_0)), \end{aligned} \tag{11.39}$$

for $\beta \in [\beta_0, \beta_0 + 2\pi)]$. The fan beam filtered back-projection algorithm can be applied to $P_{\text{intrp}}(\beta, \gamma, z_0)$ to obtain an approximate reconstruction for $\mu(x, y, z_0)$.

This approach has some problems. The most obvious problem is that the interpolated data are inconsistent. This is inevitable because for each value of β, projection data are obtained from a different function. This problem is most pronounced for the first and last scans: The interpolation scheme defines

$$\begin{aligned} P_{\text{intrp}}(\beta_0, \gamma, z_0) &= P(\beta_0 - 2\pi, \gamma, z_0 - 2\pi c) \text{ and} \\ P_{\text{intrp}}(\beta_0 + 2\pi, \gamma, z_0) &= P(\beta_0, \gamma, z_0). \end{aligned} \tag{11.40}$$

These two functions of γ may be quite different, whereas if this were projection data from a single function, they would be identical. Inconsistency in the data produces artifacts in the reconstructed images that appear as streaks. The severity of this problem is proportional to c : the faster the table moves, the more the individual measurements vary.

If the table is stationary during the acquisition of a slice, then the slice thickness, 2Δ mm, is determined by the width of the x-ray beam and size of the detector. Indeed the slice selectivity profile, defined in (11.3), is very close to a rectangle of width 2Δ mm. In a spiral scan machine, the x-ray beam and detector still have thickness 2Δ mm, but the data for a single "slice" are obtained with the x_3-coordinate of the source ranging over an interval of width $(4\pi c)$ mm. Naively, the slice thickness could be estimated to be $(2\Delta + 4\pi c)$ mm. A more careful analysis shows that the interpolation scheme weights the data for values of β close to β_0 more heavily than the data near the $\beta_0 \pm 2\pi$. Nonetheless, the intuition that the effective slice thickness increases with c is correct.

For a spiral scan machine the slice selectivity profile is no longer a rectangle. It depends on a variety of parameters, including the beam width, the interpolation scheme, as well as a new parameter called the *pitch*. If the table moves at a speed of $v\,\frac{\text{mm}}{\text{s}}$ and the rotational period of the gantry is ρ seconds, then the pitch is the dimensionless number defined to be

$$p = \frac{v}{2\Delta} \times \rho.$$

It is the multiple of the slice thickness that the source moves in one full rotation of the gantry. If the pitch is less than 1, then the anatomy involved with the measurement at β overlaps that involved with the measurement at $\beta + 2\pi$. If the pitch is greater than 1, then these slabs do not overlap. A smaller pitch leads to less inconsistency in the data and to a narrower effective slice. It also increases the time needed to scan a large region somewhat defeating the original purpose of the spiral scan machine. There are evidently trade-offs among scan time, slice thickness, and data consistency. This is discussed in detail in [132] and [94].

In our discussion of fan beam algorithms, we observed that, due to the symmetry of the Radon transform, measurements of $P(\beta, \gamma, z_0)$ for β ranging from β_0 to $\beta_0 + \pi + 2\gamma_L$ are sufficient to apply the fan beam filtered back-projection algorithm. Using this fact, interpolation schemes employing spiral scan measurements for β ranging over an interval of length $2(\pi + 2\gamma_L)$ have been developed. Using a smaller range of angles reduces both the effective slice thickness and the inconsistencies in the interpolated data. These methods require more complicated weighting schemes. Several such methods are described in [26].

We close this brief account of spiral scan machines with an observation about the placement of slices. In a conventional scanner, data are collected for a discrete sequence of slices $\{z' + m\Delta z \ : \ m = 1, 2, \ldots, M\}$ and nothing is measured "in between." In a spiral scan machine, data are acquired for a "continuously" varying collection of slices. With the interpolation methods described previously, we are free to select the center, z_0, of the slice. In this way we can reconstruct images from more slices than would be available in a conventional machine. This has been found to improve the visualization of three-dimensional structures as well as the sensitivity of the CT scanner to small high-contrast objects. This is considered in detail in [94].

11.7.2 3d-Reconstruction Formulæ

In addition to the interpolation methods described previously, which use spiral scan data to simulate conventional scanner data, there are other methods that use a three-dimensional data set directly to reconstruct an image. This approach requires entirely different reconstruction algorithms. In the most recent machines there are several rows of detector arrays set side by side on the gantry. For each position of the source-detector gantry, these machines collect samples of line integrals of μ for lines lying in a *two-dimensional* family; this is called *cone beam data*. The mathematical basis for much of the recent work on 3d-reconstruction methods is Grangeat's formula, which relates the "cone-beam" transform to the 3-dimensional Radon transform, treated in Section 6.7. We close this section by giving this formula.

The cone beam transform is a map from functions on $\mathbb{R}^3$ to functions on $\mathbb{R}^3 \times S^2$ defined by the integral:

$$Df(\boldsymbol{y}, \boldsymbol{\theta}) = \int_0^\infty f(\boldsymbol{y} + t\boldsymbol{\theta})dt. \tag{11.41}$$

Note that the integral runs from 0 to ∞. Because the x-ray source lies outside the object being imaged, the data collected in any x-ray CT machine can be modeled as samples of the cone beam transform, with source points $\boldsymbol{y}$ lying along a curve in $\mathbb{R}^3$. In the section above we considered the case where the source moves along a helix surrounding the object being imaged.

If $\boldsymbol{\theta}$ and $\boldsymbol{\omega}$ are orthogonal unit vectors, then we define

$$\nabla_{\boldsymbol{\theta}} Df(\boldsymbol{y}, \boldsymbol{\omega}) = \partial_t Df(\boldsymbol{y}, t\boldsymbol{\theta} + \sqrt{1 - t^2}\boldsymbol{\omega}) \restriction_{t=0} . \tag{11.42}$$

Recall that the 3d-Radon transform of f is the function on $\mathbb{R} \times S^2$ defined by

$$Rf(s, \boldsymbol{\omega}) = \int\limits_{\{\boldsymbol{x}:\, \boldsymbol{x}\cdot\boldsymbol{\omega}=s\}} f(\boldsymbol{x})dA. \tag{11.43}$$

The basis for the 3d-reconstruction formulæ is:

Theorem 11.7.1 (Grangeat's formula). *If f is a compactly supported function defined in $\mathbb{R}^3$, and $\boldsymbol{y} \in \mathbb{R}^3$ satisfies $\boldsymbol{y} \cdot \boldsymbol{\theta} = s$, then*

$$\partial_s Rf(s, \boldsymbol{\theta}) = \int\limits_{\boldsymbol{\theta}^{\perp}\cap S^2} \nabla_{\boldsymbol{\theta}} Df(\boldsymbol{y}, \boldsymbol{\omega})d\boldsymbol{\omega}. \tag{11.44}$$

As everything is rotationally invariant it suffices to prove this theorem for $\boldsymbol{\theta} = (0, 0, 1)$. Once this is understood, the proof of the theorem is a simple calculation, which we leave to the reader. This formula is a very good starting pointing for deriving inversion formulæ, as the inverse of the 3d-Radon transform is given by

$$f(\boldsymbol{x}) = \frac{-1}{8\pi^2} \int\limits_{S^2} \partial_s^2 Rf(\boldsymbol{x} \cdot \boldsymbol{\theta}, \boldsymbol{\theta})dA_{S^2}. \tag{11.45}$$

A discussion of these methods is beyond the scope of this text; the interested reader is directed to [77], [65], [46], [96] or [78]. The first reference is a monograph devoted to the problems of reconstruction in cone beam machines.

11.8 The Gridding Method*

The algorithms described previously are generally known as filtered back-projections. Let f denote the function we would like to reconstruct and $\hat{f}$ its Fourier transform. If the numbers of samples collected in the radial and angular directions are $O(N)$ and the reconstruction is performed on an $N \times N$ grid, then these methods require $O(N^3)$ arithmetic operations. Sampling $\hat{f}$ on a uniform rectangular grid allows the use of a *fast* direct Fourier inversion to obtain an approximation to f in $O([N \log_2 N]^2)$ operations. Neither a parallel beam, nor fan beam machine collects data that are easily converted to uniformly spaced samples of $\hat{f}$. In this section we discuss *the gridding method*, which is a family of efficient methods for passing from realistic data sets to uniformly spaced samples of a function simply related to $\hat{f}$. These methods work in any number of dimensions, so we describe this method in $\mathbb{R}^n$ for any $n \in \mathbb{N}$.

Let f denote a function defined on $\mathbb{R}^n$ that is supported in a cube

$$D = \underbrace{[-M, M] \times \cdots \times [-M, M]}_{n\text{-times}},$$

and $\hat{f}$ its Fourier transform. Suppose that samples of $\hat{f}$ are collected on a discrete set $\{\boldsymbol{y}_l : l = 1, \ldots, L\}$, and we wish to determine approximate values of f on another discrete set $\{\boldsymbol{x}_j : j = 1, \ldots, J\}$. The sort of approximation to $f(\boldsymbol{x}_j)$ we are considering uses the data to approximate the inverse Fourier transform by a sum of the form:

$$f(\boldsymbol{x}_j) \approx \frac{1}{(2\pi)^n} \sum_{l=1}^{L} \hat{f}(\boldsymbol{y}_j) e^{i\boldsymbol{x}_j \cdot \boldsymbol{y}_l} c_l. \tag{11.46}$$

The gridding method is a technique for efficiently evaluating sums like that appearing in equation (11.46). The accuracy of this sum as an approximation for $f(\boldsymbol{x}_j)$ depends, in a very *subtle* way on the choice of the coefficients $\{c_l\}$. The accuracy in going from (11.46) to its faster, gridding approximation is much easier to control.

This sum can be thought of as a Riemann sum for the inverse Fourier transform:

$$f(\boldsymbol{x}) = \frac{1}{(2\pi)^n} \int_{\mathbb{R}^n} \hat{f}(\boldsymbol{y}) e^{i\boldsymbol{y}\cdot\boldsymbol{x}} d\boldsymbol{y}. \tag{11.47}$$

This interpretation suggests that we should efficiently divide up the space containing the sample points $\{\boldsymbol{y}_l\}$ into disjoint polyhedra $\{P_l\}$, and then, for each l, define the coefficient c_l to be the n-dimensional volume of $|P_l|$. This is a reasonable thing to try, but in fact, using these "geometrically defined" coefficients does not usually give a very accurate approximation to $f(\boldsymbol{x}_j)$. The value of a Fourier integral is largely determined by the cancellations that occur due to the oscillatory nature of the integrand. If the samples of $\hat{f}$ are uniformly spaced, then, using the obvious choice of coefficients: $c_l = c$, for all l, these cancellations also occur between the terms of the finite sum, and we obtain a very accurate approximation to the integral. See Section 10.2.2.

When the samples are not uniformly spaced, then, using the geometrically defined coefficients does not allow these cancellations to occur, and the sum does not, in general, provide an accurate approximation to $f(\boldsymbol{x}_j)$. The problem of "optimally" selecting the coefficients is not, as of 2007, in any real sense, solved, see [48]. For the remainder of this chapter we simply assume that the coefficients $\{c_l\}$ are fixed. As noted above, the gridding method is just a technique for accelerating the evaluation of sums like that appearing in (11.46), for $\{\boldsymbol{x}_j\}$ lying on a uniformly spaced grid.

Suppose that w is a function that is positive on the support of f, and that its Fourier transform $\hat{w}$ is a function with small effective support. Suppose, moreover that we can efficiently evaluate both w and $\hat{w}$. The basis of the gridding method is the following observation: Define the function

$$\tilde{f}(\boldsymbol{x}) = \frac{1}{(2\pi)^n} \sum_{l=1}^{L} \hat{f}(\boldsymbol{y}_j) e^{i\boldsymbol{x}\cdot\boldsymbol{y}_l} c_l. \tag{11.48}$$

The Fourier coefficients of the function $w \cdot \tilde{f}(x)$ are given by

$$\widehat{w \cdot \tilde{f}}(\boldsymbol{k}) = \frac{1}{(2\pi)^n} \sum_{l=1}^{L} \hat{f}(\boldsymbol{y}_l) \hat{w}(\boldsymbol{k} - \boldsymbol{y}_l) c_l. \tag{11.49}$$

Using this sum to evaluate the Fourier transform $\widehat{w \cdot \tilde{f}}(\boldsymbol{\xi}_k)$, for $\{\boldsymbol{\xi}_k\}$ lying on a uniform grid, we can use the fast (inverse) Fourier transform to evaluate an accurate approximation to $\{w \cdot \tilde{f}(\boldsymbol{x}_j)\}$, for $\{\boldsymbol{x}_j\}$ on the dual, uniform grid. If the (effective) support of $\hat{w}$ is small enough then, using a small number of operations, we can accurately evaluate the sums in (11.49). Thus the total amount of computation needed to evaluate $w(\boldsymbol{x}_j) \cdot \tilde{f}(\boldsymbol{x}_j)$ can be made much smaller than that needed to directly evaluate the sums in equation (11.46) for $j = 1, \dots, J$. To determine $\{\tilde{f}(\boldsymbol{x}_j)\}$ we need to divide the computed values, $\{w(\boldsymbol{x}_j) \cdot \tilde{f}(\boldsymbol{x}_j)\}$, by $\{w(\boldsymbol{x}_j)\}$. This does not change the number of operations in a significant way, but explains why we need to assume that the values $\{w(\boldsymbol{x}_j)\}$ are bounded away from zero.

Let $S_{\text{data}} = \{\boldsymbol{y}_l : l = 1, \dots, L\}$ denote a set of points in $\mathbb{R}^n$; the data that are available are approximations to the samples $\{\hat{f}_l \approx \hat{f}(\boldsymbol{y}_l) : l = 1, \dots, L\}$. Suppose that these points lie in a cube

$$E = [-B, B] \underbrace{\times \cdots \times}_{n-\text{times}} [-B, B]$$

in Fourier space. In order for this (or any other) method to provide good results, it is necessary that the actual Fourier transform of f be small outside of E (i.e., $2B$ should equal or exceed the effective bandwidth of f). The goal is to compute the values of $\widehat{w \cdot \tilde{f}}$ on a uniformly spaced grid in E and use the FFT to reconstruct $w \cdot \tilde{f} \approx w \cdot f$.

Let $d = B/N$ denote the sample spacing (equal in all directions) in E. Use the bold letters $\boldsymbol{j}, \boldsymbol{k}$ to denote points in $\mathbb{Z}^n$. The subset $S_{\text{unif}} \subset \mathbb{Z}^n$ indexes the sample points in E with $(k_1, \dots, k_n) = \boldsymbol{k} \in S_{\text{unif}}$ if

$$-N \le k_i \le N, \quad i = 1, \dots, n,$$

then

$$S_{\text{unif}} \ni \boldsymbol{k} \leftrightarrow (dk_1, \dots, dk_n) \in E.$$

From the discussion in Section 8.2.1 it follows that the effective *field of view* of our sampled Fourier data is $2\pi/d$. That is, in order to avoid spatial aliasing, d must be chosen small enough so that the function f is (essentially) supported in

$$G = [-\frac{\pi}{d}, \frac{\pi}{d}] \underbrace{\times \cdots \times}_{n-\text{times}} [-\frac{\pi}{d}, \frac{\pi}{d}].$$

If the set of sample points $\{\boldsymbol{y}_l\}$ is a sufficiently dense subset of E, and d, the sample spacing in the Fourier domain, satisfies $M \le \pi/d$, then *spatial* aliasing does not occur. The sample points in G are also indexed by S_{unif} with

$$S_{\text{unif}} \ni \boldsymbol{k} \leftrightarrow \frac{\pi}{B}\boldsymbol{k} \in G.$$

Here we use the fact that dN equals B.

Let $\hat{w}$ be a function defined in $\mathbb{R}^n$ effectively supported in the cube

$$W = [-Kd, Kd] \underbrace{\times \cdots \times}_{n-\text{times}} [-Kd, Kd]$$

whose inverse Fourier transform does not vanish in G. As we eventually need to divide by w, it is important that $w(\boldsymbol{x})$ is actually larger than a fixed positive constant for $\boldsymbol{x}$ in the field of view. We use the sum in (11.49) to compute

$$\widehat{w \cdot \tilde{f}}(d\boldsymbol{k}) := \frac{1}{(2\pi)^n} \sum_{l=1}^{L} \hat{w}(d\boldsymbol{k} - \boldsymbol{y}_l) \hat{f}_l c_l \approx \hat{w} * \hat{f}(d\boldsymbol{k}). \tag{11.50}$$

Supposing that $\hat{w}$ decays rapidly enough so that, for the accuracy we are seeking to achieve, we can regard $\hat{w}$ as being supported in W, this calculation requires $O(K^n)$ operations for each point $\boldsymbol{k} \in S_{\text{unif}}$. The efficient evaluation of these sums requires an efficient method for computing (or the a priori determination of) the values of $\{\hat{w}(d\boldsymbol{k} - \boldsymbol{y}_l) : d\boldsymbol{k} - \boldsymbol{y}_l \in W\}$, needed for each index $\boldsymbol{k}$.

After zero padding to obtain a power of 2, the FFT can be used to compute approximate values of $g_j = w \cdot \tilde{f}(\boldsymbol{j}\frac{\pi}{B})$. The gridding method is completed by setting

$$f_j = \frac{g_j}{w(\boldsymbol{j}\frac{\pi}{B})}.$$

If N is a power of 2, then this step requires $O((N \log_2 N)^n)$ operations. As noted above, f_j is an very accurate approximation to the following trigonometric sum

$$f_j \approx \frac{1}{(2\pi)^n} \sum_{l=1}^{L} \hat{f}_l e^{i\boldsymbol{y}_l \cdot \boldsymbol{x}_j} c_j. \tag{11.51}$$

Indeed, aliasing is the only possible source of error in using the gridding method to compute this sum. [1]

Two considerations limit how large K can be taken. The amount of computation needed to do the gridding step is $O((KN)^n)$. For example, if $n = 2$, this is $O(K^2N^2)$. The amount of computation needed for the filtered back-projection algorithms is $O(N^3)$. For the gridding method to remain competitive, we see that $K << \sqrt{N}$. The second constraint comes from the requirement that $w(\boldsymbol{x}) \neq 0$ for points, $\boldsymbol{x}$ in the cube D. From our discussion of the Fourier transform, we know that the smaller the support of $\hat{w}$, the larger the set around zero in which w is nonvanishing. Thus K needs to be chosen small enough so that w is nonvanishing throughout D. A very nice discussion of these methods appears in [47]. An analysis of the errors entailed in this method is presented in [112], along with examples using different choices of $\hat{w}$.

As noted above, the main source of error in using non-uniformly spaced Fourier data, in the manner described in this section, comes from the choice of the coefficients $\{c_l\}$ in (11.46). Implicit in this approach is the idea that we should try to minimize the L^2-error, $\|f - \tilde{f}\|_{L^2(G)}$. This has been the standard approach in imaging for many years. Recent work of Donoho, Candes, and others indicates that this may not be the optimal way to use the measured data $\{\hat{f}(\boldsymbol{y}_l)\}$. They have introduced very different approaches to image

[1] I thank Jeremy Magland for pointing out this interpretation.

Figure 11.19. A three dimensional rendering of an x-ray CT micrograph made in a EVS® CT microscope. The resolution in this image is 15 microns. (This image kindly provided by Dr. Felix Wehrli. The EVS scanner was purchased by the Laboratory for Structural NMR Imaging at Hospital of the University of Pennsylvania on a shared instrumentation grant from the National Institutes of Health.)

reconstruction where, for example, one might look amongst functions h that satisfy the constraints imposed by the measurements:

$$\hat{h}(\boldsymbol{y}_l) = \hat{f}(\boldsymbol{y}_l) \text{ for } l = 1, \ldots, L, \tag{11.52}$$

for a function of minimal L^1-norm, or perhaps minimal total variation norm. Though it requires considerably more computation, using such an approach has been found, in some circumstances, to produce much better images than those obtained using an L^2-notion of error. This approach, sometimes called "compressive sampling" and "L^1-magic" is described in [18, 19].

11.9 Conclusion

At the end of the 1960s it was quite a remarkable idea that slices through a human body could be reconstructed using what amounted to a large collection of carefully measured, tiny x-ray images. The secret, of course, was mathematics. The original approach of Hounsfield was a variant of the algebraic reconstruction method, which is described in Chapter 13. It did not give very good images. An example of a reconstruction obtained using this method is given in Figure 13.7. Next, a method using (unfiltered) back-projection was tried, which also produced rather poor images. Finally, the Radon inversion formula was *rediscovered* and with it the possibility of accurately reconstructing images from projection data. The earliest images also had many artifacts caused by aliasing, the Gibbs phenomenon, miscalibration of the detectors, unstable x-ray sources, beam hardening, geometric miscalibration of the gantry, and so on. X-ray CT is now a highly developed field in medical imaging; the algorithms we discuss were mostly developed before 1980. A very nice exposition of the different sorts of practical algorithms which can be derived from exact reconstruction formulæ is given in [85].

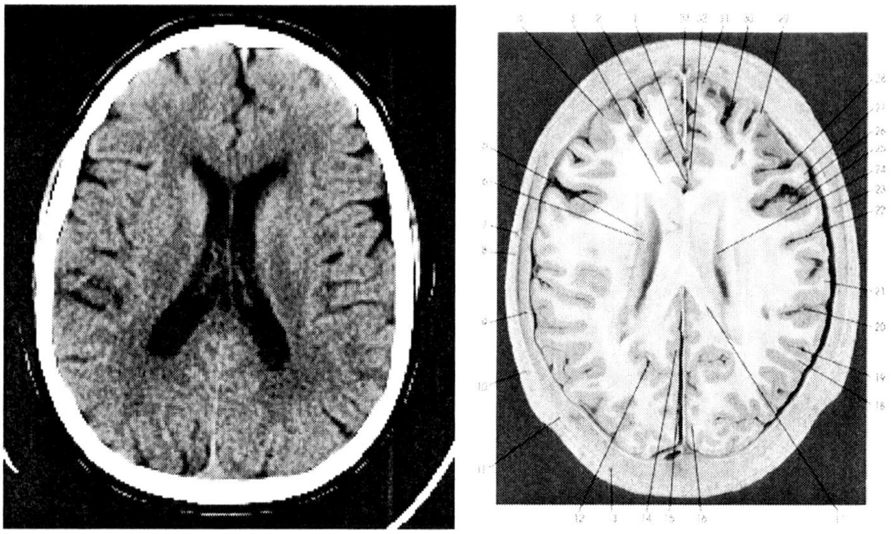

(a) Section of a human brain scanned in a GE® scanner circa 2002. (This image kindly provided by Dr. David Hackney of the Hospital of the University of Pennsylvania.)

(b) An anatomical section of the a very similar slice. (This image provided courtesy of Dr. Bergman of the Virtual Hospital. It is copyrighted material used with permission of the author and the University of Iowa's Virtual Hospital, www.vh.org.)

Figure 11.20. A slice of the human brain made with a 2002 GE® scanner and an anatomical section of a similar slice.

The filtered back-projection algorithms derived in this chapter have been standard in commercial CT-machines for over two decades. Much of the research in this field is now directed toward postprocessing the reconstructed images to locate organ boundaries and do quantitative measurements of anatomical structures. The need to reduce scan times has recently led to the development of spiral scan, cone beam, and fast electron beam machines. In a cone beam machine, each measurement consists of samples of the Radon transform from a two-dimensional family of lines. The fast electron beam machine is sometimes called a "fifth-generation scanner." It collects essentially the same data as a spiral scan machine but operates under entirely different engineering principles, leading to even shorter scan times. Electron beam CT is described in [14] and [13]. These newer machines have, in turn, led to a resurgence of interest in algorithm development with a new emphasis on fully three-dimensional algorithms.

In figures 11.19 and 11.20 we show two modern images produced using x-ray tomography. Figure 11.19 shows a three-dimensional rendering of a section of human trabecular bone made with an EVS® x-ray CT microscope. The CT microscope is a "cone beam" machine; it uses the *Feldkamp algorithm* to reconstruct images; see [40]. Figure 11.20(a) is a cross section of the human brain made using a GE® scanner. Figure 11.20(b) shows an anatomic section of a nearby slice in a human brain. These images demonstrate the remarkable capabilities of circa 2002 CT imaging.

The mathematics used to reconstruct images in x-ray CT is applicable (and actually applied) to reconstruction problems for any "nondiffracting" imaging modality (e.g., positron emission tomography and magnetic resonance imaging). Much of the mathematical technology is also used in the study of "diffracting" modalities such as ultrasound, impedance tomography, and infrared imaging. The "inverse problems" for these latter modalities are nonlinear, and to date they lack complete mathematical solutions. In consequence of this fact, these modalities have not yet come close to attaining their full potential. A challenge for tomorrow is the exact solutions of the idealized reconstruction problems for these modalities. The report of the National Academy of Sciences, *Mathematics and Physics of Emerging Biomedical Imaging* provides an excellent overview of the open problems and state of mathematical research in medical image reconstruction; see [23].

The World Wide Web provides a remarkable set of resources for the student of medical imaging. I list here a very small selection of Web sites devoted to this topic. The addresses were correct as of August 2002.

1. The **visible human project**:
`http://www.nlm.nih.gov/research/visible/visible_human.html`
2. The **radiological anatomy browser**:
`http://rad.usuhs.mil/rad/radbrowser2/index2.html`
3. The **Radiology home page** contains links to many other radiology sites:
`http://home.earthlink.net/~terrass/radiography/medradhome.html`
4. **Neuroland**, contains links to many sites with neuroradiology images: `http://neuroland.com/neuro_images/`

In the next chapter we consider a more realistic model for the measurements as averages of the Radon transform of the attenuation coefficient over the cross section of the x-ray beam. We then compute the point spread functions for both the continuum approximations to the filtered back-projection formula and their implementations on sampled data. We then analyze the effects on the reconstructed image of various sorts of systematic measurement errors, concluding with a short discussion of beam hardening.

Chapter 12

Imaging Artifacts in X-Ray Tomography

In the previous chapter we derived finite algorithms to approximately reconstruct a function of two variables with bounded support from finitely many samples of its Radon transform. For a variety of reasons this is still a highly idealized situation. In this chapter we analyze these algorithms with a more realistic model for the measurement process. The first issue we address is the fact that the x-ray beam has a finite (two-dimensional) width. There is a simple linear model for this effect as weighted averaging in the affine parameter. More careful investigation reveals that this is a nonlinear phenomenon that leads to the *nonlinear partial volume effect*. Averaging in the affine parameter is a form of lowpass filtering, which is, in turn, important for the analysis of the aliasing that results from sampling the Radon transform.

We next derive the total point spread function (PSF) for the measurement and reconstruction process first without and then with sampling. Once the measurements are sampled in the affine parameter, the reconstruction process is no longer shift invariant, so it is not described by a single point spread function. The width of the central peak of the PSF provides an indication of the resolution available in the reconstructed image. We analyze the effects of the beam width and sample spacing on the shape of the PSF. Using the PSF, we consider the consequences for the reconstructed image of various sorts of measurement errors. At the end of the chapter we consider beam hardening; this results from the fact that the x-ray beam is not monochromatic. Beam hardening is another fundamentally nonlinear phenomenon.

12.1 The Effect of a Finite Width X-Ray Beam

Up to now, we have assumed that an x-ray beam is just a line with no width and that the measurements are integrals over such lines. What is really measured is better approximated by averages of such integrals. We now consider how the finite width of the x-ray beam affects the measured data. Our treatment closely follows the discussion in [113].

12.1.1 A Linear Model for Finite Beam Width

A simple linear model for the effect of finite beam width is to replace the Radon transform of f by a weighted average of the Radon transform. For a weight function w, define

$$\mathscr{R}_W f(t, \boldsymbol{\omega}) = \int_{-\infty}^{\infty} w(u)\, \mathscr{R} f(t-u, \boldsymbol{\omega})\, du.$$

Such an average is called a *strip integral.* The weight is a nonnegative function that models both the distribution of energy *across* the x-ray beam and the detector used to make the measurements (see figure 12.1). This function is sometimes called the *beam profile*, though of course the actual beam profile must incorporate the third dimension as well.

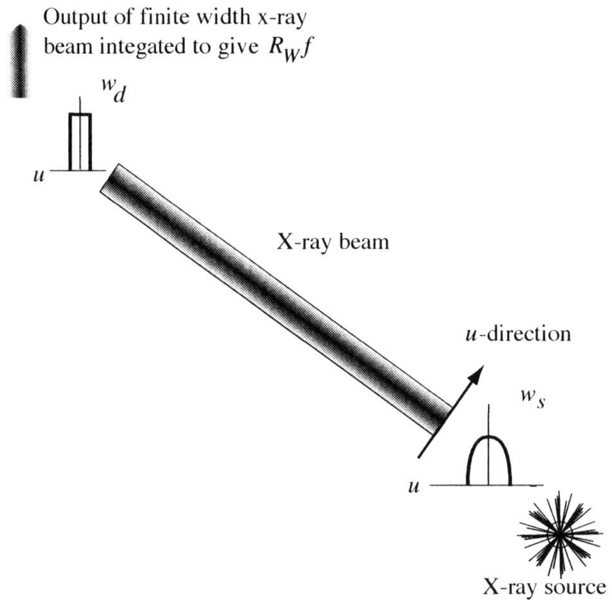

Figure 12.1. The finite size of the x-ray source and detector means real measurements are modeled by strip integrals.

The relationship between $\mathscr{R} f$ and $\mathscr{R}_W f$ is a consequence of the convolution theorem for the Radon transform. In the imaging literature it is due to Shepp and Logan.

Theorem 12.1.1 (Shepp and Logan). *The weighted Radon transform $\mathscr{R}_W f$ is the Radon transform of the convolution, $f * k$, where k is the radial function*

$$k(x, y) = -\frac{1}{\pi \rho} \partial_\rho \left[\int_\rho^\infty \frac{w(u) u}{\sqrt{u^2 - \rho^2}} du \right]_{\rho=\sqrt{x^2+y^2}}.$$

***Remark* 12.1.1.** If w has bounded support then the integrand of k is zero for sufficiently large ρ; hence $k(\sqrt{x^2+y^2})$ also has bounded support. Similarly, if k has bounded support, then so does w.

Proof. The theorem is an immediate consequence of Proposition 6.1.1. The function $\mathcal{R}_W f$ is the convolution in the t-parameter of $\mathcal{R}f$ with w. If k is a function on $\mathbb{R}^2$ such that $\mathcal{R}k = w$, then the proposition states that

$$\mathcal{R}_W f = \mathcal{R}(f * k).$$

Since w is independent of $\boldsymbol{\omega}$, it follows that k must also be a radial function. The formula for k is the Radon inversion formula for radial functions derived in Section 3.5. □

***Example* 12.1.1.** Some simple examples of (w, k)-pairs are

$$w^1(u) = \begin{cases} \frac{1}{2\delta} & u \in [-\delta, \delta], \\ 0 & |u| > \delta, \end{cases} \qquad k^1(\rho) = \begin{cases} \frac{1}{2\pi\delta}\frac{1}{\sqrt{\delta^2-\rho^2}} & 0 \leq \rho < \delta, \\ 0 & \rho \geq \delta, \end{cases}$$

and

$$w^2(u) = \begin{cases} \frac{1}{\pi\delta^2}(\delta^2 - u^2)^{1/2} & |u| < \delta, \\ 0 & |u| > \delta, \end{cases} \qquad k^2(\rho) = \begin{cases} \frac{1}{\pi\delta^2} & 0 \leq \rho < \delta, \\ 0 & \rho \geq \delta. \end{cases}$$

A consequence of finite strip width is that the actual measurements are samples of the Radon transform of $f * k$, which is a somewhat smoothed version of f. Indeed,

$$\widetilde{\mathcal{R}_W f}(r, \boldsymbol{\omega}) = \hat{w}(r)\widetilde{\mathcal{R}f}(r, \boldsymbol{\omega}), \tag{12.1}$$

and therefore the finite strip width leads to lowpass filtering of $\mathcal{R}f$ in the affine parameter. This has the desirable effect of reducing the aliasing artifacts that result from sampling. In x-ray tomography this is essentially the only way to lowpass filter the data before it is sampled. Of course, this averaging process also leads to a loss of resolution; so the properties of the averaging function w need to be matched with the sample spacing. As we saw in Section 9.1.9, the effects of such averaging can, to some extent, be removed; nonetheless algorithms are often evaluated in terms of their ability to reconstruct samples of $f * k$ rather than f itself.

As mentioned previously, the beam profile, w, models the effects of the source and detector together. This function is built out of two pieces: the detector response function, w_d, and the source function, w_s. If $I(u)$ describes the intensity of the (two-dimensional) x-ray beam incident on the (one-dimensional) detector at the point u, then the output of the detector is modeled as

$$\int_{-\infty}^{\infty} w_d(u) I(u)\, du.$$

If the function w_s models the x-ray source, then the energy of the source, in the interval $[a, b]$, is given by

$$\int_a^b w_s(u)\, du.$$

If the source and detector are fixed in space, relative to one another, then the combined effect of this source-detector pair is modeled by the pointwise product $w = w_s \cdot w_d$. This is the geometry in a third-generation, fan beam scanner. In some parallel beam scanners and fourth-generation scanners, the detectors are fixed and the source moves. In this case a model for the source-detector pair is the convolution $w = w_s * w_d$. The detector is often modeled by a simple function like w^1 or w^2 defined in Example 12.1.1, while the source is often described by a Gaussian, $ce^{-\frac{u^2}{\sigma}}$. In this case the x-ray source is said to have a *Gaussian focal spot*. In any case the *beam width* is usually taken to equal the FWHM of w.

***Remark* 12.1.2.** A thorough treatment of the problem of modeling x-ray sources and detectors is given in [6].

Exercises

***Exercise* 12.1.1.** Explain why the source-detector pair is modeled as $w_s \cdot w_d$, if the source and detector are fixed relative to one another, and as $w_d * w_s$ if they move. Assume that the x-rays are non-diverging. In particular, explain how relative motion of the source and detector leads to a convolution.

***Exercise* 12.1.2.** A real x-ray beam is three dimensional. Suppose that the third dimension is modeled as in (11.3) by the slice selectivity profile w_{ssp}. Give a linear model for what is measured, analogous to the Shepp-Logan result, that includes the third dimension.

***Exercise* 12.1.3.** What is the physical significance of the total integral of w?

12.1.2 A Nonlinear Model for Finite Beam Width

Unfortunately, the effect of finite beam width is a bit more complicated than described in the previous section. If we could produce a one-dimensional x-ray beam, then what we would measure would actually be

$$I_o = I_i \exp[-\mathscr{R}f(t, \omega)],$$

where I_i is the intensity of the x-ray source and I_o is the measured output. For a strip, what is actually measured is therefore better modeled by

$$\frac{I_o}{I_i} \approx \int_{-\infty}^{\infty} w(u) \exp[-\mathscr{R}f(t-u, \omega)]\, du.$$

Thus the measurement depends nonlinearly on the attenuation coefficient. If w is very concentrated near $u = 0$ and $\int w(u)\,du = 1$ then

$$\log \frac{I_o}{I_i} \approx \int_{-\infty}^{\infty} w(u)\,\mathcal{R}f(t-u,\boldsymbol{\omega})\,du.$$

To derive this expression, we use the Taylor expansions:

$$e^x = 1 + x + \frac{x^2}{2} + \frac{x^3}{6} + O(x^4), \quad \log(1+x) = x - \frac{x^2}{2} + \frac{x^3}{3} + O(x^4). \tag{12.2}$$

This analysis assumes that the oscillation of $\mathcal{R}f(t-u,\boldsymbol{\omega})$ over the support of w is small. We begin by factoring out $\exp(-\mathcal{R}f)$:

$$\int_{-\infty}^{\infty} w(u)\exp(-\mathcal{R}f(t-u,\boldsymbol{\omega}))\,du$$

$$= \exp(-\mathcal{R}f(t,\boldsymbol{\omega}))\int_{-\infty}^{\infty} w(u)\exp[\mathcal{R}f(t,\boldsymbol{\omega}) - \mathcal{R}f(t-u,\boldsymbol{\omega})]\,du. \tag{12.3}$$

Using the the Taylor expansion for e^x gives

$$\int_{-\infty}^{\infty} w(u)\exp(-\mathcal{R}f(t-u,\boldsymbol{\omega}))\,du$$

$$= \exp(-\mathcal{R}f(t,\boldsymbol{\omega}))\Bigg[\int_{-\infty}^{\infty} w(u)[1 + (\mathcal{R}f(t,\boldsymbol{\omega}) - \mathcal{R}f(t-u,\boldsymbol{\omega})) +$$
$$O((\mathcal{R}f(t-u,\boldsymbol{\omega}) - \mathcal{R}f(t,\boldsymbol{\omega}))^2)]\,du\Bigg]$$

$$= \exp(-\mathcal{R}f(t,\boldsymbol{\omega}))\Bigg[1 - \int_{-\infty}^{\infty} w(u)(\mathcal{R}f(t-u,\boldsymbol{\omega}) - \mathcal{R}f(t,\boldsymbol{\omega}))\,du +$$
$$O\left(\int w(u)[\mathcal{R}f(t-u,\boldsymbol{\omega}) - \mathcal{R}f(t,\boldsymbol{\omega})]^2\,du\right)\Bigg].$$

Taking $-\log$, using the Taylor expansion for $\log(1+x)$ and the assumption that $\int w(u) = 1$, gives

$$\log \frac{I_o}{I_i} \approx \int w(u)\,\mathcal{R}f(t-u,\boldsymbol{\omega})\,du + O\left(\int w(u)[\mathcal{R}f(t-u,\boldsymbol{\omega}) - \mathcal{R}f(t,\boldsymbol{\omega})]^2\,du\right). \tag{12.4}$$

The leading order error is proportional to the mean square oscillation of $\mathcal{R}f$ weighted by w.

12.1.3 The Partial Volume Effect

If the variation of $\mathcal{R}f(t, \omega)$ is large over the width of the strip, then the error term dominates in (12.4). In practice, this happens if part of the x-ray beam intercepts bone and the remainder passes through soft tissue. In imaging applications this is called the *partial volume effect*. To illustrate this we consider a simple special case. Suppose that the intensity of the x-ray beam is constant across a strip of width 1. Half the strip is blocked by a rectangular object of height 1 with attenuation coefficient 2, and half the strip is empty. If we assume that $w(u) = \chi_{[0,1]}(u)$, then

$$-\log\left[\int_{-\infty}^{\infty} w(u)\exp[-\mathcal{R}f(t-u,\omega)]\,du\right] = -\log\left[\frac{1+e^{-2}}{2}\right] \simeq 0.5662,$$

whereas

$$\int_{-\infty}^{\infty} w(u)\,\mathcal{R}f(t-u,\omega)\,du = 1.$$

In Table 12.1 we give the linear and non-linear computations for an absorbent unit square with two attenuation coefficients μ_0, μ_1 each occupying half (see figure 12.2).

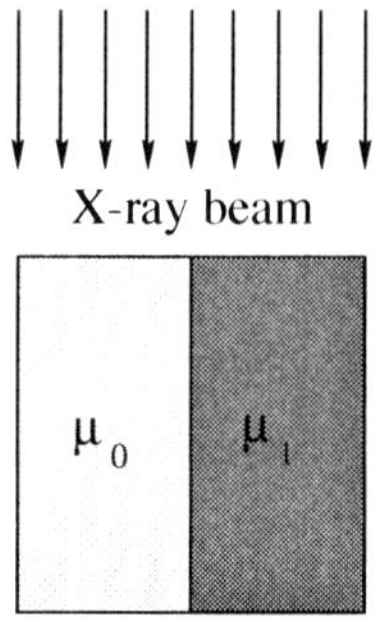

Figure 12.2. Absorbing square.

An even more realistic example is provided by a long rectangle of absorbing material with a small inclusion of more absorbent material, as shown in Figure 12.3. The graphs in Figure 12.4 show the relative errors with $\mu_0 = 1$ and $\mu_1 \in \{1.5, 2, 2.5, 3\}$. This is a model for a long stretch of soft tissue terminating at a piece of bone.

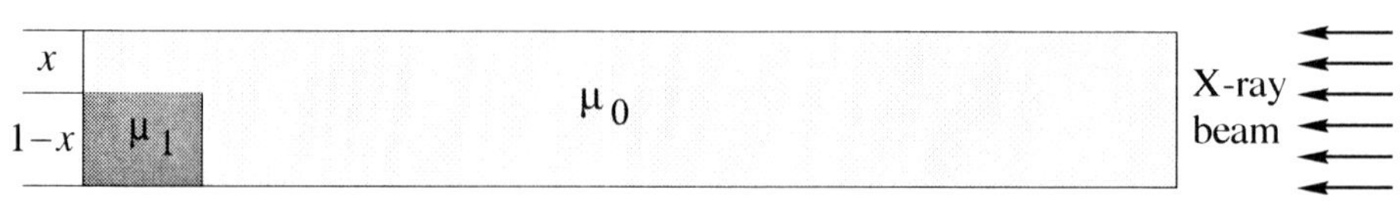

Figure 12.3. Rectangle with a small inclusion.

μ_0	μ_1	Non-linear	Linear	Relative error
0	.01	.00499	.005	2%
0	.1	.0488	.05	2.4%
0	.5	.2191	.25	12.4%
0	1	.3799	.5	24%
0	2	.5662	1	43%
.3	.4	.34875	.35	3.5%
.3	1.3	.6799	.8	15%
1	2	1.38	1.5	8%

Table 12.1. Errors due to the partial volume effect

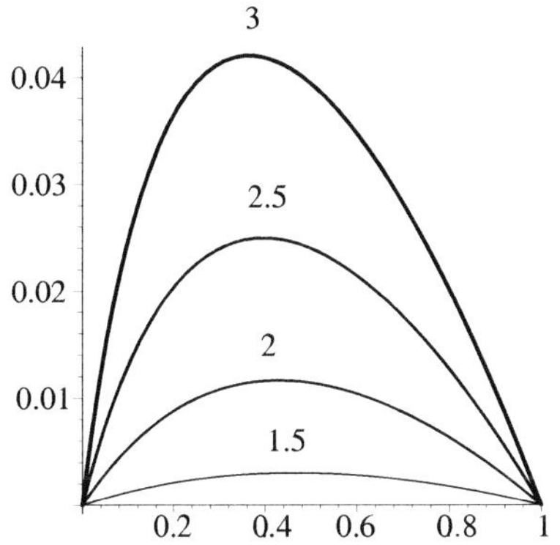

Figure 12.4. Relative errors with small inclusion. An error of 2% is significant in a medical imaging context.

The artifact caused by the partial volume effect is the result of the discrepancy between the nonlinear data that are actually collected and the linear model for the data collection, used in the derivation of the reconstruction algorithms. The algorithm assumes that what is collected are samples of $\mathcal{R}f^k(t, \omega)$; because of the nonlinear nature of the measurement process, this is not so. Even if we could measure a projection for all relevant pairs (t, ω), our algorithm would not reconstruct f^k exactly but rather some further *nonlinear* transformation applied to f. In real images the partial volume effect appears as abnormally bright spots or streaks emanating a hard object (see Figure 12.5).

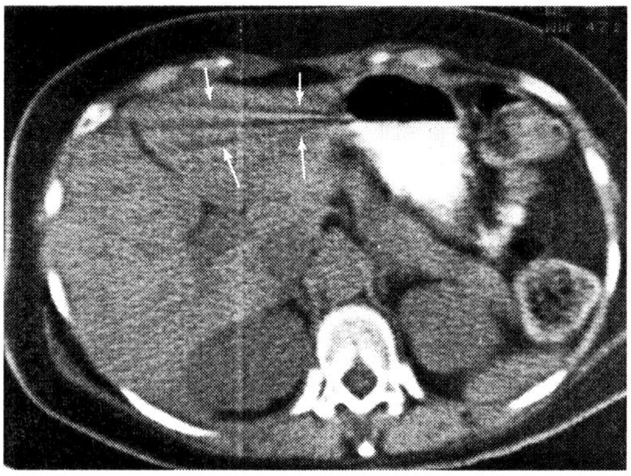

Figure 12.5. The partial volume effect is responsible for the streaks indicated by the arrows. The large white blob is highly absorbent contrast material in the patient's stomach. (This image provided by Dr. Peter Joseph.)

12.2 The Point Spread Function Without Sampling

See: 6.4.

Later in this chapter we analyze artifacts that arise in image reconstruction using realistic data and a reasonable model for the source-detector pair. The explanation for a given artifact is usually found by isolating the features of the image that produce it. At the center of this discussion are the point spread and modulation transfer functions (PSF and MTF), characterizing the measurement and reconstruction process. Once the data are sampled, the measurement process is no longer translation invariant and therefore it is not described by a single PSF. Instead, for each point (x, y) there is a function $\Psi(x, y; a, b)$ so that the reconstructed image at (x, y) is given by

$$f_{\Psi}(x, y) = \int\limits_{\mathbb{R}^2} \Psi(x, y; a, b) f(a, b)\, da\, db.$$

Because the filter it defines is not shift invariant, strictly speaking Ψ is not a point spread function. Following the standard practice in engineering, we also call Ψ a point spread function, though it must be remembered that there is a different PSF for each source location (a, b).

Our derivation of Ψ is done in two steps. First we find a PSF that incorporates a model for the source-detector pair and the filter used in the filtered back-projection step. This part is both translation invariant and isotropic. Afterward we incorporate the effects of sampling the measurements to obtain an expression for $\Psi(x, y; a, b)$. Only the parallel beam geometry is considered; our presentation follows [74]. The results for the fan beam geometry are similar but a little more complicated to derive; see [72] and [70]. In this chapter we illustrate the effects of sampling, averaging and various sorts of measurement errors by using a variety of mathematical phantoms as well as real measured data.

12.2.1 Point Sources

As a function of (x, y), $\Psi(x, y; a, b)$ is the output of the measurement-reconstruction process applied to a unit point source at (a, b), which is modeled by

$$\delta_{(a,b)}(x, y) = \delta((x, y) - (a, b)).$$

To facilitate the computation of Ψ, is it useful to determine the Radon transform of this generalized function, which should itself be a generalized function on $\mathbb{R} \times S^1$. Since $\delta_{(a,b)}$ is $\delta_{(0,0)}$ translated by (a, b), it suffices to determine $\mathcal{R}\delta_{(0,0)}$. Let φ_ϵ be a family of smooth functions converging to the $\delta_{(0,0)}$ in the sense that $\varphi_\epsilon * f$ converges uniformly to f, for f a continuous function with bounded support. The convolution theorem for the Radon transform, Proposition 6.1.1, says that

$$\mathcal{R}(\varphi_\epsilon * f)(t, \boldsymbol{\omega}) = \mathcal{R}\varphi_\epsilon *_t \mathcal{R}f(t, \boldsymbol{\omega}).$$

Since the left-hand side converges to $\mathcal{R}f$, as $\epsilon \to 0$, it follows that

$$\mathcal{R}\delta_{(0,0)}(t, \boldsymbol{\omega}) = \lim_{\epsilon \to 0} \mathcal{R}\varphi_\epsilon(t, \boldsymbol{\omega}) = \delta(t).$$

Using Proposition 6.1.2, we obtain the general formula

$$\mathcal{R}\delta_{(a,b)}(t, \boldsymbol{\omega}) = \delta(t - \langle \boldsymbol{\omega}, (a, b) \rangle). \tag{12.5}$$

Exercise

***Exercise* 12.2.1.** Derive (12.5) by using the family of functions

$$\varphi_\epsilon(x, y) = \frac{1}{\epsilon^2} \chi_{[0,\epsilon^2]}(x^2 + y^2).$$

12.2.2 The PSF without Sampling

See: A.3.3.

For the purposes of this discussion we use the simpler, linear model for a finite width x-ray beam. Let w be a nonnegative function with total integral 1. Our model for a measurement is a sample of

$$\mathcal{R}_W f(t, \boldsymbol{\omega}) = \int_{-\infty}^{\infty} w(u) \mathcal{R}f(t - u, \boldsymbol{\omega}).$$

If all the relevant data

$$\{\mathcal{R}_W f(t, \omega) \; : \; t \in [-L, L], \omega \in S^1\}$$

were available, then the filtered back-projection reconstruction, with filter function ϕ, would be

$$f_{\phi,w}(x, y) = (\mathcal{R}^* Q_\phi \mathcal{R}_W f)(x, y).$$

Here $\mathcal{R}^*$ denotes the back-projection operation and

$$\begin{aligned} Q_\phi g(t, \omega) &= \int_{-\infty}^{\infty} g(t - s, \omega)\phi(s)\, ds \\ &= \frac{1}{2\pi} \int_{-\infty}^{\infty} \tilde{g}(r, \omega)\hat{\phi}(r)e^{irt}\, dr. \end{aligned} \tag{12.6}$$

Because $\mathcal{R}_W f$ is defined by convolving $\mathcal{R} f$ with w in the t-parameter, it is a simple computation to see that

$$f_{\phi,w}(x, y) = \mathcal{R}^* Q_{\phi * w} \mathcal{R} f, \tag{12.7}$$

where $\phi * w$ is a one-dimensional convolution.

Using the central slice theorem in (12.6) gives

$$f_{\phi,w}(x, y) = \frac{1}{[2\pi]^2} \int_0^{\infty} \int_0^{2\pi} \hat{f}(r\omega)\hat{\phi}(r)\hat{w}(r)e^{ir\langle (x,y),\omega\rangle} dr\, d\omega. \tag{12.8}$$

As $\hat{\phi}(r) \approx |r|$ for small r, it is reasonable to assume that $\hat{\phi}(0) = 0$ and define $\hat{\psi}(r)$ by the equation

$$\hat{\phi}(r) = |r|\hat{\psi}(r).$$

Substituting this into (12.8) we recognize $r\, dr\, d\omega$ as the area element on $\mathbb{R}^2$, to get

$$f_{\phi,w}(x, y) = \frac{1}{[2\pi]^2} \int_{\mathbb{R}^2} \hat{f}(\xi)\hat{\psi}(\|\xi\|)\hat{w}(\|\xi\|)e^{i\langle (x,y),\xi\rangle} d\xi. \tag{12.9}$$

The MTF for the operation $f \mapsto f_{\phi,w}$ is therefore

$$\hat{\Psi}_0(\xi) = \hat{\psi}(\|\xi\|)\hat{w}(\|\xi\|). \tag{12.10}$$

It is important to keep in mind that the Fourier transforms on the right-hand side of (12.10) are *one* dimensional, while that on the left is a *two*-dimensional transform.

The PSF is obtained by applying the inverse Fourier transform to the MTF:

$$\begin{aligned}\Psi_0(x, y) &= \frac{1}{[2\pi]^2} \int_{\mathbb{R}^2} \hat{\psi}(\|\xi\|)\hat{w}(\|\xi\|)e^{i\langle(x,y),\xi\rangle}d\xi \\ &= \frac{1}{2\pi} \int_0^\infty \hat{\psi}(r)\hat{w}(r)J_0(r\rho)r\,dr,\end{aligned} \tag{12.11}$$

where $\rho = \|(x, y)\|$ denotes the radius function in the spatial variables. Sampling is not included in this model; however, if $[-B, B]$ is the effective passband of $\hat{\psi}$, then Nyquist's theorem implies that $d = \frac{\pi}{B}$ is a reasonable proxy for the sample spacing. In the following examples we examine the relationship between the beam width and this proxy for sample spacing. If we replace the beam width, δ, by $\alpha\delta$ ($w(u) \mapsto \alpha^{-1}w(\alpha^{-1}u)$) and the sample spacing, d, by βd ($\phi(t) \mapsto \phi(\beta^{-1}t)$), then the PSF becomes

$$\frac{1}{2\pi\alpha^2} \int_0^\infty \hat{\psi}(\frac{\beta}{\alpha}r)\hat{w}(r)J_0(r\frac{\rho}{\alpha})r\,dr.$$

Hence, for a given beam profile and filter function, and up to an overall scaling in ρ, the *qualitative properties* of the PSF depend only on the ratio $\frac{\delta}{d}$, which is the number of *samples per beam width*. In most of the examples that follow we use values of δ and d with ratios between .5 and 2.

***Example* 12.2.1.** For the first example we consider the result of using a sharp cutoff in frequency space. The apodizing function for the filter is $\hat{\psi}(r) = \chi_{[-\frac{\pi}{d},\frac{\pi}{d}]}(r)$, with the beam profile function

$$w_\delta = \frac{1}{2\delta}\chi_{[-\delta,\delta]}.$$

The MTF is given by

$$\hat{\Psi}_0(\xi) = \operatorname{sinc}(\delta\|\xi\|)\chi_{[-\frac{\pi}{d},\frac{\pi}{d}]}(\|\xi\|).$$

Figure 12.6(a) shows the PSFs with $d = 1$, $\delta = .5$, 1, and 2; Figure 12.6(b) shows the corresponding MTFs.

Notice the large oscillatory side lobes when the effective sample spacing is greater than the beam width (i.e., $\delta = .5$). Using a filter of this type may lead to severe Gibbs artifacts; that is, a sharp edge in the original image produces large oscillations, parallel to the edge in the reconstructed image. On the other hand, the width of the central peak in the PSF grows as the width of w increases. This is indicative of the lower resolution available in the measured data. With two samples per beam width ($\delta = 2$), the PSF is no longer peaked at zero. In the imaging literature this depression of the PSF near zero is called the *volcano effect*.

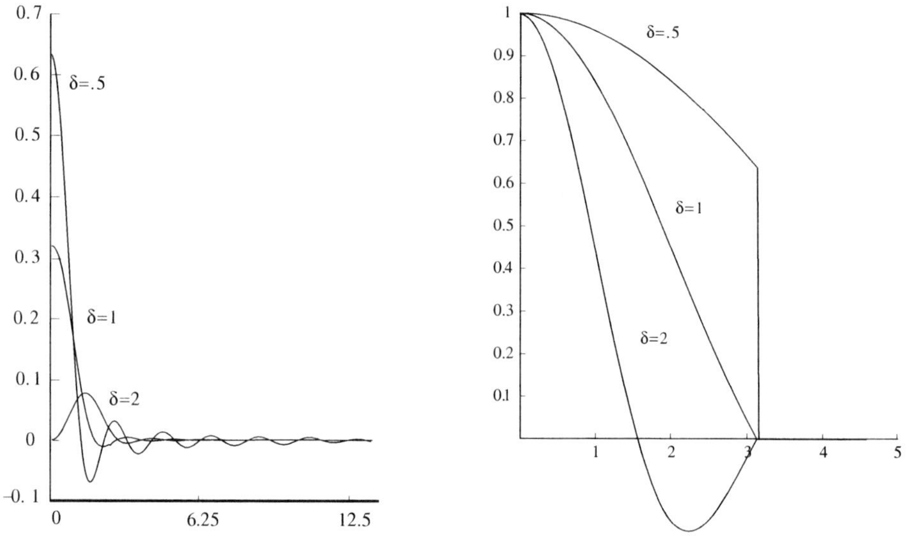

(a) Several PSFs. (b) Their MTFs.

Figure 12.6. Examples of PSF and MTF with bandlimiting regularization.

***Example* 12.2.2.** We consider the family of examples with w_δ as in the previous example and smoother apodizing function

$$\hat{\psi}_\epsilon(r) = e^{-\epsilon|r|};$$

the MTF is given by

$$\hat{\Psi}_{0(\delta,\epsilon)}(\boldsymbol{\xi}) = \operatorname{sinc}(\delta\|\boldsymbol{\xi}\|)e^{-\epsilon\|\boldsymbol{\xi}\|}.$$

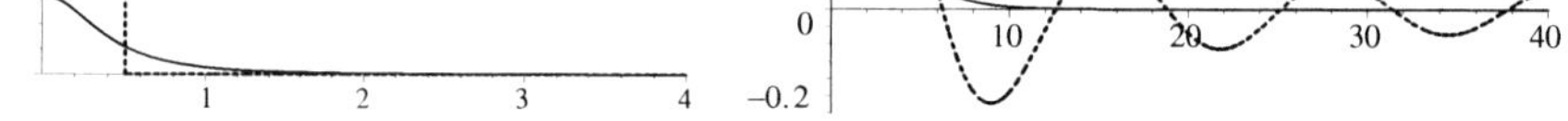

(a) Limiting PSFs. (b) Limiting MTFs.

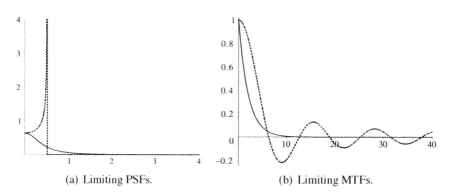

Figure 12.7. Limits for the PSF and MTF in the filtered back-projection algorithm.

If $\epsilon = 0$ (no regularizing function) or $\delta = 0$ (1-dimensional x-ray beam) then the integrals defining Ψ_0 exist as improper Riemann integrals,

$$\Psi_{0(\delta,0)}(\rho) = \frac{1}{2\pi\delta} \cdot \frac{\chi_{[0,\delta)}(\rho)}{\sqrt{\delta^2 - \rho^2}}, \quad \Psi_{0(0,\epsilon)}(\rho) = \frac{\epsilon}{2\pi} \cdot \frac{1}{[\epsilon^2 + \rho^2]^{\frac{3}{2}}}.$$

The graphs of these functions are shown Figure 12.7(a); the dotted curve shows $\Psi_{0(.5,0)}$ and the solid line is $\Psi_{0(0,.5)}$. Figure 12.7(b) shows the corresponding MTFs. The PSF in the $\epsilon = 0$ case displays an extreme version of the volcano effect.

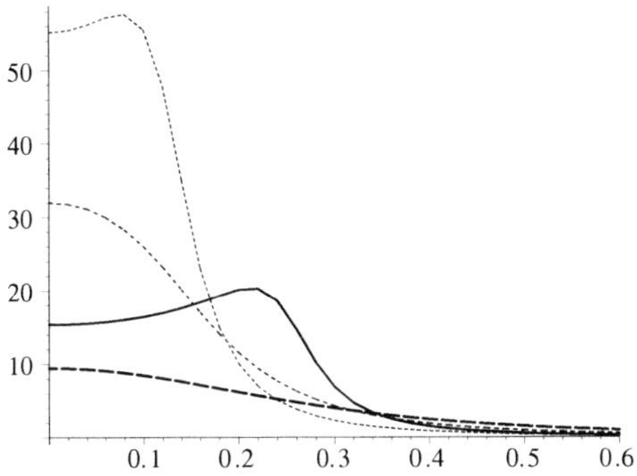

(a) Several PSFs.

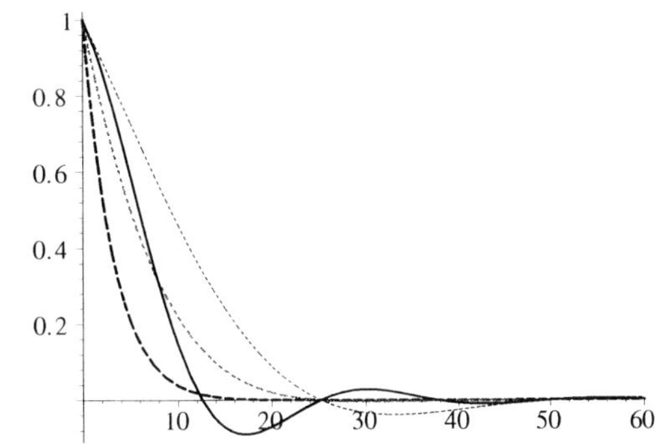

(b) Their MTFs.

Figure 12.8. Examples of PSF and MTF with exponential regularization.

Graphs of $\Psi_{0(\delta,\epsilon)}(\rho)$ for several values of (δ, ϵ) are shown in Figure 12.8(a). The MTFs are shown in Figure 12.8(b). The values used are (.25, .05), (.125, .05), (.125, .125), and (.125, .3); smaller values of ϵ produce a sharper peak in the PSF. For $\epsilon << \delta$ the PSF resembles the limiting case, $\epsilon = 0$, with a crater near $\rho = 0$.

***Example* 12.2.3.** As a final example we consider the Shepp-Logan filter. The regularizing filter has (one-dimensional) transfer function

$$\hat{\phi}(r) = |r| \left|\operatorname{sinc}\left(\frac{dr}{2}\right)\right|^3.$$

Using the same model for the source-detector pair as before gives the total MTF:

$$\hat{\Psi}_{0(\delta,d)}(\xi) = \operatorname{sinc}(\delta\|\xi\|) \left|\operatorname{sinc}\left(\frac{d\|\xi\|}{2}\right)\right|^3.$$

Recall that the Shepp-Logan filter is linearly interpolated and d represents the sample spacing. Here 2δ is the width of the source-detector pair. Graphs, in the radial variable of the PSFs and corresponding MTFs, for the pairs (.125, .05), (.125, .125), and (.125, .3) are shown in Figure 12.9. Again smaller values of d produce a more sharply peaked PSF.

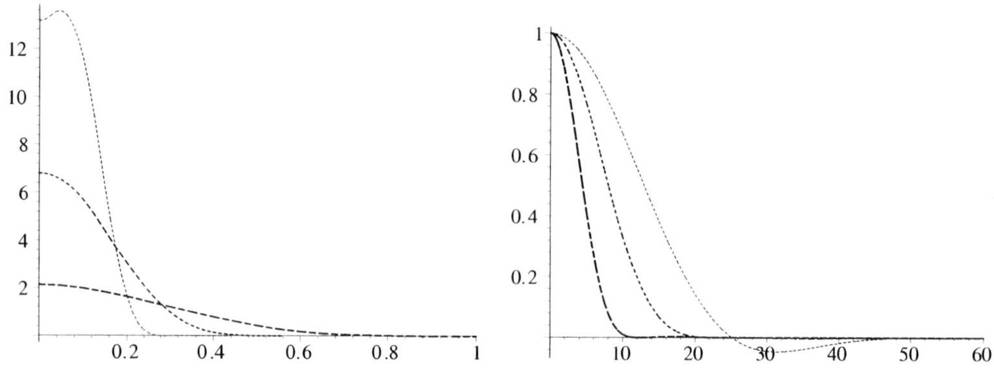

(a) Several PSFs. (b) Their MTFs.

Figure 12.9. Examples of PSF and MTF with Shepp-Logan regularization.

It is apparent in the graphs of the PSFs with exponential and Shepp-Logan regularization that these functions do not have long oscillatory tails and so the effect of convolving a piecewise continuous, bounded function with Ψ_0 should be an overall blurring, *without* oscillatory artifacts. Such artifacts are absent because the MTF decays *smoothly* and sufficiently rapidly to zero. The PSFs obtained using a sharp cutoff in frequency have long oscillatory tails, which, in turn, produce Gibbs artifacts in the reconstructed images. Oscillatory artifacts can also result from sampling. This is considered in the following section. From both (12.11) and (12.12) it is clear that the roles of the beam profile w and the filter function ϕ are entirely interchangeable in the unsampled PSF. This is no longer the case after sampling is done in the t-parameter.

Examining the graphs of the PSFs, it appears that, once the beam width is fixed, the full-width half-maximum of the PSF is not very sensitive to the sample spacing. However, smaller sample spacing produces a sharper peak, which should, in turn, lead to less blurring in the reconstructed image. From the limiting case shown in Figure 12.7(a), it is clear that the resolution is ultimately limited by the beam width. Since the PSF tends to infinity, the FWHM definition of resolution is not applicable. Half of the volume under the PSF (as a radial function on $\mathbb{R}^2$) lies in the disk of radius $d/2$, indicating that the maximum available resolution, with the given beam profile, is about half the width of the beam. This is in good agreement with experimental results, which show that two samples per beam width lead to a better reconstruction, and little improvement is seen beyond four samples per beam width; see [74] or [75]. To measure the resolution of a CT machine or reconstruction algorithm, it is customary to use a "resolution phantom." This is an array of disks of various sizes with various spacings. An example is shown in Figure 12.10.

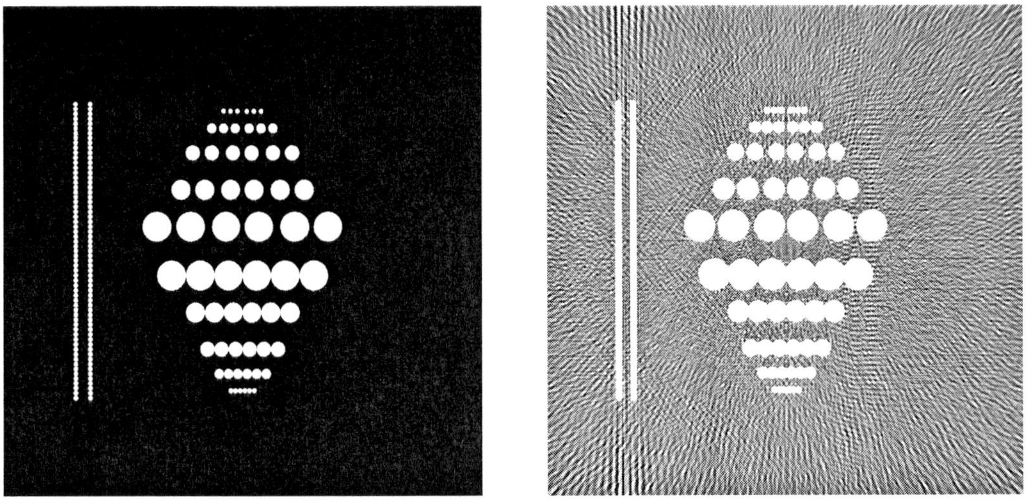

(a) A resolution phantom. (b) Its reconstruction using a fan beam algorithm.

Figure 12.10. Resolution phantoms are used to measure the actual resolution of a CT scanner and/or reconstruction algorithm.

Exercises

***Exercise* 12.2.2.** Using the formula for $\mathcal{R}\delta_{(a,b)}$, derive the alternate expression for $\Psi_0(x, y)$:

$$\Psi_0(x, y) = \frac{1}{2\pi}\int_0^{\pi}\int_{-\infty}^{\infty} w(\langle(x, y), \boldsymbol{\omega}\rangle) - s)\phi(s)ds d\boldsymbol{\omega}. \tag{12.12}$$

***Exercise* 12.2.3.** By considering the decay properties of the MTFs in Examples 12.2.2 and 12.2.3, explain why one does not expect the PSFs to have slowly decaying, oscillatory tails.

***Exercise* 12.2.4.** Using the graphs in the preceding examples, 12.2.1, 12.2.2 and 12.2.3, determine the FWHM of each of the PSFs for which it makes sense. For which cases is the FWHM definition of resolution inapplicable? What definition of resolution would be more meaningful in these cases?

***Exercise* 12.2.5.** When replacing the beam width δ by $\alpha\delta$, why do we replace $w(u)$ with $\alpha^{-1}w(\alpha^{-1}u)$? When replacing the sample spacing d by βd, why do we replace $\phi(t)$ with $\phi(\beta^{-1}t)$?

12.3 The PSF with Sampling

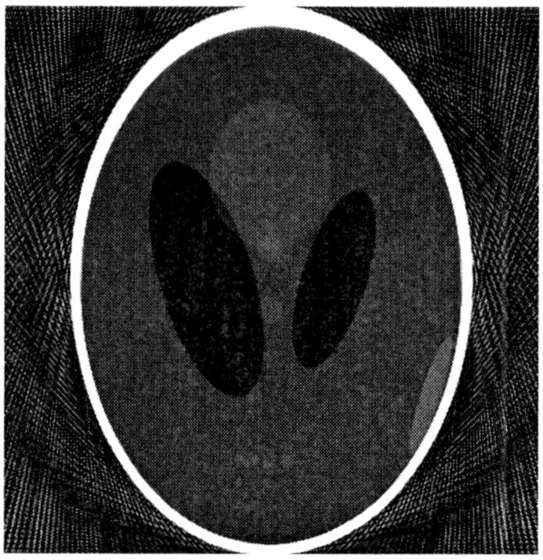

(a) Reconstruction using parallel beam data.

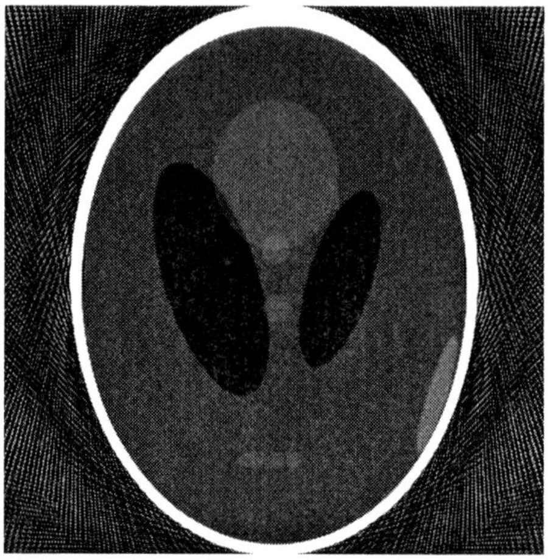

(b) Reconstruction using fan beam data.

Figure 12.11. Reconstructions of the Shepp-Logan phantom using filtered back-projection algorithms.

Real measurements entail both ray and view sampling. For a parallel beam machine, ray sampling refers to sampling in the t-parameter and view sampling to the $\boldsymbol{\omega}$- (or θ) parameter. For the sake of simplicity, these effects are usually considered separately. We take this route, first finding the kernel function incorporating ray sampling and then view sampling. Each produces distinct artifacts in the reconstructed image. As ray sampling is not a shift invariant operation, the measurement and reconstruction process can no longer be described by a single PSF but instead requires a different integrand for each point in the reconstruction grid. For the purpose of comparison, the PSFs are often evaluated at the center of reconstruction grid [i.e., $(0, 0)$], though it is also interesting to understand how certain artifacts depend on the location of the source. Both aliasing artifacts and the Gibbs phenomenon are consequences of slow decay in the Fourier transform, which is typical of functions that change abruptly. We therefore would expect to see a lot of oscillatory artifacts produced by inputs with sharp edges. To test algorithms, we typically use the characteristic functions of disks or polygons placed at various locations in the image. Figure 12.11 shows reconstructions of the Shepp-Logan phantom, Figure 11.3 made with filtered back-projection algorithms. Note the oscillatory artifacts parallel to sharp boundaries as well as the patterns of oscillations in the exterior region.

12.3.1 Ray Sampling

Suppose that d is the sample spacing in the t-parameter and that the image is reconstructed using a linearly interpolated filter. The choice of interpolation scheme is important because it affects the rate of decay for the MTF (see the appendix to Section 11.4.4). For the purposes of this section, we suppose that sampling is only done in the affine parameter so that

$$\{\mathcal{R}_W f(jd, \boldsymbol{\omega}) : j = -N, \ldots, N\}$$

is collected for all $\boldsymbol{\omega} \in S^1$. With ϕ the filter function, the reconstructed image is

$$\tilde{f}_{\phi,w}(x, y) = \frac{1}{2\pi} \int_0^{2\pi} Q_{\phi,w} \tilde{f}(\langle (x, y), \boldsymbol{\omega} \rangle, \boldsymbol{\omega})\, d\boldsymbol{\omega}, \tag{12.13}$$

where

$$Q_{\phi,w} \tilde{f}(t, \boldsymbol{\omega}) = d \sum_{j=-\infty}^{\infty} \phi(t - jd) \mathcal{R}_W f(jd, \boldsymbol{\omega}).$$

For the derivation of the PSF, we let $f = \delta_{(a,b)}$, as noted previously:

$$\mathcal{R}_W f(jd, \boldsymbol{\omega}) = w(jd - \langle (a, b), \boldsymbol{\omega} \rangle).$$

The linear interpolation is easier to incorporate in the Fourier representation. If

$$\hat{\phi}_P(r) = \sum_{j=-\infty}^{\infty} \phi(jd) e^{-ijdr},$$

then, as shown in Section 11.4.4,

$$\hat{\phi}(r) = \operatorname{sinc}^2\left(\frac{rd}{2}\right) \hat{\phi}_P(r).$$

The Fourier transform of $Q_{\phi,w} \tilde{f}$ in the t-variable is given by

$$\widetilde{Q_{\phi,w} \tilde{f}}(r, \boldsymbol{\omega}) = d \operatorname{sinc}^2\left(\frac{rd}{2}\right) \hat{\phi}_P(r) \sum_{j=-\infty}^{\infty} w(jd - \langle (a, b), \boldsymbol{\omega} \rangle) e^{-ijdr}. \tag{12.14}$$

To evaluate the last sum, we can use the dual Poisson summation formula, (8.13), obtaining

$$d \sum_{j=-\infty}^{\infty} w(jd - \langle (a, b), \boldsymbol{\omega} \rangle) e^{-ijdr} = e^{-i\langle (a,b), r\boldsymbol{\omega} \rangle} \sum_{j=-\infty}^{\infty} \hat{w}(r + \frac{2\pi j}{d}) e^{-i\frac{2\pi j}{d}\langle (a,b), \boldsymbol{\omega} \rangle}. \tag{12.15}$$

For this computation to be valid we need to assume that both w and $\hat{w}$ decay sufficiently rapidly for the Poisson summation formula to be applicable. In particular, w must be smoother than the functions, w_δ, used in Example 12.2.2.

Using the Fourier inversion formula to express $Q_{\phi,w} \tilde{f}$ in (12.13) gives the kernel function,

$$\Psi(x, y; a, b) = \frac{1}{[2\pi]^2} \int_0^{\infty} \int_0^{2\pi} \operatorname{sinc}^2\left(\frac{rd}{2}\right) \hat{\phi}_P(r) e^{i\langle (x-a, y-b), r\boldsymbol{\omega} \rangle} \times \left[\sum_{j=-\infty}^{\infty} \hat{w}(r + \frac{2\pi j}{d}) e^{-i\frac{2\pi j}{d}\langle (a,b), \boldsymbol{\omega} \rangle} \right] dr\, d\boldsymbol{\omega}. \tag{12.16}$$

It is apparent that Ψ is not a function of $(x-a, y-b)$. The symmetry in the roles played by ϕ and w has also been lost. The infinite sum in (12.16) leads to aliasing errors, a sharper beam profile producing larger errors. This infinite sum defines a $\frac{2\pi}{d}$-periodic function of r. In terms of the rate of decay, the principal difference between the integrand in (12.11) and that in (12.16) is that the decay coming from $\hat{w}(\|\xi\|)$ is lost in (12.16).

***Example* 12.3.1.** We first consider simple bandlimiting with the apodizing filter defined by $\hat{\psi}_d(r) = \chi_{[-\frac{\pi}{d},\frac{\pi}{d}]}(r)$ and the rectangular windows defined by w_δ. In this case we can find an explicit formula for $\hat{\phi}_p$. At the sample points we have

$$\phi_d(jd) = \begin{cases} \frac{\pi}{2d^2} & \text{if } j = 0, \\ \frac{(-1)^j - 1}{\pi j^2 d^2} & \text{if } j \neq 0. \end{cases}$$

From the definition it follows that

$$\hat{\phi}_p(r) = \frac{\pi}{2d^2} - \frac{4}{\pi d^2} \sum_{j=1}^{\infty} \frac{\cos(2j-1)rd}{(2j-1)^2}. \tag{12.17}$$

The function on the right-hand side of (12.17) has a simple formula: It is a $\frac{2\pi}{d}$-periodic function with

$$\hat{\phi}_p(r) = \frac{|r|}{d} \qquad \text{for } |r| < \frac{\pi}{d}. \tag{12.18}$$

Though the functions, w_δ, are too singular to apply the Poisson summation argument, the sums on the left-hand side of (12.15) are finite. The MTF for this combination of filtering, sample spacing, and beam profile is

$$\begin{aligned} \hat{\Psi}_{\delta,d}(\xi; a, b) =& d \operatorname{sinc}^2\left(\frac{\|\xi\| d}{2}\right) \hat{\phi}_p(\|\xi\|)\|\xi\|^{-1} \times \\ & \sum_{j=-\infty}^{\infty} w_\delta\left(jd - \left\langle (a,b), \frac{\xi}{\|\xi\|}\right\rangle\right) e^{-ijd\|\xi\|}. \end{aligned} \tag{12.19}$$

For example, if $d = 1$, $\delta = .5$ and $(a, b) = (0, 0)$, then only the $j = 0$ term is nonzero and this reduces to

$$\hat{\Psi}_{.5,1}(\xi; 0, 0) = \operatorname{sinc}^2\left(\frac{\|\xi\|}{2}\right) \hat{\phi}_p(\|\xi\|)\|\xi\|^{-1}.$$

The graphs in Figure 12.12(a) are radial graphs of the PSFs for $(a, b) = (0, 0)$ with $d = 1$ and $\delta = .5, 1$ and 2. The graphs in Figure 12.12(b) show the corresponding MTFs. It is important to remember that the overall filtering operation is *not* shift invariant and the kernel function $\Psi_{\delta,d}$ depends in a nontrivial way on (a, b). Contour plots illustrating this fact are shown in Figure 12.13.

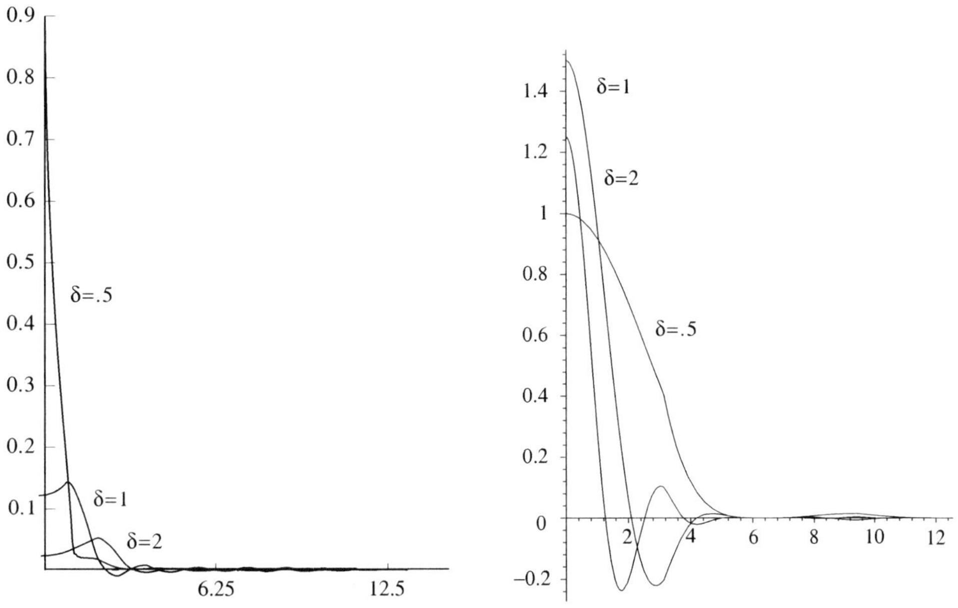

(a) The PSFs with $d = 1$, $\delta = .5, 1, 2$. (b) The MTFs with $d = 1, \delta = .5, 1, 2$.

Figure 12.12. Sampled PSFs and MTFs at $(0, 0)$ using bandlimiting, a rectangular beam profile, and linear interpolation.

The graphs in Figure 12.12 should be compared with the unsampled versions in Figure 12.6. Somewhat paradoxically, the side lobes of the sampled PSF are not much larger than those of the unsampled PSF. This is because the linear interpolation used to define ϕ between sample points leads to a smoother MTF, which decays like $\|\xi\|^{-3}$. For $\delta = .5$ the side lobes in the sampled PSF are smaller! In this case the unsampled MTF has a jump discontinuity at π, which explains the sinc-like behavior in the PSF. The linear interpolation is also evident in the graphs in Figure 12.12(a). As the beam width increases (hence more samples per beam width), the peak of the PSF broadens and it displays a more pronounced volcano effect.

We conclude this example with contour plots of $\Psi_{1,1}(x, y; a, b)$ for several values of (a, b). The plots in Figure 12.13 were computed using a finite approximation to the inverse Fourier transform and rectangular partial sums. Figure 12.13(a) is a contour plot of $\Psi_{1,1}(x, y; 0, 0)$. Near to the principal peak the contours are circles; closer to the boundary of the plot the radial symmetry starts to break down. This is a consequence of the rectangular partial sums used in the numerical computation. Such sums are used in actual reconstructions, and we also see artifacts with a similar rectangular symmetry. These artifacts are not caused by noise, measurement error, or even by sampling, per se but rather by the precise nature of the approximation to the inverse Fourier transform used in the numerical algorithms. Figures 12.13(b–d) show $\Psi_{1,1}(x, y; a, b)$ for three other values of (a, b). Each plot shows a very pronounced peak centered at (a, b). As none of these plots displays even an approximate radial symmetry, it is clear that they are not the result of translating $\Psi_{1,1}(x, y; 0, 0)$. Note finally that the Gibbs oscillation is most pronounced for $(a, b) = (0, 0)$.

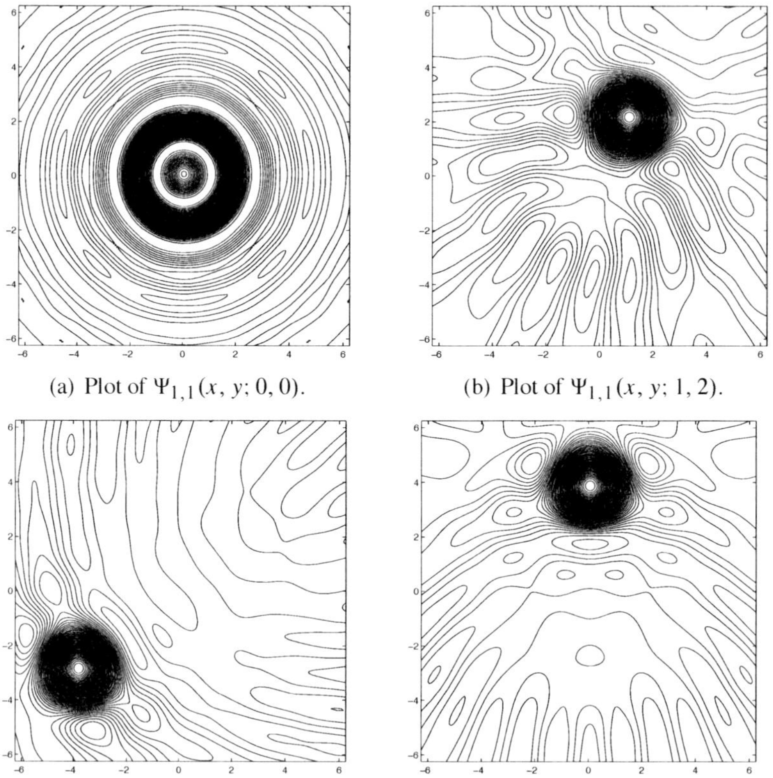

(a) Plot of $\Psi_{1,1}(x, y; 0, 0)$.

(b) Plot of $\Psi_{1,1}(x, y; 1, 2)$.

(c) Plot of $\Psi_{1,1}(x, y; -3, -4)$.

(d) Plot of $\Psi_{1,1}(x, y; 0, 4)$.

Figure 12.13. Contour plots of the ray sampled kernel function using a rectangular beam profile, and simple bandlimiting with linear interpolation. These plots show that once sampling is done in the affine parameter, the measurement-reconstruction process is no longer shift invariant.

***Example* 12.3.2.** In this example we use w_δ convolved with a Gaussian for the beam profile. The window function w_δ models the detector, while the Gaussian models the x-ray source. A Gaussian focal spot of "width" $h/\sqrt{2}$ is described by the function

$$s_h(u) = \frac{1}{\pi h^2} e^{-\left(\frac{u}{h}\right)^2}.$$

For our examples we fix $h = \frac{1}{2}$ and $\delta = 1$. The total beam profile is then

$$w(u) = \frac{1}{2} \int_{-1}^{1} s_{\frac{1}{2}}(u - v)\, dv,$$

with

$$\hat{w}(r) = \operatorname{sinc}(r)e^{-\left(\frac{r^2}{16}\right)}.$$

We use the Shepp-Logan regularizing filter, for which

$$\hat{\phi}_p(r) = |r \cdot \operatorname{sinc}\left(\frac{rd}{2}\right)|.$$

As before, the issue here is the relationship between d, the sample spacing, and the width of w. Colloquially, we ask for the number of samples per beam width. With the given parameters, the FWHM(w) is very close to 2.

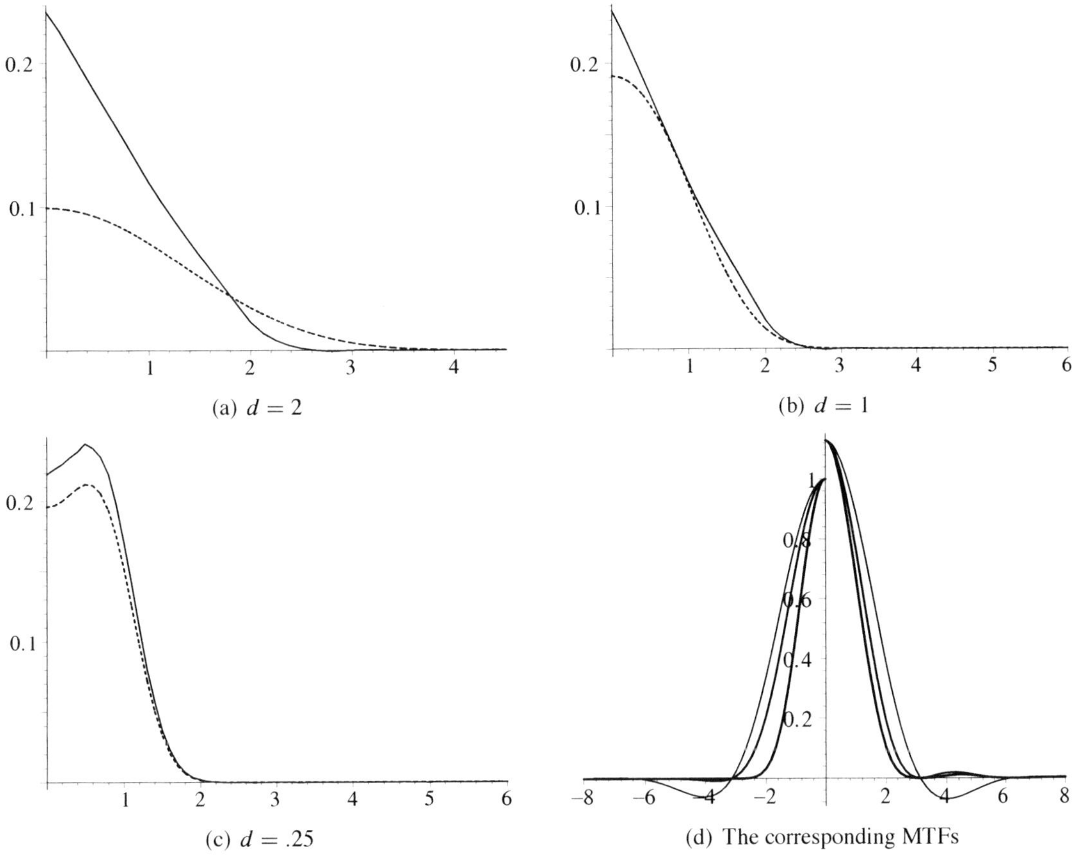

Figure 12.14. The effect of ray sampling on the PSF.

While the overall filtering operation is no longer isotropic, the function $\Psi(x, y; 0, 0)$ is radial; Figures (12.14)(a–c) show this function (of ρ) with various choices for d, with and without the effects of ray sampling. The dotted line is the unaliased PSF and the solid line the aliased. As before, smaller values of d give rise to sharper peaks in the PSF. The corresponding MTFs are shown in Figure 12.14(d).

The graphs on the right-hand side of Figure 12.14(d) include the effect of aliasing, while those on the left are the unaliased MTFs; as d decreases, the passband of the MTF broadens. With this choice of beam profile and regularizing filter, once there is at least one sample per beam width, the resolution, measured by FWHM, is not affected much by aliasing. Though it is not evident in the pictures, these PSFs also have long oscillatory tails. The small amplitude of these tails is a result of using a smooth, rapidly decaying apodizing function.

Exercises

***Exercise* 12.3.1.** When doing numerical computations of Ψ, it is sometimes helpful to use the fact that

$$d \sum_{j=-\infty}^{\infty} w(jd - \langle (a,b), \boldsymbol{\omega} \rangle) e^{-ijdr}$$

is a periodic function. Explain this observation and describe how it might figure in a practical computation. It might be helpful to try to compute this sum approximately using both representations in (12.15).

***Exercise* 12.3.2.** Verify (12.18) by a direct computation.

***Exercise* 12.3.3.** Continue the computations begun in Example 12.3.1, and draw plots of $\Psi_{\delta,d}(x, y; a, b)$ for other values of (a, b) and (δ, d). Note that $\Psi_{\delta,d}$ is no longer a radial function of (x, y), so a two-dimensional plot is required. Repeat this experiment with the beam profile and apodizing function used in Example 12.3.2.

12.3.2 View Sampling

We now turn to artifacts that result from using finitely many views and begin by considering the reconstruction of a mathematical phantom made out of constant density elliptical regions. In Figure 12.15 note the pattern of oscillations in the exterior region along lines tangent to the boundary of ellipse and the absence of such oscillations in the interior. A somewhat subtler observation is that the very pronounced, coherent pattern of oscillations does not begin immediately but rather at a definite distance from the boundary of the ellipse. This phenomenon is a consequence of sampling in the angular parameter and the filtering operations needed to invert the Radon transform approximately. Our discussion of these examples closely follows that in [114].

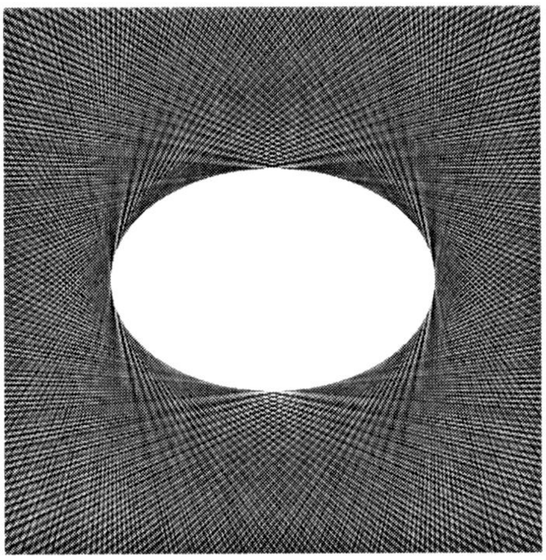

Figure 12.15. Filtered back-projection reconstruction of elliptical phantom.

***Example* 12.3.3.** Suppose the object, E, is of constant density 1 with boundary the locus of points $\frac{x^2}{a^2} + \frac{y^2}{b^2} = 1$. The line integral of $f = \chi_E$ along a line $l_{t,\theta}$ is simply the length of the intersection of the line with E (see Figure 12.16). Let $s_\pm(t, \theta)$ denote the s-parameters for the intersection points of the line $l_{t,\theta}$ with the boundary of E. The distance between these two points $|s_+(t, \theta) - s_-(t, \theta)|$ is $\mathscr{R} f(t, \boldsymbol{\omega}(\theta))$. Plugging the parametric form of the line into the equation for the ellipse and expanding gives

$$s^2\left(\frac{\sin^2\theta}{a^2} + \frac{\cos^2\theta}{b^2}\right) + 2st\sin\theta\cos\theta\left(\frac{1}{b^2} - \frac{1}{a^2}\right) + \frac{t^2\cos^2\theta}{a^2} + \frac{t^2\sin^2\theta}{b^2} - 1 = 0.$$

Rewrite the equation as

$$p(t, \theta)s^2 + q(t, \theta)s + r(t, \theta) = 0,$$

where p, q and r are the corresponding coefficients. The two roots are given by

$$s_\pm = \frac{-q \pm \sqrt{q^2 - 4ps}}{2p};$$

the distance between the two roots is therefore

$$s_+ - s_- = \frac{\sqrt{q^2 - 4ps}}{p}.$$

This gives the formula for $\mathscr{R} f$:

$$\mathscr{R} f(t, \boldsymbol{\omega}(\theta)) = \begin{cases} 2\beta(\theta)\sqrt{\alpha(\theta)^2 - t^2} & |t| \leq \alpha(\theta), \\ 0 & |t| > \alpha(\theta), \end{cases}$$

where

$$\alpha(\theta) = \sqrt{\frac{a^4\cos^2(\theta) + b^4\sin^2(\theta)}{a^2\cos^2(\theta) + b^2\sin^2(\theta)}},$$

$$\beta(\theta) = \sqrt{\frac{(a^2\cos^2(\theta) + b^2\sin^2(\theta))(b^2\cos^2(\theta) + a^2\sin^2(\theta))}{a^4\cos^2(\theta) + b^4\sin^2(\theta)}}.$$

Both α and β are smooth, nonvanishing functions of θ.

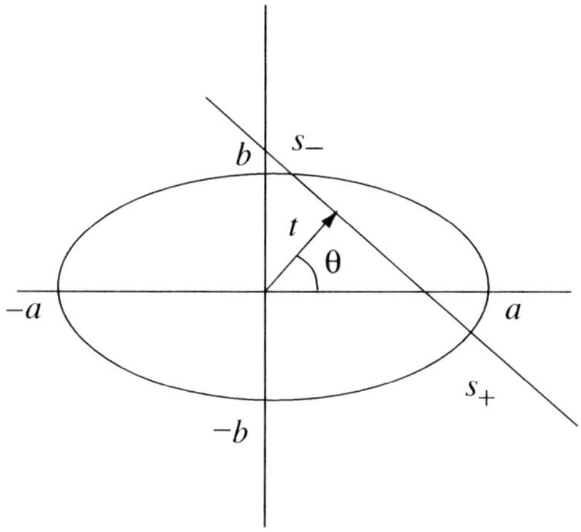

Figure 12.16. Parameters describing the Radon transform of χ_E.

Doing the exact filtration,

$$\mathcal{G}\mathcal{R}f(t, \boldsymbol{\omega}(\theta)) = \frac{1}{i}\mathcal{H}\partial_t\mathcal{R}f(t, \boldsymbol{\omega}(\theta)),$$

gives

$$\mathcal{G}\mathcal{R}f(t, \boldsymbol{\omega}(\theta)) = \begin{cases} 2\beta(\theta) & |t| \le \alpha(\theta), \\ 2\beta(\theta)\left(1 - \frac{|t|}{\sqrt{t^2 - \alpha^2(\theta)}}\right) & |t| > \alpha(\theta). \end{cases}$$

In an actual reconstruction this filter is regularized. Let $Q_\phi f$ denote the approximately filtered measurements, with filter function ϕ. Approximating the back-projection with a Riemann sum gives

$$\tilde{f}_\phi(x, y) \approx \frac{1}{2(M+1)}\sum_{k=0}^{M} Q_\phi f(\langle (x, y), \boldsymbol{\omega}(k\,\Delta\theta)\rangle, \boldsymbol{\omega}(k\,\Delta\theta)).$$

For points inside the ellipse,

$$\tilde{f}_\phi(x, y) \approx \frac{1}{2(M+1)}\sum_{k=0}^{M} \beta(k\,\Delta\theta).$$

This is well approximated by the Riemann sum because β is a smooth bounded function. For points outside the ellipse, there are three types of lines that appear in the back-projection:

1. Lines that pass through E

2. Lines that are distant from E

3. Lines outside E that pass very close to the boundary of E

The first two types of lines are not problematic. However, for any point in the exterior of the ellipse, the back-projection involves lines that are exterior to the ellipse but pass very close to it. This leads to the oscillations apparent in the reconstruction along lines tangent to the boundary of the regions. This is a combination of the Gibbs phenomenon and aliasing. To compute an accurate value for $\tilde{f}_\phi$ at an exterior point requires delicate cancellations between the moderate positive and sometimes large negative values assumed by $Q_\phi f$. For points near enough to the boundary of E, there is a sufficient density of samples to obtain the needed cancellations. For more distant points, the cancellation does not occur and the pattern of oscillations appears. In the next paragraph we derive a "far field" approximation to the reconstruction of such a phantom. This gives, among other things, a formula for the radius where the oscillatory artifacts first appear. In [72] such a formula is derived for a fan beam scanner. Using the approach of [72], we derive a similar formula for the parallel beam case. Also apparent in Figure 12.15 is an oscillation very near and *parallel* to the boundary of E; this is the usual combination of the Gibbs phenomenon and aliasing caused by ray sampling.

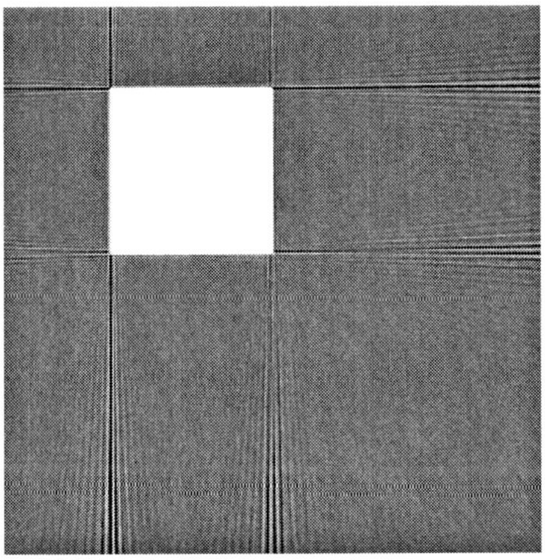

Figure 12.17. Filtered back-projection reconstruction of square phantom.

Example **12.3.4.** A striking example of this phenomenon can be seen in the reconstruction of a rectangular region. If f is the characteristic function of the square with vertices

$\{(\pm 1, \pm 1)\}$, then for $|\theta| < \frac{\pi}{4}$,

$$\mathcal{R}f(t, \boldsymbol{\omega}(\theta)) = \begin{cases} 0 & \text{if } |t| > \cos\theta + \sin\theta, \\ \frac{\cos\theta + \sin\theta - t}{\cos\theta \sin\theta} & \text{if } \cos\theta - \sin\theta < t < \cos\theta + \sin\theta, \\ \frac{2}{\cos\theta} & \text{if } \sin\theta - \cos\theta < t < \cos\theta - \sin\theta, \\ \frac{\cos\theta + \sin\theta + t}{\cos\theta \sin\theta} & \text{if } -(\cos\theta + \sin\theta) < t < \sin\theta - \cos\theta. \end{cases}$$

The function $\mathcal{R}f(t, \boldsymbol{\omega}(\theta))$ is periodic of period $\frac{\pi}{2}$ in the θ parameter. If $\theta \neq 0$, then $\mathcal{R}f(t, \theta)$ is a continuous, piecewise differentiable function in θ, whereas

$$\mathcal{R}f(t, (1, 0)) = \chi_{[-2,2]}(t)$$

has a jump discontinuity. Note the pronounced oscillatory artifact in the exterior of the square along lines tangent to the sides of the square in Figure 12.17. As before, there is also a Gibbs oscillation in the reconstructed image, parallel to the boundary of the square. This image should be compared to Figure 7.11, which illustrates the Gibbs phenomenon for two-dimensional Fourier series. Unlike Figure 7.11, the oscillatory artifact near the corner (in the "Gibbs shadow") is as large as that parallel to the edges, indicating that it is not simply the Gibbs phenomenon for the partial sums of the Fourier series.

Exercises

***Exercise* 12.3.4.** Derive the formula for $\mathcal{R}f$ and $\mathcal{G}\mathcal{R}f$ in Example 12.3.3.

***Exercise* 12.3.5.** Compute $-i\mathcal{H}\partial_t \mathcal{R}f$ for Example 12.3.4.

A Far Field Approximation for the Reconstructed Image

In this section we obtain an approximate formula for the reconstruction of a small radially symmetric object at points far from the object. Let ϕ denote a linearly interpolated filter function and w a function describing the source-detector pair. If f is the data, then the approximate, filtered back-projection reconstruction is given by

$$f_{\phi,w}(x, y) = \frac{1}{2\pi} \int_0^{\pi} Q_{\phi,w} f(\langle (x, y), \boldsymbol{\omega} \rangle, \boldsymbol{\omega})\, d\boldsymbol{\omega},$$

where

$$Q_{\phi,w} f(t, \boldsymbol{\omega}) = \int_{-\infty}^{\infty} \mathcal{R}_W f(s, \boldsymbol{\omega}) \phi(t - s)\, ds.$$

Here $\mathcal{R}_W f$ denotes the w-averaged Radon transform of f. We now consider the effect of sampling in the $\boldsymbol{\omega}$-parameter, leaving t as a continuous parameter. Equation (12.7) shows that, in this case, ϕ and w are interchangeable; the effects of finite beam width and regularizing the filter are both captured by using $\phi * w$ as the filter function. We analyze the difference between $f_{\phi,w}(x, y)$ and the Riemann sum approximation

$$\tilde{f}_{\phi,w}(x, y) = \frac{1}{4\pi} \sum_{j=0}^{M} \mathcal{Q}_{\phi,w} f(\langle (x, y), \boldsymbol{\omega}(j\, \Delta\theta)\rangle, \boldsymbol{\omega}(j\, \Delta\theta))\, \Delta\theta.$$

For simplicity, we restrict attention to functions of the form

$$f^{(a,b)}(x, y) = f((x, y) - (a, b)),$$

where f is a radial function. The averaged Radon transform of f is independent of $\boldsymbol{\omega}$; to simplify the notation, we suppress the $\boldsymbol{\omega}$-dependence, writing

$$\mathcal{R}_W f(t), \quad \mathcal{Q}_{\phi,w} f(t)$$

and therefore

$$\begin{aligned} \mathcal{R}_W f^{(a,b)}(t, \boldsymbol{\omega}) &= \mathcal{R}_W f(t - \langle (a, b), \boldsymbol{\omega}\rangle), \\ \mathcal{Q}_{\phi,w} f^{(a,b)}(t, \boldsymbol{\omega}) &= \mathcal{Q}_{\phi,w} f(t - \langle (a, b), \boldsymbol{\omega}\rangle) \end{aligned} \tag{12.20}$$

as well. From equation (12.20) we obtain

$$f_{\phi,w}^{(a,b)}(x, y) = \frac{1}{2\pi} \int_0^{\pi} \mathcal{Q}_{\phi,w} f(\langle (x, y) - (a, b), \boldsymbol{\omega}\rangle)\, d\boldsymbol{\omega}.$$

Letting $(x - a, y - b) = R(\cos\varphi, \sin\varphi)$ and $\boldsymbol{\omega}(\theta) = (\cos\theta, \sin\theta)$ gives

$$f_{\phi,w}^{(a,b)}(x, y) = \frac{1}{2\pi} \int_0^{\pi} \mathcal{Q}_{\phi,w} f(R\cos(\theta - \varphi))\, d\theta$$

with Riemann sum approximation

$$\tilde{f}_{\phi,w}^{(a,b)}(x, y) = \frac{1}{2\pi} \sum_{j=0}^{M} \mathcal{Q}_{\phi,w} f(R\cos(j\, \Delta\theta - \varphi))\Delta\theta. \tag{12.21}$$

The objects of principal interest in this analysis are small and hard. The examples presented in the previous paragraph indicate that we should concentrate on the reconstruction at points in the exterior of the object. The function $f_{\phi,w}^{(a,b)}$ is an approximation to $f * k$, where k is the inverse Radon transform of w. If w has bounded support and ϕ provides a good approximation to $-i\partial_t \mathcal{H}$, then $f_{\phi,w}^{(a,b)}$ should be very close to zero for points outside the support of $f * k$. Indeed, if w has small support, then so does k, and therefore the support of $f * k$ is a small enlargement of the support of f itself. We henceforth assume that, for points of interest,

$$f_{\phi,w}^{(a,b)}(x, y) \approx 0 \tag{12.22}$$

and therefore any significant deviation of $\tilde{f}^{(a,b)}_{\phi,w}(x, y)$ from zero is an error.

If f is the characteristic function of a disk of radius r, then $Q_{\phi,w}f(t)$ falls off rapidly for $|t| >> r$. There is a j_0 so that

$$\begin{aligned} j_0 \Delta\theta - \varphi &< \frac{\pi}{2}, \\ (j_0 + 1)\Delta\theta - \varphi &\geq \frac{\pi}{2}. \end{aligned} \tag{12.23}$$

If we let $\Delta\varphi = \frac{\pi}{2} - j_0\Delta\theta + \varphi$ then $0 < \Delta\varphi \leq \Delta\theta$ and

$$\tilde{f}^{(a,b)}_{\phi,w}(x, y) = \frac{1}{2\pi} \sum_{j=0}^{M} Q_{\phi,w} f(R \sin(j\, \Delta\theta - \Delta\varphi))\Delta\theta.$$

As the important terms in this sum are those with $|j|$ close to zero, we approximate the sine function by using

$$\sin(j\, \Delta\theta - \Delta\varphi) \approx j\, \Delta\theta - \Delta\varphi,$$

obtaining

$$\tilde{f}^{(a,b)}_{\phi,w}(x, y) \approx \frac{1}{2\pi} \sum_{j=-\infty}^{\infty} Q_{\phi,w} f(R(j\, \Delta\theta - \Delta\varphi))\Delta\theta.$$

The limits of summation have also been extended from $-\infty$ to ∞. The error this introduces is small as $\phi(t) = O(t^{-2})$.

The Poisson summation formula can be used to evaluate the last expression; it gives

$$\tilde{f}^{(a,b)}_{\phi,w}(x, y) \approx \frac{1}{2\pi R} \sum_{j=-\infty}^{\infty} \widetilde{Q_{\phi,w} f}\left(\frac{2\pi j}{R\, \Delta\theta}\right) e^{-\frac{2\pi i j\, \Delta\varphi}{\Delta\theta}}.$$

From the central slice theorem,

$$\widetilde{Q_{\phi,w} f}(\rho) = \hat{\phi}(\rho)\hat{w}(\rho)\hat{f}(\rho).$$

Assuming that $\hat{\phi}(0) = 0$ and that w is a even function gives the simpler formula

$$\tilde{f}^{(a,b)}_{\phi,w}(x, y) \approx \frac{1}{\pi R} \sum_{j=1}^{\infty} \hat{\phi} \cdot \hat{w} \cdot \hat{f}\left(\frac{2\pi j}{R\Delta\theta}\right) \cos\left(\frac{2\pi j\, \Delta\varphi}{\Delta\theta}\right). \tag{12.24}$$

In order for this sum to be negligible at a point whose distance to (a, b) is R, the angular sample spacing, $\Delta\theta$, must be chosen so that the effective support of $\hat{\phi} \cdot \hat{w} \cdot \hat{f}$ is contained in

$$\left(-\frac{2\pi}{R\Delta\theta}, \frac{2\pi}{R\Delta\theta}\right).$$

This explains why the oscillatory artifacts only appear at points that are at a definite distance from the object: For small values of R the sum itself is very small. For large enough R the product $\hat{\phi} \cdot \hat{w} \cdot \hat{f}$ is evaluated at small arguments and the sum may become large.

Suppose, for example, that $R\Delta\theta$ is such that all terms in this sum but the first are negligible. Then

$$\tilde{f}^{(a,b)}_{\phi,w}(x,y) \approx \frac{1}{\pi R}\hat{\phi}\cdot\hat{w}\cdot\hat{f}\left(\frac{2\pi}{\|(x,y)-(a,b)\|\,\Delta\theta}\right)\cos\left(\frac{2\pi\,\Delta\varphi}{\Delta\theta}\right). \tag{12.25}$$

The cosine factor produces an oscillation in the sign of the artifact whose period equals $\Delta\theta$. This is apparent in Figures 12.18 and 12.19. The amplitude of the artifact depends on the distance to the object through the product $\hat{\phi}\cdot\hat{w}\cdot\hat{f}$. This allows us to relate the angular sample spacing, needed to obtain an artifact free reconstruction in a disk of given radius, to the source-detector function w. For simplicity suppose that

$$w(u) = \frac{1}{2\delta}\chi_{[-\delta,\delta]}(u) \qquad \text{so that } \hat{w}(\rho) = \operatorname{sinc}(\rho\delta).$$

The first zero of $\hat{w}$ occurs at $\rho = \pm\frac{\pi}{\delta}$, which suggests that taking

$$\Delta\theta < \frac{2\delta}{R}$$

is a minimal requirement to get artifact-free reconstructions in the disk of radius R. This ignores the possible additional attenuation of the high frequencies resulting from $\hat{\phi}$, which is consistent with our desire to get a result that is independent of the sample spacing in the t-parameter. The estimate for $\Delta\theta$ can be rewritten as

$$\frac{\pi R}{2\delta} < \frac{\pi}{\Delta\theta}.$$

The quantity on the right-hand side is the number of samples, $M+1$, in the $\boldsymbol{\omega}$-direction. As 2δ is the width of the source, Nyquist's theorem implies that the maximum spatial frequency available in the data is about $(4\delta)^{-1}$. If we denote this by ν, then the estimate reads

$$2\pi R\nu < M+1.$$

Essentially the same result was obtained in Section 11.4.5, with much less effort! The difference in the analyses is that, in the earlier discussion, it was *assumed* that the data are effectively bandlimited to $[\frac{-\pi}{2\delta},\frac{\pi}{2\delta}]$. Here this bandlimiting is a consequence of the lowpass filtering that results from averaging over the width of the x-ray beam.

It is important to note that the artifacts that result from view sampling are present whether or not the data are sampled in the t-parameter. These artifacts can be reduced by making either ϕ or w smoother. This is in marked contrast to the result obtained for ray sampling. In that case the aliasing errors are governed solely by w and cannot be reduced by changing ϕ. If f describes a smooth object, so that $\hat{f}$ decays rapidly, then it is unlikely that view sampling aliasing artifacts will appear in the reconstruction region.

***Example* 12.3.5.** To better understand formula (12.25), we consider $(a,b) = (0,0)$ and $f(x,y) = \chi_{B_{\frac{1}{10}}}(x,y)$. For simplicity we use $w = \frac{1}{2}\chi_{[-1,1]}$ for the beam profile and the

Shepp-Logan filter with $d = .25$. We consider the right-hand side of 12.25 with $\Delta\theta = \frac{2\pi}{8}$ and $\frac{2\pi}{32}$. The three-dimensional plots in Figures 12.18(a) and 12.19(b) of the functions defined in (12.21) give an idea of how the artifacts appear in a reconstructed image. Notice that the sign of the error reverses along a circle. The two-dimensional graphs in these figures are sections of the three-dimensional plots along lines of constant φ. These graphs allow for a quantitative appreciation for the size of the errors and their dependence on $\Delta\theta$.

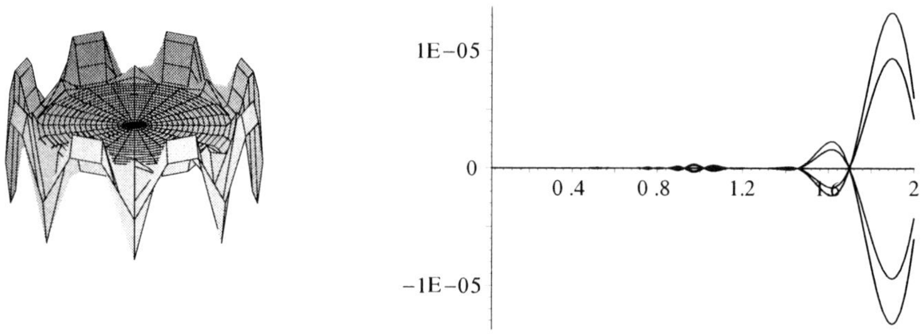

(a) 3D plot. (b) Sections along radii.

Figure 12.18. View sampling artifacts with $\Delta\theta = \frac{2\pi}{32}$.

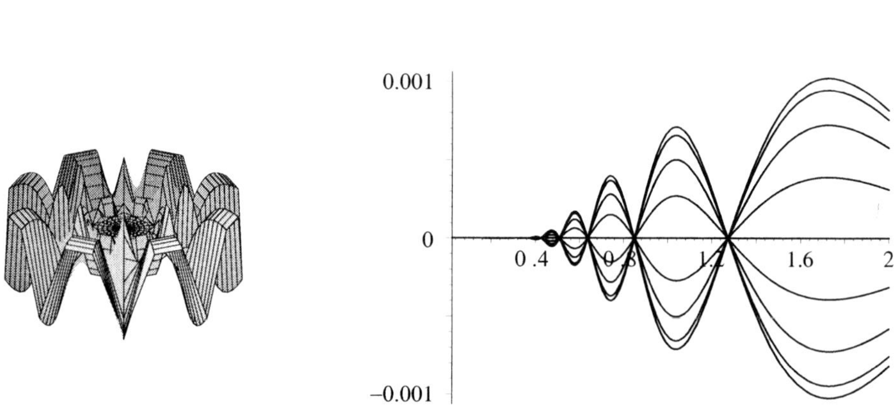

(a) 3D plot. (b) Sections along radii.

Figure 12.19. View sampling artifacts with $\Delta\theta = \frac{2\pi}{8}$.

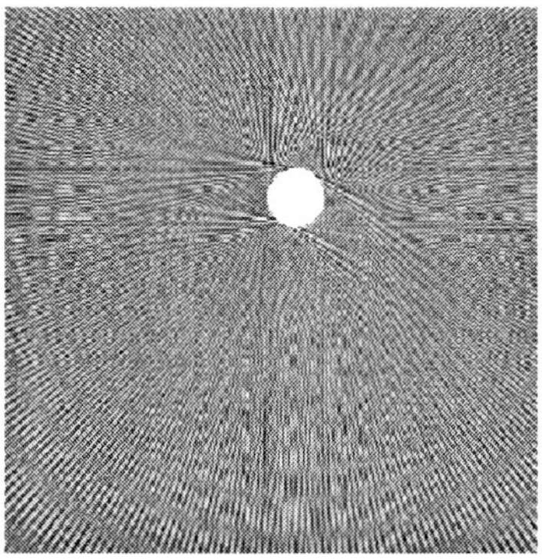

(a) Parallel beam.

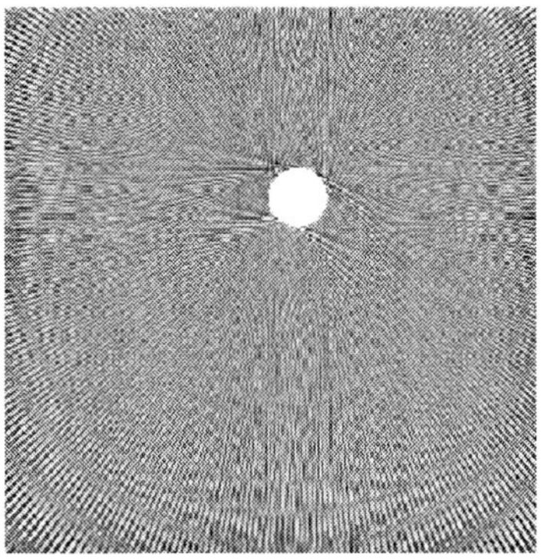

(b) Fan beam.

Figure 12.20. Examples comparing view aliasing in parallel beam and fan beam scanners. The view sampling artifact is the pattern of light and dark radial lines beginning about four diameters from the circle.

***Remark* 12.3.1.** In [72] a similar analysis is presented for a fan beam machine. The results are similar though a bit more complicated. The artifacts produced by view sampling in a fan beam machine differ in one important way: In a parallel beam machine the pattern of oscillations is circularly symmetric and depends on the distance from the center of a radially symmetric object. For a fan beam machine the pattern displays a similar circular symmetry, but the center of the circle no longer agrees, in general, with the center of the object (see Figure 12.20).

Exercises

***Exercise* 12.3.6.** Given that we use the Poisson summation formula to derive (12.24), why is it allowable to use $w(u) = (2\delta)^{-1}\chi_{[-\delta,\delta]}(u)$ in this analysis?

***Exercise* 12.3.7.** Show that the PSF for a filtered back-projection reconstruction, incorporating the beam width function w and sampling in the $\boldsymbol{\omega}$-parameter, is

$$\Psi(x, y; a, b) = \frac{\Delta\theta}{2\pi}\sum_{j=0}^{M} \phi * w(\langle(x-a, y-b), \boldsymbol{\omega}(j\,\Delta\theta)\rangle).$$

Note that this operation is shift invariant.

***Exercise* 12.3.8.** Using the result of the previous exercise, compute the MTF for a filtered back-projection algorithm, incorporating the beam width function w and sampling in the $\boldsymbol{\omega}$-parameter.

12.4 The Effects of Measurement Errors

The artifacts considered in the previous sections are algorithmic artifacts, resulting from the sampling and approximation used in any practical reconstruction method. The final class of linear artifacts we consider is the effects of *systematic* measurement errors. This should be contrasted to the analysis in Chapter 16 of the effects of *random* measurement errors or noise. Recall that the measurements made by a CT machine are grouped into views and each view is comprised of a collection of rays. We now consider the consequences of having a bad ray in a single view, a bad ray in every view, and a single bad view. These analyses illustrate a meta-principle, called the *smoothness principle*, often invoked in medical imaging:

> The filtered back-projection algorithm is very sensitive to errors that vary abruptly from ray to ray or view to view but relatively tolerant of errors that vary gradually.

This feature of the algorithm is a reflection of the fact that the filter function, $\hat{\phi}(r)$, approximates $|r|$ and therefore attenuates low frequencies and amplifies high frequencies. Our discussion is adapted from [115], [70], and [71]. Many other errors are analyzed in these references.

For this analysis we suppose that the measurements, made with a parallel beam scanner, are the samples

$$\{P(t_j, \omega(k\,\Delta\theta)) : k = 0, \dots, M, j = 1, \dots, N\}$$

of $\mathscr{R}_W f(t, \omega)$. The coordinates are normalized so that the object lies in $[-1, 1] \times [-1, 1]$. The angular sample spacing is

$$\Delta\theta = \frac{\pi}{M+1}$$

and the rays are uniformly sampled at

$$t_j = -1 + (j - \frac{1}{2})d.$$

If ϕ is the filter function, which is specified at the sample points and linearly interpolated in between, then the approximate reconstruction is given by

$$\tilde{f}_\phi(x, y) = \frac{1}{N(M+1)} \sum_{k=0}^{M} \sum_{j=1}^{N} P(t_j, \omega(k\,\Delta\theta))\phi(x\cos(k\,\Delta\theta) + y\sin(k\,\Delta\theta) - t_j). \quad (12.26)$$

For purposes of comparison, we use the Shepp-Logan filter

$$\phi(0) = \frac{4}{\pi d^2}, \quad \phi(jd) = \frac{-4}{\pi d^2(4j^2 - 1)}.$$

12.4.1 A Single Bad Ray

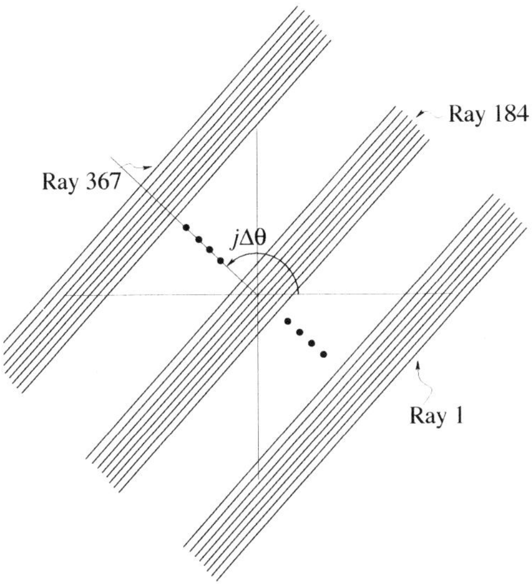

(a) Ray numbering scheme for the jth view.

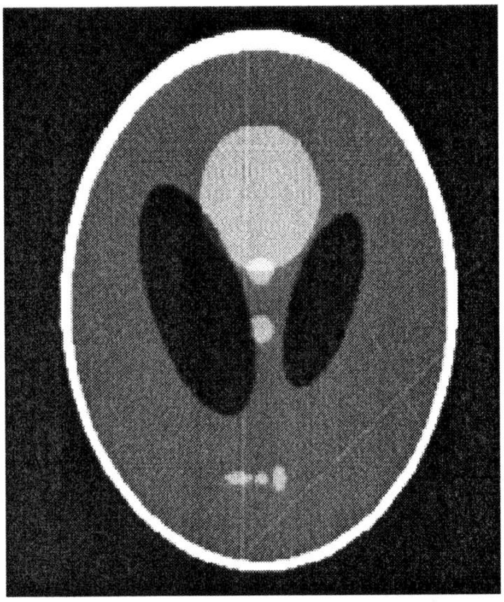

(b) Reconstruction with a few bad rays.

Figure 12.21. The numbering scheme for rays and views used in the examples in this section and a reconstruction of the Shepp-Logan phantom with a few isolated bad rays.

The first effect we consider is an isolated measurement error in a single ray, from a single view. Suppose that $P(t_{j_0}, \omega(k_0\Delta\theta))$ differs from the "true" value by ϵ. As formula (12.26) is linear, this measurement error produces a reconstruction error at (x, y) equal to

$$\Delta \tilde{f}_\phi(x, y) = \frac{\epsilon\phi(x\cos(k_0\,\Delta\theta) + y\sin(k_0\,\Delta\theta) - t_{j_0})}{N(M+1)}$$
$$= \frac{\epsilon}{N(M+1)}\phi(\mathrm{dist}((x, y), l_{t_{j_0}, \omega(k_0\Delta\theta)})). \quad (12.27)$$

The effect of this error at (x, y) only depends on the distance from (x, y) to the "bad ray," $l_{t_{j_0}, \omega(k_0\Delta\theta)}$. In light of the form of the function ϕ, the error is worst along the ray itself, where it equals

$$\frac{\epsilon N}{\pi(M+1)}.$$

***Example* 12.4.1.** In this example we use the Shepp-Logan phantom with 180 views and 367 rays in each view. The rays are numbered so that ray number 184 passes through the center of the image. The labeling of views and rays is shown in Figure 12.21(a). Figure 12.21(b) shows a reconstruction of the Shepp-Logan phantom with errors in several rays. We list rays in the form (j, θ), where j is the ray number and θ is in degrees. We introduce an error in two adjacent rays to get a clearer picture of the artifact. Rays $(220, 30)$, $(221, 30)$, $(250, 90)$, $and(251, 90)$ each have a 10% error, while rays

(190, 180), (191, 180) have a %20 error and rays (110, 135) and (111, 135) have a 30% error.

12.4.2 A Bad Ray in Each View

A bad ray might result from a momentary surge in the output of the x-ray tube. If, on the other hand, a single detector in the detector array is malfunctioning, then the same ray in each view will be in error. Let ϵ_k denote the error in $P(t_{j_0}, \omega(k\,\Delta\theta))$. In light of the linearity of (12.26), the error at (x, y) is now

$$\Delta\tilde{f}_{\phi}(x, y) = \sum_{k=0}^{M} \frac{\epsilon_k \phi(x\cos(k\,\Delta\theta) + y\sin(k\,\Delta\theta) - t_{j_0})}{N(M+1)}. \tag{12.28}$$

If $(x, y) = r(\cos\varphi, \sin\varphi)$ in polar coordinates and $\epsilon = \epsilon_k$ for all k, then

$$\begin{aligned}\Delta\tilde{f}_{\phi}(x, y) &= \sum_{k=0}^{M} \frac{\epsilon\phi(r\cos(\varphi - k\,\Delta\theta) - t_{j_0})}{N(M+1)} \\ &\approx \frac{\epsilon}{\pi N}\int_{0}^{\pi} \phi(r\cos(\varphi - s) - t_{j_0})\,ds.\end{aligned} \tag{12.29}$$

Because the function ϕ is sharply peaked at zero and

$$\int_{-\infty}^{\infty} \phi(s)\,ds = 0,$$

this artifact is worst for points where $r = t_{j_0}$ and $0 < \varphi < \pi$. At other points the integrand is either uniformly small or the integral exhibits a lot of cancellation. Due to the periodicity, of the integrand, in φ it is clear that the largest error occurs where $r = t_{j_0}$ and $\varphi = \frac{\pi}{2}$. The reason the error is only large in half the circle is that samples are only collected for $0 \le \theta \le \pi$. If data are collected over the full circle, then the result of an error ϵ in the j_0th ray is approximately

$$\Delta\tilde{f}_{\phi} \approx \frac{\epsilon}{2\pi N}\int_{0}^{2\pi} \phi(r\cos(s) - t_{j_0})\,ds. \tag{12.30}$$

If the data are collected over half the circle then $[4\pi]^{-1}\Delta\tilde{f}_{\phi}$ is the *average* error for points on the circle of radius r. Figure 12.22 shows graphs of the average error as a function of r with $t_{j_0} = 0, .25, .5,$ and $.75$. The integral in (12.30) is difficult to numerically evaluate if ϕ is a linearly interpolated function. Instead we have used the approximate filter function

$$\phi(t) = \frac{d^2 - t^2}{(d^2 + t^2)^2},$$

which was introduced in Section 6.4. These graphs bear out the prediction that the error is worst where $r = t_{j_0}$; moreover, the sharpness of the peak also depends on t_{j_0}.

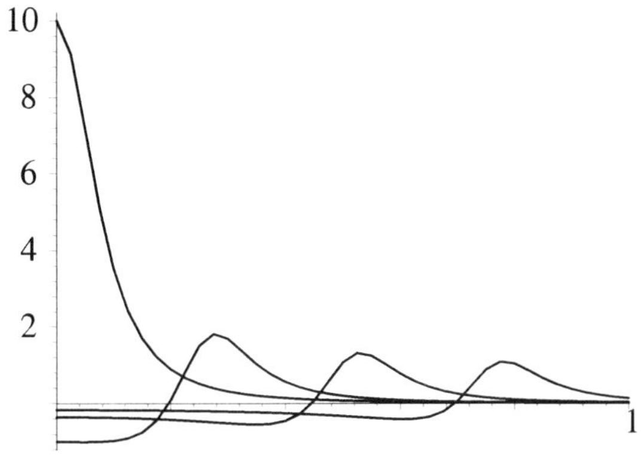

Figure 12.22. Radial graphs of the average error with systematic bad rays at different affine parameters.

Since all the bad rays have the same affine parameter, t_{j_0}, they are all tangent to the circle, centered at $(0, 0)$ of radius t_{j_0}. In [115] it is shown that the average error along this circle is given approximately by

$$\frac{\epsilon\sqrt{N}}{\pi^2\sqrt{t_{j_0}}} \text{ if } t_{j_0} >> 0 \text{ and } \frac{\epsilon N}{\pi} \text{ if } t_{j_0} = 0.$$

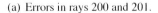

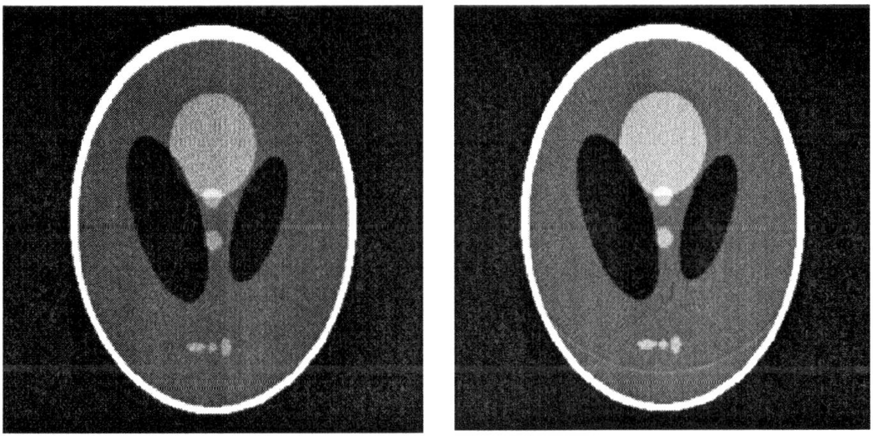

(a) Errors in rays 200 and 201. (b) Errors in rays 90 and 91.

Figure 12.23. Reconstructions of the Shepp-Logan phantom with projection data having errors in a pair of rays in every view. Note the semicircular artifacts.

***Example* 12.4.2.** The setup is the same as in Example 12.4.1. The images in Figure 12.23 show reconstructions of the Shepp-Logan phantom with an error in a pair of adjacent rays in every view. In Figure 12.23(a), rays 200 and 201 have an error that is at most 1% in every view. In Figure 12.23(b), rays 90 and 91 have an error that is at most 2% in every view.

Exercises

***Exercise* 12.4.1.** Explain why the error is only large in a semi-circle if samples are collected for $0 \leq \theta \leq \pi$.

***Exercise* 12.4.2.** Prove that $[4\pi]^{-1}\Delta\tilde{f}_\phi(r)$ represents the average error on the circle of radius r if data are collected for $0 \leq \theta \leq \pi$.

12.4.3 A Bad View

In a third-generation machine a single bad detector would result in the situation analyzed in the previous section: The measurement of the same ray would be erroneous in every view. This is because a view, for a third-generation machine, is determined by the *source* position. In a fourth-generation scanner, a miscalibrated detector could instead result in every ray from a single *view* being in error. This is because a view is determined, in a fourth-generation machine, by a *detector*.

We now analyze the effect on the reconstruction of having an error in every measurement from a single view. As before, we assume the image is reconstructed using the parallel beam algorithm. Suppose that ϵ_j is the error in the measurement $P(t_j, \boldsymbol{\omega}(k_0\,\Delta\theta))$; then the reconstruction error at (x, y) is

$$\Delta\tilde{f}_\phi(x, y) = \frac{1}{N(M+1)}\sum_{j=1}^{N}\epsilon_j\phi(\langle(x, y), \boldsymbol{\omega}(k_0\Delta\theta)\rangle - t_j).$$

If the error $\epsilon_j = \epsilon$ for all rays in the k_0th view, then the error can be approximated by an integral

$$\Delta\tilde{f}_\phi(x, y) \approx \frac{\epsilon}{2(M+1)}\int_{-1}^{1}\phi(\langle(x, y), \boldsymbol{\omega}(k_0\,\Delta\theta)\rangle - t)\,dt. \tag{12.31}$$

As before, the facts that ϕ is even and has total integral zero make this artifact most severe at the points $\langle(x, y), \omega(k_0\Delta\theta)\rangle = \pm(1-d)$ and least severe along the line

$$\langle(x, y), \boldsymbol{\omega}(k_0\Delta\theta)\rangle = 0.$$

This is the "central ray" of the k_0th view. From the properties of ϕ, we conclude that the worst error

$$\max(\Delta\tilde{f}_\phi) \propto \frac{\epsilon d}{8\pi(M+1)}\phi(0) = \frac{\epsilon N}{2\pi(M+1)}.$$

On the other hand, for points near to (0, 0) the error is approximately

$$\frac{\epsilon}{2(M+1)} \int\limits_{-1+\delta}^{1+\delta} \phi(t)\,dt,$$

where δ is the distance between (x, y) and the central ray of the bad view. The integral from $(-1+\delta)$ to $(1-\delta)$ is approximated by $\frac{2}{\pi(1-\delta)}$ and the integral from $(1-\delta)$ to $(1+\delta)$ is $O(\delta)$; hence

$$\Delta \tilde{f}_{\phi}(x, y) \approx \frac{\epsilon}{\pi(M+1)} \frac{1}{1-\delta}.$$

***Example* 12.4.3.** The setup is again as in Example 12.4.1. The values of the projections in view 0 have all been divided by 2. Indeed, the preceding estimate is overly cautious, and the change in the image produced by such errors in a single view is hardly discernible. Figure 12.24(a) shows the Shepp-Logan phantom with these systematic errors in view 0. Figure 12.24(b) is the difference of the image in (a) and the reconstruction obtained without errors in the data. It shows that the error is most pronounced where, as predicted, $\langle (x, y), \boldsymbol{\omega} \rangle \approx \pm 1$. The gray scale in Figure 12.24(b) has been compressed in order to make these vertical lines visible. This example indicates that even a large, smoothly varying error has very little effect on the reconstructed image. This is consistent with the smoothness principle.

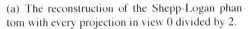

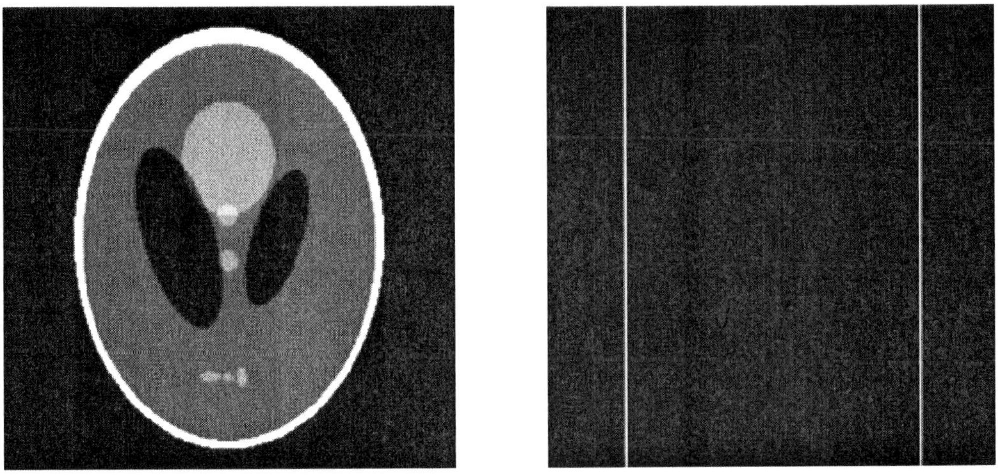

(a) The reconstruction of the Shepp-Logan phantom with every projection in view 0 divided by 2.

(b) The difference of the image in (a) and the reconstruction without systematic errors in the data.

Figure 12.24. The error produced in the reconstruction with a single bad view.

***Remark* 12.4.1.** Many other artifacts have been analyzed in [115] and [71]. We have selected examples that have a simple mathematical structure and illustrate the usage of the tools developed in the earlier chapters. Due to the success in analyzing these artifacts, they are largely absent from modern CT images.

Exercises

***Exercise* 12.4.3.** Justify the approximation as an integral in (12.31).

***Exercise* 12.4.4.** Describe the qualitative properties of the measurement error that would result from a momentary surge in the x-ray source in a fourth-generation CT-scanner.

***Exercise* 12.4.5.** Suppose that *every* measurement is off by ϵ. Show that the reconstructed image has a error

$$\Delta \tilde{f}_{\phi}(x, y) \approx \frac{\epsilon}{\pi \sqrt{1 - x^2 - y^2}}.$$

12.5 Beam Hardening

We close our discussion of imaging artifacts with a very short discussion of beam hardening. Because it is nonlinear, beam hardening is qualitatively quite different from the foregoing phenomena. It is, instead, rather similar to the partial volume effect. Beam hardening is caused by the fact that the x-ray beam is not monochromatic and the attenuation coefficient depends, in a nontrivial way, on the energy of the incident beam. Recall that an actual measurement is the ratio I_i/I_o, where I_i is the *total* energy in the incident beam and I_o is the total energy in the output. The energy content of the x-ray beam is described by its spectral function, $S(\mathcal{E})$; it satisfies

$$I_i = \int_0^{\infty} S(\mathcal{E})\, d\mathcal{E}.$$

If a (thin) x-ray beam is directed along the line $l_{t,\omega}$, then the measured output is

$$I_{o,(t,\omega)} = \int_0^{\infty} S(\mathcal{E}) \exp\left[- \int_{-\infty}^{\infty} f(s\hat{\omega} + t\omega; \mathcal{E})\, ds \right] d\mathcal{E}.$$

Here $f(\boldsymbol{x}; \mathcal{E})$ is the attenuation coefficient, with its dependence on the energy explicitly noted. A typical spectral function is shown in Figure 3.8. Due to this nonlinear distortion, the raw measurements are *not* the Radon transform of f; in the imaging literature it is often said that such measurements are *inconsistent*.

Suppose that D is a bounded object whose attenuation coefficient, $f(\mathcal{E})$, only depends on the energy. Even in this case, the function

$$\log \left[\frac{I_i}{I_{o,(t,\omega)}} \right]$$

is *not* a linear function of length of the intersection of the line $l_{t,\omega}$ with D. If T denotes the length of this line segment, then

$$\log\left[\frac{I_i}{I_{o,(t,\omega)}}\right] = H_f(T) \stackrel{d}{=} \log\left[\frac{\int S(\mathcal{E})e^{-Tf(\mathcal{E})}\,d\mathcal{E}}{\int S(\mathcal{E})\,d\mathcal{E}}\right]. \tag{12.32}$$

Because $S(\mathcal{E})$ and $f(\mathcal{E})$ are nonnegative functions, it is immediate that $H_f(T)$ is a strictly decreasing function. This implies that the inverse function, H_f^{-1}, is well defined. Thus by measuring or computing $H_f(T)$, for T in the relevant range, its inverse function can be tabulated. The attenuation coefficient of water, $f_w(\mathcal{E})$, as well as $H_w(T) - H_w'(0)T$, for a typical spectral function, are shown in Figure 12.25(a–b).

Using H_f^{-1}, the Radon transform of χ_D can be determined from x-ray attenuation measurements

$$\mathcal{R}\chi_D(t,\boldsymbol{\omega}) = H_f^{-1}\left(\log\left[\frac{I_i}{I_{o,(t,\omega)}}\right]\right). \tag{12.33}$$

The region D could now be reconstructed using the methods described previously for approximately inverting the Radon transform.

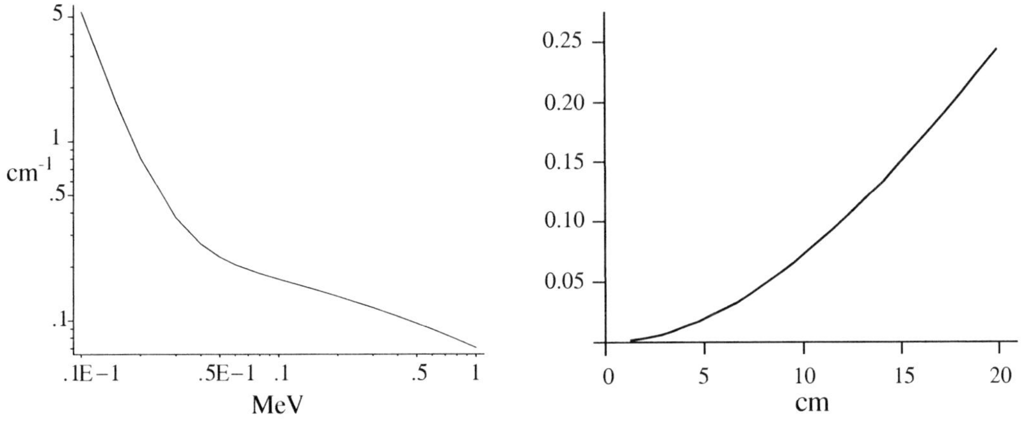

(a) Log-log plot of the attenuation coefficient of water. The unit along the horizontal axis is MeV, the vertical axis is cm^{-1}. (The data for this figure are from the National Institute of Standards and Technology Web site http://physics.nist.gov/PhysRefData/ XrayMassCoef/cover.html.)

(b) A plot of the difference between the linear and nonlinear models for the attenuation of an x-ray beam, with average energy 70 keV, caused by water. (I am very grateful to Dr. Peter Joseph for doing the computations for and providing this figure.)

Figure 12.25. Beam hardening through water.

The attenuation coefficients of the soft tissues in the human body are close to that of water, and their dependence on the energy is similar. If $f_w(\mathcal{E})$ is the attenuation coefficient of water, then this amounts to the statement that the ratio

$$\rho = \frac{f(\boldsymbol{x};\mathcal{E})}{f_w(\mathcal{E})}$$

is nearly independent of the energy. Let H_w denote the function defined in (12.32) with $f = f_w$. For slices that contain little or no bone the function, H_w^{-1} can be used as in (12.33) to correct for beam hardening. This substantially reduces the inconsistencies in the measurements and allows the usage of the Radon formalism to reconstruct ρ.

The measurement is reexpressed in terms of ρ by

$$I_{o,(t,\omega)} = \int_0^\infty S(\mathcal{E}) \exp\left[-f_w(\mathcal{E}) \int_{-\infty}^{\infty} \rho(s\hat{\omega} + t\omega)\, ds\right] d\mathcal{E}.$$

Applying H_w^{-1} to these measurements gives the Radon transform of ρ,

$$H_w^{-1}\left(\log\left[\frac{I_i}{I_{o,(t,\omega)}}\right]\right) = \int_{-\infty}^{\infty} \rho(s\hat{\omega} + t\omega)\, ds.$$

The function ρ is a nonnegative function that reflects the internal structure of the slice in much the same way as a monoenergetic attenuation coefficient. Having materials of very different densities in a slice leads to a much more difficult beam hardening problem—one that is, as of this writing, not completely solved. In x-ray CT this is principally the result of bones intersecting the slice.

Beam hardening causes dark streak artifacts, similar in appearance to those in Figure 12.5 (caused by the non-linear partial volume effect). The analysis of this problem is beyond the scope of this text. In [73] an effective algorithm is presented to substantially remove these artifacts. Another method, requiring two sets of measurements with x-ray beams having different spectral functions, is described in [3]. The discussion in this section is adapted from [73], which contains many additional illustrative examples.

Exercises

***Exercise* 12.5.1.** What is $H_f'(T)$?

***Exercise* 12.5.2.** Prove that $H_f(T)$ is a strictly monotone decreasing function.

***Exercise* 12.5.3.** Find a Taylor series expansion for H_f^{-1}.

12.6 Conclusion

We began our analysis of x-ray CT in Chapter 3 with a continuum model for the measurements as the Radon transform of a two-dimensional slice of the attenuation coefficient. With this model the reconstruction problem was simply that of inverting the Radon transform. The filtered back-projection formula derived in Chapter 6 is a complete solution to this problem. A more realistic model for the measurements was then introduced as samples of the Radon transform. We were then faced with the problem of approximately implementing the filtered back-projection formula on finitely sampled data. After studying the general problem of implementing shift invariant filters, we obtained finite reconstruction algorithms for x-ray CT in Chapter 11. As part of this discussion we saw how the actual imaging hardware determines the samples that are collected, which, in turn, dictates the structure of the reconstruction algorithm.

Assuming that the x-ray beam is one dimensional and monochromatic, the model for the measured data used in Chapter 11 is still highly idealized. In this chapter we introduced a more realistic model for the measurements as strip integrals and obtained point spread functions for the measurement and reconstruction process. We then used the reconstruction formulæ to analyze the consequences of various measurement errors on the reconstructed image. The chapter concludes with a brief discussion of beam hardening, a consequence of the fact that the x-rays available for medical tomography are not monochromatic. The introduction of the mathematical phantom and the mathematical analysis of algorithmic and modeling errors were essential steps in the reduction of imaging artifacts.

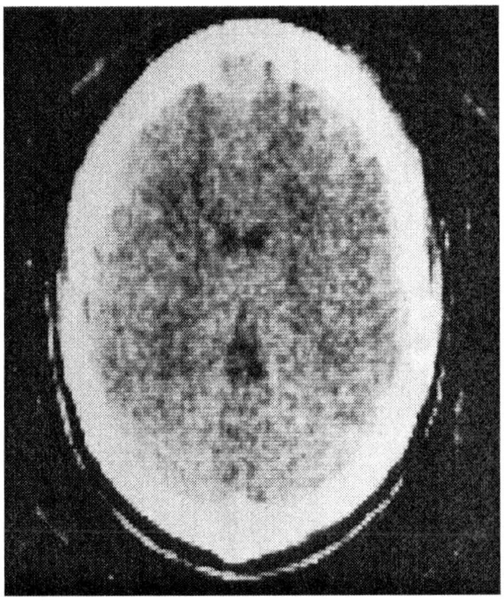

(a) A brain section made in 1976 with an EMI® scanner using a filtered back-projection algorithm. (This image provided by Dr. Peter Joseph.)

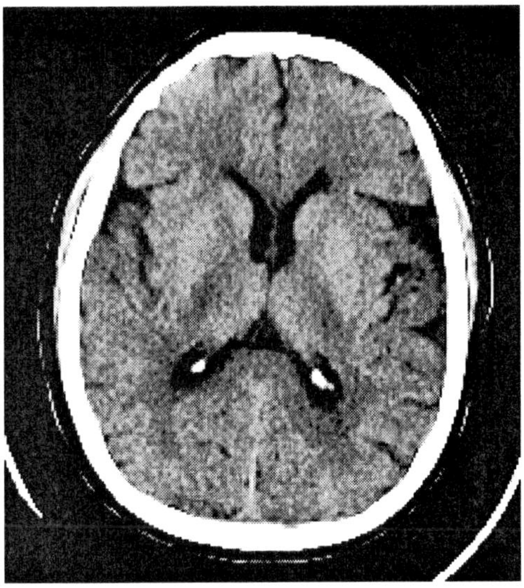

(b) A similar brain section made in 2002 with a GE® scanner also using a filtered back-projection algorithm. (This image provided by Dr. David Hackney.)

Figure 12.26. Technological improvements and mathematical analysis have led to enormous improvements in CT images.

Figure 12.26(a) shows a CT image of a slice of the head from a 1970s era machine. Note the enormous improvement in the image from 2002 of a similar slice, shown in Figure 12.26(b). The detailed anatomy of the brain, which is only hinted at in Figure 12.26(a), is readily apparent in the 2002 image. For example, we can see the folds in the cerebral cortex as well as the difference between white and gray matter. While most of the improvement in the 2002 image is due to finer sampling and better detectors, some improvement is the result of improved algorithms. For example, in the 1970s image the boundary between the brain and skull is blurred, while it is sharply delineated in the more recent image. This blurring, caused by beam hardening, is corrected in the later image.

In the next chapter we introduce algebraic reconstruction techniques that provide alternative methods for reconstructing an image. This approach employs a model for the measurements as averages of the Radon transform of a function but does not attempt to approximate the filtered back-projection formula. Rather it phrases the reconstruction problem as a very large, sparse system of linear equations. These methods are less sensitive to the details of the measurement process.

The last topic we consider is the role of random phenomena in x-ray tomography. The theory of probability and random processes provides a language to discuss noise and other nondeterministic sources of error. The main source of noise in x-ray tomography is called quantum noise; it is due to the fact that the x-ray beam is not a continuous flux of energy but instead a flux of finitely many photons. Indeed, the output of an x-ray source is also modeled as a random process, and therefore the measurements made in x-ray CT are themselves nondeterministic. Probability theory also provides a definition for the *signal-to-noise ratio,* which quantifies the amount of usable information in measurements and reconstructed images.

On the one hand, the model for x-ray CT presented thus far has been very successful: The images produced in real clinical applications provide accurate, high-resolution representations of the internal anatomy without exposing the patients to excessive levels of radiation. On the other hand, the models for the measurements used to derive the reconstruction algorithms are still somewhat metaphoric, and this makes it difficult to use the output of the reconstruction algorithm for quantitative analysis. Due to effects like beam hardening and quantum noise, it remains a challenge to describe precisely the physical quantity that is actually reconstructed in x-ray CT.

Chapter 13

Algebraic Reconstruction Techniques

Algebraic reconstruction techniques (ARTs) are techniques for reconstructing images that have no direct connection to the Radon inversion formula. Instead these methods use a linear model for the measurement process so that the reconstruction problem can be posed as a system of linear equations. Indeed the underlying mathematical concepts of ART can be applied to approximately solve many types of large systems of linear equations. This chapter contains a very brief introduction to these ideas. We present this material for historical reasons and to expose the reader to other approaches to medical image reconstruction. An extensive discussion of these methods can be found in [52]. The material presented in this chapter in not used in the sequel.

As before, boldface letters are used to denote vector or matrix quantities, while the scalar entries of a vector or matrix are denoted in regular type with subscripts. For example, $\boldsymbol{r}$ is a matrix, $\{\boldsymbol{r}_i\}$ its rows (which are vectors), and r_{ij} its entries (which are scalars).

13.1 Algebraic Reconstruction

The main features of the Radon transform of interest in ART are as follows: (1) The map $f \mapsto \mathcal{R}f$ is linear, and (2) for a function defining a simple object with bounded support, $\mathcal{R}f$ has a geometric interpretation. The first step in an algebraic reconstruction technique is the choice of a finite collection of basis functions,

$$\{b_1(x, y), \ldots, b_J(x, y)\}.$$

Certain types of a priori knowledge about the expected data and the measurement process itself can be "encoded" in the choice of the basis functions. At a minimum, it is assumed that the attenuation coefficients we are likely to encounter are well approximated by functions in the linear span of the basis functions.

If f is a two-dimensional slice of an attenuation coefficient, then for some choice of constants $\{x_j\}$, the difference

$$f - \sum_{j=1}^{J} x_j b_j$$

should be small, in an appropriate sense. For medical imaging, it is reasonable to require that the finite sum approximate f in that the gray scale images that they define "look similar" (see Section 9.4). It is also important that

$$\mathcal{R}f \approx \sum_{j=1}^{J} x_j \mathcal{R}b_j.$$

A third criterion is to choose basis function for which $\mathcal{R}b_j$ can be efficiently approximated.

The *pixel basis* is a piecewise constant family of functions often used in ART. Suppose that the support of f lies in the square $[-1, 1] \times [-1, 1]$. The square is uniformly subdivided into a $K \times K$ grid. When using ART methods, it is convenient to label the subsquares sequentially one after another, as in Figure 13.1(a). The elements of the $K \times K$ pixel basis are defined by

$$b_j^K(x, y) = \begin{cases} 1 & \text{if } (x, y) \in j\text{th} - \text{square}, \\ 0 & \text{otherwise.} \end{cases}$$

If x_j is the average of f in the jth-square, then

$$\bar{f}^K = \sum_{j=1}^{J} x_j b_j^K$$

provides an approximation to f in terms of the pixel basis. It is easy to see that the $\{b_j^K\}$ are orthogonal with respect to the usual inner product on $L^2(\mathbb{R}^2)$ and that $\bar{f}^K$ is the orthogonal projection of f into the span of the $\{b_j^K\}$.

For f a continuous function with bounded support, the sequence $< \bar{f}^K >$ converges uniformly to f as $K \to \infty$. If f represents an image, in the usual sense of the word, then as $K \to \infty$ the image defined by $\bar{f}^K$ also converges to that defined by f. The Radon transform is linear and therefore

$$\mathcal{R}\bar{f}^K = \sum_{j=1}^{J} x_j \mathcal{R}b_j^K.$$

Another advantage of the pixel basis is that $\mathcal{R}b_j^K(t, \omega)$ is, in principle, very easy to compute, being simply the length of the intersection of $l_{t,\omega}$ with the jth square.

The pixel basis is a good example to keep in mind. It has been used in many research papers on ART as well as in commercial applications. Expressing a function as a linear combination of basis functions is, in fact, the same process used in our earlier analysis of the Radon inversion formula. The only difference lies in the choice of basis functions. In ART methods we typically use a localized basis like the $\{b_j^K\}$, where each function has support in a small set. For our analysis of the Radon transform we adopted the basis provided by the exponential functions, $\{e^{i\xi\cdot x}\}$. These basis functions are well localized in frequency space but are not localized in physical space. The exponential basis is useful because it diagonalizes the linear transformations used to invert the Radon transform. On the other hand, it suffers from artifacts like the Gibbs phenomenon, which are a consequence of its non-localized, oscillatory nature. Wavelet bases are an attempt to strike a balance between these two extremes; they are localized in space but also have a fairly well-defined frequency. A good treatment of wavelets can be found in [58].

Returning now to our description of ART, assume that $\{b_j\}$ is a localized basis, though not necessarily the pixel basis. As before, the measurements are modeled as samples of $\mathcal{R}f$. The samples are labeled sequentially by $i \in \{1, \ldots, I\}$, with $\mathcal{R}f$ sampled at

$$\{(t_1, \omega_1), (t_2, \omega_2), \ldots, (t_I, \omega_I)\}.$$

Unlike the filtered back-projection (FBP) algorithm, ART methods are insensitive to the precise nature of the data set. The *measurement matrix* models the result of applying the measurement process to the basis functions. One way to define the measurement matrix is as the line integrals:

$$r_{ij} = \mathcal{R}b_j(t_i, \omega_i), \quad i = 1, \ldots, I; \tag{13.1}$$

see Figure 13.1(a). Using the definition of $\boldsymbol{r}$ in (13.1), the entries of $\boldsymbol{p}$, the vector of measurements, are defined to be

$$p_i = \mathcal{R}f(t_i, \omega_i), \quad i = 1, \ldots, I.$$

The reconstruction problem is now phrased as a system of I equations in J unknowns:

$$\sum_{j=1}^{J} r_{ij} x_j = p_i \qquad \text{for } i = 1, \ldots, I \tag{13.2}$$

or more succinctly, $\boldsymbol{rx} = \boldsymbol{p}$. A further flexibility of ART methods lies in the definition of the measurement matrix. As we shall see, simplified models are often introduced to compute its components.

The easiest type of linear system to solve is one defined by a diagonal matrix. In the Fourier approach to image reconstruction, the FFT is used to reduce the reconstruction problem to a diagonal system of equations. This explains why it was not necessary to address explicitly the problem of solving linear equations. The difficulty of using ART comes from the size of the linear system (13.2). If the square is divided into $J = 128 \times 128 \simeq 16,000$ subsquares, then, using the pixel basis, there are $16,000$ unknowns. A reasonable number of measurements is 150 samples of the Radon transform at each of 128 equally spaced angles, so that $I \simeq 19,000$. That gives a $19,000 \times 16,000$ system of equations. Even today, it is not practical to solve a system of this size directly. Indeed, as is typical in ART, this is an overdetermined system, so it is unlikely to have an exact solution. Consulting Figure 13.1, it is apparent that for each i there are about K values of j such that $r_{ij} \neq 0$. A matrix with "most" of its entries equal to zero is called a *sparse* matrix. Since $K \simeq \sqrt{I}$, r_{ij} is a sparse matrix. Localized bases are used in ART methods because it is essential for the measurement matrix to be sparse.

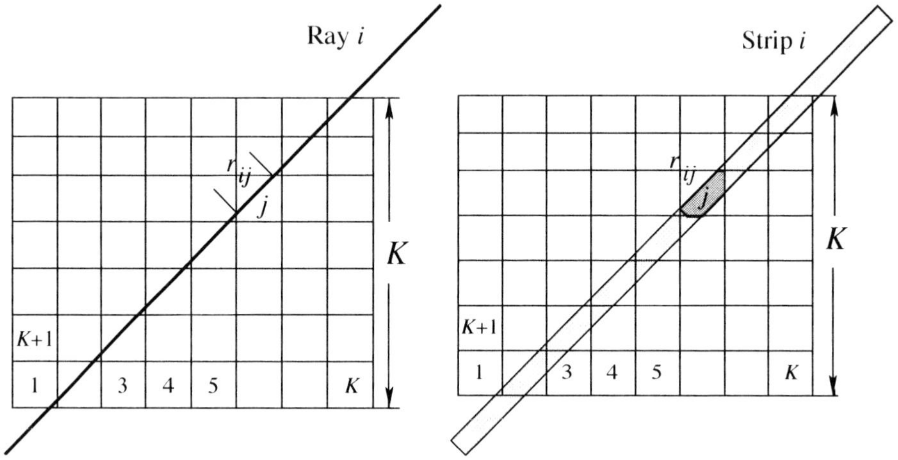

(a) Using one-dimensional rays. (b) Using two-dimensional strips.

Figure 13.1. Two definitions of the measurement matrix using a pixel basis.

With the pixel basis and a one-dimensional x-ray beam, the "exact" measurement matrix would be

$$r_{ij} = \text{length of the intersection of the } i\text{th ray with the } j\text{th pixel.}$$

If the x-ray beam has a one-dimensional *cross section*, then these rays above are replaced by strips. The value of r_{ij} could then be the area of intersection of the ith-strip with the jth-pixel [see Figure 13.1(b)]. A more complicated beam profile could also be included by weighting different parts of the strips differently. In either case, the calculation of r_{ij} requires a lot of work. Much of the literature on ART discusses the effects of using various schemes to approximate the measurement matrix. A very crude method, which was actually used in the earliest commercial machines, is to set $r_{ij} = 1$ if the center of the jth-pixel is contained in the ith-strip and 0 otherwise. For such a simple scheme, the values of r_{ij} can be computed at run time and do not have to be stored. An undesirable consequence of crudely approximating r_{ij} is that it leads to inconsistent systems of equations. If the measurement matrix is not an accurate model for the measurement process, then, given the overdetermined character of (13.2), we neither expect an actual vector of measurements to satisfy

$$\boldsymbol{rx} = \boldsymbol{p}, \tag{13.3}$$

for any choice of $\boldsymbol{x}$, nor that a solution of this equation gives a good approximation to the actual attenuation coefficient. A practical ART method needs to strike a balance between the computational load of accurately computing $\boldsymbol{r}$ and the inconsistencies that result from crude approximations.

An approach for handling inconsistent or overdetermined problems is to look for a vector $\tilde{\boldsymbol{x}}$ that minimizes an error function,

$$e(\boldsymbol{r}\tilde{\boldsymbol{x}} - \boldsymbol{p}).$$

The most common choice of error function is Euclidean (or l_2) norm of the difference,

$$e_2(\boldsymbol{r}\boldsymbol{x} - \boldsymbol{p}) = \|e(\boldsymbol{r}\boldsymbol{x} - \boldsymbol{p})\|^2.$$

In this chapter $\|\cdot\|$ refers to the Euclidean norm. Minimizing $e_2(\boldsymbol{r}\boldsymbol{x} - \boldsymbol{p})$ leads to the *least squares method.* This method is often reasonable on physical grounds, and from the mathematical standpoint it is a very simple matter to derive the *linear* equations that an optimal vector, $\tilde{\boldsymbol{x}}$, satisfies. Using elementary calculus and the bilinearity of the inner product defining the norm gives the variational equation:

$$\frac{d}{dt}\langle \boldsymbol{r}(\tilde{\boldsymbol{x}} + t\boldsymbol{v}) - \boldsymbol{p}, \boldsymbol{r}(\tilde{\boldsymbol{x}} + t\boldsymbol{v}) - \boldsymbol{p}\rangle\bigg|_{t=0} = 0 \quad \text{for all vectors } \boldsymbol{v}.$$

Expanding the inner product gives

$$t^2\langle \boldsymbol{r}\boldsymbol{v}, \boldsymbol{r}\boldsymbol{v}\rangle + 2t\langle \boldsymbol{r}\boldsymbol{v}, \boldsymbol{r}\tilde{\boldsymbol{x}} - \boldsymbol{p}\rangle + \langle \boldsymbol{r}\tilde{\boldsymbol{x}} - \boldsymbol{p}, \boldsymbol{r}\tilde{\boldsymbol{x}} - \boldsymbol{p}\rangle.$$

Hence, the derivative at $t = 0$ vanishes if and only if

$$2\langle \boldsymbol{r}\boldsymbol{v}, \boldsymbol{r}\tilde{\boldsymbol{x}} - \boldsymbol{p}\rangle = 2\langle \boldsymbol{v}, \boldsymbol{r}^t(\boldsymbol{r}\tilde{\boldsymbol{x}} - \boldsymbol{p})\rangle = 0.$$

Since $\boldsymbol{v}$ is an arbitrary vector, it follows that

$$\boldsymbol{r}^t\boldsymbol{r}\tilde{\boldsymbol{x}} = \boldsymbol{r}^t\boldsymbol{p}. \tag{13.4}$$

These are sometimes called the *normal equations.* If $\boldsymbol{r}$ has maximal rank, then $\boldsymbol{r}^t\boldsymbol{r}$ is invertible, which implies that the minimizer is unique. We might consider solving this system of equations. However, for realistic imaging data, it is about a $10^4 \times 10^4$ system, which again is too large to solve directly. Moreover, the matrix $\boldsymbol{r}^t\boldsymbol{r}$ may fail to be sparse even though $\boldsymbol{r}$ is.

Exercises

***Exercise* 13.1.1.** If f is a continuous function with bounded support, show that $\bar{f}^K$ converges uniformly to f as $K \to \infty$.

***Exercise* 13.1.2.** Let $f = \chi_{[-a,a]}(x)\chi_{[-b,b]}(y)$. By examining $\bar{f}^K$, show that there is no Gibbs phenomenon for the pixel basis. In what sense does $\bar{f}^K$ converge to f?

***Exercise* 13.1.3.** Suppose that f is a piecewise continuous function. Find norms N_1, N_2 so that $N_1(\bar{f}^K - f)$ and $N_2(\mathcal{R}\bar{f}^K - \mathcal{R}f)$ tend to zero as $K \to \infty$.

***Exercise* 13.1.4.** Prove directly that if $\boldsymbol{r}$ has maximal rank, then the normal equations have a unique solution.

13.2 Kaczmarz's Method

Most of the techniques used in numerical linear algebra are iterative. Instead of attempting to directly solve an equation like (13.4), we use an algorithm that defines a sequence, $< \boldsymbol{x}^{(k)} >$, of vectors that get closer and closer to a solution (or approximate solution). The principal method used in ART derives from the *Kaczmarz method* or *method of projections*. The idea can be explained using a very simple example. Consider the 2×2 system of equations:

$$\begin{aligned} r_{11}x_1 + r_{12}x_2 &= p_1, \\ r_{21}x_1 + r_{22}x_2 &= p_2. \end{aligned}$$

For $i = 1, 2$ $r_{i1}x_1 + r_{i2}x_2 = p_i$ defines a line l_i in the plane. The solution for this system of equations is the point of intersection of these two lines. The method of projections is very simple to describe geometrically:

1. Choose an arbitrary point and call it $\boldsymbol{x}^{(0)}$.
2. Orthogonally project $\boldsymbol{x}^{(0)}$ onto l_1, and denote the projected point by $\boldsymbol{x}^{(0,1)}$. Orthogonally project $\boldsymbol{x}^{(0,1)}$ onto l_2, denote the projected point by $\boldsymbol{x}^{(0,2)}$. This completes one iteration. Set $\boldsymbol{x}^{(1)} \stackrel{d}{=} \boldsymbol{x}^{(0,2)}$.
3. Go back to (2), replacing $\boldsymbol{x}^{(0)}$ with $\boldsymbol{x}^{(1)}$, and so on.

This algorithm is illustrated in Figure 13.2(a).

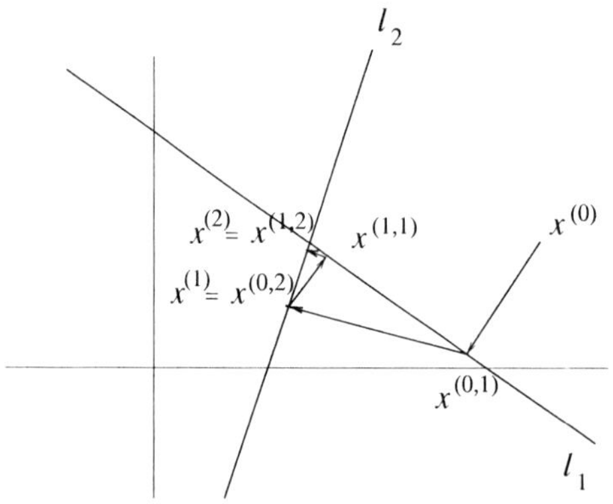

(a) Convergence in infinitely many steps.

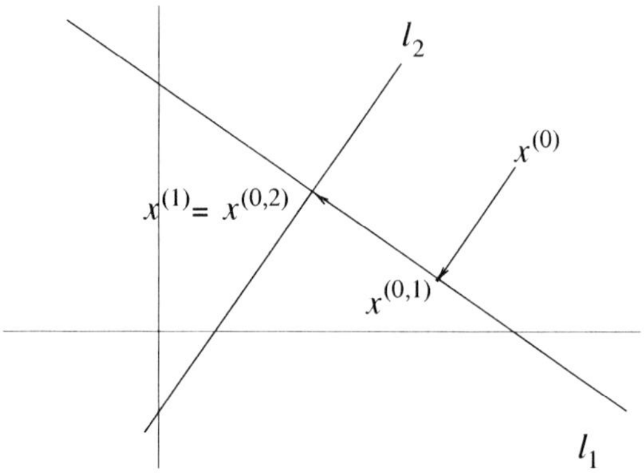

(b) Convergence in one iteration.

Figure 13.2. The method of projections for a 2×2 system.

The algorithm gives a sequence $< \boldsymbol{x}^{(j)} >$ which, in case the lines intersect, converges, as $j \to \infty$ to the solution of the system of equations. If the two lines are orthogonal, a single iteration is enough (see Figure 13.2(b)). However, the situation is not always so simple. Figure 13.3(a) shows that the algorithm does not converge for two parallel lines—this corresponds to an inconsistent system that has no solution. Figure 13.3(b) depicts an over-determined, inconsistent 3×2 system; the projections are trapped inside the triangle but do not converge as $j \to \infty$.

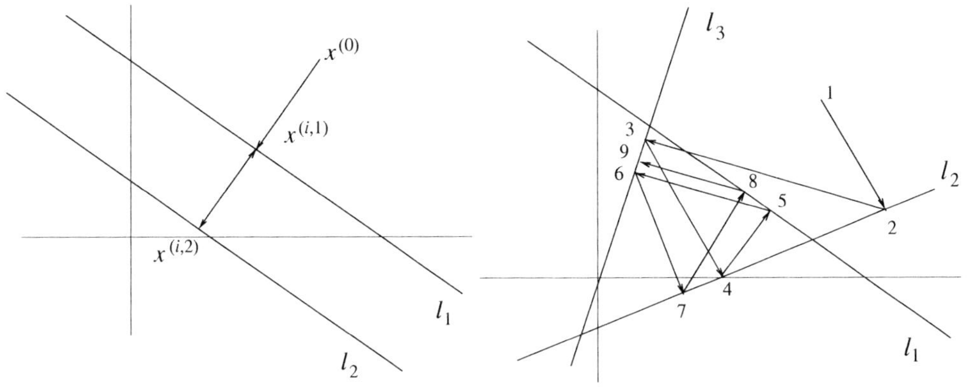

(a) Two parallel lines. (b) Overdetermined system.

Figure 13.3. Examples where the projection algorithm does not converge.

The equations that arise imaging applications can be rewritten in the form

$$\boldsymbol{r}_i \cdot \boldsymbol{x} = p_i, \quad i = 1, \ldots, I,$$

where $\boldsymbol{r}_i$ is the ith-row of the measurement matrix, $\boldsymbol{r}$. Each pair $(\boldsymbol{r}_i, p_i)$ defines a hyperplane in $\mathbb{R}^J$:

$$\{\boldsymbol{x} \ : \ \boldsymbol{r}_i \cdot \boldsymbol{x} = p_i\}.$$

Following exactly the same process used previously gives the basic Kaczmarz iteration:

1. Choose an initial vector $\boldsymbol{x}^{(0)}$.

2. Orthogonally project $\boldsymbol{x}^{(0)}$ into $\boldsymbol{r}_1 \cdot \boldsymbol{x} = p_1 \to \boldsymbol{x}^{(0,1)}$;
   orthogonally project $\boldsymbol{x}^{(0,1)}$ into $\boldsymbol{r}_2 \cdot \boldsymbol{x} = p_2 \to \boldsymbol{x}^{(0,2)}$;

   $$\vdots$$

   orthogonally project $\boldsymbol{x}^{(0,I-1)}$ into $\boldsymbol{r}_I \cdot \boldsymbol{x} = p_I \to \boldsymbol{x}^{(0,I)} \stackrel{d}{=} \boldsymbol{x}^{(1)}$.

3. Go back to (2), replacing $\boldsymbol{x}^{(0)}$ with $\boldsymbol{x}^{(1)}$, and so on.

To do these computations requires a formula for the orthogonal projection of a vector into a hyperplane. The vector $\boldsymbol{r}_i$ is orthogonal to the hyperplane $\boldsymbol{r}_i \cdot \boldsymbol{x} = p_i$. The orthogonal projection of a vector $\boldsymbol{y}$ onto $\boldsymbol{r}_i \cdot \boldsymbol{x} = p_i$ is found by subtracting a multiple of $\boldsymbol{r}_i$ from $\boldsymbol{y}$. Let $\boldsymbol{y}^{(1)} = \boldsymbol{y} - \alpha \boldsymbol{r}_i$. Then α must satisfy

$$p_i = \boldsymbol{y}^{(1)} \cdot \boldsymbol{r}_i = \boldsymbol{y} \cdot \boldsymbol{r}_i - \alpha \boldsymbol{r}_i \cdot \boldsymbol{r}_i.$$

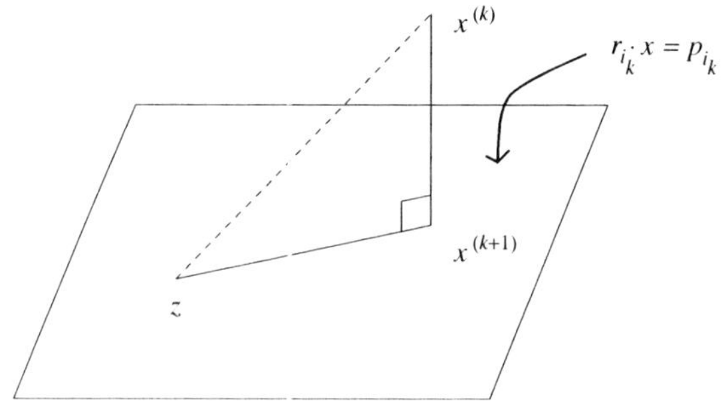

Figure 13.4. One step in the Kaczmarz algorithm.

Solving this equation gives

$$\alpha = \frac{\boldsymbol{y} \cdot \boldsymbol{r}_i - p_i}{\boldsymbol{r}_i \cdot \boldsymbol{r}_i}.$$

The explicit algorithm is therefore

$$\begin{aligned}
\boldsymbol{x}^{(k)} &\mapsto \boldsymbol{x}^{(k)} - \frac{\boldsymbol{x}^{(k)} \cdot \boldsymbol{r}_1 - p_1}{\boldsymbol{r}_1 \cdot \boldsymbol{r}_1} \boldsymbol{r}_1 = \boldsymbol{x}^{(k,1)}, \\
\boldsymbol{x}^{(k,1)} &\mapsto \boldsymbol{x}^{(k,1)} - \frac{\boldsymbol{x}^{(k,1)} \cdot \boldsymbol{r}_2 - p_2}{\boldsymbol{r}_2 \cdot \boldsymbol{r}_2} \boldsymbol{r}_2 = \boldsymbol{x}^{(k,2)}, \\
&\vdots \\
\boldsymbol{x}^{(k,I-1)} &\mapsto \boldsymbol{x}^{(k,I-1)} - \frac{\boldsymbol{x}^{(k,I-1)} \cdot \boldsymbol{r}_I - p_I}{\boldsymbol{r}_I \cdot \boldsymbol{r}_I} \boldsymbol{r}_I = \boldsymbol{x}^{(k+1)}.
\end{aligned}$$

Does the sequence $< \boldsymbol{x}^{(k)} >$ converge and, if so, to what does it converge? As was apparent in the trivial cases considered previously, the answer depends on the situation. The fundamental case to consider is when the system $\boldsymbol{r}\boldsymbol{x} = \boldsymbol{p}$ has a solution. In this case $< \boldsymbol{x}^{(k)} >$ *does* converge to a solution. This fact is very important, even though this case is unusual in imaging. For underdetermined systems, Tanabe has shown that this sequence converges to the solution $\boldsymbol{x}_s$, closest to the initial vector, $\boldsymbol{x}^{(0)}$; see [122].

Theorem 13.2.1. *Let $< \boldsymbol{r}_i >$ be a sequence of vectors in $\mathbb{R}^J$. If the system of equations*

$$\boldsymbol{r}_i \cdot \boldsymbol{x} = p_i, \quad i = 1, \dots, I,$$

has a solution, then the Kaczmarz iteration converges to a solution.

Proof. For the proof of this theorem it is more convenient to label the iterates sequentially

$$\boldsymbol{x}^{(1)}, \boldsymbol{x}^{(2)} \dots, \boldsymbol{x}^{(k)}, \dots$$

instead of

$$\boldsymbol{x}^{(0,1)}, \dots, \boldsymbol{x}^{(0,I)}, \boldsymbol{x}^{(1,0)}, \dots, \boldsymbol{x}^{(1,I)}, \dots.$$

Thus $\boldsymbol{x}^{(j,k)} \leftrightarrow \boldsymbol{x}^{(jI+k)}$.

Let z denote any solution of the system of equations. To go from $\boldsymbol{x}^{(k)}$ to $\boldsymbol{x}^{(k+1)}$ entails projecting into a hyperplane $\boldsymbol{r}_{i_k} \cdot \boldsymbol{x} = p_{i_k}$ (see Figure 13.4). The difference $\boldsymbol{x}^{(k)} - \boldsymbol{x}^{(k+1)}$ is orthogonal to this hyperplane. As both $\boldsymbol{x}^{(k+1)}$ and z lie in this hyperplane, it follows that

$$\langle \boldsymbol{x}^{(k)} - \boldsymbol{x}^{(k+1)}, \boldsymbol{x}^{(k+1)} - z \rangle = 0.$$

The Pythagorean theorem implies that

$$\|\boldsymbol{x}^{(k+1)} - z\|^2 + \|\boldsymbol{x}^{(k)} - \boldsymbol{x}^{(k+1)}\|^2 = \|\boldsymbol{x}^{(k)} - z\|^2 \Rightarrow \|\boldsymbol{x}^{(k+1)} - z\|^2 \leq \|\boldsymbol{x}^{(k)} - z\|^2. \tag{13.5}$$

The sequence $< \|\boldsymbol{x}^{(k)} - z\|^2 >$ is nonnegative and decreasing; hence, it converges to a limit. This shows that $< \boldsymbol{x}^{(k)} >$ lies in a ball of finite radius, and so the Bolzano-Weierstrass theorem implies that it has a convergent subsequence $\boldsymbol{x}^{(k_j)} \rightarrow \boldsymbol{x}^*$.

Observe that each index k_j is of the form $l_j + nI$, where $l_j \in \{0, \dots, I - 1\}$. This means that, for some l, there must be an infinite subsequence, $\{k_{j_i}\}$, so that $k_{j_i} = l + n_i I$. All the vectors $\{\boldsymbol{x}^{(k_{j_i})} : i = 1, 2, \dots\}$ lie in the hyperplane $\{\boldsymbol{r}_l \cdot \boldsymbol{x} = p_l\}$. As a hyperplane is a closed set, this implies that the limit,

$$\boldsymbol{x}^* = \lim_m \boldsymbol{x}^{(k_{j_m})},$$

also belongs to the hyperplane $\boldsymbol{r}_l \cdot \boldsymbol{x} = p_l$.

On the other hand, it follows from (13.5) and the fact that $\|\boldsymbol{x}^{(k)} - z\|$ converges that

$$\lim_{j \to \infty} \|\boldsymbol{x}^{(k_j+1)} - \boldsymbol{x}^{(k_j)}\| = 0.$$

Thus $\boldsymbol{x}^{(k_j+1)}$ also converges to $\boldsymbol{x}^*$. The definition of the Kaczmarz algorithm implies that $\boldsymbol{x}^{(k_j+1)} \in \{\boldsymbol{x} : \boldsymbol{r}_{l+1} \cdot \boldsymbol{x} = p_{l+1}\}$. As previously, this shows that $\boldsymbol{x}^*$ is in this hyperplane as well. Repeating this argument I times, we conclude that

$$\boldsymbol{x}^* \in \{\boldsymbol{r}_i \cdot \boldsymbol{x} = p_i\}, \qquad \text{for all } i = 1, \dots I.$$

That is, $\boldsymbol{x}^*$ is a solution of the original system of equations. To complete the proof, we need to show that the original sequence, $< \boldsymbol{x}^{(k)} >$, converges to $\boldsymbol{x}^*$. Recall that $\|\boldsymbol{x}^{(k)} - z\|$ tends to a limit as $k \rightarrow \infty$ for *any* solution z. Let $z = \boldsymbol{x}^*$. Then $\|\boldsymbol{x}^{(k)} - \boldsymbol{x}^*\| \rightarrow \lambda$. For the subsequence $\{k_j\}$, it follows that

$$\lim_{j \to \infty} \|\boldsymbol{x}^{(k_j)} - \boldsymbol{x}^*\| = 0.$$

Thus $\lambda = 0$ and $\lim_{k \to \infty} \boldsymbol{x}^{(k)} = \boldsymbol{x}^*$. □

As it generally requires an infinite number of iterates to find the solution, the result is largely of theoretical interest. In fact, in medical imaging applications only a few complete iterations are actually used. One reason is that the size of the system prevents using more. More important, it is an empirical fact that the quality of the reconstructed image improves for a few iterates but then begins to deteriorate. This is thought to be a consequence of noise in the data and inconsistencies introduced by approximating the measurement matrix. The image in Figure 13.5(a) is obtained using one iteration, while three iterations are used for Figure 13.5(b). Note the absence of Gibbs artifacts parallel to the sides of the squares, though view sampling artifacts are still apparent.

With an algorithm of this type it is easy to take advantage of the sparseness of $\boldsymbol{r}$. For each i, let $(j_1^i, \dots, j_{k_i}^i)$ be a list of the indices of nonzero entries in row i. Knowing the

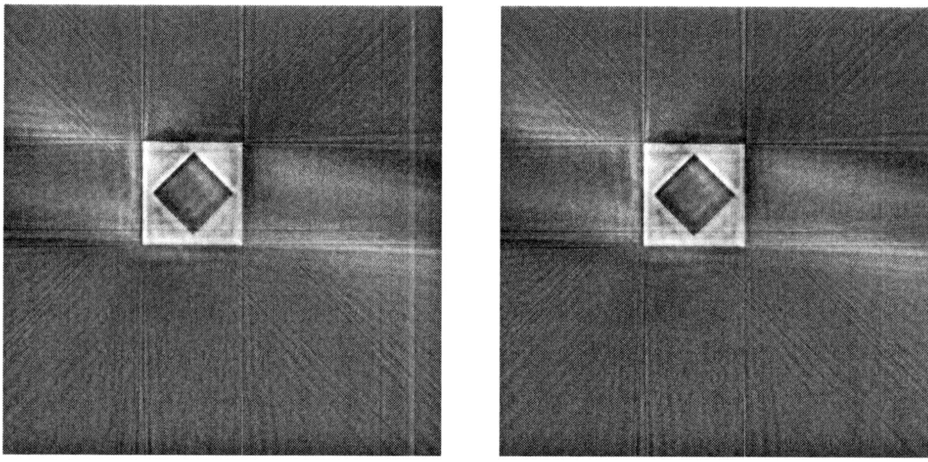

(a) One iteration. (b) Three iterations.

Figure 13.5. Examples of reconstructions using ART.

locations of the nonzero entries greatly reduces the amount of computation needed to find $\boldsymbol{r}_i \cdot \boldsymbol{x}^{(k,i)}$ as well as $\boldsymbol{r}_i \cdot \boldsymbol{r}_i$. Note also that in passing from $\boldsymbol{x}^{(k,i)}$ to $\boldsymbol{x}^{(k,i+1)}$, only entries at locations where $r_{(i+1)j} \neq 0$ are changed. This makes these methods practical even for the very large, sparse systems that arise in imaging applications.

If the equation has more than one solution, then using the Kaczmarz iteration with initial vector 0 gives the least squares solution.

Lemma 13.2.1. *If* $\boldsymbol{x}^{(0)} = 0$, *then* $\boldsymbol{x}^*$ *is the solution of* (13.2) *with minimal* l^2*-norm.*

Proof. Suppose that $A : \mathbb{R}^J \to \mathbb{R}^I$ with $I \leq J$, a matrix of maximal rank. The solution to $A\boldsymbol{x} = \boldsymbol{y}$ with minimal l^2-norm is given by $A^t\boldsymbol{u}$, where $\boldsymbol{u}$ is the unique solution to $AA^t\boldsymbol{u} = \boldsymbol{y}$. To see this, let $\boldsymbol{x}_0$ be any solution to $A\boldsymbol{x} = \boldsymbol{y}$, and let $\boldsymbol{v} \in \ker A$ be such that $\|\boldsymbol{x}_0 + \boldsymbol{v}\|^2$ is minimal. The minimal norm solution is the one perpendicular to $\ker A$ so that

$$0 = \frac{d}{dt}\langle \boldsymbol{x}_0 + \boldsymbol{v} + t\boldsymbol{w}, \boldsymbol{x}_0 + \boldsymbol{v} + t\boldsymbol{w}\rangle\bigg|_{t=0} = 2\langle \boldsymbol{x}_0 + \boldsymbol{v}, \boldsymbol{w}\rangle, \text{ for all } \boldsymbol{w} \in \ker A.$$

The assertion follows from the fact that the range of A^t is the orthogonal complement of $\ker A$ (see Theorem 2.1.2).

Suppose in our application of the Kaczmarz method we use an initial vector $\boldsymbol{x}^{(0)} = \boldsymbol{r}^t\boldsymbol{u}$ for some $\boldsymbol{u}$. From the formula for the algorithm, it follows that all subsequent iterates are also of this form. Hence $\boldsymbol{x}^* = \boldsymbol{r}^t\boldsymbol{u}$ for some $\boldsymbol{u}$ and it satisfies $\boldsymbol{r}\boldsymbol{r}^t\boldsymbol{u} = \boldsymbol{p}$. By the previous claim, $\boldsymbol{x}^*$ is the least squares solution. Taking $\boldsymbol{u} = 0$ gives $\boldsymbol{x}^{(0)} = 0$, and this completes the proof of the lemma. □

This lemma and its proof are taken from [53].

In Chapter 12 it is shown that many reconstruction artifacts appear as rapid oscillations; hence it is of interest to find a solution with the smallest possible variation. The minimal norm solution is often the minimal "variance" solution as well. Set

$$\boldsymbol{e} = (1, \ldots, 1).$$

Then

$$\mu_x = \frac{\langle \boldsymbol{e}, \boldsymbol{x} \rangle}{J}$$

is the *average value* of the coordinates of $\boldsymbol{x}$. The variance is then defined to be

$$\sigma_x^2 = \|\boldsymbol{x} - \mu_x \boldsymbol{e}\|^2.$$

Proposition 13.2.1. *If* $\boldsymbol{e} = \sum \alpha_i \boldsymbol{r}_i$ *for some* α_i, *then the minimum variance solution is also the minimum norm solution.*

Proof. If $\boldsymbol{x}$ is a solution of $\boldsymbol{r}\boldsymbol{x} = \boldsymbol{p}$, then

$$\begin{aligned}
\|\boldsymbol{x} - \frac{1}{J}\langle \boldsymbol{e}, \boldsymbol{x} \rangle \boldsymbol{e}\|^2 &= \|\boldsymbol{x}\|^2 - 2\frac{\langle \boldsymbol{e}, \boldsymbol{x} \rangle^2}{J} + \frac{\langle \boldsymbol{e}, \boldsymbol{x} \rangle^2}{J} \\
&= \|\boldsymbol{x}\|^2 - \frac{1}{J}\sum \alpha_i \langle \boldsymbol{r}_i, \boldsymbol{x} \rangle \\
&= \|\boldsymbol{x}\|^2 - \frac{1}{J}\sum \alpha_i p_i.
\end{aligned}$$

The second line follows from the fact that $\boldsymbol{x}$ is assumed to satisfy $\langle \boldsymbol{r}_i, \boldsymbol{x} \rangle = p_i$ for all i. Hence the $\|\boldsymbol{x}\|^2$ and its variance differ by a constant. This shows that minimizing the variance is equivalent to the minimizing the Euclidean norm. If $\boldsymbol{x}^{(0)} = 0$, then the lemma shows that the Kaczmarz solution has minimal Euclidean norm and therefore also minimal variance. □

Exercise

***Exercise* 13.2.1.** Prove the assertion, made in the proof, that if $\boldsymbol{x}^{(0)} = \boldsymbol{r}^t \boldsymbol{u}$, for some vector $\boldsymbol{u}$ then this is true of all subsequent iterates as well.

13.3 A Bayesian Estimate

A small modification of the ART algorithm leads to an algorithm that produces a Bayesian estimate for an optimal solution to (13.2). Without going into the details, in this approach we have prior information that the solution should be close to a known vector $\boldsymbol{v}_0$. Instead of looking for a least squares solution to the original equation, we try to find the vector that minimizes the combined error function:

$$\mathcal{B}_\rho(\boldsymbol{x}) \stackrel{d}{=} \rho\|\boldsymbol{r}\boldsymbol{x} - \boldsymbol{p}\|^2 + \|\boldsymbol{x} - \boldsymbol{v}_0\|^2.$$

Here ρ is a fixed, positive number. It calibrates the relative weight given to the measurements versus the prior information. If $\rho = 0$, then the measurements are entirely ignored; as $\rho \to \infty$, less and less weight is given to the prior information. In many different measurement schemes it is possible to use the measurements alone to compute the average value, μ_x, of the entries $\boldsymbol{x}$. If we set

$$\boldsymbol{v}_0 = \mu_x \boldsymbol{e},$$

then the prior information is the belief that the variance of the solution should be as small as possible.

The vector $\boldsymbol{x}_\rho$ that minimizes $\mathcal{B}_\rho(\boldsymbol{x})$ can be found as the minimal norm solution of a *consistent* system of linear equations. In light of Theorem 13.2.1 and Lemma 13.2.1, this vector can then be found using the Kaczmarz algorithm. The trick is to think of the error

$$\boldsymbol{u} = \boldsymbol{r}\boldsymbol{x} - \boldsymbol{p}$$

as an independent variable. Let $\begin{pmatrix} \boldsymbol{u} \\ \boldsymbol{z} \end{pmatrix}$ denote an $I + J$-column vector and E the $I \times I$ identity matrix. The system of equations we use is

$$[E \; \rho \boldsymbol{r}] \begin{pmatrix} \boldsymbol{u} \\ \boldsymbol{z} \end{pmatrix} = \rho \, [\boldsymbol{p} - \boldsymbol{r}\boldsymbol{v}_0]. \tag{13.6}$$

Theorem 13.3.1. *The system of equations* (13.6) *has a solution. If* $\begin{pmatrix} \boldsymbol{u}_\rho \\ \boldsymbol{z}_\rho \end{pmatrix}$ *is its minimal norm solution, then* $\boldsymbol{x}_\rho = \boldsymbol{z}_\rho + \boldsymbol{v}_0$ *minimizes the function* $\mathcal{B}_\rho(\boldsymbol{x})$.

Proof. That (13.6) has solutions is easy to see. For any choice of $\boldsymbol{x}$, setting

$$\boldsymbol{u} = \rho \, [\boldsymbol{p} - \boldsymbol{r}\boldsymbol{z} - \boldsymbol{r}\boldsymbol{v}_0]$$

gives a solution to this system of equations. A minimal norm solution to (13.6) is orthogonal to the null space of $[E \quad \rho \boldsymbol{r}]$. This implies that it belongs to the range of the transpose; that is,

$$\boldsymbol{z}_\rho = \rho \boldsymbol{r}^t \boldsymbol{u}_\rho. \tag{13.7}$$

On the other hand, a vector $\boldsymbol{x}_\rho$ minimizes $\mathcal{B}_\rho$ if and only if it satisfies the variational equation:

$$\rho \boldsymbol{r}^t (\boldsymbol{r}\boldsymbol{x}_\rho - \boldsymbol{p}) = \boldsymbol{v}_0 - \boldsymbol{x}\rho. \tag{13.8}$$

The relation, (13.7), between $\boldsymbol{u}_\rho$ and $\boldsymbol{z}_\rho$ implies that

$$\rho(\boldsymbol{r}\boldsymbol{z}_\rho - \boldsymbol{p}) = -(\boldsymbol{u}_\rho + \rho \boldsymbol{r}\boldsymbol{v}_0)$$

and therefore

$$\rho \boldsymbol{r}^t (\boldsymbol{r}\boldsymbol{z}_\rho - \boldsymbol{p}) = -\boldsymbol{z}_\rho - \rho \boldsymbol{r}^t \boldsymbol{r}\boldsymbol{v}_0.$$

This, in turn, shows that

$$\rho \boldsymbol{r}^t (\boldsymbol{r}\boldsymbol{x}_\rho - \boldsymbol{p}) = -\boldsymbol{z}_\rho = \boldsymbol{v}_0 - \boldsymbol{x}_\rho.$$

Thus $\boldsymbol{x}_\rho$ satisfies the variational equation (13.8) and therefore minimizes $\mathcal{B}_\rho$. □

Because (13.6) is consistent, the Kaczmarz method applied to this system, starting with the zero vector, converges to the minimum norm solution of (13.8), which therefore also minimizes $\mathcal{B}_\rho$. This algorithm is easy to describe explicitly in terms of $\boldsymbol{u}$ and $\boldsymbol{x}$. The initial vector is

$$\begin{pmatrix} \boldsymbol{u}^{(0)} \\ \boldsymbol{x}^{(0)} \end{pmatrix} = \begin{pmatrix} 0 \\ \boldsymbol{v}_0 \end{pmatrix}.$$

Suppose we have found $\boldsymbol{u}^{(k,i)}$ and $\boldsymbol{x}^{(k,i)}$. Then

$$\begin{aligned} \boldsymbol{u}^{(k,i+1)} &= \boldsymbol{u}^{(k,i)} + c^{(k,i)}\boldsymbol{e}_i, \\ \boldsymbol{x}^{(k,i+1)} &= \boldsymbol{x}^{(k,i)} + \rho c^{(k,i)}\boldsymbol{r}_i, \\ \text{where } c^{(k,i)} &= \frac{\rho(p_i - \langle \boldsymbol{r}_i, \boldsymbol{x}^{(k,i)}\rangle) - u_i^{(k,i)}}{1+\rho^2\|\boldsymbol{r}_i\|^2}. \end{aligned} \tag{13.9}$$

Here $\{\boldsymbol{e}_i, i = 1, \ldots, I\}$ is the standard basis for $\mathbb{R}^I$. This theorem and its proof are taken from [52].

Exercises

***Exercise* 13.3.1.** Suppose that the measurements are obtained using a parallel beam scanner. Explain how to compute an approximation to the average value $\mu_{\boldsymbol{x}}$. How would you try to minimize the effects of measurement error?

***Exercise* 13.3.2.** Explain (13.7) and derive the variational equation (13.8). Show that any vector that satisfies (13.8) also minimizes $\mathcal{B}_\rho$.

***Exercise* 13.3.3.** If I and J are comparable and the pixel basis is used, how does the amount of computation, per iteration, required in (13.9) compare to that required in the normal Kaczmarz algorithm?

13.4 Variants of the Kaczmarz Method

There are many ways to modify the basic Kaczmarz algorithm to obtain algorithms that give better results in a few iterations or reduce the effects of noise and modeling error. We give a small sample of this large subject.

13.4.1 Relaxation Parameters

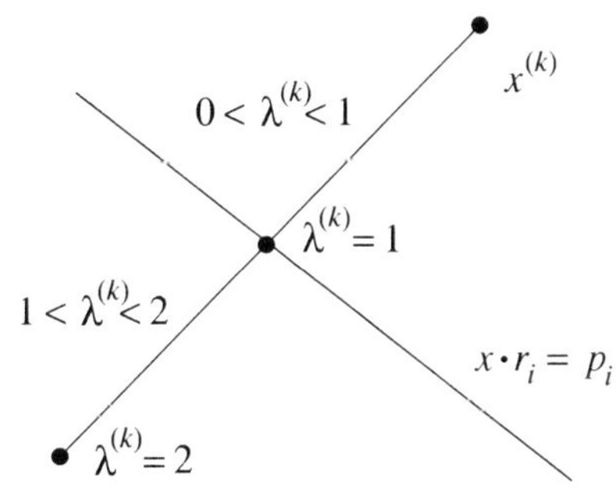

Figure 13.6. The range of values for the relaxation parameter.

The systems of equations encountered in medical imaging are often overdetermined and inconsistent because the data are noisy or the measurement matrix is only computed approximately. All these problems call for some kind of smoothing to be included in the algorithm. A common way to diminish noise and speed up convergence is to use *relaxation parameters*. Instead of applying the full correction, a scalar multiple is used. To that end, the ART algorithm is modified by putting in factors, $\{\lambda_k\}$, to obtain

$$\boldsymbol{x}^{(k,i)} \to \boldsymbol{x}^{(k,i)} - \lambda_k \frac{\boldsymbol{x}^{(k,i)} \cdot \boldsymbol{r}_i - p_i}{\boldsymbol{r}_i \cdot \boldsymbol{r}_i} \boldsymbol{r}_i.$$

The $\{\lambda_k\}$ are called *relaxation parameters*.

If $\lambda_k = 0$, then $\boldsymbol{x}^{(k,i+1)} = \boldsymbol{x}^{(k,i)}$, while $\lambda_k = 1$ gives the original algorithm. If $0 < \lambda_k < 1$, then $\boldsymbol{x}^{(k,i+1)}$ is on the same side of the hyperplane as $\boldsymbol{x}^{(k,i)}$. If $1 < \lambda_k < 2$, then $\boldsymbol{x}^{(k+1)}$ is on the opposite side of the hyperplane. If $\lambda_k = 2$, then $\boldsymbol{x}^{(k,i+1)}$ is the reflection of $\boldsymbol{x}^{(k,i)}$ in the hyperplane. So long as $0 < a \leq \lambda_k \leq b < 2$, for all k, and the system of equations has a solution, the modified algorithm also converges to a solution. If $\boldsymbol{x}^{(0)} = \boldsymbol{r}^t \boldsymbol{u}$, then the limit is again the L^2-minimum norm solution. Proofs of these facts are given in [52]. By making the sequence $< \lambda_k >$ tend to zero, a limit can be obtained even though the system of equations has no solution. Using algorithms of this type, we can find approximate solutions that are optimal for several different criteria; see [20] and [52]. As explained in these papers, the choice of relaxation parameters is largely empirical, with some choices suppressing noise and other choices improving the contrast in the final image.

Another trick used in actual reconstruction algorithm stems from the following observation. Two adjacent rays produce measurement vectors $\boldsymbol{r}_i, \boldsymbol{r}_{i+1}$ that are very nearly parallel and therefore

$$\frac{\langle \boldsymbol{r}_i, \boldsymbol{r}_{i+1} \rangle}{\|\boldsymbol{r}_i\| \cdot \|\boldsymbol{r}_{i+1}\|} \approx 1.$$

Going from i to $i+1$ will, in general, lead to a very small change in the approximate solution; small corrections often get lost in the noise and round-off error. To speed up the convergence, the successive hyperplanes are ordered to be as close to orthogonal as possible. The quality of the image produced by a few iterations is therefore likely to be improved by ordering the hyperplanes so that successive terms of the iteration come from hyperplanes that are not close to parallel. This is sometimes accomplished by "randomly" ordering the measurements, so that the expected correlation between successive measurements is small.

13.4.2 Other Related Algorithms

There are many variants of the sort of iteration used in the Kaczmarz method. For example, we can think of

$$\Delta x_j^{(k,i)} = x_j^{(k,i)} - x_j^{(k,i-1)} = \frac{p_i - \boldsymbol{r}_i \cdot \boldsymbol{x}^{(k,i-1)}}{\boldsymbol{r}_i \cdot \boldsymbol{r}_i} r_{ij}$$

as a correction that is applied to the jth-entry of our vector but defer applying the corrections until we have cycled once through all the equations. Define

$$\delta x_j^{(k,i)} = \frac{p_i - \boldsymbol{r}_i \cdot \boldsymbol{x}^{(k)}}{\boldsymbol{r}_i \cdot \boldsymbol{r}_i} r_{ij}.$$

After this quantity is computed for all pairs $1 \leq i \leq I$, and $1 \leq j \leq J$, the approximate solution is updated:

$$\begin{aligned} x_j^{(k+1)} &= x_j^{(k)} + \frac{1}{N_j} \sum_i \delta x_j^{(k,i)} \\ &= x_j^{(k)} + \frac{1}{N_j} \sum_i \frac{p_i - \boldsymbol{r}_i \cdot \boldsymbol{x}^{(k)}}{\boldsymbol{r}_i \cdot \boldsymbol{r}_i} r_{ij}. \end{aligned}$$

Here N_j is the number of i for which $r_{ij} \neq 0$. This is the number of iterates in which the value of x_j actually changes. In this algorithm we use the average of the corrections. This type of an algorithm is sometimes called a *simultaneous iteration reconstruction technique* or *SIRT*. A slight variant of the last algorithm that is used in real applications is to set

$$x_j^{(k+1)} = x_j^{(k)} + \sum_i \frac{\left[r_{ij} \frac{p_i - \boldsymbol{r}_i \cdot \boldsymbol{x}^{(k)}}{\sum_{j=1}^N r_{ij}} \right]}{\sum_{j=1}^N r_{ij}}.$$

The denominator, $\sum_j r_{ij}$, equals the length of the intersection of the ith ray with the image region. Using $\sum_{j=1}^N r_{ij}$ instead of $\sum_{j=1}^N r_{ij}^2$ is done for dimensional reasons and because it appears to give superior results.

Note, finally, that a density function is normally assumed to be nonnegative. Bases used in ART methods usually consist of nonnegative functions, and therefore the coefficients of a density function should also be nonnegative. This observation can be incorporated into ART algorithms in various ways. The simplest approach is to replace the final coefficients with the maximum of the computed value and 0. It is also possible to do this at each step of the algorithm, replacing the entries of $\boldsymbol{x}^{(k,i)}$ with the maximum of $x_l^{(k,i)}$ and 0 for each $1 \leq l \leq J$. Beyond this, ART methods can be used to find vectors that satisfy a collection of *inequalities*, rather than equations. Such algorithms are described in [52].

13.5 Conclusion

ART algorithms provide a flexible alternative to filtered back-projection. Unlike FBP, they are insensitive to the details of the data set. Through the usage of relaxation parameters, noisy or inconsistent data can also be effectively handled. Though most present day machines use some form of filtered back-projection, Godfrey Hounsfield used an ART algorithm in the first commercial CT scanner. Figure 13.7 is an 80×80 reconstruction of a brain section made with the original EMI scanner using an ART algorithm; it should be compared to Figure 11.19(a). A comprehensive discussion of these methods, along with references to the extensive literature, is given in [52].

In the next three chapters we introduce the language and concepts of probability theory. These concepts are applied to study a variety of problems in medical imaging. Among them we derive Beer's law and give probabilistic models for x-ray sources and detectors.

We also estimate the signal-to-noise ratio in an image reconstructed using the filtered back-projection algorithm. In the last section of Chapter 16 we describe the maximum likelihood method for reconstructing images in positron emission tomography. The iteration used in this approach is quite similar to that used in the Kaczmarz method. A description of the usage of ART techniques in this context is given in [54].

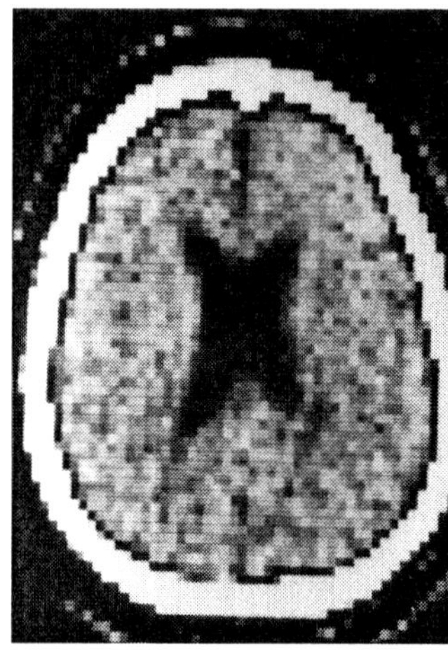

Figure 13.7. A very early x-ray CT image made using the EMI scanner and an ART algorithm. Note the dark edge just inside the skull. This is an artifact of the reconstruction algorithm called the *false subarachnoid space*. The controversy over whether this dark edge was anatomical or an artifact lasted for several years. (This image provided by Dr. Peter Joseph.)

Chapter 14

Magnetic Resonance Imaging

14.1 Introduction

Nuclear magnetic resonance (NMR) is a subtle quantum mechanical phenomenon that has played a major role in the revolution in medical imaging over the last 30 years. Before being used in imaging, NMR was employed by chemists to do spectroscopy, and remains a very important technique for determining the structure of complex chemical compounds like proteins. There are many points of contact between these technologies, and problems of interest to mathematicians. Spectroscopy is an applied form of spectral theory. NMR imaging is connected to Fourier analysis, and more general Fourier Integral Operators. The problem of *selective excitation* in NMR is easily translated into a classical inverse scattering problem, and in this formulation, is easily solved.

In this chapter, we give a rapid introduction to nuclear magnetic resonance imaging. While the physical basis for this technique is purely quantum mechanical, i.e., the fact that the protons in hydrogen atoms are spin-$\frac{1}{2}$ particles, there is an entirely classical model that is adequate to understand most applications of NMR in medical imaging. This is called the Bloch phenomenological equation, and it forms the foundation of our presentation. X-ray tomography is based upon a much simpler physical phenomenon: the scattering and absorption of x-ray photons, and therefore offers very few mechanisms for creating contrast between different tissue types, and essentially no mechanism for imaging metabolic processes. NMR is based on subtle and complex physical phenomena, which allows for a great variety of different contrast mechanisms.

In our treatment, little attention is paid to either NMR spectroscopy, or the quantum description of NMR. Those seeking a more complete introduction to these subjects should consult the monographs of Abragam, [1], Levitt [84], or Ernst, Bodenhausen and Wokaun, [38], for spectroscopy, and those of Callaghan, [17], Haacke, et al. [50], and Bernstein, et al. [9], for imaging. All six books consider the quantum mechanical description of the these phenomena.

14.2 Nuclear Magnetic Resonance

A proton is a spin-$\frac{1}{2}$ particle. This means that it has both an *intrinsic* angular momentum, $\boldsymbol{J}_p$ and magnetic moment, $\boldsymbol{\mu}_p$. These are $\mathbb{R}^3$-valued quantum mechanical observables, which transform, under the action of $SO(3)$, by the standard 3-dimensional representation. The Wigner-Eckert theorem therefore implies that these observables are proportional, i.e., there is a constant, γ_p, such that

$$\boldsymbol{\mu}_p = \gamma_p \boldsymbol{J}_p. \tag{14.1}$$

This constant is called the *gyromagnetic ratio*. If one could somehow contrive to have a stationary proton in a uniform magnetic field, $\boldsymbol{B}_0$, then the quantum mechanical expectation, $< \boldsymbol{J}_p >$, of the angular momentum would satisfy the ordinary differential equation:

$$\frac{d < \boldsymbol{J}_p >}{dt} = < \boldsymbol{\mu}_p > \times \boldsymbol{B}_0. \tag{14.2}$$

In light of equation (14.1), this implies that the expectation of $\boldsymbol{\mu}_p$ satisfies

$$\frac{d < \boldsymbol{\mu}_p >}{dt} = \gamma_p < \boldsymbol{\mu}_p > \times \boldsymbol{B}_0. \tag{14.3}$$

This equation shows that $< \boldsymbol{\mu}_p >$ precesses about the direction of $\boldsymbol{B}_0$ with a characteristic angular frequency, $\omega_0 = \gamma_p \|\boldsymbol{B}_0\|$. A vector $\boldsymbol{v}$ precesses about a vector $\boldsymbol{w}$ if $\boldsymbol{v}$ rotates at a fixed angle about the axis determined by $\boldsymbol{w}$, see Figure 14.1. This precessional motion, at a precisely determined angular frequency, is the resonance phenomenon that characterizes NMR. The frequency ω_0 is called the *Larmor frequency*; it depends on the strength of the background field.

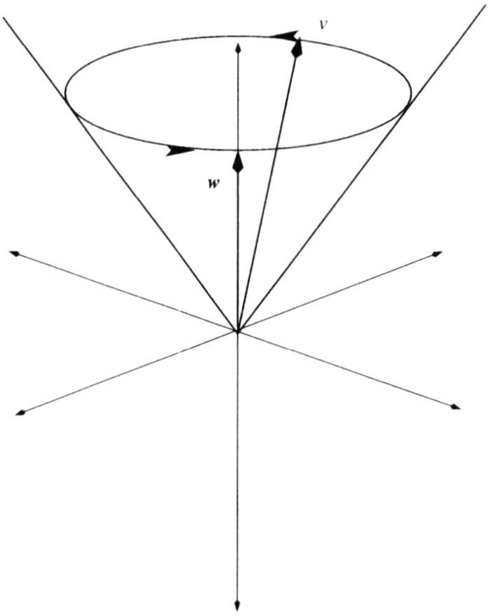

Figure 14.1. Precession of the vector $\boldsymbol{v}$ about the vector $\boldsymbol{w}$.

Of course one cannot obtain a stationary isolated proton, so this constant has never been measured experimentally. What can actually be measured are resonances for protons contained in nuclei of molecules. The electron cloud in a molecule affects the magnetic field at the nuclei, leading to small shifts in the observed resonances. This phenomenon is one of the basic ingredients needed to use NMR spectroscopy to determine molecular structure. For hydrogen protons in water molecules

$$\gamma \approx 42.5764 \times 10^6 \frac{\text{cycles}}{\text{sec} \cdot \text{Tesla}}. \tag{14.4}$$

The Tesla is a unit of magnetic induction; for purposes of comparison, the strength of the earth's magnetic field is about 5×10^{-5} Tesla. For hydrogen protons in other molecules, the gyromagnetic ratio is expressed in the form $(1 - \sigma)\gamma$. The coefficient σ is called the *chemical shift*. It is typically between 10^{-6} and 10^{-4}. In the sequel we use γ to denote the gyromagnetic ratio of a hydrogen proton in a water molecule.

The strength of a standard magnet used in a hospital MR imaging device is in the 1-3 Tesla range, and spectrometers typically use magnets in the 5-15 Tesla range. For imaging magnets, the resonance frequencies are in the 40-120 MHz range. That this is the standard FM band turns out to be a great piece of luck for medical applications. The quantum mechanical energy ($E = h\nu$) at these frequencies is too small to break chemical bonds, and so the radiation used in MR is fundamentally "safe" in ways that X-rays are not. Technologically, it is also a relatively simple frequency band to work in.

14.3 The Bloch Phenomological Equation

In most NMR imaging applications one is attempting to determine the distribution of water molecules in an extended object. Let (x, y, z) denote orthogonal coordinates in the region occupied by the sample, and $\rho(x, y, z)$ denote the density of water molecules at (x, y, z). The nuclear spins in a complex, extended object interact with one another. If the object is placed within a magnetic field $\boldsymbol{B}_0$ (no longer assumed to be homogeneous) then the spins become *polarized* leading to a net bulk *equilibrium magnetization* $\boldsymbol{M}_0$. The strength of $\boldsymbol{M}_0$ is determined by thermodynamic considerations: there is a universal constant C so that

$$\boldsymbol{M}_0(x, y, z) = \frac{C}{T_K} \rho(x, y, z) \boldsymbol{B}_0(x, y, z). \tag{14.5}$$

Here T_K is the absolute (or Kelvin) temperature. At room temperature, in a 1 Tesla field, roughly 1 in 10^6 moments are aligned with $\boldsymbol{B}_0$. Thus $\boldsymbol{M}_0$ is a tiny perturbation of $\boldsymbol{B}_0$, which would be very difficult to directly detect.

Felix Bloch introduced a phenomenological equation, which describes the interactions of the bulk magnetization, resulting from the nuclear spins, with one another, and with an applied field. If $\boldsymbol{B}$ is a magnetic field of the form $\boldsymbol{B}_0(x, y, z) + \widetilde{\boldsymbol{B}}(x, y, z; t)$, with the time dependent parts much smaller than $\boldsymbol{B}_0$, then the bulk magnetization, $\boldsymbol{M}$, satisfies the equation

$$\begin{aligned}\frac{d\boldsymbol{M}(x,y,z;t)}{dt} =&\gamma\,\boldsymbol{M}(x,y,z;t)\times\boldsymbol{B}(x,y,z;t)-\frac{1}{T_2}\boldsymbol{M}^{\perp}(x,y,z;t)+\\ &\frac{1}{T_1}(\boldsymbol{M}_0(x,y,z)-\boldsymbol{M}^{\|}(x,y,z;t)).\end{aligned}\tag{14.6}$$

Here $\boldsymbol{M}^{\perp}$ is the component of $\boldsymbol{M}$, perpendicular to $\boldsymbol{B}_0$, (called the *transverse component*), and $\boldsymbol{M}^{\|}$ is the component of $\boldsymbol{M}$, parallel to $\boldsymbol{B}_0$, (called the *longitudinal component*). The first term $\gamma\,\boldsymbol{M}(x,y,z;t)\times\boldsymbol{B}(x,y,z;t)$ comes from the direct interaction of the spins with the magnetic field. If $\boldsymbol{B}(x,y,z;t)$ is actually time independent, then this equation predicts that the magnetization vector, $\boldsymbol{M}(x,y,z;t)$ will precess about $\boldsymbol{B}(x,y,z)$ at frequency $\gamma\,\|\boldsymbol{B}(x,y,z)\|$.

The other terms in equation (14.6) are called relaxation terms. The term with coefficient $\frac{1}{T_1}$ describes how the spins become *polarized* by the fixed background field $\boldsymbol{B}_0$, and the term with coefficient $\frac{1}{T_2}$ describes the decays of components of $\boldsymbol{M}$ perpendicular to $\boldsymbol{B}_0$. If $\boldsymbol{B}$ has no time dependent component, then this equation predicts that the sample becomes polarized with the transverse part of $\boldsymbol{M}$ decaying as $e^{-\frac{t}{T_2}}$, and the longitudinal component approaching the equilibrium field as $1-e^{-\frac{t}{T_1}}$. In Bloch's model, the spins at different points do not directly interact. Instead, the relaxation terms describe averaged interactions with "spin baths." This simple model is adequate for most imaging applications. Indeed, for many purposes, it is sufficient to use the Bloch equation without the relaxation terms. See [11] and [126].

Typically, $\boldsymbol{B}_0$, the *background field*, is assumed to be a strong uniform field, $\boldsymbol{B}_0 = (0,0,b_0)$, and $\boldsymbol{B}$ takes the form

$$\boldsymbol{B}=\boldsymbol{B}_0+\boldsymbol{G}+\boldsymbol{B}_1,\tag{14.7}$$

where $\boldsymbol{G}$ is a *gradient field*. Usually the gradient fields are "piecewise time independent" fields, small relative to $\boldsymbol{B}_0$. By piecewise time independent field we mean a collection of static fields that, in the course of the experiment, are turned on, and off. The $\boldsymbol{B}_1$ component is a time dependent radio frequency field, nominally at right angles to $\boldsymbol{B}_0$. It is usually taken to be spatially homogeneous, with time dependence of the form:

$$\boldsymbol{B}_1(t)=\begin{pmatrix}\mathrm{Re}[(\alpha(t)+i\beta(t))e^{i\omega_0 t}]\\ \mathrm{Im}[(\alpha(t)+i\beta(t))e^{i\omega_0 t}]\\ 0\end{pmatrix},\tag{14.8}$$

here $\omega_0=\gamma\,b_0$.

If $\boldsymbol{B}_1=\boldsymbol{G}=0$ and $\alpha(t)=\beta(t)\equiv 0$, then the solution operator for the Bloch equation, without relaxation terms, is

$$U(t)=\begin{bmatrix}\cos\omega_0 t & -\sin\omega_0 t & 0\\ \sin\omega_0 t & \cos\omega_0 t & 0\\ 0 & 0 & 1\end{bmatrix}.\tag{14.9}$$

This is just the "resonance rotation" at time t, about the $\boldsymbol{B}_0$-field.

14.4 The Rotating Reference Frame

In light of (14.9) it is convenient to introduce the *rotating reference frame*. Let

$$U(t)^{-1} = W(t) = \begin{bmatrix} \cos\omega_0 t & \sin\omega_0 t & 0 \\ -\sin\omega_0 t & \cos\omega_0 t & 0 \\ 0 & 0 & 1 \end{bmatrix}; \tag{14.10}$$

to define the rotating reference frame we replace $\boldsymbol{M}$ with $\boldsymbol{m}$ where

$$\boldsymbol{m}(x, y, z; t) = W(t)\boldsymbol{M}(x, y, z; t). \tag{14.11}$$

It is a classical result of Larmor, that if $\boldsymbol{M}$ satisfies (14.6), then $\boldsymbol{m}$ satisfies

$$\begin{aligned} \frac{d\boldsymbol{m}(x, y, z; t)}{dt} =& \gamma\, \boldsymbol{m}(x, y, z; t) \times \boldsymbol{B}_{\text{eff}}(x, y, z; t) - \frac{1}{T_2}\boldsymbol{m}^{\perp}(x, y, z; t) + \\ & \frac{1}{T_1}(\boldsymbol{M}_0(x, y, z) - \boldsymbol{m}^{\|}(x, y, z; t)), \end{aligned} \tag{14.12}$$

where

$$\boldsymbol{B}_{\text{eff}} = W(t)\boldsymbol{B} - (0, 0, \frac{\omega_0}{\gamma}).$$

If $\boldsymbol{G}$, in (14.7), is much smaller than $\boldsymbol{B}$ and quasi-static, then one may usually ignore the components of $\boldsymbol{G}$ that are orthogonal to $\boldsymbol{B}_0$. In this case we often use the notation $\boldsymbol{G} = (*, *, l(x, y, z))$, with $*$s indicating field components that may be ignored. In imaging applications one usually assumes that the components of $\boldsymbol{G}$ depend linearly on (x, y, z) with the $\hat{z}$-component given by $\langle (x, y, z), (g_1, g_2, g_3)\rangle$. The constant vector $\boldsymbol{g} = (g_1, g_2, g_3)$ is called the *gradient vector*. With $\boldsymbol{B}_0 = (0, 0, b_0)$ and $\boldsymbol{B}_1$ given by (14.8), we see that $\boldsymbol{B}_{\text{eff}}$ can be taken to equal $(0, 0, \langle (x, y, z), \boldsymbol{g}\rangle) + (\alpha(t), \beta(t), 0)$. In the remainder of this chapter we work in the rotating reference frame and assume that $\boldsymbol{B}_{\text{eff}}$ takes this form.

If $\boldsymbol{G} = 0$ and $\beta(t) \equiv 0$, then, in the rotating reference frame, the solution operator for the Bloch equation, without relaxation terms, is

$$V(t) = \begin{bmatrix} 1 & 0 & 0 \\ 0 & \cos\theta(t) & \sin\theta(t) \\ 0 & -\sin\theta(t) & \cos\theta(t) \end{bmatrix}, \tag{14.13}$$

where

$$\theta(t) = \int_0^t \alpha(s)ds. \tag{14.14}$$

This is simply a rotation about the x-axis through the angle $\theta(t)$. If $\boldsymbol{B}_1 \neq 0$ for $t \in [0, \tau]$, then the magnetization is rotated through the angle $\theta(\tau)$. Thus RF-excitation can be used to move the magnetization out of its equilibrium state. As we shall soon see, this is crucial for obtaining a measurable signal. The equilibrium magnetization is a tiny perturbation of the very large $\boldsymbol{B}_0$-field and is therefore in practice not directly measurable. Only the precessional motion of the transverse components of $\boldsymbol{M}$ produces a measurable signal. More general $\boldsymbol{B}_1$-fields, i.e. with both α and β non-zero, have more complicated effects on the magnetization. In general the angle between $\boldsymbol{M}$ and $\boldsymbol{M}_0$ at the conclusion of the RF-excitation is called the *flip angle*.

Now suppose that $\boldsymbol{B}_1 = 0$ and $\boldsymbol{G} = (*, *, l(x, y, z))$, where $l(\cdot)$ is a function. As noted above, the $*$s indicate field components that may be ignored. The solution operator V now depends on (x, y, z), and is given by

$$V(x, y, z; t) = \begin{bmatrix} \cos \gamma l(x, y, z)t & -\sin \gamma l(x, y, z)t & 0 \\ \sin \gamma l(x, y, z)t & \cos \gamma l(x, y, z)t & 0 \\ 0 & 0 & 1 \end{bmatrix} \tag{14.15}$$

This is precession about $\boldsymbol{B}_0$ at an angular frequency that depends on the local field strength $b_0 + l(x, y, z)$. If both $\boldsymbol{B}_1$ and $\boldsymbol{G}$ are simultaneously non-zero, then, starting from equilibrium, the solution of the Bloch equation, at the conclusion of the RF-pulse, has a nontrivial spatial dependence. In other words, the flip angle becomes a function of the spatial variables. We return to this in Section 14.6.

Exercises

1. Prove Larmor's theorem: If $\boldsymbol{M}(t)$ satisfies (14.6) and $\boldsymbol{m}(t)$ is given by (14.11), then $\boldsymbol{m}(t)$ satisfies (14.12).

2. Suppose that $\boldsymbol{B}(x, y, z) = (0, 0, b_0) + \boldsymbol{G}$, where $\boldsymbol{G} = (l_x, l_y, l_z)$, with coordinates depending on (x, y, z). Use Taylor's theorem to expand $\|\boldsymbol{B}\| = \|\boldsymbol{B}_0 + \boldsymbol{G}\|$. Explain why it is reasonable to ignore components of $\boldsymbol{G}$ orthogonal to $\boldsymbol{B}_0$, if $b_0 = 1.5$T and $\|\boldsymbol{G}\| = .001$T.

14.5 A Basic Imaging Experiment

With these preliminaries we can describe the basic measurements in magnetic resonance imaging. The sample is polarized, and then an RF-field, of the form given in (14.8), (with $\beta \equiv 0$) is turned on for a finite time T. This is called an *RF-excitation*. For the purposes of this discussion we suppose that the time is chosen so that $\theta(T) = 90°$. As $\boldsymbol{B}_0$ and $\boldsymbol{B}_1$ are spatially homogeneous, the magnetization vectors within the sample remain parallel throughout the RF-excitation. At the conclusion of the RF-excitation, $\boldsymbol{M}$ is orthogonal to $\boldsymbol{B}_0$. With the RF turned off, the vector field $\boldsymbol{M}(t)$ precesses about $\boldsymbol{B}_0$, *in phase*, with angular velocity ω_0. The transverse component of $\boldsymbol{M}$ decays exponentially. If we normalize the time so that $t = 0$ corresponds to the conclusion of the RF-pulse, then

$$\boldsymbol{M}(x, y, z; t) = C\omega_0 \rho(x, y, z)[e^{-\frac{t}{T_2}} e^{i(\omega_0 t + \phi)}, (1 - e^{-\frac{t}{T_1}})]. \tag{14.16}$$

In this formula and in the sequel we follow the standard practice in MR of expressing the magnetization as a pair consisting of a complex and a real number: $[M_x + iM_y, M_z]$. The complex part, $M_x + iM_y$, is the transverse magnetization and M_z is the longitudinal magnetization. Here ϕ is a fixed real phase.

Recall Faraday's Law: A changing magnetic field induces an electro-motive-force (EMF) in a loop of wire according to the relation

$$\text{EMF}_{\text{loop}} = \frac{d\Phi_{\text{loop}}}{dt}. \tag{14.17}$$

Here Φ_{loop} denotes the flux of the field through the loop of wire. The transverse components of $\boldsymbol{M}$ are a rapidly varying magnetic field, which, according to Faraday's law, induce a current in a loop of wire. In fact, by placing several coils near to the sample we can measure a complex valued signal of the form:

$$S_0(t) = C'\omega_0^2 e^{i\omega_0 t} \int\limits_{\text{sample}} \rho(x, y, z) e^{-\frac{t}{T_2(x,y,z)}} b_{1\text{rec}}(x, y, z)\,dx\,dy\,dz. \tag{14.18}$$

Here $b_{1\text{rec}}(x, y, z)$ quantifies the sensitivity of the detector to the precessing magnetization located at (x, y, z). The fact that the transverse relaxation rate depends on spatial location is the source of a very important contrast mechanism in MR-imaging called T_2-weighting.

For the moment we ignore the spatial dependence of T_2 and assume that T_2 is just a constant. With this simplifying assumption we see that it is easy to obtain a measurement of the integral of $\rho b_{1\text{rec}}$. By using a carefully designed detector, $b_{1\text{rec}}$ can be taken to be a constant, and therefore we can determine the total spin density within the object of interest. For the rest of this section we assume that $b_{1\text{rec}}$ is a constant. Note that the size of the measured signal is proportional to ω_0^2, which is, in turn, proportional to $\|\boldsymbol{B}_0\|^2$. This explains, in part, why it is so useful to have a very strong background field. Even with a 1.5T magnet the measured signal is still quite small, in the micro-watt range, see [62, 63].

Let $l(x, y, z) = \langle (x, y, z), (g_x, g_y, g_z)\rangle$. Suppose that at the end of the RF-excitation, we instead turn on $\boldsymbol{G}$. It follows from (14.15) that, in this case, the measured signal would be approximately:

$$S_l(t) \approx C' b_{1\text{rec}} e^{-\frac{t}{T_2}} \omega_0^2 e^{i\omega_0 t} \int\limits_{\text{sample}} \rho(x, y, z) e^{it\gamma\langle (x,y,z),(g_x,g_y,g_z)\rangle}\,dx\,dy\,dz. \tag{14.19}$$

Up to a constant, $e^{-i\omega_0 t} e^{\frac{t}{T_2}} S_l(t)$ is simply the Fourier transform of ρ at $\boldsymbol{k} = -t\gamma\,(g_x, g_y, g_z)$. By sampling in time and using a variety of different linear functions l, we can sample the Fourier transform of ρ in neighborhood of 0. This suffices to reconstruct an approximation to ρ.

The signal in MR is usually sampled at uniformly spaced sample times, $j\Delta t : j = 1, \ldots, N$. If $g = \|(g_x, g_y, g_z)\|$, then this translates into uniformly spaced samples in frequency space, with sample spacing given by:

$$\Delta k = \gamma g \Delta t. \tag{14.20}$$

Since we are sampling in the frequency domain, the sample spacing is determined, according to Nyquist's theorem, by the size of the object we are trying to image. As follows from the Poisson summation formula, if the sample spacing is Δk, then the maximum size of object that can be reconstructed, without aliasing errors, is

$$L = \frac{2\pi}{\Delta k}. \tag{14.21}$$

This is an idealized model for the measurement process. The fact that the human body is electrically conductive leads to Johnson noise in the measurements, with the size of the noise signal proportional to the bandwidth of the measured data and the volume of the sample. The measurement apparatus itself also produces noise. From the Riemann Lebesgue lemma, it follows that $e^{\frac{t}{T_2}} S_l(t)$ is decaying. On the other hand, the noise is of constant mean amplitude. This places a practical limit on how long the signal can be sampled, which translates into a maximum absolute frequency in Fourier space that can reliably be measured. Finally this limits the resolution attainable in the reconstructed image.

The approach to imaging described above captures the spirit of the methods used in real applications. It is however, not representative in several particulars. It is unusual to sample the 3-dimensional Fourier transform of ρ. Rather, a specially shaped RF-pulse is used in the presence of nontrivial field gradients, to excite the spins in a thin 2-dimensional slice of the sample. The spins slightly outside this slice remain in the equilibrium state. This means that any measured signal comes predominately from the excited slice. This process is called *selective excitation.* In the next section we explain a technique used to design RF-pulses that produce a selective excitation. It is also much more common to sample the Fourier domain in a rectilinear fashion, rather than in the radial fashion described above. This makes it possible to use the "fast Fourier transform" algorithm to reconstruct the image, vastly reducing the computational requirements of the reconstruction step.

Imagine that we collect samples of $\hat{\rho}(\boldsymbol{k})$ on a uniform rectangular grid

$$\{(j_x \Delta k_x, j_y \Delta k_y, j_z \Delta k_z) : -\frac{N_x}{2} \le j_x \le \frac{N_x}{2}, -\frac{N_y}{2} \le j_y \le \frac{N_y}{2}, -\frac{N_z}{2} \le j_z \le \frac{N_z}{2}\}.$$

Since we are sampling in the Fourier domain, the Nyquist sampling theorem implies that the sample spacings determine the spatial field-of-view from which we can reconstruct an artifact free image: in order to avoid aliasing artifacts, the support of ρ must lie in a rectangular region with side lengths $[\Delta k_x^{-1}, \Delta k_y^{-1}, \Delta k_z^{-1}]$, see [50, 5]. In typical medical applications the support of ρ is much larger in one dimension than the others and so it turns out to be impractical to use the simple data collection technique described above. Instead the RF-excitation takes place in the presence of non-trivial gradient fields, which allows for a spatially selective excitation: the magnetization in one region of space obtains a transverse component, while that in the complementary region is left in the equilibrium state. In this we can way collect data from an essentially two dimensional slice. This is described in the next section.

Exercises

1. Why is the measured signal in (14.18) proportional to ω_0^2.

2. Explain why the sample spacing determines the field of view, and why the maximum frequency sampled limits the resolution available in the measured data.

14.6 Selective Excitation

As remarked above, practical imaging techniques do not excite all the spins in an object and directly measure samples of the 3d Fourier transform. Rather the spins lying in a slice are excited and samples of the 2d Fourier transform are then measured. This process is called selective excitation and may be accomplished by applying the RF-excitation with a gradient field turned on. With this arrangement, the strength of the static field, $\boldsymbol{B}_0 + \boldsymbol{G}$, varies with spatial position, hence the response to the RF-excitation does as well. Suppose that $\boldsymbol{G} = (*, *, \langle (x, y, z), \boldsymbol{g} \rangle)$ and set $f = [2\pi]^{-1}\gamma \langle (x, y, z), \boldsymbol{g} \rangle$. This is called the *offset frequency* as it is the amount by which the local resonance frequency differs from the resonance frequency, ω_0 of the unperturbed $\boldsymbol{B}_0$-field. The result of a selective RF-excitation is described by a magnetization profile $\boldsymbol{m}^{\infty}(f)$, this is a unit 3-vector valued function of the offset frequency. A typical case would be

$$\boldsymbol{m}^{\infty}(f) = \begin{cases} [0, 0, 1] & \text{for } f \notin [f_0 - \delta, f_1 + \delta] \\ [\sin\theta, 0, \cos\theta] & \text{for } f \in [f_0, f_1]. \end{cases} \tag{14.22}$$

Here $\delta > 0$ allows for a smooth transition between the spins which are flipped and those left in their equilibrium state. The magnetization is flipped through an angle θ, in regions of space where the offset frequency lies in $[f_0, f_1]$, and is left in the equilibrium state otherwise.

Typically the excitation step takes a few milliseconds and is much shorter than either T_1 or T_2; therefore one generally uses the Bloch equation, without relaxation, in the analysis of selective excitation. In the rotating reference frame the Bloch equation, without relaxation, takes the form

$$\frac{d\boldsymbol{m}(f;t)}{dt} = \begin{bmatrix} 0 & 2\pi f & -\gamma\beta \\ -2\pi f & 0 & \gamma\alpha \\ \gamma\beta & -\gamma\alpha & 0 \end{bmatrix} \boldsymbol{m}(f;t). \tag{14.23}$$

The problem of designing a selective pulse is non-linear. Indeed, the selective excitation problem can be rephrased as a classical inverse scattering problem: one seeks a function $\alpha(t) + i\beta(t)$ with support in an interval $[t_0, t_1]$ so that, if $\boldsymbol{m}(f;t)$ is the solution to (14.23) with $\boldsymbol{m}(f;t_0) = [0, 0, 1]$, then $\boldsymbol{m}(f;t_1) = \boldsymbol{m}^{\infty}(f)$. If one restricts attention to flip angles close to 0, then there is a simple linear model that can be used to find approximate solutions.

If the flip angle is close to zero, then $m_3 \approx 1$ throughout the excitation. Using this approximation, we derive the low flip angle approximation to the Bloch equation, without relaxation:

$$\frac{d(m_1 + im_2)}{dt} = -2\pi i f(m_1 + im_2) + i\gamma(\alpha + i\beta). \tag{14.24}$$

From this approximation we see that

$$\alpha(t) + i\beta(t) \approx \frac{\mathcal{F}(m_1^\infty + i m_2^\infty)(t)}{\gamma i}, \quad \text{where } \mathcal{F}(h)(t) = \int_{-\infty}^{\infty} h(f) e^{-2\pi i f t} df. \tag{14.25}$$

For an example like that in (14.22), with θ close to zero, and $f_0 = -f_1$, we obtain

$$\alpha + i\beta \approx \frac{i \sin\theta \sin f_1 t}{\pi \gamma t}. \tag{14.26}$$

A pulse of this sort is called a sinc-pulse. A sinc-pulse is shown in Figure 14.2(a), the result of applying it in Figure 14.2(b).

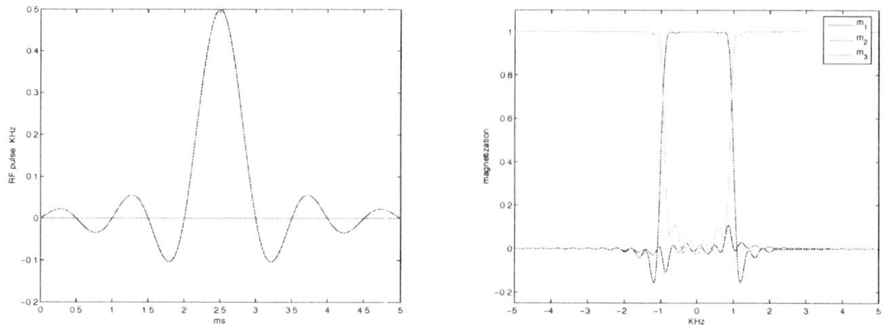

(a) Profile of a 90°-sinc pulse.

(b) The magnetization profile produced by the pulse in (a).

Figure 14.2. A selective 90°-pulse and profile designed using the linear approximation.

A more accurate pulse can be designed using the Shinnar-Le Roux algorithm; see [101, 117]. The problem of selective pulse design can be reformulated as a classical inverse scattering problem for the Zakharov-Sabat, or AKNS 2×2 system. This was discovered in the late 1980s by Alberto Grünbaum, J.F. Carlson, David Rourke, and others. For a modern treatment of this approach and an efficient, accurate algorithm see [34, 92]. An inverse scattering 90°-pulse is shown in Figure 14.3(a) and the response in Figure 14.3(b).

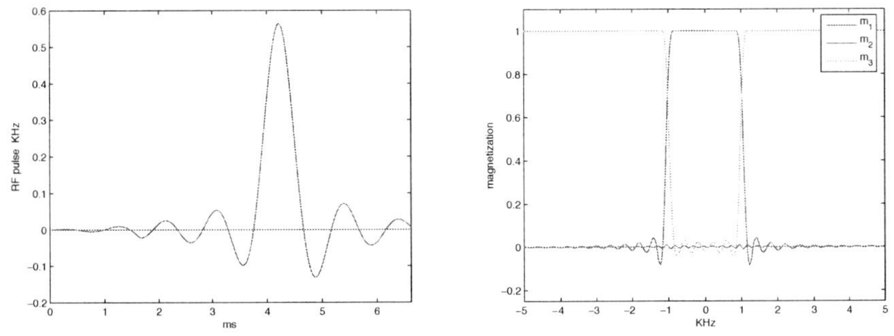

(a) Profile of a 90° inverse scattering pulse. (b) The magnetization profile produced by the pulse in (a).

Figure 14.3. A selective 90°-pulse and profile designed using the inverse scattering approach.

14.7 Spin-warp Imaging

In Section 14.5 we showed how NMR measurements could be used to measure the 3d Fourier transform of ρ. In this section we consider a more practical technique, that of measuring the 2d Fourier transform of a "slice" of ρ. Applying a selective RF-pulse, as described in the previous section, we can flip the magnetization in a region of space $z_0 - \Delta z < z < z_0 + \Delta z$, while leaving it in the equilibrium state outside a slightly larger region. Observing that a signal near the resonance frequency is only produced where the magnetization has a non-zero transverse component, we can now measure samples of the 2d Fourier transform of the slice averaged spin density:

$$\overline{\rho}_{z_0}(x, y) = \frac{1}{2\Delta z} \int\limits_{z_0-\Delta z}^{z_0+\Delta z_0} \rho(x, y, z)dz. \tag{14.27}$$

If Δz is sufficiently small, then $\overline{\rho}_{z_0}(x, y) \approx \rho(x, y, z_0)$.

In order to be able to use the fast Fourier transform algorithm (FFT) to do the reconstruction, it is very useful to sample $\widehat{\overline{\rho}}_{z_0}$ on a uniform grid. To that end we use the gradient fields as follows: After the RF-excitation we apply a gradient field of the form $\boldsymbol{G}_{\text{ph}} = (*, *, -g_2 y + g_1 x)$, for a certain period of time, T_{ph}. This is called a *phase encoding* gradient. At the conclusion of the phase encoding gradient, the transverse components of the magnetization from the excited spins has the form

$$\boldsymbol{m}^{||}(x, y) \propto e^{-2\pi i(k_y y - k_x x)}\overline{\rho}_{z_0}(x, y), \tag{14.28}$$

where $(k_x, k_y) = [2\pi]^{-1}\gamma T_{\text{ph}}(-g_1, g_2)$. At time T_{ph} we turn off the y-component of $\boldsymbol{G}_{\text{ph}}$ and reverse the polarity of the x-component. At this point we begin to measure the signal, obtaining samples of $\widehat{\overline{\rho}}(k, k_y)$ where k varies from $-k_{x\max}$ to $k_{x\max}$. By repeating this process with the strength of the y-phase encoding gradient being stepped through a sequence of uniformly spaced values, $g_2 \in \{n\Delta g_y\}$, and collecting samples at a uniformly spaced set of times, we collect the set of samples

$$\{\widehat{\overline{\rho}}_{z_0}(m\Delta k_x, n\Delta k_y) : -\frac{N_x}{2} \leq m \leq \frac{N_x}{2}, -\frac{N_y}{2} \leq n \leq \frac{N_y}{2}\}. \tag{14.29}$$

The reconstruction algorithm is simply the inverse FFT applied to the sampled data; it produces an approximation to $\overline{\rho}_{z_0}$, within the field of view. Because the reconstruction algorithm is unitary it is intrinsically stable.

The gradient $\boldsymbol{G}_{\text{fr}} = (*, *, -g_1 x)$, left on during signal acquisition, is called a *frequency encoding gradient.* While there is no difference mathematically between the phase encoding and frequency encoding steps, there are significant practical differences. This approach to sampling is known as *spin-warp imaging*; it was introduced in [32]. The steps of this experiment are summarized in a pulse sequence timing diagram, shown in Figure 14.4. This graphical representation for the steps followed in a magnetic resonance imaging experiment is ubiquitous in the literature.

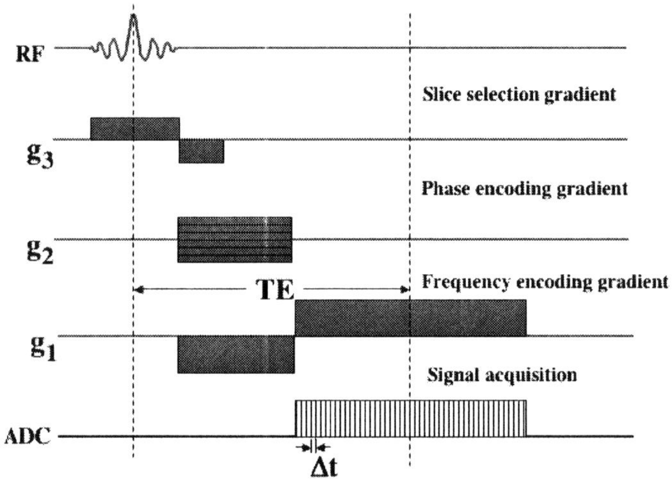

Figure 14.4. Pulse timing diagram for spin-warp imaging. During the positive lobe of the frequency-encoding gradient, the analog-to-digital converter (ADC) collects samples of the signal produced by the rotating transverse magnetization.

As noted above, to avoid aliasing artifacts, the sample spacings Δk_x and Δk_y must be chosen so that the excited portion of the sample is contained in a region of size $\Delta k_x^{-1} \times \Delta k_y^{-1}$. This is called the *field-of-view* or *FOV.* Since we can only collect the signal for a finite period of time, the Fourier transform $\widehat{\rho}(k_x, k_y)$ is sampled at frequencies lying in a rectangle with vertices $(\pm k_{x\,\max}, \pm k_{y\,\max})$, where

$$k_{x\,\max} = \frac{N_x \Delta k_x}{2}, \quad k_{y\,\max} = \frac{N_y \Delta k_y}{2}. \tag{14.30}$$

The maximum frequencies sampled effectively determine the resolution available in the reconstructed image. Heuristically, this resolution limit equals half the shortest measured wavelength:

$$\Delta x \approx \frac{1}{2k_{x\,\max}} = \frac{FOV_x}{N_x} \quad \Delta y \approx \frac{1}{2k_{y\,\max}} = \frac{FOV_y}{N_y}. \tag{14.31}$$

Whether one can actually resolve objects of this size in the reconstructed image depends on other factors such as the available contrast and the signal-to-noise ratio (SNR).

Spin-warp imaging is a simple example of the many possible approaches to collecting data in an MR-scanner. Different ways of sampling $\boldsymbol{k}$-space lead to different types of contrast in the reconstructed image. This flexibility makes it possible to examine many different aspects of anatomy and physiology using MR-imaging. A thorough discussion of this topic is given in [9].

14.8 Contrast and Resolution

The single most distinctive feature of MRI is its extraordinarily large *innate contrast*. For two soft tissues, it can be on the order of several hundred percent. By comparison, contrast in X-ray imaging is a consequence of differences in the attenuation coefficients for two adjacent structures and is typically on the order of a few percent.

We have seen in the preceding sections that the physical principles underlying MRI are radically different from those of X-ray computed tomography in that the signal elicited is generated by the spins themselves in response to an external perturbation. The contrast between two regions, A and B, with signals S_A, S_B respectively, is defined as

$$C_{AB} = \frac{S_A - S_B}{S_A}. \tag{14.32}$$

If the only contrast mechanism were differences in the proton spin density of various tissues, then contrast would be on the order of 5-20%. In reality, it can be several hundred percent. The reason for this discrepancy is that the MR signal is acquired under *non-equilibrium conditions*. At the time of excitation, the spins have typically not recovered from the effect of the previous cycle's RF pulses, nor is the signal usually detected immediately after its creation. indexcontrast to noise ratio

Typically, in spin-warp imaging, a *spin-echo* is detected as a means to alleviate spin coherence losses from static field inhomogeneity. A spin echo is the result of applying an RF-pulse that has the effect of taking (m_1, m_2, m_3) to $(m_1, -m_2, -m_3)$. As such a pulse effects a 180° rotation of the $\hat{z}$-axis, it is also called a π-pulse. If, after such a pulse, the spins continue to evolve in the same environment then, following a certain period of time, the transverse components of the magnetization vectors throughout the sample become aligned. Hence a pulse of this type is also called a refocusing pulse. The time when all the transverse components are rephased is called the *echo time*, TE.

The spin-echo signal amplitude for an RF pulse sequence $\frac{\pi}{2} - \tau - \pi - \tau$, repeated every T_R, seconds is approximately given by:

$$S(t = 2\tau) \approx \rho(1 - e^{-T_R/T_1})e^{-TE/T_2}. \tag{14.33}$$

This is a good approximation as long as $TE << T_R$ and $T_2 << T_R$ in which case the transverse magnetization decays essentially to zero between successive pulse sequence cycles. In equation (14.33), ρ is voxel spin density and the echo time $TE = 2\tau$. Empirically, it is known that tissues differ in at least one of the intrinsic quantities, T_1, T_2 or ρ. It therefore suffices to acquire images in such a manner that contrast is sensitive to one particular parameter. For example, a "T_2-weighted" image would be acquired with $TE \sim T_2$ and $TR >> T_1$ and, similarly, a "T_1-weighted" image with $TR < T_1$ and $TE << T_2$, with T_1, T_2 representing typical tissue proton relaxation times. Figure 14.5 shows two images obtained with the same scan parameters except for TR and TE illustrating the fundamentally different image contrasts that are achievable.

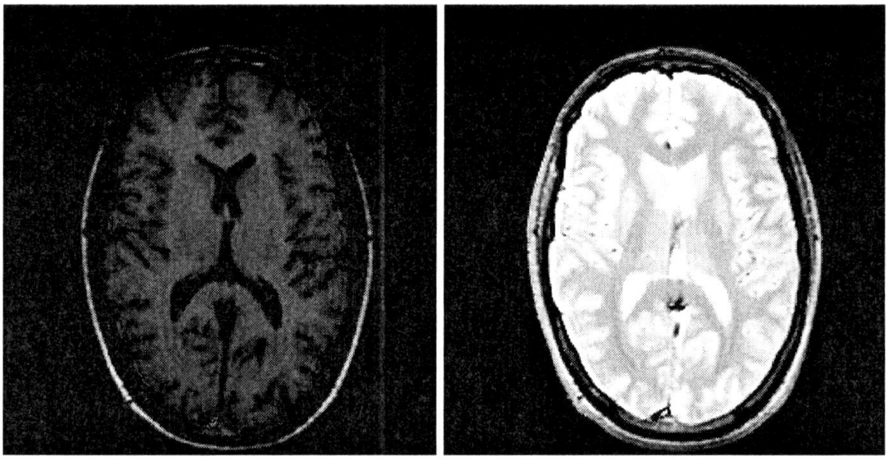

(a) (b)

Figure 14.5. Dependence of image contrast on pulse sequence timing parameters: a) T_1-weighted; b) proton density-weighted.

It is noteworthy that object visibility is not just determined by the contrast between adjacent structures but is also a function of the noise. It is therefore useful to define the contrast-to-noise ratio as

$$CNR_{AB} = \frac{C_{AB}}{\sigma_{\text{eff}}} \tag{14.34}$$

where σ_{eff} is the effective standard deviation of the signal. We have previously shown that the limiting resolution is given by $k_{\max}$, the largest spatial frequency sampled; see (14.31).

In reality, however, the actual resolution is always lower. For example spin-spin (T_2) relaxation causes the signal to decay during the acquisition. In spin-warp imaging this causes the high spatial frequencies to be further attenuated.

A further consequence of finite sampling is a ringing or Gibbs artifact that is most prominent at sharp intensity discontinuities. In practice, these artifacts are mitigated by applying an appropriate apodizing filter to the data. Figure 14.6 shows a portion of a brain image obtained at two different resolutions. In Figure 14.6(b), the total $\boldsymbol{k}$-space area covered was 16 times larger than for the acquisition of the image in b). Artifacts from finite sampling and blurring of fine detail such cortical blood vessels are clearly visible in the low-resolution image. SNR, according to equation (16.36), is reduced in the latter image by a factor of 4.

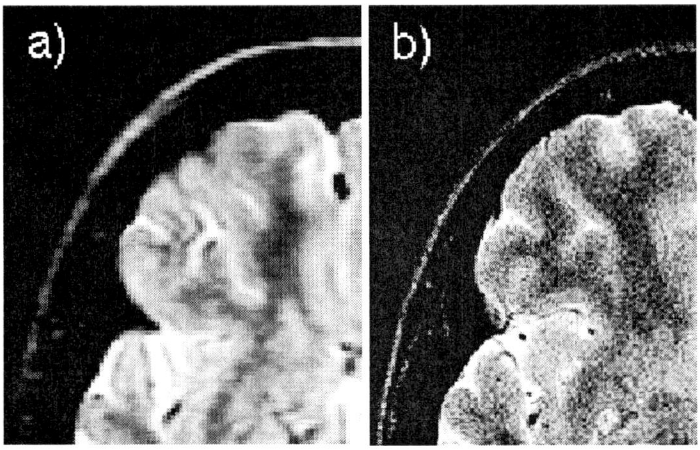

Figure 14.6. Effect of $\boldsymbol{k}$-space coverage on spatial resolution in axial image of the brain: the field of view in both images 20 cm and all scan parameters were the same except that a) was acquired with $N_x = N_y = 128$ and $N_x = N_y = 512$.

14.9 Conclusion

This section gives the merest hint of the remarkable richness of the phenomenon of nuclear magnetic resonance and its applications in medicine, chemistry and physics. The interested reader is referred to the excellent texts, listed at the end of introduction to this chapter, for more complete discussions of various aspects of this subject. This chapter was adapted from [37] and [33].

Chapter 15

Probability Theory and Random Variables

Up to this point, we have only considered deterministic models. These are models where a known input produces a definite output. We now begin to discuss probability theory, which is the language of noise analysis. But what is noise? There are two essentially different sources of noise in the mathematical description of a physical system or measurement process. The first step in building a mathematical model is to isolate a physical system from the world in which it sits. Once such a separation is fixed, the effects of the outside world on the state of the system are often modeled, a posteriori, as noise. In our model for CT imaging it is assumed that every x-ray that is detected is produced by our x-ray source. In reality, there are many other sources of x-rays that might also impinge on our detectors. Practically speaking, it is not possible to give a complete description of all such external sources, but sometimes it is possible to describe them probabilistically.

Many physical processes are inherently probabilistic, and that is the second source of noise. In CT imaging the interaction of an x-ray "beam" with an object is a probabilistic phenomenon. The beam is, in fact, a collection of discrete photons. Whether or not a given photon, entering an object on one side, reemerges on the other side, traveling in the same direction, depends on the very complicated interactions this photon has with the microscopic structure of the object it passes through. These individual interactions cannot be modeled in a useful way. If μ is the attenuation coefficient and $I(t)$ is the incident flux of photons at t, then Beer's law says that the change in the flux, $I(t + \Delta t) - I(t)$, over a small distance Δt is

$$I(t + \Delta t) - I(t) \approx -\mu \, \Delta t I(t)$$

or

$$I(t + \Delta t) = (1 - \mu \, \Delta t) I(t).$$

This can be interpreted as the statement that an incident photon has probability $(1 - \mu \, \Delta t)$ of being emitted. Beer's law describes the average behavior of a photon and is useful only if the x-ray beam is composed of a large number of photons.

For a given measurement it is not possible to quantify the exact discrepancy between the measured value and that predicted by Beer's law. Instead we have a probabilistic description that describes the *statistics* of this discrepancy. Suppose that N transmitted photons are measured; we can think of the measurement as having two parts:

$$N = N_d + \upsilon,$$

where N_d is a deterministic quantity (i.e., the part predicted by Beer's law) and υ is a random "variable" that models the noise in the system. A good probabilistic model for the noise component aids in interpretation of the measurements. Even the model of an x-ray beam as a constant flux of photons is an approximation to the truth; the actual output of an x-ray source also has a useful probabilistic description.

In this chapter we review some of the basic concepts of measure theory and probability theory. It is not intended as a development of either subject *ab ovo*, but merely a presentation of the main ideas so that we can discuss noise in image reconstruction and later random processes. These concepts are presented using the mathematical framework provided by measure theory. Historically, probability theory preceded measure theory by many decades but was found, in the early twentieth century, to lack rigorous foundations. Kolmogorov discovered that measure theory provided the missing foundations. It is not necessary to have a background in measure theory; we use it as a language to give precise definitions of the concepts used in probability theory and later to pass to random processes. Many of the examples in this chapter are elementary so that readers not interested in technical details can still get a flavor of these subjects. Those with a good background in probability theory can safely skip to Section 15.3, where we introduce the probability distributions needed in imaging applications. An introduction to probability can be found in [27], an introduction to random processes in [30] or [57], and an introduction to measure theory in [42] or [109].

15.1 Measure Theory*

Mathematical probability theory is a subset of measure theory. This is the branch of mathematics that gives a framework in which to study integration. For concreteness we discuss probability theory from the point of view of doing an "experiment." We are then interested in quantifying the likelihood that the experiment has a particular outcome.

15.1.1 Allowable Events

In measure theory we work with a measure space. This is a pair $(X, \mathcal{M})$, where X is the underlying space and $\mathcal{M}$ is a collection of subsets of X called a σ-algebra. From the point view of probability theory, X is called the *sample space*; it is the set of all possible outcomes of the experiment. Subsets of X are collections of possible outcomes, which in probability theory are called *events*. The subsets of X belonging to $\mathcal{M}$ are *allowable* events. These are the events that can be assigned a well-defined probability of occurring. A simple physical example serves to explain, in part, why it may not be possible to assign a probability of occurrence to every event (i.e., to every subset of X).

***Example* 15.1.1.** Suppose that the experiment involves determining where a particle strikes a line. The sample space X is the real line $\mathbb{R}$. For each $n \in \mathbb{Z}$, the measuring device can determine whether on not the particle fell in $[n, n+1)$ but not where in this interval the particle fell. For each n, the event

$$A_n = \text{ the particle fell in } [n, n+1)$$

is therefore an allowable event. On the other hand, the event "the particle fell in $[.3, .7)$" is not an allowable event, as the measurements cannot determine whether or not this occurs. For the set of allowable events, we take arbitrary unions of the subsets $\{A_n : n \in \mathbb{Z}\}$.

***Example* 15.1.2.** Perhaps the simplest, interesting experiment is that of "flipping a coin." This is an experiment that has two possible outcomes, which we label H and T. The sample space is

$$X_1 = \{H, T\}.$$

The possible *events* are

- We get a head.
- We get a tail.
- We get a head or a tail.

These events correspond to the subsets of X_1:

$$\{H\}, \{T\}, \{H, T\}.$$

***Example* 15.1.3.** Suppose that instead of flipping a coin once, the experiment involves flipping a coin two times. The possible outcomes are sequences of length two in the symbols H and T:

$$X_2 = \{(H, H), (H, T), (T, H), (T, T)\}.$$

If the experiment involves flipping a coin N times, then the sample space consists of all sequences of length N using the symbols H and T. We denote this space by X_N; for each N the sample space is a finite set. The set of allowable events is usually taken to be *all* the subsets of X_N; we denote this collection by $\mathcal{M}_N$.

As suggested by Example 15.1.1, the collection, $\mathcal{M}$, of allowable events provides a mathematical framework for describing the operation of a measuring apparatus. It has the following axiomatic properties:

1. $X \in \mathcal{M}$, this is the statement that X is the collection of *all* possible outcomes.

2. If $A \in \mathcal{M}$ and $B \in \mathcal{M}$, then $A \cup B \in \mathcal{M}$; in other words, if the events A and B are each allowable, then the event "A or B" is also allowable.

3. If $A \in \mathcal{M}$, then $X \setminus A \in \mathcal{M}$; in other words, if the event "A occurs" is allowable, then the event "A does not occur" is also allowable.

4. If we have a countable collection of allowable events $A_j \in \mathcal{M}$, then

$$\bigcup_{j=1}^{\infty} A_j \in \mathcal{M}$$

as well.

Applying (2) recursively, we conclude that if $\{A_1, \ldots, A_n\}$ belong to $\mathcal{M}$, then so does their union $A_1 \cup \cdots \cup A_n$. Condition (4) is the limit of this statement as n goes to infinity. It is a technical condition that is essential to have a good mathematical theory of integration. It is important when taking limits of sequences. Subsets belonging to $\mathcal{M}$ are called allowable events or *measurable sets*. A collection of subsets of a space that satisfies these axioms is called a *σ-algebra*.

As a consequence of these axioms, we need to introduce the notion of the *empty set*. It is the subset of X that contains *no* elements and is denoted by $\emptyset$. The empty set is a subset of any set. The need for the empty set arises because (2) and (3) imply that if A and B are allowable events, then so is $A \cap B$. However, if A and B have no points in common, then $A \cap B = \emptyset$. It is a linguistic device, encapsulating the idea that the experiment has no outcome. As we shall see, this "event" always has probability zero of occurring.

***Definition* 15.1.1.** If A and B are allowable events such that $A \cap B = \emptyset$, then A and B are said to be *mutually exclusive* events.

***Example* 15.1.4.** The list given in Example 15.1.2 is not quite all of $\mathcal{M}_1$, but rather

$$\mathcal{M}_1 = \{\emptyset, \{H\}, \{T\}, \{H, T\}\}.$$

***Example* 15.1.5.** Some examples of allowable events for the case X_N defined previously are as follows:

- All N flips produce heads.
- Half the flips are tails.
- If $k \leq N$, then the kth flip is a head.
- At most $N/3$ flips are heads.

***Example* 15.1.6.** Suppose that the result of our experiment is a real number; then X is $\mathbb{R}$. Allowable events might include the following:

- $\{x\}$ for an $x \in \mathbb{R}$, the outcome of the experiment is the number x.
- $[0, \infty)$, the outcome of the experiment is a nonnegative number.
- $(1, 2)$, the outcome of the experiment is a number between 1 and 2.

In many practical situations the sample space is the real line $\mathbb{R}$. In these cases the σ-algebra is often taken to be the Borel sets.

***Definition* 15.1.2.** The smallest collection of subsets of $\mathbb{R}$ that includes all intervals and is a σ-algebra is called the *Borel sets*.

The Borel sets are discussed in [42].

Exercises

***Exercise* 15.1.1.** Show that the collection of allowable events, $\mathcal{M}$, in Example 15.1.1 has the following description: For A in $\mathcal{M}$, either $A = \emptyset$ or there are, possibly bi-infinite, sequences $< a_i >$ and $< b_i >$ of integers so that

$$\cdots < a_i < b_i < a_{i+1} < b_{i+1} < \cdots$$

and

$$A = \bigcup_{j=-\infty}^{\infty} [a_j, b_j).$$

***Exercise* 15.1.2.** Show that the collection of sets defined in Example 15.1.1 along with the empty set is a σ-algebra.

***Exercise* 15.1.3.** Show that the space X_N in Example 15.1.3 has 2^N elements. How many allowable events are there; that is, how large is $\mathcal{M}_N$?

***Exercise* 15.1.4.** Let $A \subset X$. Then *the complement of* A *in* X is the subset of X defined by

$$A^c = X \setminus A = \{x \in X \ : \ x \notin A\}.$$

Show that if $A, B \subset X$, then

$$(A \cup B)^c = A^c \cap B^c.$$

Conclude that if $A, B \in \mathcal{M}$, then $A \cap B \in \mathcal{M}$ as well.

***Exercise* 15.1.5.** Suppose that X is a set and $\mathcal{M}$ is a σ-algebra of subsets of X. Show that the empty set belongs to $\mathcal{M}$.

***Exercise* 15.1.6.** Suppose that $\mathcal{M}$ is a σ-algebra of subsets of $\mathbb{R}$ that contains all closed intervals $\{[a, b]\}$. Show that it also contains all open and half-open intervals

$$\{(a, b), (a, b], [a, b) \ : \ a \leq b\}.$$

15.1.2 Measures and Probability

See: B.6, B.7.

So far we have no notion of the probability that an event occurs. Since events are subsets of X, what is required is a way to measure the size of a subset. Mathematically, this is described by a real-valued function ν defined on the set of allowable events. This function is called a *probability measure* if it has the following properties:

1. $\nu(A) \geq 0$ for all $A \in \mathcal{M}$, that is an allowable event occurs with nonnegative probability.

2. $\nu(X) = 1$, so that X is the list of *all* possible outcomes.

3. If $A, B \in \mathcal{M}$ and $A \cap B = \emptyset$, then

$$\nu(A \cup B) = \nu(A) + \nu(B).$$

This is called additivity for mutually exclusive events.

3′. If we have a list of events $A_i \in \mathcal{M}$, for $i \in \mathbb{N}$ such that $A_i \cap A_j = \emptyset$ for $i \neq j$, then

$$\nu\left(\bigcup_{j=1}^{\infty} A_j\right) = \sum_{j=1}^{\infty} \nu(A_j).$$

Conditions (2) and (3) imply that $\nu(A) \leq 1$ for any $A \in \mathcal{M}$. A triple $(X, \mathcal{M}, \nu)$ consisting of a space X, with a σ-algebra of subsets $\mathcal{M}$, and a probability measure $\nu : \mathcal{M} \to [0, 1]$, is called a *probability space*. In the sequel we often use

$$\text{Prob}(A) \stackrel{d}{=} \nu(A)$$

to denote the *probability of the event* A. Note that (2) and (3) imply that $\text{Prob}(\emptyset) = 0$.

Applying (3) recursively shows that if $\{A_1, \ldots, A_n\}$ are mutually exclusive events, then

$$\nu(A_1 \cup \cdots \cup A_n) = \nu(A_1) + \cdots + \nu(A_n).$$

Condition (3′) is the limit of this condition as n tends to infinity; as such it can be seen as a continuity condition imposed on ν. It is called *countable additivity.* This condition is needed in order to have a good mathematical theory of integration. It turns out that many interesting functions that satisfy these conditions cannot be extended to *all* the subsets of X in a reasonable way. For example, there is *no* way to extend the naive notion of length to every subset of the real line *and* satisfy these axioms. Subsets that do not belong to $\mathcal{M}$ are called *nonmeasurable*. A simple example of a nonmeasurable set is given in Example 15.1.1. In most cases nonmeasurable sets are very complicated and do not arise naturally in applications.

***Example* 15.1.7.** For the case of a single coin toss, the allowable events are

$$\mathcal{M}_1 = \{\emptyset, \{H\}, \{T\}, \{H, T\}\}.$$

The function $\nu_1 : \mathcal{M}_1 \to [0, 1]$ is fixed once we know the probability of H. For say that $\nu_1(H) = p$. Because $\nu_1(H \cup T) = 1$ and H and T are mutually exclusive events, it follows that $\nu_1(T) = 1 - p$. Clearly, $\nu(\emptyset) = 0$.

***Example* 15.1.8.** For the case of general N, the allowable events are all collections of sequences $\boldsymbol{a} = (a_1, \ldots, a_N)$, where $a_j \in \{H, T\}$ for $j = 1, \ldots, N$. The most general probability function on $\mathcal{M}_N$ is defined by choosing numbers $\{p_{\boldsymbol{a}} : \boldsymbol{a} \in X_N\}$ so that

$$0 \leq p_a \leq 1$$

and

$$\sum_{\boldsymbol{a} \in X_N} p_a = 1. \tag{15.1}$$

If $A \in \mathcal{M}_N$ is an event, then

$$\nu(A) = \sum_{a \in A} p_a.$$

In most instances we use a much simpler measure to define the probability of events in $\mathcal{M}_N$. Instead of directly assigning a probability to each sequence of length N, we use the assumption that an H occurs (at any position in the sequence) with probability p and a T with probability $1 - p$. We also assume that outcomes of the various flips are independent of one another. With these assumptions we can show that

$$\nu_{p,N}(\{\boldsymbol{a}\}) = p^{m_a}(1-p)^{N-m_a}, \tag{15.2}$$

where $m_{\boldsymbol{a}}$ is the number of Hs in the sequence $\boldsymbol{a}$.

***Example* 15.1.9.** Suppose that $X = \mathbb{R}$, with $\mathcal{M}$ the Borel sets. For any $a < b$, the half-open interval $[a, b)$ is an allowable event. Let f be a nonnegative, continuous function defined on X with total integral 1:

$$\int_{-\infty}^{\infty} f(t)\,dt = 1.$$

Define the probability of the event $[a, b)$ to be

$$\text{Prob}([a, b)) = \int_a^b f(t)\,dt.$$

We can show that this defines a function on $\mathcal{M}$ satisfying the properties enumerated previously, see [42] or [57].

***Example* 15.1.10.** In the situation described in Example 15.1.1, X equals $\mathbb{R}$. Choose a bi-infinite sequence $\{a_n : n \in \mathbb{Z}\}$ of nonnegative numbers such that

$$\sum_{n=-\infty}^{\infty} a_n = 1$$

and define

$$\nu([n, n+1)) = a_n \qquad \text{for } n \in \mathbb{Z}.$$

This means that a_n is the probability that the particle fell in $[n, n+1)$. This function is easily extended to define a measure on the allowable events defined in Example 15.1.1. Again note that it is not possible to assign a probability to the event "the particle fell in $[.3, .7)$" as our measurements are unable to decide whether or not this happens.

***Example* 15.1.11.** Suppose that X is the unit disk in the plane, and let $\mathcal{M}$ be the Lebesgue measurable subsets of X. These are the subsets whose surface area is well defined. The outcome of the experiment is a point in X. For a measurable set A, define the probability of the event "the point lies in A" to be

$$\text{Prob}(A) = \frac{\text{area}(A)}{\pi}.$$

In a reasonable sense, this means that each point in X is equally likely to be the outcome of the experiment. Note that if the set A consists of a single point $\{(x, y)\}$, then $\text{Prob}(A) = 0$.

How is the notion of probability introduced previously related to the outcomes of experiments? Consider the case of flipping a coin; what does it mean to say that heads occurs with probability p and tails occurs with probability $1 - p$? Suppose we flip a coin N times, and let $H(N)$ and $T(N)$ be the number of heads and tails, respectively. One possible interpretation of the statement $\text{Prob}(\{H\}) = p$ is that

$$\lim_{N\to\infty} \frac{H(N)}{N} = p.$$

This is sometimes called a *time average*: We perform a sequence of identical, independent experiments and take the average of the results. We expect that the average over time will approach the theoretical probability. Another way to model the connection between the outcomes of experiments and probability theory is to consider *ensemble averages*. For the example of coin tossing, imagine that we have N *identical* coins. We flip each coin once; again let $H(N)$ be the number of heads. If

$$\lim_{N\to\infty} \frac{H(N)}{N} = p,$$

then we say that heads occurs, for this coin, with probability p.

The existence of either of these limits means that the process of flipping coins has some sort of statistical regularity. It is a simple matter to concoct sequences $< H(N) >$ for which the ratio $H(N)/N$ does not converge to a limit. It is, in essence, an experimental fact that such sequences are very unlikely to occur. The relationship between experimental science and the theory of probability is a complex philosophical issue. E.T. Jaynes beautifully described the two major schools of thought in his seminal paper[1]:

[1]Quotation taken from: *Information Theory and Statistical Mechanics*, by E. T. Jaynes, in *Physical Review*, vol. 106 (1957), p. 622.

The "objective" school of thought regards the probability of an event as an objective property of that event, always capable in principle of empirical measurement by observation of frequency ratios in a random experiment. In calculating a probability distribution the objectivist believes that he is making predictions which are in principle verifiable in every detail, just as those of classical mechanics. The test of a good probability distribution $p(x)$ is: does it correctly represent the observable fluctuations of x?

On the other hand the "subjective" school of thought regards probabilities as expressions of human ignorance; the probability of an event is merely a formal expression of our expectation that the event will or did occur, based on whatever information is available. To the subjectivist, the purpose of probability is to help us in forming plausible conclusions in cases where there is not enough information available to lead to certain conclusions; thus detailed verification is not expected. The test of a good subjective probability distribution is does it correctly represent our state of knowledge as to the value of x?

Exercises

***Exercise* 15.1.7.** Prove that if $(X, \mathcal{M}, \nu)$ is any probability space, then $\nu(\emptyset) = 0$.

***Exercise* 15.1.8.** In Example 15.1.11, explain why $\text{Prob}(\{(x, y)\}) = 0$.

***Exercise* 15.1.9.** Show that the probability measure defined in (15.2) satisfies (15.1).

***Exercise* 15.1.10.** In Example 15.1.11, let A_r denote the circle of radius r centered at 0. Compute $\text{Prob}(A_r)$.

***Exercise* 15.1.11.** In Section 3.4 we defined a "set of measure zero." What is the probabilistic interpretation of such a set?

15.1.3 Integration

See: B.6.

Using the measure defined on a probability space, $(X, \mathcal{M}, \nu)$, we can define a notion of integration. The first step is to define a class of functions that might be integrable. Such a function is called a measurable function.

***Definition* 15.1.3.** Let $(X, \mathcal{M}, \nu)$ be a probability space. A real-valued function, f, defined on X is *measurable* if, for every $t \in \mathbb{R}$, the set $\{x \in X : f(x) \leq t\}$ belongs to $\mathcal{M}$.

A fundamental consequence of axiom (3′) for a σ-algebra is the fact that a pointwise convergent sequence of measurable functions is also measurable.

For an arbitrary subset A, define the *indicator function of* A to be

$$\chi_A(x) = \begin{cases} 1 \text{ if } x \in A, \\ 0 \text{ if } x \notin A. \end{cases}$$

Earlier in the book this function would have been called the characteristic function of the set A. In probability theory, *characteristic function* has a different meaning, so here we use (the also standard terminology), indicator function.

***Example* 15.1.12.** If A is an element of $\mathcal{M}$, then χ_A is a measurable function.

***Example* 15.1.13.** If X is the real line and $\mathcal{M}$ is the collection of Borel sets, then any continuous function is measurable.

For the indicator function of a set A in $\mathcal{M}$, it is clear that the only reasonable way to define its integral is to let

$$\int_X \chi_A(x)\,d\nu(x) \stackrel{d}{=} \nu(A).$$

As $\nu(X) = 1$, this defines $\int \chi_A\,d\nu$ as the average of χ_A over X.

***Definition* 15.1.4.** A function f is called a *simple function* if there are sets $A_j \in \mathcal{M}$, $j = 1, \ldots, m$, and real constants $\{a_j\}$ so that

$$f = \sum_{j=1}^{N} a_j \chi_{A_j}.$$

It is not difficult to show that a simple function is measurable. Since the integral should be linear, it is clear that we must define

$$\int_X \sum_{j=1}^{N} a_j \chi_{A_j}(x)\,d\nu(x) \stackrel{d}{=} \sum_{j=1}^{N} a_j \nu(A_j).$$

While this formula is intuitively obvious, it requires proof that it makes sense. This is because a simple function can be expressed as a linear combination of indicator functions in different ways. If, as functions,

$$\sum_{j=1}^{N} a_j \chi_{A_j} = \sum_{k=1}^{M} b_k \chi_{B_k}, \tag{15.3}$$

then it is necessary to show that

$$\sum_{j=1}^{N} a_j \nu(A_j) = \sum_{k=1}^{M} b_k \nu(B_k).$$

This is left as an exercise for the reader.

Suppose that f is a bounded, measurable function. Fix a positive integer N. For each $j \in \mathbb{Z}$, define the measurable set

$$A_{N,j} \stackrel{d}{=} f^{-1}\left([\frac{j}{N}, \frac{j+1}{N})\right). \tag{15.4}$$

Since f is a bounded function, the function

$$F_N = \sum_{j=-\infty}^{\infty} \frac{j}{N} \chi_{A_{N,j}} \tag{15.5}$$

is a simple function with the following properties:

1. $$0 \leq f(x) - F_N(x) \leq N^{-1}, \text{ for all } x,$$

2. $$\int_X F_N(x) d\nu(x) = \sum_{j=-\infty}^{\infty} \frac{j}{N} \nu(A_{N,j}),$$

3. $$F_N(x) \leq F_{N+1}(x), \text{ for all } x.$$

In other words, a bounded measurable function can be approximated by simple functions for which the integral is defined. This explains, in part, why we introduce the class of measurable functions. For a bounded, nonnegative, measurable function, the ν-integral is defined by

$$\int_X f(x)\, d\nu(x) = \lim_{N \to \infty} \int_X F_N(x)\, d\nu(x).$$

Condition (3) implies that $\int F_N\, d\nu$ is an increasing function of N so this limit exists. By approximating nonnegative, measurable functions f in a similar way, the definition of the integral can be extended to this class. If f is a nonnegative measurable function, then its integral over X is denoted by

$$\int_X f(x)\, d\nu(x).$$

The integral may equal $+\infty$. If f is measurable, then the functions

$$f_+(x) = \max\{0, f(x)\}, \quad f_-(x) = \max\{0, -f(x)\}$$

are also measurable. If either $\int_X f_\pm d\nu$ is finite, then the integral of f is defined to be

$$\int_X f(x)\, d\nu(x) \stackrel{d}{=} \int_X f_+(x)\, d\nu(x) - \int_X f_-(x)\, d\nu(x).$$

***Definition* 15.1.5.** Let $(X, \mathcal{M}, \nu)$ be a probability space. A measurable function f is *integrable* if both of the integrals

$$\int_X f_\pm(x)\, d\nu(x)$$

are finite. In this case

$$\int_X f(x)\,d\nu(x) \stackrel{d}{=} \int_X f_+(x)\,d\nu(x) - \int_X f_-(x)\,d\nu(x).$$

This the *Lebesgue* integral.

A more complete discussion of Lebesgue integration can be found in [42] or [109]. For our purposes, certain formal properties of the integral are sufficient. Let f, g be measurable functions and a a real number. Then the integral defined by ν is *linear* in the sense that

$$\begin{aligned} \int_X (f+g)\,d\nu &= \int_X f\,d\nu + \int_X g\,d\nu \text{ and} \\ \int_X af\,d\nu &= a\int_X f\,d\nu. \end{aligned} \tag{15.6}$$

These conditions imply that if $\{A_j\}$ is a collection of pairwise disjoint subsets belonging to $\mathcal{M}$ and $\{a_j\}$ is a bounded sequence of numbers, then the function

$$f(x) = \sum_{j=1}^{\infty} a_j \chi_{A_j}(x)$$

is an integrable function with

$$\int_X f d\nu = \sum_{j=1}^{\infty} a_j \nu(A_j).$$

Note that here we consider infinite sums, whereas previously we only considered finite sums.

***Example* 15.1.14.** Let $(X_N, \mathcal{M}_N, \nu_{p,N})$ be the probability space introduced in the Example 15.1.8. Since $\mathcal{M}_N$ contains all subsets of X_N, any function on X_N is measurable. Using the properties of the integral listed previously, it is not difficult to show that if f is a function on X_N, then

$$\int_{X_N} f(x)\,d\nu_{p,N}(x) = \sum_{\boldsymbol{a}\in X_N} f(\boldsymbol{a})\nu_{p,N}(\boldsymbol{a}).$$

For a finite probability space, an integral reduces to a finite sum.

***Example* 15.1.15.** If f is a nonnegative, continuous function on $\mathbb{R}$ with

$$\int_{\mathbb{R}} f(x)\,dx = 1,$$

then we can define a probability measure on the Borel sets by setting

$$\nu(A) = \int_A f(x)\,dx = \int_{\mathbb{R}} \chi_A(x) f(x)\,dx.$$

With this definition, it is not difficult to show that for any bounded, measurable function g we have

$$\int_{\mathbb{R}} g(x)\,d\nu(x) = \int_{\mathbb{R}} g(x) f(x)\,dx.$$

Example* 15.1.16. Suppose that F is a nonnegative function defined on $\mathbb{R}$ that satisfies the following conditions:

1. F is a monotone nondecreasing function: $x < y$ implies that $F(x) \leq F(y)$.

2. F is continuous from the right: For all x, $F(x) = \lim_{y \to x^+} F(y)$.

3. $\lim_{x \to -\infty} F(x) = 0, \quad \lim_{x \to \infty} F(x) = 1.$

Such a function defines a measure on the Borel subsets of $\mathbb{R}$. The measure of a half-ray is defined to be

$$\nu_F((-\infty, a]) \stackrel{d}{=} F(a),$$

and the measure of an interval $(a, b]$ is defined to be

$$\nu_F((a, b]) \stackrel{d}{=} F(b) - F(a).$$

Note that if $(a, b]$ is written as a disjoint union,

$$(a, b] = \bigcup_{j=1}^{\infty} (a_j, b_j],$$

then

$$F(b) - F(a) = \sum_{j=1}^{\infty} F(b_j) - F(a_j). \tag{15.7}$$

This condition shows that the ν_F-measure of an interval $(a, b]$ is well defined.

If $a_1 < b_1 \leq a_2 < b_2 \leq \cdots$, then the measure of the union of intervals is defined, by countable additivity, to be

$$\nu_F\left(\bigcup_{j=1}^{\infty} (a_j, b_j]\right) = \sum_{j=1}^{\infty} F(b_j) - F(a_j).$$

The measure of an arbitrary Borel set is defined by approximating it by intersections of unions of intervals. The measure ν_F is called the *Lebesgue-Stieltjes measure* defined by F. The ν_F-integral of a function g is usually denoted

$$\int_{-\infty}^{\infty} g(x)\,dF(x).$$

It is called the *Lebesgue-Stieltjes integral* defined by F.

A similar construction can be used to define measures on $\mathbb{R}^n$. Here F is assumed to be a nonnegative function that is monotone, nondecreasing, and continuous from the right in each variable separately, satisfying

$$\lim_{x_j \to -\infty} F(x_1, \ldots, x_j, \ldots, x_n) = 0 \qquad \text{for } j = 1, \ldots, n.$$

The measure of $(-\infty, a_1] \times \cdots \times (-\infty, a_n]$ is defined to be

$$\nu_F((-\infty, a_1] \times \cdots \times (-\infty, a_n]) = F(a_1, \ldots, a_n).$$

By appropriately adding and subtracting, we define the measure of a finite rectangle. For example, if $n = 2$, then

$$\nu_F((a_1, b_1] \times (a_2, b_2]) = F(b_1, b_2) + F(a_1, a_2) - F(a_2, b_1) - F(a_1, b_2). \tag{15.8}$$

A discussion of Lebesgue-Stieltjes integrals can be found in [29].

***Example* 15.1.17.** Let X equal $\mathbb{R}$ and let $< a_n >$ be a bi-infinite sequence of nonnegative numbers such that

$$\sum_{j=-\infty}^{\infty} a_j = 1.$$

We let $\mathcal{M}$ be the collection of all subsets of $\mathbb{R}$ and define the measure ν by letting

$$\nu(A) = \sum_{\{n\,:\,n \in A\}} a_n.$$

Any function g is measurable and

$$\int_{\mathbb{R}} g(x)\,d\nu(x) = \sum_{n=-\infty}^{\infty} a_n g(n).$$

Exercises

***Exercise* 15.1.12.** Suppose that $(X, \mathcal{M}, \nu)$ is a probability space and $A, B \in \mathcal{M}$ are two allowable events. Show that

$$\nu(A \cup B) = \nu(A) + \nu(B) - \nu(A \cap B).$$

Explain why this is reasonable from the point of view of probability.

***Exercise* 15.1.13.** Show that if $A \in \mathcal{M}$, then χ_A is a measurable function.

***Exercise* 15.1.14.** For the σ-algebra $\mathcal{M}$ defined in Example 15.1.1, what are the measurable functions?

***Exercise* 15.1.15.** If $X = \mathbb{R}$ and $\mathcal{M}$ is the collection of Borel sets, show that any continuous function on X is measurable.

***Exercise* 15.1.16.** Prove that if f is a measurable function, then the sets $A_{N,j}$ defined in (15.4) belong to $\mathcal{M}$ for every j and N.

***Exercise* 15.1.17.** For F_N defined in (15.5), show that

$$\int_X F_N(x)\,d\nu(x) \leq \int_X F_{N+1}(x)\,d\nu(x).$$

***Exercise* 15.1.18.** If f is a bounded function, then for each N, there exists an M_N so that

$$A_{N,j} = \emptyset$$

if $|j| \geq M_N$.

***Exercise* 15.1.19.** Show that if f is a measurable function such that $|f(x)| \leq M$ for all $x \in X$, then

$$\int_X f(x)\,d\nu(x) \leq M.$$

What can be concluded from the equality condition in this estimate?

***Exercise* 15.1.20.** With $\mathcal{M}$ defined in Example 15.1.1 and ν defined in Example 15.1.10, which functions are integrable and what is the integral of an integrable function?

***Exercise* 15.1.21.** Give the details for the extension of the definition of the integral to nonnegative measureable functions, which may not be bounded.

***Exercise* 15.1.22.** If $(X, \mathcal{M}, \nu)$ is a probability space, $\{A_j\} \subset \mathcal{M}$ are pairwise disjoint subsets, and $< a_j >$ is a bounded sequence, then show that

$$\sum_{j=1}^{\infty} |a_j|\nu(A_j) < \infty.$$

***Exercise* 15.1.23.** Suppose that F is a differentiable function with nonnegative derivative f. Show that F defines a Lebesgue-Stieltjes measure and

$$\int_{-\infty}^{\infty} g(x)\,dF(x) = \int_{-\infty}^{\infty} g(x)f(x)\,dx. \tag{15.9}$$

***Exercise* 15.1.24.** Suppose that F has a weak derivative f. Does (15.9) still hold?

***Exercise* 15.1.25.** Let $F(x) = 0$ for $x < 0$ and $F(x) = 1$ for $x \geq 0$. For a continuous function, g, what is

$$\int_{-\infty}^{\infty} g(x)\,dF(x)?$$

15.1.4 Independent Events

Suppose that a coin is flipped several times in succession. The outcome of one flip should not affect the outcome of a successive flip; nor is it affected by an earlier flip. They are *independent events*. This is a general concept in probability theory.

***Definition* 15.1.6.** Let $(X, \mathcal{M}, \nu)$ be a probability space. Two allowable events, A and B, are called *independent* if

$$\text{Prob}(A \cap B) = \text{Prob}(A)\,\text{Prob}(B). \tag{15.10}$$

Earlier we said that two events A and B were mutually exclusive if $A \cap B = \emptyset$. In this case,

$$\text{Prob}(A \cup B) = \text{Prob}(A) + \text{Prob}(B).$$

Note the difference between these concepts.

***Example* 15.1.18.** Let $X_N = \{(a_1, a_2, \ldots, a_N) : a_i \in \{H, T\}\}$ be the sample space for flipping a coin N times. Suppose that $\text{Prob}(H) = p$ and $\text{Prob}(T) = 1 - p$. If successive flips are independent, then formula (15.10) implies that

$$\text{Prob}((a_1, a_2, \ldots, a_N)) = \prod_{i=1}^{N} \text{Prob}(a_i) = p^{m_a}(1-p)^{N-m_a}.$$

To prove this, observe that the event $a_1 = H$ is the set $\{(H, a_2, \ldots, a_N) : a_j \in \{H, T\}\}$; it has probability p because it evidently only depends on the outcome of the first flip. Similarly, the event $a_1 = T$ has probability $1 - p$. Indeed, for any fixed j,

$$\text{Prob}(a_j = H) = p \text{ and } \text{Prob}(a_j = T) = 1 - p.$$

The event $A_k = \{a_i = H, \quad i = 1, \ldots k \text{ and } a_j = T, \quad j = k+1, \ldots, N\}$ can be expressed as an intersection:

$$\left[\bigcap_{i=1}^{k} \{a_i = H\}\right] \bigcap \left[\bigcap_{j=k+1}^{N} \{a_j = T\}\right].$$

Since this is an intersection of independent events, it follows that

$$\text{Prob}(A_k) = p^k(1-p)^{N-k}.$$

A similar argument applies if we permute the order in which the heads and tails arise.

For each integer $0 \le k \le N$, define the event

$$H_k = \{\boldsymbol{a} : k \text{ of the } a_i \text{ are } H\}.$$

An element of H_k is a sequence of length N consisting of k H's and $(N-k)$ T's. Since every sequence in H_k occurs with this probability, we can compute $\text{Prob}(H_k)$ by determining the number of different sequences that contain k H's and $(N - k)$ T's. This equals the number of ways to choose k numbers out of $\{1, \dots, N\}$, which is

$$\binom{N}{k} = \frac{N!}{k!(N-k)!};$$

hence,

$$\text{Prob}(H_k) = \binom{N}{k} p^k (1-p)^{N-k}.$$

Exercises

***Exercise* 15.1.26.** Suppose that we perform the experiment of flipping a coin N times. The outcome is the number of heads. The sample space for this experiment is $X = \{0, 1, \dots, N\}$. Show that for each $0 \leq p \leq 1$ the function defined on X by

$$\text{Prob}(\{k\}) = \binom{N}{k} p^k (1-p)^{N-k}$$

defines a probability measure on X.

***Exercise* 15.1.27.** Suppose that we perform the experiment of flipping a coin N times. Suppose that the probability that the ith flip is a head equals p_i and that all the flips are independent. What is the probability of getting exactly k heads in N flips? Find a plausible physical explanation for having different probabilities for different flips while maintaining the independence of the successive flips.

***Exercise* 15.1.28.** Suppose that the probability that a coin lands on heads is p. Describe an experiment to decide whether or not successive flips are independent.

15.1.5 Conditional Probability

Another important notion in probability theory is called *conditional probability*. Here we suppose that we have a probability space $(X, \mathcal{M}, \nu)$ and that $B \in \mathcal{M}$ is an event such that $\text{Prob}(B) > 0$. Assume we *know* that the event B occurs. The conditional probability of an event "$A \in \mathcal{M}$ given B" is defined to be

$$\text{Prob}(A|B) = \frac{\text{Prob}(A \cap B)}{\text{Prob}(B)}.$$

Notice $\text{Prob}(B|B) = 1$, as it must since we *know* that B occurs. We can also define the probability of B given A:

$$\text{Prob}(B|A) = \frac{\text{Prob}(A \cap B)}{\text{Prob}(A)}.$$

They are related by *Bayes's law*

$$\text{Prob}(B|A) = \frac{\text{Prob}(A|B)\,\text{Prob}(B)}{\text{Prob}(A)}.$$

If A and B are independent events, then

$$\text{Prob}(A|B) = \text{Prob}(A).$$

This shows that the terminology is consistent, for if A is independent of B, then the fact that B occurs can have no bearing on the probability of A occurring.

***Example* 15.1.19.** In our coin tossing experiment, we could consider the following question: Suppose that we know that k heads turn up in N flips. What is the probability that $a_1 = T$? That is, what is $\text{Prob}(a_1 = T\,|\,H_k)$? It is clear that

$$\begin{aligned}\text{Prob}(\{a_1 = T\} \cap H_k) &= \text{Prob}(k \text{ of } (a_2, \dots a_N) \text{ are } H \text{ and } a_1 = T)\\ &= (1-p)\binom{N-1}{k} p^k (1-p)^{N-1-k}.\end{aligned} \tag{15.11}$$

On the other hand,

$$\text{Prob}(H_k) = \binom{N}{k} p^k (1-p)^{N-k};$$

hence,

$$\text{Prob}(a_1 = T\,|\,H_k) = 1 - \frac{k}{N}.$$

***Example* 15.1.20.** Let X equal $\mathbb{R}$ and $\mathcal{M}$ be the Borel sets. A nonnegative, measurable function f with

$$\int_{-\infty}^{\infty} f(t)\,dt = 1$$

defines a probability measure

$$\nu_f(A) = \int_{-\infty}^{\infty} f(t)\chi_A(t)\,dt.$$

Let $B \in \mathcal{M}$ be a set for which $\nu_f(B) > 0$. Then the conditional probability of A given B is given by

$$\text{Prob}(A|B) = \frac{\nu_f(A \cap B)}{\nu_f(B)} = \frac{\int_{A\cap B} f(t)\,dt}{\int_B f(t)\,dt}.$$

Exercises

***Exercise* 15.1.29.** In the setup of Example 15.1.19, calculate the conditional probability $\text{Prob}(a_1 = H|H_k)$.

***Exercise* 15.1.30.** Let $(X, \mathcal{M}, \nu)$ be a probability space and B an element of $\mathcal{M}$ such that $\nu(B) > 0$. Define a function on $\mathcal{M}$ by setting

$$\nu_B(A) = \text{Prob}(A|B).$$

Show that ν_B defines a probability measure on $(X, \mathcal{M})$.

***Exercise* 15.1.31.** What is the significance of $\text{Prob}(A|B) = 0$?

15.2 Random Variables*

The sample space for the coin tossing experiment is X_N. For many questions X_N contains more information than is needed. For example, if we are interested only in the number of heads, then we could use the simpler sample space $\{0, 1, \dots, N\}$. The event $\{k\} \subset \{0, 1, \dots, N\}$ corresponds to the event $H_k \subset X_N$. The sample space $\{0, \dots, N\}$ contains strictly less information than X_N, but for the purpose of counting the number of heads, it is sufficient. It is often useful to employ the simplest possible sample space.

Another way to think of the sample space $\{0, \dots, N\}$ is as the range of a function defined on the "full" sample space X_N. Define the function χ^H on $X_1 = \{H, T\}$ by

$$\chi^H(H) = 1, \chi^H(T) = 0.$$

Similarly, on X_N define

$$\chi_N^H(\boldsymbol{a}) = \sum_{i=1}^{N} \chi^H(a_i),$$

where $\boldsymbol{a} = (a_1, \dots, a_N)$. The set $\{0, \dots, N\}$ is the range of this function. The event that k heads arise is the event $\{\boldsymbol{a} : \chi_N^H(\boldsymbol{a}) = k\}$; its probability is

$$\text{Prob}(H_k) = \nu(\{\boldsymbol{a} : \chi_N^H(\boldsymbol{a}) = k\}) = \binom{N}{k} p^k (1-p)^{N-k}. \tag{15.12}$$

The expression on the right-hand side of (15.12) can be thought of as defining a probability measure on the space $\{0, \dots, N\}$ with

$$\text{Prob}(\{k\}) = \binom{N}{k} p^k (1-p)^{N-k}.$$

Let $(X, \mathcal{M}, \text{Prob})$ be a probability space. Recall that a real-valued function f is measurable if for every $t \in \mathbb{R}$ the set $f^{-1}((-\infty, t])$ belongs to $\mathcal{M}$. A complex-valued function is measurable if its real and imaginary parts are measurable.

***Definition* 15.2.1.** A real-valued, measurable function defined on the sample space is called a *random variable*. A complex-valued, measurable function is a *complex random variable*.

***Example* 15.2.1.** The function χ_N^H is a random variable on X_N.

***Example* 15.2.2.** Let X be the unit interval $[0, 1]$, $\mathcal{M}$ the Borel sets in $[0, 1]$, and $\nu([a, b]) = b - a$. The function $f(x) = x$ is a measurable function. It is therefore a random variable on X. For each $k \in \mathbb{Z}$ the function $e^{2\pi ikx}$ is measurable and is therefore a complex random variable.

Thinking of the sample space X as all possible outcomes of an experiment, its points give a complete description of the possible states of the system under study. With this interpretation, we would not expect to be able to completely determine which $x \in X$ is the outcome of an experiment. Instead we expect to be able to measure some function of x. This is the way that random variables enter in many practical applications.

***Example* 15.2.3.** Consider a system composed of a very large number, N, of gas particles contained in a fixed volume. The sample space, $X = \mathbb{R}^{6N}$, describes the position and momentum of every particle in the box. To each configuration in X we associate a temperature T and a pressure P. These are real-valued, random variables defined on the sample space. In a realistic experiment we can measure T and P, though not the actual configuration of the system at any given moment.

Given the probabilistic description, $(X, \mathcal{M}, \nu)$, for the state of a system and a random variable χ, which can be measured, it is reasonable to enquire what value of χ we should *expect* to measure. Because $\nu(X) = 1$, the integral of a function over X is a weighted average. In probability theory this is called the *expected value.*

***Definition* 15.2.2.** Let $(X, \mathcal{M}, \nu)$ be a probability space and χ a random variable. Define the *expected value* or *mean* of the random variable χ by setting

$$\mu_\chi = E[\chi] \stackrel{d}{=} \int_X \chi(x)\, d\nu(x).$$

If either $\chi_\pm$ has integral $+\infty$, then χ does *not* have an expected value.

***Remark* 15.2.1.** In the literature $< \chi >$ is often used to denote the expected value of χ. We avoid this notation as we already use $< \cdot >$ to denote sequences. In this text μ is also used to denote the attenuation coefficient. The meaning should be clear from the context.

Because the expected value is an integral and an integral depends linearly on the integrand, the expected value does as well.

Proposition 15.2.1. *Suppose that $(X, \mathcal{M}, \nu)$ is a probability space and the random variables χ and ψ have finite expected values. Then so does their sum and*

$$E[\chi + \psi] = E[\chi] + E[\psi].$$

***Example* 15.2.4.** We can ask how many heads will occur, on average, among N tosses. This is the expected value of the function χ_N^H:

$$E[\chi_N^H] = \sum_{k=0}^{N} k \operatorname{Prob}(\{\chi_N^H = k\}) = \sum_{k=0}^{N} k \binom{N}{k} p^k (1-p)^{N-k} = pN.$$

This expected value can be also be expressed as the integral over X_N:

$$E[\chi_N^H] = \int_{X_N} \chi_N^H(\boldsymbol{a})\, d\nu_{p,N}(\boldsymbol{a}).$$

***Example* 15.2.5.** Suppose we play a game: We get one dollar for each head and lose one dollar for each tail. What is the expected outcome of this game? Note that the number of tails in a sequence $\boldsymbol{a}$ is $\chi_N^T(\boldsymbol{a}) = N - \chi_N^H(\boldsymbol{a})$. The expected outcome of this game is the expected value of $\chi_N^H - \chi_N^T$. It is given by

$$E[\chi_N^H - \chi_N^T] = E[2\chi_N^H - N] = 2E[\chi_N^H] - E[N] = 2pN - N = (2p-1)N.$$

If $p = \frac{1}{2}$, then this is a fair game: The expected outcome is 0. If $p > \frac{1}{2}$, we expect to make money from this game.

***Example* 15.2.6.** Suppose that X is the unit disk and the probability measure is dA/π. The distance from the origin, $r = \sqrt{x^2 + y^2}$, is a measurable function. Its expected value is

$$E[r] = \int_X r\frac{dA}{\pi} = \frac{1}{\pi}\int_0^{2\pi}\int_0^1 r^2\, dr\, d\theta = \frac{2}{3}.$$

In other words, if a point is picked randomly in the unit disk, its expected distance from $(0, 0)$ is $2/3$.

Exercises

***Exercise* 15.2.1.** Derive the result found in Example 15.2.4.

***Exercise* 15.2.2.** Instead of making or losing one dollar for each toss, we could adjust the amount of money for each outcome to make the game in Example 15.2.5 into a fair game. For a given p, find the amount we should receive for each head and pay for each tail to make this a fair game.

***Exercise* 15.2.3.** In Example 15.2.6, what are $E[x]$ and $E[y]$?

***Exercise* 15.2.4.** Let X be the unit circle in $\mathbb{R}^2$ with

$$\nu(A) = \int_A \frac{d\theta}{2\pi}.$$

The exponential functions $\{e^{2\pi ik\theta}\}$ are complex random variables. What is $E[e^{2\pi ik\theta}]$ for $k \in \mathbb{Z}$?

15.2.1 Cumulative Distribution Function

Associated to a real-valued, random variable is a probability measure on the *real line*. The *cumulative distribution function* for χ is defined to be

$$P_\chi(t) \stackrel{d}{=} \text{Prob}(\{x : \chi(x) \leq t\}).$$

A cumulative distribution function has several basic properties:

1. It is monotone increasing and continuous from the right.
2. $\lim_{t\to-\infty} P_\chi(t) = 0.$
3. $\lim_{t\to\infty} P_\chi(t) = 1.$

A function satisfying these conditions defines a Lebesgue-Stieltjes probability measure on $\mathbb{R}$ with

$$\nu_\chi((a, b]) = P_\chi(b) - P_\chi(a);$$

see Example 15.1.16. This measure is defined on the Borel subsets.

Often the cumulative distribution function can be expressed as the integral of a nonnegative function

$$P_\chi(t) = \int_{-\infty}^{t} p_\chi(s)\, ds.$$

The function p_χ is called the *density* or *distribution function* for χ. In terms of the distribution function,

$$\text{Prob}(a \leq \chi \leq b) = \int_a^b p_\chi(t)\, dt.$$

Heuristically $p_\chi(t)$ is the "infinitesimal" probability that the value of χ lies between t and $t + dt$. Since probabilities are nonnegative, this implies that

$$p_\chi(t) \geq 0 \text{ for all } t.$$

The third property of the cumulative distribution implies that

$$\int_{-\infty}^{\infty} p_\chi(t)\, dt = 1.$$

The expected value of χ can be computed from the distribution function:

$$E[\chi] = \int_X \chi(x)\, d\nu(x) = \int_{-\infty}^{\infty} t p_\chi(t)\, dt.$$

Notice that we have replaced an integration over the probability space X by an integration over the *range* of the random variable χ. Often the sample space X and the probability measure ν on X are not explicitly defined. Instead we just speak of a random variable with a given distribution function. The "random variable" can then be thought of as the coordinate on the real line; its cumulative distribution defines a Lebesgue-Stieltjes measure on $\mathbb{R}$.

***Example* 15.2.7.** A random variable χ is said to be *Gaussian* with mean zero if

$$P_{\chi}(t) = \frac{1}{\sqrt{2\pi}\,\sigma}\int_{-\infty}^{t} \exp\left[-\frac{x^2}{2\sigma^2}\right] dx.$$

If χ describes the outcome of an experiment, then the probability that the outcome lies in the interval $[a, b]$ is

$$\text{Prob}(a \leq \chi \leq b) = \frac{1}{\sqrt{2\pi}\,\sigma}\int_{a}^{b} \exp\left[-\frac{x^2}{2\sigma^2}\right] dx.$$

Notice that we have described the properties of this random variable *without* defining the space X on which it is defined.

Let χ be a random variable. The kth *moment* of χ exists if

$$\int_X |\chi|^k \, d\nu < \infty.$$

If χ has a distribution function p_χ, then this is equivalent to the condition that

$$\int_{-\infty}^{\infty} |t^k| p_\chi(t)\, dt < \infty.$$

The kth moment of χ is then defined to be

$$E[\chi^k] = \int_X \chi^k(x)\, d\nu(x).$$

In terms of the distribution function,

$$E[\chi^k] = \int_{-\infty}^{\infty} t^k p_\chi(t) dt.$$

A more useful quantity is the kth-*centered moment* given by $E[(\chi - \mu_\chi)^k]$. The centered moments measure the deviation of a random variable from its mean value. The moments of a random variable may not be defined.

***Example* 15.2.8.** Suppose that a real-valued random variable, χ, has cumulative distribution

$$P_\chi(t) = \frac{1}{\pi}\int_{-\infty}^{t} \frac{dx}{1+x^2}.$$

Neither the expected value of χ nor of $|\chi|$ exists because

$$\frac{1}{\pi}\int_{-\infty}^{\infty} \frac{|x|}{1+x^2}dx = \infty.$$

Exercises

***Exercise* 15.2.5.** Let χ be a random variable. Prove that P_χ is a monotone increasing function.

***Exercise* 15.2.6.** For each N, what is the cumulative distribution function for the random variables χ_N^H defined on X_N? Do these random variables have distributions functions?

***Exercise* 15.2.7.** Suppose that $(X, \mathcal{M}, \nu)$ is a probability space and χ is a nonnegative random variable. Show that $E[\chi] \geq 0$.

***Exercise* 15.2.8.** Suppose that $(X, \mathcal{M}, \nu)$ is a probability space and χ is a random variable for which there exists a number c so that

$$\text{Prob}(\chi = c) > 0.$$

Show that χ does not have a distribution function.

***Exercise* 15.2.9.** Suppose that $(X, \mathcal{M}, \nu)$ is a probability space and χ is a nonnegative random variable with $E[\chi] = \alpha$. Show that for $t > 0$,

$$\text{Prob}(\chi \geq t) \leq \frac{\alpha}{t}. \tag{15.13}$$

The estimate in (15.13) is called the Chebyshev inequality.

***Exercise* 15.2.10.** Give a heuristic justification for the following statement: "If χ is a random variable with distribution function $\frac{1}{\pi(1+x^2)}$, then its expected value is zero." Why is this statement not strictly correct?

***Exercise* 15.2.11.** Suppose that χ is a random variable with distribution function p_χ. For a function f such that $f(\chi)$ is also a random variable with a finite expected value, show that

$$E[f(\chi)] = \int_{-\infty}^{\infty} f(t)p_\chi(t)\,dt.$$

15.2.2 The Variance

Of particular interest in applications is the second centered moment, or *variance* of a random variable. It is defined by

$$\sigma_\chi^2 \stackrel{d}{=} E[(\chi - \mu_\chi)^2].$$

The variance is a useful measure of how frequently a random variable differs from its mean. It can be expressed in terms of the expectations of χ and χ^2:

$$\begin{aligned} \sigma_\chi^2 &= E[(\chi - E[\chi])^2] \\ &= E[\chi^2] - 2E[\chi]^2 + E[\chi]^2 \\ &= E[\chi^2] - E[\chi]^2. \end{aligned} \tag{15.14}$$

As the expected value of nonnegative random variable, the variance is always nonnegative. The positive square root of the variance σ_χ is called the *standard deviation*. Zero standard deviation implies that, with probability one, χ is equal to its mean. In experimental applications the variance is used to quantify the relative uncertainty in the data.

***Definition* 15.2.3.** If a measurement is described by a random variable χ, then the *signal-to-noise ratio* in the measurement is defined to be

$$\text{SNR}(\chi) \stackrel{d}{=} \frac{\mu_\chi}{\sigma_\chi}.$$

A larger signal-to-noise ratio indicates a more reliable result.

The expectation $E[|\chi - \mu_\chi|^k]$, for any positive number k, could also be used as a measure of the deviation of a random variable from its mean. The variance occurs much more frequently in applications because it is customary and computations involving the variance are much simpler than those for k not equal to 2.

***Example* 15.2.9.** In the coin tossing example,

$$E[(\chi_N^H)^2] = \sum_{k=0}^{N} k^2 \binom{N}{k} p^k (1-p)^{N-k} = pN[p(N-1)+1]$$

Using (15.14), the variance is

$$E[(\chi_N^H - E[\chi_N^H])^2] = pN[p(N-1)+1] - p^2N^2 = p(1-p)N.$$

If $p = 0$, the standard deviation is zero and the coin always falls on tails. A coin is *fair* if $p = \frac{1}{2}$. A fair coin has the largest standard deviation, $\frac{1}{4}N$.

***Example* 15.2.10.** Suppose that χ is a random variable with mean μ and variance σ^2. Then

$$\text{Prob}(|\chi - \mu| \geq t) \leq \frac{\sigma^2}{t^2}. \tag{15.15}$$

This is also called Chebyshev's inequality. The proof uses the observation

$$\{x : |\chi(x) - \mu| \geq t\} = \{x : |\chi(x) - \mu|^2 \geq t^2\}$$

and therefore

$$\begin{aligned}\text{Prob}(|\chi - \mu|) \geq t) &= \int\limits_{\{x : |\chi(x)-\mu|^2 \geq t^2\}} d\nu \\ &\leq \int\limits_{\{x : |\chi(x)-\mu|^2 \geq t^2\}} \frac{|\chi - \mu|^2}{t^2} d\nu \\ &\leq \frac{\sigma^2}{t^2}.\end{aligned} \tag{15.16}$$

This indicates why the variance is regarded as a measure of the uncertainty in the value of a random variable.

Exercises

***Exercise* 15.2.12.** Why does $E[\chi - \mu_\chi]$ provide a poor measure of the deviation of χ from its mean?

***Exercise* 15.2.13.** Let χ be a Gaussian random variable with distribution function

$$p_\chi(t) = \frac{1}{\sqrt{2\pi}\,\sigma} \exp\left[-\frac{(t-a)^2}{2\sigma^2}\right].$$

What are $E[\chi]$ and σ_χ?

***Exercise* 15.2.14.** In (15.16), justify the transitions from the first to the second and the second to the third lines.

Exercise* 15.2.15. Deduce from (15.15) that

$$\text{Prob}(|\chi - \mu| < t) \geq 1 - \frac{\sigma^2}{t^2}.$$

15.2.3 The Characteristic Function

Another important function of a random variable χ is the expected value of $e^{-2\pi i \lambda \chi}$.

***Definition* 15.2.4.** Let χ be a random variable. The function of $\lambda \in \mathbb{R}$ defined by

$$M_\chi(\lambda) = E[e^{-2\pi i \lambda \chi}] = \int\limits_X e^{-2\pi i \lambda \chi(x)} d\nu(x)$$

is called the *characteristic function* of χ.

If the cumulative distribution for χ has a density function p_χ, Then

$$M_\chi(\lambda) = \int_{-\infty}^{\infty} e^{-2\pi i\lambda t} p_\chi(t)\,dt.$$

Up to the factor of 2π, this is the Fourier transform of the density function. As the density function is a nonnegative integrable function, its Fourier transform is continuous and, by the Riemann-Lebesgue lemma, it tends to zero as $|\lambda| \to \infty$.

As we saw in Chapter 4, rapid decay of the density function at infinity makes the characteristic function differentiable. Its derivatives at $\lambda = 0$ determine the moments of χ. They are given by

$$\left[\frac{-\partial_\lambda}{2\pi i}\right]^k M_\chi(\lambda)\Big|_{\lambda=0} = E[\chi^k].$$

Using the Taylor series for the exponential and computing *formally* gives

$$\begin{aligned} E[e^{-2\pi i\lambda\chi}] &= \sum_{j=0}^{\infty} \frac{[-2\pi i\lambda]^j}{j!} \int_{-\infty}^{\infty} t^j p_\chi(t)\,dt \\ &= \sum_{j=0}^{\infty} \frac{[-2\pi i\lambda]^j}{j!} E[\chi^j]. \end{aligned} \tag{15.17}$$

For this reason the $E[e^{-2\pi i\lambda\chi}]$ is sometimes called the *generating function* for the moments of χ. Note that the expected value of $e^{-2\pi i\lambda\chi}$ always exists, while the moment themselves may not.

***Example* 15.2.11.** In the coin tossing example, the characteristic function is

$$\begin{aligned} E[e^{-2\pi i\lambda\chi_N^H}] &= \sum_{k=0}^{N} e^{-2\pi i\lambda k} \binom{N}{k} p^k(1-p)^{N-k} \\ &= (1-p)^N \sum_{k=0}^{N} \binom{N}{k} [e^{-2\pi i\lambda} \frac{p}{1-p}]^k \\ &= (1-p)^N \left(1 + \frac{e^{-2\pi i\lambda} p}{1-p}\right)^N = (1 - p(1 - e^{-2\pi i\lambda}))^N. \end{aligned}$$

Notice again that we do not integrate over the space X_N, where the random variable χ_N^H is defined, but rather over the range of χ_N^H. This again shows the utility of replacing a complicated sample space by a simpler one when doing calculations with random variables.

What happens to the distribution function if a random variable is shifted or rescaled? Suppose χ is a random variable and a new random variable is defined by

$$\psi = \frac{\chi - \mu}{\sigma}.$$

This is often done for convenience; for example, if $\mu = \mu_\chi$ and $\sigma = \sigma_\chi$, then ψ is random variable with mean zero and variance one. The cumulative distribution function is

$$\text{Prob}(\psi \leq t) = \text{Prob}(\frac{\chi - \mu}{\sigma} \leq t) = \text{Prob}(\chi \leq \sigma t + \mu).$$

If p_χ is the distribution function for χ, then the distribution function for ψ is

$$p_\psi(t) = \sigma p_\chi(\sigma t + \mu) \text{ and } M_\psi(\lambda) = e^{\frac{2\pi i \lambda \mu}{\sigma}} M_\chi(\frac{\lambda}{\sigma}). \tag{15.18}$$

Exercises

***Exercise* 15.2.16.** Show that $E[e^{-2\pi i \lambda \chi}]$ is defined for any random variable χ.

***Exercise* 15.2.17.** What is the characteristic function of the Gaussian random variable defined in Exercise 15.2.13?

***Exercise* 15.2.18.** Derive the formulæ in (15.18).

15.2.4 A Pair of Random Variables

Often we have more than one random variable. It is then important to understand how these random variables are related. Suppose that χ_1 and χ_2 are random variables defined on the same space X. By analogy to the cumulative distribution for a single random variable, we define the *joint cumulative distribution function* of χ_1 and χ_2 by

$$\text{Prob}(\chi_1 \leq s \text{ and } \chi_2 \leq t) = \nu(\{\chi_1^{-1}(-\infty, s]\} \cap \{\chi_2^{-1}(-\infty, t]\}).$$

This function is monotone nondecreasing and continuous from the right in each variable and therefore defines a Lebesgue-Stieltjes measure ν_{χ_1,χ_2} on $\mathbb{R}^2$ (see Example 15.1.16). The measure of a rectangle is given by the formula

$$\begin{aligned}\nu_{\chi_1,\chi_2}((a, b] \times (c, d]) = &\text{Prob}(\chi_1 \leq b, \text{ and } \chi_2 \leq d) + \text{Prob}(\chi_1 \leq a, \text{ and } \chi_2 \leq c) - \\ &\text{Prob}(\chi_1 \leq a, \text{ and } \chi_2 \leq d) - \text{Prob}(\chi_1 \leq b, \text{ and } \chi_2 \leq c)\end{aligned} \tag{15.19}$$

If there is a function p_{χ_1,χ_2} defined on $\mathbb{R}^2$ such that

$$\text{Prob}(\chi_1 \leq s, \text{ and } \chi_2 \leq t) = \int_{-\infty}^{s} \int_{-\infty}^{t} p_{\chi_1,\chi_2}(x, y)\, dy\, dx,$$

then we say that p_{χ_1,χ_2} is the *joint distribution function* for the pair of random variables (χ_1, χ_2).

It is clear that

$$\begin{aligned}\text{Prob}(\chi_1 \leq s, \text{ and } \chi_2 \leq \infty) &= \text{Prob}(\chi_1 \leq s) \text{ and} \\ \text{Prob}(\chi_2 \leq s, \text{ and } \chi_1 \leq \infty) &= \text{Prob}(\chi_2 \leq s).\end{aligned} \tag{15.20}$$

This is reasonable because the condition $\chi_i \leq \infty$ places no restriction on χ_i. This is expressed in terms of the distribution functions by the relations

$$\begin{aligned}\int_{-\infty}^{s}\int_{-\infty}^{\infty} p_{\chi_1,\chi_2}(x, y)\, dy\, dx &= \int_{-\infty}^{s} p_{\chi_1}(x)\, dx,\\ \int_{-\infty}^{s}\int_{-\infty}^{\infty} p_{\chi_1,\chi_2}(x, y)\, dx\, dy &= \int_{-\infty}^{s} p_{\chi_2}(y)\, dy.\end{aligned} \tag{15.21}$$

The joint distribution function therefore, is not independent of the distribution functions for individual random variables. It must satisfy the consistency conditions:

$$p_{\chi_2}(y) = \int_{-\infty}^{\infty} p_{\chi_1,\chi_2}(x, y)\, dx \text{ and } p_{\chi_1}(x) = \int_{-\infty}^{\infty} p_{\chi_1,\chi_2}(x, y)\, dy.$$

Recall that two events A and B are independent if

$$\text{Prob}(A \cap B) = \text{Prob}(A)\,\text{Prob}(B),$$

Similarly, two random variables, χ_1 and χ_2, are *independent* if

$$\text{Prob}(\chi_1 \leq s \text{ and } \chi_2 \leq t) = \text{Prob}(\chi_1 \leq s)\,\text{Prob}(\chi_2 \leq t).$$

In terms of their distribution functions, this is equivalent to

$$p_{\chi_1,\chi_2}(x, y) = p_{\chi_1}(x) p_{\chi_2}(y).$$

The expected value of a product of random variables, having a joint distribution function, is given by

$$E[\chi_1\chi_2] = \int_{-\infty}^{\infty}\int_{-\infty}^{\infty} xy \cdot p_{\chi_1,\chi_2}(x, y)\, dx\, dy \tag{15.22}$$

Whether or not χ_1 and χ_2 have a joint distribution function, this expectation is an integral over the sample space, and therefore $E[\chi_1\chi_2]$ satisfies the Cauchy-Schwarz inequality.

Proposition 15.2.2. *Let χ_1 and χ_2 be a pair of random variables defined on the same sample space with finite second moments. Then*

$$|E[\chi_1\chi_2]| \leq \sqrt{E[|\chi_1|^2]E[|\chi_2|^2]}. \tag{15.23}$$

It is useful to have a simple way to quantify the degree of independence of a pair of random variables.

***Definition* 15.2.5.** The *covariance* of χ_1 and χ_2 is defined by

$$\mathrm{Cov}(\chi_1, \chi_2) \stackrel{d}{=} E[(\chi_1 - \mu_{\chi_1})(\chi_2 - \mu_{\chi_2})] = E[\chi_1\chi_2] - E[\chi_1]E[\chi_2],$$

and *correlation coefficient* by

$$\rho_{\chi_1\chi_2} \stackrel{d}{=} \frac{\mathrm{Cov}(\chi_1, \chi_2)}{\sigma_{\chi_1}\sigma_{\chi_2}}.$$

These are the fundamental measures of independence. If χ_1 is measured in units u_1 and χ_2 is measured in units u_2, then the covariance has units $u_1 \cdot u_2$. The correlation coefficient is a more useful measure of the independence of χ_1 and χ_2 because it is dimensionally independent; it is a pure number taking values between ± 1 (see Exercise 15.1.3).

If χ_1, χ_2 are independent, then

$$\begin{aligned} E[\chi_1\chi_2] &= \int_{-\infty}^{\infty} y p_{\chi_2}(y)\, dy \int_{-\infty}^{\infty} x p_{\chi_1}(x)\, dx \\ &= E[\chi_1]E[\chi_2]. \end{aligned} \tag{15.24}$$

In this case, the covariance, $\mathrm{Cov}(\chi_1, \chi_2)$, is equal to zero. This is a necessary but not sufficient condition for two random variables to be independent.

***Example* 15.2.12.** Zero covariance does *not* imply the independence of two random variables. We illustrate this point with a simple example. Let $X = [0, 1]$ and $d\nu = dx$. Two random variables are defined by

$$\chi_1 = \cos 2\pi x, \quad \chi_2 = \sin 2\pi x.$$

Their means are clearly zero, $E[\chi_1] = E[\chi_2] = 0$. They are also uncorrelated:

$$\mathrm{Cov}(\chi_1, \chi_2) = \int_0^1 \cos 2\pi x \sin 2\pi x\, dx = 0.$$

On the other hand, we compute the probability

$$\mathrm{Prob}\left(0 \le |\sin 2\pi x| \le \frac{1}{\sqrt{2}} \text{ and } \frac{1}{\sqrt{2}} \le |\cos 2\pi x| \le 1\right).$$

From the identity $\cos^2\theta + \sin^2\theta = 1$, the first condition is equivalent to the second one. Using the graph of $\sin 2\pi x$, we can easily check that

$$\begin{aligned} \mathrm{Prob}\left(0 \le |\sin 2\pi x| \le \frac{1}{\sqrt{2}} \text{ and } \frac{1}{\sqrt{2}} \le |\cos 2\pi x| \le 1\right) &= \mathrm{Prob}\left(0 \le |\sin 2\pi x| \le \frac{1}{\sqrt{2}}\right) \\ &= \frac{1}{2}. \end{aligned} \tag{15.25}$$

But the product is

$$\text{Prob}(0 \leq |\sin 2\pi x| \leq \frac{1}{\sqrt{2}}) \, \text{Prob}(\frac{1}{\sqrt{2}} \leq |\cos 2\pi x| \leq 1) = \frac{1}{2} \cdot \frac{1}{2} = \frac{1}{4}.$$

Hence these are not independent variables.

Random variables are, in many ways, like ordinary variables. The properties of a single real-valued, random variable are entirely specified by its cumulative distribution, which, in turn, defines a measure on the real line. Indeed nothing is lost by thinking of a single random variable χ as *being* the coordinate on $\mathbb{R}$. If f is a function, then $f(\chi)$ is a new random variable and it bears the same relation to χ as $f(x)$ bears to x; for example,

$$E[f(\chi)] = \int_{-\infty}^{\infty} f(x) p_\chi(x) \, dx$$

and

$$\text{Prob}(a \leq f(\chi) \leq b) = \int_{f^{-1}([a,b])} p_\chi(x) \, dx.$$

So long as we are only interested in random variables that are functions of χ, we can think of our sample space as being the real line.

Two functions defined on the plane are thought of as being independent if they behave like x, y-coordinates. That is, one is in no way a function of the other. This is the essence of the meaning of independence for random variables: One is not a function of the other, with probability 1. Independence is a very strong condition. If χ_1, χ_2 are independent, then

$$E[f(\chi_1)g(\chi_2)] = E[f(\chi_1)]E[g(\chi_2)] \tag{15.26}$$

for any functions f and g such that this makes sense. To work with a pair of random variables, we use $\mathbb{R}^2$ as the underlying sample space. If the variables are independent, then ν_{χ_1,χ_2} is the induced product measure, $\nu_{\chi_1} \times \nu_{\chi_2}$ (see Appendix B.8). As with ordinary functions, there are degrees of dependence between two random variables. If χ_1 and χ_2 are random variables that are *not* independent, then it does not mean that one is a function of the other or that there is a third random variable, χ_3, so that $\chi_1 = f(\chi_3)$ and $\chi_2 = g(\chi_3)$. When working with random variables, it is often useful to replace them by coordinates or functions of coordinates on a Euclidean space.

***Example* 15.2.13.** Let X_N be the sample space for N coin tosses. As noted previously, we usually assume that the results of the different tosses in the sequence are independent of one another. In Example 15.1.18 we showed that the probability that a sequence has k heads and $N - k$ tails is $p^k(1-p)^{N-k}$. The corresponding measure on X_N is denoted $\nu_{p,N}$. To translate this example into the language of random variables, we define the functions $\chi_j : X_N \to \{0, 1\}$ by letting

$$\chi_j(\boldsymbol{a}) = \begin{cases} 1 & \text{if } a_j = H, \\ 0 & \text{if } a_j = T. \end{cases}$$

With the probability defined by $\nu_{p,N}$, these random variables are pairwise independent. Observe that χ_j is only a function of a_j so the various χ_j are also *functionally* independent of one another. Using a different probability measure, we could arrange to have $\sigma_{\chi_i\chi_j} \neq 0$, so that the $\{\chi_j\}$ are no longer independent as random variables.

***Example* 15.2.14.** As a nice application of the characteristic function formalism, introduced in Section 15.2.3, we compute the distribution function for the sum of a pair of independent random variables. Suppose χ_1, χ_2 are independent random variables with distribution functions p_{χ_1}, p_{χ_2}, respectively. What is the distribution function for $\chi_1 + \chi_2$? It is calculated as follows:

$$\begin{aligned} M_{\chi_1+\chi_2}(\lambda) &= E[e^{-2\pi i\lambda(\chi_1+\chi_2)}] = E[e^{-2\pi i\lambda\chi_1}e^{-2\pi i\lambda\chi_2}] \\ &= E[e^{-2\pi i\lambda\chi_1}]E[e^{-2\pi i\lambda\chi_2}] \\ &= M_{\chi_1}(\lambda)M_{\chi_2}(\lambda). \end{aligned}$$

The second line comes from the fact that χ_1 and χ_2 are independent. On the other hand, $M_\chi(\lambda) = \hat{p}_\chi(2\pi\lambda)$, hence $\hat{p}_{\chi_1+\chi_2} = \hat{p}_{\chi_1}\hat{p}_{\chi_2}$. This implies that

$$p_{\chi_1+\chi_2} = p_{\chi_1} * p_{\chi_2}$$

and therefore

$$\begin{aligned} \text{Prob}(\chi_1 + \chi_2 \leq t) &= \int_{-\infty}^{t} p_{\chi_1+\chi_2}(s)\,ds \\ &= \int_{-\infty}^{t}\int_{-\infty}^{\infty} p_{\chi_1}(s-y)p_{\chi_2}(y)\,dy\,ds \qquad (15.27) \\ &= \int_{-\infty}^{\infty}\int_{-\infty}^{t-y} p_{\chi_1}(x)\,dx\,p_{\chi_2}(y)\,dy. \end{aligned}$$

Exercises

***Exercise* 15.2.19.** Suppose that $(X, \mathcal{M}, \nu)$ is a probability space and χ_1 and χ_2 are random variables. Express $E[\chi_1\chi_2]$ as an integral over X.

***Exercise* 15.2.20.** Give a geometric explanation for formula (15.19). When a joint distribution function exists, show that $\nu_{\chi_1,\chi_2}((a, b] \times (c, d])$ reduces to the expected integral.

***Exercise* 15.2.21.** Suppose that χ_1 and χ_2 are random variables with finite mean and variance. Show that $-1 \leq \rho_{\chi_1\chi_2} \leq 1$.

***Exercise* 15.2.22.** In the situation of the previous exercise, show that $|\rho_{\chi_1\chi_2}| = 1$ if and only if $\chi_2 = a\chi_1 + b$ for some constants a, b. More precisely,

$$\text{Prob}(\chi_2 = a\chi_1 + b) = 1.$$

***Exercise* 15.2.23.** Prove the expectation version of the Cauchy-Schwarz inequality, (15.2.2).

***Exercise* 15.2.24.** Prove the statement in Example 15.2.13, that χ_j and χ_k are independent random variables if $j \neq k$.

***Exercise* 15.2.25.** Suppose that χ is a random variable with distribution function p_χ. Let f and g be functions. Show that $\text{Prob}(f(\chi) \leq a, g(\chi) \leq b)$ can be expressed in the form

$$\text{Prob}(f(\chi) \leq a, g(\chi) \leq b) = \int_{E_{f,g}} p_\chi(x)\, dx,$$

where $E_{f,g}$ is a subset of $\mathbb{R}$.

***Exercise* 15.2.26.** Suppose that χ_1 and χ_2 are random variables on X. Recast the independence of χ_1 and χ_2 as a statement about the independence of certain events.

***Exercise* 15.2.27.** Suppose that χ_1 and χ_2 are independent random variables and f, g are functions. Show that $f(\chi_1)$ and $g(\chi_2)$ are also independent random variables.

***Exercise* 15.2.28.** Suppose that χ_1 and χ_2 are random variables and that f and g are functions. Does

$$E[\chi_1\chi_2] = E[\chi_1] \cdot E[\chi_2]$$

imply that

$$E[f(\chi_1)g(\chi_2)] = E[f(\chi_1)] \cdot E[g(\chi_2)]?$$

Give a proof or counterexample.

***Exercise* 15.2.29.** A probability measure is defined on $\mathbb{R}^2$ by

$$\text{Prob}(A) = \frac{1}{\pi} \iint_A \exp[-(x^2 + y^2)]\, dx\, dy.$$

Are the functions $x + y$ and $x - y$ independent random variables? How about x and $x + y$?

***Exercise* 15.2.30.** Suppose that (χ_1, χ_2) is a pair of independent random variables with means (μ_1, μ_2) and variances (σ_1^2, σ_2^2). Show that

$$\mu_{\chi_1+\chi_2} = \mu_1 + \mu_2 \text{ and } \sigma_{\chi_1+\chi_2}^2 = \sigma_1^2 + \sigma_2^2. \tag{15.28}$$

***Exercise* 15.2.31.** Suppose that (χ_1, χ_2) is a pair of random variables with means (μ_1, μ_2), variances (σ_1^2, σ_2^2), and covariance σ_{12}. Find formulæ for $\mu_{\chi_1+\chi_1}$ and $\sigma_{\chi_1+\chi_2}^2$.

***Exercise* 15.2.32.** In Example 15.2.13, find a probability measure on the space of N flips, X_N so that $\text{Cov}(\chi_i, \chi_j) \neq 0$.

15.2.5 Several Random Variables

The concepts introduced in the previous section can be generalized to cover the case of two or more random variables. Suppose that $\{\chi_1, \ldots, \chi_m\}$ is a collection of m real-valued, random variables. Their *joint cumulative distribution* is a function on $\mathbb{R}^m$ defined by

$$\begin{aligned} P_{\chi_1,\ldots,\chi_m}(t_1, \ldots, t_m) &= \text{Prob}(\chi_1 \leq t_1 \text{ and } \ldots \text{ and } \chi_m \leq t_m) \\ &= \nu\left(\bigcap_{j=1}^{m}\{\chi_j^{-1}((-\infty, t_j])\}\right). \end{aligned} \tag{15.29}$$

This function is monotone nondecreasing and continuous from the right in each variable separately and therefore defines a Lebesgue-Stieltjes measure on $\mathbb{R}^m$. If there is a function $p_{\chi_1,\ldots,\chi_m}(t_1, \ldots, t_m)$ so that

$$P_{\chi_1,\ldots,\chi_m}(t_1, \ldots, t_m) = \int_{-\infty}^{t_1} \cdots \int_{-\infty}^{t_m} p_{\chi_1,\ldots,\chi_m}(s_1, \ldots, s_m) ds_1 \cdots ds_m,$$

then $p_{\chi_1,\ldots,\chi_m}$ is called the *joint distribution function* for this collection of random variables.

***Definition* 15.2.6.** A collection $\{\chi_1, \ldots, \chi_m\}$ of random variables is independent if

$$\text{Prob}(\chi_1 \leq t_1 \text{ and } \ldots \text{ and } \chi_m \leq t_m) = \prod_{j=1}^{m} \text{Prob}(\chi_j \leq t_j).$$

If there is a joint distribution function, $p_{\chi_1,\ldots,\chi_m}$, then this is equivalent to

$$p_{\chi_1,\ldots,\chi_m}(t_1, \ldots, t_m) = \prod_{j=1}^{m} p_{\chi_j}(t_j).$$

Once again, it is useful to have a statistical measure of independence. The expected value of the products $E[\chi_j \chi_k]$ is an $m \times m$ matrix called the *correlation matrix*, and the difference

$$\text{Cov}(\chi_j, \chi_k) = E[\chi_j \chi_k] - E[\chi_j]E[\chi_k]$$

is called the *covariance matrix*. The dimensionless version is the normalized correlation matrix defined by

$$\rho_{\chi_j,\chi_k} = \frac{\text{Cov}(\chi_j, \chi_k)}{\sigma_{\chi_j}\sigma_{\chi_k}}.$$

If $\{\chi_1, \ldots, \chi_m\}$ are random variables, then there is a nice formalism for computing certain conditional probabilities. For example, suppose we would like to compute the probability of the event

$$\chi_1 \leq t_1, \ldots, \chi_k \leq t_k \text{ given that } \chi_{k+1} = s_{k+1}, \ldots, \chi_m = s_m. \tag{15.30}$$

To find the distribution function, the event $\{\chi_{k+1} = s_{k+1}, \ldots, \chi_m = s_m\}$ is thought of as the limit of the events $\{|\chi_{k+1} - s_{k+1}| \leq \epsilon, \ldots, |\chi_m - s_m| \leq \epsilon\}$ as $\epsilon \to 0$. The limiting distribution function does not exist unless the event $\{\chi_j = s_j$ for $j = k+1, \ldots, m\}$ has nonzero "infinitesimal probability;" that is,

$$p_{\chi_{k+1},\ldots,\chi_m}(s_{k+1}, \ldots, s_m) \neq 0.$$

To obtain a simple formula, we also require that the density function $p_{\chi_{k+1},\ldots,\chi_m}$ be continuous at $(s_{k+1}, \ldots, s_m)$. The probability in the limiting case is then given by

$$P(\chi_1 \leq t_1, \ldots, \chi_k \leq t_k | \chi_{k+1} = s_{k+1}, \ldots, \chi_m = s_m) = \frac{\int\limits_{-\infty}^{t_k} \cdots \int\limits_{-\infty}^{t_1} p_{\chi_1,\ldots,\chi_m}(x_1, \ldots, x_k, s_{k+1}, \ldots, s_m)\, dx_1 \cdots dx_k}{p_{\chi_{k+1},\ldots\chi_m}(s_{k+1}, \ldots, s_m)}. \tag{15.31}$$

The joint density for the random variables $\{\chi_1, \ldots, \chi_k\}$, given that

$$\chi_j = s_j \qquad \text{for } j = k+1, \ldots, m,$$

is therefore

$$\frac{p_{\chi_1,\ldots,\chi_m}(x_1, \ldots, x_k, s_{k+1}, \ldots, s_m)}{p_{\chi_{k+1},\ldots,\chi_m}(s_{k+1}, \ldots, s_m)}. \tag{15.32}$$

Exercises

The properties of a finite collection of random variables are developed in the following exercises.

***Exercise* 15.2.33.** Show that if $\{\chi_1, \ldots, \chi_m\}$ are independent random variables, then for each pair $i \neq j$ the variables χ_i and χ_j are independent. Is the converse statement true (i.e., does pairwise independence imply that a collection of random variables are independent)?

***Exercise* 15.2.34.** Suppose that the random variables $\{\chi_1, \ldots, \chi_m\}$ are pairwise independent. Show that $\text{Cov}(\chi_j, \chi_k) = 0$.

***Exercise* 15.2.35.** Let $\{\chi_1, \ldots, \chi_m\}$ be random variables with joint distribution function $p_{\chi_1,\ldots,\chi_m}$. Show that

$$p_{\chi_1,\ldots,\chi_{m-1}}(t_1, \ldots, t_{m-1}) = \int\limits_{-\infty}^{\infty} p_{\chi_1,\ldots,\chi_m}(t_1, \ldots, t_{m-1}, s)\, ds.$$

***Exercise* 15.2.36.** Show that if $\{\chi_1, \ldots, \chi_m\}$ have a joint distribution function and $1 \leq i_1 < \cdots < i_k \leq m$, then $\{\chi_{i_1}, \ldots, \chi_{i_k}\}$ also have a joint distribution function. Give a formula for the joint distribution function of $\{\chi_{i_1}, \ldots, \chi_{i_k}\}$.

***Exercise* 15.2.37.** Suppose that $\{\chi_1, \ldots, \chi_m\}$ are independent random variables with means $\{\mu_1, \ldots, \mu_m\}$ and variances $\{\sigma_1^2, \ldots, \sigma_m^2\}$. Let

$$\bar{\chi} = \frac{\chi_1 + \cdots + \chi_m}{m}.$$

Show that

$$\mu_{\bar{\chi}} = \frac{\mu_1 + \cdots + \mu_m}{m} \text{ and } \sigma_{\bar{\chi}}^2 = \frac{\sigma_1^2 + \cdots + \sigma_m^2}{m^2}. \tag{15.33}$$

Does this formula remain valid if we only assume that $\operatorname{Cov}(\chi_i, \chi_j) = 0$?

***Exercise* 15.2.38.** Suppose that $\{\chi_1, \ldots, \chi_m\}$ are independent random variables with distribution functions $\{p_{\chi_1}, \ldots, p_{\chi_m}\}$. What is the distribution function of $\chi_1 + \cdots + \chi_m$? *Hint:* Show that the characteristic function of the sum is the product $M_{\chi_1} \cdots M_{\chi_m}$.

***Exercise* 15.2.39.** Suppose that $\{\chi_1, \ldots, \chi_m\}$ are random variables on a probability space $(X, \mathcal{M}, \nu)$ and let $c_{ij} = E[\chi_i \chi_j]$ be their correlation matrix. Show that this matrix is nonnegative definite; that is, if $(x_1, \ldots, x_m) \in \mathbb{R}^m$, then

$$\sum_{i=1}^{m} \sum_{j=1}^{m} c_{ij} x_i x_j \geq 0.$$

Hint: Express this as the expectation of a nonnegative random variable.

***Exercise* 15.2.40.** Fill in the details in the derivation of the formula (15.32) for the density function of the conditional probability,

$$P(\chi_1 \leq t_1, \ldots, \chi_k \leq t_k | \chi_{k+1} = s_{k+1}, \ldots, \chi_m = s_m).$$

15.3 Some Important Random Variables

In medical imaging and in physics there are three fundamental probability distributions. We now introduce them and discuss some of their properties.

15.3.1 Bernoulli Random Variables

A Bernoulli random variable is specified by two parameters, $p \in [0, 1]$ and $N \in \mathbb{N}$. The variable χ assumes the values $\{0, 1, \ldots, N\}$ with probabilities given by

$$\operatorname{Prob}(\chi = k) = \binom{N}{k} p^k (1-p)^{N-k}. \tag{15.34}$$

The number of heads in N independent coin tosses is an example of a Bernoulli random variable. Sometimes these are called *binomial random variables*.

Recall that in the coin tossing experiment we defined a function χ_N^H such that

$$\chi_N^H((a_1, \ldots, a_N)) = \text{number of heads in } \boldsymbol{a}.$$

There is a similar model for a γ-ray detector. The model is summarized by the following axioms:

- Each photon incident on the detector is detected with probability p.
- Independence axiom: The detection of one photon is independent of the detection of any other.

Let χ denote the number of photons detected out of N arriving at the detector. The probability that k out of N incident photons are detected is given by (15.34). We see that

$$\begin{aligned} \text{Expected value:} \quad & E[\chi] = pN, \\ \text{Variance:} \quad & \sigma^2 = E[(\chi - Np)^2] = p(1-p)N. \end{aligned}$$

If $p = 1$, then we have a perfect detector; hence there is no variance. There is also no variance if $p = 0$. In the latter case the detector is turned off.

Suppose we know the detector (i.e., p is known from many experiments). The number N characterizes the intensity of the source. We would like to know how many photons were emitted by the source. If we measure M photons, a reasonable guess for N is given by $pN = M$. Of course, we do not really believe this because the variance is, in general, not zero. What this means is that, if all our assumptions are satisfied, *and* we repeat the measurement many times, then the average value of the measurements should approach pN.

15.3.2 Poisson Random Variables

A Poisson random variable χ assumes the values $\{0, 1, 2, \ldots\}$ and is characterized by the following probability distribution: $\text{Prob}(\chi = k) = \frac{\lambda^k}{k!}e^{-\lambda}$, where λ is a positive number. This defines a probability measure on the nonnegative integers since

$$\sum_{k=0}^{\infty} \frac{\lambda^k}{k!} e^{-\lambda} = e^{\lambda} e^{-\lambda} = 1.$$

The expected value is given by

$$E[\chi] = \sum_{k=0}^{\infty} k \frac{\lambda^k}{k!} e^{-\lambda} = \lambda.$$

The constant λ is called the *intensity*. Poisson random variables are used to model many different situations. Some examples are as follows:

- The arrival of patients at a doctor's office
- The number of telephone calls passing through a switch
- The number of radioactive decays occurring in a large quantity of a radioactive element, in a fixed amount of time
- The generation of x-rays

The variance is given by

$$\sigma_\chi^2 = E[(\chi - E[\chi])^2] = \lambda.$$

Notice that the variance is equal to the expected value. This has an interesting consequence: The signal-to-noise ratio for a Poisson random variable is given by

$$\frac{\text{expected value}}{\text{standard deviation}} = \frac{\lambda}{\sqrt{\lambda}} = \sqrt{\lambda}.$$

Hence, the intensity of a Poisson random variable measures the relative noise in the system.

Exercise

***Exercise* 15.3.1.** Derive the formulæ for the mean and standard deviation of a Poisson random variable.

15.3.3 Gaussian Random Variables

The final class of distributions we discuss is Gaussian random variables, which have already been discussed briefly. Gaussian random variables are determined by their first and second moments and have many special properties as a consequence of this fact. They are very important in the context of measurement because the average of a large collection of independent random variables is approximately Gaussian, almost no matter how the individual variables are distributed. This fact, known as the central limit theorem, is treated in the next section.

A random variable χ is Gaussian if and only if its cumulative distribution is given by

$$\text{Prob}(\chi \leq t) = \frac{1}{\sqrt{2\pi}\,\sigma} \int_{-\infty}^{t} \exp\left[-\frac{(x-\mu)^2}{2\sigma^2}\right] dx$$

for μ a real number and σ a positive real number. Integrating gives formulæ for the mean and variance:

$$E[\chi] = \mu, \quad E[(\chi - \mu)^2] = \sigma^2.$$

For a Gaussian random variable the probability that χ lies between $\mu - \sigma$ and $\mu + \sigma$ is about $2/3$. From the definition it is clear that the distribution function of a Gaussian random variable is determined by its mean and variance. A Gaussian random variable is said to be *normalized* if its mean is 0 and its variance 1.

The characteristic function of a Gaussian random variable is

$$M_\chi(\lambda) = e^{2\pi i \mu \lambda} e^{-\frac{\sigma^2 (2\pi\lambda)^2}{2}}.$$

Higher moments are easily computed using the fact that if

$$f(t) = \frac{1}{\sqrt{t}} = \frac{1}{\sqrt{2\pi}\,\sigma} \int_{-\infty}^{\infty} \exp\left[-t\frac{(x-\mu)^2}{2\sigma^2}\right] dx,$$

then

$$E[(\chi-\mu)^{2k}] = (-1)^k 2^k \sigma^{2k} f^{[k]}(1).$$

Thus

$$E[(\chi-\mu)^k] = \begin{cases} 0 & \text{if } k \text{ is odd,} \\ 1\cdot 3\cdots(k-1)\sigma^k & \text{if } k \text{ is even.} \end{cases}$$

Two random variables, χ_1, χ_2, are jointly Gaussian if their joint density function is given by

$$p_{\chi_1,\chi_2}(x,y) = \frac{1}{2\pi\sigma_{\chi_1}\sigma_{\chi_2}} \exp\left[\frac{1}{1-\rho^2}\left[\left(\frac{x-\mu_{\chi_1}}{\sigma_{\chi_1}}\right)^2 - 2\rho\left(\frac{x-\mu_{\chi_1}}{\sigma_{\chi_1}}\right)\left(\frac{y-\mu_{\chi_2}}{\sigma_{\chi_2}}\right) + \left(\frac{y-\mu_{\chi_2}}{\sigma_{\chi_2}}\right)^2\right]\right]; \tag{15.35}$$

where

$$\begin{aligned} \mu_{\chi_i} &= E[\chi_i], \text{ are real numbers,} \\ \sigma_{\chi_i} &= \text{are positive numbers,} \\ \rho &= \frac{E[(\chi_1-\mu_{\chi_1})(\chi_2-\mu_{\chi_2})]}{\sigma_{\chi_1}\sigma_{\chi_2}}, \quad \text{is a real number between } -1 \text{ and } +1. \end{aligned}$$

The number ρ is called the normalized correlation coefficient; it is a dimensionless measure of the independence of χ_1 and χ_2. Once again, the joint distribution function of a pair of Gaussian random variables is determined by the second-order statistics of the pair of variables, $\{E[\chi_i], E[\chi_i\chi_j] : i, j = 1, 2\}$.

Proposition 15.3.1. *Let χ_1 and χ_2 be Gaussian random variables. Then they are independent if and only if they are uncorrelated.*

Proof. If a pair of random variables has a joint density function, then they are independent if and only if

$$p_{\chi_1,\chi_2}(x,y) = p_{\chi_1}(x)p_{\chi_2}(y). \tag{15.36}$$

From the form of the joint density of a pair of Gaussian variables, it is clear that (15.36) holds if and only if $\rho = 0$. □

More generally, a collection of m random variables, $\{\chi_1, \cdots, \chi_m\}$, is Gaussian if and only if the density of the joint distribution function has the form

$$p_{\chi_1,\ldots,\chi_m}(t_1,\ldots,t_m) = \sqrt{\frac{\det a_{ij}}{[2\pi]^m}} \exp\left[-\frac{1}{2}\sum_{i,j=1}^m a_{ij}(t_i-\mu_i)(t_j-\mu_j)\right]. \tag{15.37}$$

Here (a_{ij}) is assumed to be a symmetric, positive definite matrix. That is, for some $c > 0$

$$a_{ij} = a_{ji} \text{ and } \sum_{i,j=1}^m a_{ij}x_ix_j \geq c\sum_{j=1}^m x_j^2.$$

Proposition 15.3.2. *If* $\{\chi_1, \ldots, \chi_m\}$ *are jointly Gaussian random variables with the joint density function given in* (15.37)*, then*

$$E[\chi_i] = \mu_i,$$

where a_{ij} *is the inverse of* $\mathrm{Cov}(\chi_i, \chi_j)$.

Evidently, the joint distribution of a collection of Gaussian random variables is again determined by the second-order statistics. This proposition can also be viewed as an existence theorem. Given a collection of numbers $\{\mu_1, \ldots, \mu_m\}$ and a positive definite matrix (r_{ij}), there is a collection of Gaussian random variables $\{\chi_1, \ldots, \chi_m\}$ with

$$E[\chi_i] = \mu_i \text{ and } \mathrm{Cov}(\chi_i, \chi_j) = r_{ij}. \tag{15.38}$$

In practical situations, if all that is known about a collection of random variables is their means and covariance, then it is often assumed that they are Gaussian.

Suppose that $\{\chi_1, \cdots, \chi_m\}$ is a collection of independent Gaussian random variables. If $\{a_j\}$ are constants, then the linear combination

$$\chi_a = \sum_{j=1}^{m} a_j \chi_j$$

is also a Gaussian random variable. The easiest way to see this is to compute the characteristic function of χ_a. From Exercise 15.2.38 it follows that

$$M_a(\lambda) = (a_1 \cdots a_m) M_{\chi_1}(\lambda) \cdots M_{\chi_m}(\lambda). \tag{15.39}$$

Because a sum of quadratic functions is a quadratic function, this implies that

$$p_{\chi_a} = \frac{1}{\sqrt{2\pi}\,\sigma_a} \exp\Big[-\frac{(x-\mu_a)^2}{2\sigma_a^2}\Big],$$

for some constants μ_a, σ_a.

Exercises

***Exercise* 15.3.2.** Suppose that χ_1 and χ_2 are jointly Gaussian random variables. Show that for any constants a, b, the linear combination $a\chi_1 + b\chi_2$ is also a Gaussian random variable. Compute its mean and standard deviation.

***Exercise* 15.3.3.** Using formula (15.39), find an expression for the mean and variance of χ_a in terms of the means and variances of the $\{\chi_j\}$.

***Exercise* 15.3.4.** Suppose that χ_1 and χ_2 are jointly Gaussian random variables. Show that there is an invertible matrix

$$\begin{pmatrix} a & b \\ c & d \end{pmatrix}$$

so that $a\chi_1 + b\chi_2$ and $c\chi_1 + d\chi_2$ are independent, Gaussian random variables.

***Exercise* 15.3.5.** Suppose that real numbers, $\{\mu_1, \dots, \mu_m\}$, and a positive definite $m \times m$ matrix, (r_{ij}) are given. Find a sample space X, a probability measure on X, and random variables $\{\chi_1, \dots, \chi_m\}$ defined on X that are jointly Gaussian, satisfying (15.38).

***Exercise* 15.3.6.** Suppose that $\{\chi_1, \dots, \chi_m\}$ are jointly Gaussian random variables with means $\{\mu_1, \dots, \mu_m\}$. Show that they are pairwise independent if and only if

$$\operatorname{Cov}(\chi_i, \chi_j) = \delta_{ij}\sigma_{\chi_i}^2.$$

***Exercise* 15.3.7.** Suppose that $\{\chi_1, \dots, \chi_m\}$ are jointly Gaussian random variables. Show that they are independent if and only if they are pairwise independent.

15.4 Limits of Random Variables

Often we take limits of random variables as some parameter tends to infinity. In this section we consider several basic examples. We do not treat the problem of convergence of the random variables themselves but only the behavior, under limits, of their distribution functions. The former, more difficult, problem is treated in [30] or [41]. We begin our discussion with the most important general convergence result, the central limit theorem.

15.4.1 The Central Limit Theorem

In applications random variables are often assumed to be Gaussian. This is, of course, not always true, but the following theorem explains why it is often a reasonable approximation.

Theorem 15.4.1 (Central limit theorem). *Let $\{\chi_1, \chi_2, \dots, \}$ be a sequence of independent identically distributed random variables with mean μ and variance σ^2. Let*

$$Z_n = \frac{\chi_1 + \cdots + \chi_n - n\mu}{\sigma\sqrt{n}} = \sqrt{n}\frac{\bar{\chi}_n - \mu}{\sigma},$$

where $\bar{\chi}_n = (\chi_1 + \cdots + \chi_n)/n$. The sequence of distribution functions for the variables $< Z_n >$ tends to a normalized Gaussian as $n \to \infty$. That is,

$$\lim_{n\to\infty} \operatorname{Prob}(Z_n \leq t) = \int_{-\infty}^{t} e^{-\frac{x^2}{2}} \frac{dx}{\sqrt{2\pi}}.$$

The hypothesis that the variables are independent is quite important. To get a sensible limit we must subtract the mean of $\chi_1 + \cdots + \chi_n$ and divide by $\sqrt{n}$. Also notice that there is an implicit hypothesis: The second moments of the $\{\chi_i\}$ are assumed to exist. Notwithstanding these hypotheses, this is a remarkable result. It says that the distribution function of the average of a large collection of independent random variables approaches a Gaussian, *no matter* how the individual random variables are distributed.

Before proving the theorem, we derive an interesting consequence. Is there any reason to expect that the average of a collection of measurements will converge to the theoretical mean value? Assume that the individual measurements are independent random variables $\{\chi_i\}$, each with finite mean μ and variance σ. Since they are the result of performing the same experiment over and over, they can also be assumed to be identically distributed. As before, we set

$$\bar{\chi}_n = \frac{\chi_1 + \cdots + \chi_n}{n}$$

and observe that

$$\begin{aligned}\text{Prob}(|\bar{\chi}_n - \mu| \leq \epsilon) &= \text{Prob}(|\frac{\bar{\chi}_n - \mu}{\sigma/\sqrt{n}}| \leq \frac{\epsilon\sqrt{n}}{\sigma}) \\ &\geq \text{Prob}(|Z_n| < N)\end{aligned}$$

for any N such that $\epsilon\sqrt{n}/\sigma > N$. Hence

$$\lim_{n\to\infty} \text{Prob}(|\bar{\chi}_n - \mu| \leq \epsilon) \geq \int_{-N}^{N} \frac{e^{-\frac{x^2}{2}}}{\sqrt{2\pi}}.$$

As this holds for any N, we conclude that $\lim_{n\to\infty} \text{Prob}(|\bar{\chi}_n - \mu| \leq \epsilon) = 1$. This is called the *weak law of large numbers*. If we recall that each χ_j is a function on the sample space $(X, \mathcal{M}, \nu)$, the weak law of large numbers says that for any $\epsilon > 0$, the average,

$$\frac{\chi_1(x) + \cdots + \chi_n(x)}{n},$$

will eventually be within ϵ of μ, for almost every $x \in X$. That is, if we do many independent trials of the same experiment and average the results, there is good reason to expect that this average will approach the theoretical mean value. This explains, in part, the importance of the assumption that the individual trials of an experiment are independent of one another.

The weak law of large numbers says that, in a weak sense, the sequence of random variables

$$\bar{\chi}_n = \frac{\chi_1 + \cdots + \chi_n}{n}$$

converges to the random variable that is a constant equal to the common mean value. There are many theorems of this type, where we consider different ways of measuring convergence. The central limit theorem itself is such a statement; it asserts that the sequence of random variables $\{Z_n\}$ converges to a normalized Gaussian *in distribution*. That is, the cumulative distribution for Z_n converges to the cumulative distribution of the normalized Gaussian. Other results of this type can be found in [30] and [29].

We now turn to the proof of the central limit theorem.

Proof. Let p be the (common) density function for $\{\chi_i - \mu\}$. The hypotheses imply that

$$\int_{-\infty}^{\infty} p(x)\,dx = 1, \quad \int_{-\infty}^{\infty} xp(x)\,dx = 0, \quad \int_{-\infty}^{\infty} x^2 p(x)\,dx = \sigma^2. \tag{15.40}$$

For simplicity, we assume that the characteristic function of $\chi_i - \mu$ has two derivatives at the origin. Using the Taylor expansion and the relations (15.40) gives

$$\hat{p}(\xi) = 1 - \frac{\sigma^2\xi^2}{2} + o(\xi^2).$$

Let q_n be the density function for the shifted and scaled random variables, $\{(\chi_i - \mu)/\sigma\sqrt{n}\}$. It is given by $q_n(x) = \sigma\sqrt{n} \cdot p(\sigma\sqrt{n}x)$, and therefore

$$\text{Prob}(\frac{\chi_j - \mu}{\sigma\sqrt{n}} \leq t) = \int_{-\infty}^{t} q_n(x)\,dx.$$

The Fourier transform of q_n and its Taylor expansion are

$$\begin{aligned} \hat{q}_n(\xi) &= \hat{p}(\frac{\xi}{\sigma\sqrt{n}}), \\ \hat{q}_n(\xi) &= 1 - \frac{\xi^2}{2n} + o(\frac{\xi^2}{n}). \end{aligned}$$

Since $\{(\chi_i - \mu)/(\sigma\sqrt{n})\}$ are independent random variables, the characteristic function of their sum,

$$Z_n = \frac{\chi_1 - \mu}{\sigma\sqrt{n}} + \frac{\chi_2 - \mu}{\sigma\sqrt{n}} + \cdots + \frac{\chi_n - \mu}{\sigma\sqrt{n}},$$

is just the product of the characteristic functions of each:

$$\begin{aligned} \hat{p}_{Z_n}(\xi) &= E[e^{-i\xi Z_n}] = E[e^{-i\xi\frac{\chi_1-\mu}{\sigma\sqrt{n}}}]\cdots E[e^{-i\xi\frac{\chi_n-\mu}{\sigma\sqrt{n}}}] \\ &= [\hat{q}_n(\xi)]^n = \big[1 - \frac{\xi^2}{2n} + o(\frac{\xi^2}{n})\big]^n. \end{aligned}$$

The last term is negligible, as $n \to \infty$; therefore,

$$\hat{p}_{Z_n}(\xi) = \big[1 - \frac{\xi^2}{2n} + o(\frac{\xi^2}{n})\big]^n \to e^{-\frac{\xi^2}{2}}.$$

Thus, by the Fourier inversion formula, the density function of Z_n converges to the Gaussian:

$$\mathscr{F}^{-1}[\hat{p}_{Z_n}] \to \int_{-\infty}^{\infty} e^{i\xi x} e^{-\frac{\xi^2}{2}} \frac{dx}{\sqrt{2\pi}} = \frac{1}{\sqrt{2\pi}} e^{-\frac{x^2}{2}};$$

see Section 4.2.3. □

15.4.2 Other Examples of Limiting Distributions

See: A.3.3.

The central limit theorem is often applied to obtain approximate formulæ for a (non-Gaussian) distribution function as a parameter gets large. In this section we begin with such an example and then consider limiting distributions for other limits of random variables.

***Example* 15.4.1.** Denote by X_∞ the set of all infinite sequences of heads and tails. Let

$$\text{Prob}(H) = p, \quad \text{Prob}(T) = 1 - p.$$

We assume that this holds for all flips, and each flip is independent of the others. Let χ_i be the random variable defined by

$$\chi_i((a_1, \dots, a_n, \dots)) = \begin{cases} 1 & \text{if } a_i = H, \\ 0 & \text{if } a_i = T. \end{cases}$$

These are independent identically distributed random variables with expected value and variance given by

$$E[\chi_i] = p, \quad \sigma^2_{\chi_i} = p(1-p).$$

The central limit theorem implies that

$$\text{Prob}(Z_n \le t) \to \int_{-\infty}^{t} e^{-\frac{x^2}{2}} \frac{dx}{\sqrt{2\pi}} \quad \text{as} \quad n \to \infty,$$

where

$$Z_n = \frac{\chi_1 + \chi_2 + \dots + \chi_n - np}{\sqrt{np(1-p)}}.$$

We use this fact to approximate the Bernoulli distribution. The probability for the Bernoulli distribution is given by

$$\text{Prob}(\chi_1 + \dots + \chi_n \le k) = \sum_{j=0}^{k} \binom{n}{j} p^j (1-p)^{n-j}.$$

Let $k = [\sqrt{np(1-p)}t + np]$. The central limit theorem implies that

$$\text{Prob}(\chi_1 + \dots + \chi_n \le k) \approx \int_{-\infty}^{t} e^{-\frac{x^2}{2}} \frac{dx}{\sqrt{2\pi}}.$$

The combinatorial quantity, on the left, is rather complicated to compute, whereas the right side is the integral of a very smooth, rapidly decaying function. It is often more useful to have a rapidly convergent integral approximation rather than an exact combinatorial formula. Note that this approximation is useful even for moderately sized n. The graphs in Figures 15.1 and 15.2 show the distribution functions for Bernoulli distributions with $p = .1, .5$ and $n = 10, 30, 60$ along with the Gaussians having the same mean and variance. The Bernoulli distribution is only defined for integral values; for purposes of comparison it has been linearly interpolated in the graphs. Note the more rapid convergence for $p = .5$.

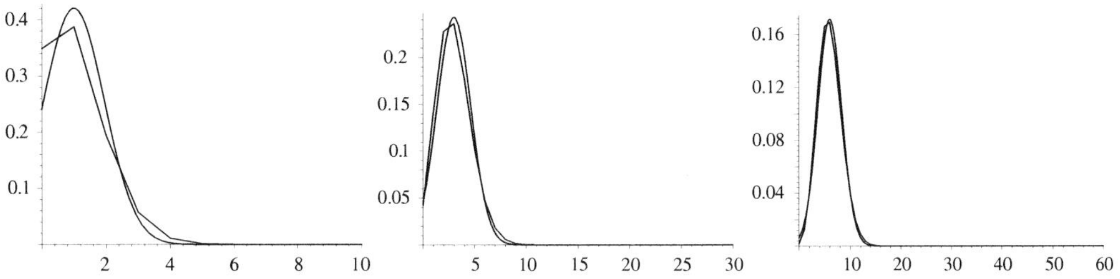

(a) $p = .1, n = 10$ (b) $p = .1, n = 30$ (c) $p = .1, n = 60$

Figure 15.1. Comparisons of Bernoulli and Gaussian distribution functions with $p = .1$.

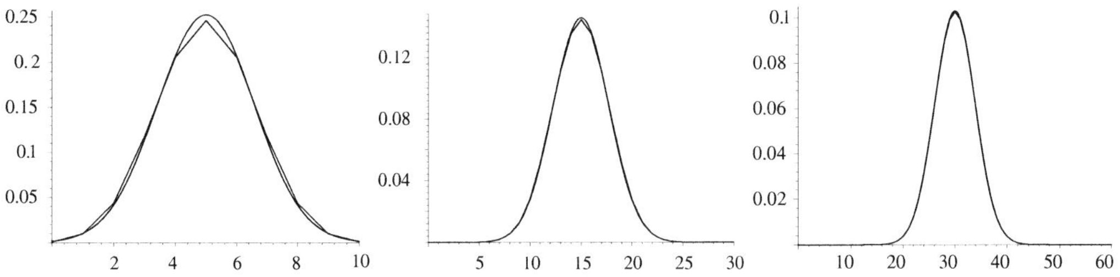

(a) $p = .5, n = 10$ (b) $p = .5, n = 30$ (c) $p = .5, n = 60$

Figure 15.2. Comparisons of Bernoulli and Gaussian distribution functions with $p = .5$.

***Example* 15.4.2.** We now consider a different limit of the Bernoulli distribution. This distribution is used to model the number of radioactive decays occurring in a fixed time interval. Suppose there are N particles and each has a probability p of decaying in a fixed time interval, $[0, T]$. If we suppose that the decay of one atom is independent of the decay of any other and let χ denote the number of decays occurring in $[0, T]$, then

$$\text{Prob}(\chi = k) = \binom{N}{k} p^k (1 - p)^{N-k}.$$

The number of decays is therefore a Bernoulli random variable. An actual sample of any substance contains $O(10^{23})$ atoms. In other words, N is a huge number, which means that p must be a very small number. Suppose that we let $N \to \infty$ and $p \to 0$ in such a way that $Np \to \bar{\lambda}$ for some constant $\bar{\lambda} > 0$. It is not difficult to find the limit of the Bernoulli distribution under these hypotheses. Assuming that $Np = \bar{\lambda}$, we get that

$$\binom{N}{k}\left(\frac{\bar{\lambda}}{N}\right)^k\left(1-\frac{\bar{\lambda}}{N}\right)^{N-k} = \frac{N(N-1)\cdots(N-(k-1))}{k!}\left(1-\frac{\bar{\lambda}}{N}\right)^N/\left(\frac{N}{\bar{\lambda}}-1\right)^k$$
$$= \frac{1}{k!}\frac{N^k(1-1/N)\cdots(1-(k-1)/N)}{N^k(1/\bar{\lambda}-1/N)^k}\left(1-\frac{\bar{\lambda}}{N}\right)^N.$$

Since $(1-\frac{\alpha}{N})^N \to e^{-\alpha}$, as $N \to \infty$, we have that

$$\binom{N}{k}\left(\frac{\bar{\lambda}}{N}\right)^k\left(1-\frac{\bar{\lambda}}{N}\right)^{N-k} \to \frac{\bar{\lambda}^k e^{-\bar{\lambda}}}{k!}.$$

This explains why the Poisson process provides a good model for radioactive decay. The parameter $\bar{\lambda}$ is a measure of the intensity of the radioactive source.

***Example* 15.4.3.** As a final example, we consider the behavior of a Poisson random variable as the intensity gets to be very large. The probability density for a Poisson random variable χ with intensity λ is the generalized function

$$p_\chi = \sum_{k=0}^{\infty}\delta(x-k)\frac{\lambda^k e^{-\lambda}}{k!}.$$

This density function has a variety of defects from the point of view of applications, among them the following: (1) It is combinatorially complex; (2) it is not a function, but rather a generalized function. We are interested in the behavior of $\text{Prob}(\chi = k)$ as λ becomes very large with $|k-\lambda|$ reasonably small compared to λ. Recall that the mean of χ is λ and the standard deviation is $\sqrt{\lambda}$. This means that the

$$\text{Prob}(\lambda - m\sqrt{\lambda} \le \chi \le \lambda + m\sqrt{\lambda})$$

is very close to 1 for reasonably small values of m (e.g., $m = 3, 4, 5$, etc.). Let

$$p_\lambda(k) = \frac{\lambda^k e^{-\lambda}}{k!}.$$

To obtain an asymptotic form for $p_\lambda(k)$, we use Stirling's formula. It implies that for large values of k, $k! = \sqrt{2\pi k}k^k e^{-k}(1 + O(\frac{1}{k}))$; see Section A.3.3 or [130]. Using this approximation for large k gives

$$p_\lambda(k) \approx \frac{1}{\sqrt{2\pi\lambda}}\left[\frac{\lambda}{k}\right]^{k+\frac{1}{2}} e^{k-\lambda}. \tag{15.41}$$

To find a useful asymptotic formula, we set $k = \lambda + x$ with the understanding that $x\lambda^{-1} < < 1$. In terms of x,

$$\begin{aligned}\left[\frac{\lambda}{k}\right]^{k+\frac{1}{2}} &= e^{(\lambda+x+\frac{1}{2})\log(1+\frac{x}{\lambda})}\\ &\approx e^{-[x+\frac{x^2}{2\lambda}]}.\end{aligned} \tag{15.42}$$

In the second line we use the Taylor polynomial for $\log(1+y)$. Putting this into the formula for p_λ shows that

$$p_\lambda(k) \approx \frac{e^{-\frac{(k-\lambda)^2}{2\lambda}}}{\sqrt{2\pi\lambda}}, \tag{15.43}$$

provided that $|\lambda - k|/\sqrt{\lambda}$ remains bounded and λ is large.

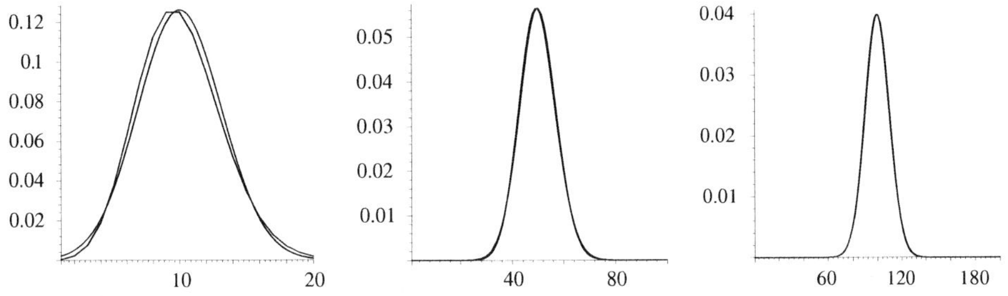

(a) $\lambda = 10$ (b) $\lambda = 50$ (c) $\lambda = 100$

Figure 15.3. Comparisons of Poisson and Gaussian distribution functions.

Once again, we see that the Gaussian distribution provides a limiting form for a random variable. Using formula (15.43), we can approximately compute expected values for functions of χ. If λ is large and f is a reasonably well-behaved function, then

$$E[f(\chi)] \approx \frac{1}{\sqrt{2\pi\lambda}} \int_{-m\sqrt{\lambda}}^{m\sqrt{\lambda}} f(t)e^{-\frac{(t-\lambda)^2}{2\lambda}}\,dt.$$

Here m is chosen to make the error as small as needed provided only that it remains small compared to $\lambda^{\frac{1}{4}}$. Figure 15.3 shows the graphs of the Poisson and Gaussian distributions for $\lambda = 10, 50$, and 100.

Exercise

***Exercise* 15.4.1.** Use Example 15.2.11 to show that if χ is a Poisson random variable with $E[\chi] = \lambda$, then

$$E[e^{-2\pi i\chi\xi}] = e^{-\lambda(1-e^{-2\pi i\xi})}.$$

15.5 Statistics and Measurements

We close our discussion of probability theory by considering how these ideas apply in a simple practical situation. Suppose that χ is a real-valued, random variable that describes the outcome of a experiment. By describing the outcome of the experiment in these terms, we are acknowledging that the measurements involved in the experiment contain errors. At the same time, we are asserting that the experimental errors have a statistical regularity in that they are distributed according to a definite but a priori unknown law. Let p_χ denote the density function for χ so that, for any $a < b$,

$$\text{Prob}(a \leq \chi \leq b) = \int_a^b p_\chi(x)\,dx. \tag{15.44}$$

Often we know that p_χ belongs to a family of distributions. For example, if χ is the number of radioactive decays that occur in a fixed time interval, then we know χ is a Poisson random variable and is therefore determined by its intensity, $\lambda = E[\chi]$. On the other hand, the general type of distribution may not be known in advance. For most practical applications we would be satisfied with estimates for the mean and variance of χ:

$$\mu_\chi = E[\chi] \text{ and } \sigma_\chi^2 = E[(\chi - \mu_\chi)^2].$$

The mean represents the idealized outcome of the experiment, while the variance quantifies the uncertainty in the measurements themselves. Let $< \chi_i >$ denote a sequence of independent random variables that are all distributed according to (15.44). This is a model for independent trials of an experiment. If the experiment is performed N times, then the probability that the results lie in a rectangle $[a_1, b_1] \times \cdots [a_N, b_N]$ is

$$\text{Prob}(a_1 < \chi_1 \leq b_1, \dots, a_N < \chi_N \leq b_N) = \int_{a_1}^{b_1} \cdots \int_{a_N}^{b_N} p_\chi(x_1) \cdots p_\chi(x_N)\,dx_1 \cdots dx_N. \tag{15.45}$$

Our discussion of experimental errors is strongly predicated on the assumption that the various trials of the experiment are independent.

Let $\bar{\chi}_N$ denote the average of the first N trials: $\bar{\chi}_N = \frac{\chi_1 + \cdots + \chi_N}{N}$. Because the trials are independent, formula (15.33) gives the mean and variance of $\bar{\chi}_N$:

$$\mu_{\bar{\chi}_N} = \mu \text{ and } \sigma_{\bar{\chi}_N}^2 = \frac{\sigma^2}{N}. \tag{15.46}$$

The various laws of large numbers imply that $\bar{\chi}_N$ converges to the constant function, μ in a variety of different senses. Since the variables in question have finite variance, the Chebyshev inequality, (15.15), gives the estimate

$$\text{Prob}(|\bar{\chi}_N - \mu| < \epsilon) = 1 - \text{Prob}(|\bar{\chi}_N - \mu| \geq \epsilon) \leq 1 - \frac{\sigma^2}{N\epsilon^2}. \tag{15.47}$$

This explains a sense in which the variance is a measure of experimental error. Indeed, the central limit theorem implies that, for large N,

$$\text{Prob}(-\epsilon \leq \bar{\chi}_N - \mu \leq \epsilon) \approx \frac{1}{\sqrt{2\pi}} \int_{-\frac{\sqrt{N}\epsilon}{\sigma}}^{\frac{\sqrt{N}\epsilon}{\sigma}} e^{-\frac{x^2}{2}}\,dx.$$

This is all very interesting, but it does not really address the question of how to estimate μ and σ^2 using the actual outcomes $\{x_1, \dots, x_N\}$ of N trials of our experiment. An estimate for μ is an estimate for the idealized outcome of the experiment, whereas an estimate for σ^2 provides an estimate for the uncertainty in our results. These questions are properly questions in estimation theory, which is a part of statistics, a field distinct from probability per se. We consider only very simple answers to these questions. Our answers are motivated by (15.46) and (15.47). The results of N trials defines a point in $\boldsymbol{x} \in \mathbb{R}^N$, so our answer is phrased in terms of functions on $\mathbb{R}^N$. As an estimate for μ we use the *sample mean*,

$$m(\boldsymbol{x}) = \frac{x_1 + \cdots + x_N}{N},$$

and for σ^2, the *sample variance*,

$$S^2(\boldsymbol{x}) = \frac{1}{N-1}\sum_{i=1}^{N}(x_i - m(\boldsymbol{x}))^2.$$

The sample mean is exactly what we would expect; the sample variance requires some explanation. We might expect that we should define S^2 by subtracting μ from the measurements, but of course, we do not know μ, and that is why we subtract m instead. We also might have expected a factor $1/N$ instead of $1/(N-1)$; however, due to the nonlinearity of S^2, this would lead to an estimate of σ^2 whose expected value is *not* σ^2. An estimate with the wrong expected value is called a *biased estimate*.

Since $\mathbb{R}^N$ parameterizes the outcomes of N-independent trials, we think of it as a probability space, with the cumulative distribution defined in (15.45). With this interpretation, the expected value and variance of m are given by

$$E[m] = \mu \text{ and } E[(m-\mu)^2] = \frac{\sigma^2}{N}. \tag{15.48}$$

Hence an estimate for σ^2 leads to an estimate for the error in asserting that $m(\boldsymbol{x})$ equals μ. For example, Chebyshev's inequality gives

$$\text{Prob}(|m(\boldsymbol{x}) - \mu| < \epsilon) \geq 1 - \frac{\sigma^2}{N\epsilon^2}.$$

It is important to recognize that in a situation of this sort, the best that we can hope for is a statement to the effect that $|m(\boldsymbol{x}) - \mu| < \epsilon$ with a specified probability.

Expanding the square to compute $E[S^2]$ gives

$$E[S^2] = \frac{1}{N-1}E\left[\sum_{i=1}^{N}x_i^2 - 2m\sum_{i=1}^{N}x_i + m^2\right] = \frac{1}{N-1}\sum_{i=1}^{N}E[x_i^2 - m^2]. \tag{15.49}$$

In the Exercises it is shown that

$$E[m^2] = \frac{\sigma^2}{N} + \frac{(N-1)\mu^2}{N}, \tag{15.50}$$

from which it follows that

$$E[S^2] = \frac{1}{N-1}[N(\sigma^2 + \mu^2) - (\sigma^2 + (N-1)\mu^2)] = \sigma^2.$$

This explains the factor $N - 1$: With this factor the expected valued of S^2 is the true variance. Finally, we would like to compute the variance in S^2,

$$E[(S^2 - \sigma^2)^2] = \frac{(N-1)^2}{N^3} E[(\chi - \mu)^4] - \frac{(N-1)(N-3)\sigma^4}{N^3}. \tag{15.51}$$

The variance of S^2 depends on the fourth moment of the original random variable χ. If χ is a Gaussian random variable, then $E[(\chi - \mu)^4] = 3\sigma^4$ and therefore

$$E[(S^2 - \sigma^2)^2] = \frac{2\sigma^4(N-1)}{N^2}.$$

If the variance is large, then m provides a much better estimate for μ than S^2 for σ^2.

In the final analysis, these formulæ are unsatisfactory in that they all involve σ^2, which we do not know. To use these formulæ with any confidence therefore requires an a priori estimate for σ^2. By this we mean an upper bound for σ^2 derived from principles unconnected to the measurements themselves.

We have only made a tiny scratch in the surface of estimation theory, a subject with a very interesting interplay of probability and empirical experience. Underlying its application to real experiments are deep, philosophical assumptions about the nature of the universe we live in. Our treatment of this subject is adapted from [4] and [30]. The former reference is a very good and complete introduction to the application of probability and statistics to experimental science.

Exercises

***Exercise* 15.5.1.** Show that

$$E[m^2] = \frac{\sigma^2}{N} + \frac{(N-1)\mu^2}{N}.$$

***Exercise* 15.5.2.** Derive the formula for $E[(S^2 - \sigma^2)^2]$.

***Exercise* 15.5.3.** What are $E[(m-\mu)^2]$ and $E[(S^2-\sigma^2)^2]$ if χ is a Poisson random variable with intensity λ?

15.6 Conclusion

Probability theory is an essential tool in the interpretation of experimental results and measure theory provides a language to make the intuitively appealing concepts of probability theory meaningful. The following chart summarizes the correspondence between the basic vocabularies of the two subjects:

	Measure Theory	**Probability Theory**
X	space	sample space
$\mathcal{M}$	σ – algebra	allowable events
ν	measure	probability
χ	measurable function	random variable
M_{χ}	Fourier transform	characteristic function.

Probability theory has manifold applications to medical imaging. The basic physical processes, such as radioactive decay, x-ray production, and attenuation, are most usefully described as random variables. In our discussion of imaging up to this point, we have used a deterministic model for the measurement process. In truth, each measurement is a sample of a random variable. Thus far we have largely ignored this fact, focusing instead on reconstruction errors that result from the practical necessity of sampling functions of continuous variables.

Randomness in a measurement process is usually called noise. The signal-to-noise ratio is a probabilistically defined assessment of the quality of measured data. The presence of noise in the x-ray tomography limits the amount of useful information available in the reconstructed image. Probability theory allows for quantitative descriptions of uncertainty and information content. In the next chapter we use this language to assess the consequences of the randomness of the physical processes involved in x-ray tomography on the quality and interpretation of the reconstructed image.

Chapter 16

Applications of Probability in Medical Imaging

In this chapter we consider several applications of the theory of probability to medical imaging. We begin with probabilistic models for x-ray generation and detection and a derivation of Beer's law. In the next section we analyze the propagation of noise through the filtered back-projection algorithm. This leads to relations between the signal-to-noise ratio and the sample spacing. Using the connection between the sample spacing and spatial resolution, this, in turn, gives a relationship between the dosage of radiation and the resolution at a fixed signal-to-noise ratio. In the last section we briefly describe positron emission tomography and the maximum likelihood algorithm for reconstructing images. This algorithm is a probabilistic alternative to filtered back-projection that may give superior results if the measurements are very noisy.

16.1 Applications of Probability Theory to X-Ray Imaging

In this section we present simple probabilistic models for an x-ray detector and a source-detector pair. We also give a probabilistic derivation of Beer's law.

16.1.1 Modeling a Source-Detector Pair

Suppose that we have a Poisson source with intensity λ and a *Bernoulli detector*. This means that each incident photon has probability p, $0 \leq p \leq 1$ of being detected and each detection event is independent of any other. What is the statistical description of the output of such a source-detector pair? The probability of observing k photons, given that N photons arrive, is a conditional probability:

$$P_d(k|N) = \begin{cases} \binom{N}{k} p^k (1-p)^{N-k} & k = 0, \cdots, N, \\ 0 & k > N. \end{cases}$$

On the other hand, the source is described by

$$P_s(\chi = N) = \frac{\bar{\lambda}^N e^{-\bar{\lambda}}}{N!}.$$

The probability that the detector observes k photons, $P_o(d = k)$, is therefore

$$\begin{aligned}
P_o(d = k) &= \sum_{N=k}^{\infty} P_s(N) P_d(k|N) \\
&= \sum_{N=k}^{\infty} \binom{N}{k} p^k (1-p)^{N-k} \frac{\bar{\lambda}^N e^{-\bar{\lambda}}}{N!} \\
&= e^{-\bar{\lambda}} \sum_{N=k}^{\infty} \frac{1}{k!(N-k)!} (\bar{\lambda} p)^k (\bar{\lambda}(1-p))^{N-k} \\
&= \frac{(\bar{\lambda} p)^k}{k!} e^{-\bar{\lambda}} e^{-\bar{\lambda}(1-p)} = \frac{(\bar{\lambda} p)^k}{k!} e^{-\bar{\lambda} p}.
\end{aligned} \tag{16.1}$$

Hence the source-detector pair is again a Poisson random variable with the intensity scaled by the probability of detection. This is a general feature of Poisson and Bernoulli random variables: If a Poisson random variable is the "input" to a Bernoulli random variable, then the output is again a Poisson random variable.

16.1.2 Beer's Law

It has been asserted several times that Beer's law is essentially a prescription for the behavior of the *mean value* of a random variable. In this section we consider Beer's law from this perspective. In our analysis we consider a "beam" of N photons traveling, through a material, along an interval $[a, b]$ contained in a line l. Let χ_N be a random variable that equals the numbers of photons that are emitted. The attenuation coefficient μ is a non-negative function defined along this line. Suppose that Δs is a very small number (really an infinitesimal). We assume that the individual photons are independent and the transmission of a particle through a thin slab of material is a Bernoulli random variable with the following properties:

1. A single particle that is incident upon the material at point s has a probability $(1 - \mu(s)\,\Delta s)$ of being emitted and therefore probability $\mu(s)\,\Delta s$ of being absorbed.

2. Each particle is independent of each other particle.

3. Disjoint subintervals of $[a, b]$ are independent.

To derive Beer's law, we subdivide $[a, b]$ into m subintervals

$$J_k = [a + \frac{(k-1)(b-a)}{m}, a + \frac{k(b-a)}{m}), \quad k = 1, \dots, m.$$

In order for a particle incident at a to emerge at b, it must evidently pass through every subinterval. The probability that a particle passes through J_k is approximately

$$p_{k,m} \approx (1 - \mu(a + \frac{k(b-a)}{m})\frac{b-a}{m}).$$

By hypothesis (3) it follows that the probability that a particle incident at a emerges at b is the product of these probabilities:

$$p_{ab,m} \approx \prod_{k=1}^{m} p_{k,m}. \tag{16.2}$$

This is an approximate result because we still need to let m tend to infinity.

If μ is a constant, μ_0, then it is an elementary result that the limit of this product, as $m \to \infty$, is $e^{-\mu_0(b-a)}$. Hence a single particle incident at a has a probability $e^{-\mu_0(b-a)}$ of emerging at b. The independence of the individual photons implies that the probability that k out of N photons emerge is

$$P(k, N) = \binom{N}{k} e^{-k\mu_0(b-a)}(1 - e^{-\mu_0(b-a)})^{N-k}.$$

If N photons are incident, then number of photons expected to emerge is therefore

$$E[\chi_N] = e^{-\mu_0(b-a)}N;$$

see Example 15.2.4. The variance, computed in Example 15.2.9, is

$$\sigma_{\chi_N}^2 = Ne^{-\mu_0(b-a)}(1 - e^{-\mu_0(b-a)}).$$

For this experiment the signal-to-noise ratio of χ_N is

$$\text{SNR}(\chi_N) = \sqrt{N\left(\frac{e^{-\mu_0(b-a)}}{1 - e^{-\mu_0(b-a)}}\right)}.$$

This is an important result: The quality of the measurements can be expected to increase as the $\sqrt{N}$. Moreover, the greater the fraction of photons absorbed, the less reliable the measurements. This has an important consequence in imaging: Measurements corresponding to rays that pass through more (or harder) material have a lower SNR than those passing through less (or softer) material.

In general, the attenuation coefficient is not a constant, and we therefore need to compute the limit in (16.2). After taking the logarithm, this is easily done:

$$\log p_{ab,m} = \sum_{k=1}^{m} \log\left((1 - \mu(a + \frac{k(b-a)}{m})\frac{b-a}{m}\right). \tag{16.3}$$

The Taylor expansion for the logarithm implies that

$$\log p_{ab,m} = -\sum_{k=1}^{m}\left[\mu(a+\frac{k(b-a)}{m})\frac{b-a}{m}\right] + O(m^{-1}). \tag{16.4}$$

As m tends to infinity, the right-hand side of (16.4) converges to

$$-\int_a^b \mu(s)\,ds.$$

Hence the probability that a particle incident at a emerges at b is

$$p_\mu = \exp\left[-\int_a^b \mu(s)\,ds\right].$$

Arguing exactly as before, we conclude that if N photons are incident then the probability that $k \leq N$ emerge is

$$P(k, N) = \binom{N}{k} p_\mu^k (1-p_\mu)^{N-k} \tag{16.5}$$

and therefore the expected number to emerge is

$$E[\chi_N] = N\exp\left[-\int_a^b \mu(s)\,ds\right], \tag{16.6}$$

exactly as predicted by Beer's law! The variance is

$$\text{Var}(\chi_N) = p_\mu(1-p_\mu)N \tag{16.7}$$

so the signal-to-noise ratio is

$$\text{SNR}(\chi_N) = \sqrt{N\left(\frac{p_\mu}{1-p_\mu}\right)}.$$

In medical imaging, N, the number of incident photons, is also a random variable. It is usually assumed to satisfy a Poisson distribution. In the previous section it is shown that having a Poisson random variable as the "input" to a Bernoulli process leads to Poisson random variable. This is considered in Exercise 16.1.3.

Exercises

***Exercise* 16.1.1.** In Example 15.4.2 we considered a different limit for a Bernoulli distribution from that considered in this section and got a different result. What is the underlying physical difference in the two situations?

***Exercise* 16.1.2.** Suppose that the probability that k out of N photons are emitted is given by (16.5) and that each emitted photon is detected with probability q. Assuming that there are N (independent) incident photons, show that the probability that k are detected is

$$P_{\mathrm{det}}(k, N) = \binom{N}{k} (p_\mu q)^k (1 - p_\mu q)^{N-k}. \tag{16.8}$$

***Exercise* 16.1.3.** Suppose that the number of x-ray photons emitted is a Poisson random variable with intensity N, and the Bernoulli detector has a probability q of detecting each photon. Show that the overall system of x-ray production, attenuation, and detection is a Poisson random variable with intensity $p_\mu q N$.

***Exercise* 16.1.4.** Suppose that the process of absorption of x-ray photons through a slab is modeled as a Poisson random variable. If the expected number of emitted photons is given by Beer's law, what is the variance in the number of emitted photons? Is this a reasonable model?

16.2 Noise in the Filtered Back-Projection Algorithm

In Chapter 11 we determined the point spread function of the measurement and reconstruction process for a parallel beam scanner. It is shown in examples that if there are two samples per beam width, then the resolution in the reconstructed image is essentially equal to the full-width half-maximum of the beam profile function. A second conclusion of that analysis is that the effects of aliasing, resulting from ray sampling, are well controlled by using the Shepp-Logan filter and a Gaussian focal spot. Decreased sample spacing sharpens the peak of the PSF and does not produce oscillatory side lobes. Finally, the effect of view sampling is an oscillatory artifact, appearing at a definite distance from a hard object. The distance is proportional to $\Delta\theta^{-1}$. This analysis, and considerable empirical evidence, shows that by decreasing $\Delta\theta$ we can obtain an "artifact-free" region of any desired size.

In this section we consider how noise in the measurements obtained in a CT scanner propagates through a reconstruction algorithm. The principal source of noise in CT imaging is *quantum noise*. This is a consequence of the fact that x-rays "beams" are really composed of discrete photons, whose number fluctuates in a random way. The output of an x-ray source is usually modeled as Poisson random variable, with the intensity equal to the expected number of photons per unit time. Recall that the signal-to-noise ratio for a Poisson random variable is the square root of the intensity. Because each x-ray photon carries a lot of energy, safety considerations limit the number of photons that can be used to form an image. This is why quantum noise is an essentially unavoidable problem in x-ray imaging. Matters are further complicated by the fact that the x-ray photons have different energies, but we do not consider this effect; see [6].

In most of this section we assume that the absorption and detection processes are Bernoulli. As shown in Section 16.1.2 (especially Exercise 16.1.3), the complete system of x-ray production, absorption, and detection is modeled as a Poisson random variable. In x-ray CT, the actual measurement is the number of photons emerging from an object along a finite collection of rays, $N_{\text{out}}(t_j, \omega(k\,\Delta\theta))$. This number is compared to the number of photons that entered, $N_{\text{in}}(t_j, \omega(k\,\Delta\theta))$. Both of these numbers are Poisson random variables, and Beer's law is the statement that

$$E[N_{\text{out}}(t_j, \omega(k\Delta\theta))] = E[N_{\text{in}}(t_j, \omega(k\Delta\theta))]\exp\left[-\mathcal{R}_W f(t_j, \omega(k\Delta\theta))\right].$$

Here f is the attenuation coefficient of the object and $\mathcal{R}_W f$ is the Radon transform averaged with the beam profile. We use the signal-to-noise ratio as a measure of the useful information in the reconstructed image.

16.2.1 A Simple Model with Sampled Data

In this section we consider sampled data with uncertainty in the measurements. The variance in the value of the reconstruction of each pixel is estimated in terms of the variance of the noise. Let $\{P(t_k, \theta_j)\}$ denote approximate samples of the Radon transform of an attenuation coefficient f. In this section the spatial coordinates are normalized so that f is supported in $[-1, 1] \times [-1, 1]$. The Radon transform is sampled at $M + 1$ equally spaced angles,

$$\{j\Delta\theta \,:\, j = 0, \ldots, M\} \qquad \text{with } \Delta\theta = \frac{\pi}{M+1}.$$

The sample spacing in the affine parameter is denoted by d. Given the normalization of the spatial coordinates,

$$N = \frac{2}{d}$$

is the number of samples in the t-direction. With ϕ a choice of filter function, the filtered back-projection formula gives

$$\tilde{f}_\phi(x, y) = \frac{d}{2(M+1)} \sum_{j=0}^{M} \sum_{k=-\infty}^{\infty} P(t_k, \theta_j)\phi(\langle (x, y), \boldsymbol{\omega}(j\Delta\theta)\rangle - t_k). \tag{16.9}$$

The basic constraint on the filter function ϕ is that

$$\hat{\phi}(\xi) \approx |\xi| \qquad \text{for } |\xi| < \Omega, \tag{16.10}$$

where Ω represents the effective bandwidth of the measured data.

The measurement is modeled as the "true value" plus noise. The noise is modeled as a collection of random variables $\{\eta_{kj}\}$, so that the measurements are

$$Q_{kj} = P(t_k, \boldsymbol{\omega}(j\Delta\theta)) + \sigma\,\eta_{kj}.$$

The statistical assumptions made on the noise are

$$E[\eta_{kj}] = 0, \quad E[\eta_{kj}\eta_{lm}] = \delta_{kl}\delta jm. \tag{16.11}$$

The condition on the mean implies that there are no systematic errors in the measurements. The second assumption asserts that the errors made in different measurements are uncorrelated. This is a reasonable assumption, though is not realistic to assume that variance is the same for all measurements. The mean and variance of the individual measurements are

$$E[Q_{kj}] = P(t_k, \omega(j\Delta\theta)) \text{ and } E[(Q_{kj} - < Q_{kj} >)^2] = \sigma^2.$$

Given these measurements, the reconstructed image is

$$\breve{f}_\phi(x, y) = \frac{d}{2(M+1)} \sum_{j=0}^{M} \sum_{k=-\infty}^{\infty} Q_{kj} \phi(\langle (x, y), \omega(j\Delta\theta)\rangle - t_k). \tag{16.12}$$

***Remark* 16.2.1 (Important notational remark).** In the remainder of this section, the notation $\breve{f}_\phi$ refers to a reconstruction using noisy data, as in (16.12). This allows us to distinguish such approximate reconstructions from the approximate reconstruction, $\tilde{f}_\phi$, made with "exact" data.

Since $E[Q_{kj}] = P(t_k, \omega(j\Delta\theta))$, the expected value of the output is

$$E[\breve{f}_\phi(x, y)] = \frac{d}{2(M+1)} \sum_{j=0}^{M} \sum_{k=-\infty}^{\infty} E[Q_{kj}] \phi(\langle (x, y), \omega(j\Delta\theta)\rangle - t_k) = \tilde{f}_\phi(x, y).$$

The variance at (x, y) is

$$E[(\breve{f}_\phi(x, y) - \tilde{f}_\phi(x, y))^2] = E[\breve{f}_\phi^2(x, y)] - \tilde{f}_\phi^2(x, y).$$

Expanding the square in the reconstruction formula gives

$$\begin{aligned}
\breve{f}_\phi^2 &= \left(\frac{d}{2(M+1)}\right)^2 \left[\sum_{j=0}^{M} \sum_{k=-\infty}^{\infty} (P(t_k, \omega(j\Delta\theta)) + \sigma\eta_{kj}) \phi(\langle (x, y), \omega(j\Delta\theta)\rangle - t_k)\right]^2 \\
&= \left(\frac{d}{2(M+1)}\right)^2 \left[\sum_{j=0}^{M} \sum_{k=-\infty}^{\infty} P(t_k, \omega(j\Delta\theta)) \phi(\langle (x, y), \omega(j\Delta\theta)\rangle - t_k)\right]^2 \\
&+ 2\sigma \left(\frac{d}{2(M+1)}\right)^2 \sum_{j,k} \sum_{l,m} P(t_k, \omega(j\Delta\theta)) \phi(\langle (x, y), \omega(j\Delta\theta)\rangle - t_k) \times \\
&\qquad \phi(\langle (x, y, \omega(m\Delta\theta)\rangle - t_l) \eta_{l,m} \\
&+ \sigma^2 \left(\frac{d}{2(M+1)}\right)^2 \sum_{j,k} \sum_{l,m} \phi(\langle (x, y), \omega(j\Delta\theta)\rangle - t_k) \times \\
&\qquad \phi(\langle (x, y), \omega(m\Delta\theta)\rangle - t_l) \eta_{j,k} \eta_{l,m}.
\end{aligned}$$

Using the hypothesis that the noise for different measurements is uncorrelated leads to

$$E[\breve{f}_\phi^2] = \tilde{f}_\phi^2 + \sigma^2 \left[\frac{d}{2(M+1)}\right]^2 \sum_{j=0}^{M} \sum_{k=-\infty}^{\infty} \phi^2(\langle (x, y), \omega(j\Delta\theta)\rangle - t_k), \tag{16.13}$$

and therefore

$$\sigma_\phi^2 = \sigma^2 \left[\frac{d}{2(M+1)}\right]^2 \sum_{j=0}^{M} \sum_{k=-\infty}^{\infty} \phi^2(\langle (x, y), \omega(j\Delta\theta)\rangle - t_k). \tag{16.14}$$

For each fixed j, the sum on k is an approximation to the integral of ϕ^2,

$$d \sum_{k=-\infty}^{\infty} \phi^2(\langle (x, y), \omega(j\Delta\theta)\rangle - t_k) \approx \int_{-\infty}^{\infty} \phi^2(t)\, dt.$$

Therefore,

$$\sigma_\phi^2 \approx \sigma^2 \frac{d}{4(M+1)} \int_{-\infty}^{\infty} \phi^2(t)\, dt = \frac{\sigma^2}{2N(M+1)} \int_{-\infty}^{\infty} \phi^2(t)\, dt. \tag{16.15}$$

The noise variance *per pixel* is therefore proportional to the integral of the square of the filter function. Note that this variance is independent of the point in image. This is a consequence of assuming that the variance in the number of measured photons is constant.

Parseval's theorem says that

$$\int_{-\infty}^{\infty} \phi^2(t)\, dt = \frac{1}{2\pi} \int_{-\infty}^{\infty} |\hat{\phi}(\xi)|^2\, d\xi,$$

and therefore if ϕ satisfies (16.10), then

$$\int_{-\infty}^{\infty} \phi^2(t)\, dt \approx \frac{1}{\pi} \int_0^{\Omega} \xi^2\, d\xi = \frac{\Omega^3}{3\pi}.$$

The ratio σ_ϕ/σ is called the *noise amplification factor*. Its square is given approximately by

$$\frac{\sigma_\phi^2}{\sigma^2} \approx \frac{\Omega^3}{6\pi N(M+1)}.$$

From Nyquist's theorem, $\Omega \approx N/2$, so we see that

$$\frac{\sigma_\phi^2}{\sigma^2} \approx \frac{N^2}{48\pi(M+1)}.$$

In order to have the resolution in the angular direction equal that in the radial direction, we need to take $(M+1) \approx 2\pi d^{-1}$ and therefore

$$\frac{\sigma_\phi^2}{\sigma^2} \approx \frac{N}{48\pi} = \frac{1}{24\pi^2 d} \tag{16.16}$$

This is an estimate of the noise amplification for a single pixel. Our discussion is adapted from [114].

16.2.2 A Computation of the Variance in the Measurements

In the previous section we considered the effect of additive noise, assuming that the variance is the same in all the measurements. This is not a reasonable assumption because the variance in the number of photons counted is proportional to the number of *measured* photons. This is true whether the number of detected photons is modeled as a Bernoulli random variable (deterministic source) or a Poisson random variable (Poisson source). These numbers can vary quite a lot due to difference thicknesses and absorbencies encountered along different rays. Using the same geometry as in the previous calculation (a parallel beam scanner with sample spacing d for the affine parameter), we derive an estimate for $\text{Var}(Q_{kj})$ from the assumption that the number of photons counted is a Poisson random variable. The computation of the variance is complicated by the fact that the input to the reconstruction algorithm is not the *number* of measured photons, but rather

$$\log\left(\frac{N_{\text{in}}}{N_{\text{out}}}\right).$$

The nonlinearity of the log renders the estimation of the variance in Q_{kj} a nontrivial calculation.

Let $N_\theta(kd)$ denote the number of photons measured for the ray $l_{kd,\omega(\theta)}$. The idealized measurement would give

$$P_\theta(kd) = \int\limits_{l_{kd,\omega(\theta)}} f ds.$$

For each ray the number of measured photons is a Poisson random variable. Let $\bar{N}_\theta(kd)$ denote the expected value $E[N_\theta(kd)]$. For simplicity we assume that N_{in}, the number of incident photons in each beam, is a deterministic fixed, large number. Beer's law is the statement that

$$\bar{N}_\theta(kd) = N_{\text{in}} e^{-P_\theta(kd)}.$$

Because $N_\theta(kd)$ is a Poisson random variable, its probability distribution is determined by its expected value,

$$\text{Prob}(N_\theta(kd) = l) = \frac{[\bar{N}_\theta(kd)]^l e^{-\bar{N}_\theta(kd)}}{l!}, \tag{16.17}$$

and its variance is

$$\text{Var}(N_\theta(kd)) = \bar{N}_\theta(kd). \tag{16.18}$$

The SNR of an individual measurement is therefore

$$\frac{E[N_\theta(kd)]}{\sigma_{N_\theta(kd)}} = \sqrt{\bar{N}_\theta(kd)}.$$

This is characteristic of Poisson random variables: The signal-to-noise ratio is proportional the square root of the expected value.

16.2.3 The Variance of the Radon Transform

Let $P_\theta^m(kd)$ denote the measured value of $P_\theta(kd)$:

$$P_\theta^m(kd) = \log\left(\frac{N_{\text{in}}}{N_\theta(kd)}\right).$$

The expected value of the measurement is given by

$$E[P_\theta^m(kd)] = E[\log N_{\text{in}} - \log N_\theta(kd)] = E[N_{\text{in}}] - E[\log N_\theta(kd)], \tag{16.19}$$

where

$$E[\log N_\theta(kd)] = \sum_{l=0}^{\infty} \frac{\text{Ln}(l)[\bar{N}_\theta(kd)]^l e^{-\bar{N}_\theta(kd)}}{l!}.$$

Because $\log 0$ is infinity, we define $\text{Ln}(0) = 0$ in this summation. Unfortunately, there is no simple closed form for this expression. Since the logarithm is not a linear function,

$$E[\log N_\theta(kd)] \neq \log E[N_\theta(kd)].$$

Using Taylor's formula, we derive an expression for the difference,

$$E[\log N_\theta(kd)] - \log E[N_\theta(kd)].$$

Let y be a nonnegative random variable with density function p, with

$$\bar{y} = \int_{-\infty}^{\infty} yp(y)\,dy \text{ and } \sigma^2 = \int_{-\infty}^{\infty} (y-\bar{y})^2 p(y)\,dy.$$

Assuming that $\bar{y}$ is a large number and that p is sharply peaked around its mean,

$$\begin{aligned}
E[\log y] &= \int_0^\infty (\log y)p(y)\,dy \\
&= \int_{-\bar{y}}^\infty \log(x+\bar{y})p(x+\bar{y})\,dx \\
&= \int_{-\bar{y}}^\infty \left[\log\bar{y} + \log(1+\frac{x}{\bar{y}})\right]p(x+\bar{y})\,dx \\
&\approx \log\bar{y} + \int_{-\bar{y}}^{\bar{y}} \left[\frac{x}{\bar{y}} - \frac{1}{2}(\frac{x}{\bar{y}})^2 + \cdots\right]p(x+\bar{y})\,dx \\
&\approx \log\bar{y} - \frac{1}{2\bar{y}^2}\sigma^2.
\end{aligned} \tag{16.20}$$

To apply this computation, we approximate the distribution function for a Poisson random variable, which is, in fact, a sum of δ-functions, by a smooth Gaussian distribution.

As shown in Section 15.4.3, the distribution of a Poisson random variable with intensity $\lambda >> 1$ is well approximated by

$$p_\lambda(x) = \frac{1}{\sqrt{2\pi\lambda}} e^{-\frac{(x-\lambda)^2}{2\lambda}}.$$

For large λ, the standard deviation, $\sqrt{\lambda}$, is much smaller than λ. So long as λ is large, the approximation of $E[\log y]$ in (16.20) is applicable.

Using our assumption that $\bar{N}_\theta(kd)$ is a large number and that

$$\sigma_{\theta,k}^2 = \bar{N}_\theta(kd),$$

the foregoing computation gives the estimate

$$E[\log N_\theta(kd)] \approx \log E[N_\theta(kd)] - \frac{\sigma_{\theta,k}^2}{2E[N_\theta(kd)]} = \log \bar{N}_\theta(kd) - \frac{1}{2\bar{N}_\theta(kd)}.$$

Using this approximation in equation (16.19) gives

$$E[P_\theta^m(kd)] \approx \log\left(\frac{N_{\text{in}}}{\bar{N}_\theta(kd)}\right) = P_\theta(kd). \tag{16.21}$$

The variance is

$$\begin{aligned} \text{Var}(P_\theta^m(kd)) &= E[(P_\theta^m(kd) - P_\theta(kd))^2] \\ &= E[(\log\left(\frac{N_\theta(kd)}{N_{\text{in}}}\right) - \log\left(\frac{\bar{N}_\theta(kd)}{N_{\text{in}}}\right))^2] \\ &= E[(\log\left(\frac{N_\theta(kd)}{\bar{N}_\theta(kd)}\right))^2]. \end{aligned} \tag{16.22}$$

Assuming that the x-ray source is deterministic, the variance in the measurements is independent of the source intensity. The variance is approximated as before by

$$\text{Var}(P_\theta^m(kd)) \approx \int_{-\bar{y}}^{\bar{y}} \left[\log\left(1 + \frac{x}{\bar{y}}\right)\right]^2 p(x+\bar{y})\, dx, \tag{16.23}$$

which is easily seen to give

$$\text{Var}(P_\theta^m(kd)) \approx \frac{1}{\bar{N}_\theta(kd)}. \tag{16.24}$$

This verifies the claim that the variance in a measurement of $\mathcal{R}f$ is inversely proportional to the number of photons measured. This computation assumes that the number of incident photons N_{in} is a fixed number.

Exercises

***Exercise* 16.2.1.** Compute the variance in $P_\theta^m(kd)$ assuming that the source is also a Poisson random variable with intensity N_{in}.

***Exercise* 16.2.2.** Derive (16.24) from (16.23).

16.2.4 The Variance in the Reconstructed Image

Assuming that the measurement errors in different rays are uncorrelated, we now find a more accurate computation for the variance in the reconstructed image. The reconstructed image is given by (16.9) with the actual measurements in place of the idealized values:

$$\breve{f}_\phi = \frac{\pi d}{(M+1)} \sum_{j=0}^{M} \sum_{k} P_{\theta_i}^{m}(kd)\phi(\langle (x,y), \boldsymbol{\omega}(j\Delta\theta)\rangle - kd).$$

As the errors have mean zero, the linearity of the reconstruction formula implies that

$$E[\breve{f}_\phi(x,y)] = \tilde{f}_\phi(x,y).$$

The variance of the reconstructed image is given by

$$\operatorname{Var}(\breve{f}_\phi(x,y)) = \left(\frac{\pi d}{M+1}\right)^2 \sum_{j=0}^{M} \sum_{k} \frac{1}{\bar{N}_{\theta_i}(kd)} \phi^2(\langle (x,y), \boldsymbol{\omega}(j\Delta\theta)\rangle - kd).$$

This is quite similar to what was obtained before with the small modification that the contribution to the variance of each projection is weighted according to the expected number of measured photons, $1/\bar{N}_{\theta_i}$. The thicker parts of the object contribute more to the variance.

Using the formula

$$\bar{N}_{\theta_i}(kd) = N_{in} e^{-P_\theta(kd)},$$

the variance can be rewritten

$$\operatorname{Var}(\breve{f}_\phi(x,y)) = \left(\frac{\pi d}{M+1}\right)^2 \frac{1}{N_{in}} \sum_{j=0}^{M} \sum_{k} e^{P_{\theta_j}(kd)} \phi^2(\langle (x,y), \boldsymbol{\omega}(j\Delta\theta)\rangle - kd).$$

At the center of the image, the variance is therefore

$$\operatorname{Var}(\breve{f}_\phi(0,0)) \approx \left(\frac{\pi d}{M+1}\right)^2 \sum_{k} \sum_{j=0}^{M} \frac{\phi^2(-kd)}{\bar{N}_{\theta_i}(kd)}.$$

Assuming that the object is radially symmetric, with constant attenuation coefficient m, implies that

$$\bar{N}_{\theta_j}(kd) = N_{\text{in}} e^{-2m\sqrt{1-(kd)^2}}$$

for all j. Replacing the sum with an integral gives

$$\operatorname{Var}(\breve{f}_\phi(0,0)) \approx \frac{\pi^2 d}{(M+1)N_{\text{in}}} \int_{-1}^{1} \phi^2(t) e^{2m\sqrt{1-t^2}} dt. \tag{16.25}$$

The graphs in Figure 16.1 show $\operatorname{Var}(\breve{f}_\phi(0,0))$ as functions of m and $1/d$. In Figure 16.1(a) formula 16.25 is used, while in Figure 16.1(b) formula 16.15 is used, with

$$\sigma^2 = \int_0^1 e^{2m\sqrt{1-t^2}}\,dt$$

the average. Note that for larger m the constant variance estimate is much smaller.

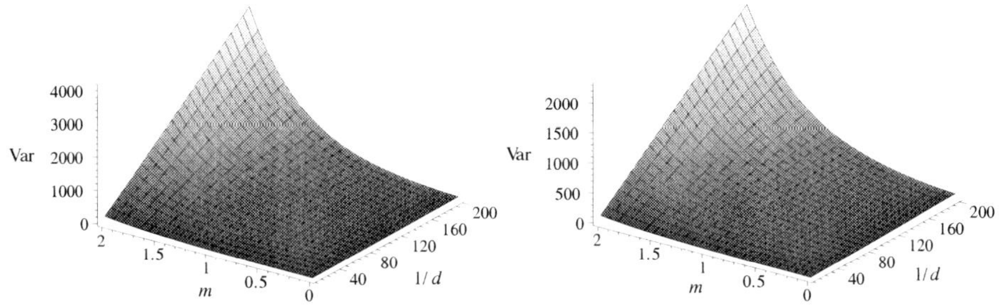

(a) Image variance from 16.25 as a function of m and $1/d$.

(b) Image variance from 16.15 as a function of m and $1/d$.

Figure 16.1. Comparison of the image variance using different models for the variance in the measurements.

Exercises

***Exercise* 16.2.3.** Find a formula for $\operatorname{Var}(\breve{f}_\phi(x,y))$ for other points in the disk. Graph the result for fixed values of (d,m).

***Exercise* 16.2.4.** Find a formula for $\operatorname{Cov}(\breve{f}_\phi(0,0), \breve{f}_\phi(x,y))$. Graph the result for fixed values of (d,m).

16.2.5 Signal-to-Noise Ratio, Dosage and Contrast

We now compute the signal-to-noise ratio at the center of a homogeneous disk of radius R and attenuation coefficient m. From the previous section we have

$$\begin{aligned} \bar{N}_0 &= N_{\text{in}} e^{-2mR}, \\ E[\breve{f}_\phi(0,0)] &\approx \tilde{f}_\phi(0,0). \end{aligned}$$

Approximating the integral in (16.25) gives

$$\operatorname{Var}(\tilde{f}_\phi(0,0)) \approx \frac{\pi^2 d}{M N_{\text{in}} e^{-2mR}} \int_{-\infty}^{\infty} \phi^2(t)\,dt.$$

Using the Parseval formula and the assumption that $\hat{\phi}(\xi) \approx \chi_{[0,\Omega]}(|\xi|)|\xi|$, we obtain

$$\operatorname{Var}(\tilde{f}_\phi(0,0)) \approx \frac{\pi d \Omega^3}{6 M N_{\text{in}} e^{-2mR}}.$$

The resolution $\delta \propto \Omega^{-1}$; hence the signal-to-noise ratio is

$$SNR \propto \sqrt{\delta^2 M N_{\text{in}} e^{-mR}} m e^{-\frac{1}{2}mR}.$$

Let D denote the dosage of radiation absorbed by the center pixel in units of rad/cm^3. The photon density passing through the center pixel is proportional to $MN_{\text{in}}e^{-\mu_0 R}$. Assuming that the pixel size is proportional to the resolution, the number photons absorbed is proportional to $\delta MN_{\text{in}}e^{-\mu_0 R}$. If the thickness of the slice is also proportional to the resolution, then

$$D \propto \frac{\delta M N_{\text{in}} e^{-mR}}{\delta^3}.$$

Using this in the formula for the signal-to-noise ratio leads to

$$\text{SNR} \propto \sqrt{\delta^4 D} m e^{-\frac{1}{2}mR}. \tag{16.26}$$

This shows that the signal-to-noise ratio has a harsh dependence on the thickness and density of the object (i.e., $me^{-\frac{1}{2}mR}$ decays rapidly as mR increases). It also demonstrates the "fourth-power law" relating resolution to dosage: In order to increase the resolution by a factor of 2 (i.e., $\delta \to \frac{1}{2}\delta$), keeping the signal-to-noise ratio constant, we need to increase the dosage by a factor of 16!

What is the importance of the signal-to-noise ratio? In a real physical measurement such as that performed in x-ray CT, the measured quantity assumes values in a definite range. In medical applications the attenuation coefficient, quoted in Hounsfield units, takes values between -1000 (air) and 1000 (bone) (see Table 3.1). The structures of interest are usually soft tissues, and these occupy a tiny part of this range, about -50 to 60, or 5%. The signal-to-noise ratio in the measurements determines the *numerical resolution* or accuracy in the reconstructed attenuation coefficient. In imaging this is called *contrast*. Noise in the measurements interferes with the discrimination of low-contrast objects (that is, contiguous objects with very similar attenuation coefficients). In clinical applications the image is usually viewed on a monitor and the range of grays or colors available in the display is mapped to a certain part of the dynamic range of the reconstructed attenuation coefficient. If, for example, the features of interest lie between 0 and 100 Hounsfield units, then everything below 0 is mapped to white and everything above 100 to black. If the attenuation coefficient is reconstructed with an accuracy of $\frac{1}{2}$%, then a difference of 10 Hounsfield unit is meaningful and, by scaling the range of displayed values, should be discernible in the output. If, on the other hand, the measured values are only accurate to 2% or 40 Hounsfield units, then the scaled image will have a mottled appearance and contain little useful information.

The *accuracy* of the reconstruction should not be confused with the *spatial resolution*. In prosaic terms the accuracy is the number of significant digits in the values of the reconstructed attenuation coefficient. The reconstructed values approximate spatial averages of the actual attenuation coefficient over a pixel or, if the slice thickness is included, voxel of a certain size. The spatial resolution is a function of the dimensions of a voxel. This, in turn, is largely determined by the beam width, sample spacing in the affine parameter, and

FWHM of the reconstruction algorithm. Increasing the resolution is essentially the same thing as decreasing the parameter δ in (16.26). If the dosage is fixed, then this leads to a decrease in the SNR and a consequent decrease in the contrast available in the reconstructed image. Joseph and Stockham give an interesting discussion of the relationship of contrast and resolution in CT images; see [75].

***Remark* 16.2.2.** Our discussion of SNR is adapted from [6].

16.3 Signal-to-Noise in Magnetic Resonance Imaging

In this brief section we discuss the computation of the signal-to-noise ratio in magnetic resonance imaging. In this modality the measurements are a collection of complex numbers, nominally samples of the Fourier transform of the spin-density weighted by a variety of contrast mechanisms, $\{\hat{\rho}(m\Delta k_x, n\Delta k_y)\}$. The measurements of the real and imaginary parts of $\hat{\rho}$ are independent; the noise in each measurement is modeled as white noise, filtered by the measuring harware. Thus we model the measured signal as

$$S(t) = \hat{\rho}(k_x(t), k_y(t)) + \nu_x(t) + i\nu_y(t), \tag{16.27}$$

where ν_x, ν_y are independent Gaussian random variables, of mean zero.

At a given spatial resolution, image quality is largely determined by the signal-to-noise ratio (SNR), and the contrast between the different materials making up the imaging object. SNR in MRI is defined as the voxel signal amplitude, in the reconstructed image, divided by the noise standard deviation. The reconstruction process in MRI is essentially the inverse Fourier transform, and is modeled as a unitary transformation, U, from the finite dimensional space of measurements to the space of discretized images. This means that the noise in the measurements is transformed by the reconstruction process to a noise process $n_x + in_y$ in the reconstructed image:

$$n_x + in_y = U(\nu_x + i\nu_y) \tag{16.28}$$

Because the reconstruction algorithm is unitary n_x and n_y are again independent Gaussian random variables with the same variance and mean as ν_x and ν_y respectively.

Ignoring contributions from quantization, for example, due to limitations of the analog-to-digital converter, the noise in the measured signal can be ascribed to random thermal fluctuations in the receive circuit, see [31]. The variance is given by

$$\sigma^2_{thermal} = 4k_B T R \Delta f, \tag{16.29}$$

where k is Boltzmann's constant, T is the absolute temperature, R is the effective resistance (resulting from both receive coil, R_c and object, R_o), and Δf is the receive bandwidth. Both R_c and R_o are frequency dependent, with $R_c \propto \omega^{\frac{1}{2}}$, and $R_o \propto \omega$; recall that $\omega = \gamma \|\boldsymbol{B}_0\|$ is the Larmor frequency. Their relative contributions to overall circuit resistance depend in a complicated manner on coil geometry, and the imaging object's shape, size and electrical conductivity; see [21]. Hence, at high magnetic field, and for large objects, as in most medical applications, the resistance from the object dominates and the noise scales linearly with frequency. Since the signal is proportional to ω^2, in MRI, the SNR increases in proportion to the field strength.

As the reconstructed image is complex valued, it is customary to display the magnitude rather than the real component. Doing so, however, has some consequences on the noise properties. In regions where the signal is much larger than the noise, the Gaussian approximation is valid. However, in regions where the signal is low, rectification causes the noise to assume a Raleigh distribution. Mean and standard deviation can be calculated from the joint probability distribution:

$$P(n_x, n_y) = \frac{1}{2\pi\sigma^2} e^{-(n_x^2+n_y^2)/2\sigma^2}, \tag{16.30}$$

where n_x and n_y are the noise in the real and imaginary parts of ρ. When the signal is large compared to noise, one finds that the variance $\sigma_m^2 = \sigma^2$. In the other extreme of nearly zero signal, one obtains for the mean:

$$\widehat{S} = \sigma\sqrt{\pi/2} \cong 1.253\sigma \tag{16.31}$$

and, for the variance:

$$\sigma_m^2 = 2\sigma^2(1 - \pi/4) \cong 0.655\sigma^2. \tag{16.32}$$

Of particular practical significance is the SNR dependence on the imaging parameters. The voxel noise variance is reduced by the total number of samples collected during the data acquisition process, i.e.,

$$\sigma_m^2 = \sigma_{thermal}^2/N, \tag{16.33}$$

where $N = N_x N_y$ in a 2d spin-warp experiment. Incorporating the contributions to thermal noise variance, other than bandwidth, into a constant

$$u = 4k_B T R, \tag{16.34}$$

we obtain for the noise variance:

$$\sigma_m^2 = \frac{u\Delta f}{N_x N_y N_{avg}}. \tag{16.35}$$

Here N_{avg} is the number of signal averages collected at each phase-encoding step. We obtain a simple formula for SNR per voxel of volume ΔV :

$$SNR = C\widetilde{\rho}\Delta V\sqrt{\frac{N_x N_y N_{avg}}{u\Delta f}} = C\widetilde{\rho}\Delta x\Delta y d_z\sqrt{\frac{N_x N_y N_{avg}}{u\Delta f}}, \tag{16.36}$$

where Δx, Δy are defined in (14.31), d_z is the thickness of the slab selected by the slice-selective RF pulse, and $\widetilde{\rho}$ denotes the spin density weighted by effects determined by the (spatially varying) relaxation times T_1 and T_2 and the pulse sequence timing parameters.

Figure 16.2 shows two images of the human brain obtained from the same anatomic location but differing in SNR.

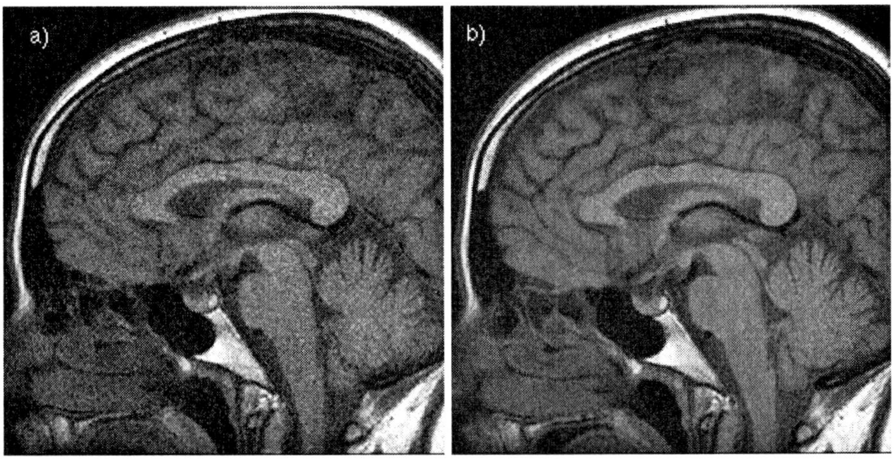

Figure 16.2. T_1-weighted sagittal images through the midline of the brain: Image b) has twice the SNR of image a) showing improved conspicuity of small anatomic and low-contrast detail. The two images were acquired at 1.5 Tesla field strength using 2D spin-warp acquisition and identical scan parameters, except for N_{avg}, which was 1 in a) and 4 in b).

***Remark* 16.3.1.** Section 16.3 is adapted from [37]. An extensive discussion of noise analysis in MR-imaging can be found in [61].

16.4 Image Reconstruction in Positron Emission Tomography

In this section we consider the image reconstruction problem in positron emission tomography or PET. This modality has many features in common with x-ray tomography. Line integrals of a nonnegative function again provide a simple model for the data that are collected. Indeed a filtered back-projection algorithm is often used to reconstruct images in this modality. The actual measurements are again better described as Poisson random variables. In PET the number of photons that are used to form an image is much smaller than in x-ray CT, and therefore quantum noise is a bigger problem. Shepp and Vardi introduced a probabilistic method for image reconstruction in PET using the *maximum likelihood* algorithm. We begin with a brief description of the physics underlying PET and then describe the likelihood function and the maximum likelihood algorithm. This material is a nice application of probabilistic concepts. Our treatment closely follows the original paper of Shepp and Vardi, [116]. The review article, [86] gives an up-to-date, *circa* 2002, account of image reconstruction in emission tomography as well as a very extensive list of references.

16.4.1 Positron Emission Physics

A positron is an *anti-electron*, or the anti-particle of the electron. It has the same mass as the electron and opposite charge. When an electron and a positron meet, they annihilate each other, producing two 511-keV γ-ray photons; this is called a *annihilation event*. If the total momentum of the incoming pair is zero, then the outgoing γ-rays travel in opposite directions. If the incoming pair of particles has nonzero total momentum, then, in order to conserve momentum, the trajectories of the outgoing γ-rays are not quite oppositely directed (see Figure 16.3).

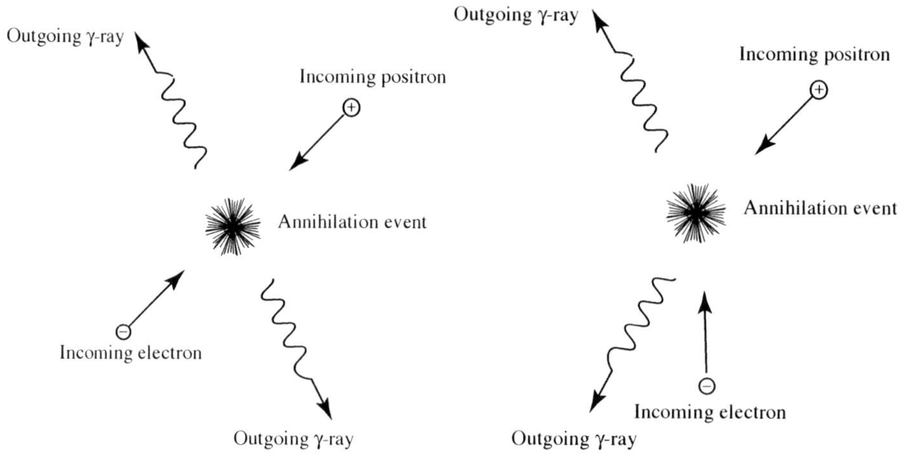

(a) The annihilation of an electron-positron pair with zero momentum. (b) The annihilation of an electron-positron pair with nonzero momentum

Figure 16.3. The annihilation of an electron-positron pair.

In medical applications the total momentum of the incoming electron-positron pairs is small; we assume that the total incoming momentum is zero. The outgoing γ-rays therefore travel in opposite directions along a line where the annihilation event took place. The direction taken by the outgoing γ-rays is a uniformly distributed random variable. As $\boldsymbol{\omega}(\theta)$ and $\boldsymbol{\omega}(\theta + \pi) = -\boldsymbol{\omega}(\theta)$ define the same direction, the correspondence

$$\theta \leftrightarrow \pm\boldsymbol{\omega}(\theta), \quad \theta \in [0, \pi),$$

conveniently parameterizes the directions of lines in the plane. If χ_θ denotes this random variable, then for $0 \leq a < b < \pi$

$$\text{Prob}(a \leq \chi_\theta \leq b) = \frac{b - a}{\pi}.$$

Several organic elements (oxygen, carbon, nitrogen) have short-lived isotopes that decay via positron emission. We call this a *decay event*. In medical applications a positron emitting organic substance is injected into the patient. This is metabolized and taken up selectively by certain structures in the body. After a short time (and for a short time) the distribution of the radioactive substance in the patient is described by a nonnegative function ρ. The goal of PET is to reconstruct this function by counting γ-ray photons outside the patient's body. Because the distribution of positron emitters is a result of metabolism, PET can be used to study metabolic processes, in marked contrast to x-ray CT. As with x-ray CT, a slicing method is often used to reconstruct ρ. We now describe several models for the measurements made in a single two-dimensional slice. We use coordinates (x, y) for this slice.

The number of decays in a fixed time, originating from a small region B inside the object, is a Poisson random variable with intensity proportional to

$$\iint_B \rho(x, y)\, dx\, dy.$$

Ideally, the object would be encircled by (a continuum of) γ-ray detectors (see Figure 16.4). As the annihilation event resulting from a decay event produces a pair of oppositely directly γ-rays, the detector is designed to look for pairs of incoming γ-rays that arrive at essentially the same time. We call these *coincidence events*. The γ-rays move at the speed of light; if r, the radius of the detector circle, is 1 meter, then the difference in arrival times at the detector circle for a pair of γ-rays, coming from a single decay event, is less than 4 nanoseconds.

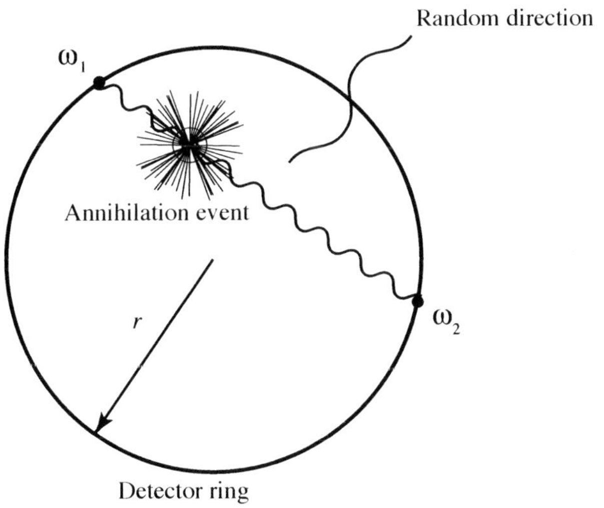

Figure 16.4. An idealized positron emission detector.

Suppose, for given pair of directions (ω_1, ω_2), the measuring apparatus counts $n(\omega_1, \omega_2)$ coincidence events. We use the notation ℓ_{ω_1,ω_2} to denote the line segment joining the points $r\omega_1$ and $r\omega_2$ on the detector circle. Forgetting attenuation effects, the number of decays should be proportional to the integrated intensity of the source along l_{ω_1,ω_2}, which is given by

$$r\int_0^1 \rho(sr\omega_1 + (1-s)r\omega_2)\,ds. \tag{16.37}$$

In this simple model $n(\omega_1, \omega_2)$ provides a measurement of the Radon transform of ρ. Several physical phenomena serve to confound this interpretation. As we know from our study of x-ray tomography, the human body attenuates radiation by scattering and absorption. The integral in (16.37) should therefore be modified to account for attenuation. If $l_{t,\omega}$ is a line passing through the object and meeting the detector at $r\omega_1$ and $r\omega_2$, then the probability that a decay at a point $s\hat{\omega} + t\omega$, traveling along $l_{t,\omega}$, is detected at $r\omega_1$ is

$$\exp\left[-\int_s^r \mu(u\hat{\omega} + t\omega)\,du\right].$$

The probability that the oppositely directly γ -ray is detected at $r\omega_2$ is

$$\exp\left[-\int_{-r}^s \mu(u\hat{\omega} + t\omega)\,du\right].$$

These two probabilities are different, indicating that many decay events do *not* produce coincidence events at the detector ring.

Assuming that the two detection events are independent, the probability that the coincidence event is detected is the product

$$\begin{aligned}\exp\left[-\int_s^r \mu(u\hat{\omega} + t\omega)\,du\right] \times \exp\left[-\int_{-r}^s \mu(u\hat{\omega} + t\omega)\,du\right] &= \exp\left[-\int_{-r}^r \mu(u\hat{\omega} + t\omega)\,du\right] \\ &= \exp[-\mathcal{R}\mu(t,\omega)].\end{aligned} \tag{16.38}$$

This is independent of the location of the decay event along $l_{t,\omega}$. Using the nonstandard notation $\mathcal{R}\mu(\omega_1, \omega_2)$ to denote the integral of the μ along the line from $r\omega_1$ to $r\omega_2$, we see that the number of coincidence events observed at this pair of detectors is predicted to be

$$\exp[-\mathcal{R}\mu(\omega_1, \omega_2)] \int_{\ell_{\omega_1,\omega_2}} \rho\,ds. \tag{16.39}$$

If μ is known, then, at least in principle, ρ could be reconstructed from this collection of measurements. In actual practice a filtered back-projection is often used to reconstruct an approximation to ρ. The difficulty with this approach is that the data are very noisy. A different approach would be to embrace the probabilistic character of both radioactive decay and γ -ray detection. A model of this type is described in the next section.

16.4.2 A Probabilistic Model for PET

As before, we have a nonnegative function ρ defined in a bounded region D in the plane. Imagine that D is divided into a collection of disjoint boxes $\{B_j : j = 1, \ldots, N\}$. The number of decay events occurring (in a fixed time) in box j is a Poisson random variable n_j with intensity

$$\lambda_j = \iint_{B_j} \rho(x, y)\, dx\, dy. \tag{16.40}$$

The random variables $\{n_1, \ldots, n_N\}$ are independent. These random variables are, in principle, unobservable and

$$E[n_j] = \lambda_j. \tag{16.41}$$

Define the vector of intensities

$$\boldsymbol{\lambda} = (\lambda_1, \ldots, \lambda_N).$$

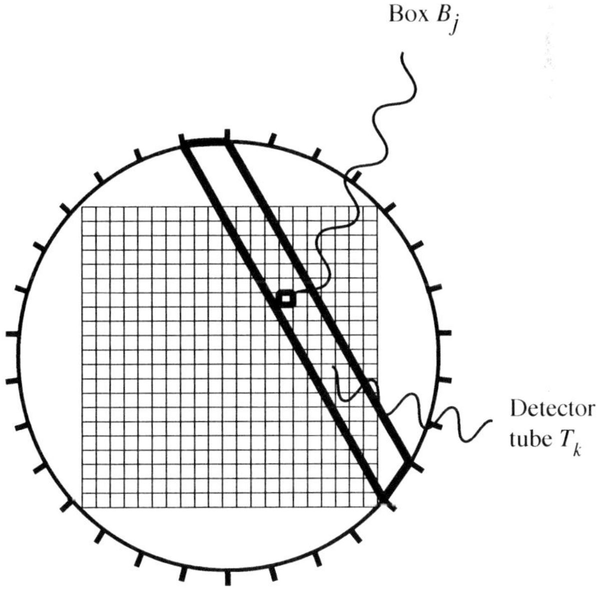

Figure 16.5. The object is divided into boxes, and each pair of detectors defines a tube.

Instead of a continuum of (infinitesimal) detectors, we imagine a finite collection of detectors of finite size placed in a ring surrounding D. Each pair of detectors defines the tube bounded by the lines joining their outer edges. Let $\{T_k : k = 1, \ldots, M\}$ be a listing of these tubes (see Figure 16.5). For each pair (j, k), let p_{jk} be the probability that a decay event in box B_j results in a coincidence event detected in tube T_k. The *transition matrix* (p_{jk}) is computed from the geometry of the detector array; it can incorporate models for attenuation and other physical phenomena. Observe that

$$p_j = \sum_{k=1}^{M} p_{jk}$$

is the probability that a decay in box j is detected at all. As noted previously, many decays may not produce a coincidence event, and therefore generally each p_j is less than 1.

Let n_k^* be the number of coincidence events counted in detector tube k with

$$\boldsymbol{n}^* = (n_1^*, \ldots, n_M^*)$$

the vector of measurements. The reconstruction problem in PET is to estimate $\boldsymbol{\lambda}$, the vector of intensities, from $\boldsymbol{n}^*$. For each choice of $\boldsymbol{\lambda}$ there is a well-defined probability of observing $\boldsymbol{n}^*$. This can be thought of as a conditional probability, which is denoted by

$$\mathcal{L}(\boldsymbol{\lambda}) = P(\boldsymbol{n}^*|\boldsymbol{\lambda}).$$

This is called the *likelihood function*; we give an explicit formula for it in the next section. A *maximum likelihood estimate* is a vector $\hat{\boldsymbol{\lambda}}$ such that

$$\mathcal{L}(\hat{\boldsymbol{\lambda}}) = \max_{\{\boldsymbol{\lambda}:\lambda_j \geq 0 \forall j\}} \{\mathcal{L}(\boldsymbol{\lambda})\}.$$

Among the models we are considering, $\hat{\boldsymbol{\lambda}}$ provides the model that is most consistent with our measurements. An algorithm for finding $\hat{\boldsymbol{\lambda}}$ is called a *maximum likelihood algorithm.* This approach to image reconstruction is likely to succeed if $\boldsymbol{n}^*$ lies close to the mean, which is likely if the actual distribution function is highly peaked. For Poisson random variables the (relative) width of the peak decreases as the intensity increases.

***Example* 16.4.1.** Let n be a single Poisson random variable whose (unknown) intensity we would like to estimate. If the intensity of the variable is λ, then the likelihood function is

$$P(n|\lambda) = \frac{\lambda^n e^{-\lambda}}{n!}.$$

If we measure n^* counts, then the maximum likelihood estimate for λ is

$$\hat{\lambda} = n^*. \tag{16.42}$$

It simplifies the discussion to assume that $p_j = 1$ for all j. In fact, this has little effect on the final result. If we let n'_j denote the number of decays in box j *that are detected,* then, as is shown in Section 16.1.1, this is again a Poisson random variable and

$$p'_{jk} = \frac{p_{jk}}{p_j}$$

is the conditional probability that a decay in box j is detected in tube k. Indeed if λ'_j denotes the intensity of the random variable defined as "decays in box j that are detected," then

$$\lambda'_j = p_j \lambda_j \tag{16.43}$$

and

$$P(\boldsymbol{n}^*|\boldsymbol{\lambda}) = P'(\boldsymbol{n}^*|\boldsymbol{\lambda}').$$

This shows that if $\hat{\boldsymbol{\lambda}}$ is a maximum likelihood estimate for $\boldsymbol{\lambda}$, then $\hat{\boldsymbol{\lambda}}'$, a maximum likelihood estimate for $\boldsymbol{\lambda}'$, satisfies (16.43). Since we assume that the $\{p'_j\}$ are known, finding $\hat{\boldsymbol{\lambda}}'$ also determines $\hat{\boldsymbol{\lambda}}$. The details of this argument can be found in [118]. Henceforth we assume that $p_j = 1$ for all j. We now turn to a description of the maximum likelihood algorithm.

16.4.3 The Maximum Likelihood Algorithm

Let n_{jk} denote the number of events counted in tube k resulting from a decay event in box j. These are mutually independent Poisson random variables. If λ_j is the intensity of the Poisson source in box j, then

$$E[n_{jk}] = \lambda_{jk} = \lambda_j p_{jk}.$$

Indeed the observed counts $\{n_k^*\}$ are themselves independent Poisson random variables. This follows from the fact that the $\{n_{jk}\}$ are mutually independent Poisson random variables and therefore

$$n_k^* = \sum_{j=1}^{N} n_{jk}$$

are as well; see Exercise 16.4.1. The expected value of n_k^* is given by

$$\lambda_k^* = E[n_k^*] = \sum_{j=1}^{N} \lambda_j p_{jk}. \tag{16.44}$$

This suggests one possible approach to estimating $\hat{\boldsymbol{\lambda}}$: Use the measured values $\{n_k^*\}$ for $\{\lambda_k^*\}$ to obtain the system of linear equations

$$\sum_{j=1}^{N} \lambda_j p_{jk} = n_k^*, \quad k = 1, \ldots, M.$$

Unfortunately, this is usually an ill-conditioned system and the data $\boldsymbol{n}^*$ are noisy, so a less direct approach is often preferable.

With these preliminaries we can write down the likelihood of observing $\boldsymbol{n}^*$ for given $\boldsymbol{\lambda}$. Let $\mathcal{A}_{\boldsymbol{n}^*}$ consists of all $N \times M$ matrices (n_{jk}) with

$$n_k^* = \sum_{j=1}^{N} n_{jk}.$$

For a fixed $\boldsymbol{\lambda}$ and (n_{jk}) in $\mathcal{A}_{\boldsymbol{n}^*}$, the product

$$\prod_{\substack{j=1,\ldots,N \\ k=1,\ldots,M}} \frac{e^{-\lambda_{jk}} \lambda_{jk}^{n_{jk}}}{n_{jk}!}$$

is the probability that this collection of decay events produced the measured data. The likelihood function is the sum of these probabilities over all possible ways of getting the measured data:

$$\mathscr{L}(\boldsymbol{\lambda}) = \sum_{(n_{jk}) \in \mathscr{A}_{n^*}} \prod_{\substack{j=1,\dots,N \\ k=1,\dots,M}} \frac{e^{-\lambda_{jk}} \lambda_{jk}^{n_{jk}}}{n_{jk}!}. \tag{16.45}$$

While this is a very complicated function, it is nonetheless possible to compute explicitly the first and second derivatives of $l(\boldsymbol{\lambda}) = \log \mathscr{L}(\boldsymbol{\lambda})$. The first partial derivatives are given by

$$\frac{\partial l}{\partial \lambda_j} = -1 + \sum_{k=1}^{M} \left[\frac{n_k^* p_{jk}}{\sum_{n=1}^{N} \lambda_n p_{nk}} \right].$$

A derivation using the properties of Poisson random variables is given in [116]. It is now a simple computation to show that

$$\frac{\partial^2 l}{\partial \lambda_i \partial \lambda_j} = -\sum_{k=1}^{M} \frac{n_k^* p_{ik} p_{jk}}{\left[\sum_{n=1}^{N} \lambda_n p_{nk}\right]^2}.$$

From this formula we conclude that

$$\sum_{i,j=1}^{N} \frac{\partial^2 l}{\partial \lambda_i \partial \lambda_j} x_i x_j = -\sum_{k=1}^{M} n_k^* c_k^2, \tag{16.46}$$

where

$$c_k = \frac{\sum_{j=1}^{N} x_j p_{jk}}{\sum_{j=1}^{N} \lambda_j p_{jk}}.$$

As the components of $\boldsymbol{n}^*$ are nonnegative, this shows that the Hessian of l is a negative semidefinite quadratic form; hence l is a concave function. This implies that all maxima of l are global maxima.

There are many ways to find the maximum of a concave function. The algorithm proposed by Shepp and Vardi is the following. We begin with an initial guess $\hat{\boldsymbol{\lambda}}^0$ with all components positive. If $\hat{\boldsymbol{\lambda}}^{\text{old}}$ denotes our current guess, then define

$$\hat{\lambda}_j^{\text{new}} = \hat{\lambda}_j^{\text{old}} \sum_{k=1}^{M} \left[\frac{n_k^* p_{jk}}{\sum_{n=1}^{N} \hat{\lambda}_n^{\text{old}} p_{nk}} \right] \quad \text{for } j = 1, \dots, N. \tag{16.47}$$

We multiply each component of $\hat{\boldsymbol{\lambda}}^{\text{old}}$ by 1 plus the corresponding component of the gradient of l. Notice that the denominator in (16.47) cannot vanish unless the numerator does as well. In this case we define $\frac{0}{0}$ to be zero. A nonnegative input leads to a nonnegative output; moreover,

$$\sum_{j=1}^{N} \hat{\lambda}_j^{\text{new}} = \sum_{k=1}^{M} n_k^*.$$

This means that the true number of counts is preserved in the iterative step, which agrees well with (16.44). Shepp and Vardi go on to prove that

$$\mathscr{L}(\hat{\boldsymbol{\lambda}}^{\text{new}}) \geq \mathscr{L}(\hat{\boldsymbol{\lambda}}^{\text{old}}),$$

with strict inequality unless $\mathscr{L}(\hat{\boldsymbol{\lambda}}^{\text{old}})$ is the maximum. This iteration has a limit $\hat{\boldsymbol{\lambda}}^{\infty}$ that provides a maximum likelihood estimate for $\boldsymbol{\lambda}$. For the details of these arguments the reader is referred to [116].

Exercises

***Exercise* 16.4.1.** Suppose that χ_1 and χ_2 are independent Poisson random variables with intensities λ_1 and λ_2, respectively. Show that $\chi_1 + \chi_2$ is also a Poisson random variable and compute $E[\chi_1 + \chi_2]$. *Hint:* See Exercise 15.4.1.

***Exercise* 16.4.2.** Suppose that we make a measurement of a single Poisson random variable and are attempting to determine its intensity using maximum likelihood. If the intensity is λ, then the probability of measuring n counts is

$$P(n|\lambda) = \frac{\lambda^n e^{-\lambda}}{n!}.$$

If we measure n^* counts, prove that $\hat{\lambda} = n^*$.

Exercise* 16.4.3. Derive the formulæ for the first and second partial derivatives of l. The derivation of the formula for the first derivatives should not be done as a direct calculation.

***Exercise* 16.4.4.** Deduce (16.46) from the formula for the second derivatives of l.

***Exercise* 16.4.5.** Suppose that f is a concave function defined on $\mathbb{R}^n$. Show that all maxima of f are global maxima and that if f assumes it maximum at both $\boldsymbol{x}_0$ and $\boldsymbol{x}_1$, then it also assumes its maximum at the points $\{t\boldsymbol{x}_0 + (1-t)\boldsymbol{x}_1 \; : \; t \in [0, 1]\}$.

***Exercise* 16.4.6.** Show that if $\hat{\boldsymbol{\lambda}}^{\text{new}} = \hat{\boldsymbol{\lambda}}^{\text{old}}$, then

$$\frac{\partial l}{\partial \lambda_j}(\hat{\boldsymbol{\lambda}}^{\text{new}}) = 0.$$

16.4.4 Determining the Transition Matrix

To complete our discussion of the maximum likelihood approach to image reconstruction in PET, we need to specify the transition matrix. This is something of an art. Conceptually simple definitions for (p_{jk}) may be difficult to compute, while simple but unintuitive choices are found, empirically, to give satisfactory results.

Recall that p_{jk} is the probability that a decay in box j is detected in tube k. This means that the two detectors that define this tube both observe a γ-ray at the "same" time. We first describe a continuum model. The "detectors" are parameterized by points $(\boldsymbol{\omega}_1, \boldsymbol{\omega}_2)$ in $S^1 \times S^1$. These points also parameterize the oriented line segments lying in D. The segment $\ell_{\boldsymbol{\omega}_1,\boldsymbol{\omega}_2}$ is oriented *from* $r\boldsymbol{\omega}_1$ *to* $r\boldsymbol{\omega}_2$. We have to exclude the subset of $S^1 \times S^1$, where $\boldsymbol{\omega}_1 = \boldsymbol{\omega}_2$. A subset C of $S^1 \times S^1$ is *even* if

$$(\boldsymbol{\omega}_1, \boldsymbol{\omega}_2) \in C \Longrightarrow (\boldsymbol{\omega}_2, \boldsymbol{\omega}_1) \in C.$$

Let B be a subset of D and C an even subset of $S^1 \times S^1$. Let $\mathscr{A}_{BC}$ denote the set of lines passing though B with endpoints in C. We could define the probability that a decay occurring in D is detected by a pair of detectors lying in C as

$$p_{BC}^{\text{bad}} = \frac{\text{area}(\mathscr{A}_{BC})}{4\pi^2}.$$

This is not a good model because it does not incorporate the hypothesis that the direction taken by the outgoing γ-rays is uniformly distributed.

To incorporate the uniform distribution of directions, we proceed somewhat differently. Let $\boldsymbol{x}$ be a point in D. For an even subset of pairs C, let $\mathscr{I}_{C,\boldsymbol{x}}$ be the subset of $[0, \pi)$ defined by

$$\theta \in \mathscr{I}_{C,\boldsymbol{x}} \text{ if the line with direction } \boldsymbol{\omega}(\theta) \text{ passing through } \boldsymbol{x} \text{ is in } C. \tag{16.48}$$

The probability that a decay event at $\boldsymbol{x}$ is detected by a pair of detectors in C is defined to be

$$p^{\text{unif}}(\boldsymbol{x}, C) = \frac{1}{\pi r^2}\frac{|\mathscr{I}_{C,\boldsymbol{x}}|}{\pi},$$

where $|\mathscr{I}_{C,\boldsymbol{x}}|$ is the Lebesgue measure of $\mathscr{I}_{C,\boldsymbol{x}}$. This model incorporates the uniform distribution but ignores attenuation effects. To include attenuation, each line through $\boldsymbol{x}$ is weighted according to the likelihood that *both* γ-rays are detected. As shown in (16.38), the probability that a decay at $\boldsymbol{x}$ is detected by a pair in C should therefore be

$$p^{\text{unif},\mu}(\boldsymbol{x}, C) = \frac{1}{\pi r^2}\left[\frac{1}{\pi}\int\limits_{\mathscr{I}_{C,\boldsymbol{x}}} \exp[-\mathscr{R}\mu(\langle \boldsymbol{x}, \boldsymbol{\omega}(\theta)\rangle, \boldsymbol{\omega}(\theta))]\, d\theta\right].$$

By integrating these distributions, either model can be used to define a transition matrix. These matrices are given by

$$p_{jk} = \int\limits_{B_j} p^{\text{unif}}(\boldsymbol{x}, T_k) d\boldsymbol{x} \text{ and } p_{jk}^{\mu} = \int\limits_{B_j} p^{\text{unif},\mu}(\boldsymbol{x}, T_k)\, d\boldsymbol{x}.$$

Here T_k is identified with the even subset of $S^1 \times S^1$ corresponding to the endpoints of lines lying in T_k. The functions p^{unif} and $p^{\text{unif},\mu}$ are difficult to compute and therefore simpler models are often employed. For example, we could set p_{jk} equal to the probability that a decay at the *center* of B_j is detected in T_k. We could even incorporate a probabilistic description of the incoming momenta of the pairs into the definition of the transition matrix. Empirically it was found in [116] that the final result is not too sensitive to the definition of (p_{jk}) in that several reasonable choices produced similar reconstructions.

***Remark* 16.4.1.** *Single Photon Emission CT* or *SPECT* is a related technique in medical imaging. Instead of two collinear photons, the radioactive compounds used in this modality produce a single γ-ray which is then detected by a collimated detector. The reconstruction problem in this modality is considerably more difficult and involves inverting the attenuated x-ray transform. The mathematical solution of this problem was only achieved by Roman Novikov in 2001! The interested reader should see [86] for a discussion of SPECT and the papers of Novikov for a discussion of the attenuated x-ray transform, [98], [99].

Exercises

***Exercise* 16.4.7.** What line $l_{t,\omega}$ corresponds to ℓ_{ω_1,ω_2}?

***Exercise* 16.4.8.** For a point $x \in D$, what is the distribution function for the direction in which a decay occurs implied by p^{bad}?

***Exercise* 16.4.9.** How should (p_{jk}) be defined if the incoming momenta are Gaussian distributed with expected mean magnitude $p_m \neq 0$?

16.5 Conclusion

In this chapter we presented a few simple applications of probability theory to problems in medical imaging. Beer's law, the basic physical principle underlying the measurement model in x-ray tomography, was shown to have a probabilistic origin. We also replaced our earlier deterministic description of the measurements with a probabilistic model. By describing the measurements as random variables, we were able to establish relationships between noise in the measurements and uncertainty in the reconstructed images. Of particular importance is the relationship between the dosage of radiation the patient receives and the resolution of the reconstructed image.

Because the variance in the measurements is inversely proportional to the number of measured photons, reducing the width of the beam has an adverse effect on the signal-to-noise ratio in the measurements. This, in turn, reduces the contrast available in the reconstructed image. A consequence of these analyses is that an assessment of the resolution in an image is only meaningful if it is coupled with an estimate of the signal-to-noise ratio. Because of safety considerations, the ultimate image quality achievable in x-ray tomography is probably limited by the noise inherent in the measurements themselves.

In the next and final chapter we define random processes that are parameterized families of random variables. This extension of the formalism of probability theory is necessary when studying continuum models. Among other things it provides a method to analyze the output of a linear filter when the input is contaminated by noise. It is also useful for the design of "optimal" filters. These are filters that remove noise as well as possible, under given assumptions about the signal and the noise.

Chapter 17

Random Processes*

To model noise in measurement and filtering processes requires concepts more general than that of a random variable. This is because we need to discuss the results of passing a noisy signal through a linear filter. As was the case in chapter 9, it is easier to present this material in terms of functions of a single variable, though the theory is easily adapted to functions of several variables. After giving the basic definitions, we consider some important examples of random processes and analyze the effect of a linear shift invariant filter on random inputs. The chapter concludes with an analysis of noise in the continuum model for filtered back-projection. Our discussion of random processes is very brief, aimed squarely at the goal of analyzing the effects of noise in the filtered back-projection algorithm.

17.1 Random Processes in Measurements

To motivate this discussion, we think of the familiar example of a radio signal. A radio station broadcasts a signal $s_b(t)$ as an electromagnetic wave. A receiver detects a signal $s_r(t)$, which we model as a sum

$$s_r(t) = F(s_b)(t) + n(t).$$

Here F is a filtering operation that describes the propagation of the broadcast signal. For the purposes of this discussion we model F as attenuation; that is, $F(s_b) = \lambda s_b$ for a $0 < \lambda < 1$. The other term is noise. In part, the noise records "random" aspects of the life history of the broadcast signal that are not modeled by F. Beyond that it is an aggregation of other signals that happen to be present at the carrier frequency of the broadcast signal. The existence of the second part is easily verified by tuning an (old) radio to a frequency for which there is no local station. Practically speaking, it is not possible to give a formula for the noise. Because we cannot give an exact description of the noise, we instead describe it in terms of its statistical properties.

We begin with the assumption that the noise is a bounded function of t. What can be done to specify the noise further? Recall the ensemble average definition of the probability of an event as the average of the results of many "identical experiments." In the radio example, imagine having many different radios, labeled by a set $\mathscr{A}$. For each $\alpha \in \mathscr{A}$ we let $s_{r,\alpha}(t)$ be the signal received by radio α at time t. The collection of radios $\mathscr{A}$ *is* the sample space. The value of $s_{r,\alpha}$ at a time t is a function on the sample space—in other words, a random variable. From the form given previously we see that

$$s_{r,\alpha}(t) = \lambda s_b(t) + n_\alpha(t).$$

The noise can then be described in terms of the statistical properties of the random variables $\{n_\alpha(t)\}$ for different values of t. We emphasize that the sample space is $\mathscr{A}$, the collection of different receivers; the time parameter simply labels different random variables defined on the sample space. A family of random variables, defined on the same sample space, parameterized by a real variable, is called a *random process* or, more precisely, a continuous parameter random process.

Once the sample space $\mathscr{A}$ is equipped with a σ-algebra $\mathscr{M}$ and a probability measure ν, we can discuss the statistics of the noise. For example, at each time t the random variable $n_\alpha(t)$ has an expected value

$$E[n_\alpha(t)] = \int_{\mathscr{A}} n_\alpha(t)\, d\nu(\alpha).$$

In many applications we assume that the noise has mean zero (i.e., $E[n_\alpha(t)] = 0$ for all t). This means that if we make many different independent measurements and average them, we should get a good approximation to $\lambda s_r(t)$. The correlation between the noise at one moment in time and another is given by

$$E[n_\alpha(t_1)n_\alpha(t_2)] = \int_{\mathscr{A}} n_\alpha(t_1)n_\alpha(t_2)\, d\mu(\alpha).$$

How should the sample space be described mathematically? In an example like this, the usual thing is to use the space of all bounded functions as the index set. That is, *any* bounded function is a candidate for the noise in our received signal. In principle, the probabilistic component of the theory could then be encoded in the choice of a σ-algebra and probability measure on the space of bounded functions. The probability measure is rarely made explicit. Instead we specify the cumulative joint distributions of the noise process at finite collections of times. This means that for any $k \in \mathbb{N}$, any k times $(t_1, \ldots, t_k)$, and any k values $(v_1, \ldots, v_k)$, the joint probability that

$$n_\alpha(t_j) \leq v_j \qquad \text{for } j = 1, \ldots, k$$

is specified. If the joint distributions satisfy the usual consistency conditions (see Exercise 15.2.35), then a result of Kolmogorov states that there is a probability measure on $\mathscr{A}$, inducing the joint distributions, with σ-algebra chosen so that all the sets

$$\{\alpha \in \mathscr{A} \,:\, n_\alpha(t_j) \leq v_j\}$$

are measurable. In this chapter we give a brief introduction to the basic concepts of random processes. Our treatment, though adequate for our applications to imaging, is neither complete nor rigorous. In particular we do not establish the existence of random processes as outlined previously. Complete treatments can be found in [29] or [128].

17.2 Basic Definitions

See: B.8.

Let $(X, \mathcal{M}, \nu)$ be a probability space; as noted above a random process is an indexed family of random variables defined on a fixed probability space. There are two main types of random processes. If the index set is a subset of integers, for examplethe natural numbers $\mathbb{N}$, then the process is a *discrete parameter random process*. Such a random process is a sequence $\{\chi_1, \chi_2, \ldots, \}$ of random variables defined on X. A *continuous parameter random process* is a collection of random variables, $\{\chi_t\}$ indexed by a continuous parameter. Often the whole real line is used as the index set, though one can also use a finite interval or a half ray. For each t, χ_t is a random variable, that is a measurable function on X. We can think of χ as a function of the pair $(w; t)$ where $w \in X$. For a fixed $w \in X$, the map $t \mapsto \chi_t(w)$ is called a *sample path* for this random process. Depending on the context, sample paths are denoted using either standard functional notation $\chi(t)$, or subscript notation χ_t. The dependence on the point in X is suppressed unless it is required for clarity. In the continuous parameter case, a rigorous treatment of this subject requires hypotheses about the regularity properties of the random variables as functions of t, see [29] or [128].

It would appear that the first step in the discussion of a random process should be the definition of the measure space and a probability measure defined on it. As noted previously, this is rarely done. Instead the random process is defined in terms of the properties of the random variables themselves. In the continuous time case, for each k and every k-tuple of times $t_1 \leq \cdots \leq t_k$ the cumulative distributions

$$P_{t_1,\ldots,t_k}(s_1, \ldots, s_k) = \operatorname{Prob}(\chi_{t_1} \leq s_1, \chi_{t_2} \leq s_2, \ldots, \chi_{t_k} \leq s_k)$$

are specified. In favorable circumstances, these distributions are given by integrals of density functions:

$$P_{t_1,\ldots,t_k}(s_1, \ldots, s_k) = \int\limits_{-\infty}^{s_1} \cdots \int\limits_{-\infty}^{s_k} p_{t_1,\ldots,t_k}(x_1, \cdots, x_k)\, dx_1 \cdots dx_k,$$

The distributions must satisfy the usual consistency conditions:

$$\begin{aligned} p_{t_1,\ldots,t_k}(x_1, \ldots, x_k) &= \int\limits_{-\infty}^{\infty} p_{t_1,\ldots,t_{k+1}}(x_1, \ldots, x_k, x_{k+1})\, dx_{k+1}, \\ p_{t_{\tau(1)},\ldots,t_{\tau(k)}}(x_{\tau(1)}, \ldots, x_{\tau(k)}) &= p_{t_1,\ldots,t_k}(x_1, \ldots, x_k), \end{aligned} \tag{17.1}$$

where τ is any permutation of $\{1, \ldots, k\}$.

In the discrete case the joint distribution functions are specified for any finite subset of the random variables. That is, for each $k \in \mathbb{N}$ and each k-multi-index $\boldsymbol{i} = (i_1, \ldots, i_k)$, the cumulative distribution

$$P_{\boldsymbol{i}}(s_1, \ldots, s_k) = \operatorname{Prob}(\chi_{i_1} \leq s_1, \chi_{i_2} \leq s_2, \ldots, \chi_{i_k} \leq s_k)$$

is specified. The $\{P_i\}$ also need to satisfy the consistency conditions for joint cumulative distributions. If $\{\chi_i\}$ is a discrete parameter random process, we say that the terms of the sequence are *independent* if for any choice of distinct indices $\{i_1, \ldots, i_k\}$ the random variables $\{\chi_{i_1}, \ldots, \chi_{i_k}\}$ are independent.

The cumulative distribution functions $P_{t_1,\ldots,t_k}(s_1, \ldots, s_k)$ (or $P_i(s_1, \ldots, s_k)$) are called the *finite-dimensional distributions* of the random process. A basic result of Kolmogorov states that if finite-dimensional distributions are specified that satisfy the compatibility conditions, then there is a probability space $(X, \mathcal{M}, \nu)$ and a random process χ_t defined on it that induces the given finite-dimensional distributions. We take this result for granted; for a proof see [29] or [128].

Some examples of random processes should help to clarify these ideas.

***Example* 17.2.1.** Let X be the set of *all* bounded sequences of real numbers, that is,

$$X = \{\boldsymbol{a} = (a_1, a_2, \ldots), \quad a_i \in \mathbb{R} \quad \text{with} \limsup_{i\to\infty} |a_i| < \infty\}.$$

Define a discrete parameter, random process $\{\chi_1, \chi_2, \ldots\}$ by setting

$$\chi_i(\boldsymbol{a}) = a_i.$$

To describe the measure theoretic aspects of this process, we choose a probability measure, ν on $\mathbb{R}$. For all i define

$$\text{Prob}(\chi_i \leq t) = \int_{-\infty}^{t} d\nu.$$

Supposing further that the $\{\chi_i\}$ are independent random variables, we can compute the joint distributions: For each $k \in \mathbb{N}$, multi-index $\{i_1, \ldots, i_k\}$ and $(s_1, \ldots, s_k) \in \mathbb{R}^k$, we have

$$\text{Prob}(\chi_{i_1} \leq s_1, \ldots, \chi_{i_k} \leq s_k) = \int_{-\infty}^{t_1} d\nu \cdots \int_{-\infty}^{t_k} d\nu.$$

These properties serve to characterize the random process, though the proof that a σ-algebra and measure exist on X inducing these joint distribution functions is by no means trivial.

***Example* 17.2.2.** Another example is the set of infinite sequences of coin flips. The sample space X is a set of all possible infinite sequences of heads and tails. As previously, define

$$\chi_i(\boldsymbol{a}) = \begin{cases} 1 & \text{if } a_i = H, \\ 0 & \text{if } a_i = T. \end{cases}$$

The $\{\chi_i\}$ are then taken to be independent random variables with

$$\text{Prob}(\chi_i = 0) = 1 - p, \quad \text{Prob}(\chi_i = 1) = p.$$

Such a process is called a *Bernoulli random process* because each χ_i is Bernoulli random variable.

***Example* 17.2.3.** An example of a continuous time random process is provided by setting $X = \mathscr{C}_0(\mathbb{R}_+)$, the set of continuous functions on $\mathbb{R}_+$ that vanish at 0. For each $t \in \mathbb{R}_+$ we have the random variable χ_t defined at $w \in X$ by

$$\chi_t(w) = w(t);$$

χ_t is the evaluation of the function w at time t. As before, it is difficult to give a direct description of the σ-algebra and measure on X. Instead the process is described in terms of its joint distribution functions. We need to specify

$$\text{Prob}(\chi_{t_1} \leq s_1, \chi_{t_2} \leq s_2, \ldots, \chi_{t_k} \leq s_k)$$

for all $k \in \mathbb{N}$ and all pairs of real k-tuples, $((t_1, \ldots, t_k), (s_1, \ldots, s_k))$.

An important special case of this construction is given by

$$\text{Prob}(\chi_{t_1} \leq s_1, \chi_{t_2} \leq s_2, \ldots, \chi_{t_k} \leq s_k) =$$
$$\int_{-\infty}^{s_k} \cdots \int_{-\infty}^{s_1} \frac{e^{-\frac{x_1^2}{2t_1}}}{\sqrt{2\pi t_1}} \prod_{j=2}^{k} \frac{e^{-\frac{(x_j - x_{j-1})^2}{2(t_j - t_{j-1})}}}{\sqrt{2\pi(t_j - t_{j-1})}} dx_1 \cdots dx_k, \tag{17.2}$$

if $t_1 < t_2 < \cdots < t_k$. This process is called *Brownian motion*. For each t the random variable χ_t is Gaussian, and for any finite set of times, $(t_1, \ldots, t_k)$ the random variables $\{\chi_{t_1}, \ldots, \chi_{t_k}\}$ are jointly Gaussian. We return to this example later.

Exercises

***Exercise* 17.2.1.** Show that the cumulative distributions defined in Example 17.2.1 satisfy the consistency conditions.

***Exercise* 17.2.2.** In many applications we need to approximate a random process by a finite-dimensional sample space. Suppose that in Example 17.2.1 we consider finite, real sequences of length N. The sample space is then $\mathbb{R}^N$. The random variables $\{\chi_1, \ldots, \chi_N\}$ are defined on this space. Find a probability measure on $\mathbb{R}^N$ that gives the correct joint distributions functions for these variables. Are there others that would also work?

17.2.1 Statistical Properties of Random Processes

In practical applications a random process is described by its statistical properties. The simplest are the mean

$$\mu_\chi(t) = E[\chi_t]$$

and variance

$$\sigma_\chi^2(t) = E[(\chi_t - \mu_\chi(t))^2].$$

A measure of the relationship of χ at two different times is the *autocorrelation function*:

$$R_\chi(t_1, t_2) = < \chi_{t_1} \chi_{t_2} > .$$

As before, the *covariance* is defined by

$$\begin{aligned}\operatorname{Cov}(\chi_{t_1}, \chi_{t_2}) &= R_\chi(t_1, t_2) - <\chi_{t_1}><\chi_{t_2}> \\ &= E[(\chi_{t_1} - \mu_\chi(t_1))(\chi_{t_2} - \mu_\chi(t_2))].\end{aligned} \tag{17.3}$$

The normalized correlation coefficient is

$$\rho(t_1, t_2) = \frac{\operatorname{Cov}(\chi_{t_1}, \chi_{t_2})}{\sigma_\chi(t_1)\sigma_\chi(t_2)}.$$

If the cumulative joint distribution for χ_{t_1} and χ_{t_2} has a distribution function, then

$$R_\chi(t_1, t_2) = \int_{-\infty}^{\infty}\int_{-\infty}^{\infty} xyp_{t_1,t_2}(x, y)\,dx\,dy.$$

Using the Cauchy-Schwarz inequality and (17.1), we deduce that

$$\begin{aligned}R_\chi(t_1, t_2) &= \iint_{\mathbb{R}^2} xyp_{t_1,t_2}(x, y)\,dx\,dy \\ &\leq \Big[\iint_{\mathbb{R}^2} x^2 p_{t_1,t_2}(x, y)\,dx\,dy\Big]^{1/2}\Big[\iint_{\mathbb{R}^2} y^2 p_{t_1,t_2}(x, y)\,dx\,dy\Big]^{1/2} \\ &= \Big[\int_{-\infty}^{\infty} x^2 p_{t_1}(x)\,dx\Big]^{1/2}\Big[\int_{-\infty}^{\infty} y^2 p_{t_2}(y)\,dy\Big]^{1/2}.\end{aligned}$$

Hence, we have the estimate

$$|R_\chi(t_1, t_2)| \leq \sqrt{E[\chi_{t_1}^2]E[\chi_{t_2}^2]}. \tag{17.4}$$

Exercise

***Exercise* 17.2.3.** Show how to derive (17.4) using the formalism of a probability space (i.e., by integrating over X with respect to $d\nu$).

17.2.2 Stationary Random Processes

A important notion is that of a *stationary random process*. Heuristically a noise process is stationary if it does not matter when you start looking; the noise is always "the same."

***Definition* 17.2.1.** Let χ_t be a continuous parameter random process. It is a stationary process if

1. $\operatorname{Prob}(\chi_t \leq \lambda)$ is independent of t.

2. For any $\tau \in \mathbb{R}$,

$$\operatorname{Prob}(\chi_{t_1} \leq r, \chi_{t_2} \leq s) = \operatorname{Prob}(\chi_{(t_1+\tau)} \leq r, \chi_{(t_2+\tau)} \leq s).$$

2′. Similarly, for any collection of $(t_1, \dots, t_k)$, $(s_1, \dots, s_k)$ and τ we have

$$\operatorname{Prob}(\chi_{t_1} \leq s_1, \dots, \chi_{t_k} \leq s_k) = \operatorname{Prob}(\chi_{(t_1+\tau)} \leq s_1, \dots, \chi_{(t_k+\tau)} \leq s_k).$$

If χ_t is a stationary random process, then

$$R_\chi(t_1, t_2) = E[\chi_{t_1}\chi_{t_2}] = E[\chi_0\chi_{(t_2-t_1)}]. \tag{17.5}$$

Setting $r_\chi(\tau) = E[\chi_0\chi_\tau]$, it follows that

$$R_\chi(t_1, t_2) = r_\chi(t_2 - t_1).$$

On the other hand, the fact that $R_\chi(t_1, t_2)$ is a function of $t_2 - t_1$ does *not* imply that χ_t is a stationary process. A process satisfying this weaker condition is called a *weak sense stationary random process*. For a weak sense stationary process,

$$r_\chi(\tau) \leq E[\chi_0^2] = r_\chi(0). \tag{17.6}$$

This coincides well with intuition. If something is varying in a "random" but stationary way, then it is unlikely to be better correlated at two different times than at a given time.

17.2.3 Spectral Analysis of Stationary Processes*

The reason for introducing the autocorrelation function is that it allows the use of Fourier theory to study weak sense stationary processes.

***Definition* 17.2.2.** If χ_t is a weak sense stationary random process and r_χ is integrable (or square integrable), then its Fourier transform,

$$S_\chi(\xi) = \int_{-\infty}^{\infty} r_\chi(\tau)e^{-i\tau\xi}\,d\tau,$$

is called the *spectral density function* for the process χ.

The autocorrelation function is not always integrable but, as shown in the proof of Proposition 17.2.1, it is a "nonnegative definite function." If it is also continuous, then it follows from a theorem of Bochner that its Fourier transform is well defined as a measure on $\mathbb{R}$. This means that, while S_χ may not be well defined at points, for any $[a, b]$ the integral

$$\frac{1}{2\pi}\int_a^b S_\chi(\xi)\,d\xi$$

is meaningful. This measure is called the spectral density measure. The integral *defines* the power contained in the process in the interval $[a, b]$. A sufficient condition for r_χ to be continuous is that $t \mapsto \chi_t$ is continuous in the L^2-sense:

$$\lim_{t-s\to 0} E[(\chi_t - \chi_s)^2] = 0. \tag{17.7}$$

The proposition enumerates the important properties of r_χ and S_χ.

Proposition 17.2.1. *Let r_χ, be the autocorrelation function of a real, weak sense stationary random process. Suppose that r_χ is continuous and has Fourier transform S_χ. Then the following statements hold:*

1. *r_χ is real valued.*
2. *The autocorrelation is an even function: $r_\chi(\tau) = r_\chi(-\tau)$.*
3. *S_χ is a real-valued even function.*
4. *S_χ is nonnegative.*
5. *The total power of the process is the variance:*

$$r_\chi(0) = E[\chi_t^2] = \frac{1}{2\pi}\int_{-\infty}^{\infty} S_\chi(\xi)\,d\xi.$$

Proof. By definition, we have

$$r_\chi(\tau) = E[\chi_t\chi_{(t+\tau)}] = E[\chi_{(t-\tau)}\chi_t] = r_\chi(-\tau).$$

Hence r_χ is even; Proposition 4.2.6 then implies that S_χ is also even. To show that the spectral density function is real valued, take the conjugate of S_χ:

$$\bar{S}_\chi(\xi) = \int_{-\infty}^{\infty} r_\chi(\tau)e^{i\xi\tau}\,d\tau = \int_{-\infty}^{\infty} r_\chi(-\tau)e^{i\xi\tau}\,d\tau = S_\chi(\xi).$$

The fourth fact is not obvious from the definition. This follows because the autocorrelation function r_χ is a *nonnegative definite function.* This means that for any vectors $(x_1, \dots, x_N)$ and $(\tau_1, \dots, \tau_N)$, we have that

$$\sum_{i,j=1}^{N} r_\chi(\tau_i - \tau_j)x_i x_j \geq 0.$$

In other words, the symmetric matrix $a_{ij} = r_\chi(\tau_i - \tau_j)$ is nonnegative definite.

This inequality is a consequence of the fact that the expected value of a nonnegative random variable is nonnegative. If χ_t is any real-valued, continuous time random process with finite mean and covariance, then

$$0 \leq E[\left|\sum_{1}^{N} x_i\chi_{\tau_i}\right|^2] = \sum_{i,j=1}^{N} x_i x_j E[\chi_{\tau_i}\chi_{\tau_j}] = \sum_{i,j=1}^{N} x_i x_j R_\chi(\tau_i, \tau_j).$$

Hence, $\sum_{i,j=1}^{N} R_\chi(\tau_i, \tau_j)x_i x_j$ is nonnegative. For a weak sense stationary process, $R_\chi(\tau_1, \tau_2)$ equals $r_\chi(\tau_1 - \tau_2)$ and thus

$$\sum_{i,j=1}^{N} r_\chi(\tau_i - \tau_j)x_i x_j \geq 0.$$

The Bochner theorem states that a continuous function is the Fourier transform of a positive measure if and only if it is nonnegative definite. Hence, the fact that r_χ is nonnegative definite implies that $S_\chi d\xi$ is a nonnegative measure; that is,

$$\int_a^b S_\chi(\xi)\, d\xi = \int_a^b \int_{-\infty}^{\infty} r_\chi(\tau)e^{-i\tau\xi} d\tau \geq 0;$$

see [79]. Fact (5) follows from the Fourier inversion formula. □

If f is a real-valued, bounded, integrable function, then we see that its "autocorrelation function,"

$$r_f(t) = \int_{-\infty}^{\infty} f(t)f(t+\tau)\, dt$$

is well defined. Computing the Fourier transform of r_f gives

$$\hat{r}_f(\xi) = \int_{-\infty}^{\infty}\int_{-\infty}^{\infty} f(t)f(t+\tau)e^{-i\xi\tau} dt\, d\tau = |\hat{f}(\xi)|^2.$$

These computations are not immediately applicable to a random process but hold in a time-averaged, probabilistic sense.

Let χ_t be a real-valued, weak sense stationary random process. For each $T > 0$ define

$$\widehat{\chi_T}(\xi) = \int_{-T}^{T} \chi_t e^{-it\xi} dt.$$

We compute the expected value of $|\widehat{\chi_T}(\xi)|^2$,

$$\begin{aligned} E[|\widehat{\chi_T}(\xi)|^2] &= E\left[\int_{-T}^{T} \chi_t e^{-it\xi} dt \int_{-T}^{T} \chi_s e^{is\xi} ds\right] \\ &= \int_{-T}^{T}\int_{-T}^{T} r_\chi(t-s)e^{-i(t-s)\xi} dt\, ds. \end{aligned} \tag{17.8}$$

Letting $\tau = t - s$, we obtain

$$\begin{aligned} E[|\widehat{\chi}_T(\xi)|^2] &= \int_{-2T}^{2T} (2T - |\tau|) r_\chi(\tau) e^{-i\tau\xi}\, d\tau \\ &= (2T) \int_{-2T}^{2T} \left(1 - \frac{|\tau|}{2T}\right) r_\chi(\tau) e^{-i\tau\xi}\, d\tau. \end{aligned} \tag{17.9}$$

Proposition 17.2.2. *If χ_t is a real-valued, weak sense stationary random process and r_χ is integrable, then*

$$S_\chi(\xi) = \lim_{T\to\infty} \frac{1}{2T} E[|\widehat{\chi}_T(\xi)|^2]. \tag{17.10}$$

Proof. Under the hypotheses, the Lebesgue-dominated convergence theorem applies to show that the limit can be taken inside the integral in the second line of (17.9), giving the result. □

This proposition justifies the description of S_χ as the power spectral density of the process. Given a point w in the sample space, $t \mapsto \chi_t(w)$ is a sample path. A reasonable definition for the autocorrelation function of a single sample path is

$$r_w(\tau) = \lim_{T\to\infty} \frac{1}{2T} \int_{-T}^{T} \chi_t(w)\chi_{(t+\tau)}(w)\, dt.$$

This is sometimes called the *time autocorrelation function*. For a given choice of w, this limit may or may not exist. It turns out that for many interesting classes of random processes this time average exists, defines a random variable on X, and, with probability 1, does not depend on the choice of path. In fact,

$$\text{Prob}(r_w = r_\chi) = 1.$$

Exercises

***Exercise* 17.2.4.** Derive (17.5).

***Exercise* 17.2.5.** Prove (17.6).

***Exercise* 17.2.6.** Derive the first line of (17.9).

***Exercise* 17.2.7.** Suppose that χ_t is a random process so that $E[|\chi_t|]$ is independent of t. Show either $\chi_t \equiv 0$, with probability 1, or

$$\int_{-\infty}^{\infty} |\chi_t|\, dt = \infty,$$

with probability 1.

17.2.4 Independent and Stationary Increments

Many processes encountered in imaging applications are not themselves stationary but satisfy the weaker hypothesis of having *stationary increments.*

***Definition* 17.2.3.** Let χ_t be a continuous parameter random process such that for any finite sequence of times $t_1 < t_2 < \cdots < t_n$, $n \geq 3$ the random variables

$$\chi_{t_2} - \chi_{t_1}, \chi_{t_3} - \chi_{t_2}, \ldots, \chi_{t_n} - \chi_{t_{n-1}}$$

are independent. Such a process is said to have *independent increments.* If, moreover, the probability $\text{Prob}(\chi_t - \chi_s \leq \lambda)$ depends only on $t - s$, then the process has *stationary increments.*

A weaker condition is that a process have uncorrelated increments; that is,

$$E[(\chi_{t_2} - \chi_{s_2})(\chi_{t_1} - \chi_{s_1}) = E[(\chi_{t_2} - \chi_{s_2})]E[(\chi_{t_1} - \chi_{s_1})],$$

provided that $[t_2, s_2] \cap [t_1, s_1] = \emptyset$. If $E[|\chi_t - \chi_s|^2]$ only depends on $t - s$, then the process is said to have *wide sense stationary increments.*

***Example* 17.2.4.** Brownian motion is a random process, parameterized by $[0, \infty)$, that describes, among other things, the motion of tiny particles in a fluid. It is defined as a process χ_t with independent increments, such that for every s, t the increment $\chi_t - \chi_s$ is a Gaussian random variable with

$$E[\chi_t - \chi_s] = 0 \text{ and } E[(\chi_t - \chi_s)^2] = \sigma^2|t - s|^2.$$

This process is often normalized by fixing $\chi_0 = a \in \mathbb{R}$, with probability 1. A very important fact about Brownian motion is that it is essentially the only process with independent increments whose sample paths are continuous, with probability 1. Brownian motion is frequently called the *Wiener process.*

Exercise

***Exercise* 17.2.8.** Show that if a Gaussian process has $E[\chi_t] = 0$ and uncorrelated increments, then it has independent increments.

17.3 Examples of Random Processes

For many applications in medical imaging a small collection of random processes suffice. Several have already been defined; a few more are described in this section.

17.3.1 Gaussian Random Process

A Gaussian random process is a family $\{\chi_t\}$ (or sequence of random variables $\{\chi_i\}$) that, for each t (or i), is a Gaussian random variable. The finite-dimensional distributions are also assumed to be Gaussian. As we saw in Section 15.3.3, the joint distributions for Gaussian random variables are determined by their means and covariance matrix. This remains true of Gaussian processes, and again the converse statement is also true. Suppose that T is the parameter space for a random process and that there are real-valued functions $\mu(t)$ defined on T and $r(s,t)$ defined on $T \times T$. The function r is assumed to satisfy the following conditions:

1. For any pair $s, t \in T$ $r(s,t) = r(t,s)$, and
2. If $\{t_1, \dots, t_m\} \subset T$, then the matrix $[r(t_i, t_j)]$ is nonnegative definite.

There exists a Gaussian random process $\{\chi_t \, : \, t \in T\}$ such that

$$E[\chi_t] = \mu(t) \text{ and } E[\chi_s \chi_t] = r(s,t).$$

If we are only concerned with the second-order statistics of a random process, then we are free to assume that the process is Gaussian.

Brownian motion, defined in Example 17.2.4, is an important example of a Gaussian process. As remarked there, we can fix $\chi_0 = 0$ with probability 1. Since $E[\chi_t - \chi_s] = 0$ for all t, s it follows that

$$E[\chi_t] = 0 \qquad \text{for all } t \in [0, \infty).$$

The autocorrelation function can now be computed using the hypothesis

$$E[(\chi_t - \chi_s)^2] = \sigma^2 |t - s|. \tag{17.11}$$

Let $0 < s < t$. Then, as $\chi_0 = 0$ with probability 1,

$$E[\chi_s \chi_t] = E[(\chi_s - \chi_0)(\chi_t - \chi_0)].$$

This can be rewritten as

$$\begin{aligned} E[\chi_s \chi_t] &= E[(\chi_s - \chi_0)^2] + E[(\chi_s - \chi_0)(\chi_t - \chi_s)] \\ &= \sigma^2 |s| = \sigma^2 \min\{|s|, |t|\}. \end{aligned} \tag{17.12}$$

In passing from the first line to the second, we use the independence of the increments and (17.11). Thus Brownian motion is not a weak sense stationary process.

17.3.2 The Poisson Counting Process

The Poisson counting process is another example of a process with independent, stationary increments. This process is a family of random variables $\{\chi_t\}$ defined for $t \geq 0$ that take values in $\{0, 1, 2, \dots\}$. The Poisson counting process has a nice axiomatic characterization. For convenience, let

$$P(k,t) \stackrel{d}{=} \text{Prob}(\chi_t = k).$$

Here are the axioms phrased in terms of counting emitted particles:

INDEPENDENT INCREMENTS: The number of particles emitted in $[t_1, t_2]$ is independent of the number in $[t_3, t_4]$ if $[t_1, t_2] \cap [t_3, t_4] = \emptyset$.

SHORT TIME BEHAVIOR: The probability that one particle is emitted in a very short time interval is given by $P(1, \Delta t) = \lambda \Delta t + o(\Delta t)$ for some constant λ, where $o(\Delta t)$ denotes a term such that $\lim_{\Delta t \to 0} o(\Delta t)/\Delta t = 0$. As a consequence, $P(1, 0) = 0$.

STATIONARY INCREMENTS: The process has stationary increments:

$$\text{Prob}(\chi_t - \chi_s = k) = \text{Prob}(\chi_{(t+\tau)} - \chi_{(s+\tau)} = k), \quad \forall \tau \geq 0, 0 \leq s \leq t.$$

We can now estimate the probability that two particles are emitted in a short interval $[0, \Delta t]$: In order for this to happen, there must be a $0 < \tau < \Delta t$ such that one particle is emitted in $[0, \tau]$ and one is emitted in $(\tau, \Delta]$. The hypothesis of independent stationary increments and the short time behavior imply that

$$P(2, \Delta t) \leq \max_{\tau \in (0, \Delta t)} P(1, \tau) P(1, \Delta t - \tau) = O((\Delta t)^2).$$

From the independent, stationary increments axiom, we have that

$$P(0, t + \Delta t) = P(0, t) P(0, \Delta t).$$

For any time Δt it is clear that

$$\sum_{k=0}^{\infty} P(k, \Delta t) = 1.$$

Arguing as before we can show that, for sufficiently small Δt, we have the estimates $P(k, \Delta t) \leq [P(1, \Delta t)]^k$. Combining these observations leads to

$$P(0, \Delta t) + P(1, \Delta t) = 1 + o(\Delta t). \tag{17.13}$$

Hence

$$P(0, \Delta t) = 1 - \lambda \Delta t + o(\Delta t), \tag{17.14}$$
$$P(0, t + \Delta t) = P(0, t) P(0, \Delta t) = P(0, t)[1 - \lambda \Delta t + o(\Delta t)]. \tag{17.15}$$

Letting $\Delta t \to 0$, we have

$$\lim_{\Delta t \to 0} \frac{P(0, t + \Delta t) - P(0, t)}{\Delta t} = \frac{P(0, t)(-\lambda \Delta t + o(\Delta t))}{\Delta t} = -\lambda P(0, t).$$

This provides a differential equation for $P(0, t)$:

$$\frac{dP}{dt}(0, t) = -\lambda P(0, t), \text{ with } P(0, 0) = 1.$$

The solution of this equation is

$$P(0, t) = e^{-\lambda t}.$$

The probabilities $\{P(k,t)\}$ for $k > 1$ are obtained recursively. For each $t \geq 0$ and $j \leq k$ suppose that

$$P(j,t) = \frac{(\lambda t)^j}{j!}e^{-\lambda t}.$$

The hypothesis of independent, stationary increments implies that

$$\begin{aligned} &P(k+1,t+\Delta t) \\ &\quad = P(k+1,t)P(0,\Delta t) + P(k,t)P(1,\Delta t) + \sum_{j=0}^{k-1} P(j,t)P(k+1-j,\Delta t) \\ &\quad = P(k,t)P(1,\Delta t) + P(k+1,t)P(0,\Delta t) + o(\Delta t). \end{aligned} \tag{17.16}$$

We use the induction hypothesis to conclude that the sum is $o(\Delta t)$ as Δt tends to zero. From equations (17.14) and (17.16) we obtain

$$P(k+1,t+\Delta t) - P(k+1,t) = P(k,t)(\lambda\Delta t + o(\Delta t)) + P(k+1,t)[P(0,\Delta t) - 1] + o(\Delta t),$$

which leads to

$$\frac{dP}{dt}(k+1,t) = \lambda P(k,t) - \lambda P(k+1,t). \tag{17.17}$$

As $P(k,0)$ is zero, we obtain that

$$P(k+1,t) = \frac{(\lambda t)^{k+1}}{(k+1)!}e^{-\lambda t}.$$

For each t, $\{\chi_t\}$ is a Poisson random variable with intensity λt. As the intensity changes with time it follows that this cannot be a stationary random process. The expected value and the variance are

$$E[\chi_t] = \lambda t, \tag{17.18}$$
$$E[(\chi_t - \lambda t)^2] = \lambda t, \tag{17.19}$$

as follows from the formulæ in Section 15.3.2.

Suppose we know that one particle is emitted in an interval $[0, T]$. What is the probability distribution for the time the particle is emitted? This is a question about conditional probability. We formulate it as follows, for $0 \leq t \leq T$:

$$\begin{aligned} \text{Prob}(\chi_t = 1 | \chi_T = 1) &= \frac{\text{Prob}(\chi_t = 1, \text{ and } \chi_T = 1)}{\text{Prob}(\chi_T = 1)} \\ &= \frac{\text{Prob}(\chi_t = 1, \chi_T - \chi_t = 0)}{\text{Prob}(\chi_T = 1)}. \end{aligned}$$

Using the distributions obtained previously, we see that this equals

$$\text{Prob}(\chi_t = 1 | \chi_T = 1) = \frac{P(1,t)P(0,T-t)}{P(1,T)} = \frac{\lambda t e^{-\lambda t} e^{-\lambda(T-t)}}{\lambda T e^{-\lambda T}} = \frac{t}{T}.$$

This says that each time in $[0, T]$ is equally probable. The Poisson counting process is used to describe radioactive decay. If it is known that one decay was observed in a certain interval, then the time of decay is uniformly distributed over the interval. This is why it is said that the time of decay is completely random.

Next we compute the autocorrelation function $E[\chi_t \chi_s]$ using the identity

$$\begin{aligned} E[(\chi_t - \chi_s)^2] &= E[\chi_t^2 - 2\chi_t\chi_s + \chi_s^2] \\ &= E[\chi_t^2] - 2E[\chi_t\chi_s] + E[\chi_s^2]. \end{aligned}$$

From the stationary increments property and $\chi_0 = 0$, it follows that

$$E[(\chi_t - \chi_s)^2] = E[\chi_{(t-s)} - \chi_0)^2] = E[\chi_{(t-s)}^2].$$

Assume that $t \geq s$. Then

$$\begin{aligned} E[\chi_{(t-s)}^2] &= \sum_{k=0}^{\infty} k^2 \operatorname{Prob}(\chi_{(t-s)} = k) \\ &= \sum_{k=0}^{\infty} \frac{1}{k!} k^2 [(\lambda(t-s))^k e^{-\lambda(t-s)}] = \lambda(t-s) + \lambda^2(t-s)^2. \end{aligned}$$

The autocorrelation is now easily obtained:

$$E[\chi_t\chi_s] = \frac{1}{2}(E[\chi_t^2] + E[\chi_s^2] - E[(\chi_t - \chi_s)^2]) = \lambda \min(t, s) + \lambda^2 ts.$$

The Poisson counting process is also not a weak sense stationary process. Substituting $E[\chi(t)] = \lambda t$ gives

$$\operatorname{Cov}(\chi_t, \chi_s) = \lambda \min(t, s).$$

Exercises

***Exercise* 17.3.1.** Deduce from the axioms that $P(k, \Delta t) \leq [P(1, \Delta t)]^k$ for small enough Δt. Using the final form of $P(k, t)$, give an estimate, in terms of λ, for how small Δt must be taken for these estimates to hold for all $k > 1$.

***Exercise* 17.3.2.** Prove that $P(k, t)$ satisfies (17.17).

17.3.3 Poisson Arrival Process

Let χ_t be a continuous parameter, Poisson counting process. Strictly speaking, a continuous parameter Poisson process is a function of two variables (t, w), where w is a point in the sample space. Several interesting processes, with the same underlying sample space, can be built out of the counting process. We now describe the Poisson arrival process. Let $T_1(w)$ be the time the first particle arrives, and $T_2(w)$ the second arrival time and, recursively,

$T_n(w)$ the nth arrival time. This is called the Poisson arrival process. Taking differences gives a further process:

$$\begin{aligned} Z_1 &= T_1, \\ Z_2 &= T_2 - T_1, \\ &\vdots \\ Z_i &= T_i - T_{i-1}. \end{aligned}$$

The hypothesis that the original process $\{\chi_t\}$ has independent increments implies that $\{Z_i\}$ are independent random variables. They are identically distributed because the counting process has stationary increments. The original process is a function of a continuous parameter that takes integer values. The arrival process and its increments are sequences of random variables indexed by positive integers taking continuous values.

We now work out the distribution functions for these two processes. The probability that the first particle arrives *after* time t equals the probability that χ_t is zero:

$$\text{Prob}(Z_1 > t) = \text{Prob}(\chi_t = 0) = e^{-\lambda t}.$$

Hence,

$$\text{Prob}(Z_1 \leq t) = 1 - e^{-\lambda t} = \int_{-\infty}^{t} \lambda e^{-\lambda s} \chi_{[0,\infty]}(s)\, ds. \tag{17.20}$$

The density function of Z_1, hence that of Z_i for each i, is $\lambda e^{-\lambda t}\chi_{[0,\infty]}(t)$. The expected value of Z_1 is

$$E[Z_1] = \int_{0}^{\infty} t\lambda e^{-\lambda t} dt = \frac{1}{\lambda}.$$

For radioactive decay this says that the expected length of time before the first decay is $1/\lambda$, which agrees well with intuition. The more intense a radioactive source is, the less time we expect to wait for the first decay. The variance and standard deviation are

$$E[Z_1^2] = \int_{0}^{\infty} t^2\lambda e^{-\lambda t} dt = \frac{2}{\lambda^2} \Rightarrow \sigma_{Z_1}^2 = E[Z_1^2] - E[Z_1]^2 = \frac{1}{\lambda^2}.$$

The arrival time, T_n, is the sum of the differences of arrival times (i.e., $T_n = Z_1 + \cdots + Z_n$). Thus each T_n is a sum of independent, identically distributed random variables. The

$E[e^{-i\xi Z_n}]$ is

$$\hat{p}(\xi) = \int_{-\infty}^{\infty} e^{-i\xi t} \lambda e^{-\lambda t} \chi_{[0,\infty]}(t)\, dt$$
$$= \lambda \int_{0}^{\infty} e^{-t(i\xi+\lambda)} dt$$
$$= \frac{\lambda}{\lambda + i\xi}.$$

It therefore follows from Exercise 15.2.38 that $E[e^{-i\xi T_n}]$ is given by

$$E[e^{-i\xi T_n}] = E[e^{-i\xi Z_1}] \cdots E[e^{-i\xi Z_n}] = [\hat{p}(\xi)]^n.$$

Using a complex contour integral, the Fourier inversion of $[\hat{p}(\xi)]^n$ is obtained:

$$\int_{-\infty}^{\infty} \frac{\lambda^n}{(\lambda + i\xi)^n} e^{i\xi t} d\xi = \begin{cases} 0 & t < 0, \\ \frac{1}{(n-1)!} \lambda e^{-\lambda t} (\lambda t)^{n-1} & t \geq 0. \end{cases}$$

The probability distribution for T_n is therefore

$$\text{Prob}(T_n \leq t) = \int_{0}^{t} \frac{\lambda e^{-\lambda t} (\lambda t)^{n-1}}{(n-1)!} = \frac{1}{(n-1)!} \int_{0}^{\lambda t} e^{-\tau} \tau^{n-1} d\tau.$$

Recall that the Gamma function is defined by

$$\Gamma(x) = \int_{0}^{\infty} e^{-s} s^{x-1} ds.$$

That the Γ-function satisfies

$$\Gamma(n+1) = n! \qquad \text{for any } n > 0, \tag{17.21}$$

implies that

$$\lim_{t \to \infty} \text{Prob}(T_n \leq t) = 1.$$

Indeed this gives another proof of (17.21)! Since the $\{Z_i\}$ are identically distributed, independent random variables, the central limit theorem applies to show

$$\lim_{n \to \infty} \text{Prob}\left(\frac{T_n - n/\lambda}{\sqrt{n}/\lambda} \leq t \right) \to \int_{-\infty}^{t} e^{-\frac{x^2}{2}} \frac{dx}{\sqrt{2\pi}}.$$

Exercises

***Exercise* 17.3.3.** Prove that the $\{Z_n\}$ are independent, identically distributed random variables.

***Exercise* 17.3.4.** What is $E[T_n]$ for each $n \in \mathbb{N}$?

17.3.4 Fourier Coefficients for Periodic Processes*

Suppose that χ_t is a weak sense stationary random process with finite variance and correlation function r_χ. A process is said to be *mean square T-periodic* if

$$E[|\chi_{t+T} - \chi_t|^2] = 0 \qquad \text{for all } t.$$

Since χ_t has finite variance, with probability 1, the sample paths have finite square integral over the interval $[0, T]$, with probability 1:

$$E\left[\int_0^T |\chi_t|^2\,dt\right] = \left[\int_0^T E[|\chi_t|^2]\,dt\right] = Tr_\chi(0).$$

Because the process is weak sense stationary, $E[|\chi_t|^2]$ is constant. In light of these estimates, the Fourier coefficients

$$\zeta_k = \frac{1}{T}\int_0^T \chi_t e^{-\frac{2\pi ikt}{T}}\,dt, \qquad k \in \mathbb{Z}$$

are defined, with probability 1. These are *complex-valued* functions defined on the same sample space as χ_t. The integral can be understood as a limit of finite sums; using standard measure theory we can show that the $\{\zeta_k\}$ are measurable functions on the sample space. Hence $\{\zeta_k\}$ defines a complex, discrete parameter random process.

We first consider the autocorrelation function of the original process.

Proposition 17.3.1. *If χ_t is a mean square T-periodic, weak sense stationary random process, then $r_\chi(\tau)$ is T-periodic.*

Proof. We need to show that $r_\chi(\tau + T) = r_\chi(\tau)$ for any τ. The proof is a simple computation:

$$\begin{aligned} r_\chi(\tau + T) &= E[\chi_0\chi_{\tau+T}] \\ &= E[\chi_T\chi_{\tau+T}] + E[(\chi_0 - \chi_T)\chi_{\tau+T}] \\ &= r_\chi(\tau) + E[(\chi_0 - \chi_T)\chi_{\tau+T}]. \end{aligned} \tag{17.22}$$

The Cauchy-Schwarz inequality, (15.2.2) gives the estimate

$$E[(\chi_0 - \chi_T)\chi_{\tau+T}] \leq \sqrt{E[(\chi_0 - \chi_T)^2]E[(\chi_{\tau+T})^2]} = 0.$$

The right hand side is zero because χ_t is mean square T-periodic. This completes the proof of the proposition. □

Since r_χ is T-periodic and bounded by $r_\chi(0)$, it has a Fourier series expansion

$$r_\chi = \sum_{k=-\infty}^{\infty} r_k e^{\frac{2\pi i k\tau}{T}}.$$

Here equality is in the "limit in the mean" sense; see Proposition 7.3.1 and

$$r_k = \frac{1}{T}\int_0^T r_\chi(\tau) e^{-\frac{2\pi i k\tau}{T}}\, d\tau.$$

According to the Parseval formula, the coefficients satisfy

$$\sum_{k=-\infty}^{\infty} |r_k|^2 = \frac{1}{T}\int_0^T |r_\chi(\tau)|^2\, d\tau.$$

Using the definition of the autocorrelation function gives the formula

$$r_k = E[\chi_0 \zeta_k], \tag{17.23}$$

which brings us to the main result of this section:

Proposition 17.3.2. *The random variables* $\{\zeta_k\}$ *are pairwise uncorrelated (i.e.,* $E[\zeta_k \overline{\zeta_l}] = 0$ *if* $k \neq l$*) and*

$$E[|\zeta_k|^2] = r_k. \tag{17.24}$$

***Remark* 17.3.1.** It is natural to use $E[f\bar{g}]$ when working with complex-valued random variables so that this quadratic form defines an hermitian inner product.

Proof. Once again, the proof is a simple computation interchanging an expectation with integrals over $[0, T]$:

$$\begin{aligned} E[\zeta_k \overline{\zeta_l}] &= \frac{1}{T^2} E\left[\int_0^T \chi_t e^{-\frac{2\pi i k t}{T}}\, dt \int_0^T \chi_s e^{\frac{2\pi i l s}{T}}\, ds\right] \\ &= \frac{1}{T^2}\int_0^T\int_0^T \left[e^{-\frac{2\pi i k t}{T}} e^{\frac{2\pi i k s}{T}} r_\chi(t-s)\right] dt\, ds \\ &= \frac{r_k}{T}\int_0^T e^{\frac{2\pi i (l-k) t}{T}}\, dt. \end{aligned} \tag{17.25}$$

The passage from the second to the third lines is a consequence of the T-periodicity of r_χ. The proposition follows as the integral in last line of (17.25) is $\delta_{kl} T$. □

Thus a mean square T-periodic, weak sense stationary process leads to a discrete parameter process of uncorrelated variables. The final question we address is the convergence of the Fourier series

$$\sum_{k=-\infty}^{\infty} \zeta_k e^{\frac{2\pi ikt}{T}}. \tag{17.26}$$

Given that the χ_t is *mean square* periodic, it is reasonable to examine the convergence in $L^2([0,T])$.

Proposition 17.3.3. *The Fourier series in* (17.26) *converges to* χ_t *in* $L^2([0,T])$ *with probability 1 if and only if*

$$\lim_{N\to\infty} \sum_{k=-N}^{N} r_k = r_\chi(0).$$

Proof. We need to compute the expected value of

$$\int_0^T \left| \chi_t - \sum_{k=-\infty}^{\infty} \zeta_k e^{\frac{2\pi ikt}{T}} \right|^2 dt.$$

Once again, this reduces to interchanging an expectation with an ordinary integral. We begin by considering a finite partial sum:

$$\begin{aligned} E\left[\int_0^T \left| \chi_t - \sum_{k=-N}^{N} \zeta_k e^{\frac{2\pi ikt}{T}} \right|^2 dt\right] &= E\left[\int_0^T |\chi_t|^2 dt - T\sum_{k=-N}^{N} |\zeta_k|^2\right] \\ &= T(r_\chi(0) - \sum_{k=-N}^{N} r_k). \end{aligned} \tag{17.27}$$

The definition of r_χ and equation (17.24) are used to go from the first to the second line. As

$$\int_0^T \left| \chi_t - \sum_{k=-N}^{N} \zeta_k e^{\frac{2\pi ikt}{T}} \right|^2 dt$$

is a decreasing, nonnegative sequence of functions on the sample space, the limit, as N tends to infinity, can be taken inside the expectation to give:

$$E\left[\int_0^T \left| \chi_t - \sum_{k=-\infty}^{\infty} \zeta_k e^{\frac{2\pi ikt}{T}} \right|^2 dt\right] = T(r_\chi(0) - \sum_{k=-\infty}^{\infty} r_k). \tag{17.28}$$

The statement of the proposition follows easily from this equation. □

Another way to state the conclusion of the proposition is that the series in (17.26) represents χ_t, in the mean, with probability 1 if and only if r_χ is represented pointwise by its Fourier series at 0. This, in turn, depends on the regularity of $r_\chi(\tau)$ for τ near to zero.

***Remark* 17.3.2.** This material is adapted from [30], where a thorough treatment of eigenfunction expansions for random processes can be found.

Exercises

***Exercise* 17.3.5.** Prove (17.23).

***Exercise* 17.3.6.** Provide the details of the computations in (17.27).

17.3.5 White Noise*

See: A.4.5, A.4.6.

In applications it is often assumed that the noise component is modeled by *white noise*. This is a mean zero random process that is uncorrelated from time to time. The continuous parameter version of this process turns out to be rather complicated. In this section we use elementary facts about generalized functions. We begin with a discussion of the discrete parameter case.

A random process $\{\chi_n\}$, indexed by $\mathbb{Z}$, is called a *white noise* process if

$$\begin{aligned} E[\chi_n] = 0, \quad E[|\chi_n|^2] < \infty \text{ for all } n, \text{ and} \\ E[\chi_m \chi_n] = 0 \text{ if } m \neq n. \end{aligned} \tag{17.29}$$

A white noise process is simply an orthogonal collection of random variables on the sample space where the inner product is defined by $E[f\bar{g}]$. The Fourier coefficients of a weak sense stationary, mean square periodic process therefore define a (complex-valued) discrete, white noise process.

In the continuous parameter case we would like to do the same thing. White noise should be defined to be a random process, W_t, which is weak sense stationary and satisfies

$$E[W_t] = 0, \quad r_W(\tau) = E[W_t W_{t+\tau}] = \sigma^2 \delta(\tau).$$

These properties are intuitively appealing because they imply that the noise is completely uncorrelated from one time to another. On the other hand, its variance, $\sigma_W = \sqrt{r_W(0)}$, is infinite. The power spectral density is given by

$$S_W(\xi) = \int_{-\infty}^{\infty} \sigma^2 \delta(\tau) e^{-i\xi\tau} d\tau = \sigma^2.$$

Thus white noise has the same amount of power at every frequency. The problem is that there is no real-valued random process with these properties. For example, it makes no sense to ask for the value of $\text{Prob}(W_t \leq \lambda)$.

However, this concept is constantly used, so what does it mean? White noise is, in a sense, a random process whose sample paths are generalized functions. We cannot speak of the value of the process at any given time, in much the same way that the δ-function does not have a well-defined value at 0. On the other hand, if f is continuous, then

$$\int_{-\infty}^{\infty} f(t)\delta(t)\,dt = f(0)$$

makes perfect sense. Similarly, we can give a precise meaning to time averages of white noise. If f is a continuously differentiable function, then

$$W_f = \int_a^b f(t) W_t\,dt$$

makes sense as a random variable; it has the "expected" mean and variance:

$$E[W_f] = \int_a^b f(t)E[W_t]\,dt = 0 \text{ and } E[W_f^2] = \sigma^2 \int_a^b f^2(t)\,dt.$$

In a similar way ,it makes sense to pass white noise through an sufficiently smoothing, linear filter. It is much more complicated to make sense of nonlinear operations involving white noise.

The sample paths for a white noise process are usually described as the derivatives of the sample paths of an ordinary continuous time random process. Of course, the sample paths of a random process are not classically differentiable, with probability 1, so these derivatives must be interpreted as weak derivatives. We close this section by *formally* constructing a white noise process as the distributional derivative of Brownian motion.

Let $\{\chi_t : t \geq 0\}$ denote Brownian motion and recall that

$$E[\chi_t] = 0 \text{ and } E[\chi_s \chi_t] = \sigma^2 \min\{s, t\}.$$

Formally, we set $W_t = \partial_t \chi_t$. Commuting the derivative and the integral defining the expectation gives

$$E[W_t] = E[\partial_t \chi_t] = \partial_t E[\chi_t] = 0;$$

hence W_t has mean zero. To compute the variance, we again commute the derivatives and the expectation to obtain

$$E[W_t W_s] = \partial_t \partial_s E[\chi_t \chi_s] = \partial_t \partial_s \sigma^2 \min\{s, t\}.$$

The rightmost expression is well defined as the weak derivative of a function of two variables:

$$\partial_t \partial_s \sigma^2 \min\{s, t\} = 2\sigma^2 \delta(t - s). \tag{17.30}$$

To prove (17.30), we let φ be a smooth function with bounded support in $[0, \infty) \times [0, \infty)$. The weak derivative in (17.30) is *defined* by the condition

$$\int_0^\infty \int_0^\infty \partial_t \partial_s \min\{s, t\} \varphi(s, t)\, ds\, dt = \int_0^\infty \int_0^\infty \min\{s, t\} \partial_t \partial_s \varphi(s, t)\, ds\, dt,$$

for every test function φ. Writing out the integral on the right-hand side gives

$$\begin{aligned} \int_0^\infty \int_0^\infty \min\{s, t\} \partial_t \partial_s \varphi(s, t)\, ds\, dt &= \int_0^\infty \int_s^\infty s \partial_t \partial_s \varphi(s, t)\, dt\, ds + \int_0^\infty \int_t^\infty t \partial_s \partial_t \varphi(s, t)\, ds\, dt \\ &= -\int_0^\infty s \partial_s \varphi(s, s)\, ds - \int_0^\infty t \partial_t \varphi(t, t)\, dt \\ &= 2 \int_0^\infty \varphi(s, s)\, ds. \end{aligned} \tag{17.31}$$

The last line follows by integration by parts, using the bounded support of φ to eliminate the boundary terms. At least formally, this shows that the first (weak) derivative of Brownian motion is white noise.

Exercises

***Exercise* 17.3.7.** Give a detailed justification for the computations in (17.31).

***Exercise* 17.3.8.** Show that the first derivative of the Poisson counting process is also, formally, a white noise process. What do the sample paths for this process look like?

17.4 Random Inputs to Linear Systems

See: B.8.

In the analysis of linear filters, the input and output are often interpreted as the sum of a deterministic part plus noise. The noise is modeled as a random process, and therefore the output is also a random process. We often wish to understand the statistical properties of the output in terms of those of the input. In this connection it is useful to think of a continuous parameter, random process as a function of two variables $\chi(t; w)$, with w a point in the sample space, X, and t a time. In this section we use functional notation, $\chi(t)$ [or $\chi(t; w)$], for sample paths and let ν denote a probability measure on X inducing the finite-dimensional distributions of the random process χ.

The output of a shift invariant filter H with the impulse response function h, for such a random input, is again a random process on the same sample space. It is given *formally* by

$$\Upsilon(t; w) = H(\chi(t; w)) = \int_{-\infty}^{\infty} h(t-s)\chi(s; w)\, ds.$$

Such an expression is useful if it makes sense with probability 1. The statistics of the output process Υ are determined by the impulse response and the statistics of the input process. The expected value of the output of a linear system is an iterated integral:

$$\begin{aligned} E[\Upsilon(t)] &= \int_X \Upsilon(t; w)\, d\nu(w) \\ &= \int_X \int_{-\infty}^{\infty} h(t-s)\chi(s; w)\, ds\, d\nu(w). \end{aligned}$$

Under reasonable hypotheses (e.g., χ is bounded and h is integrable), the order of the two integrations can be interchanged. Interchanging the order of the integrations gives

$$\begin{aligned} E[\Upsilon(t)] &= \int_{-\infty}^{\infty} \int_X h(t-s)\chi(s; w)\, d\nu(w)\, ds \\ &= \int_{-\infty}^{\infty} h(t-s) E[\chi(s)]\, ds. \end{aligned}$$

The expected output of a linear filter applied to a random process is the result of applying the linear filter to the expected value of the random process. Some care is necessary, even at this stage. If $E[\chi(s)]$ is a constant μ_χ, then the expected value of Υ is

$$E[\Upsilon(t)] = \mu_\chi \int_{-\infty}^{\infty} h(t-s)\, ds = \mu_\chi \int_{-\infty}^{\infty} h(s)\, ds = \mu_\chi \hat{h}(0).$$

For this to make sense, we should assume that $\int |h(s)| < \infty$. If the input random process χ is stationary, then the output process $H\chi$ is as well.

Exercise

***Exercise* 17.4.1.** Suppose that χ is a stationary random process and H is a linear shift invariant filter for which $H\chi$ makes sense (with probability 1). Show that $H\chi$ is also a stationary process.

17.4.1 The Autocorrelation of the Output

To analyze shift invariant, linear systems we used the Fourier transform. In this case, it cannot be used directly since a random process does not have a Fourier transform in the ordinary sense. Observe that for any positive number k

$$E[\int_{-\infty}^{\infty} |\chi(s)|^k \, ds] = \int_{-\infty}^{\infty} E[|\chi(s)|^k] \, ds.$$

For a stationary process, these integrals diverge unless $E[|\chi(s)|^k] \equiv 0$, which would imply that the process equals zero, with probability 1! The sample paths of a nontrivial, stationary process do have Fourier transforms (which are also functions) with probability 1. To get around this difficulty, we consider the autocorrelation function. It turns out that the autocorrelation function for a stationary process is frequently square integrable.

For a continuous time process, the autocorrelation function R_χ is defined by

$$R_\chi(t_1, t_2) = E[\chi(t_1)\chi(t_2)].$$

Recall that the process is weak sense stationary if there is a function r_χ so that the autocorrelation function is given by

$$R_\chi(t_1, t_2) = r_\chi(t_1 - t_2).$$

Given two random processes, χ, Υ, on the same underlying probability space, the *cross-correlation function* is defined to be

$$R_{\chi,\Upsilon}(t_1, t_2) = E[\chi(t_1)\Upsilon(t_2)].$$

For two stationary processes, $R_{\chi,\Upsilon}$ is only a function of $t_2 - t_1$. If we define

$$r_{\chi,\Upsilon}(\tau) = E[\chi(t)\Upsilon(t + \tau)],$$

then

$$R_{\chi,\Upsilon}(t_1, t_2) = r_{\chi,\Upsilon}(t_1 - t_2).$$

Now suppose that H is a linear shift invariant filter, with impulse response h, and that χ is a random process for which $H\chi$ makes sense, with probability 1. The autocorrelation of the output process is

$$\begin{aligned} R_{H\chi}(t_1, t_2) &= E[H\chi(t_1)H\chi(t_2)] \\ &= E[\int_{-\infty}^{\infty} h(t_1 - s_1)\chi(s_1)\, ds_1 \int_{-\infty}^{\infty} h(t_2 - s_2)\chi(s_2)\, ds_2]. \end{aligned} \tag{17.32}$$

The expected value is itself an integral; interchanging the order of the integrations leads to

$$\begin{aligned} R_{H\chi}(t_1, t_2) &= E[\int_{-\infty}^{\infty} h(t_1 - s_1)\chi(s_1; w)\, ds_1 \int_{-\infty}^{\infty} h(t_2 - s_2)\chi(s_2; w)\, ds_2] \\ &= \int_{-\infty}^{\infty}\int_{-\infty}^{\infty} h(t_1 - s_1)h(t_2 - s_2)E[\chi(s_1; w)\chi(s_2; w)]\, ds_1\, ds_2 \\ &= \int_{-\infty}^{\infty}\int_{-\infty}^{\infty} h(t_1 - s_1)h(t_2 - s_2)R_\chi(s_1, s_2)\, ds_1\, ds_2 \\ &= [h^{(2)} * R_\chi](t_1, t_2), \end{aligned}$$

where $h^{(2)}(x, y) \stackrel{d}{=} h(x)h(y)$. Hence, $R_{H\chi}$ is expressible as a two-dimensional convolution with R_χ.

For the case of a weak sense stationary process, the result is simpler. Recall that

$$R_\chi(t_1, t_2) = r_\chi(t_1 - t_2).$$

Letting $\tau_i = t_i - s_i,\ i = 1, 2$, we obtain

$$R_{H\chi}(t_1, t_2) = \int_{-\infty}^{\infty}\int_{-\infty}^{\infty} h(\tau_1)h(\tau_2)r_\chi(\tau_1 - \tau_2 + t_2 - t_1)\, d\tau_1\, d\tau_2.$$

Thus the output is also weak sense stationary with

$$r_{H\chi}(\tau) = \int_{-\infty}^{\infty}\int_{-\infty}^{\infty} h(s + t)h(t)r_\chi(s - \tau)\, dt\, ds.$$

In Proposition 17.2.1, the properties of the power spectral density of a stationary random process are enumerated. Using the formula for $r_{H\chi}$, we compute the spectral power density of the output in terms of the spectral power density of the input, obtaining

$$S_{H\chi}(\xi) = |\hat{h}(\xi)|^2 S_\chi(\xi). \tag{17.33}$$

This is consistent with the "determinate" case, for if x is a finite energy signal, with $y = Hx$ and $\hat{y} = \hat{h}\hat{x}$, we have

$$|\hat{y}(\xi)|^2 = |\hat{h}(\xi)|^2|\hat{x}(\xi)|^2. \tag{17.34}$$

Note that the total power of the input is given by

$$E[\chi^2] = r_\chi(0) = \frac{1}{2\pi}\int_{-\infty}^{\infty} S_\chi(\xi)\, d\xi,$$

which we compare with the power in the output,

$$E[(H\chi)^2] = r_{H\chi}(0) = \frac{1}{2\pi}\int_{-\infty}^{\infty} |\hat{h}(\xi)|^2 S_\chi(\xi)\, d\xi.$$

The input and output variances are given by

$$\begin{aligned} \sigma_\chi^2 &= \frac{1}{2\pi}\int_{-\infty}^{\infty} S_\chi(\xi)\, d\xi - \mu_\chi^2 \\ \frac{1}{2\pi}\sigma_{H\chi}^2 &= \int_{-\infty}^{\infty} |\hat{h}(\xi)|^2 S_\chi(\xi)\, d\xi - |\hat{h}(0)|^2\mu_\chi^2. \end{aligned} \tag{17.35}$$

To compute the power or variance of the output requires a knowledge of both the spectral density function S_χ of the process as well as the transfer function of the filter.

Exercises

***Exercise* 17.4.2.** If χ and Υ are stationary processes, show that $R_{\chi,\Upsilon}(t_1, t_2)$ only depends on $t_2 - t_1$.

***Exercise* 17.4.3.** Derive (17.33).

17.4.2 Thermal or Johnson Noise

Current is the flow of electrons through a conductor. The electrons can be thought of as discrete particles that, at normal room temperature, move in a random way through the conductor. Even with no applied voltage, the randomness of this motion produces fluctuations in the voltage measured across the conductor. The thermal motion of electrons produces noise, known as *Johnson noise*, in essentially any electrical circuit. While not an important source of noise in CT imaging, Johnson noise is the main source of noise in MRI. The intensity of this noise is related to the impedance of the electrical circuit. To understand this dependence, we examine the result of using a white noise voltage source as the input to the simple electrical circuit shown in Figure 17.1.

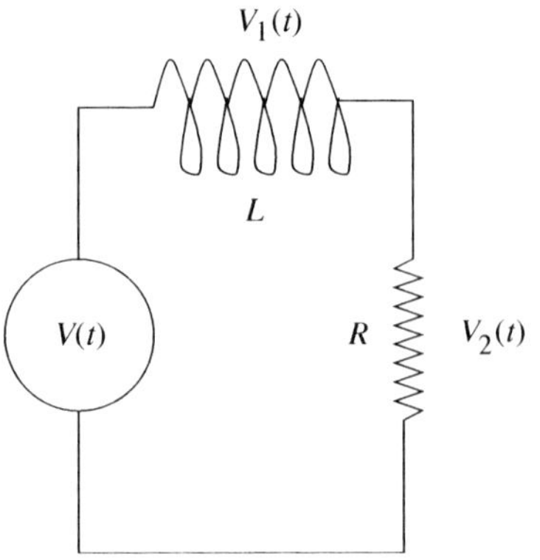

Figure 17.1. An RL circuit.

Thermodynamic considerations show that the expected power through the circuit, due to the thermal fluctuations of the electrons, is

$$E[\frac{LI^2}{2}] = \frac{kT}{2}, \tag{17.36}$$

where T is the absolute temperature and k is Boltzmann's constant. The voltage source $V(t)$ is a white noise process with intensity σ. This means that the spectral density of the noise is constant with

$$S_V(\xi) = \sigma^2.$$

Using Kirchoff's laws, it is not difficult to show that the transfer function for the current through this circuit is

$$\hat{h}(\xi) = \frac{1}{R + iL\xi};$$

see [8]. Since the input is a random process, the output current, $I(t)$, is also a random process. According to (17.33), its spectral density function is

$$S_I(\xi) = \sigma^2 \frac{1}{R^2 + (L\xi)^2}.$$

This allows the computation of $E[I^2]$,

$$E[I^2] = \frac{1}{2\pi} \int_{-\infty}^{\infty} \frac{\sigma^2 \, d\xi}{R^2 + (L\xi)^2} = \frac{\sigma^2}{2RL}. \tag{17.37}$$

Comparing this result with the (17.36) gives the intensity of the white noise process:

$$\sigma^2 = 2RkT.$$

This result and its generalizations are also known as Nyquist's theorem. At room temperature (about 300°K) with a resistance $R = 10^6$ Ohms, the intensity of the Johnson noise process is

$$\sigma^2 \approx 4 \times 10^{-15} (\text{volt})^2 \text{s}.$$

Of course, in a real physical system the spectrum of the thermal noise cannot be flat, for this would imply that the noise process contains an infinite amount of energy. It is an empirical fact that the spectrum is essentially flat up to a fairly high frequency. Instead Johnson noise is describes as a random process, χ, with

$$S_\chi(\xi) = \sigma^2 \chi_{[0,B]}(|\xi|),$$

or briefly, as *bandlimited white noise*. The integral over $\mathbb{R}$ in (17.37) is then replaced by an integral from $-B$ to B. If B is reasonably large, then the result is nearly the same. The total power of the (bandlimited) Johnson noise is therefore

$$S_{\text{tot}} \approx \frac{RkTB}{\pi}.$$

In many practical applications, the spectrum of the noise is bandlimited because the data are bandlimited. The formula for S_{tot} shows that any attempt to increase the bandwidth of the data increases the total power of the Johnson noise commensurately.

***Remark* 17.4.1.** Our treatment of Johnson noise is adapted from [28].

Exercise

***Exercise* 17.4.4.** If the input is B-bandlimited white noise, what is the exact formula for $E[I^2]$?

17.4.3 Optimal Filters

As a final application of these ideas, we consider the design of a noise reducing filter that is "optimal" in some sense. Let the signal x be modeled as

$$x(t) = s(t) + n(t),$$

where s stands for the signal and n, the noise. The signal and noise are assumed to be uncorrelated, weak sense stationary, finite variance random processes:

$$E[s(t_1)n(t_2)] = 0 \qquad \text{for all } t_1 \text{ and } t_2. \tag{17.38}$$

We would like to design a linear, shift invariant filter H that minimizes the error in the detected signal in the sense that expected mean square error, $E[|s - Hx|^2(t)]$, is minimized. In this case both s and Hx are weak sense stationary processes, and therefore the value of the error is independent of t.

The solution of the minimization problem is characterized by an orthogonality condition,

$$E[(Hx - s)(t_1)x(t_2)] = 0, \text{ for all } t_1, t_2. \tag{17.39}$$

We give a formal derivation of this condition. Suppose that h is the impulse response of an optimal filter and that k is an "arbitrary" impulse response. The optimality condition implies that

$$\frac{d}{d\lambda} E[|(s - (h + \lambda k) * x)(t_1)|^2]\Big|_{\lambda=0} = 0, \text{ for any } t_1.$$

Expanding the square and differentiating in t gives

$$E[((h * x - s)k * x)(t_1)] = 0 \text{ for any } k \text{ and } t_1. \tag{17.40}$$

Given that the various convolutions make sense, the derivation up to this point has been fairly rigorous. Choose a smooth, nonnegative function φ with bounded support and total integral 1. For any t_2, taking

$$k_\epsilon(t) = \frac{1}{\epsilon}\varphi\left(\frac{t - (t_1 - t_2)}{\epsilon}\right)$$

gives a sequence of very smooth test functions that "converge" to $\delta(t - (t_1 - t_2))$. Assuming that the limit makes sense, (17.40) implies that

$$0 = \lim_{\epsilon \downarrow 0} E[((h * x - s)k_\epsilon * x)(t_1)] = E[(h * x - s)(t_1)x(t_2)], \tag{17.41}$$

which is the desired orthogonality condition. By using finite sums to approximate $k * x$, the condition in (17.40) is easily deduced from (17.39).

Using (17.38), the orthogonality condition can be rewritten in terms of s and n as

$$\begin{aligned}
0 &= E[(h * s(t_1) + h * n(t_1) - s(t_1))(s(t_2) + n(t_2))] \\
&= E[h * s(t_1)s(t_2) + h * n(t_1)n(t_2) - s(t_1)s(t_2)] \\
&= \int_{-\infty}^{\infty} E[s(\tau)s(t_2)]h(t_1 - \tau)\,d\tau + \int_{-\infty}^{\infty} E[n(\tau)n(t_2)]h(t_1 - \tau)\,d\tau - E[s(t_1)s(t_2)].
\end{aligned}$$

Let r_s and r_n be the autocorrelation functions for the signal and noise, respectively. Letting $t = \tau - t_2$ and $\sigma = t_1 - t_2$ gives

$$\int_{-\infty}^{\infty} r_s(t)h(\sigma - t)\,dt + \int_{-\infty}^{\infty} r_n(t)h(\sigma - t)\,dt = r_s(\sigma).$$

This is a convolution equation, so taking the Fourier transform gives the relation

$$S_s(\xi)\hat{h}(\xi) + S_n(\xi)\hat{h}(\xi) = S_s(\xi).$$

Recalling that the spectral density function is nonnegative, we may divide to obtain the transfer function for the optimal filter:

$$\hat{h}(\xi) = \frac{S_s(\xi)}{S_s(\xi) + S_n(\xi)} = \frac{1}{1 + S_n(\xi)/S_s(\xi)} \approx \begin{cases} 1 & S_n(\xi)/S_s(\xi) << 1, \\ 0 & S_n(\xi)/S_s(\xi) >> 1. \end{cases}$$

This shows how we can use the power spectrum of the noise and a probabilistic description of the signal to design an optimal filter. This example is called the *Wiener filter*; it is a very simple example of an optimal filter. There are many variants on this approach using different classes of filters and different kinds of random processes. Kalman and Bucy found a different approach to the problem of optimal filtering. More complete treatments of this subject can be found in [30] or [16].

Exercises

***Exercise* 17.4.5.** Prove that if H defines the optimal filter, then

$$E[|Hx|^2] = E[sHx]. \tag{17.42}$$

***Exercise* 17.4.6.** Using (17.42), compute the expected mean squared error for the optimal filter, H:

$$E[|Hx - s|^2] = r_s(0) - \int_{-\infty}^{\infty} h(t) r_s(t)\, dt. \tag{17.43}$$

***Exercise* 17.4.7.** Using the Parseval formula and (17.43) prove that

$$E[|Hx - s|^2] = \frac{1}{2\pi} \int_{-\infty}^{\infty} \frac{S_s(\xi) S_n(\xi)\, d\xi}{S_s(\xi) + S_n(\xi)}.$$

***Exercise* 17.4.8.** If the signal and noise have nonzero cross-correlation of the form

$$r_{sn}(\tau) = E[s(t)n(t+\tau)],$$

show that the orthogonality condition for the optimal filter becomes

$$\int_{-\infty}^{\infty} r_s(t) h(\sigma - t)\, dt + \int_{-\infty}^{\infty} r_n(t) h(\sigma - t)\, dt = r_s(\sigma) + r_{sn}(\sigma).$$

Find the transfer function for the optimal filter in this case.

17.5 Noise in the Continuum Model for Filtered Back-Projection

Using the techniques introduced in this chapter, we study the propagation of noise through the continuum model for the filtered back-projection algorithm. In Section 16.2 we did analogous computations for the finite filtered back-projection algorithm. Computations in a continuum model are often easier and provide a good cross check for the earlier discrete computations. Let f be a function supported in the disk of radius L. We begin our analysis by assuming that $\mathcal{R}f$ can be measured for all $(t, \omega) \in [-L, L] \times S^1$ and that f is approximately reconstructed using filtered back-projection,

$$f_\phi(x, y) = \frac{1}{2\pi} \int_0^\pi \int_{-L}^{L} \mathscr{R} f(\langle (x, y), \omega\rangle - s, \omega)\phi(s)\, ds\, d\omega. \tag{17.44}$$

To simplify the notation in this section, we omit explicit reference to the beam width function. The results in Section 12.2.2 show that this does not reduce the generality of our results.

The uncertainty in the measurements can be modeled in two different ways. On the one hand, we can imagine that f itself is corrupted by noise, so that the measurements are of the form $\mathscr{R}(f + n_i)$. Here n_i is a random process, represented by functions on $\mathbb{R}^2$, that models the uncertainty in the input f. On the other hand, f can be considered to be determinate, but the measurements themselves are corrupted by noise. In this case the measurements are modeled as $\mathscr{R} f + n_m$. Here n_m is a random process, represented by functions on $\mathbb{R} \times S^1$, that models the uncertainty in the measurements. Of course, the real situation involves a combination of these effects. We analyze these sources of error, assuming that f itself is zero.

The first case is easy to analyze, as we can simply apply the results in Section 17.4. The map from f to f_ϕ is a shift invariant linear filter with MTF given by

$$\hat{\Psi}(\xi) = \hat{\psi}(\|\xi\|),$$

where

$$\hat{\phi}(r) = |r|\hat{\psi}(r).$$

Assume that $n_i(x, y)$ is a stationary random process with mean zero for each (x, y) in $\mathbb{R}^2$. Denote the autocorrelation function by

$$r_i(x, y) = E[n_i(0, 0)n_i(x, y)].$$

Its Fourier transform $S_i(\xi)$ is the power spectral density in the input noise process. The power spectral density of the output is given by (17.33),

$$S_o(\xi) = S_i(\xi)|\hat{\Psi}(\xi)|^2.$$

The total noise power in the output is therefore

$$S_{\text{tot}} = \frac{1}{[2\pi]^2} \int_0^{2\pi} \int_0^{\infty} S_i(r\omega)|\hat{\psi}(r)|^2 r\, dr\, d\omega. \tag{17.45}$$

A useful, though not too realistic example, is to assume that n_i is a white noise process with $S_i(\xi) = \sigma^2$. If ϕ is the Shepp-Logan filter with

$$\hat{\phi}(r) = |r| \left|\operatorname{sinc}\left(\frac{dr}{2}\right)\right|^3,$$

then the total noise in the output, which equals the variance $r_o(0)$, is

$$S_{\text{tot}} = C\frac{\sigma^2}{d^2}, \tag{17.46}$$

where C is a positive constant. The reconstruction algorithm amplifies the uncertainty in f by a factor proportional to d^{-2}.

The other possibility is that the noise is measurement noise. In this case n_m is a function on $\mathbb{R} \times S^1$. Using an angular coordinate, we can think of n_m as a function of (t, θ) that is 2π-periodic. We assume that the noise process is weak sense stationary, so that

$$E[n_m(t_1, \theta_1)n_m(t_2, \theta_2)] = r_m(t_1 - t_2, \theta_1 - \theta_2),$$

where $r_m(\tau, \theta)$ is also 2π-periodic in θ. The filtered back-projection algorithm applied to the noise gives

$$n_{m\phi}(x, y) = \frac{1}{4\pi}\int_0^{\pi}\int_{-\infty}^{\infty} n_m(\langle(x, y), \boldsymbol{\omega}\rangle - s, \boldsymbol{\omega})\phi(s)\, ds\, d\boldsymbol{\omega}.$$

For convenience, we have replaced the finite limits of integration with infinite limits. Because the noise is bounded and the Shepp-Logan filter is absolutely integrable, this does not significantly affect the outcome.

The autocorrelation of the noise in the output is

$$E[n_{m\phi}(x, y)n_{m\phi}(0, 0)] =$$
$$\frac{1}{[2\pi]^2}\int_0^{\pi}\int_0^{\pi}\int_{-\infty}^{\infty}\int_{-\infty}^{\infty} r_m(\langle(x, y), \boldsymbol{\omega}(\theta_1)\rangle + s_2 - s_1, \theta_1 - \theta_2)\phi(s_1)\phi(s_2)\, ds_1\, ds_2\, d\theta_1\, d\theta_2. \tag{17.47}$$

Without further information this expression is difficult to evaluate. We make the hypothesis that the measurement noise is white (i.e., the errors in one ray are uncorrelated with the errors in another). This means that

$$r_m(\tau, \theta) = \sigma^2\delta(\tau)\delta(\theta),$$

where, strictly speaking θ should be understood in this formula as $\theta \bmod 2\pi$. That the errors from ray to ray are weakly correlated is not an unreasonable hypothesis; however, the analysis in Section 16.1.2, particularly equation (16.7), shows that the variance is unlikely to be constant. These assumptions give

$$E[n_{m\phi}(x, y)n_{m\phi}(0, 0)] = \frac{\sigma^2}{[2\pi]^2}\int_0^{\pi}\int_{-\infty}^{\infty} \phi(s_1)\phi(s_1 - \langle(x, y), \boldsymbol{\omega}(\theta_1)\rangle)\, ds_1\, d\theta_1. \tag{17.48}$$

Because ϕ is an even function and $\hat{\phi}(0) = 0$, this can be reexpressed as a two-dimensional inverse Fourier transform

$$E[n_{m\phi}(x, y)n_{m\phi}(0, 0)] = \frac{\sigma^2}{[2\pi][4\pi]^2} \int_0^{\pi} \int_{-\infty}^{\infty} \frac{|\hat{\phi}(r)|^2}{|r|} e^{ir\langle(x,y),\omega\rangle} |r| \, dr \, d\omega. \tag{17.49}$$

The power spectral density in the output is therefore

$$S_o(\xi) = \frac{\sigma^2}{8\pi} \frac{|\hat{\phi}(\|\xi\|)|^2}{\|\xi\|}.$$

Using the same filter as before, the total noise power in the output is

$$S_{\text{tot}} = C' \frac{\sigma^2}{d^3}, \tag{17.50}$$

where again C' is a positive constant. The total noise power in the measurements is amplified by a factor proportional to d^{-3}. Recalling that the resolution is proportional to d, it follows that, as the resolution increases, errors in measurement have a much greater affect on the reconstructed image than uncertainty in f itself. In either case the noise is assumed to have mean zero so a nonzero f would only change the variance computations by a bounded function of d. As the number of pixels in the reconstruction grids is $O(d^{-2})$, this result agrees with equation (16.16), where we computed the noise power per pixel. Note finally that with either sort of noise, the variance tends to infinity as d goes to zero. This substantiates our claim that noise necessitates the use of regularization in the reconstruction process. This discussion is adapted in part from [69].

Exercise

***Exercise* 17.5.1.** Repeat the computations in this section with f a nonzero input.

17.6 Conclusion

We have now completed our journey, starting with very simple mathematical models of very simple physical systems and ending with a rather comprehensive model for the measurement and reconstruction processes in x-ray tomography. Along the way we have introduced many of the mathematical techniques that are applied in the full spectrum of imaging modalities. Of necessity, much has been omitted and some topics have only been outlined. Nonetheless, it is my hope that a good command of the material in this book be adequate preparation for reading much of the current research literature in medical imaging. Let me know what *you* think!

Appendix A

Background Material

In applied subjects, mathematics needs to be appreciated in three rather distinct ways: (1) in the abstract context of perfect and complete knowledge generally employed in mathematics itself; (2) in a less abstract context of fully specified, but incompletely known functions—this is the world of mathematical approximation; and (3) in a realistic context of partially known functions and noisy, approximate data, which is closer to the real world of measurements. With these different perspectives in mind, we discuss some of the mathematical concepts underlying image reconstruction and signal processing. The bulk of this material is usually presented in undergraduate courses in linear algebra, analysis, and functional analysis. Instead of a giving the usual development, which emphasizes mathematical rigor and proof techniques, we present this material from an engineering perspective. Many of the results are proved in exercises, and examples are given to illustrate general phenomena. This material is intended to provide background material and recast familiar material in a more applied framework; it should be referred to as needed.

A.1 Numbers

We begin by discussing numbers, beginning with the abstract concepts of numbers and their arithmetic properties. Representations of numbers are then considered, leading to a comparison between abstract numbers and the way numbers are actually used in computation.

A.1.1 Integers

Mathematicians think of numbers as a set that has two operations, addition and multiplication, that satisfy certain properties. The mathematical discussion of this subject always begins with the integers. We denote the set of integers by $\mathbb{Z}$ and the set of positive integers (the whole or natural numbers) by $\mathbb{N}$. There are two operations defined on the integers: addition, $+$, and multiplication, $\times$. Associated to each of these operations is a special number: For addition that number is 0 it is *defined* by the property

$$n + 0 = 0 + n = n \qquad \text{for every integer } n.$$

For multiplication that number is 1 and it is *defined* by the property

$$n \times 1 = 1 \times n = n \qquad \text{for every integer } n.$$

The important axiomatic properties of addition and multiplication are as follows:

COMMUTATIVE LAW: $n + m = m + n, \quad n \times m = m \times n$, for every $m, n \in \mathbb{Z}$

ASSOCIATIVE LAW: $(n + m) + p = n + (m + p), \quad (n \times m) \times p = n \times (m \times p)$, for every $m, n, p \in \mathbb{Z}$

DISTRIBUTIVE LAW: $(m + n) \times p = m \times p + n \times p$ for every $m, n, p \in \mathbb{Z}$.

These rules are familiar from grade school, and we use them all the time when we do computations by hand.

In mathematics numbers are treated in an axiomatic way. Neither a *representation* of numbers nor an *algorithm* to perform addition and multiplication has yet to be considered. We normally use the decimal representation, when working with numbers "by hand." To define a representation of numbers, we first require some special symbols; for the decimal representation we use the symbols $0, 1, 2, 3, 4, 5, 6, 7, 8, 9$, which represent the numbers zero through nine. We also introduce an additional symbol, $-$, to indicate that a number is smaller than zero. The decimal representation of a positive integer is a string of numbers

$$a_m a_{m-1} \dots a_1 a_0 \text{ where } 0 \le a_j \le 9, \text{ for } j = 0, \dots, m.$$

What does this string of numbers mean? By definition,

$$a_m a_{m-1} \dots a_1 a_0 \stackrel{d}{=} \sum_{j=0}^{m} a_j 10^j.$$

What appears on the right-hand side of this formula is a mathematical number; what appears on the left is its decimal or base 10 representation. A negative number is represented by prepending the minus sign $-a_m \dots a_0$. For each positive integer $k > 1$ there is an analogous representation for integers called the base-k or k-ary expansion.

Algorithms for addition and multiplication require addition and multiplication tables. To do addition in base 10, we need to *know* how to do the sums $a + b$ for $0 \le a, b \le 9$; then we use "carrying" to add larger numbers. To do multiplication in base 10, we need to *know* how to do the products $a \times b$ for $0 \le a, b \le 9$. The normal human mind has no difficulty remembering these base 10 addition and multiplication tables. In the early days of computing this was a large burden to place on a machine. It was found to be much easier to build a machine that uses a base 2 or binary representation to store and manipulate numbers. In the binary representation an integer is represented by a string of zeros and ones. By definition,

$$b_m b_{m-1} \dots b_1 b_0 = \sum_{j=0}^{m} b_j 2^j, \quad \text{where } b_j \in \{0, 1\} \text{ for } j = 0, \dots, m.$$

The analogous algorithms for adding and multiplying in base 2 only require a knowledge of $a + b, a \times b$ for $0 \leq a, b \leq 1$, which is a lot less to remember. On the other hand, you need to do a lot more carrying to add or multiply numbers of a given size.

Even in this very simple example we see that there is a trade-off in efficiency of computation between the amount of memory utilized and the number of steps needed to do a certain computation. There is a second reason why binary representations are preferred for machine computation. For a machine to evaluate a binary digit, it only needs to distinguish between two possible states. This is easy to do, even with inexpensive hardware. To evaluate a decimal digit, a machine would need to distinguish between 10 different possible states. This would require a much more expensive machine. Finally, there is the issue of tradition. It might be cheaper and more efficient to use base 3 for machine computation, but the mere fact that so many base 2 machines already exist make it highly unlikely that we will soon have to learn to do arithmetic in base 3.

Because we have a conceptual basis for numbers, there is no limit to size of the numbers we can work with. Could a given number N be the largest number we can "handle?" It would be hard to see why, because if we could handle N, then we could certainly $N + 1$. In fact, this is essentially the mathematical proof that there is no largest integer. The same cannot be said of a normally programmed computer; it has numbers of maximum and minimum size with which it can work.

Exercises

***Exercise* A.1.1.** Write algorithms to do addition and multiplication using the decimal representation of numbers.

***Exercise* A.1.2.** Adding the symbols A, B, C, D, E, F to represent the decimal numbers 10, 11, 12, 13, 14, 15 leads to the base 16 or *hexadecimal* representation of numbers. Work out the relationship between the binary and hexadecimal representations. Write out the addition and multiplication tables in hexadecimal.

A.1.2 Rational Numbers

The addition operation also has an inverse operation which we call subtraction: Given a number n, there is a number $-n$ that has the property $n + (-n) = 0$. We are so used to this that it is difficult to see this as a "property," but note that, if we are only permitted to use integers, then the multiplication operation does not have an inverse. This can be thought of in terms of solving equations: Any equation of the form

$$x + m = n,$$

where $m, n \in \mathbb{Z}$, has an integer solution $x = n - m$. On the other hand, for many choices of $m, n \in \mathbb{Z}$ the equation

$$n \times x = m \tag{A.1}$$

does not have an integer solution.

Again we learned in grade school how to handle this problem: We introduce *fractions* and then (A.1) has the solution

$$x = \frac{m}{n}.$$

This is just a *symbolic* formula, and its meaning is a good deal more subtle than the meaning of $x = n - m$. First, if $n = 0$, then it means nothing. If $n \neq 0$ and p is another nonzero integer, then the solution of the equation

$$p \times n \times x = p \times m \tag{A.2}$$

is the same as the solution to (A.1). This means that the *number* represented by the symbol $\frac{p \times m}{p \times n}$ is the same as the number represented by the symbol $\frac{m}{n}$. We now introduce rational numbers, $\mathbb{Q}$, as the set of symbols

$$\{\frac{m}{n} : m, n \in \mathbb{Z}\},$$

with the understanding that

1. The denominator n is nonzero and
2. As numbers
$$\frac{m}{n} = \frac{p}{q}$$
if
$$m \times q = p \times n. \tag{A.3}$$

We have defined the set of rational numbers and now have to define the operations of addition and multiplication on them. Thus far, all we know is how to add and multiply integers. Our definitions for addition and multiplication of rational numbers have to be given in terms of these operations. Multiplication is relatively easy:

$$\frac{m}{n} \times \frac{p}{q} \overset{d}{=} \frac{m \times p}{n \times q}.$$

To define addition we use the familiar concept of a common denominator and set

$$\frac{m}{n} + \frac{p}{q} \overset{d}{=} \frac{m \times q + n \times p}{n \times q}. \tag{A.4}$$

The formula only involves operations that we have already defined, though it is not immediately obvious that this is actually an operation on *numbers* and not merely an operation on *symbols*. Equation (A.1) can now be solved for any $m, n \in \mathbb{Z}$ as long as $n \neq 0$. In fact, we get a little more for our effort; the equations

$$p \times x = q$$

can be solved for any rational numbers q and $p \neq 0$.

There are two different ways to represent rational numbers: (1) as fractions or (2) as k-ary expansions analogous to those used for integers. Decimal representations of the form

$$a_m \dots a_0 a_{-1} \dots a_{-n} \stackrel{d}{=} \sum_{j=-n}^{m} a_j 10^{-j}, \text{ where } 0 \le a_j \le 9$$

represent rational numbers. It is easy to see that only fractions of the form

$$\frac{n}{10^k} \text{ for } n, k \in \mathbb{N},$$

have such a finite decimal representation. For some purposes the representation as fractions is more useful; it is certainly more efficient. For example, using a fraction, we have an exact representation of the number $1/3$; using long division, we find that

$$\frac{1}{3} = \sum_{j=1}^{\infty} \frac{3}{10^j}.$$

In other words, to *exactly* represent $1/3$ as a decimal requires infinitely many decimal places. Thus far we have not even defined infinite sums, but from the engineering point of view it is clear what this means.

Because the representation as fractions is not unique and because of the need to find common denominators for addition, fractions are not well adapted to machine computation. In a computer rational numbers are represented as strings of zeros and ones. Such a string of zeros and ones is called a *binary string*. Depending on the application, different numbers can be assigned to a given binary string. One way to assign a number to a binary string with $2N + 2$ entries or *bits* is to set

$$a_{N+1} a_N \dots a_{-N} \stackrel{d}{=} (-1)^{a_{N+1}} 2^N \sum_{j=-N}^{N} a_j 2^j.$$

This called a *fixed-point* representation. With this choice, the spacing between consecutive numbers is 1 and the maximum and minimum numbers that can be represented are $\pm(2^{2N+1} - 1)$. This allows the representation of large numbers but sacrifices accuracy. If we knew in advance that all our numbers would lie between -1 and $+1$ then we could use the same $2N + 2$ bits to get more accurate representations for a smaller range of numbers by instead assigning the number

$$(-1)^{a_{N+1}} \frac{1}{2^N} \sum_{j=-N}^{N} a_j 2^j$$

to this binary string. Here the minimum spacing between numbers is 2^{-2N}.

Floating-point numbers represents a compromise between these two extremes. The string of binary digits is divided into two parts, an exponent and a fractional part. Writing the string as $b_s e_s e_0 \dots e_m f_1 \dots f_n$, with $m + n = 2N$, the corresponding number is

$$(-1)^{b_s} 2^{\left[(-1)^{e_s} \sum_{j=0}^{m} e_j 2^j\right]} \sum_{k=1}^{n} f_k 2^{-k}.$$

Using a floating-point representation, we can represent a much larger range of numbers. If, for example, we let $m = n = N$, then with $2N + 2$ bits we can represent numbers between $\pm 2^{2^N}$. The accuracy of the representation is *proportional* to the size of the number. For numbers between 2^{k-1} and 2^k the minimum spacing is 2^{k-N}. In applications this is a reasonable choice to make. Suppose a number x is the result of a measurement and its value is determined within Δx. The number Δx is called the *absolute error*; usually it is not a very interesting number. The ratio $\frac{\Delta x}{x}$ is more useful; it is is called the *relative error*. In a floating-point representation the relative accuracy of the representation is constant throughout the range of representable numbers. On the other hand, it places subtle constraints on the kinds of computations that can accurately be done. For example, subtracting numbers of vastly different sizes does not usually give a meaningful result.

Since we only have finitely many digits, computations done in a computer are essentially never exact. It is therefore very important to use algorithms that are not sensitive to repeatedly making small errors of approximation. In image reconstruction this is an important issue as the number of computations used to reconstruct a single image is usually in the millions. For a thorough discussion of treatment of numbers in machine computation, see [127].

Exercises

***Exercise* A.1.3.** Show that the condition in (A.3) is the correct condition to capture the elementary concept that two fractions represent the same number.

***Exercise* A.1.4.** Show that formula (A.4) defines an operation on rational numbers. That is, if $\frac{m}{n} = \frac{m'}{n'}$ and $\frac{p}{q} = \frac{p'}{q'}$, then

$$\frac{m \times q + n \times p}{n \times q} = \frac{m' \times q' + n' \times p'}{n' \times q'}$$

as rational numbers.

***Exercise* A.1.5.** Find the exact binary representation of $1/3$.

***Exercise* A.1.6.** What would it mean to represent a number in base 1? What numbers can be represented this way? Find as many problems with base 1 as you can. (Thanks to Dr. Fred Villars for suggesting this question.)

***Exercise* A.1.7.** Describe binary algorithms for addition, subtraction, multiplication, and division.

A.1.3 Real Numbers

In practice we can never use anything beyond rational numbers; indeed for machine computation we have at most a finite collection of numbers at our disposal. We could take the attitude that there is no point in considering numbers beyond rational numbers. Some people do, but it vastly limits the mathematical tools at our disposal. From a mathematical perspective, the rational numbers are inadequate. For example, there is no rational number solving the equation

$$x^2 = 2.$$

In other words, there are "holes" in the rational numbers. Calculus relies on the concept of a *continuum*, so it is necessary to fill these holes. It is well beyond the scope of this book to give an axiomatic development for the real numbers. Instead we assume that the real numbers exist and describe the essential difference between the real numbers and the rational numbers: The real numbers are *complete*. To define this concept this we need to define the limit of a sequence of numbers. Recall the absolute value function

$$|x| = \begin{cases} x & \text{for } x \geq 0, \\ -x & \text{for } x < 0. \end{cases}$$

The distance between two numbers x and y is defined to be

$$d(x, y) \stackrel{d}{=} |x - y|.$$

It is easy to see that this has the three basic properties of a distance:

NONDEGENERACY: $d(x, y) \geq 0$ and $d(x, y) = 0$ if and only if $x = y$

SYMMETRY: $d(x, y) = d(y, x)$

THE TRIANGLE INEQUALITY: $d(x, y) \leq d(x, z) + d(z, y)$

The third condition is called the triangle inequality by analogy with the familiar fact from Euclidean geometry: The length of one side of a triangle is less than the sum of the lengths of the other sides. We use the standard notation $\mathbb{R}$ to denote the set of real numbers. The following definitions are useful when discussing sets of real numbers.

***Definition* A.1.1.** A subset S of the real numbers is bounded from below if there is a number m so that

$$m \leq x, \text{ for all } x \in S,$$

and bounded from above if there is a number M such that

$$x \leq M, \text{ for all } x \in S.$$

If a set is bounded from above and below, then we say it is bounded.

***Definition* A.1.2.** If a subset S of $\mathbb{R}$ is bounded from below then, we define the *infimum* of S, $\inf S$, as the largest number m such that $m \leq x$ for all $x \in S$. If S is bounded from above, we define the *supremum* of S, $\sup S$ to be the smallest number M such that $x \leq M$ for all $x \in S$.

Sequences

A sequence of real numbers is an ordered list of numbers. A sequence can be either finite or infinite. In this section we consider only infinite sequences. Frequently the terms of a sequence are labeled or *indexed* by the positive integers $x_1, x_2, x_3, \ldots$. The notation $< x_n >$ refers to a sequence indexed by n. A sequence is *bounded* if there is a number M so that

$$|x_n| \leq M$$

for all choices of the index n. It is *monotone increasing* if $x_n \leq x_{n+1}$ for all n. The definition of *limit* and the completeness axiom for the real numbers follow:

LIMITS: If $< x_n >$ is a sequence of real numbers, then we say that $< x_n >$ converges to x if the distances, $d(x_n, x)$ can be made arbitrarily small by taking the index sufficiently large. More technically, *given* a positive number $\epsilon > 0$ we can *find* an integer N so that

$$d(x_n, x) < \epsilon \text{ provided } n > N.$$

In this case we say the "limit of the sequence $< x_n >$ is x" and write

$$\lim_{n \to \infty} x_n = x.$$

COMPLETENESS AXIOM: If $< x_n >$ is a monotone increasing, bounded sequence of real numbers, then $< x_n >$ converges to limit; that is, there exists a real number x such that $\lim_{n \to \infty} x_n = x$.

From the completeness axiom it is easy to show that bounded, monotone decreasing sequences also converge. The completeness axiom is what distinguishes the real numbers from the rational numbers. For example, it is not difficult to construct a bounded, monotone sequence of rational numbers $< x_n >$ that get closer and closer to $\sqrt{2}$; see Exercise A.1.10. That is, $d(x_n, \sqrt{2})$ can be made as small as we like by taking n sufficiently large. But $\sqrt{2}$ is not a rational number, showing that $< x_n >$ cannot converge to a rational number. The rational numbers are not complete!

Using the completeness axiom, it is not difficult to show that every real number has a decimal expansion. That is, given a positive real number x, we can find a (possibly infinite) sequence $< a_m, a_{m-1}, \cdots >$ of numbers such that $0 \leq a_j \leq 9$ and

$$x = \lim_{N \to \infty} \left[\sum_{j=-N}^{m} a_j 10^j \right].$$

In this context the index set for the sequence $< a_j >$ is the set of integers less than or equal to m. If x has only finitely many nonzero terms in its decimal expansion, then, by convention, we set all the remaining digits to zero. To study infinite decimal expansions, it is useful to have a formula for the sum of a geometric series.

Proposition A.1.1. *If $r \in \mathbb{R}$ and $N \in \mathbb{N}$, then*

$$\sum_{j=0}^{N} r^j = \frac{r^{N+1} - 1}{r - 1}. \tag{A.5}$$

If $|r| < 1$, then the limit of this sum exists as $N \to \infty$; it is given by

$$\sum_{j=0}^{\infty} r^j = \frac{1}{1 - r}. \tag{A.6}$$

Because the digits in the decimal expansion are restricted to lie between zero and nine, we can estimate the error in replacing x by a finite part of its decimal expansion:

$$0 \leq x - \sum_{j=-N}^{m} a_j 10^j \leq \sum_{j=N+1}^{\infty} \frac{9}{10^j} = \frac{1}{10^N},$$

which agrees with our intuitive understanding of decimal representations. It tells us that real numbers can be approximated, with arbitrary accuracy by rational numbers. The addition and multiplication operations can therefore be extended by continuity to all real numbers: Suppose that $< x_n >$ and $< y_n >$ are sequences of rational numbers converging to real numbers x and y. Then

$$x + y \stackrel{d}{=} \lim_{n \to \infty} (x_n + y_n) \text{ and } x \times y \lim_{n \to \infty} \stackrel{d}{=} x_n \times y_n.$$

Arguing in a similar way, we can show that any positive number x has a binary representation; this is a (possibly infinite) binary sequence $< b_n, b_{n-1}, \cdots >$ such that

$$x = \lim_{N \to \infty} \left[\sum_{j=-N}^{n} b_j 2^j \right].$$

A finite part of the binary expansion gives an approximation for x that satisfies

$$0 \leq x - \sum_{j=-N}^{n} b_j 2^j \leq \frac{1}{2^N}.$$

This introduction to real numbers suffices for our applications; a very good and complete introduction to this subject can be found in [27].

***Remark* A.1.1.** As it serves no further pedagogical purpose to use $\times$ to indicate multiplication of numbers, we henceforth follow the standard notation of indicating multiplication of numbers by juxtaposition: If a, b are numbers, then ab is the product of a and b.

Exercises

***Exercise* A.1.8.** Prove (A.5).

***Exercise* A.1.9.** Show that completeness axiom is equivalent to the following statement: If S is a subset of $\mathbb{R}$ that is bounded from above, then sup S exists.

***Exercise* A.1.10.** Define the sequence by letting $x_0 = 2$ and

$$x_{j+1} = \frac{x_j}{2} + \frac{1}{x_j} \qquad \text{for } j > 0.$$

Show that $< x_n >$ is a bounded, monotone decreasing sequence of rational numbers and explain why its limit must be $\sqrt{2}$. **Extra credit:** Show that there is a constant C such that

$$|x_j - \sqrt{2}| < C2^{-2^j}.$$

This shows that $< x_n >$ converges very quickly to $\sqrt{2}$.

***Exercise* A.1.11.** Show that the definitions of addition and multiplication of real numbers as limits make sense. That is, if $< x_n >$ and $< x'_n >$ both converge to x and $< y_n >$ and $< y'_n >$ both converge to y, then

$$\lim_{n\to\infty} x_n + y_n = \lim_{n\to\infty} x'_n + y'_n \text{ and } \lim_{n\to\infty} x_n \times y_n = \lim_{n\to\infty} x'_n \times y'_n.$$

A.1.4 Cauchy Sequences

In the previous section we discussed the properties of convergent sequences of numbers. Suppose that $< x_n >$ is a sequence of numbers. How do we decide if it has a limit or not? The definition of completeness only considers bounded monotone sequences; many convergent sequences are not monotone. In light of this it would be useful to have a more flexible criterion for a sequence to have a limit. If $< x_n >$ converges to x^*, then, as n gets large, x_n gets closer to x^*. As an inevitable consequence, the distances between the terms of the sequence, $\{|x_n - x_m|\}$, must become small as *both* m and n get large. In order to converge, the terms of the sequence must cluster closer and closer to *each other* as the index gets large. A sequence with this latter property is called a Cauchy sequence.

***Definition* A.1.3.** A sequence of real numbers $< x_n >$ is called a *Cauchy sequence* if given $\epsilon > 0$ there is an N so that

$$|x_n - x_m| < \epsilon \qquad \text{whenever both } m \text{ and } n \text{ are larger than } N. \tag{A.7}$$

The fundamental importance of this concept is contained in the following theorem. This is sometimes called the *Cauchy criterion*.

Theorem A.1.1 (Cauchy criterion). *A sequence of real numbers converges if and only if it is a Cauchy sequence.*

The proof can be found in [27].

If we imagine "observing" a convergent sequence of numbers, then it seems unlikely that one could directly observe its limit. On the other hand, the clustering described in the Cauchy criterion is something that is readily observed.

***Example* A.1.1.** Let $x_n = n^{-1}$. If $n < m$, then

$$|x_n - x_m| \leq \frac{1}{n}.$$

This shows that x_n is a Cauchy sequence.

***Example* A.1.2.** Suppose that $< x_n >$ is a sequence and it is known that for any $\epsilon > 0$ there is an N so that $|x_n - x_{n+1}| < \epsilon$ if $n > N$. This does not imply that the sequence converges. For the sequence defined by

$$x_n = \sum_{j=1}^{n} \frac{1}{j},$$

the differences $x_{n+1} - x_n = (n+1)^{-1}$ go to zero as n tends to infinity. However, $< x_n >$ is unbounded as n tends to infinity. This shows that it is not enough for the *successive* terms of a sequence to be close together. The Cauchy criterion requires that the differences $|x_n - x_m|$ be small for *all* sufficiently large values of m and n.

Exercises

***Exercise* A.1.12.** Suppose that $< x_n >$ is a sequence of real numbers such that

$$\lim_{N\to\infty} \sum_{j=1}^{N} |x_j - x_{j+1}| < \infty.$$

Show that $\lim_{n\to\infty} x_j$ exists.

***Exercise* A.1.13.** Show that a convergent sequence satisfies the Cauchy convergence criterion.

A.2 Vector Spaces

We now discuss the linear structure of Euclidean space, linear transformations, and different ways to measure distances and angles. Geometrically, $\mathbb{R}$ is usually represented by a straight line; the numbers are coordinates for this line. We can specify coordinates on a plane by choosing two intersecting straight lines, and coordinates in space are determined by choosing three lines that intersect in a point. Of course, we can continue in this way. We denote the set of ordered pairs of real numbers by

$$\mathbb{R}^2 = \{(x, y) \,:\, x, y \in \mathbb{R}\}$$

and the set of ordered triples by

$$\mathbb{R}^3 = \{(x, y, z) \ : \ x, y, z \in \mathbb{R}\}.$$

These are known as the Euclidean 2-space and 3-space, respectively. From a mathematical perspective, there is no reason to stop at 3; for each $n \in \mathbb{N}$ we let $\mathbb{R}^n$ denote the set of ordered n-tuples $(x_1, x_2, \ldots, x_n)$ of real numbers. This is called the Euclidean n-space, or just n-space for short. From a physical perspective, we can think of n-space as giving (local) coordinates for a system with n-degrees of freedom. The physical space we occupy is 3-space; if we include time, then this gives us 4-space. If we are studying the weather, then we would want to know the temperature, humidity, and barometric pressure at each point in space-time, so this requires 7 parameters (x, y, z, t, T, H, P). More complicated and complete physical models are often described as spaces with more dimensions.

A.2.1 Euclidean n-Space

All the Euclidean n-spaces have the structure of *linear or vector spaces*. This means that we know how to add two n-tuples of real numbers

$$(x_1, \ldots, x_n) + (y_1, \ldots, y_n) = (x_1 + y_1, \ldots, x_n + y_n)$$

and multiply an n-tuple of real numbers by a real number

$$a \cdot (x_1, \ldots, x_n) = (ax_1, \ldots ax_n).$$

These two operations are compatible in that

$$\begin{aligned} a \cdot (x_1, \ldots, x_n) + a \cdot (y_1, \ldots, y_n) &= a \cdot (x_1 + y_1, \ldots, x_n + y_n) \\ &= (a(x_1 + y_1), \ldots, a(x_n + y_n)). \end{aligned}$$

An ordered n-tuple of numbers is called an *n-vector* or *vector*. The first operation is called vector addition (or just addition), and the second operation is called *scalar* multiplication. For most values of n there is no way to define a compatible notion of vector multiplication. There are five cases where this can be done: if $n = 1$ (real numbers), $n = 2$ (complex numbers), $n = 3$ (cross product), $n = 4$ (quaternions), and $n = 8$ (Cayley numbers). It is often convenient to use a single letter to denote an n-tuple of numbers. In this book boldface, Roman letters are used to denote vectors; that is,

$$\boldsymbol{x} = (x_1, \ldots, x_n).$$

Provisionally we also use $a \cdot \boldsymbol{x}$ to denote scalar multiplication. The compatibility of vector addition and scalar multiplication is then written as

$$a \cdot (\boldsymbol{x} + \boldsymbol{y}) = a \cdot \boldsymbol{x} + a \cdot \boldsymbol{y}.$$

There is a special vector all of whose entries are zero denoted by $\mathbf{0} = (0, \ldots, 0)$. It satisfies

$$\boldsymbol{x} + \mathbf{0} = \boldsymbol{x} = \mathbf{0} + \boldsymbol{x}$$

for any vector $\boldsymbol{x}$. It is also useful to single out a collection of n *coordinate vectors*. Let $\boldsymbol{e}_j \in \mathbb{R}^n$ denote the vector with all entries zero but for the jth-entry, which equals 1. For example, if $n = 3$, then the coordinate vectors are

$$\boldsymbol{e}_1 = (1, 0, 0), \quad \boldsymbol{e}_2 = (0, 1, 0), \quad \boldsymbol{e}_3 = (0, 0, 1).$$

These are called coordinate vectors because we can express any vector as a sum of these vectors. If $\boldsymbol{x} \in \mathbb{R}^n$, then

$$\boldsymbol{x} = x_1 \cdot \boldsymbol{e}_1 + \cdots + x_n \cdot \boldsymbol{e}_n = \sum_{j=1}^{n} x_j \cdot \boldsymbol{e}_j. \tag{A.8}$$

The n-tuple of numbers $(x_1, \ldots, x_n)$ are then the *coordinates* for the vector $\boldsymbol{x}$. The set of vectors $\{\boldsymbol{e}_1, \ldots, \boldsymbol{e}_n\}$ is also called the *standard basis* for $\mathbb{R}^n$.

The linear structure singles out a special collection of real-valued functions.

***Definition* A.2.1.** A function $f : \mathbb{R}^n \to \mathbb{R}$ is linear if it satisfies the following conditions: For any pair of vectors $\boldsymbol{x}, \boldsymbol{y} \in \mathbb{R}^n$ and $a \in \mathbb{R}$,

$$\begin{aligned} f(\boldsymbol{x} + \boldsymbol{y}) &= f(\boldsymbol{x}) + f(\boldsymbol{y}), \\ f(a \cdot \boldsymbol{x}) &= af(\boldsymbol{x}). \end{aligned} \tag{A.9}$$

In light of (A.8), it is clear that a linear function on $\mathbb{R}^n$ is completely determined by the n values $\{f(\boldsymbol{e}_1), \ldots, f(\boldsymbol{e}_n)\}$. For an arbitrary $\boldsymbol{x} \in \mathbb{R}^n$, (A.8) and (A.9) imply

$$f(\boldsymbol{x}) = \sum_{j=1}^{n} x_j f(\boldsymbol{e}_j).$$

On the other hand, it is easy to see that given n numbers $\{a_1, \ldots, a_n\}$, we can *define* a linear function on $\mathbb{R}^n$ by setting

$$f(\boldsymbol{x}) = \sum_{j=1}^{n} a_j x_j.$$

We therefore have an explicit knowledge of the collection of linear functions.

What measurements are required to determine a linear function? While it suffices, it is not actually necessary to measure $\{f(\boldsymbol{e}_1), \ldots, f(\boldsymbol{e}_n)\}$. To describe what is needed requires a definition.

***Definition* A.2.2.** If $\{\boldsymbol{v}_1, \ldots, \boldsymbol{v}_n\}$ is a collection of n vectors in $\mathbb{R}^n$ with the property that every vector $\boldsymbol{x}$ can be represented as

$$\boldsymbol{x} = \sum_{j=1}^{n} a_j \cdot \boldsymbol{v}_j, \tag{A.10}$$

for a set of scalars $\{a_1, \ldots, a_n\}$, then we say that these vectors are a *basis* for $\mathbb{R}^n$. The coefficients are called the coordinates of $\boldsymbol{x}$ with respect to this basis.

Note that the standard bases, defined previously, satisfy (A.10).

***Example* A.2.1.** The standard basis for $\mathbb{R}^2$ is $\boldsymbol{e}_1 = (1, 0)$, $\boldsymbol{e}_2 = (0, 1)$. The vectors $\boldsymbol{v}_1 = (1, 1)$, $\boldsymbol{v}_2 = (0, 1)$ also define a basis for $\mathbb{R}^2$. To see this, we observe that

$$\boldsymbol{e}_1 = \boldsymbol{v}_1 - \boldsymbol{v}_2 \text{ and } \boldsymbol{e}_2 = \boldsymbol{v}_2;$$

therefore, if $\boldsymbol{x} = x_1 \cdot \boldsymbol{e}_1 + x_2 \cdot \boldsymbol{e}_2$, then

$$\boldsymbol{x} = x_1 \cdot (\boldsymbol{v}_1 - \boldsymbol{v}_2) + x_2 \cdot \boldsymbol{v}_2 = x_1 \cdot \boldsymbol{v}_1 + (x_2 - x_1) \cdot \boldsymbol{v}_2.$$

Proposition A.2.1. *A collection of n vectors, $\{\boldsymbol{v}_1, \dots, \boldsymbol{v}_n\}$ in $\mathbb{R}^n$ defines a basis if and only if the only n-tuple, $(a_1, \dots, a_n)$ for which*

$$\sum_{j=1}^{n} a_j \cdot \boldsymbol{v}_j = 0$$

is the zero vector.

The proposition implies that the coefficients appearing in (A.10) are uniquely determined by $\boldsymbol{x}$.

If f is a linear function, then the proposition shows that the values

$$\{f(\boldsymbol{v}_1), \dots, f(\boldsymbol{v}_n)\},$$

for any basis $\{\boldsymbol{v}_1, \dots, \boldsymbol{v}_n\}$, suffice to determine f. On the other hand, given any set of n numbers, $\{a_1, \dots, a_n\}$, we can define a linear function f by setting

$$f(\boldsymbol{v}_j) = a_j \qquad \text{for } 1 \leq j \leq n \tag{A.11}$$

and extending *by linearity*. This means that if

$$\boldsymbol{x} = \sum_{j=1} b_j \cdot \boldsymbol{v}_j,$$

then

$$f(\boldsymbol{x}) = \sum_{j=1}^{n} b_j a_j. \tag{A.12}$$

From the standpoint of measurement, how are vectors in $\mathbb{R}^n$ distinguished from one another? Linear functions provide an answer to this question. Let $\{\boldsymbol{v}_1, \dots, \boldsymbol{v}_n\}$ be a basis and for each $1 \leq j \leq n$ we define the linear function f_j by the conditions

$$f_j(\boldsymbol{v}_j) = 1, \quad f_j(\boldsymbol{v}_i) = 0 \qquad \text{for } i \neq j.$$

Suppose that for each j we can build a machine whose output is $f_j(\boldsymbol{x})$. Two vectors $\boldsymbol{x}$ and $\boldsymbol{y}$ are equal if and only if $f_j(\boldsymbol{x}) = f_j(\boldsymbol{y})$ for $1 \leq j \leq n$.

Exercises

***Exercise* A.2.1.** Prove Proposition A.2.1.

***Exercise* A.2.2.** Show that the function defined in (A.11) and (A.12) is well defined and linear.

***Exercise* A.2.3.** Let $f : \mathbb{R}^n \to \mathbb{R}$ be a nonzero linear function. Show that there is a basis $\{\boldsymbol{v}_1, \ldots, \boldsymbol{v}_n\}$ for $\mathbb{R}^n$ such that

$$f(\boldsymbol{v}_1) = 1 \text{ and } f(\boldsymbol{v}_j) = 0 \qquad \text{for } 2 \leq j \leq n.$$

A.2.2 General Vector Spaces

As is often the case in mathematics, it is useful to introduce an abstract concept that encompasses many special cases. The Euclidean spaces introduced in the previous sections are examples of vector spaces.

***Definition* A.2.3.** Let V be a set; it is a *real vector space* if it has two operations:

ADDITION: Addition is a map from $V \times V \to V$. If $(\boldsymbol{v}_1, \boldsymbol{v}_2)$ is an element of $V \times V$, then we denote this by $(\boldsymbol{v}_1, \boldsymbol{v}_2) \mapsto \boldsymbol{v}_1 + \boldsymbol{v}_2$.

SCALAR MULTIPLICATION: Scalar multiplication is a map from $\mathbb{R} \times V \to V$. If $a \in \mathbb{R}$ and $\boldsymbol{v} \in V$, then we denote this by $(a, \boldsymbol{v}) \mapsto a \cdot \boldsymbol{v}$.

The operations have the following properties:

COMMUTATIVE LAW: $\boldsymbol{v}_1 + \boldsymbol{v}_2 = \boldsymbol{v}_2 + \boldsymbol{v}_1$

ASSOCIATIVE LAW: $(\boldsymbol{v}_1 + \boldsymbol{v}_2) + \boldsymbol{v}_3 = \boldsymbol{v}_1 + (\boldsymbol{v}_2 + \boldsymbol{v}_3)$

DISTRIBUTIVE LAW: $a \cdot (\boldsymbol{v}_1 + \boldsymbol{v}_2) = a \cdot \boldsymbol{v}_1 + a \cdot \boldsymbol{v}_2$

Finally, there is a special element $\boldsymbol{0} \in V$ such that

$$\boldsymbol{v} + \boldsymbol{0} = \boldsymbol{v} = \boldsymbol{0} + \boldsymbol{v} \text{ and } \boldsymbol{0} = 0 \cdot \boldsymbol{v};$$

this vector is called the *zero vector.*

We consider some examples of vector spaces.

***Example* A.2.2.** For each $n \in \mathbb{N}$, the space $\mathbb{R}^n$ with the addition and scalar multiplication defined previously is a vector space.

***Example* A.2.3.** The set of real-valued functions defined on $\mathbb{R}$ is a vector space. We define addition of functions by the rule $(f+g)(x) = f(x)+g(x)$; scalar multiplication is defined by $(a \cdot f)(x) = af(x)$. We denote the space of functions on $\mathbb{R}$ with these operations by $\mathcal{F}$.

***Example* A.2.4.** If f_1 and f_2 are linear functions on $\mathbb{R}^n$, then define $f_1 + f_2$ as previously:

$$(f_1 + f_2)(\boldsymbol{x}) = f_1(\boldsymbol{x}) + f_2(\boldsymbol{x}) \qquad \text{for all } \boldsymbol{x} \in \mathbb{R}^n$$

and $(a \cdot f)(\boldsymbol{x}) = af(\boldsymbol{x})$. A sum of linear functions is a linear function, as is a scalar multiple. Thus the set of linear functions on $\mathbb{R}^n$ is also a vector space. This vector space is called the *dual* vector space; it is denoted by $(\mathbb{R}^n)'$.

***Example* A.2.5.** For each $n \in \mathbb{N} \cup \{0\}$, let $\mathcal{P}_n$ denote the set of real-valued polynomial functions, of degree at most n. Since the sum of two polynomials of degree at most n is again a polynomial of degree at most n, as is a scalar multiple, it follows that $\mathcal{P}_n$ is a vector space.

Many natural mathematical objects have a vector space structure. Often a vector space is subset of a larger vector space.

***Definition* A.2.4.** Let V be a vector space; a subset $U \subset V$ is a *subspace* if, whenever $\boldsymbol{u}_1, \boldsymbol{u}_2 \in U$, then $\boldsymbol{u}_1 + \boldsymbol{u}_2 \in U$ and for every $a \in \mathbb{R}$, $a \cdot \boldsymbol{u}_1 \in U$ as well. Briefly, a subset U is a subspace if it is a vector space with the addition and scalar multiplication it inherits from V.

***Example* A.2.6.** The subset of $\mathbb{R}^2$ consisting of the vectors $\{(x, 0) : x \in \mathbb{R}\}$ is a subspace.

***Example* A.2.7.** Let $f : \mathbb{R}^n \to \mathbb{R}$ be a linear function; the set $\{\boldsymbol{v} \in \mathbb{R}^n : f(\boldsymbol{v}) = 0\}$ is a subspace. This subspace is called the null space of the linear function f.

***Example* A.2.8.** The set of polynomials of degree at most 2 is a subspace of the set of polynomials of degree at most 3.

***Example* A.2.9.** The set of vectors $\{\boldsymbol{v} \in \mathbb{R}^n : f(\boldsymbol{v}) = 1\}$ in *not* a subspace.

***Example* A.2.10.** If $g : (x, y) \to \mathbb{R}$ is defined by $g(x, y) = x^2 - y$, then the set of vectors $\{(x, y) \in \mathbb{R}^2 : g(x, y) = 0\}$ is *not* a subspace.

***Definition* A.2.5.** Let $\{\boldsymbol{v}_1, \ldots, \boldsymbol{v}_m\}$ be a collection of vectors in a vector space V. A vector of the form

$$\boldsymbol{v} = a_1 \cdot \boldsymbol{v}_1 + \cdots + a_m \cdot \boldsymbol{v}_m$$

is called a *linear combination* of the vectors $\{\boldsymbol{v}_1, \ldots, \boldsymbol{v}_m\}$. The *linear span* of these vectors is the set of all linear combinations

$$\operatorname{span}(\boldsymbol{v}_1, \ldots, \boldsymbol{v}_m) \stackrel{d}{=} \{a_1 \cdot \boldsymbol{v}_1 + \cdots + a_m \cdot \boldsymbol{v}_m : a_1, \ldots a_m \in \mathbb{R}\}.$$

***Example* A.2.11.** The linear span of a collection of vectors $\{\boldsymbol{v}_1, \ldots, \boldsymbol{v}_m\} \subset V$ is a subspace of V.

A basic feature of a vector space is its dimension. This is a precise mathematical formulation of the number of degrees of freedom. The vector space $\mathbb{R}^n$ has dimension n. The general concept of a *basis* is needed to define the dimension.

***Definition* A.2.6.** Let V be a vector space; a set of vectors $\{\boldsymbol{v}_1, \ldots, \boldsymbol{v}_n\} \subset V$ is said to be *linearly independent* if

$$\sum_{j=1}^{n} a_j \cdot \boldsymbol{v}_j = 0$$

implies that $a_j = 0$ for $j = 1, \ldots, n$. This is another way of saying that it is not possible to write one of these vectors as a linear combination of the others. A finite set of vectors $\{\boldsymbol{v}_1, \ldots, \boldsymbol{v}_n\} \subset V$ is a *basis* for V if

1. The vectors are linearly independent.

2. Every vector in V is a linear combination of these vectors; that is,

$$\text{span}(\boldsymbol{v}_1, \ldots, \boldsymbol{v}_n) = V.$$

The definition of a basis given earlier for the vector spaces $\mathbb{R}^n$ is a special case of this definition. If a vector space V has a basis then every basis for V has the same number of elements. This fact makes it possible to define the dimension of a vector space.

***Definition* A.2.7.** If a vector space V has a basis consisting of n vectors, then the *dimension* of V is n. We write

$$\dim V = n.$$

If $\{\boldsymbol{v}_1, \ldots, \boldsymbol{v}_n\}$ is a basis for V, then for every vector $\boldsymbol{v} \in V$ there is a unique point $(x_1, \ldots, x_n) \in \mathbb{R}^n$ such that

$$\boldsymbol{v} = x_1 \cdot \boldsymbol{v}_1 + \cdots + x_n \cdot \boldsymbol{v}_n. \tag{A.13}$$

A vector space V of dimension n has exactly the same number of degrees of freedom as $\mathbb{R}^n$. In fact, by choosing a basis we define an *isomorphism* between V and $\mathbb{R}^n$. This is because if $\boldsymbol{v} \leftrightarrow (x_1, \ldots, x_2)$ and $\boldsymbol{v}' \leftrightarrow (y_1, \ldots, y_n)$ in (A.13), then

$$\boldsymbol{v} + \boldsymbol{v}' = (x_1 + y_1) \cdot \boldsymbol{v}_1 + \cdots + (x_n + y_n) \cdot \boldsymbol{v}_n$$

and for $a \in \mathbb{R}$

$$a \cdot \boldsymbol{v} = (ax_1) \cdot \boldsymbol{v}_1 + \cdots + (ax_n) \cdot \boldsymbol{v}_n.$$

From this point of view, all vector spaces of dimension n are the same. The abstract concept is still useful. Vector spaces often do not come with a *natural* choice of basis. Indeed the possibility of changing the basis—that is changing the identification of V with $\mathbb{R}^n$—is a very powerful tool. In applications we try to choose a basis that is well adapted to the problem at hand. It is important to note that many properties of vector spaces are independent of the choice of basis.

***Example* A.2.12.** The vector space $\mathcal{F}$ of all functions on $\mathbb{R}$ does not have a basis; that is, we cannot find a finite collection of functions such that any function is a linear combination of these functions. The vector space $\mathcal{F}$ is infinite dimensional. The study of infinite-dimensional vector spaces is called functional analysis; we return to this subject in Section A.3.

***Example* A.2.13.** For each n the set $\{1, x, \ldots, x^n\}$ is a basis for the $\mathcal{P}_n$. Thus the $\dim \mathcal{P}_n = n + 1$.

Exercises

***Exercise* A.2.4.** Show that $\mathcal{F}$, defined in Example A.2.3, is a vector space.

***Exercise* A.2.5.** Show that the set of polynomials $\{x^j : 0 \le j \le n\}$ is a basis for $\mathcal{P}_n$.

***Exercise* A.2.6.** Show that the set of polynomials $\{x^j(1-x)^{n-j} \ : \ 0 \leq j \leq n\}$ is a basis for $\mathcal{P}_n$.

***Exercise* A.2.7.** Show that if a vector space V has a basis, then any basis for V has the same number of vectors.

***Exercise* A.2.8.** Let V be a vector space with $\dim V = n$ and let V' denote the set of linear functions on V. Show that V' is also a vector space with $\dim V' = n$.

A.2.3 Linear Transformations and Matrices

The fact that both $\mathbb{R}^n$ and $\mathbb{R}^m$ have linear structures allows us to single out a special class of maps between these spaces.

***Definition* A.2.8.** A map $F : \mathbb{R}^n \rightarrow \mathbb{R}^m$ is called a *linear transformation* if for all pairs $\boldsymbol{x}, \ \boldsymbol{y} \in \mathbb{R}^n$ and $a \in \mathbb{R}$ we have

$$\begin{aligned} F(\boldsymbol{x}+\boldsymbol{y}) &= F(\boldsymbol{x}) + F(\boldsymbol{y}), \\ F(a \cdot \boldsymbol{x}) &= a \cdot F(\boldsymbol{x}). \end{aligned} \tag{A.14}$$

Comparing the definitions, we see that a linear function is just the $m = 1$ case of a linear transformation. For each $n \in \mathbb{N}$ there is a special linear transformation of $\mathbb{R}^n$ to itself, called the identity map. It is defined by $\boldsymbol{x} \mapsto \boldsymbol{x}$ and denoted by Id_n.

If $\{\boldsymbol{v}_1, \ldots, \boldsymbol{v}_n\}$ is a basis for $\mathbb{R}^n$, then a linear transformation is determined by the values $\{F(\boldsymbol{v}_1), \ldots, F(\boldsymbol{v}_n)\}$. If $\boldsymbol{x} = a_j \cdot \boldsymbol{v}_1 + \cdots + a_n \cdot \boldsymbol{v}_n$, then (A.14) implies that

$$F(\boldsymbol{x}) = \sum_{j=1}^{n} a_j \cdot F(\boldsymbol{v}_j).$$

In this section, linear transformations are denoted by bold, uppercase, Roman letters (e.g., $\boldsymbol{A}, \boldsymbol{B}$). The action of the linear transform $\boldsymbol{A}$ on the vector $\boldsymbol{x}$ is denoted $\boldsymbol{Ax}$. Connected to a linear transformation $\boldsymbol{A} : \mathbb{R}^n \rightarrow \mathbb{R}^m$ are two natural subspaces.

***Definition* A.2.9.** The set of vectors $\{\boldsymbol{x} \in \mathbb{R}^n \ : \ \boldsymbol{Ax} = 0\}$ is called the kernel or null space of the linear transformation $\boldsymbol{A}$; we denote this subspace by $\ker \boldsymbol{A}$.

***Definition* A.2.10.** The set of vectors $\{\boldsymbol{Ax} \in \mathbb{R}^m \ : \ \boldsymbol{x} \in \mathbb{R}^n\}$ is called the *image* of the linear transformation $\boldsymbol{A}$; we denote this $\mathrm{Im}\,\boldsymbol{A}$.

The kernel and image of a linear transformation are basic examples of subspaces of a vector space that are defined *without reference* to a basis. There is, in general, no natural choice of a basis for either subspace.

Bases and Matrices

As previously, let $\{\boldsymbol{v}_1, \ldots, \boldsymbol{v}_n\}$ be a basis for $\mathbb{R}^n$. If we also choose a basis $\{\boldsymbol{u}_1, \ldots, \boldsymbol{u}_m\}$ for $\mathbb{R}^m$, then there is a collection of mn-numbers $\{a_{ij}\}$ so that for each j,

$$\boldsymbol{A}(\boldsymbol{v}_j) = \sum_{i=1}^{m} a_{ij} \boldsymbol{u}_i.$$

Such a collection of numbers, labeled with two indices, is called a *matrix*. Once bases for the domain and range of $\boldsymbol{A}$ are fixed, the matrix determines and is determined by the linear transformation. If

$$\boldsymbol{x} = x_1 \cdot \boldsymbol{v}_1 + \cdots + x_n \cdot \boldsymbol{v}_n,$$

then the coefficients $(y_1, \dots, y_m)$ of $\boldsymbol{A}\boldsymbol{x}$ with respect to $\{\boldsymbol{u}_k\}$ are given by

$$y_i = \sum_{j=1}^{n} a_{ij} x_j \qquad \text{for } i = 1, \dots, m.$$

If $\boldsymbol{A} : \mathbb{R}^n \to \mathbb{R}^n$, then we usually select a single basis $\{\boldsymbol{v}_j\}$ and use it to represent vectors in both the domain and range of $\boldsymbol{A}$. Often it is implicitly understood that the bases are the standard bases.

***Example* A.2.14.** If $\{\boldsymbol{v}_1, \dots, \boldsymbol{v}_n\}$ is a basis for $\mathbb{R}^n$, then $\mathrm{Id}_n(\boldsymbol{v}_j) = \boldsymbol{v}_j$. The matrix for Id_n, with respect to any basis is denoted by

$$\delta_{ij} = \begin{cases} 1 & \text{if } i = j, \\ 0 & \text{fi } i \neq j. \end{cases}$$

Once a pair of bases is fixed, then we can identify the set of linear transformations from $\mathbb{R}^n$ to $\mathbb{R}^m$ with the collection of $m \times n$-arrays (read m by n) of numbers. If we think of (a_{ij}) as a rectangular array of numbers, then the first index, i, labels the rows and the second index, j, labels the columns.

$$\begin{pmatrix} a_{11} & \dots & a_{1n} \\ \vdots & & \vdots \\ a_{m1} & \dots & a_{mn} \end{pmatrix} \tag{A.15}$$

A vector in $\mathbb{R}^n$ can be thought of as either a row vector—that is, an $1 \times n$ matrix—or a column vector—that is, an $n \times 1$ matrix. An $m \times n$ matrix has n columns consisting of $m \times 1$ vectors,

$$\boldsymbol{a} = (\boldsymbol{a}^1 \dots \boldsymbol{a}^n),$$

or m rows consisting of $1 \times n$ vectors

$$\boldsymbol{a} = \begin{pmatrix} \boldsymbol{a}_1 \\ \vdots \\ \boldsymbol{a}_m \end{pmatrix}.$$

Precisely how we wish to think about a matrix depends on the situation at hand.

We can define a notion of multiplication between column vectors and matrices.

***Definition* A.2.11.** Let $\boldsymbol{a}$ be an $m \times n$ matrix with entries a_{ij}, $1 \le i \le m$, $1 \le j \le n$, and $\boldsymbol{x}$ be an n-vector with entries x_j, $1 \le j \le n$. Then we define the product $\boldsymbol{a} \cdot \boldsymbol{x}$ to be the m-vector $\boldsymbol{y}$ with entries

$$y_i = \sum_{j=1}^{n} a_{ij} x_j, \quad i = 1, \dots, m.$$

Concisely this is written $\boldsymbol{y} = \boldsymbol{a} \cdot \boldsymbol{x}$. In this section we use lowercase, bold Roman letters to denote matrices (e.g., $\boldsymbol{a}$, $\boldsymbol{b}$).

Proposition A.2.2. *Let $\boldsymbol{a}$ be an $m \times n$ matrix, $\boldsymbol{x}_1, \boldsymbol{x}_2$ two n-vectors, and a a real number. Then*

$$\boldsymbol{a} \cdot (\boldsymbol{x}_1 + \boldsymbol{x}_2) = \boldsymbol{a} \cdot \boldsymbol{x}_1 + \boldsymbol{a} \cdot \boldsymbol{x}_2 \text{and } \boldsymbol{a} \cdot (a \cdot \boldsymbol{x}) = a \cdot (\boldsymbol{a} \cdot \boldsymbol{x}).$$

These conditions show that the map $\boldsymbol{x} \mapsto \boldsymbol{a} \cdot \boldsymbol{x}$ is a linear transformation of $\mathbb{R}^n$ to $\mathbb{R}^m$.

Matrix Multiplication

We can also define multiplication between matrices with compatible dimensions. Let $\boldsymbol{a}$ be an $m \times n$ matrix and $\boldsymbol{b}$ be an $l \times m$ matrix. If (a_{ij}) are the entries of $\boldsymbol{a}$ and (b_{pq}) the entries of $\boldsymbol{b}$, then the entries of their product $\boldsymbol{c} = \boldsymbol{b} \cdot \boldsymbol{a}$ are given by

$$c_{pj} \stackrel{d}{=} \sum_{i=1}^{m} b_{pi} a_{ij}.$$

This shows that we can multiply an $l \times m$ matrix by an $m \times n$ matrix and the result is an $l \times n$ matrix. If $\boldsymbol{a}$ and $\boldsymbol{b}$ are both $n \times n$ matrices, then both products $\boldsymbol{a} \cdot \boldsymbol{b}$, $\boldsymbol{b} \cdot \boldsymbol{a}$ are defined. In general, they are *not* equal. We say that matrix multiplication is *noncommutative*. The product of an $m \times n$ matrix and an n-vector is the special case of multiplying an $m \times n$ matrix by $n \times 1$ matrix; as expected, the result if an $m \times 1$ matrix or an m-vector.

The matrix product is associative; that is, if $\boldsymbol{a}$ is an $m \times n$ matrix, $\boldsymbol{b}$ an $l \times m$ matrix, and $\boldsymbol{c}$ a $k \times l$ matrix, then

$$(\boldsymbol{c} \cdot \boldsymbol{b}) \cdot \boldsymbol{a} = \boldsymbol{c} \cdot (\boldsymbol{b} \cdot \boldsymbol{a}).$$

If $\boldsymbol{x}$ is an n-vector, then $\boldsymbol{a} \cdot \boldsymbol{x}$ is an m-vector so $\boldsymbol{b} \cdot (\boldsymbol{a} \cdot \boldsymbol{x})$ is also defined. The matrix product $\boldsymbol{b} \cdot \boldsymbol{a}$ defines a linear transformation from $\mathbb{R}^n$ to $\mathbb{R}^l$. The associative law for the matrix product shows that

$$(\boldsymbol{b} \cdot \boldsymbol{a}) \cdot \boldsymbol{x} = \boldsymbol{b} \cdot (\boldsymbol{a} \cdot \boldsymbol{x}).$$

The Change-of-Basis Formula

Suppose that $\{\boldsymbol{v}_1, \ldots, \boldsymbol{v}_n\}$ and $\{\boldsymbol{u}_1, \ldots, \boldsymbol{u}_n\}$ are both bases for $\mathbb{R}^n$. The definition of a basis implies that there are $n \times n$ matrices $\boldsymbol{a} = (a_{ij})$ and $\boldsymbol{b} = (b_{ij})$ so that

$$\boldsymbol{v}_i = \sum_{j-1}^{n} a_{ji} \cdot \boldsymbol{u}_j \text{ and } \boldsymbol{u}_i = \sum_{j=1}^{n} b_{ji} \cdot \boldsymbol{v}_j.$$

These are called *change-of-basis* matrices. If $\boldsymbol{x} \in \mathbb{R}^n$, then there are vectors $(a_1, \ldots, a_n)$ and $(b_1, \ldots, b_n)$ so that

$$\boldsymbol{x} = \sum_{j=1}^{n} a_j \cdot \boldsymbol{v}_j \text{ and also } \boldsymbol{x} = \sum_{j=1}^{n} b_j \cdot \boldsymbol{u}_j.$$

Substituting our expression for the $\{v_j\}$ in terms of the $\{u_j\}$ gives

$$\begin{aligned} x &= \sum_{j=1}^{n} a_j \cdot \left[\sum_{k=1}^{n} a_{kj} \cdot u_k \right] \\ &\sum_{k=1}^{n} \left[\sum_{j=1}^{n} a_{kj} a_j \right] \cdot u_k. \end{aligned} \tag{A.16}$$

Comparing (A.16) with our earlier formula, we see that

$$b_k = \sum_{j=1}^{n} a_{kj} a_j \qquad \text{for } k = 1, \ldots, n.$$

This explains why a is called the change-of-basis matrix.

Suppose that $A : \mathbb{R}^n \to \mathbb{R}^m$ is a linear transformation and we select bases $\{v_1, \ldots, v_n\}$ and $\{u_1, \ldots, u_m\}$ for $\mathbb{R}^n$ and $\mathbb{R}^m$, respectively. Let (a_{ij}) denote the matrix of this linear transformation with respect to this choice of bases. How does the matrix change if these bases are replaced by a different pair of bases? We consider what it means for "(a_{ij}) to be the matrix representing A with respect to the bases $\{v_j\}$ and $\{u_i\}$" by putting into words the computations performed previously: Suppose that x is a vector in $\mathbb{R}^n$ with coordinates $(x_1, \ldots, x_n)$ with respect to the basis $\{v_j\}$. Then the coordinates of $y = Ax$ with respect to $\{u_i\}$ are

$$y_i = \sum_{j=1}^{n} a_{ij} x_j, \quad i = 1, \ldots, m.$$

The fact to keep in mind is that we are dealing with different representations of fixed (abstract) vectors x and Ax.

Suppose that $\{v'_j\}$ and $\{u'_i\}$ are new bases for $\mathbb{R}^n$ and $\mathbb{R}^m$, respectively, and let (b_{lj}) and (c_{ki}) be change-of-basis matrices; that is,

$$v'_j = \sum_{l=1}^{n} b_{lj} \cdot v_l \text{ and } u_i = \sum_{k=1}^{m} c_{ki} \cdot u'_k.$$

Let a'_{ij} be the matrix of A with respect to $\{v'_j\}$ and $\{u'_i\}$. If $(x'_1, \ldots, x'_n)$ are the coordinates of x with respect to $\{v'_j\}$ and $(y'_1, \ldots, y'_m)$ the coordinates of Ax with respect to $\{u'_i\}$, then

$$y'_i = \sum_{j=1}^{n} a'_{ij} x'_j.$$

Formula (A.16) tells us that

$$x_j = \sum_{l=1}^{n} b_{jl} x'_l$$

and therefore

$$\begin{aligned} y_i &= \sum_{j=1}^{n} a_{ij} \left[\sum_{l=1}^{n} b_{jl} x'_l \right] \\ &= \sum_{l=1}^{n} \left[\sum_{j=1}^{n} a_{ij} b_{jl} \right] x'_l \end{aligned} \tag{A.17}$$

gives the expression for $\boldsymbol{A}\boldsymbol{x}$ with respect to the $\{\boldsymbol{u}_i\}$. To complete our computation, we only need to reexpress $\boldsymbol{A}\boldsymbol{x}$ with respect to the basis $\{\boldsymbol{u}'_i\}$. To that end, we apply (A.16) one more time to obtain that

$$y'_i = \sum_{k=1}^{m} c_{ik} y_k.$$

Putting this into (A.17) and reordering the sums, we obtain that

$$y'_i = \sum_{j=1}^{n} \left[\sum_{k=1}^{m} \sum_{l=1}^{n} c_{ik} a_{kl} b_{lj} \right] x'_j.$$

This shows that

$$a'_{ij} = \sum_{k=1}^{m} \sum_{l=1}^{n} c_{ik} a_{kl} b_{lj}.$$

Using $\boldsymbol{a}$, $\boldsymbol{a}'$, $\boldsymbol{b}$, $\boldsymbol{c}$ to denote the matrices defined previously, we can rewrite these expressions more concisely as

$$\begin{pmatrix} x_1 \\ \vdots \\ x_n \end{pmatrix} = \boldsymbol{b} \cdot \begin{pmatrix} x'_1 \\ \vdots \\ x'_n \end{pmatrix}, \quad \begin{pmatrix} y_1 \\ \vdots \\ y_m \end{pmatrix} = \boldsymbol{c} \cdot \begin{pmatrix} y'_1 \\ \vdots \\ y'_m \end{pmatrix}, \quad \boldsymbol{a}' = \boldsymbol{c} \cdot \boldsymbol{a} \cdot \boldsymbol{b}. \tag{A.18}$$

The reader should be aware that this formula differs slightly from that usually given in textbooks; this is because $\boldsymbol{b}$ changes from $(x'_1, \ldots, x'_n)$ to $(x_1, \ldots, x_n)$ whereas $\boldsymbol{c}$ changes from $(y'_1, \ldots, y'_m)$ to $(y_1, \ldots, y_m)$.

Exercises

***Exercise* A.2.9.** Show that that if $\boldsymbol{A} : \mathbb{R}^n \to \mathbb{R}^m$ and $\boldsymbol{B} : \mathbb{R}^m \to \mathbb{R}^l$ are linear transformations, then the composition $\boldsymbol{B} \circ \boldsymbol{A}(\boldsymbol{x}) \stackrel{d}{=} \boldsymbol{B}(\boldsymbol{A}(\boldsymbol{x}))$ is a linear transformation from $\mathbb{R}^n$ to $\mathbb{R}^l$.

***Exercise* A.2.10.** Let $\boldsymbol{A} : \mathbb{R}^n \to \mathbb{R}^m$ be a linear transformation. Show that $\ker \boldsymbol{A}$ is a subspace of $\mathbb{R}^n$.

***Exercise* A.2.11.** Let $\boldsymbol{A} : \mathbb{R}^n \to \mathbb{R}^m$ be a linear transformation. Show that $\operatorname{Im} \boldsymbol{A}$ is a subspace of $\mathbb{R}^m$.

***Exercise* A.2.12.** Suppose that we use a basis $\{\boldsymbol{v}_1, \ldots, \boldsymbol{v}_n\}$ for the domain and $\{\boldsymbol{u}_1, \ldots, \boldsymbol{u}_n\}$ for the range. What is the matrix for Id_n?

***Exercise* A.2.13.** Prove Proposition A.2.2.

***Exercise* A.2.14.** If

$$\boldsymbol{a} = \begin{pmatrix} 0 & 0 \\ 1 & 0 \end{pmatrix} \text{ and } \boldsymbol{b} = \begin{pmatrix} 0 & 1 \\ 0 & 0 \end{pmatrix},$$

then show that $\boldsymbol{a} \cdot \boldsymbol{b} \neq \boldsymbol{b} \cdot \boldsymbol{a}$.

***Exercise* A.2.15.** Show that if $\boldsymbol{a}$ is the matrix of a linear transformation $\boldsymbol{A} : \mathbb{R}^n \to \mathbb{R}^m$ and $\boldsymbol{b}$ is the matrix of a linear transformation $\boldsymbol{B} : \mathbb{R}^m \to \mathbb{R}^l$, then $\boldsymbol{b} \cdot \boldsymbol{a}$ is the matrix of their composition $\boldsymbol{B} \circ \boldsymbol{A} : \mathbb{R}^n \to \mathbb{R}^l$.

***Exercise* A.2.16.** Show that $\partial_x : \mathcal{P}_n \to \mathcal{P}_n$ is a linear transformation. It is defined without reference to a basis. Find the basis for ∂_x in terms of the basis $\{1, x, \ldots, x^n\}$. Find bases for $\ker \partial_x$ and $\operatorname{Im} \partial_x$.

***Exercise* A.2.17.** Show that the space of linear transformations from $\mathbb{R}^n$ to $\mathbb{R}^m$ is a vector space with addition defined by

$$(\boldsymbol{A} + \boldsymbol{B})\boldsymbol{x} \stackrel{d}{=} \boldsymbol{A}\boldsymbol{x} + \boldsymbol{B}\boldsymbol{x} \qquad \text{for all } \boldsymbol{x} \in \mathbb{R}^n$$

and scalar multiplication defined by

$$(a \cdot \boldsymbol{A})(\boldsymbol{x}) \stackrel{d}{=} a \cdot (\boldsymbol{A}\boldsymbol{x}).$$

Let $\{\boldsymbol{v}_1, \ldots, \boldsymbol{v}_n\}$ and $\{\boldsymbol{u}_1, \ldots, \boldsymbol{u}_m\}$ be bases for $\mathbb{R}^n$ and $\mathbb{R}^m$, respectively. For $1 \leq i \leq m$ and $1 \leq j \leq n$ define the linear transformations $\boldsymbol{l}_{ij}$ by letting

$$\boldsymbol{l}_{ij}(\boldsymbol{v}_j) = \boldsymbol{u}_i \text{ and } l_{ij}(\boldsymbol{v}_k) = 0 \text{ if } k \neq j.$$

Show that the $\{\boldsymbol{l}_{ij} : 1 \leq i \leq m, \quad 1 \leq j \leq n\}$ is a basis for this vector space. This shows that the space of linear transformations from $\mathbb{R}^n$ to $\mathbb{R}^m$ is isomorphic to $\mathbb{R}^{mn}$.

***Remark* A.2.1 (Important notational remark).** From this point on, we no longer use "$\cdot$" to denote the operations of scalar multiplication or multiplication of a vector by a matrix. That is, for $\boldsymbol{x} \in \mathbb{R}^n$, $a \in \mathbb{R}$, the notation $a\boldsymbol{x}$ indicates scalar multiplication and, for $\boldsymbol{a}$ and $\boldsymbol{b}$ matrices, $\boldsymbol{ab}$ is the matrix product of $\boldsymbol{a}$ and $\boldsymbol{b}$.

A.2.4 Norms and Metrics

In the previous section we concentrated on algebraic properties of vector spaces. In applications of linear algebra to physical problems, it is also important to have a way to quantify errors. To that end, we now introduce notions of distance for vector spaces. Measurement of distance in a vector space usually begins with a notion of *length*. Taking advantage of the underlying linear structure, the distance between two vectors $\boldsymbol{x}$ and $\boldsymbol{y}$ is then defined as the length of $\boldsymbol{x} - \boldsymbol{y}$.

There are many reasonable ways to measure length in $\mathbb{R}^n$. The most common way to define the length of $(x_1, \ldots, x_n)$ is to set

$$\|(x_1, \ldots, x_n)\|_2 = \sqrt{\sum_{j=1}^{n} x_j^2}.$$

This is called the *Euclidean length*. Two other reasonable definitions of length are

$$\begin{aligned} \|(x_1, \ldots, x_n)\|_1 &= \sum_{j=1}^{n} |x_j|, \\ \|(x_1, \ldots, x_n)\|_\infty &= \max\{|x_1|, \ldots, |x_n|\}. \end{aligned} \tag{A.19}$$

What makes a notion of length reasonable? There are three basic properties that a reasonable notion of length should have. Let N denote a real-valued function defined on a vector space V; it defines a reasonable notion of length if it satisfies the following conditions:

NONDEGENERACY: For every $\boldsymbol{v} \in V$, $N(\boldsymbol{v}) \geq 0$ and $N(\boldsymbol{v}) = 0$ if and only if $\boldsymbol{v} = \boldsymbol{0}$. In other words, every vector has nonnegative length and only the zero vector has zero length.

HOMOGENEITY: If $a \in \mathbb{R}$ and $\boldsymbol{v} \in V$, then $N(a\boldsymbol{v}) = |a|N(\boldsymbol{v})$. If we scale a vector, its length gets multiplied by the scaling factor.

THE TRIANGLE INEQUALITY: If $\boldsymbol{v}, \boldsymbol{v}' \in V$, then

$$N(\boldsymbol{v} + \boldsymbol{v}') \leq N(\boldsymbol{v}) + N(\boldsymbol{v}'). \tag{A.20}$$

***Definition* A.2.12.** A function $N : V \to \mathbb{R}$ that satisfies these three conditions is called a *norm*. A vector space V with a norm is called a *normed vector space*.

The functions defined in (A.19) satisfy these conditions and therefore define norms.

***Example* A.2.15.** If p is a real number, larger than or equal to 1, then the function

$$\|(x_1, \ldots, x_n)\|_p \overset{d}{=} \left[\sum_{j=1}^{n} |x_j|^p\right]^{\frac{1}{p}} \tag{A.21}$$

defines a norm on $\mathbb{R}^n$. If $p = \infty$ then we define

$$\|(x_1, \ldots, x_n)\|_\infty \overset{d}{=} \max\{|x_1|, \ldots, |x_n|\}. \tag{A.22}$$

This is called the *sup norm*.

Using a norm N, we can define a notion of distance between two vectors by setting

$$d(\boldsymbol{x}, \boldsymbol{y}) = N(\boldsymbol{x} - \boldsymbol{y}).$$

For any choice of norm, this function has the following properties:

NONDEGENERACY: $d(\boldsymbol{x}, \boldsymbol{y}) \geq 0$ with equality if and only if $\boldsymbol{x} = \boldsymbol{y}$.

SYMMETRY: $d(\boldsymbol{x}, \boldsymbol{y}) = d(\boldsymbol{y}, \boldsymbol{x})$.

THE TRIANGLE INEQUALITY: For any 3 points $\boldsymbol{x}, \boldsymbol{y}, \boldsymbol{z}$, we have that

$$d(\boldsymbol{x}, \boldsymbol{z}) \leq d(\boldsymbol{x}, \boldsymbol{y}) + d(\boldsymbol{y}, \boldsymbol{z}).$$

Any function $d : \mathbb{R}^n \times \mathbb{R}^n \to \mathbb{R}$ with these properties is called a *metric*. While any norm defines a metric, there are metrics on $\mathbb{R}^n$ that are not defined by norms.

A metric gives a way to measure distances and therefore a way to define the convergence of sequences.

***Definition* A.2.13.** Suppose that $d(\cdot, \cdot)$ is a metric defined by a norm and that $< \boldsymbol{x}_j > \subset \mathbb{R}^n$ is a sequence of vectors. The sequence converges to $\boldsymbol{x}$ in the d-sense if

$$\lim_{j \to \infty} d(\boldsymbol{x}_j, \boldsymbol{x}) = 0.$$

Given a notion of distance, it is also possible to define Cauchy sequences.

***Definition* A.2.14.** Suppose that d is a metric on $\mathbb{R}^n$ and $< \boldsymbol{x}_n >$ is a sequence. It is a *Cauchy sequence* with respect to d if, for any $\epsilon > 0$, there exists an N so that

$$d(\boldsymbol{x}_n, \boldsymbol{x}_m) < \epsilon \qquad \text{provided that } m \text{ and } n > N.$$

The importance of this concept is contained in the following theorem.

Theorem A.2.1. *Let d be a metric on $\mathbb{R}^n$ defined by a norm. A sequence $< \boldsymbol{x}_n >$ converges in the d-sense if and only if it is a Cauchy sequence.*

If a norm is used to define a distance function, then it is reasonable to enquire if the convergence properties of sequences depend on the choice of norm. The next proposition implies that they do *not*.

Proposition A.2.3. *Suppose that $\| \cdot \|$ and $\| \cdot \|'$ are two norms on $\mathbb{R}^n$. Then there is a positive constant C so that*

$$C^{-1}\|\boldsymbol{x}\|' \leq \|\boldsymbol{x}\| \leq C\|\boldsymbol{x}\|' \qquad \text{for all } \boldsymbol{x} \in \mathbb{R}^n.$$

The proof is left as an exercise.

The choice of which norm to use in a practical problem is often dictated by physical considerations. For example, if we have a system whose state is described by a point in $\mathbb{R}^n$ and we allow the same uncertainty in each of our measurements, then it would be reasonable to use the sup norm (i.e., $\| \cdot \|_\infty$). If, on the other hand, we can only tolerate a certain fixed aggregate error, but it is not important how this error is distributed among the various measurements, then it would be reasonable to use $\| \cdot \|_1$ to define the norm. If the errors are expected to follow a Gaussian distribution, then we would usually use the Euclidean norm.

There are also computational considerations that can dictate the choice of a norm. If $\boldsymbol{a}$ is an $m \times n$ matrix with $m > n$, then the system of linear equations $\boldsymbol{ax} = \boldsymbol{y}$ is overdetermined. For most choices of $\boldsymbol{y}$ it has no solution. A way to handle such equations is to look for a vector such that the "size" of the error $\boldsymbol{ax} - \boldsymbol{y}$ is minimized. To do this, we need to choose a norm on $\mathbb{R}^m$ to measure the size of the error. It turns out that among all possible choices the Euclidean norm leads to the simplest minimization problems. The vector $\bar{\boldsymbol{x}}$ such that $\|\boldsymbol{a}\bar{\boldsymbol{x}} - \boldsymbol{y}\|_2$ is minimal is called the *least squares solution.*

Matrix and Operator Norms

In Exercise A.2.17 it is shown that the space of linear transformations from $\mathbb{R}^n$ to $\mathbb{R}^m$ is a vector space. When discussing numerical methods for solving linear equations, it is very useful to have a way to measure the size of a linear transformation that is connected to its geometric properties as a map. We can use norms on the domain and range to define a notion of size for a linear transformation $\boldsymbol{A} : \mathbb{R}^n \to \mathbb{R}^m$. Let $\|\cdot\|$ be a norm on $\mathbb{R}^n$ and $\|\cdot\|'$ be a norm on $\mathbb{R}^m$. The *operator norm* of $\boldsymbol{A}$ (with respect to $\|\cdot\|$ and $\|\cdot\|'$), denoted by $|||\boldsymbol{A}|||$, is defined by

$$|||\boldsymbol{A}||| \overset{d}{=} \max_{\boldsymbol{x} \in \mathbb{R}^n \setminus \{0\}} \frac{\|\boldsymbol{Ax}\|'}{\|\boldsymbol{x}\|}. \tag{A.23}$$

This norm gives a measure of how much $\boldsymbol{A}$ changes the lengths of vectors. For all $\boldsymbol{x} \in \mathbb{R}^n$ we have the estimate

$$\|\boldsymbol{Ax}\|' \leq |||\boldsymbol{A}||| \|\boldsymbol{x}\|. \tag{A.24}$$

The estimate (A.24) implies that a linear transformation from $\mathbb{R}^n$ to $\mathbb{R}^m$ is always continuous. This is because $\boldsymbol{Ax}_1 - \boldsymbol{Ax}_2 = \boldsymbol{A}(\boldsymbol{x}_1 - \boldsymbol{x}_2)$. Thus we see that

$$\|\boldsymbol{Ax}_1 - \boldsymbol{Ax}_2\|' = \|\boldsymbol{A}(\boldsymbol{x}_1 - \boldsymbol{x}_2)\| \leq |||\boldsymbol{A}||| \|\boldsymbol{x}_1 - \boldsymbol{x}_2\|. \tag{A.25}$$

There are other ways to define norms on linear transformations. If we fix bases in the domain and range,then a norm defined $\mathbb{R}^{mn}$ can be used to define a norm on the set of linear transformations from $\mathbb{R}^n$ to $\mathbb{R}^m$. If (a_{ij}) is the matrix of a linear transformation $\boldsymbol{A}$, then we can, for example, define

$$|||\boldsymbol{A}|||_p = \left[\sum_{i=1}^{m} \sum_{j=1}^{n} |a_{ij}|^p \right]^{\frac{1}{p}}.$$

These norms are not as closely connected to the geometric properties of the map. If $p \neq 2$, then it is *not* generally true that $\|\boldsymbol{Ax}\|_p \leq |||\boldsymbol{A}|||_p \|\boldsymbol{x}\|_p$. Results like that in Exercise A.2.23 are also generally false for these sorts of norms. In physical applications a linear transformation or matrix often models a measurement process: If $\boldsymbol{x}$ describes the state of a system, then $\boldsymbol{Ax}$ is the result of performing measurements on the system. The appropriate notion of size for $\boldsymbol{A}$ may then be determined by the sorts of errors which might arise in the model.

Exercises

***Exercise* A.2.18.** Suppose that $\boldsymbol{x}, \boldsymbol{y}, \boldsymbol{z} \in \mathbb{R}^n$ and that $d(\boldsymbol{x}, \boldsymbol{y}) = \|\boldsymbol{x} - \boldsymbol{y}\|_2$. If

$$d(\boldsymbol{x}, \boldsymbol{z}) = d(\boldsymbol{x}, \boldsymbol{y}) + d(\boldsymbol{y}, \boldsymbol{z}),$$

then show that the three points lie along a line in the indicated order. Is this true if we use $\| \cdot \|_p$ with $p \neq 2$ to define the metric?

***Exercise* A.2.19.** Prove Proposition A.2.3. *Hint*: Use the fact that $\|a\boldsymbol{x}\| = |a|\|\boldsymbol{x}\|$.

***Exercise* A.2.20.** Let $\|\cdot\|$ and $\|\cdot\|'$ be two norms on $\mathbb{R}^n$ and d, d' the corresponding metrics. Show that a sequence $< \boldsymbol{x}_j >$ converges to $\boldsymbol{x}$ in the d-sense if and only if it converges in the d'-sense. This shows that the notion of limits on Euclidean spaces is independent of the choice of norm. *Hint*: Use Proposition A.2.3.

***Exercise* A.2.21.** Suppose that w_1, w_2 are positive numbers. Show that

$$N_w((x_1, x_2)) = \sqrt{w_1 x_1^2 + w_2 x_2^2}$$

defines a norm on $\mathbb{R}^2$. What physical considerations might lead to using a norm like N_w instead of the standard Euclidean norm?

***Exercise* A.2.22.** Use estimate (A.25) to show that if a sequence $< \boldsymbol{x}_n >$ converges to $\boldsymbol{x}$ in $\mathbb{R}^n$ then $< \boldsymbol{A}\boldsymbol{x}_n >$ also converges to $\boldsymbol{A}\boldsymbol{x}$ in $\mathbb{R}^m$. This is just the statement that $\boldsymbol{A} : \mathbb{R}^n \to \mathbb{R}^m$ is continuous.

***Exercise* A.2.23.** Let $\boldsymbol{A} : \mathbb{R}^n \to \mathbb{R}^m$ and $\boldsymbol{B} : \mathbb{R}^m \to \mathbb{R}^l$. Choose norms $\| \cdot \|$, $\| \cdot \|'$, and $\| \cdot \|''$ for $\mathbb{R}^n$, $\mathbb{R}^m$, and $\mathbb{R}^l$, respectively and let $|||\cdot|||_{n\to m}$, $|||\cdot|||_{m\to l}$ and $|||\cdot|||_{n\to l}$ denote the operator norms they define. Show that

$$|||\boldsymbol{B} \circ \boldsymbol{A}|||_{n\to l} \leq |||\boldsymbol{A}|||_{n\to m} |||\boldsymbol{B}|||_{m\to l}.$$

***Exercise* A.2.24.** Let $A : \mathbb{R}^n \to \mathbb{R}^n$ have matrix (a_{ij}) with respect to the standard basis. Show that

$$\|\boldsymbol{A}\boldsymbol{x}\|_2 \leq |||\boldsymbol{a}|||_2 \|\boldsymbol{x}\|_2.$$

A.2.5 Inner Product Structure

The notions of distance considered in the previous section do not allow for the measurement of angles between vectors. Recall the formula for the dot product in $\mathbb{R}^2$,

$$\boldsymbol{x} \cdot \boldsymbol{y} = x_1 y_1 + x_2 y_2 = \|\boldsymbol{x}\|_2 \|\boldsymbol{y}\|_2 \cos\theta,$$

where θ is the angle between $\boldsymbol{x}$ and $\boldsymbol{y}$. We can generalize the notion of the dot product to n dimensions by setting

$$\boldsymbol{x} \cdot \boldsymbol{y} = \sum_{j=1}^{n} x_j y_j.$$

In this book the dot product is usually called an *inner product* and is denoted by $\langle \boldsymbol{x}, \boldsymbol{y} \rangle$.

Proposition A.2.4. *If x, y, z are vectors in $\mathbb{R}^n$ and $a \in \mathbb{R}$, then*

$$\begin{aligned}\langle \boldsymbol{x}, \boldsymbol{y} \rangle &= \langle \boldsymbol{y}, \boldsymbol{x} \rangle, \\ \langle (\boldsymbol{x} + \boldsymbol{y}), \boldsymbol{z} \rangle &= \langle \boldsymbol{x}, \boldsymbol{z} \rangle + \langle \boldsymbol{y}, \boldsymbol{z} \rangle \text{ and} \\ \langle a\boldsymbol{x}, \boldsymbol{y} \rangle &= a\langle \boldsymbol{x}, \boldsymbol{y} \rangle = \langle \boldsymbol{x}, a\boldsymbol{y} \rangle.\end{aligned} \tag{A.26}$$

The inner product is connected with the Euclidean norm by the relation

$$\langle \boldsymbol{x}, \boldsymbol{x} \rangle = \|\boldsymbol{x}\|_2^2.$$

Most of the special properties of the Euclidean norm stem from this fact. There is a very important estimate which also connects these two objects called the *Cauchy-Schwarz inequality*:

$$|\langle \boldsymbol{x}, \boldsymbol{y} \rangle| \leq \|\boldsymbol{x}\|_2 \|\boldsymbol{y}\|_2. \tag{A.27}$$

It is proved in Exercise A.2.26. It implies that

$$-1 \leq \frac{\langle \boldsymbol{x}, \boldsymbol{y} \rangle}{\|\boldsymbol{x}\|_2 \|\boldsymbol{y}\|_2} \leq 1. \tag{A.28}$$

In light of (A.28) we can define the angle θ between two nonzero vectors in $\mathbb{R}^n$ by the formula

$$\cos\theta = \frac{\langle \boldsymbol{x}, \boldsymbol{y} \rangle}{\|\boldsymbol{x}\|_2 \|\boldsymbol{y}\|_2}.$$

An important special case is an angle of $90°$, which is the case if $\langle \boldsymbol{x}, \boldsymbol{y} \rangle = 0$. The vectors $\boldsymbol{x}$ and $\boldsymbol{y}$ are said to be *orthogonal*.

Suppose that $\{\boldsymbol{v}_1, \dots, \boldsymbol{v}_n\}$ is a basis for $\mathbb{R}^n$. In order to make practical use of this basis, it is necessary to be able to determine the coordinates of a vector with respect to it. Suppose that $\boldsymbol{x}$ is a vector. We would like to find scalars $\{a_j\}$ so that

$$\boldsymbol{x} = \sum_{j=1}^{n} a_j \boldsymbol{v}_j.$$

Expressing the basis vectors and $\boldsymbol{x}$ in terms of the standard basis, $\boldsymbol{x} = (x_1, \dots, x_n)$ and $\boldsymbol{v}_j = (v_{1j}, \dots, v_{nj})$, this can be reexpressed as a system of linear equations,

$$\sum_{j=1}^{n} v_{ij} a_j = x_i \qquad \text{for } i = 1, \dots, n.$$

In general, this system can be quite difficult to solve; however, there is a special case when it is easy to write down a formula for the solution.

Suppose that the basis vectors are of Euclidean length 1 and pairwise orthogonal; that is,

$$\|\boldsymbol{v}_j\|_2 = 1 \text{ for } j = 1, \dots, n \text{ and } \langle \boldsymbol{v}_i, \boldsymbol{v}_j \rangle = 0 \text{ if } i \neq j.$$

Such a basis is called an *orthonormal basis*. The standard basis is an orthonormal basis. If $\{\boldsymbol{v}_j\}$ is an orthonormal basis, then the coordinates of $\boldsymbol{x}$ with respect to $\{\boldsymbol{v}_j\}$ can be computing by simply evaluating inner products,

$$a_j = \langle \boldsymbol{x}, \boldsymbol{v}_j \rangle;$$

hence

$$\boldsymbol{x} = \sum_{j=1}^{n} \langle \boldsymbol{x}, \boldsymbol{v}_j \rangle \boldsymbol{v}_j. \tag{A.29}$$

A consequence of (A.29) is that, in any orthonormal basis $\{\boldsymbol{v}_j\}$, the Pythagorean theorem holds:

$$\|\boldsymbol{x}\|_2^2 = \sum_{j=1}^{n} |\langle \boldsymbol{x}, \boldsymbol{v}_j \rangle|^2. \tag{A.30}$$

An immediate consequence of (A.30) is that the individual coordinates of a vector, with respect to an orthonormal basis, are bounded by the Euclidean length of the vector. Orthonormal bases are often preferred in applications because they display stability properties not shared by arbitrary bases.

***Example* A.2.16.** If $\epsilon \neq 0$, then the vectors $\boldsymbol{v} = (1, 0)$ and $\boldsymbol{u}_\epsilon = (1, \epsilon)$ are a basis for $\mathbb{R}^2$. If ϵ is small, then the angle between these vectors is very close to zero. The representation of $(0, 1)$ with respect to $\{\boldsymbol{v}, \boldsymbol{u}_\epsilon\}$ is

$$(0, 1) = \frac{1}{\epsilon}\boldsymbol{u}_\epsilon - \frac{1}{\epsilon}\boldsymbol{v}.$$

The coefficients blow up as ϵ goes to zero. For non-orthonormal bases, it can be difficult to estimate the sizes of the coefficients in terms of the length of the vector.

The Gram-Schmidt Method

The problem then arises of how to construct orthonormal bases; this problem is solved using the *Gram-Schmidt* method. Beginning with an arbitrary basis $\{\boldsymbol{u}_1, \ldots, \boldsymbol{u}_n\}$, the Gram-Schmidt method produces an orthonormal basis. It has the special property that for each $1 \leq j \leq n$ the linear span of $\{\boldsymbol{u}_1, \ldots, \boldsymbol{u}_j\}$ is the same as the linear span of $\{\boldsymbol{v}_1, \ldots, \boldsymbol{v}_j\}$. This method is important for both theoretical and practical applications.

We describe the Gram-Schmidt method as an algorithm:

STEP 1: Replace $\boldsymbol{u}_1$ with the vector

$$\boldsymbol{v}_1 = \frac{\boldsymbol{u}_1}{\|\boldsymbol{u}_1\|_2}.$$

The span of $\boldsymbol{v}_1$ clearly agrees with that of $\boldsymbol{u}_1$.

STEP 2: For a $1 \leq j < n$, suppose that we have found orthonormal vectors $\{\boldsymbol{v}_1, \ldots, \boldsymbol{v}_j\}$ such that the linear span of $\{\boldsymbol{v}_1, \ldots, \boldsymbol{v}_j\}$ is the same as that of $\{\boldsymbol{u}_1, \ldots, \boldsymbol{u}_j\}$. Set

$$\boldsymbol{v}'_{j+1} = \boldsymbol{u}_{j+1} + \sum_{k=1}^{j} \alpha_k \boldsymbol{v}_k,$$

where $\alpha_k = -\langle \boldsymbol{u}_{j+1}, \boldsymbol{v}_k \rangle$. A calculation shows that

$$\langle \boldsymbol{v}'_{j+1}, \boldsymbol{v}_k \rangle = 0 \qquad \text{for } 1 \leq k \leq j.$$

STEP 3: Since $\{\boldsymbol{u}_i\}$ is a basis and $\{\boldsymbol{v}_1, \ldots, \boldsymbol{v}_j\}$ are in the linear span of $\{\boldsymbol{u}_1, \ldots, \boldsymbol{u}_j\}$, it follows that $\boldsymbol{v}'_{j+1} \neq 0$; thus we can set

$$\boldsymbol{v}_{j+1} = \frac{\boldsymbol{v}'_{j+1}}{\|\boldsymbol{v}'_{j+1}\|_2}.$$

STEP 4: If $j = n$, we are done; otherwise, return to Step 2.

This algorithm shows that there are many orthonormal bases.

Linear Functions and Inner Products

The inner product also gives a way to represent linear functions. A vector $\boldsymbol{y} \in \mathbb{R}^n$ defines a linear function $l_{\boldsymbol{y}}$ by the rule

$$l_{\boldsymbol{y}}(\boldsymbol{x}) = \langle \boldsymbol{x}, \boldsymbol{y} \rangle. \tag{A.31}$$

Proposition A.2.5. *The linear function $l_{\boldsymbol{y}}$ is zero if and only if $\boldsymbol{y} = \boldsymbol{0}$. Moreover,*

$$l_{\boldsymbol{y}_1 + \boldsymbol{y}_2} = l_{\boldsymbol{y}_1} + l_{\boldsymbol{y}_2},$$

and if $a \in \mathbb{R}$, then $l_{a\boldsymbol{y}_1} = a l_{\boldsymbol{y}_1}$.

The proposition shows that the map $\boldsymbol{y} \mapsto l_{\boldsymbol{y}}$ defines an isomorphism between $\mathbb{R}^n$ and $(\mathbb{R}^n)'$. The map is clearly linear; because $l_{\boldsymbol{y}} = 0$ if and only if $\boldsymbol{y} = 0$, it follows [from (A.34)] that the image of the map is all of $(\mathbb{R}^n)'$. In other words, every linear function on $\mathbb{R}^n$ has a representation as $l_{\boldsymbol{y}}$ for a unique $\boldsymbol{y} \in \mathbb{R}^n$.

Let $\boldsymbol{A} : \mathbb{R}^n \to \mathbb{R}^m$ be a linear transformation. If $\boldsymbol{y} \in \mathbb{R}^m$, then $\boldsymbol{x} \mapsto \langle \boldsymbol{A}\boldsymbol{x}, \boldsymbol{y} \rangle$ is a linear function on $\mathbb{R}^n$. This means that there is a vector $\boldsymbol{z} \in \mathbb{R}^n$ such that

$$\langle \boldsymbol{A}\boldsymbol{x}, \boldsymbol{y} \rangle = \langle \boldsymbol{x}, \boldsymbol{z} \rangle, \text{ for all } \boldsymbol{x} \in \mathbb{R}^n.$$

We denote this vector by $\boldsymbol{A}^t \boldsymbol{y}$. It is not difficult to show that the map $\boldsymbol{y} \mapsto \boldsymbol{A}^t \boldsymbol{y}$ is a linear transformation from $\mathbb{R}^m$ to $\mathbb{R}^n$.

Proposition A.2.6. *If $A : \mathbb{R}^n \to \mathbb{R}^m$ has matrix (a_{ij}) with respect to the standard bases, then A^t has matrix (a_{ji}) with respect to the standard bases,*

$$(A^t y)_i = \sum_{j=1}^{m} a_{ji} y_j.$$

The linear transformation $A^t : \mathbb{R}^m \to \mathbb{R}^n$ is called the *transpose* (or *adjoint*) of A. Note that while the matrices representing A and its transpose are simply related, the transpose is defined without reference to a choice of basis:

$$\langle Ax, y \rangle_m = \langle x, A^t y \rangle_n$$

for all $x \in \mathbb{R}^n$ and $y \in \mathbb{R}^m$. In order to avoid confusion, we have used $\langle \cdot, \cdot \rangle_n$ (respectively, $\langle \cdot, \cdot \rangle_m$) to denote the inner product on $\mathbb{R}^n$ (respectively, $\mathbb{R}^m$).

We close this section by placing these considerations in a slightly more abstract framework.

***Definition* A.2.15.** Let V be a vector space, a function $b : V \times V \to \mathbb{R}$, that satisfies the conditions $b(v, v) \geq 0$ with $b(v, v) = 0$ if and only if $v = 0$, and for all $v, w, z \in V$ and $a \in \mathbb{R}$

$$\begin{aligned} b(v, w) &= b(w, v), \\ b(v + w, z) &= b(v, z) + b(w, z) \text{ and} \\ b(av, w) &= ab(v, w) = b(v, aw) \end{aligned} \tag{A.32}$$

defines an *inner product* on V. A function with the properties in (A.32) is called a *bilinear function.*

***Example* A.2.17.** Let $A : \mathbb{R}^n \to \mathbb{R}^n$ be a linear transformation with $\ker A = \{0\}$. Then

$$\langle x, y \rangle_{\mathcal{A}} = \langle Ax, Ay \rangle$$

defines an inner product on $\mathbb{R}^n$.

***Example* A.2.18.** Let $\mathcal{P}_n$ be the real-valued polynomials of degree at most n. Then

$$b_n(p, q) = \int_{-1}^{1} p(x) q(x)\, dx$$

defines an inner product on $\mathcal{P}_n$.

Exercises

***Exercise* A.2.25.** Suppose that $A : \mathbb{R}^n \to \mathbb{R}^m$ and $B : \mathbb{R}^m \to \mathbb{R}^l$. Show that

$$(B \circ A)^t = A^t \circ B^t.$$

Express this relation in terms of the matrices for these transformations with respect to the standard bases.

***Exercise* A.2.26.** Calculus can be used to proved (A.27). Let $\boldsymbol{x}$ and $\boldsymbol{y}$ be vectors in $\mathbb{R}^n$ and define the function

$$f(t) = \langle \boldsymbol{x} + t\boldsymbol{y}, \boldsymbol{x} + t\boldsymbol{y} \rangle = \|\boldsymbol{x} + t\boldsymbol{y}\|_2^2.$$

This function satisfies $f(t) \geq 0$ for all $t \in \mathbb{R}$. Use calculus to locate the value of t where f assumes it minimum. By evaluating f at its minimum and using the fact that $f(t) \geq 0$, show that (A.27) holds.

***Exercise* A.2.27.** Let b be an inner product on a vector space V. Using the idea outlined in Exercise A.2.26, show that

$$|b(\boldsymbol{v}_1, \boldsymbol{v}_2)| \leq \sqrt{b(\boldsymbol{v}_1, \boldsymbol{v}_1) b(\boldsymbol{v}_2, \boldsymbol{v}_2)}.$$

***Exercise* A.2.28.** Show that the Gram-Schmidt procedure can be applied to an arbitrary vector space with an inner product.

***Exercise* A.2.29.** Apply the Gram-Schmidt process to the basis $\{1, x, x^2\}$ with the inner product given in Example A.2.18 to find an orthonormal basis for $\mathcal{P}_2$.

***Exercise* A.2.30.** Prove Proposition A.2.4.

***Exercise* A.2.31.** If $\boldsymbol{a}$ is an $m \times n$ matrix and $\boldsymbol{x} \in \mathbb{R}^n$, then we can use the inner product to express the matrix product $\boldsymbol{a}\boldsymbol{x}$. Show that if we write $\boldsymbol{a}$ in terms of its rows

$$\boldsymbol{a} = \begin{pmatrix} \boldsymbol{a}_1 \\ \vdots \\ \boldsymbol{a}_m \end{pmatrix},$$

then

$$\boldsymbol{a}\boldsymbol{x} = \begin{pmatrix} \langle \boldsymbol{a}_1, \boldsymbol{x} \rangle \\ \vdots \\ \langle \boldsymbol{a}_m, \boldsymbol{x} \rangle \end{pmatrix}. \tag{A.33}$$

***Exercise* A.2.32.** If $\boldsymbol{a}$ is an $m \times n$ matrix and $\boldsymbol{b}$ an $n \times l$ matrix, then we can use the inner product to express the matrix product $\boldsymbol{a}\boldsymbol{b}$. Show that if we write $\boldsymbol{a}$ in terms of its rows

$$\boldsymbol{a} = \begin{pmatrix} \boldsymbol{a}_1 \\ \vdots \\ \boldsymbol{a}_m \end{pmatrix},$$

and $\boldsymbol{b}$ in terms of its columns

$$\boldsymbol{b} = (\boldsymbol{b}^1, \ldots, \boldsymbol{b}^l),$$

then the ij entry of $\boldsymbol{a}\boldsymbol{b}$ is given by $\langle \boldsymbol{a}_i, \boldsymbol{b}^j \rangle$.

***Exercise* A.2.33.** Prove formula (A.29).

***Exercise* A.2.34.** Prove Proposition A.2.5.

***Exercise* A.2.35.** Prove Proposition A.2.6.

***Exercise* A.2.36.** Show that $\langle \boldsymbol{x}, \boldsymbol{y} \rangle_{\boldsymbol{A}}$ is an inner product. Why do we need to assume that $\ker \boldsymbol{A} = \{0\}$?

***Exercise* A.2.37.** Prove that b_n defined in Example A.2.18 is an inner product.

A.2.6 Linear Transformations and Linear Equations

Linear transformations give a geometric way to think about linear equations. A system of m linear equations in n unknowns is given by

$$\sum_{j=1}^{n} a_{ij}x_j = y_i \qquad \text{for } i = 1, \ldots, m.$$

The matrix $\boldsymbol{a}$ defines a linear transformation $\boldsymbol{A} : \mathbb{R}^n \to \mathbb{R}^m$. The null space of $\boldsymbol{A}$ is none other than the set of solutions to the *homogeneous equation*

$$\boldsymbol{a}\boldsymbol{x} = \boldsymbol{0}.$$

The system of equations

$$\boldsymbol{a}\boldsymbol{x} = \boldsymbol{y}$$

has a solution if and only if $\boldsymbol{y}$ belongs to the image of $\boldsymbol{A}$. Theorem 2.1.2 relates the dimensions of the null space and image of $\boldsymbol{A}$; they satisfy the relation

$$\dim \ker \boldsymbol{A} + \dim \operatorname{Im} \boldsymbol{A} = n. \tag{A.34}$$

If $\boldsymbol{A} : \mathbb{R}^n \to \mathbb{R}^n$ and $\dim \ker \boldsymbol{A} = 0$, then formula (A.34) implies that $\dim \operatorname{Im} \boldsymbol{A} = n$ and therefore for every $\boldsymbol{y} \in \mathbb{R}^n$ there is a *unique* $\boldsymbol{x} \in \mathbb{R}^n$ such that

$$\boldsymbol{A}\boldsymbol{x} = \boldsymbol{y}.$$

A linear transformation with this property is called *invertible;* we let $\boldsymbol{A}^{-1}$ denote the *inverse* of $\boldsymbol{A}$. It is also a linear transformation. A linear transformation and its inverse satisfy the relations

$$\boldsymbol{A}^{-1}\boldsymbol{A} = \operatorname{Id}_n = \boldsymbol{A}\boldsymbol{A}^{-1}.$$

If (a_{ij}) is the matrix of $\boldsymbol{A}$ with respect to a basis and (b_{ij}) is the matrix for $\boldsymbol{A}^{-1}$, then these relations imply that

$$\sum_{j=1}^{n} b_{ij}a_{jk} = \delta_{ik} = \sum_{j=1}^{n} a_{ij}b_{jk}. \tag{A.35}$$

From a purely mathematical standpoint, the problem of solving the linear equation $\boldsymbol{A}\boldsymbol{x} = \boldsymbol{y}$ is simply a matter of computing the matrix representing $\boldsymbol{A}^{-1}$. Cramer's rule gives an explicit formula for $\boldsymbol{A}^{-1}$, though it is very unusual to solve linear equations this way. The direct computation of $\boldsymbol{A}^{-1}$ is computationally expensive and usually unstable. Less direct, computationally more stable and efficient methods are usually employed.

***Definition* A.2.16.** An $n \times n$ matrix (a_{ij}) is called *upper triangular* if

$$a_{ij} = 0 \qquad \text{if } j < i.$$

A system of equations is upper triangular if its matrix of coefficients is upper triangular.

Upper triangular systems are easy to solve. Suppose that (a_{ij}) is an upper triangular matrix with all of its diagonal entries $\{a_{ii}\}$ nonzero. The system of equations $\boldsymbol{ax} = \boldsymbol{y}$ becomes

$$\sum_{j=i}^{n} a_{ij} x_j = y_i \text{ for } i = 1, \ldots, n.$$

It is easily solved using the *back-substitution* algorithm:

STEP 1: Let $x_n = a_{nn}^{-1} y_n$.

STEP 2: For a $1 < j < n$, assume we know $(x_{j+1}, \ldots, x_n)$ and let

$$x_{j+1} = \frac{y_{j+1} - \sum_{k=j+1}^{n} a_{(j+1)k} x_k}{a_{(j+1)(j+1)}}.$$

STEP 3: If $j = n$, we are done; otherwise, return to Step 2.

Another important class of matrices has orthonormal rows (and columns).

***Definition* A.2.17.** A matrix $\boldsymbol{a} = (a_{ij})$ is *orthogonal* if

$$\sum_{j=1}^{n} a_{ij} a_{kj} = \begin{cases} 1 & \text{if } i = k, \\ 0 & \text{if } i \neq k. \end{cases}$$

In terms of matrix multiplication, this condition is expressed by

$$\boldsymbol{a}\boldsymbol{a}^t = \mathrm{Id}_n = \boldsymbol{a}^t \boldsymbol{a}.$$

Hence a matrix is orthogonal if $\boldsymbol{a}^t$ is the inverse of $\boldsymbol{a}$.

Let $\boldsymbol{a}$ be an orthogonal matrix and let $\{\boldsymbol{a}_j : i = 1, \ldots n\}$ denote its columns thought of $n \times 1$ vectors. The solution to the equation $\boldsymbol{ax} = \boldsymbol{y}$ is given by

$$x_j = \langle \boldsymbol{a}_j, \boldsymbol{y} \rangle \text{ for } j = 1, \ldots, n.$$

We have found two classes of linear equations that are computationally simple to solve. Using the Gram-Schmidt algorithm, we can prove the following statement:

Theorem A.2.2. *Suppose that $\boldsymbol{a}$ is an invertible $n \times n$ matrix. Then there exists an upper triangular matrix $\boldsymbol{r}$ and an orthogonal matrix $\boldsymbol{q}$ such that*

$$\boldsymbol{a} = \boldsymbol{q}\boldsymbol{r}.$$

Once a matrix is expressed in this form, the system of equations $\boldsymbol{ax} = \boldsymbol{y}$ is easily solved in two steps: Multiplying by $\boldsymbol{q}^t$ gives the upper triangular system $\boldsymbol{rx} = \boldsymbol{q}^t \boldsymbol{y}$, which is then solved by back substitution. There is an enormous literature devoted to practical implementations of this and similar results. A good starting point for further study is [127].

Exercises

***Exercise* A.2.38.** Show that if a_{ij} is an upper triangular matrix with $a_{ii} = 0$ for some i, then there is a nonzero vector $(x_1, \ldots, x_n)$ such that

$$\sum_{j=i}^{n} a_{ij} x_j = 0.$$

In other words, the homogeneous equation has a nontrivial solution.

***Exercise* A.2.39.** Let $\boldsymbol{a}$ be an invertible upper triangular matrix. Show that $\boldsymbol{a}^{-1}$ is also upper triangular.

***Exercise* A.2.40.** Show that if $\boldsymbol{a}$ and $\boldsymbol{b}$ are upper triangular matrices, then so is $\boldsymbol{ab}$.

***Exercise* A.2.41.** Prove Theorem A.2.2.

A.3 Functions, Theory, and Practice

The idea of a function is familiar from calculus. A real-valued function on $\mathbb{R}$ is a *rule* for assigning to each $x \in \mathbb{R}$ a unique value $y \in \mathbb{R}$. Usually we write something like $y = f(x)$. In this context, what is meant by a "rule?" The simplest functions are described by explicit formulæ involving a variable and arithmetic operations. For example,

$$\begin{aligned} f_1(x) &= 1, \\ f_2(x) &= 2 + x^3, \\ f_3(x) &= \frac{7 + 3x + 6x^3 + 17x^9}{3 + 4x^2 + 5x^4}. \end{aligned} \tag{A.36}$$

The functions we get this way are called *rational functions*; these are functions that can be expressed as ratios of polynomials. These functions have the following considerable virtue: If we "know" what x is and we can "do" arithmetic, then we can actually compute (in finite time) the value of $f(x)$. Other than expressing the numbers themselves, no *infinite processes* are required to evaluate a rational function. This is a concept we consider in some detail, so we give it a name.

***Definition* A.3.1.** A function f is a *real computable function* if its value can be determined for any real number by doing a finite number of *feasible* operations.

For more on this concept, see [39].

What are the "feasible operations?" Feasible operations are those that require only the ability to do arithmetic and to determine if a number is nonnegative. These are the operations that can be done *approximately* by a computer. We can give an analogous definition for computable functions defined on $\mathbb{R}^n$ or on subsets of $\mathbb{R}^n$, $n \geq 1$.

Rational functions are computable functions, but there are other types of computable functions. If $[a, b] \subset \mathbb{R}$ is an *interval*, that is,

$$[a, b] = \{x \in \mathbb{R} : a \leq x \leq b\},$$

then we define the characteristic function of an interval by the rule

$$\chi_{[a,b]}(x) = \begin{cases} 1 \text{ if} & x \in [a,b], \\ 0 \text{ if} & x \notin [a,b]. \end{cases}$$

Again, if we know the exact value of x, then to compute $\chi_{[a,b]}(x)$ we only need to perform feasible operations: checking if $0 \leq x - a$ and $0 \leq b - x$.

Proposition A.3.1. *Suppose that f and g are computable functions. Then $f + g$, fg, $f - g$, f/g and $f \circ g$, are also computable functions.*

The set of computable functions is, in essence, the set of functions that are actually available for computational purposes. They are the functional analogue of floating-point numbers. However, it is very easy to define functions, quite explicitly, that do not fall into this class. The function $f(x) = x^3$ is a computable function and it is one to one. That is, $f(x_1) = f(x_2)$ implies that $x_1 = x_2$. For every real number y there is a unique x so that $f(x) = y$. This means there is a function g that inverts f; that is, $g(f(x)) = x$. Of course, this is just the cube root function. Much less evident is how to compute $g(y)$ for a given value of y; g is not a computable function.

A function can also be defined implicitly via a functional relation. For example, we can think of y as a function of x defined by the relation

$$x^2 + y^2 = 1.$$

Evaluating y as a function of x entails solving this equation. Formally, we can write

$$y_{\pm}(x) = \pm\sqrt{1 - x^2}.$$

The relation actually defines two functions, which is not a serious difficulty; however, to compute either $y_+(x)$ or $y_-(x)$ requires the ability to calculate a square root. In this case there is a trick that effectively avoids the computation of the square root. If

$$x(t) = \frac{1 - t^2}{1 + t^2} \text{ and } y(t) = \frac{2t}{1 + t^2},$$

then

$$x(t)^2 + y(t)^2 = 1.$$

Both of the functions $x(t)$ and $y(t)$ are computable and so we see that, at the expense of expressing both x and y in terms of an auxiliary variable, t we are able to solve $x^2 + y^2 = 1$. For only slightly more complicated equations in two variables it is known that no such trick exists. Solving nonlinear equations in one or several variables usually leads to noncomputable functions.

Probably the most important examples of noncomputable functions are the solutions of linear, ordinary differential equations. For example, the $\sin x$, $\cos x$, $\exp x$ all arise in this context as well as the Bessel functions, Legendre functions, and so on. Such functions are called transcendental functions. For many purposes these functions are regarded as completely innocuous. They are, however, not computable, except for very special values of x. The reason that these functions are not greeted with horror is that they are all well approximated by computable functions in a precise sense: For each of these functions there are computable approximations and estimates for the differences between the actual functions and their approximations. In fact, as machine computation is always approximate, it is not necessary (or even desirable) to evaluate functions exactly. It is only necessary to be able to evaluate functions to within a *specified* error.

Exercises

***Exercise* A.3.1.** Prove Proposition A.3.1.

***Exercise* A.3.2.** Give a definition for computable functions of several variables. Show that linear functions are computable functions. Show, moreover, that the solution $\boldsymbol{x}$ of a system of linear equations

$$\boldsymbol{a}\boldsymbol{x} = \boldsymbol{y}$$

is a computable function of $\boldsymbol{y}$.

A.3.1 Power Series

Many of the functions encountered in applications can be represented as infinite sums of computable functions. Power series is the most important general class of such functions. This class is the infinite sum generalization of polynomials.

***Definition* A.3.2.** Let $< a_j >$ be a sequence of complex numbers. The *power series* with these coefficients is the infinite series

$$\sum_{j=0}^{\infty} a_j z^j; \tag{A.37}$$

z is a complex number.

As it stands, a power series is a formal expression. The theory of convergence of power series is relatively simple. Roughly speaking, a power series converges for a complex argument z provided that $\lim_{j\to\infty} |a_j z^j| = 0$. The exact result is given in the following theorem.

Theorem A.3.1. *Suppose that* $r \geq 0$ *and*

$$\lim_{j\to\infty} |a_j| r^j = 0. \tag{A.38}$$

Then the power series (A.37) *converges absolutely for all complex numbers* z *with* $|z| < r$.

The supremum of the numbers that satisfy (A.38) is called the *radius of convergence* of the power series; we denote it by r_{conv}. For values of z with $|z| < r_{\text{conv}}$, the power series converges absolutely; if $|z| = r_{\text{conv}}$, then the question of convergence or divergence of the series is again quite subtle.

***Example* A.3.1.** If $a_j = j^{-1}$, then $r_{\text{conv}} = 1$. For $|z| < 1$ the series

$$\sum_{j=1}^{\infty} \frac{z^j}{j}$$

converges absolutely. If $z = 1$, then the series diverges, while if $z = -1$, the series converges.

***Example* A.3.2.** Suppose that $a_j = j^j$. Then for any number $r > 0$ we have that

$$a_j r^j = (jr)^j.$$

If $jr > 2$, then $a_j r^j > 2^j$, and this shows that the radius of convergence of the power series with these coefficients is 0. In general, if the coefficients grow too quickly, then the series does not converge for any nonzero value of z. While such series do not, strictly speaking, define functions, they often appear in applications as *asymptotic expansions* for functions; see [25].

In the set

$$B_{\text{conv}} = \{z : |z| < r_{\text{conv}}\}$$

the series (A.37) defines a function of z with many of the properties of polynomials. Let

$$f(z) = \sum_{j=0}^{\infty} a_j z^j, \tag{A.39}$$

and suppose that r_{conv}, the radius of convergence is positive. Formally differentiating gives a new power series,

$$f_1(z) = \sum_{j=1}^{\infty} j a_j z^{j-1}.$$

It is not difficult to show that the radius of convergence of this series is also r_{conv} and, in fact, $f'(z) = f_1(z)$, see [2]. This can, of course, be repeated over and over. These observations are summarized in the following theorem.

Theorem A.3.2. *Suppose that $r_{\text{conv}} > 0$, the radius of convergence of the power series* (A.39), *is positive. The function it defines in B_{conv} is infinitely differentiable. For each $k \geq 0$,*

$$f^{[k]}(z) = \sum_{j=k}^{\infty} a_j j(j-1)\ldots(j-k+1)z^{j-k}$$

also has radius of convergence r_{conv}. Note in particular that

$$f^{[k]}(0) = k! a_k.$$

***Example* A.3.3.** The functions $\sin(z)$, $\cos(z)$, $\exp(z)$ are defined as the solutions of differential equations. The sine and cosine satisfy

$$f'' + f = 0,$$

while the exponential solves

$$f' - f = 0.$$

Assuming that these functions have power series expansions, we find, by substituting into the differential equations, that

$$\begin{aligned} \sin(x) &= \sum_{j=0}^{\infty} \frac{(-1)^j x^{2j+1}}{(2j+1)!}, \\ \cos(x) &= \sum_{j=0}^{\infty} \frac{(-1)^j x^{2j}}{(2j)!}, \\ \exp(x) &= \sum_{j=0}^{\infty} \frac{x^j}{j!}. \end{aligned} \tag{A.40}$$

Here we have use the facts that $\sin(0) = 0$ and $\cos(0) = 1$. From these formulæ it is not difficult to see that the radii of convergence of these series are infinite and that each of these functions satisfies the appropriate differential equation. A power series is defined for complex numbers; substituting $z = ix$ into the series for exp, we learn that

$$\exp(ix) = \cos(x) + i\sin(x). \tag{A.41}$$

This is a very useful fact both theoretically and for computation; it is called *Euler's formula*.

A shown in Exercise A.3.5 the exponential is positive and monotone increasing on the real line. Thus exp has an inverse function $l(y)$, defined for positive real numbers y; it satisfies

$$\exp(l(y)) = y \text{ and } l(\exp(x)) = x.$$

Note that $l(1) = 0$. This function is called the logarithm (or natural logarithm). Following standard practice, we use the notation $\log(y)$ for $l(y)$. As the derivative of exp is non-vanishing, its inverse is also differentiable. Using the chain rule, we obtain that

$$\log'(y) = \frac{1}{y}. \tag{A.42}$$

This differential equation and $\log(1) = 0$ imply that $\log(y)$ can be expressed as an integral:

$$\log(y) = \int_1^y \frac{ds}{s}. \tag{A.43}$$

Because the log is not defined at 0, this function does not have a convergent power series expansion about $x = 0$. Nonetheless, the log has a representation as the definite integral of a computable function. Using numerical integration, it is therefore not difficult to evaluate $\log y$, with any specified accuracy.

While a function defined as an infinite sum is not in general a computable function, a power series is computable to any *given precision*. Suppose that f is a power series with a positive radius of convergence r_{conv}. If we specify an $\epsilon > 0$ and an argument z with $|z| < r_{\text{conv}}$, then there is a N such that

$$|f(z) - \sum_{j=0}^{N} a_j z^j| < \epsilon.$$

If we have a formula for the coefficients $\{a_j\}$, as is usually the case, then we can compute N as a function of ϵ and $|z|/r_{\text{conv}}$.

***Example* A.3.4.** The series defining the sine and cosine are alternating series; this means that

$$\left|\sin(x) - \sum_{j=0}^{N} \frac{(-1)^j x^{2j+1}}{(2j+1)!}\right| \leq \frac{|x|^{2N+3}}{(2N+3)!} \text{ and } \left|\cos(x) - \sum_{j=0}^{N} \frac{(-1)^j x^{2j}}{(2j)!}\right| \leq \frac{|x|^{2N+2}}{(2N+2)!}; \tag{A.44}$$

see (B.3). Because $\sin(x) = (-1)^k \sin(x + k\pi)$, it suffices to consider values of x between $-\pi/2$ and $\pi/2$. For such x, to compute $\sin(x)$ with a error less than $\epsilon > 0$ requires an N that satisfies

$$\frac{\pi^{2N+3}}{2^{2N+3}(2N+3)!} < \epsilon.$$

Using this formula we obtain Table A.1. This gives an effective algorithm to compute the sine (or cosine) to a given accuracy. In actual applications, computing the partial sums of the power series is not used because much faster algorithms can be obtained by using the multiple angle formulæ.

N	Maximum error
4	10^{-4}
6	10^{-8}
8	10^{-12}
10	10^{-16}

Table A.1. Errors approximating $\sin(x)$ by partial sums of its Taylor series

Exercises

***Exercise* A.3.3.** Using Euler's formula, deduce that

$$\cos(x) = \frac{e^{ix} + e^{-ix}}{2}, \quad \sin(x) = \frac{e^{ix} - e^{-ix}}{2i}. \tag{A.45}$$

***Exercise* A.3.4.** Using the uniqueness theorem for ordinary differential equations, prove that for real numbers x and y,

$$\exp(x+y) = \exp(x)\exp(y), \quad \exp(-x) = [\exp(x)]^{-1}. \tag{A.46}$$

The function $g(x) = \exp(ix)$ satisfies the ODE $g' - ig = 0$, and therefore the relations (A.46) hold with x and y replaced by ix and iy. Deduce the multiple angle formulæ

$$\cos(x+y) = \cos(x)\cos(y) - \sin(x)\sin(y), \quad \sin(x+y) = \sin(x)\cos(y) + \sin(y)\cos(x). \tag{A.47}$$

***Exercise* A.3.5.** Show that $\exp(x)$ is positive if x is real, and conclude that the derivative of exp is also positive on the real axis. Show that exp has an inverse function defined on $(0, \infty)$.

***Exercise* A.3.6.** Euler's formula shows that a complex number has a *polar representation* in the form $z = re^{i\theta}$, where r and θ are real numbers; compare with (2.24). If $w = \rho e^{i\phi}$, show that

$$zw = r\rho e^{i(\theta+\phi)}. \tag{A.48}$$

***Exercise* A.3.7.** Using the integral formula, (A.43), prove that

$$\log(xy) = \log(x) + \log(y). \tag{A.49}$$

***Exercise* A.3.8.** Show that the log has a convergent power series expansion about $y = 1$ given by

$$\log(1+t) = -\sum_{j=1}^{\infty} \frac{(-t)^j}{j}.$$

For what values of t does this series converge?

A.3.2 The Binomial Formula

The elementary binomial formula gives the expansion for $(x+y)^n$, where n is a positive integer,

$$(x+y)^n = \sum_{j=0}^{n} \binom{n}{j} x^j y^{n-j}. \tag{A.50}$$

The coefficients are the *binomial* coefficients given by

$$\binom{n}{j} = \frac{n!}{j!(n-1)!}.$$

One of the earliest uses of power series was the generalization of this formula to arbitrary values of n. If n is not a positive integer, the result is an infinite series. For a real number α we have the formula

$$(x+y)^{\alpha} = y^{\alpha}\left[1+\alpha\left(\frac{x}{y}\right)+\frac{\alpha(\alpha-1)}{2!}\left(\frac{x}{y}\right)^2 + \cdots + \frac{\alpha(\alpha-1)\dots(\alpha-k+1)}{k!}\left(\frac{x}{y}\right)^k + \dots\right]. \tag{A.51}$$

The infinite sum converges so long as $|x/y| < 1$. This formula can be used to compute approximations to the roots of numbers. Choose y to be the smallest number of the form k^n, $k \in \mathbb{N}$ that is larger than x. The general formula then gives

$$x^{\frac{1}{n}} = (k^n + x - k^n)^{\frac{1}{n}} = k\left[1 - \frac{1}{n}\left(1-\frac{x}{k^n}\right) + \frac{n-1}{2n^2}\left(1-\frac{x}{k^n}\right)^2 - \dots\right].$$

Again, in principle, we have an usable algorithm for computing the roots of positive numbers to any desired accuracy. In practice, there are more efficient algorithms than those arising from the power series representation.

By directly multiplying the power series for the exponential function, we can show that (A.46) holds for any pair of complex numbers. This gives a way to compute roots of complex numbers. Let $z = re^{i\theta}$. Then (A.46) and (A.48) imply that

$$\zeta_n = r^{\frac{1}{n}} e^{i\frac{\theta}{n}}$$

is an nth root of z. Using (A.41), we can rewrite this as

$$\zeta_n = r^{\frac{1}{n}}\left[\cos\left(\frac{\theta}{n}\right) + i\sin\left(\frac{\theta}{n}\right)\right].$$

This reduces the problem of approximating roots of complex numbers to problems we have already solved. Note that if $r = \exp(x)$ for x a real number, then

$$r^{\frac{1}{n}} = \exp\left(\frac{x}{n}\right),$$

gives another way to approximate roots of real numbers. It reduces the problem of approximating roots to that of approximating the log function.

If x is a large real number, then $\exp(-x)$ is a very small, positive number; it is given as an infinite sum by

$$\exp(-x) = 1 - x + \frac{x^2}{2!} - \frac{x^3}{3!} + \frac{x^4}{4!} - \dots,$$

whereas

$$\exp(x) = 1 + x + \frac{x^2}{2!} + \frac{x^3}{3!} + \frac{x^4}{4!} + \dots.$$

Note that the numbers that appear in these two sums are identical; only the signs are different. The first sum is a very small positive number, and the second a very large positive number. This means that there is a lot of subtle cancellation occurring in the first sum. Because of the cancellations, it is difficult to use floating-point arithmetic to accurately compute such a sum. A more accurate computation of $\exp(-x)$ is obtained by first computing an approximation $y \simeq \exp(x)$ and then setting $\exp(-x) \simeq y^{-1}$. We can compute the relative error. First suppose

$$y = e^x + \epsilon;$$

then a calculation shows that

$$\frac{1}{y} - e^{-x} = \frac{\epsilon e^{-x}}{y}.$$

This shows that the *relative* error we make in setting e^{-x} equal to y^{-1} is

$$\frac{|y^{-1} - e^{-x}|}{e^{-x}} = \frac{|\epsilon|}{y} \simeq |\epsilon| e^{-x} = \frac{|y - e^x|}{e^x}.$$

Thus if we compute e^x with a given relative error, then the relative error in using y^{-1} for e^{-x} is the same.

A.3.3 Some Higher Transcendental Functions

Functions not expressible as rational functions of powers, the exponential and logarithm functions, are called *higher transcendental functions*. The Γ-function is probably the most important such function. We briefly describe its properties and then consider the J-Bessel functions. This section uses elementary ideas from the theory of analytic functions; see [2].

The Gamma Function

Perhaps the most important higher transcendental function is the *Gamma function*. For complex numbers z with $\operatorname{Re} z > 0$ it is defined by the formula

$$\Gamma(z) = \int_0^\infty e^{-t} t^{z-1} dt. \tag{A.52}$$

From the formula it is clear that Γ is an analytic function in the right half-plane. The Γ-function satisfies a functional equation.

Proposition A.3.2. *For any z with* $\operatorname{Re} z > 0$ *the Gamma function satisfies the relation*

$$\Gamma(z+1) = z\Gamma(z). \tag{A.53}$$

The proof is a simple integration by parts. Using the functional equation recursively, it follows that for any $n \in \mathbb{N}$ the Γ-function satisfies

$$\Gamma(z+n) = z(z+1)\cdots(z+n-1)\Gamma(z).$$

Using these relations, the Γ-function can be extended to be a meromorphic function on the whole complex plane with poles at the non-positive integers. For z with $-n < \operatorname{Re} z$, $\Gamma(z)$ is defined by

$$\Gamma(z) = \frac{\Gamma(z+n)}{z(z+1)\cdots(z+n-1)}.$$

In many applications it is important to understand the behavior of Γ for real arguments tending to infinity. Stirling's formula states that for x in $\mathbb{R}$

$$\Gamma(x) = \sqrt{2\pi} x^{x-\frac{1}{2}} e^{-x}(1 + O(\frac{1}{x})). \tag{A.54}$$

We give an outline of the derivation of Stirling's formula. It is a special case of *Laplace's method* for obtaining asymptotics for functions of the form

$$f(x) = \int e^{x\phi(s)} \psi(s)\, ds.$$

The idea is very simple: Only the global maxima of the function in the exponent ϕ contribute, asymptotically, to $f(x)$ as x tends to infinity. A fuller account of this method can be found in [104].

We begin by setting $t = s(x-1)$ in (A.52) to obtain

$$\Gamma(x) = (x-1)^x \int_0^\infty e^{(x-1)(\log s - s)} ds.$$

The function in the exponent $\log s - s$ has a unique maximum at $s = 1$ where it assumes the value -1. This implies that for any small $\delta > 0$ we have the asymptotic formula

$$\int_0^\infty e^{(x-1)(\log s - s)} ds = \int_{1-\delta}^{1+\delta} e^{(x-1)(\log s - s)} ds + O(e^{-(x-1)(1+\frac{\delta^2}{2})}). \tag{A.55}$$

The second derivative of $\log s - s$ at $s = 1$ is -1, which means that the function

$$v = \begin{cases} \sqrt{(2(u - \log(1+u))} \text{ for } u > 0, \\ -\sqrt{(2(u - \log(1+u))} \text{ for } u < 0 \end{cases}$$

is smooth and invertible in an open interval around $u = 0$. Using v as the variable of integration, the integral becomes

$$\int_{1-\delta}^{1+\delta} e^{(x-1)(\log s - s)} ds = e^{-(x-1)} \int_{-\delta'}^{\delta''} e^{-(x-1)v^2} h'(v)\, dv.$$

Here δ' and δ'' are positive numbers.

As $h'(v)$ is a smooth function and $h'(0) = 1$, it is not difficult to prove that as x tends to infinity,

$$\int_{-\delta'}^{\delta''} e^{-(x-1)v^2} h'(v)\, dv = \frac{1}{\sqrt{x-1}} \int_{-\infty}^{\infty} e^{-\tau^2} d\tau (1 + O(\frac{1}{x})). \tag{A.56}$$

Collecting the pieces gives

$$\begin{aligned} \Gamma(x) &= x^x \left[1 - \frac{1}{x}\right]^x \frac{\sqrt{2\pi}\, e^{1-x}}{\sqrt{x-1}} (1 + O(\frac{1}{x})) \\ &= \sqrt{2\pi}\, x^{(x-\frac{1}{2})} e^{-x} (1 + O(\frac{1}{x})). \end{aligned} \tag{A.57}$$

As a special case, Stirling's formula gives an asymptotic formula for $n!$ as n tends to infinity:

$$n! = \Gamma(n+1) \approx \sqrt{2\pi n} \left[\frac{n}{e}\right]^n.$$

Exercises

***Exercise* A.3.9.** Prove the functional equation (A.53). Deduce that

$$\Gamma(z+n) = z(z+1)\cdots(z+n-1)\Gamma(z).$$

Show that for a positive integer n, $\Gamma(n+1) = n!$

***Exercise* A.3.10.** For m a non-positive integer, compute the limit

$$\lim_{z\to m} (z-m)\Gamma(z).$$

***Exercise* A.3.11.** Prove formula (A.55).

***Exercise* A.3.12.** Prove that v is a smooth, invertible function of u for u in an interval about 0 and that if $u = h(v)$, then $h'(0) = 1$.

***Exercise* A.3.13.** Fill in the details of the last step in the derivation of Stirling's formula.

Exercise* A.3.14. Prove that if x and y are positive real numbers, then

$$\int_0^1 t^{x-1}(1-t)^{y-1} dt = \frac{\Gamma(x)\Gamma(y)}{\Gamma(x+y)}. \tag{A.58}$$

Bessel Functions

For a complex number ν, a *Bessel function* of *order* ν is any nonzero solution of the ordinary differential equation

$$\frac{d^2 f}{dz^2} + \frac{1}{z}\frac{df}{dz} + \left[1 - \frac{\nu^2}{z^2}\right] f = 0. \tag{A.59}$$

The *J-Bessel* functions, of integral and half-integral orders, are important in Fourier analysis. If $\nu \in \mathbb{C} \setminus \{-2, -3, \dots\}$, then $J_\nu(z)$ is defined by the power series

$$J_\nu(z) = \left[\frac{z}{2}\right]^\nu \sum_{k=0}^{\infty} (-1)^k \frac{z^{2k}}{2^{2k} k! \Gamma(\nu + k + 1)}. \tag{A.60}$$

The infinite sum converges in the whole complex plane. If ν is not an integer, then $J_\nu(z)$ is defined for z with $|\arg(z)| < \pi$ by setting

$$z^\nu = e^{\nu \log z} \qquad \text{for } z \in \mathbb{C} \setminus (-\infty, 0];$$

here $\log z$ is taken to be real for positive real values of z. Graphs of J_0 and $J_{\frac{1}{2}}$ are shown in Figure A.1.

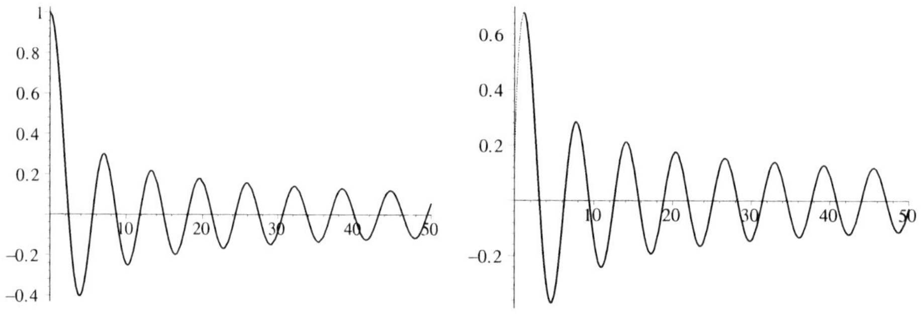

(a) J_0 (b) $J_{\frac{1}{2}}$

Figure A.1. Some J-Bessel functions.

The J-Bessel functions have a variety of integral representations. Their connection with Fourier analysis is a consequence of the formula

$$J_\nu(z) = \frac{\left[\frac{z}{2}\right]^\nu}{\Gamma\left(\nu + \frac{1}{2}\right)\Gamma\left(\frac{1}{2}\right)} \int_0^\pi e^{iz\cos\theta} \sin^{2\nu}\theta d\theta, \tag{A.61}$$

valid if $\text{Re}(\nu) > -\frac{1}{2}$. Up to a multiplicative factor, the J-Bessel function is also the one-dimensional, Fourier transform of a function:

$$J_\nu(z) = \frac{\left[\frac{z}{2}\right]^\nu}{\Gamma\left(\nu+\frac{1}{2}\right)\Gamma\left(\frac{1}{2}\right)} \int_{-1}^{1} (1-x^2)^{\nu-\frac{1}{2}} e^{izx}\,dx \qquad \text{for } \operatorname{Re}\nu > -\frac{1}{2}. \tag{A.62}$$

Each J-Bessel function has a simple asymptotic formula as $|z|$ tends to infinity:

$$J_\nu(z) = \sqrt{\frac{2}{\pi z}}\left[\cos(z - \frac{\pi\nu}{2} - \frac{\pi}{4}) - (\nu^2 - \frac{1}{4})\sin(z - \frac{\pi\nu}{2} - \frac{\pi}{4}) + O(\frac{1}{z^2})\right], \ |\arg z| < \pi. \tag{A.63}$$

This formula is proved using (A.61) and the method of stationary phase. Indeed the Bessel functions have complete asymptotic expansions. Several additional facts are outlined in the exercises. A thorough treatment of Bessel functions is given in [129]. Many useful relations and definite integrals involving Bessel functions can be found in [45].

Exercises

***Exercise* A.3.15.** Show that the function defined in (A.60) satisfies (A.59).

***Exercise* A.3.16.** Show that the function defined in (A.61) satisfies (A.59).

***Exercise* A.3.17.** Show how to deduce (A.62) from (A.61).

***Exercise* A.3.18.** Derive the power series expansion, (A.60), from (A.62). *Hint*: Write the exponential as a power series and use (A.58).

***Exercise* A.3.19.** Bessel functions with half-integral order can be expressed in terms of trigonometric functions. Show that

1. $J_{\frac{1}{2}}(z) = \sqrt{\frac{2}{\pi z}} \sin z$

2. $J_{\frac{3}{2}}(z) = \sqrt{\frac{2}{\pi z}} \left[\frac{\sin z}{z} - \cos z\right]$

***Exercise* A.3.20.** Show that

$$\frac{d}{dz} J_0(z) = -J_1(z).$$

***Exercise* A.3.21.** Use the Parseval formula to compute

$$\int_{-\infty}^{\infty} |J_\nu(x)|^2 |x|^{-2\nu}\,dx$$

for ν a nonnegative integer or half-integer.

***Exercise* A.3.22.** By considering the differential equation (A.59), for large real z, explain the asymptotic expansion of the Bessel function. In particular, why is the rate of decay of J_ν independent of ν?

A.4 Spaces of Functions*

In mathematics functions are usually grouped together into vector spaces, according to their smoothness properties or rates of decay. Norms or metrics are defined on these vector spaces which incorporate these properties.

A.4.1 Examples of Function Spaces

A basic example is the space $\mathscr{C}^0([0,1])$; it is the set of continuous functions defined on the interval $[0,1]$. A function f belongs to $\mathscr{C}^0([0,1])$ if, for every $x \in [0,1]$ $\lim_{y\to x} f(y) = f(x)$, at the endpoints we need to use one-sided limits:

$$\lim_{y\to 0^+} f(y) = 0, \qquad \lim_{y\to 1^-} f(y) = f(1).$$

A scalar multiple of a continuous function is continuous, as is a sum of two continuous functions. Thus the set $\mathscr{C}^0([0,1])$ is a vector space. Define a norm on this vector space by setting

$$\|f\|_{\mathscr{C}^0} \overset{d}{=} \sup_{x\in[0,1]} |f(x)|.$$

Notice that the expression on the right is defined for any bounded function defined on $[0,1]$.

A sequence of functions $< f_n >$ converges in this norm to a function f provided that

$$\lim_{n\to\infty} \|f_n - f\|_{\mathscr{C}^0} = 0.$$

It is a nontrivial result in analysis that if $< f_n >$ converges to f, in this sense, then f is also a continuous function. This norm is called the *uniform norm* or *sup norm*; convergence in this norm is called *uniform convergence*. The vector space $\mathscr{C}^0([0,1])$ is *complete* with respect to this norm.

For each $k \in \mathbb{N}$, we let $\mathscr{C}^k([0,1])$ denote the space of functions defined on $[0,1]$ with k continuous derivatives. We define a norm on this space by setting

$$\|f\|_{\mathscr{C}^k} = \sum_{j=0}^{k} \|f^{[j]}\|_{\mathscr{C}^0}. \tag{A.64}$$

This norm defines a notion of convergence for k-times differentiable functions. As before, if a sequence $< f_n >$ converges to f in this sense, then f is also a function with k continuous derivatives. The basic result in analysis used to study these function spaces is as follows:

Theorem A.4.1. *Let* $< f_n >$ *be a sequence of* k*-times differentiable functions defined on* $[a,b]$. *If* $< f_n >$ *converges* uniformly *to* f *and for each* j *between* 1 *and* k *the sequence of derivatives* $< f_n^{[j]} >$ *converges* uniformly *to a function* g_j, *then* f *is* k*-times continuously differentiable and*

$$f^{[j]} = g_j \qquad \textit{for } j = 1, \ldots, k.$$

A proof of this theorem is given in [111].

Let $\mathscr{C}^\infty([0,1])$ denote the vector space of functions, defined on $[0,1]$ with infinitely many continuous derivatives. The expression on the right-hand side of (A.64) makes no sense if $k=\infty$. In fact, there is no way to define a norm on the vector space $\mathscr{C}^\infty([0,1])$. We can, however, define a metric on $\mathscr{C}^\infty([0,1])$ by setting

$$d(f,g)=\sum_{j=0}^{\infty}2^{-j}\frac{\|f-g\|_{\mathscr{C}^j}}{1+\|f-g\|_{\mathscr{C}^j}}.$$

A sequence of functions $<f_n>\subset\mathscr{C}^\infty([0,1])$ converges to f if

$$\lim_{n\to\infty}d(f_n,f)=0.$$

Analogous spaces are defined with $[0,1]$ replaced by other sets; for example, $\mathscr{C}^0(\mathbb{R}^n)$ or $\mathscr{C}^\infty(S^1\times\mathbb{R})$.

The foregoing examples are defined by considering the smoothness properties of functions. Other types of spaces are defined by considering rates of decay at infinity and blow-up at finite points. For example, the space $L^2([0,1])$ consists of functions for which

$$\|f\|_2=\sqrt{\int_0^1|f(x)|^2\,dx}<\infty.$$

Such a function is said to be *square integrable*. It is not necessary for f to be continuous in order for it to belong to $L^2([0,1])$, only that this integral makes sense. The function $|x-\frac{1}{2}|^{-\frac{1}{4}}$ is not even bounded, but it belongs to $L^2([0,1])$ because

$$\int\frac{1}{\sqrt{|x-\frac{1}{2}|}}dx<\infty.$$

***Example* A.4.1.** Let $0\le a<b<1$. Then the functions $\chi_{[a,b]}$ belong to $L^2([0,1])$. Note the following:

$$\chi_{[a,b]}(x)-\chi_{(a,b)}(x)=\begin{cases}1 & \text{if } x=a \text{ or } b,\\ 0 & \text{otherwise.}\end{cases}$$

The L^2-norm cannot detect the difference between these two functions

$$\|\chi_{[a,b]}-\chi_{(a,b)}\|_2=0.$$

This is a general feature of norms defined by integrals: They do not distinguish functions that differ on very small sets. The technical term for these very small sets is *sets of measure zero*. This property does not create significant difficulties but is important to keep in mind. Indeed, this is also a feature of physical measurements and explains, in part, the relevance of integral norms in practical applications.

Once again, if $< f_n > \subset L^2([0,1])$ is a sequence of functions that converge to a function f in the sense that

$$\lim_{n\to\infty} \|f_n - f\|_2 = 0,$$

then f also belongs to $L^2([0,1])$. This type of convergence is often called *convergence in the mean*; we use the special notation

$$\operatorname*{LIM}_{n\to\infty} f_n = f.$$

The behavior of L^2-convergent sequences is quite different from that of $\mathscr{C}^0$-convergent sequences. Examples best illustrate this point.

***Example* A.4.2.** Let $f_n(x) = x^n$, if $x \in [0,1)$. Then

$$\lim_{n\to\infty} f_n(x) = 0,$$

whereas $\lim_{n\to\infty} f_n(1) = 1$. For each x, $f_n(x)$ converges, but the limit function

$$f(x) = \begin{cases} 0 & \text{if } 0 \le x < 1, \\ 1 & \text{if } x = 1 \end{cases}$$

is *not* continuous. This means that $\|f_n - f\|_{\mathscr{C}^0}$ cannot go to zero as $n \to \infty$. On the other hand,

$$\int_0^1 |x^n|^2\, dx = \frac{1}{2n+1},$$

and therefore $\lim_{n\to\infty} \|f_n - 0\|_2 = 0$. So the sequence $< f_n >$ converges in the L^2-norm to the function that is identically zero. The pointwise limit, f, cannot be distinguished from the zero function by the L^2-norm. Note also the related fact: The L^2-convergence of a sequence $< f_n >$ to a function f does *not* require that $\lim_{n\to\infty} f_n(x) = f(x)$ for all x.

***Example* A.4.3.** Define a sequence of functions

$$f_n(x) = \begin{cases} 0 & \text{for } x \in [0, \frac{n-1}{2n}], \\ nx - \frac{n-1}{2} & \text{for } x \in (\frac{n-1}{2n}, \frac{n+1}{2n}), \\ 1 & \text{for } x \in [\frac{n+1}{2n}, 1]. \end{cases}$$

Each of these functions is continuous, and it is not difficult to show that

$$f(x) = \lim_{n\to\infty} f_n(x) = \begin{cases} 0 & \text{for } x \in [0, \frac{1}{2}), \\ 1 & \text{for } x \in (\frac{1}{2}, 1]. \end{cases}$$

Once again, the limit function is not continuous, and it is easy to see that

$$\|f_n - f\|_{\mathscr{C}^0} = \frac{1}{2}$$

for every $n \in \mathbb{N}$. On the other hand, it is also not hard to show that

$$\lim_{n\to\infty} \int_0^1 |f_n(x) - f(x)|^2\,dx = 0.$$

Spaces of functions are generally infinite dimensional. Introducing a basis for a *finite-dimensional*, real vector space establishes an isomorphism between that vector space and $\mathbb{R}^n$, for some n. In Exercise A.2.3 it is shown that the notion of convergence for a sequence in $\mathbb{R}^n$ is independent of the choice of norm. This is *not* true for infinite-dimensional vector spaces. There are many non-isomorphic vector spaces, and different norms lead to different convergent sequences. By analogy to the norms, $\|\cdot\|_p$ defined on $\mathbb{R}^n$, the L^p-norms are defined for functions defined on $[0, 1]$ by setting

$$\|f\|_{L^p} = \left[\int_0^1 |f(x)|^p\,dx\right]^{\frac{1}{p}}. \tag{A.65}$$

If $1 \leq p$, then this defines a norm; the restriction on p is needed to establish the triangle inequality,

$$\|f + g\|_p \leq \|f\|_p + \|g\|_p. \tag{A.66}$$

We can also let $p = \infty$ by defining

$$\|f\|_\infty = \max\{|f(x)| \; : \; x \in [0, 1]\}.$$

Define the vector space $L^p([0, 1])$ to be those locally integrable functions f such that $\|f\|_p < \infty$. The various L^p-spaces are related by a fundamental inequality.

Theorem A.4.2 (Hölder's inequality). *Let $1 \leq p \leq \infty$, and define q by*

$$q = \begin{cases} \frac{p}{p-1} & \text{if } p \neq 1, \infty, \\ 1 & \text{if } p = \infty, \\ \infty & \text{if } p = 1. \end{cases} \tag{A.67}$$

If $f \in L^p([0, 1])$ and $g \in L^q([0, 1])$, then

$$\int_0^1 |f(x)g(x)|\,dx \leq \|f\|_{L^p}\|g\|_{L^q}. \tag{A.68}$$

In particular, the product fg belongs to $L^1([0, 1])$.

The analogous result holds with $[0, 1]$ replaced by $\mathbb{R}$.

Exercise A.4.2 shows that a sequence of functions $< f_n >$ that belongs to $L^2([0, 1])$ also belongs to $L^1([0, 1])$. The next example shows that a bounded sequence in $L^2([0, 1])$ need not have a limit in L^2-norm even though it does have a limit in the L^1-norm.

***Example* A.4.4.** Define a sequence of functions

$$f_n(x) = \begin{cases} n & \text{if } x \in [0, \frac{1}{n^2}], \\ 0 & \text{if } n \in (\frac{1}{n^2}, 1]. \end{cases}$$

Note that if $x \neq 0$, then $\lim_{n\to\infty} f_n(x) = 0$; on the other hand, for all $n \in \mathbb{N}$ we have

$$\|f_n\|_2 = 1.$$

This shows that this sequence is bounded in $L^2([0, 1])$ but does not converge to anything. Note that $\lim_{n\to\infty} \|f_n\|_1 = 0$ and therefore $< f_n >$ does converge to zero in the L^1-norm.

Exercises

***Exercise* A.4.1.** In Example A.4.2, find the maximum value of the difference $|f(x) - f_n(x)|$ for each n and show that this does not go to zero as $n \to \infty$.

***Exercise* A.4.2.** Use Hölder's inequality to show that if $f \in L^p([0, 1])$ and $1 \leq p' < p$, then $f \in L^{p'}([0, 1])$ as well. *Hint*: Take $g = 1$.

***Exercise* A.4.3.** Show that the function

$$f_\alpha(x) = \frac{1}{x^\alpha}$$

belongs to $L^p([0, 1])$ if $\alpha < p^{-1}$ and does *not* belong to $L^p([0, 1])$ if $\alpha \geq p^{-1}$.

A.4.2 Completeness

In the finite-dimensional case we introduced the concept of a Cauchy sequence as a way of describing which sequences *should* converge. The real power of this idea only becomes apparent in the infinite-dimensional context.

***Definition* A.4.1.** Let $(V, \|\cdot\|)$ be a normed vector space. A sequence $< v_n > \subset V$ is a *Cauchy sequence* if, for any $\epsilon > 0$, there exists an N so that

$$\|v_n - v_m\| < \epsilon \qquad \text{provided that } m \text{ and } n > N.$$

Reasoning by analogy, the Cauchy sequences are the ones that "should converge." Because there are many different norms that can be used on an infinite-dimensional space, this is a subtle question.

***Example* A.4.5.** Let V be the continuous functions on $[0, 1]$ and use for a norm

$$\|f\|_1 = \int_0^1 |f(x)|\,dx.$$

Define a sequence $< f_n > \subset V$ by setting

$$f_n(x) = \begin{cases} 0 & \text{for } 0 \le x \le \frac{1}{2} - \frac{1}{n}, \\ n(x - \frac{1}{2}) & \text{for } \frac{1}{2} - \frac{1}{n} \le x \le \frac{1}{2}, \\ 1 & \text{for } \frac{1}{2} \le x \le 1. \end{cases}$$

The distances between the terms of the sequence satisfy the estimates

$$\| f_n - f_m \|_1 \le \frac{1}{2}(\frac{1}{n} + \frac{1}{m}).$$

This implies that $< f_n >$ is a Cauchy sequence. Pointwise, $< f_n >$ converges to

$$f = \begin{cases} 0 & \text{for } 0 \le x < \frac{1}{2}, \\ 1 & \text{for } \frac{1}{2} \le x \le 1. \end{cases}$$

Indeed it is not difficult to show that

$$\lim_{n \to \infty} \| f_n - f \|_1 = 0.$$

The only difficulty is that f is not a continuous function. This is an example of a Cauchy sequence that does not converge.

This sort of example leads to the following definition.

***Definition* A.4.2.** A normed vector space $(V, \| \cdot \|)$ is said to be *complete* if every Cauchy sequence $< v_n > \subset V$ converges to a limit *in* V.

Note that completeness is a property of a *normed vector* space. It makes no sense to say "the set of continuous functions on [0, 1] is complete." Rather, we must say that "the set of continuous functions on [0, 1], with the sup norm is complete." Completeness is an important property for a normed linear space, and most of the spaces we consider have this property.

Theorem A.4.3. *For* $1 \le p \le \infty$, *the normed linear spaces* $L^p([0, 1])$ *(or* $L^p(\mathbb{R}^n)$*) are complete. For any nonnegative integer* k, *the normed linear spaces* $\mathcal{C}^k([0, 1])$ *(or* $\mathcal{C}^k(\mathbb{R}^n)$*) are complete.*

Exercise

***Exercise* A.4.4.** In Example A.4.5, show that $< f_n >$ is not a Cauchy sequence in the sup-norm.

A.4.3 Linear Functionals

For finite-dimensional vector spaces the concept of a linear function is given by purely algebraic conditions (A.9). For infinite-dimensional vector spaces more care is required because linear functions may not be continuous.

***Example* A.4.6.** Let V be the set of once differentiable functions on $[0, 1]$. Instead of using the usual $\mathcal{C}^1$-norm, we use the $\mathcal{C}^0$-norm. With this choice of norm a sequence of functions $< f_n >\subset V$ converges to $f \in V$ if

$$\lim_{n\to\infty} \|f_n - f\|_{\mathcal{C}^0} = 0.$$

Suppose $< f_n >$ is a sequence that converges to 0 in this sense and that $l : V \to \mathbb{R}$ is a linear function. If l is continuous, then

$$\lim_{n\to\infty} l(f_n) = 0.$$

Define a function on V by setting

$$l(f) = f'(\frac{1}{2}).$$

The usual rules of differentiation show that this is a linear function. It is, however, not continuous. Define a sequence of functions in V by letting

$$f_n(x) = \begin{cases} 0 & \text{if } x \notin (\frac{n-1}{2n}, \frac{n+3}{2n}), \\ \frac{1}{\sqrt{n}}(1 - [n(x - \frac{1}{2n})]^2)^2 & \text{if } x \in (\frac{n-1}{2n}, \frac{n+3}{2n}). \end{cases}$$

It is not difficult to show that $f_n \in V$ for each n and that

$$f_n(x) \leq \frac{1}{\sqrt{n}} \qquad \text{for } x \in [0, 1].$$

This shows that $< f_n >$ converges to $f \equiv 0$ in the sense defined previously. However, a calculation gives that

$$l(f_n) = f_n'(\frac{1}{2}) = -\frac{3}{2}\sqrt{n}.$$

In other words, $\lim_{n\to\infty} l(f_n) = -\infty$, even though $< f_n >$ converges to zero. If we use the $\mathcal{C}^1$-norm instead, then l is indeed a continuous linear function. The reason this does not contradict the previous example is that the sequence $< f_n >$ does *not* converge to zero in the $\mathcal{C}^1$-norm.

In light of this example, it is clear that additional care is needed in the study of linear functions on infinite-dimensional vector spaces.

***Definition* A.4.3.** Let V be a vector space with norm $\|\cdot\|$. A linear function $l : V \to \mathbb{R}$ is called a *linear functional* if it is continuous with respect to the norm. That is, if $< f_n >$ is sequence in V and, for an f in V,

$$\lim_{n\to\infty} \|f - f_n\| = 0,$$

then

$$\lim_{n\to\infty} l(f_n) = l(f).$$

We denote the set of linear functionals by V'; as before, it is a vector space called the *dual* vector space. It has a naturally defined norm given by

$$\|l\|' = \sup_{V \ni f \neq 0} \frac{|l(f)|}{\|f\|}. \tag{A.69}$$

For the normed vector spaces of greatest interest there is a complete description of the dual vector space. Let $1 \leq p \leq \infty$, and let q be defined by (A.67). Choose a function $g \in L^q([0,1])$; then Hölder's inequality implies that for every $f \in L^p([0,1])$ the function fg is integrable and

$$\left| \int_0^1 f(x)g(x)\,dx \right| \leq \|f\|_{L^p}\|g\|_{L^q}.$$

The real-valued function

$$l_g(f) = \int_0^1 f(x)g(x)\,dx \tag{A.70}$$

is therefore well defined for all $f \in L^p([0,1])$. The elementary properties of the integral imply that it is linear and Hölder's inequality implies that it is continuous. Suppose that $< f_n > \subset L^p([0,1])$, which converges in the L^p-sense to f. We see that

$$|l_g(f_n) - l(f)| = |l_g(f_n - f)| \leq \|f_n - f\|_{L^p}\|g\|_{L^q}.$$

This shows that $\lim_{n\to\infty} l_g(f_n) = l_g(f)$.

In fact, all linear functionals on these normed vector spaces are of this form.

Theorem A.4.4 (Riesz representation theorem 1). *If* $1 \leq p < \infty$ *and* q *is given by* (A.67), *then* $L^q([0,1])$ *is the dual space to* $L^p([0,1])$. *That is, every continuous linear function on* $L^p([0,1])$ *is given by* l_g *for some* $g \in L^q([0,1])$.

***Remark* A.4.1.** Note that the case $p = \infty$ is excluded in Theorem A.4.4. The space $L^\infty([0,1])$ turns out to be considerably more complicated, as a normed vector space, than $L^p([0,1])$ for $1 \leq p < \infty$. As the details of this space are not needed in the sequel, we do not pursue the matter further.

Starting with n-tuples of numbers and the $\|\cdot\|_p$-norm defined in (A.21) leads to yet another collection of infinite-dimensional spaces.

***Definition* A.4.4.** For $1 \leq p \leq \infty$, let l^p denote the collection of sequences $< a_j >$ such that

$$\| < a_j > \|_p = \left[\sum_{j=1}^{\infty} |a_j|^p \right]^{\frac{1}{p}} < \infty. \tag{A.71}$$

These are *complete* normed vector spaces.

***Example* A.4.7.** The space l^1 consists of sequences that have absolutely convergent sums; that is, $< a_j > \in l^1$ if and only if

$$\sum_{j=1}^{\infty} |a_j| < \infty.$$

If $p < p'$, then it is clear that $l^p \subset l^{p'}$. There is also a version of the Hölder inequality. Let $1 \leq p \leq \infty$ and q by given by (A.67); for $< a_j > \in l^p$ and $< b_j > \in l^q$ the sequence $< a_j b_j > \in l^1$ and

$$\sum_{j=1}^{\infty} |a_j b_j| \leq \| < a_j > \|_p \| < b_j > \|_q. \tag{A.72}$$

This inequality shows that if $\boldsymbol{b} =< b_j > \in l^q$, then we can define a bounded linear functional on l^p by setting

$$l_{\boldsymbol{b}}(\boldsymbol{a}) = \sum_{j=1}^{\infty} a_j b_j.$$

This again gives all bounded functionals provided p is finite.

Theorem A.4.5 (Riesz representation theorem 2). *If $1 \leq p < \infty$ and q is given by* (A.67) *then l^q is the dual space to l^p. That is, every continuous linear function on l^p is given by $l_{\boldsymbol{b}}$ for some $\boldsymbol{b} \in l^q$.*

Exercise

***Exercise* A.4.5.** Prove that $l^p \subset l^{p'}$.

A.4.4 Measurement, Linear Functionals, and Weak Convergence

Suppose that the state of a system is described by a function $f \in L^p([0, 1])$. In this case the measurements that we can make are often modeled as the evaluation of linear functionals. That is, we have a collection of functions $\{g_1, \ldots, g_k\} \subset L^q([0, 1])$ and our measurements are given by

$$m_j(f) = \int_0^1 f(x) g_j(x)\, dx, \quad j = 1, \ldots, k.$$

From the point of view of measurement, this suggests a different, perhaps more reasonable, notion of convergence. Insofar as these measurements are concerned, a sequence of states $< f_n >$ would appear to converge to a state f if

$$\lim_{n\to\infty} \int_0^1 f_n(x)g_j(x)\,dx = \int_0^1 f(x)g_j(x)\,dx, \text{ for } j = 1,\ldots k. \tag{A.73}$$

Since we are only considering finitely many measurements on an infinite-dimensional state space, this is clearly a much weaker condition than the condition that $< f_n >$ converge to f in the L^p-sense.

Of course, if $< f_n >$ converges to f in the L^p-sense, then, for any $g \in L^q([0, 1])$, $\lim_{n\to\infty} l_g(f_n) = l_g(f)$. However, the L^p-convergence is not required for (A.73) to hold. It is an important observation that the condition

$$\lim_{n\to\infty} \int_0^1 f_n(x)g(x)\,dx = \int_0^1 f(x)g(x)\,dx \text{ for } \textit{every} \text{ function } g \in L^q([0, 1])$$

is a weaker condition than L^p-convergence.

***Definition* A.4.5.** Suppose that $(V, \|\cdot\|)$ is a normed vector space and $< \boldsymbol{v}_n >$ is a sequence of vectors in V. If there exists a vector $\boldsymbol{v} \in V$ such that for *every* continuous linear function l we have that

$$\lim_{n\to\infty} l(\boldsymbol{v}_n) = l(\boldsymbol{v}),$$

then we say that $\boldsymbol{v}_n$ *converges weakly* to $\boldsymbol{v}$. This is sometimes denoted by

$$\boldsymbol{v}_n \rightharpoonup \boldsymbol{v}.$$

From the point of view of measurement, weak convergence is often the appropriate notion. Unfortunately, it *cannot* be defined by a norm, and a sequence does not exert very much control over the properties of its weak limit. For example, it is not in general true that

$$\lim_{n\to\infty} \|\boldsymbol{v}_n\| = \|\boldsymbol{v}\|$$

for a weakly convergent sequence. This is replaced by the statement

$$\text{If } \boldsymbol{v}_n \rightharpoonup \boldsymbol{v}, \text{ then } \limsup_{n\to\infty} \|\boldsymbol{v}_n\| \geq \|\boldsymbol{v}\|. \tag{A.74}$$

***Example* A.4.8.** The sequence of functions $< f_n >$ defined in Example A.4.4 is a sequence with

$$\|f_n\|_{L^2} = 1$$

for all n. On the other hand, if $x \in (0, 1]$, then

$$\lim_{n\to\infty} f_n(x) = 0.$$

These two facts allow the application of standard results from measure theory to conclude that

$$\lim_{n\to\infty}\int_0^1 f_n(x)g(x)\,dx = 0,$$

for every function $g \in L^2([0, 1])$. In other words, the sequence $< f_n >$ converges *weakly* to zero even though it does not converge to anything in the L^2-sense.

***Example* A.4.9.** Let $< \boldsymbol{a}_n > \subset l^2$ be the sequence defined by

$$\boldsymbol{a}_n(j) = \begin{cases} 1 & \text{if } j = n, \\ 0 & \text{if } j \neq n. \end{cases}$$

Since $\boldsymbol{a}_n(j) = 0$ if $j < n$, it is clear that if $< \boldsymbol{a}_n >$ were to converge to $\boldsymbol{a}$, in the l^2-sense, then $\boldsymbol{a} = \mathbf{0}$. On the other hand, $\|\boldsymbol{a}_n\|_{l^2} = 1$ for all n, and this shows that $\boldsymbol{a}_n$ cannot converge in the l^2-sense. Finally, if $\boldsymbol{b} \in l^2$, then

$$\langle \boldsymbol{a}_n, \boldsymbol{b}\rangle_{l^2} = \boldsymbol{b}(n).$$

Because $\|\boldsymbol{b}\|_{l^2} < \infty$, it is clear that

$$\lim_{n\to\infty} \boldsymbol{b}(n) = 0$$

and therefore $\boldsymbol{a}_n$ converges weakly to $\mathbf{0}$.

Exercise

***Exercise* A.4.6.** Suppose that $< f_n > \subset L^2([0, 1])$ and $< f_n >$ has a weak limit. Show that it is unique.

A.4.5 Generalized Functions on $\mathbb{R}$

Within mathematics and also in its applications, the fact that many functions are not differentiable can be a serious difficulty. Within the context of *linear analysis, generalized functions* or *distributions* provide a very comprehensive solution to this problem. Though it is more common in the mathematics literature, we avoid the term distribution, because there are so many other things in imaging that go by this name. In this section we outline the theory of generalized functions and give many examples. The reader wishing to attain a degree of comfort with these ideas is strongly urged to do the Exercises at the end of the section.

Let $\mathcal{C}_c^\infty(\mathbb{R})$ denote infinitely differentiable functions defined on $\mathbb{R}$ that vanish outside of bounded sets. These are sometimes called *test functions*.

Definition **A.4.6.** A generalized function on $\mathbb{R}$ is a linear function, l, defined on the set of test functions such that there is a constant C and an integer k so that, for every $f \in \mathscr{C}_c^\infty(\mathbb{R})$, we have the estimate

$$|l(f)| \leq C \sup_{x \in \mathbb{R}} \left[(1+|x|)^k \sum_{j=0}^{k} |\partial_x^j f(x)| \right]. \tag{A.75}$$

These are linear functions on $\mathscr{C}_c^\infty(\mathbb{R})$ that are, in a certain sense, continuous. The constants C and k in (A.75) depend on l but do not depend on f. The expression on the right-hand side defines a norm on $\mathscr{C}_c^\infty(\mathbb{R})$. For convenience, we let

$$\|f\|_k = \sup_{x \in \mathbb{R}} \left[(1+|x|)^k \sum_{j=0}^{k} |\partial_x^j f(x)| \right].$$

If $f \in \mathscr{C}_c^\infty(\mathbb{R})$, then it easy to show that $\|f\|_k$ is finite for every $k \in \mathbb{N} \cup \{0\}$.

A few examples of generalized functions should help clarify the definition.

Example **A.4.10.** The most famous generalized function of all is the Dirac δ-function. If is defined by

$$\delta(f) = f(0).$$

It is immediate from the definition that $f \mapsto \delta(f)$ is linear and

$$|\delta(f)| \leq \|f\|_0,$$

so the δ-function *is* a generalized function. For $j \in \mathbb{N}$ define

$$\delta^{(j)}(f) = \partial_x^j f(0).$$

Since differentiation is linear, these also define linear functions on $\mathscr{C}_c^\infty(\mathbb{R})$ that satisfy the estimates

$$|\delta^{(j)}(f)| \leq \|f\|_j.$$

Hence these are also generalized functions.

Example **A.4.11.** Let φ be a function that is integrable on any finite interval and such that

$$C_\varphi = \int_{-\infty}^{\infty} |\varphi(x)|(1+|x|)^{-k} \leq \infty$$

for some nonnegative integer k. Any such function defines a generalized function

$$l_\varphi(f) = \int_{-\infty}^{\infty} f(x)\varphi(x)\,dx.$$

Because f has bounded support, the integral converges absolutely. The linearity of the integral implies that $f \mapsto l_\varphi(f)$ is linear. To prove the estimate, we observe that

$$|f(x)| \leq \frac{\|f\|_k}{(1+|x|)^k}$$

and therefore

$$|l_\varphi(f)| \leq \int_{-\infty}^{\infty} \|f\|_k \frac{|\varphi(x)|}{(1+|x|)^k} dx = C_\varphi \|f\|_k.$$

Thus l_φ is also a generalized function. In particular, every function in $\mathscr{C}_c^\infty(\mathbb{R})$ defines a generalized function, so that, in a reasonable sense, a generalized function is a generalization of a function!

***Example* A.4.12.** Recall that the Cauchy principal value integral of g is defined, when the limit exists, by

$$\text{P.V.} \int_{-\infty}^{\infty} g(x)\,dx = \lim_{\epsilon \downarrow 0} \left[\int_{-\infty}^{-\epsilon} g(x)\,dx + \int_{\epsilon}^{\infty} g(x)\,dx \right],$$

A generalized function is defined by

$$l_{1/x}(f) = \text{P.V.} \int_{-\infty}^{\infty} \frac{f(x)\,dx}{x}.$$

Because $1/x$ is not integrable in any neighborhood of 0, the ordinary integral $f(x)/x$ is not defined. The principal value is well defined for any test function and defines a generalized function. To prove this, observe that, for any $\epsilon > 0$,

$$\int_{-1}^{-\epsilon} \frac{f(x)\,dx}{x} + \int_{\epsilon}^{1} \frac{f(x)\,dx}{x} = \int_{-1}^{-\epsilon} \frac{(f(x) - f(0))\,dx}{x} + \int_{\epsilon}^{1} \frac{(f(x) - f(0))\,dx}{x}.$$

This is because $1/x$ is an odd function and the region of integration is symmetric about 0. The ratio $(f(x) - f(0))/x$ is a smooth bounded function in a neighborhood of 0 and therefore the limit exists as $\epsilon \to 0$. This shows that

$$l_{1/x}(f) = \int_{-1}^{1} \frac{(f(x) - f(0))\,dx}{x} + \int_{|x| \geq 1} \frac{f(x)\,dx}{x}.$$

It is left as an exercise to show that

$$|l_{1/x}(f)| \leq C\|f\|_1. \tag{A.76}$$

As noted in Example A.4.11, the map $f \mapsto l_f$ identifies every function in $\mathscr{C}_c^\infty(\mathbb{R})$ with a unique generalized function. If the world were very simple, then *every* generalized function would be of this form for some locally integrable function. But this is not true! It is not hard to show that the δ-function is not of this form: Suppose that $\delta = l_\varphi$ for some locally integrable function φ. We can show that φ must vanish for all $x \neq 0$; this is because $\delta(f)$ only depends on $f(0)$. But an integrable function supported at one point has integral 0 so

$$l_\varphi(f) = 0$$

for all $f \in \mathscr{C}_c^\infty(\mathbb{R})$.

Recall that our goal is to extend the notion of differentiability. The clue to how this should be done is given by the integration by parts formula. Let f and g be test functions. Then

$$\int_{-\infty}^{\infty} \partial_x f(x) g(x)\, dx = -\int_{-\infty}^{\infty} f(x) \partial_x g\, dx. \tag{A.77}$$

Thinking of f as a generalized function, this formula can be rewritten as

$$l_{\partial_x f}(g) = l_f(-\partial_x g). \tag{A.78}$$

The right-hand side of (A.77) *defines* a generalized function, which we identify as the derivative of the l_f,

$$\left[\partial_x l_f\right](g) \stackrel{d}{=} -l_f(\partial_x g). \tag{A.79}$$

This equation is really just notation, but the underlying idea can be used to define the derivative of any generalized function.

Proposition A.4.1. *If l is a generalized function defined on $\mathbb{R}$, then*

$$l'(f) = l(\partial_x f)$$

is also a generalized function.

Proof. Because ∂_x maps $\mathscr{C}_c^\infty(\mathbb{R})$ to itself, the linear function l' is well defined; we only need to prove that it satisfies an estimate of the form (A.75). As l is a generalized function, there is a C and k so that

$$|l(f)| \leq C\|f\|_k.$$

From the definition of l' it is clear that

$$|l'(f)| \leq C\|\partial_x f\|_k.$$

The proof is completed by showing that

$$\|\partial_x f\|_k \leq \|f\|_{k+1}.$$

This is left as an exercise. □

With this proposition we can now define the derivative of a generalized function. It is very important to keep in mind that the derivative of a generalized function is another generalized function! To distinguish this concept of derivative from the classical one, the derivative of a generalized function is called a *weak derivative.*

***Definition* A.4.7.** Let l be a generalized function. The *weak derivative* of l is the generalized function $l^{[1]}$ defined by

$$l^{[1]}(f) \stackrel{d}{=} -l(\partial_x f). \tag{A.80}$$

If $l = l_f$ for a smooth function f, then $l^{[1]} = l_{\partial_x f}$, so this definition *extends* the usual definition of derivative. Because every generalized function is differentiable and its weak derivative is another generalized function, it follows that every generalized function is twice differentiable. Indeed, arguing recursively, it follows that every generalized function is *infinitely* differentiable. Let $\{l^{[j]}\}$ denote the successive weak derivatives of l. It is left as an exercise for the reader to prove the general formula

$$l^{[j]}(f) \stackrel{d}{=} (-1)^j l(\partial_x^j f). \tag{A.81}$$

***Example* A.4.13.** The weak derivative of the δ-function is just $-\delta^{(1)}$, as already defined in Example A.4.10. The definition states that

$$\delta^{[1]}(f) = -\delta(\partial_x f) = -\partial_x f(0).$$

It is clear that the weak derivative of $\delta^{[1]}$ is $\delta^{(2)}$ and so on.

***Example* A.4.14.** Let $\varphi = \chi_{[0,\infty)}$. Since φ is bounded and piecewise continuous, it defines a generalized function. This function also has a classical derivative away from 0 but is not even continuous at 0. Nonetheless, it has a weak derivative as a generalized function. To find it, we apply the definition

$$\begin{aligned} l^{[1]}_{\chi_{[0,\infty)}}(f) &= -l_{\chi_{[0,\infty)}}(\partial_x f) \\ &= -\int_0^\infty \partial_x f(x)\,dx = f(0). \end{aligned} \tag{A.82}$$

This shows that $l^{[1]}_{\chi_{[0,\infty)}} = \delta$. This is an example of an ordinary function, whose weak derivative, as a generalized function, is not represented by an ordinary function. However, we have accomplished exactly what we set out to do, because now $\chi_{[0,\infty)}$ has a derivative.

***Example* A.4.15.** If f is a smooth function, with bounded support, then the previous example generalizes to shows that

$$l^{[1]}_{f\chi_{[0,\infty)}} = f(0)\delta + l_{(\partial_x f)\chi_{[0,\infty)}}. \tag{A.83}$$

The set of generalized functions is a vector space. If l and k are generalized functions, then so is the sum

$$(l+k)(f) \stackrel{d}{=} l(f) + k(f)$$

as well as scalar multiples

$$(al)(f) \stackrel{d}{=} a(l(f)) \qquad \text{for } a \in \mathbb{R}.$$

Differentiation is a linear operation with respect to this vector space structure; that is,

$$(l+k)^{[1]} = l^{[1]} + k^{[1]} \text{ and } (al)^{[1]} = al^{[1]}.$$

The notion of weak convergence is perfectly adapted to generalized functions.

***Definition* A.4.8.** A sequence $\{l_n\}$ of generalized functions converges weakly to a generalized function l if, for every test function f,

$$\lim_{n\to\infty} l_n(f) = l(f).$$

Weak derivatives of generalized functions behave very nicely under weak limits.

Proposition A.4.2. *If $< l_n >$ is a sequence of generalized functions that converge weakly to a generalized function l, then, for every $j \in \mathbb{N}$, the sequence of generalized functions $< l_n^{[j]} >$ converges weakly to $l^{[j]}$.*

Generalized functions seem to have many nice properties, and they provide a systematic way to define derivatives of all functions, though the derivatives are, in general, *not* functions. Multiplication is the one basic operation that cannot be done with generalized functions. Indeed, it is a theorem that there is no way to define a product on generalized functions so that $l_f \cdot l_g = l_{fg}$. However, if f is a test function and l is a generalized function, then the product $f \cdot l$ is defined; it is

$$(f \cdot l)(g) \stackrel{d}{=} l(fg).$$

This product satisfies the usual Leibniz formula

$$(f \cdot l)^{[1]} = f \cdot l^{[1]} + \partial_x f \cdot l. \tag{A.84}$$

This is generalized slightly in Exercise A.4.16.

We close this brief introduction to the idea of a generalized function with a proposition that gives a fairly concrete picture of the "general" generalized function as a limit of simple examples.

Proposition A.4.3. *If l is a generalized function, then there is a sequence of test functions $< f_n >$ such that l is the weak limit of the sequence of generalized functions $< l_{f_n} >$.*

In other words, any generalized function is the weak limit of generalized functions defined by integration.

***Example* A.4.16.** Let φ be a smooth nonnegative function with support in $(-1, 1)$ normalized so that

$$\int_{-\infty}^{\infty} \varphi(x)\, dx = 1.$$

For each $n \in \mathbb{N}$, define $\varphi_n(x) = n\varphi(nx)$; then δ is the weak limit of l_{φ_n}.

The generalized functions considered in this section are usually called *tempered distributions* in the mathematics literature. This is because they have "tempered growth" at infinity. A more systematic development and proofs of the results in this section can be found in [7]. A very complete treatment of this subject including its higher-dimensional generalizations is given in [59].

Exercises

***Exercise* A.4.7.** Show that if $\varphi = e^{|x|}$, then l_φ is not a generalized function.

***Exercise* A.4.8.** Prove (A.76).

***Exercise* A.4.9.** Suppose that $f \in \mathcal{C}_c^\infty(\mathbb{R})$. Show that

$$\|\partial_x f\|_k \leq \|f\|_{k+1}.$$

***Exercise* A.4.10.** Prove (A.81).

***Exercise* A.4.11.** Compute the derivative of $l_{1/x}$.

***Exercise* A.4.12.** Let $\varphi(x) = (1 - |x|)\chi_{[-1,1]}(x)$ and, for $n \in \mathbb{N}$, set

$$\varphi_n(x) = n\varphi(nx).$$

Prove that l_{φ_n} converges to δ. Show by direct computation that $l_{\varphi_n}^{[1]}$ converges to $\delta^{[1]}$.

***Exercise* A.4.13.** Prove (A.83).

***Exercise* A.4.14.** Prove Proposition A.4.2.

***Exercise* A.4.15.** Prove (A.84).

***Exercise* A.4.16.** Let l be a generalized function and $f \in \mathcal{C}^\infty(\mathbb{R})$ a function with *tempered growth*. This means that there is a $k \in \mathbb{N}$ and constants $\{C_j\}$ so that

$$|\partial_x^j f(x)| \leq C_j(1 + |x|)^k.$$

Show that $(f \cdot l)(g) \stackrel{d}{=} l(fg)$ defines a generalized function.

***Exercise* A.4.17.** Show that any polynomial is a function of tempered growth. Show that a smooth periodic function is a function of tempered growth.

A.4.6 Generalized Functions on $\mathbb{R}^n$

The theory of generalized function extends essentially verbatim to functions of several variables. We give a very brief sketch. For each nonnegative integer k, define a norm on $\mathcal{C}_c^\infty(\mathbb{R}^n)$ by setting

$$\|f\|_k = \sup_{\boldsymbol{x}\in\mathbb{R}^n} \left[(|1 + \|\boldsymbol{x}\|)^k \sum_{\boldsymbol{\alpha}\ |\boldsymbol{\alpha}|\leq k} |\partial_{\boldsymbol{x}}^{\boldsymbol{\alpha}} f(\boldsymbol{x})| \right].$$

Here $\boldsymbol{\alpha}$ is an n-multi-index—that is an n-tuple of nonnegative integers, $\boldsymbol{\alpha} = (\alpha_1, \ldots, \alpha_n)$ with

$$\partial_{\boldsymbol{x}}^{\boldsymbol{\alpha}} \stackrel{d}{=} \partial_{x_1}^{\alpha_1} \cdots \partial_{x_n}^{\alpha_n} \text{ and } |\boldsymbol{\alpha}| \stackrel{d}{=} \alpha_1 + \cdots + \alpha_n.$$

***Definition* A.4.9.** A linear function $l : \mathscr{C}_c^\infty(\mathbb{R}^n) \to \mathbb{R}$ is a generalized function if there exists a $k \in \mathbb{N} \cup \{0\}$ and a constant C so that

$$|l(f)| \leq C\|f\|_k.$$

As before, the set of generalized functions is a vector space. A sequence of generalized functions, $< l_n >$, converges weakly to a generalized function l provided that

$$\lim_{n\to\infty} l_n(f) = l(f) \text{ for every } f \in \mathscr{C}_c^\infty(\mathbb{R}^n).$$

***Example* A.4.17.** The Dirac δ-function is defined in n dimensions by

$$\delta(f) = f(0).$$

It satisfies the estimate $|\delta(f)| \leq \|f\|_0$ and is therefore a generalized function.

***Example* A.4.18.** If φ is a locally integrable function of *tempered growth*, that is, there is a $k \geq 0$ and a constant so that

$$|\varphi(\boldsymbol{x})| \leq C(1 + \|\boldsymbol{x}\|)^k,$$

then

$$l_\varphi(f) = \int_{\mathbb{R}^n} \varphi(\boldsymbol{x}) f(\boldsymbol{x})\, d\boldsymbol{x}$$

satisfies

$$|l_\varphi(f)| \leq C'\|f\|_{n+1+k}. \tag{A.85}$$

This shows that l_φ is a generalized function.

If $\boldsymbol{\alpha}$ is an n-multi-index, and $f \in \mathscr{C}_c^\infty(\mathbb{R}^n)$ then $\partial_{\boldsymbol{x}}^{\boldsymbol{\alpha}} f$ is also in $\mathscr{C}_c^\infty(\mathbb{R}^n)$ and satisfies the estimates

$$\|\partial_{\boldsymbol{x}}^{\boldsymbol{\alpha}} f\|_k \leq \|f\|_{k+|\boldsymbol{\alpha}|}.$$

As before, this allows us to extend the notion of partial derivatives to generalized functions.

***Definition* A.4.10.** If l is a generalized function, then, for $1 \leq j \leq n$, the weak jth partial derivative of l is the generalized function defined by

$$[\partial_{x_j} l](f) \stackrel{d}{=} (-1) l(\partial_{x_j} f). \tag{A.86}$$

Since $\partial_{x_j} l$ is a generalized function as well, it also has partial derivatives. To make a long story short, for an arbitrary multi-index $\boldsymbol{\alpha}$ the weak $\boldsymbol{\alpha}$th partial derivative of the generalized function l is defined by

$$[\partial_{\boldsymbol{x}}^{\boldsymbol{\alpha}} l](f) \stackrel{d}{=} (-1)^{|\boldsymbol{\alpha}|} l(\partial_{\boldsymbol{x}}^{\boldsymbol{\alpha}} f). \tag{A.87}$$

If $f, g \in \mathscr{C}_c^\infty(\mathbb{R}^n)$, then the n-dimensional integration by parts formula states that

$$\int_{\mathbb{R}^n} [\partial_{x_j} f(\boldsymbol{x})] g(\boldsymbol{x})\, d\boldsymbol{x} = -\int_{\mathbb{R}^n} [\partial_{x_j} g(\boldsymbol{x})] f(\boldsymbol{x})\, d\boldsymbol{x}. \tag{A.88}$$

Applying this formula recursively gives the integration by parts for higher-order derivatives:

$$\int_{\mathbb{R}^n} [\partial_{\boldsymbol{x}}^{\boldsymbol{\alpha}} f(\boldsymbol{x})] g(\boldsymbol{x})\, d\boldsymbol{x} = (-1)^{|\boldsymbol{\alpha}|} \int_{\mathbb{R}^n} [\partial_{\boldsymbol{x}}^{\boldsymbol{\alpha}} g(\boldsymbol{x})] f(\boldsymbol{x})\, d\boldsymbol{x}. \tag{A.89}$$

It therefore follows that if $f \in \mathscr{C}_c^\infty(\mathbb{R}^n)$, then the definition of the weak partial derivatives of l_f is consistent with the classical definition of the partial derivatives of f in that

$$\partial_{\boldsymbol{x}}^{\boldsymbol{\alpha}} l_f = l_{\partial_{\boldsymbol{x}}^{\boldsymbol{\alpha}} f} \qquad \text{for all } \boldsymbol{\alpha}. \tag{A.90}$$

Finally, we remark that every generalized function on $\mathbb{R}^n$ is a weak limit of "nice" generalized functions.

Proposition A.4.4. *If l is a generalized function on $\mathbb{R}^n$, then there is a sequence of functions $< f_n > \subset \mathscr{C}_c^\infty(\mathbb{R}^n)$ so that*

$$l(g) = \lim_{n\to\infty} l_{f_n}(g) \qquad \textit{for all } g \in \mathscr{C}_c^\infty(\mathbb{R}^n).$$

Exercises

***Exercise* A.4.18.** Prove that if $f \in \mathscr{C}_c^\infty(\mathbb{R}^n)$ and $j \leq k$, then

$$|\partial_{\boldsymbol{x}}^{\boldsymbol{\alpha}} f(\boldsymbol{x})| \leq \frac{\|f\|_k}{(1 + \|\boldsymbol{x}\|)^k}$$

provided $|\boldsymbol{\alpha}| \leq k$.

***Exercise* A.4.19.** Prove (A.85).

***Exercise* A.4.20.** Let $f \in \mathscr{C}_c^\infty(\mathbb{R}^2)$. Show that

$$l(f) = \int_{-\infty}^{\infty} f(x, 0)\, dx$$

defines a generalized function.

***Exercise* A.4.21.** Prove that (A.86) defines a generalized function.

***Exercise* A.4.22.** Show that the right-hand side of (A.87) defines a generalized function.

***Exercise* A.4.23.** By writing the integrals over $\mathbb{R}^n$ as iterated one-dimensional integrals, prove (A.88). Deduce (A.89).

***Exercise* A.4.24.** Prove (A.90).

***Exercise* A.4.25.** Let $\varphi(x, y) = \chi_{[0,\infty)}(x) \cdot \chi_{[0,\infty)}(y)$. Show that

$$\partial_x \partial_y l_\varphi = \delta.$$

***Exercise* A.4.26.** Let $\varphi(x, y) = \frac{1}{4}\chi_{[-1,1]}(x)\chi_{[-1,1]}(y)$ and $\varphi_n(x, y) = n^2\varphi(nx, ny)$. Prove that

$$\lim_{n\to\infty} l_{\varphi_n}(f) = \delta(f) \qquad \text{for every } f \in \mathscr{C}_c^\infty(\mathbb{R}^2).$$

A.5 Functions in the Real World

In Section A.4 we considered functions from the point of view of a mathematician. In this approach, functions are described by abstract *properties* such as differentiability or integrability. Using these properties, functions are grouped together into normed vector spaces. The principal reason for doing this is to study the mapping properties of linear transformations. It is a very abstract situation because we do not, even in principle, have a way to compute most of the functions under consideration: They are described by their properties and not defined by rules or formulæ. This level of abstraction even leads to difficulties in the mathematical development of the subject. In practice we can only *approximately* measure a function, in most circumstances, at a finite collection of arguments. What mediates between these two very different views of functions? This is a question with many different answers, but the basic ideas involve the concepts of *approximation*, *sampling*, and *interpolation*.

A.5.1 Approximation

The basic problem of approximation theory is to begin with a function from an abstract class (for example, continuous functions) and approximate it, in an appropriate sense, by functions from a more concrete class (for example, polynomials). We begin with the basic theorem in this subject.

Theorem A.5.1 (The Weierstrass approximation theorem). *Given a function $f \in \mathscr{C}^0([0, 1])$ and an $\epsilon > 0$, there is a polynomial p such that*

$$\|f - p\|_{\mathscr{C}^0} < \epsilon. \tag{A.91}$$

The relationship of polynomial functions to continuous functions is analogous to that of finite decimal expansions to real numbers. For the purposes of approximate computations (even with a specified error), it suffices to work with polynomials. This theorem is the prototype for many other such results.

A very useful result for L^p-spaces uses approximation by *step functions*. Recall that if E is an subset of $\mathbb{R}$, then its *characteristic function* is defined by

$$\chi_E(x) = \begin{cases} 1 & \text{if } x \in E, \\ 0 & \text{if } x \notin E. \end{cases}$$

***Definition* A.5.1.** A function f is called a *step function* if there is a finite collection of intervals $\{[a_i, b_i) : i = 1, \ldots, N\}$ and constants $\{c_i\}$ so that

$$f(x) = \sum_{i=1}^{N} c_i \chi_{[a_i,b_i)}(x).$$

Step functions are computable functions.

Theorem A.5.2 (L^p-Approximation theorem). *Suppose that $1 \leq p < \infty$, $f \in L^p(\mathbb{R})$ and $\epsilon > 0$ is given. There exists a step function F such that*

$$\|f - F\|_{L^p} < \epsilon. \tag{A.92}$$

Note that $p = \infty$ is excluded; the theorem is false in this case. The proof of this theorem uses the definition of the Lebesgue integral and the structure of Lebesgue measurable sets. It is beyond the scope of this text but can be found in [42]. Theorem A.5.2 has a very useful corollary.

Corollary A.5.1. *Suppose that $1 \leq p < \infty$, $f \in L^p(\mathbb{R})$ and $\epsilon > 0$ is given. There exists a continuous function G such that*

$$\|f - G\|_{L^p} < \epsilon. \tag{A.93}$$

Proof. Theorem A.5.2 gives the existence of a step function F so that $\|f - F\|_{L^p} < \epsilon/2$. This means that it suffices to find a continuous function G so that

$$\|F - G\|_{L^p} < \frac{\epsilon}{2}.$$

In light of Exercise A.5.1, there is a sequence $a = a_0 < a_1 < \cdots < a_m = b$ and constants $\{c_j\}$ so that

$$F(x) = \sum_{j=1}^{m} c_j \chi_{[a_{j-1},a_j)}(x).$$

Such a function is easily approximated, in the L^p-norm, by continuous, piecewise linear functions. Fix an $\eta > 0$ so that

$$2\eta < \min\{a_i - a_{i-1} : i = 1, \ldots, m\}.$$

For each $1 \geq j < m$ define the piecewise linear function

$$l_j(x) = \begin{cases} 0 & \text{if } |x - a_j| > \eta, \\ c_j \frac{x-(a_j-\eta)}{2\eta} + c_{j-1}\frac{(a_j+\eta-x)}{2\eta} & \text{if } |x - a_j| \leq \eta. \end{cases}$$

For $j = 0$ or m we let

$$l_0(x) = \begin{cases} 0 & \text{if } |x - a_0| > \eta, \\ c_j \frac{x-(a_0-\eta)}{2\eta} & \text{if } |x - a_0| \leq \eta, \end{cases} \quad l_m(x) = \begin{cases} 0 & \text{if } |x - a_m| > \eta, \\ c_m \frac{(a_m+\eta-x)}{2\eta} & \text{if } |x - a_m| \leq \eta. \end{cases}$$

A continuous, piecewise linear function is defined by

$$G = \sum_{j=0}^{m} l_j(x) + \sum_{j=1}^{m} \chi_{[a_{j-1}+\eta, a_j-\eta)}(x).$$

The L^p-norm of the difference $F - G$ is estimated by

$$\|F - G\|_{L^p}^p \leq 2\eta \sum_{j=0}^{m+1} |c_j - c_{j-1}|^p.$$

Here $c_{-1} = c_{m+1} = 0$. As η can be made arbitrarily small, this proves the corollary. □

While the precise statements of these three results are quite different, their structures are identical: In each case we have a normed vector space $(V, \|\cdot\|)$ and a subspace $S \subset V$ consisting of computable functions. The theorems assert that, if the norm $\|\cdot\|$ is used to measure the error, then the set S is dense in V. Neither theorem addresses the problem of finding the approximating function, though the usual proofs of these theorems provide algorithms, at least in principle. Analogous statement hold with $\mathbb{R}$ replaced by $\mathbb{R}^n$ or finite intervals.

Polynomial Approximation

Let us return to the problem of polynomial approximation for continuous functions. We can ask a more precise question: How well can a given continuous function f be approximated by a polynomial of degree n? Let $\mathcal{P}_n$ denote the polynomials of degree at most n and define

$$E_n(f) \stackrel{d}{=} \min\{\|f - p\|_{\mathcal{C}^0} \; : \; p \in \mathcal{P}_n\}. \tag{A.94}$$

Weierstrass's theorem implies that for any $f \in \mathcal{C}^0([0, 1])$,

$$\lim_{n\to\infty} E_n(f) = 0.$$

This suggests two questions:

1. Is there an element $p_n \in \mathcal{P}_n$ for which

$$\|f - p_n\|_{\mathcal{C}^0} = E_n(f)?$$

2. Is there an estimate for the rate at which $E_n(f)$ goes to zero?

The answer to the first question yes: If f is a continuous function, then there is a unique polynomial $p_n \in \mathcal{P}_n$ such that

$$\|f - p_n\|_{\mathcal{C}^0} = E_n(f).$$

The answer to the second question turns out to depend on the smoothness of f.

Theorem A.5.3 (Jackson's theorem). *If $f \in \mathscr{C}^k([0,1])$ for a $k \in \mathbb{N}$, then*

$$E_n(f) \leq \frac{C}{n^k}.$$

The smoother the function, the faster the sequence of "best" approximating polynomials converges to it. These facts suggest another question: Can the "best" approximating polynomial be found? The answer, at present, is no. We might then ask if another approximation $q_n \in \mathscr{P}_n$ can be found such that $\|f - q_n\|_{\mathscr{C}^0}$ goes to zero at about same rate as $E_n(f)$. The answer to this question is yes, but it is generally quite complicated to do. The interested reader is referred to [108]. Later we give an effective method for finding a sequence $\{q_n\}$ such that $\|f - q_n\|_{\mathscr{C}^0}$ goes to zero at nearly the optimal rate.

We close our discussion of approximation in the $\mathscr{C}^0$-norm with a formula for an approximating sequence of polynomials that works for any continuous function.

Theorem A.5.4 (Bernstein's formula). *Let $f \in \mathscr{C}^0([0,1])$ and define the polynomial of degree n by*

$$B_n(f;x) = \sum_{j=0}^{n} f\left(\frac{j}{n}\right)\binom{j}{n} x^j(1-x)^{n-j}.$$

This sequence converges to f in the $\mathscr{C}^0$-norm; that is,

$$\lim_{n\to\infty} \|B_n(f) - f\|_{\mathscr{C}^0} = 0.$$

If f is once differentiable, then there is a constant M so that

$$|B_n(f;x) - f(x)| \leq \frac{M}{\sqrt{n}}.$$

Bernstein's formula gives a sequence of polynomials that always works and gives somewhat better results if the function is smoother. However, even for very smooth functions, the Bernstein polynomials do not behave like the best approximants.

Infinite-Dimensional Gram-Schmidt

The difficulties encountered with finding the best polynomial approximations are mostly a result of using the $\mathscr{C}^0$-norm to measure the error. A much easier approximation problem results from measuring the error in the L^2-norm. Indeed, allowing a slightly more general norm does not introduce any additional difficulties. Let w be a nonnegative, integrable function defined on $[0,1]$, and define the L^2-norm *with weight* w to be

$$\|f\|_{2,w} = \int_0^1 |f(x)|^2 w(x)\, dx.$$

An infinite-dimensional generalization of the Gram-Schmidt method leads to a very simple solution to the following problem:

$$\text{Find } p_n \in \mathcal{P}_n \text{ such that } \|f - p_n\|_{2,w} = \min\{\|f - q\|_{2,w} : q \in \mathcal{P}_n\}.$$

Let $\langle\cdot,\cdot\rangle_w$ denote the inner product associated with this weighted L^2-norm

$$\langle f, g\rangle_w = \int_0^1 f(x)g(x)w(x)\,dx.$$

We use the linearly independent functions $\{1, x, x^2, \dots\}$ to define a sequence of polynomials $\{P_n\}$ by the properties

1. $\deg P_n = n$
2. $\langle P_n, P_m\rangle_w = \delta_{mn}$

The algorithm is exactly the same as the finite-dimensional case:

STEP 1: Let $P_0 = [\langle 1, 1\rangle_w]^{-\frac{1}{2}}$.

STEP 2: Suppose that we have found $\{P_0, \dots, P_j\}$. Let

$$\widetilde{P}_{j+1} = x^{j+1} + \sum_{i=0}^{j} \alpha_i P_i,$$

where

$$\alpha_i = -\langle x^{j+1}, P_i\rangle_w.$$

This function is orthogonal to $\{P_0, \dots, P_j\}$.

STEP 3: Set

$$P_{j+1} = \frac{\widetilde{P}_{j+1}}{\sqrt{\|\widetilde{P}_{j+1}\|_{2,w}}}.$$

Because the set $\{x^j : j \geq 0\}$ is infinite, this procedure does not terminate. Observe that any polynomial p of degree n has a unique expression in the form

$$p = \sum_{j=0}^{n} a_j P_j$$

and that

$$\|p\|_{2,w}^2 = \sum_{j=0}^{n} |a_j|^2.$$

Theorem A.5.5. *If f is function with $\|f\|_{2,w} < \infty$, then the polynomial*

$$p_n = \sum_{j=0}^{n} \langle f, P_j \rangle_w P_j$$

satisfies

$$\|f - p_n\|_{2,w} = \min\{\|f - q\|_{2,w} \ : \ q \in \mathcal{P}_n\}.$$

This function is called the best, weighted least squares approximation to f of degree n. Thus the best polynomial approximations with respect to these weighted L^2-norms are easy to find. If f is continuous, how does $\|f - p_n\|_{\mathcal{C}^0}$ behave? Do these give better approximations if the function f is smoother? The answers to these questions depend, in part, on the weight function. For the case

$$w(x) = [x(1-x)]^{-\frac{1}{2}},$$

the answer happens to be very simple.

Theorem A.5.6. *Suppose that $f \in \mathcal{C}^0([0, 1])$ and p_n is the best, weighted least squares approximation of degree n. Then*

$$\|f - p_n\|_{\mathcal{C}^0} \leq \left[4 + \frac{4}{\pi^2} \log n\right] E_n(f).$$

The proof of this theorem can be found in [108]. Combining this result with Jackson's theorem, we see that if $f \in \mathcal{C}^k([0, 1])$, then there is a constant C so that

$$\|f - p_n\|_{\mathcal{C}^0} \leq \frac{C \log n}{n^k}.$$

In other words, the easily found least squares approximants give almost the optimal rate of decrease for the error *measured in the $\mathcal{C}^0$-norm.* This is a remarkable and useful fact.

Exercises

***Exercise* A.5.1.** Suppose that F is a step function. Show that there exist a finite increasing sequence $a_0 < a_1 < \cdots < a_m$ and constants $\{c_1, \ldots, c_m\}$ so that

$$F = \sum_{j=1}^{m} c_j \chi_{[a_{j-1}, a_j)}.$$

***Exercise* A.5.2.** For simplicity, we work on the interval $[-1, 1]$. For each $k \in \mathbb{N}$, show that there is a polynomial

$$T_k(x) = \sum_{j=0}^{k} a_{kj} x^j$$

so that

$$\cos(k\theta) = T_k(\cos(\theta)).$$

Show that the polynomials $\{T_k\}$ satisfy the relations

$$\int_{-1}^{1} \frac{T_k(x)T_l(x)\,dx}{\sqrt{1-x^2}} = 0 \qquad \text{if } j \neq k.$$

Hint: Use the change of variables $x = \cos(\theta)$. These polynomials are called the Chebyshev polynomials. They have many remarkable properties and are often used in approximation theory; see [107]. Note that setting $y = \frac{1}{2}(1+x)$ maps $[-1, 1]$ into $[0, 1]$ and

$$y(1-y) = \frac{1-x^2}{4}.$$

A.5.2 Sampling and Interpolation

Suppose that we have a system whose state is described by a function f of a variable t. The simplest way to model a measurement is as *evaluation* of this function. That is, we have a sequence of "times" $< t_j >$ and the measurement consists of evaluating f at these times. The sequence $< f(t_j) >$ is called the *samples* of f at the times $< t_j >$. The sample times are usually labeled in a monotone fashion; that is,

$$t_j < t_{j+1}.$$

The differences $\Delta t_j = t_j - t_{j-1}$ are called the *sample spacings*. If they are all equal to a single value Δt, then we say f is *uniformly sampled with sample spacing* Δt. In a real application we can measure at most finitely many samples, and of course we can only measure them with finite precision. Nonetheless, in analyzing measurement processes it is often useful to assume that we can evaluate f along an infinite sequence and that the measurements are exact.

A question of primary interest is to decide what the samples $< f(t_j) >$ tell us about the value of f for times t not in our sample set. The answer depends on the sample spacing and a priori knowledge of the smoothness of f. Such information is usually incorporated implicitly into a model. Suppose that we sample a *differentiable* function $f(t)$ at the points $\{t_j\}$. Let t lie between t_j and t_{j+1}. Then the mean value theorem implies that there is a point $\tau \in (t_j, t_{j+1})$ such that

$$f(t) = f(t_j) + f'(\tau)(t - t_j).$$

If the points are close together and the derivative is continuous, then

$$f'(\tau) \simeq \frac{f(t_j) - f(t_{j+1})}{t_j - t_{j+1}}.$$

Defining

$$F(t) = f(t_j) + \left[\frac{f(t_j) - f(t_{j+1})}{t_j - t_{j+1}} \right] (t - t_j), \text{ for } t \in [t_j, t_{j+1}], \tag{A.95}$$

gives a continuous, piecewise linear function with $F(t_j) = f(t_j)$, for all j. In general, F is not differentiable. We say that F is a piecewise linear function, interpolating f at the points $< t_j >$. For a smooth function the error $\| f - F \|_{\mathscr{C}^0}$ goes to zero as the sample spacing goes to zero. However, the approximating function is not differentiable. This means that we cannot use F effectively to compute approximate values of f', and the graph of F has "corners" even though the graph of f is smooth.

If we know that f is a polynomial, then a finite number of samples determines f completely. If f is a polynomial of degree 0—in other words, a constant—then a single sample determines f. If the degree is 1, then 2 samples are required, and if the degree is n, then $n + 1$ samples suffice. Indeed there are simple explicit formulæ to reconstruct a polynomial from such data. For example, if the sample points are $< t_1, t_2 >$, then a linear polynomial is reconstructed as follows:

$$f(t) = f(t_1)\frac{t - t_2}{t_1 - t_2} + f(t_2)\frac{t - t_1}{t_2 - t_1}.$$

More generally, if f is of degree n and the sample points are $< t_1, \ldots, t_{n+1} >$, then

$$f(t) = \sum_{j=1}^{n+1} f(t_j) \frac{\prod_{k \neq j}(t - t_k)}{\prod_{k \neq j}(t_j - t_k)}. \tag{A.96}$$

The expression on the right-hand side of (A.96) is called the *Lagrange interpolation formula*.

If f is a continuous function that we sample at the $n + 1$ points $< t_j >$, then we can define an nth-degree polynomial using (A.96):

$$F(t) = \sum_{j=1}^{n+1} f(t_j) \frac{\prod_{k \neq j}(t - t_k)}{\prod_{k \neq j}(t_j - t_k)}. \tag{A.97}$$

This polynomial has the property that $F(t_j) = f(t_j)$ for $j = 1, \ldots, n+1$. We say that F is the nth-degree polynomial interpolant for f at the points $< t_1, \ldots, t_{n+1} >$. The question of principal interest is how well $F(t)$ approximates $f(t)$ for $t \neq t_j$. Perhaps somewhat surprisingly, the answer to this question is that, in general, F does a very poor job. Figure A.2 shows graphs of the function $f(t) = |t - \frac{1}{2}|$ along with degree 2, 6, and 12 polynomial interpolants found using equally spaced samples. Note that, as the degree increases, the polynomial provides a worse and worse approximation to f, away from the sample points. For this reason it is unusual to use a high-degree polynomial to interpolate the values of a function. This does not contradict the results of the previous subsection on the existence of accurate, high degree polynomial approximations to continuous functions. It only demonstrates that such approximations cannot, in general, be found by simply interpolating.

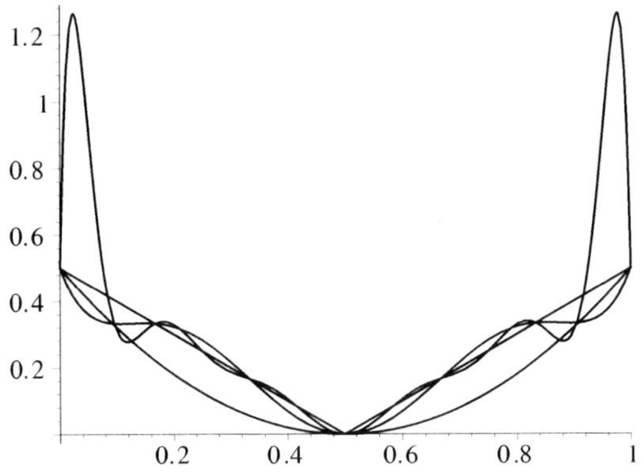

Figure A.2. Polynomial interpolants for $|x - \frac{1}{2}|$.

How then can good approximations to sampled functions be found? One answer lies in using functions which are piecewise polynomials of low degree. We consider only the simplest case. Suppose that f is a differentiable function on $[0, 1]$ and that we sample it at the points $T_n =< 0 = t_0, \dots, t_n = 1 >$. Using a piecewise cubic polynomial, we can find a function F that interpolates f at the sample points and is itself twice differentiable.

***Definition* A.5.2.** For T_n a set of points as previously, define $S(T_n)$ to be the subset of $\mathscr{C}^2([0, 1])$ consisting of functions g with the property that, for each $i \in \{0, n - 1\}$, the restriction of g to the interval $[t_i, t_{i+1}]$ is given by a cubic polynomial. Such a function is called a *cubic spline* with *nodes* $t_0 < \cdots < t_n$.

The basic approximation result is the following.

Theorem A.5.7. *Given* $n + 3$ *numbers* $\{f_0, \dots, f_n\}$ *and* $\{\alpha_0, \alpha_1\}$, *there is a unique cubic spline* $F \in S(T_n)$ *such that*

$$\begin{aligned} F(t_i) &= f_i \qquad \textit{for } i = 0, \dots, n, \\ F'(0) &= \alpha_0 \qquad \textit{and } F'(1) = \alpha_1. \end{aligned} \tag{A.98}$$

The theorem tells us that once we fix values for the derivatives at the points 0 and 1, there is a unique cubic spline that interpolates f at the given sample points. The values $< f(t_j) >$ do not determine the cubic spline interpolant for f; the numbers α_0, α_1 also need to be specified. If we know or can reasonably approximate $f'(0)$ and $f'(1)$, then these give reasonable choices for α_0 and α_1. If these data are not known, then another common way to pick a cubic spline, interpolating f, is to require that $F''(0) = F''(1) = 0$. This is sometimes called the *natural cubic spline* interpolating $< f(t_i) >$.

The problem of finding cubic splines is easily reduced to a system of linear equations. We give this reduction for the case considered in the theorem with the additional assumption that $t_i - t_{i-1} = h$ for all i. Let $f_i = f(t_i)$. To define the basic building blocks, set

$$\begin{aligned} c_i(t) = & \left[\frac{(t-t_{i+1})^2}{h^2} + \frac{2(t-t_i)(t-t_{i+1})^2}{h^3}\right] f_i + \\ & \left[\frac{(t-t_i)^2}{h^2} - \frac{2(t-t_{i+1})(t-t_i)^2}{h^3}\right] f_{i+1} + \\ & \frac{(t-t_i)(t-t_{i+1})^2}{h^2} a_i + \frac{(t-t_{i+1})(t-t_i)^2}{h^2} a_{i+1}. \end{aligned} \tag{A.99}$$

Evaluating this function gives

$$\begin{aligned} c_i(t_{i+1}) &= c_{i+1}(t_{i+1}) = f_{i+1}, \\ c_i'(t_{i+1}) &= c_{i+1}'(t_{i+1}) = a_{i+1}. \end{aligned} \tag{A.100}$$

In other words, for any choice of values $\{a_1, \ldots, a_{n-1}\}$, these functions piece together to define a continuously differentiable function, interpolating the values of f. To find the spline with these properties, we need to select these coefficients so that the resultant function also has a continuous second derivative. Evaluating the second derivatives and comparing at the adjacent endpoints, we derive the relations

$$a_i + 4a_{i+1} + a_{i+2} = \frac{3}{h}(f_{i+2} - f_i), \text{ for } i = 0, \ldots, n-2.$$

This is an invertible, "tridiagonal"system of linear equations for $\{a_1, \ldots, a_{n-1}\}$. After solving for these coefficients, setting

$$F(t) = c_i(t) \text{ for } t \in [t_i, t_{i+1}], \quad i = 0, n-1,$$

gives the desired spline.

Splines have many desirable properties. They are the interpolants that have, in a certain sense, the minimal oscillation among all twice differentiable functions that interpolate the given values. If f is twice differentiable, then the first derivative of the spline derived previously is also a good approximation to f'. This discussion is adapted from that given in [108] and [87].

Exercise

***Exercise* A.5.3.** Explain why formula (A.96) gives the correct answer if f is known to be a polynomial of degree n.

A.6 Numerical Techniques for Differentiation and Integration

In calculus we learn a variety of rules for computing derivatives and integrals of functions given by formulæ. In applications we need to have ways to approximate these operations for measured data. We briefly review some elementary method for numerically approximating integrals and derivatives. First we consider the problems of approximating integration and differentiation for exactly known functions and then the same questions for noisy sampled data.

If f is a Riemann integrable function defined on [0, 1], then its integral can be defined as the following limit:

$$\int_0^1 f(t)\,dt = \lim_{N\to\infty} \sum_{j=1}^{N} \frac{1}{N} f\left(\frac{j}{N}\right).$$

Using this sum for a fixed value of N gives a way to approximate an integral, called a *Riemann sum* approximation. For functions with more smoothness, there are better approximations. If f is differentiable, then its derivative is also defined as a limit

$$f'(t) = \lim_{\Delta t\to 0} \frac{f(t+\Delta t) - f(t)}{\Delta t}.$$

Using this formula with positive values of Δt leads to approximations for the derivative called *finite differences.*

If f is a continuous function and we set

$$m_N = \max_{0\le j\le N-1} \max_{t\in[\frac{j}{N},\frac{j+1}{N}]} |f(t) - f(j/N)|,$$

then

$$\left| \int_0^1 f(t)\,dt - \sum_{j=0}^{N-1} \frac{1}{N} f\left(\frac{j}{N}\right) \right| \le m_N.$$

To find the analogous estimate for the approximate derivative, we let

$$M_N = \max_{|s-t|\le\frac{1}{N}} |f'(t) - f'(s)|;$$

then, applying the mean value theorem, we see that

$$\left| \frac{f(t+\frac{1}{N}) - f(t)}{N^{-1}} - f'(t) \right| \le M_N.$$

Comparing these formulæ we see that the accuracy of the approximate integral is controlled by the size of m_N, while that of the approximate derivative is controlled by M_N. If f is differentiable, then $m_N \propto N^{-1}$; whereas this implies no estimate for M_N. In order to know that $M_N \propto N^{-1}$, we would need to know that f is *twice* differentiable. This indicates why, in principle, it is harder to approximate derivatives than integrals.

For real data, approximate integration is usually simpler and more accurate than approximate differentiation. The reason for this lies in the nature of noise. Suppose that f represents the "actual" state of our system. We often aggregate various (possibly unknown) sources of error and uncertainty into a single function n, which we call *noise*. What are then measured are actually samples of $f + n$.

Intuitively, noise is a random process, so it goes up and down unpredictably; such a function is typically not differentiable. This means that $|n(t + \Delta t) - n(t)|$ may well be large compared to Δt. On the other hand, if we have a good enough model, then the noise term should be equally likely to be positive or negative, even on small time scales. This implies that averages of n over small intervals should often be small. Symbolically,

$$\left|\frac{n(t+\Delta t) - n(t)}{\Delta t}\right| >> 1, \text{ whereas}$$
$$\left|\frac{1}{\Delta}\int_{t}^{t+\Delta t} n(s)\,ds\right| << 1. \tag{A.101}$$

A.6.1 Numerical Integration

In addition to the (right) Riemann sum formula,

$$R_N(f) = \frac{1}{N}\sum_{j=1}^{N} f(\frac{j}{N}),$$

there are two other commonly used formulæ for numerical integration: the trapezoidal rule and Simpson's rule. Suppose that f is a continuous function on $[0, 1]$; the trapezoidal approximation to the integral of f with $N + 1$ points is given by

$$T_N(f) = \frac{1}{2N}(f(0) + f(1)) + \frac{1}{N}\sum_{j=1}^{N-1} f\left(\frac{j}{N}\right). \tag{A.102}$$

If f is twice differentiable, then we have the error estimate

$$\left|T_N(f) - \int_0^1 f(t)\,dt\right| \leq \frac{\max_{t\in[0,1]}|f''(t)|}{12N^2}. \tag{A.103}$$

Simpson's rule also uses the midpoints; it is given by

$$S_N(f) = \frac{1}{6N}\left[f(0) + f(1) + 2\sum_{j=1}^{N-1} f\left(\frac{j}{N}\right) + 6\sum_{j=0}^{N-1} f\left(\frac{2j+1}{2N}\right)\right]. \tag{A.104}$$

Though Simpson's rule only requires about twice as much computation, it gives a much smaller error if f is *four* times differentiable. The error estimate is

$$\left|S_N(f) - \int_0^1 f(t)\,dt\right| \leq \frac{\max_{t\in[0,1]}|f^{[iv]}(t)|}{2880N^4}. \tag{A.105}$$

N	$\lvert S_N - \frac{\pi}{8}\rvert$	$\lvert R_N - \frac{\pi}{8}\rvert$
16	.0025504	.0064717
64	.0003176	.0008113
128	.0001122	.0002870
144	.0000940	.0002405

Table A.2. Comparison of absolute errors in different approximations to the integral of $\sqrt{x(1-x)}$

It is important to note that the error estimate for the trapezoidal rule assumes that f is twice differentiable and for Simpson's rule that f is four times differentiable. If this is not true, then these "higher-order" integration schemes do not produce such precise results. For real data, higher-order schemes are often not used because the noise present in the data has the effect of making the integrand nondifferentiable. This means that the higher rates of convergence are not realized. These approximate integration techniques are examples of relatively elementary methods. Each method for approximating functions leads, via integration, to a method for approximating integrals. These go under the general rubric of *quadrature methods*. What distinguishes these methods, at least theoretically, is the set of functions for which the approximate formula gives the correct answer. The Riemann sum is correct for constant functions; the trapezoidal rule gives an exact result for linear functions, and Simpson's rule gives the exact answer for cubic polynomials. A more complete discussion of this rich and important subject can be found in [56].

***Example* A.6.1.** We consider the results of using Riemann sums and Simpson's rule to approximately compute

$$\int_0^1 \sqrt{x(1-x)}\,dx = \frac{\pi}{8}.$$

This function is not differentiable at the endpoints of the interval. Table A.2 shows the absolute errors in approximating this integral by Riemann sums and Simpson's rule for several values of N. Since this function vanishes at the endpoints, the trapezoidal rule and the Riemann sum give the same result. While Simpson's rule gives a better result for a given number of samples, it improves much more slowly than it would if the integrand had four derivatives. The ratios of the error and the expected rates of decay (i.e., N^{-4} for Simpson's rule and N^{-1} for Riemann sums) are summarized in Table A.3. As this table shows, the order of convergence for Simpson's rule is much slower than it would be if f had the requisite four derivatives.

***Example* A.6.2.** We now consider these integration schemes applied to a "random" piecewise linear function taking values between -1 and $+1$. The graph of such a function is shown in Figure A.3.

N	$N^4\lvert S_N - \frac{\pi}{8}\rvert$	$N\lvert R_N - \frac{\pi}{8}\rvert$
16	167	.104
48	2596	.06
80	9304	.05
112	21571	.04
144	40426	.034

Table A.3. Comparison of rates of convergence to the "expected" rates of decay in different approximations to the integral of $\sqrt{x(1-x)}$

N	$\lvert R_N - I\rvert$	$\lvert R_N - I\rvert N$	$\lvert T_N - I\rvert$	$\lvert T_N - I\rvert N^2$
16	.254	.41	.063	16
40	.017	.69	.068	109
56	.021	1.15	.069	216
80	.007	.64	.069	441

Table A.4. Rates of convergence and errors for approximations to the integral of a random function

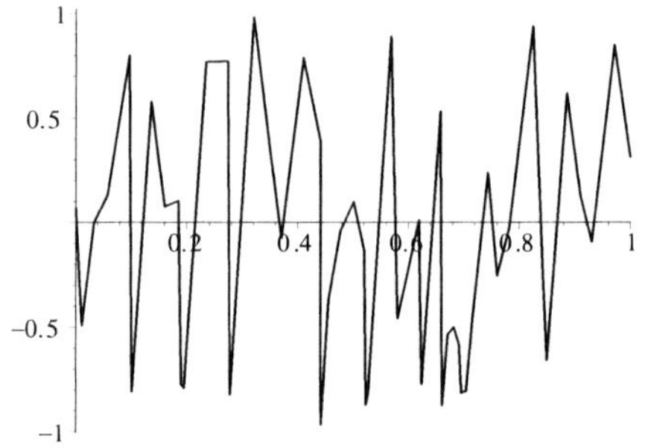

Figure A.3. A random piecewise linear function.

In Table A.3 we give the errors made using Riemann sums and the trapezoidal rule for various values of N, as well as these errors rescaled by the theoretical rate of decrease for sufficiently differentiable data. Here $I = -.06972...$ is the actual value of the integral. Note that in absolute terms, the result from the trapezoidal rule does not improve as we increase N. Relative to the expected error, things are consequently degenerating rapidly.

Exercises

***Exercise* A.6.1.** The log function is defined by an integral

$$\log(x) = \int_1^x \frac{ds}{s}.$$

By using numerical integration techniques, approximate values for $\log(x)$ can be obtained. For the Riemann sum, trapezoidal rule, and Simpson's rule, how large a value of N is needed to compute $\log(2)$ with 10 digits of accuracy?

***Exercise* A.6.2.** Use the functional equation $\log(xy) = \log(x) + \log(y)$ to devise an efficient method for approximately computing the logarithms of numbers between .01 and 100.

A.6.2 Numerical Differentiation

If f is a differentiable function, defined on $[0, 1]$ sampled at the points $\{\frac{j}{N} : j = 0, \ldots, N\}$, then we can use a finite difference formula to approximate the derivatives of f at the points in the sample set. We could use the left difference, right difference, or centered difference to obtain approximations to $f'(\frac{j}{N})$; if $j \neq 0, N$, then

$$\text{LEFT DIFFERENCE:}\quad D_l f(\tfrac{j}{N}) = \frac{f(\frac{j}{N}) - f(\frac{j-1}{N})}{N^{-1}}$$

$$\text{RIGHT DIFFERENCE:}\quad D_r f(\tfrac{j}{N}) = \frac{f(\frac{j+1}{N}) - f(\frac{j}{N})}{N^{-1}}$$

$$\text{CENTERED DIFFERENCE:}\quad D_c f(\tfrac{j}{N}) = \frac{f(\frac{j+1}{N}) - f(\frac{j-1}{N})}{2N^{-1}}$$

For a twice differentiable function, Taylor's formula gives the error estimates

$$|D_r f(\frac{j}{N}) - f'(\frac{j}{N})| \leq \frac{M_2}{N}, \quad |D_l f(\frac{j}{N}) - f'(\frac{j}{N})| \leq \frac{M_2}{N}, \tag{A.106}$$

where M_2 is proportional to the maximum of $|f^{[2]}(x)|$. If f is three time differentiable, then

$$|D_c f(\frac{j}{N}) - f'(\frac{j}{N})| \leq \frac{M_3}{N^2}, \tag{A.107}$$

where M_3 is proportional to the maximum of $|f^{[3]}(x)|$.

There are other ways to assign approximate values to $< f'(\frac{j}{N}) >$ given sampled data $< f(\frac{j}{N}) >$. For example, a cubic spline F_N, interpolating these values, is a twice differentiable function. A reasonable way to approximate the derivatives of f is to use $< F_N'(\frac{j}{N}) >$. Since F_N is defined for all $x \in [0, 1]$, we can even use $F_N'(x)$ as an approximate value for $f'(x)$ for any $x \in [0, 1]$. If f is twice differentiable and F_N is the cubic spline defined using the endpoint data,

$$a_0 = D_r f(0), \quad a_N = D_l f(1),$$

then $< F_N >$ converges to f in $\mathscr{C}^1([0, 1])$. That is,

$$\lim_{N \to \infty} [\|f - F_N\|_{\mathscr{C}^0} + \|f' - F_N'\|_{\mathscr{C}^0}] = 0.$$

For sufficiently large N, $F_N'(x)$ is a good approximation to $f'(x)$ for any $x \in [0, 1]$. The rate of convergence depends on the size of the second derivatives of f. Fourier series and integrals provide other ways to approximate the derivatives of a function.

N	.25	.5	.75
10	-7.41	16.36	3.35
100	-68.8	82.7	-8.3
1000	296.2	143.4	-280.5

Table A.5. Finite differences for a random piecewise linear function

We close this section by considering approximation of derivatives in a more realistic situation. Suppose that we are trying to sample a function f. Ordinarily, we model the actual data as samples of $f + \epsilon n$, where n is a "random" noise function. Here we scale things so the $|n(t)| \leq 1$ for all t; ϵ is then the amplitude of the noise. The noise is a random function in two different senses. In the first place we cannot say with certainty what the function is, so we usually think of this function as being *randomly selected* from some family. It is also random in the sense that functions in these families do not vary smoothly. Thus for n a fixed member of the family, the value of $n(t)$ at a given time t is itself *random*. For instance, we could use the family of piecewise constant functions, or the family of all piecewise linear functions as models for the noise. The graphs of such a function is shown in Figure A.4.

If the sampled data is of the form $< f(\frac{j}{N}) + \epsilon n(\frac{j}{N}) >$, then the right difference approximation to the first derivative is

$$f'(\frac{j}{N}) \simeq D_r f(\frac{j}{N}) + \epsilon N[n(\frac{j+1}{N}) - n(\frac{j}{N})].$$

Due to the random nature of the noise (in the second sense), there is no reason why the difference $n(\frac{j+1}{N}) - n(\frac{j}{N})$ should be small. The contribution of noise to the error in approximating $f'(\frac{j}{N})$ can only be bounded by $2\epsilon N$. In order to get a good approximate value for f', using a finite difference, it would be necessary to choose N, so that $N\epsilon$ remains small.

***Example* A.6.3.** Table A.3 shows the finite differences for the random, piecewise linear function shown in Figure A.4. The differences are evaluated at the points $x \in \{.25, .5, .75\}$ with the indicated values of N.

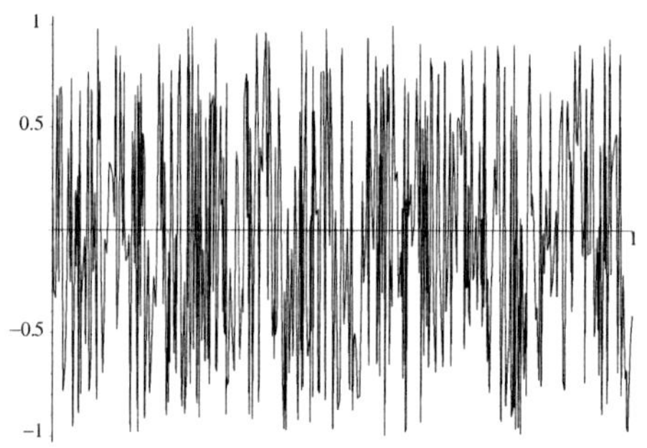

Figure A.4. A fairly random function.

In the foregoing example, the measurement process is modeled as functional evaluation. Actual measurements are always some sort of an average; to measure the value of a function at a single moment of time would require infinite energy. Random functions often have small averages, so this fact actually works in our favor when trying to approximate derivatives. The simplest model for an average is a uniform average: Instead of evaluating a function at the arguments $\{\frac{j}{N}\}$, we actually measure its average over an interval $[\frac{j}{N}, \frac{j+\delta}{N}]$. Our samples are therefore

$$f_j = \delta^{-1} \int_{\frac{j}{N}}^{\frac{j+\delta}{N}} [f(s) + n(s)]\, ds.$$

The finite difference then gives

$$\frac{f_{j+1} - f_j}{N^{-1}} = \delta^{-1} \int_{\frac{j}{N}}^{\frac{j+\delta}{N}} \frac{(f(s+N^{-1}) - f(s))\, ds}{N^{-1}} + \frac{\epsilon N}{\delta} \int_{\frac{j}{N}}^{\frac{j+\delta}{N}} [n(s+N^{-1}) - n(s)]\, ds. \quad \text{(A.108)}$$

Using the mean value theorem, it follows that for each $s \in [\frac{j}{N}, \frac{j+\delta}{N}]$ there is an ξ_s in this interval so that

$$\frac{f(s+N^{-1}) - f(s)}{N^{-1}} = f'(\xi_s).$$

Thus the first term in (A.108) is a weighted average of values of f' over the interval $[\frac{j}{N}, \frac{j+\delta}{N}]$.

Because of the randomness of n, there is no reason for the differences $[n(s + N^{-1}) - n(s)]$ to be small; however, for a large enough δ, the individual averages,

$$\frac{1}{\delta} \int_{\frac{j}{N}}^{\frac{j+\delta}{N}} n(s)\, ds, \quad \frac{1}{\delta} \int_{\frac{j+1}{N}}^{\frac{j+1+\delta}{N}} n(s)\, ds,$$

should themselves be small. This would, in turn, make the second term in (A.108) small. This illustrates a familiar dichotomy between noise reduction and resolution: By increasing δ we can diminish the effect of the noise, both in the measured values of f and in the finite difference approximations to f'. On the other hand, increasing δ also smears out the values of f'. The price for reducing the noise component of a measurement is decreasing its resolution.

***Example* A.6.4.** Using the same function considered in Example A.6.3, we compute the finite differences for averaged data. In order to be able to make comparisons we fix $N^{-1} = .1$ and consider the results obtained with $\delta \in \{.1, .02, .01, .005\}$. The data in Table A.6 bear out the prediction that averaging the data diminishes the effects of noise on the computation of finite differences, with longer averaging intervals generally producing a larger effect.

However, there is also some failure of this to occur. This is because the experiment is performed on a "random" piecewise linear function, which is, in some sense, not especially random.

δ	.25	.5	.75
0	-7.41	16.36	3.35
.1	.269	.593	2.57
.02	5.97	1.64	.281
.01	3.99	.174	5.2
.005	6.74	-1.129	5.50

Table A.6. Finite differences for an averaged, random piecewise linear function

Exercises

***Exercise* A.6.3.** Show how to use Taylor's formula to derive (A.106) and (A.107) these error estimates and give formulæ for M_2 and M_3.

***Exercise* A.6.4.** Show that for the function $f(t) = |t|$ and all t, the centered differences converge as $N \to \infty$. Is this limit always equal to $f'(t)$?

Appendix B

Basic Analysis

This appendix contains some of the basic definitions and results from analysis that are used in this book. Many good treatments of this material are available; for example, [27], [111], or [121].

B.1 Sequences

A *sequence* is an ordered list of objects. As such it can be described by a function defined on a subset of the integers of the form $\{M, M+1, \ldots, N-1, N\}$. This set is called the *index set*. If both M and N are finite, then the sequence is finite. If $M = -\infty$ and N is finite or M is finite and $N = \infty$, then it is an infinite sequence. If the index set equals $\mathbb{Z}$, then the sequence is *bi-infinite*. In this section and throughout most of the book, the term *sequence* usually refers to an infinite sequence. In this case the index set is usually taken to be the positive integers $\mathbb{N}$. For example, a sequence of real numbers is specified by a function

$$x : \mathbb{N} \longrightarrow \mathbb{R}.$$

With this notation the nth term of the sequence would be denoted $x(n)$. It is *not* customary to use functional notation but rather to use subscripts to label the terms of a sequence. The nth term is denoted by x_n and the totality of the sequence by $< x_n >$. This distinguishes a sequence from the *unordered* set consisting of its elements, which is denoted $\{x_n\}$.

Given a sequence $< x_n >$, a *subsequence* is defined by selecting a subset of $\{x_n\}$ and keeping them in the same order as they appear in $< x_n >$. In practice, this amounts to defining a monotone increasing function from $\mathbb{N}$ to itself. We denote the value of this function at j by n_j. In order to be monotone increasing, $n_j < n_{j+1}$, for every j in $\mathbb{N}$. The jth term of the subsequence is denoted by x_{n_j}, and the totality of the subsequence by $< x_{n_j} >$. As an example, consider the sequence $x_n = (-1)^n n$; setting $n_j = 2j$ defines the subsequence $x_{n_j} = (-1)^{2j} 2j$.

***Definition* B.1.1.** A sequence of real numbers, $< x_n >$, has a *limit* if there is a number L such that, given any $\epsilon > 0$, there exists an integer $N > 0$, such that

$$|x_n - L| < \epsilon \qquad \text{whenever } n > N.$$

A sequence with a limit is called a *convergent sequence*: we then write

$$\lim_{n\to\infty} x_n = L.$$

When a limit exists it is unique. A sequence may fail to have limit, but it may have a subsequence that converges. In this case the sequence is said to have a *convergent subsequence.* For example, $x_n = (-1)^n$ is not convergent, but the subsequence defined by $n_j = 2j$ is.

The computation of limits is facilitated by the rules for computing limits of algebraic combinations of convergent sequences.

Theorem B.1.1 (Algebraic rules for limits). *Suppose that* $< x_n >, < y_n >$ *are convergent sequences of real numbers. Then*

$$\begin{aligned} \lim_{n\to\infty} ax_n \textit{ exists and equals } a \lim_{n\to\infty} x_n, \textit{ for all } a \in \mathbb{R}, \\ \lim_{n\to\infty} (x_n + y_n) \textit{ exists and equals } \lim_{n\to\infty} x_n + \lim_{n\to\infty} y_n, \\ \lim_{n\to\infty} (x_n y_n) \textit{ exists and equals } (\lim_{n\to\infty} x_n)(\lim_{n\to\infty} y_n), \\ \textit{provided } \lim_{n\to\infty} y_n \neq 0, \textit{ then } \lim_{n\to\infty} \frac{x_n}{y_n} \textit{ exists and equals } \frac{\lim_{n\to\infty} x_n}{\lim_{n\to\infty} y_n}. \end{aligned} \tag{B.1}$$

In this theorem the nontrivial claim is that the limits exist; once this is established it is easy to show what the limits must be.

A problem of fundamental importance is to decide whether or not a sequence has a limit. A sequence $< x_n >$ is bounded if there is a number M such that

$$|x_n| < M \qquad \text{for all } n.$$

A sequence is non-increasing if

$$x_n \geq x_{n+1} \qquad \text{for all } n,$$

and nondecreasing if

$$x_n \leq x_{n+1} \qquad \text{for all } n.$$

The *completeness axiom* of the real numbers states that *a bounded non-increasing or non-decreasing sequence of real numbers has a limit.* If a bounded sequence is neither non-decreasing nor non-increasing, then the only general theorem about convergence is the Bolzano-Weierstrass theorem.

Theorem B.1.2 (Bolzano-Weierstrass theorem). *A bounded sequence of real numbers has a convergent subsequence.*

Note that this theorem does *not* assert that any bounded sequence converges but only that any bounded sequence has a convergent subsequence.

The following two lemmas are very useful for studying limits of sequences.

Lemma B.1.1 (Squeeze lemma). *If* $< x_n >, < y_n >, < z_n >$ *are sequences of real numbers such that*

$$x_n \le y_n \le z_n$$

and x_n *and* z_n *are convergent with*

$$L = \lim_{n\to\infty} x_n = \lim_{n\to\infty} z_n,$$

then y_n *converges with*

$$\lim_{n\to\infty} y_n = L.$$

Lemma B.1.2. *If* $x_n \ge 0$ *for* n *in* $\mathbb{N}$ *and* $< x_n >$ *is a convergent sequence, then*

$$\lim_{n\to\infty} x_n \ge 0.$$

In the preceding discussion it is assumed that the limit is known in advance. There is a criterion, due to Cauchy, that states that a given sequence has a limit but makes no reference to the limit itself.

Theorem B.1.3 (Cauchy criterion for sequences). *If* $< x_n >$ *is a sequence of real numbers such that, given* $\epsilon > 0$ *there exists an* N *for which*

$$|x_n - x_m| < \epsilon \qquad \text{whenever both } n \text{ and } m \text{ are greater than } N,$$

then the sequence is convergent.

A sequence satisfying this condition is called a *Cauchy sequence.*

B.2 Series

A series is the sum of an infinite sequence; it is denoted by

$$\sum_{n=1}^{\infty} x_n.$$

***Definition* B.2.1.** A *series* converges if the *sequence* of partial sums

$$s_k = \sum_{n=1}^{k} x_n,$$

converges. In this case

$$\sum_{n=1}^{\infty} x_n \stackrel{d}{=} \lim_{k\to\infty} s_k.$$

If a series does not converge ,then it diverges.

***Definition* B.2.2.** A series converges *absolutely* if the sum of the absolute values

$$\sum_{n=1}^{\infty} |x_n|$$

converges.

The following theorem lists the elementary properties of convergent series.

Theorem B.2.1 (Theorem on series). *Suppose that* $< x_n >, < y_n >$ *are sequences. If the series* $\sum_{n=1}^{\infty} x_n, \sum_{n=1}^{\infty} y_n$ *converge, then*

$$\sum_{n=1}^{\infty}(x_n + y_n) \text{ converges and } \sum_{n=1}^{\infty}(x_n + y_n) = \sum_{n=1}^{\infty} x_n + \sum_{n=1}^{\infty} y_n,$$

$$\text{If } a \in \mathbb{R} \sum_{n=1}^{\infty} a x_n = a \sum_{n=1}^{\infty} x_n, \tag{B.2}$$

$$\text{If } x_n \geq 0 \text{ for all } n, \text{ then } \sum_{n=1}^{\infty} x_n \geq 0.$$

There are many criteria, used to determine if a given series converges. The most important is the comparison test.

Theorem B.2.2 (Comparison Test). *Suppose that* $< x_n >, < y_n >$ *are sequences such that, for all* n, $|x_n| \leq y_n$. *If* $\sum_{n=1}^{\infty} y_n$ *converges, then so does* $\sum_{n=1}^{\infty} x_n$. *If* $0 \leq y_n \leq x_n$ *and* $\sum_{n=1}^{\infty} y_n$ *diverges, then so does* $\sum_{n=1}^{\infty} x_n$.

Since this test calls for a comparison, we need to have examples of series that are known to converge or diverge. The simplest case is a geometric series. This is because there is a formula for the partial sums:

$$\sum_{n=0}^{k} a^k = \frac{a^{k+1} - 1}{a - 1}.$$

From this formula we immediately conclude the following:

Theorem B.2.3 (Convergence of geometric series). *A geometric series converges if and only if* $|a| < 1$.

The root and ratio tests are special cases of the comparison test where the series is comparable to a geometric series.

Theorem B.2.4 (Ratio test). *If* $< x_n >$ *is a sequence with*

$$\limsup_{n\to\infty} \left| \frac{x_{n+1}}{x_n} \right| = \alpha,$$

then the series

$$\sum_{n=1}^{\infty} x_n \begin{cases} \textit{converges if} & \alpha < 1 \\ \textit{diverges if} & \alpha > 1. \end{cases}$$

The test gives no information if $\alpha = 1$.

We also have the following:

Theorem B.2.5 (Root test). *If* $< x_n >$ *is a sequence with*

$$\limsup_{n\to\infty} |x_n|^{\frac{1}{n}} = \alpha,$$

then the series

$$\sum_{n=1}^{\infty} x_n \begin{cases} \textit{converges if} & \alpha < 1 \\ \textit{diverges if} & \alpha > 1. \end{cases}$$

The test gives no information if $\alpha = 1$.

If $\alpha < 1$ in the ratio or root tests, then the series converge absolutely.

Another test is obtained by comparing a series to an integral.

Theorem B.2.6 (Integral test). *If* f *is a non-negative, monotone decreasing, integrable function defined for* $x \geq 1$, *then*

$$\sum_{n=1}^{\infty} f(n) \textit{ converges if and only if } \lim_{n\to\infty} \int_1^n f(x)\,dx \textit{ exists.}$$

From this test it follows that the series $\sum_{n=1}^{\infty} \frac{1}{n^p}$ converges if and only if $p > 1$. If a series is shown to converge using any of the foregoing tests, then the series converges absolutely. We give a final test that sometimes gives the convergence of a non-absolutely convergent series.

Theorem B.2.7 (Alternating series test). *Suppose that* $< x_n >$ *is a sequence such that the sign alternates, the* $\lim_{n\to\infty} x_n = 0$, *and* $|x_{n+1}| \leq |x_n|$. *Then*

$$\sum_{n=1}^{\infty} x_n$$

converges and

$$\left| \sum_{n=1}^{\infty} x_n - \sum_{n=1}^{N} x_n \right| \leq |x_{N+1}|. \tag{B.3}$$

Note that this test requires that the signs alternate, the sequence of absolute values is monotonely decreasing, and the sequence tends to zero. If any of these conditions are not met, the series may fail to converge.

A useful tool for working with integrals is the integration by parts formula; see Proposition B.6.1. This formula has a discrete analogue, called the *summation by parts formula*, which is important in the study of non-absolutely convergent series.

Proposition B.2.1 (Summation by parts formula). *Let* $< x_n >$ *and* $< y_n >$ *be sequences of numbers. For each* n, *let*

$$Y_n = \sum_{k=1}^{n} y_k;$$

then

$$\sum_{n=1}^{N} x_n y_n = x_N Y_N - \sum_{n=1}^{N-1} (x_{n+1} - x_n) Y_n. \tag{B.4}$$

Using this formula, it is often possible to replace a conditionally convergent sum by an absolutely convergent sum.

***Example* B.2.1.** Let $\alpha = e^{2\pi i x}$, where $x \notin \mathbb{Z}$, so that $\alpha \neq 1$. For any such α, the series

$$\sum_{n=1}^{\infty} \frac{\alpha^n}{\sqrt{n}}$$

converges. To prove this, observe that

$$B_n = \sum_{k=1}^{n} \alpha^k = \frac{\alpha^{n+1} - 1}{\alpha - 1}$$

is a uniformly bounded sequence and

$$\frac{1}{\sqrt{n}} - \frac{1}{\sqrt{n+1}} \leq \frac{1}{n^{\frac{3}{2}}}.$$

The summation by parts formula gives

$$\sum_{n=1}^{N} \frac{\alpha^n}{\sqrt{n}} = \left(\frac{1}{\sqrt{N}} - \frac{1}{\sqrt{N+1}}\right) B_N - \sum_{n=1}^{N-1} \left(\frac{1}{\sqrt{n}} - \frac{1}{\sqrt{n+1}}\right) B_n.$$

The boundary term on the right goes to zero as $N \to \infty$, and the sum is absolutely convergent. This shows how the summation by parts formula can be used to convert a conditionally convergent sum into an absolutely convergent sum.

B.3 Limits of Functions and Continuity

The next thing we consider is the behavior of functions defined on intervals in $\mathbb{R}$. Suppose that f is defined for $x \in (a, c) \cup (c, b)$. This is called a *punctured or deleted neighborhood* of c.

***Definition* B.3.1.** We say that the function f has limit L, as x approaches c if, given $\epsilon > 0$, there exists $\delta > 0$ such that

$$|f(x) - L| < \epsilon \qquad \text{provided } 0 < |x - c| < \delta;$$

we write

$$\lim_{x \to c} f(x) = L.$$

Note that in this definition nothing is said about the value of f at c. This has no bearing at all on whether or not the limit exists.

***Definition* B.3.2.** If $f(c)$ is defined and we have that

$$\lim_{x \to c} f(x) = f(c),$$

then we say that f is continuous at c. If f is continuous for all $x \in (a, b)$, then we say that f is continuous on (a, b).

In addition to the ordinary limit, we also define one-sided limits. If f is defined in (a, b) and there exists an L such that, given $\epsilon > 0$ there exists $\delta > 0$ such that

$$|f(x) - L| < \epsilon \text{ provided } 0 < x - a < \delta, \text{ then } \lim_{x \to a^+} f(x) = L.$$

If instead

$$|f(x) - L| < \epsilon \text{ provided } 0 < b - x < \delta, \text{ then } \lim_{x \to b^-} f(x) = L.$$

The rules for dealing with limits of functions are very similar to the rules for handling limits of sequences

Theorem B.3.1 (Algebraic rules for limits of functions). *Suppose that f, g are defined in a punctured neighborhood of c and that*

$$\lim_{x \to c} f(x) = L, \quad \lim_{x \to c} g(x) = M.$$

Then

$$\begin{aligned} &\lim_{x \to c} (af(x)) \text{ exists and equals } aL \text{ for all } a \in \mathbb{R}, \\ &\lim_{x \to c} (f(x) + g(x)) \text{ exists and equals } L + M, \\ &\lim_{x \to c} (f(x)g(x)) \text{ exists and equals } LM, \\ &\text{provided } M \neq 0, \lim_{x \to c} \frac{f(x)}{g(x)} \text{ exists and equals } \frac{L}{M}. \end{aligned} \tag{B.5}$$

From this we deduce the following results about continuous functions:

Theorem B.3.2 (Algebraic rules for continuous functions). *If f, g are continuous at c, then so are af, $f+g$, fg. If $g(c) \neq 0$, then f/g is also continuous at c.*

For functions there is one further operation that is very important: composition.

Theorem B.3.3 (Continuity of compositions). *Suppose that $f(x)$, $g(y)$ are two functions such that $f(x)$ is continuous at $x = c$ and $g(y)$ is continuous at $y = f(c)$. Then the composite function, $g \circ f(x)$, is continuous at $x = c$.*

***Definition* B.3.3.** A function defined on an interval $[a, b]$ is said to be *uniformly continuous* if, given $\epsilon > 0$, there exists $\delta > 0$ such that

$$|f(x) - f(y)| < \epsilon, \text{ for all } x, y \in [a, b] \text{ with } |x - y| < \delta.$$

The basic proposition is as follows:

Proposition B.3.1. *A continuous function on a closed, bounded interval is uniformly continuous.*

Using similar arguments, we can also prove the following:

Proposition B.3.2 (Max-min theorem for continuous functions). *If f is continuous on a closed bounded interval, $[a, b]$, then there exists $x_1 \in [a, b]$ and $x_2 \in [a, b]$ that satisfy*

$$f(x_1) = \sup_{x \in [a,b]} f(x), \quad f(x_2) = \inf_{x \in [a,b]} f(x).$$

As a final result on continuous functions, we have the Intermediate value theorem

Theorem B.3.4 (Intermediate value theorem). *Suppose that f is continuous on $[a, b]$ and $f(a) < f(b)$. Then for each $y \in (f(a), f(b))$ there exists $c \in (a, b)$ such that $f(c) = y$.*

B.4 Differentiability

A function defined in a neighborhood of a point c is said to be *differentiable* at c if the function

$$g(x) = \frac{f(x) - f(c)}{x - c},$$

defined in a punctured neighborhood of c, has a limit as $x \to c$. This limit is called the derivative of f at c; we denote it by $f'(c)$. A function that is differentiable at every point of an interval is said to be differentiable in the interval. If the derivative is itself continuous, then the function is said to be *continuously differentiable*. As with continuous functions, we have algebraic rules for differentiation.

Proposition B.4.1 (Rules for differentiation). *Suppose that f, g are differentiable at c; then so are af, $(f+g)$, fg. If $g(c) \neq 0$, then so is f/g. The derivatives are given by*

$$\begin{aligned}(af)'(c) &= a(f'(c)),\\ (f+g)'(c) &= f'(c)+g'(c),\\ (fg)'(c) &= f'(c)g(c)+f(c)g'(c),\\ \left(\frac{f}{g}\right)'(c) &= \frac{f'(c)g(c)-f(c)g'(c)}{g(c)^2}.\end{aligned} \tag{B.6}$$

We can also differentiate a composition.

Proposition B.4.2 (The chain rule). *If $f(x)$ is differentiable at $x = c$ and $g(y)$ is differentiable at $y = f(c)$, then $g \circ f(x)$ is differentiable at $x = c$; the derivative is*

$$g \circ f'(c) = g'(f(c))f'(c).$$

It is often useful to be able to compare the sizes of two functions f, g near a point c without being too specific. The *big O* and *little o* notations are often used for this purpose.

***Definition* B.4.1.** The notation

$$f(x) = O(g(x)) \qquad \text{near to } c$$

means that there exists an M and an $\epsilon > 0$ such that

$$|f(x)| < Mg(x) \qquad \text{provided } |x-c| < \epsilon,$$

whereas

$$f(x) = o(g(x)) \qquad \text{near to } c$$

means that

$$\lim_{x \to c} \frac{|f(x)|}{g(x)} = 0.$$

For example, a function f is differentiable at c if and only if there exists a number L for which

$$f(x) = f(c) + L(x-c) + o(|x-c|).$$

Of course, $L = f'(c)$. This implies that a function that is differentiable at a point is also continuous at that point. The converse statement is false: A function may be continuous at a point without being differentiable; for example, $f(x) = |x|$ is continuous at 0 but not differentiable.

B.5 Higher-Order Derivatives and Taylor's Theorem

If the first derivative of function, f', is also differentiable, then we say that f is twice differentiable. The second derivative is denoted by f''. Inductively, if the kth derivative is differentiable, then we say that f is $(k+1)$ times differentiable. The kth derivative of f is denoted by $f^{[k]}$. For a function that has n derivatives, there is a polynomial that agrees with f to order $n-1$ at a point.

Theorem B.5.1 (Taylor's Theorem). *Suppose that f has n derivatives in an interval $[a,b]$. If $x, c \in (a,b)$, then*

$$f(x) = \sum_{j=0}^{n-1} \frac{f^{[j]}(c)(x-c)^j}{j!} + R_n(x), \tag{B.7}$$

where

$$R_n(x) = O(|x-c|^n).$$

Formula (B.7) is called *Taylor's formula with remainder term*. There are many different formulæ for the *remainder term* R_n. One from which all the others can be derived is given by

$$R_n(x) = \frac{1}{(n-1)!} \int_c^x f^{[n]}(t)(x-t)^{n-1} dt. \tag{B.8}$$

The $n = 1$ case of Taylor's theorem is the mean value theorem.

Theorem B.5.2 (Mean value theorem). *Suppose that f is continuous on $[a,b]$ and differentiable on (a,b). Then there exists a $c \in (a,b)$ such that*

$$f(b) - f(a) = f'(c)(b-a).$$

B.6 Integration

The inverse operation to differentiation is integration. Suppose that f is a bounded function defined on a finite interval $[a,b]$. An increasing sequence $P = \{a = x_0 < x_1 < \cdots < x_N = b\}$ defines a *partition* of the interval. The mesh size of the partition is defined to be

$$|P| = \max\{|x_i - x_{i-1}| \, : \, i = 1, \ldots, N\}.$$

To each partition we associate two approximations of the area under the graph of f, by the rules

$$\begin{aligned} U(f,P) &= \sum_{j=1}^{N} \sup_{x \in [x_{j-1}, x_j]} f(x)(x_j - x_{j-1}), \\ L(f,P) &= \sum_{j=1}^{N} \inf_{x \in [x_{j-1}, x_j]} f(x)(x_j - x_{j-1}). \end{aligned} \tag{B.9}$$

These are called the *upper and lower Riemann sums*. Observe that for any partition P we have the estimate

$$U(f, P) \geq L(f, P). \tag{B.10}$$

If P and P' are partitions with the property that every point in P is also a point in P', then we say that P' is a *refinement* of P and write $P < P'$. If P_1 and P_2 are two partitions, then, by using the union of the points in the two underlying sets, we can define a new partition P_3 with the property that

$$P_1 < P_3 \text{ and } P_2 < P_3.$$

A partition with this property is called a *common refinement* of P_1 and P_2. From the definitions it is clear that if $P < P'$, then

$$U(f, P) \geq U(f, P') \text{ and } L(f, P) \leq L(f, P'). \tag{B.11}$$

***Definition* B.6.1.** A bounded function f defined on an interval $[a, b]$ is *Riemann integrable* if

$$\inf_P U(f, P) = \sup_P L(f, P).$$

In this case we denote the common value, called the *Riemann integral*, by

$$\int_a^b f(x)\,dx.$$

Most "nice" functions are Riemann integrable. For example we have the following basic result.

Theorem B.6.1. *Suppose that f is a piecewise continuous function defined on $[a, b]$. Then f is Riemann integrable and*

$$\int_a^b f(x)\,dx = \lim_{N\to\infty} \sum_{j=1}^{N} f(a + \frac{j}{N}(b-a))\frac{b-a}{N}.$$

The proof of this theorem is not difficult and relies primarily on the uniform continuity of a continuous function on a closed, bounded interval and (B.11). The sums appearing in this theorem are called *right Riemann sums*, because the function is evaluated at the right endpoint of each interval. The *left Riemann sums* are obtained by evaluating at the left endpoints. The formula for the integral holds for any Riemann integrable function but is more difficult to prove in this generality. The integral is a linear map from integrable functions to the real numbers.

Theorem B.6.2. *Suppose that f and g are Riemann integrable functions. Then $f+g$ and fg are integrable as well. If $c \in \mathbb{R}$, then*

$$\int_a^b (f(x)+g(x))\,dx = \int_a^b f(x)\,dx + \int_a^b g(x)\,dx \text{ and } \int_a^b cf(x)\,dx = c\int_a^b f(x)\,dx.$$

Theorem B.6.3. *Suppose that f is Riemann integrable on $[a,b]$ and that $c \in [a,b]$. Then f is Riemann integrable on $[a,c]$ and $[c,b]$; moreover,*

$$\int_a^b f(x)\,dx = \int_a^c f(x)\,dx + \int_c^b f(x)\,dx. \tag{B.12}$$

There is also a mean value theorem for the integral, similar to Theorem B.5.2.

Theorem B.6.4. *Suppose that f is a continuous function and w is a nonnegative integrable function. There exists a point $c \in (a,b)$ so that*

$$\int_a^b f(x)w(x)\,dx = f(c)\int_a^b w(x)\,dx.$$

The mean theorem provides a different formula for the remainder term in Taylor's theorem.

Corollary B.6.1. *Suppose that f has n-derivatives on an interval $[a,b]$ and x, c are points in (a,b). Then there exists a number d, between x and c, so that*

$$f(x) = \sum_{j=0}^{n-1} \frac{f^{[j]}(c)(x-c)^j}{j!} + \frac{f^{[n]}(d)(x-c)^n}{n!}. \tag{B.13}$$

Most elementary methods for calculating integrals come from the fundamental theorem of calculus. To state this result we need to think of the integral in a different way. As described previously, the integral associates a number to a function defined on a fixed interval. Suppose instead that f is defined and Riemann integrable on $[a,b]$. Theorem B.6.3 states that, for each $x \in [a,b]$, f is Riemann integrable on $[a,x]$. The new idea is to use the integral to define a new *function* on $[a,b]$ by setting

$$F(x) = \int_a^x f(y)\,dy.$$

This function is called the *indefinite integral* or anti-derivative of f. In this context we often refer to $\int_a^b f(x)\,dx$ as the *definite integral* of f.

Theorem B.6.5 (The Fundamental theorem of calculus). *If f is a continuous function on $[a, b]$ then F is differentiable and $F' = f$. If f is differentiable and f' is Riemann integrable then*

$$\int_a^b f'(x)dx = f(b) - f(a).$$

There are two further basics tools needed to compute and manipulate integrals. The first is called *integration by parts*, it is a consequence of the product rule for derivatives; see Proposition B.4.1.

Proposition B.6.1 (Integration by parts). *If $f, g \in \mathscr{C}^1([a, b])$ then*

$$\int_a^b f'(x)g(x)\,dx = f(b)g(b) - f(a)g(a) - \int_a^b f(x)g'(x)\,dx.$$

The other formula follows from the chain rule, Proposition B.4.2.

Proposition B.6.2 (Change of variable). *Let g be a monotone increasing, differentiable function defined $[a, b]$ with $g(a) = c$, $g(b) = d$ and let f be a Riemann integrable function on $[c, d]$. The following formula holds:*

$$\int_{g(a)}^{g(b)} f(y)\,dy = \int_a^b f(g(x))g'(x)\,dx.$$

B.7 Improper Integrals

In the previous section we defined the Riemann integral for *bounded* functions on *bounded* intervals. In applications both of these restrictions need to be removed. This leads to various notions of *improper integrals*. The simplest situation is that of a function f defined on $[0, \infty)$ and integrable on $[0, R]$ for every $R > 0$. We say that the improper integral,

$$\int_0^\infty f(x)\,dx$$

exists if the limit,

$$\lim_{R\to\infty} \int_0^R f(x)\,dx, \tag{B.14}$$

exists. In this case the improper integral is given by the limiting value. By analogy with the theory of infinite series, there are two distinct situations in which the improper integral exists. If the improper integral of $|f|$ exists, then we say that f is *absolutely integrable* on $[0, \infty)$.

***Example* B.7.1.** The function $(1+x^2)^{-1}$ is absolutely integrable on $[0, \infty)$. Indeed we see that if $R < R'$, then

$$
\begin{aligned}
0 \leq \int_0^{R'} \frac{dx}{1+x^2} - \int_0^{R} \frac{dx}{1+x^2} &= \int_R^{R'} \frac{dx}{1+x^2} \\
&\leq \int_R^{R'} \frac{dx}{x^2} \\
&\leq \frac{1}{R}.
\end{aligned} \tag{B.15}
$$

This shows that

$$\lim_{R\to\infty} \int_0^{R} \frac{dx}{1+x^2}$$

exists.

***Example* B.7.2.** The function $\frac{\sin x}{x}$ is a bounded continuous function; it is integrable on $[0, \infty)$ but not absolutely integrable. The integral of $\frac{\sin x}{x}$ over any finite interval is finite. Using integration by parts, we find that

$$\int_1^{R} \frac{\sin x dx}{x} = \frac{\cos x}{x}\Big|_1^R - \int_1^{R} \frac{\cos x dx}{x^2}.$$

Using this formula and the previous example, it is not difficult to show that

$$\lim_{R\to\infty} \int_0^{R} \frac{\sin x dx}{x}$$

exists. On the other hand because

$$\int_1^{R} \frac{dx}{x} = \log R,$$

it is not difficult to show that

$$\int_0^{R} \frac{|\sin x| dx}{x}$$

grows like $\log R$ and therefore diverges as R tend to infinity.

There are similar definitions for the improper integrals

$$\int_{-\infty}^{0} f(x)\,dx \text{ and } \int_{-\infty}^{\infty} f(x)\,dx.$$

The only small subtlety is that we say that the improper integral exists in the second case only when both the improper integrals,

$$\int_{-\infty}^{0} f(x)\,dx \text{ and } \int_{0}^{\infty} f(x)\,dx,$$

exist separately. Similar definitions apply to functions defined on bounded intervals (a, b) that are integrable on any subinterval $[c, d]$. We say that the improper integral

$$\int_{a}^{b} f(x)\,dx$$

exists if the limits

$$\lim_{c \to a^+} \int_{c}^{e} f(x)\,dx \text{ and } \lim_{c \to b^-} \int_{e}^{c} f(x)\,dx$$

both exist. Here e is any point in (a, b); the existence or nonexistence of these limits is clearly independent of which (fixed) point we use. Because improper integrals are defined by limits of proper integrals, they have the same linearity properties as integrals. For example,

Proposition B.7.1. *Suppose that f and g are improperly integrable on $[0, \infty)$. Then $f+g$ is as well and*

$$\int_{0}^{\infty} (f(x) + g(x))\,dx = \int_{0}^{\infty} f(x)\,dx + \int_{0}^{\infty} g(x)\,dx,$$

for $a \in \mathbb{R}$, af is improperly integrable and

$$\int_{0}^{\infty} af(x)\,dx = a \int_{0}^{\infty} f(x)\,dx.$$

The final case that requires consideration is that of a function f defined on a punctured interval $[a, b) \cup (b, c]$ and integrable on subintervals of the form $[a, e]$ and $[f, c]$, where $a \leq e < b$ and $b < f \leq c$. If both limits

$$\lim_{e \to b^-} \int_{a}^{e} f(x)\,dx \text{ and } \lim_{f \to b^+} \int_{f}^{c} f(x)\,dx$$

exist, then we say that f is improperly integrable on $[a, b]$. For example, the function $f(x) = x^{-\frac{1}{3}}$ is improperly integrable on $[-1, 1]$. On the other hand, the function $f(x) = x^{-1}$ is not improperly integrable on $[-1, 1]$ because

$$\lim_{e \to 0^+} \int_e^1 \frac{dx}{x} = \infty \text{ and } \lim_{f \to 0^-} \int_{-1}^f \frac{dx}{x} = -\infty.$$

There is a further extension of the notion of integrability that allows us to assign a meaning to

$$\int_{-1}^{1} \frac{dx}{x}.$$

This is called the *principal value integral* or *Cauchy principal value integral*. The observation is that for any $\epsilon > 0$

$$\int_{-1}^{-\epsilon} \frac{dx}{x} + \int_{\epsilon}^{1} \frac{dx}{x} = 0,$$

so the limit of this sum of integrals exists as ϵ goes to zero.

***Definition* B.7.1.** Suppose that f is defined on the punctured interval $[a, b) \cup (b, c]$ and is integrable on any subinterval $[a, e]$, $a \leq e < b$ or $[f, c]$, $b < f \leq c$. If the limit

$$\lim_{\epsilon \to 0} \int_a^{b-\epsilon} f(x)\,dx + \int_{b+\epsilon}^{c} f(x)\,dx$$

exists, then we say that f has a *principal value integral* on $[a, c]$. We denote the limit by

$$\text{P.V.} \int_a^c f(x)\,dx.$$

For a function that is not (improperly) integrable on $[a, b]$, the principal value integral exists because of cancellation between the divergences of the two parts of the integral. The approach to the singular point is symmetric; both the existence of the limit and its value depend crucially on this fact.

***Example* B.7.3.** We observed that the function x^{-1} has a principal value integral on $[-1, 1]$ and its value is zero. To see the importance of symmetry in the definition of the principal value integral, observe that

$$\int_{-1}^{-\epsilon} \frac{dx}{x} + \int_{2\epsilon}^{1} \frac{dx}{x} = -\log 2$$

and

$$\int_{-1}^{-\epsilon} \frac{dx}{x} + \int_{\epsilon^2}^{1} \frac{dx}{x} = -\log \epsilon.$$

In the first case we get a different limit and in the second case the limit does not exist.

The material in this chapter is usually covered in an undergraduate course in mathematical analysis. The proofs of these results and additional material can be found in [27], [111], and [121].

B.8 Fubini's Theorem and Differentiation of Integrals*

This section contains two results of a more advanced character than those considered in the previous sections of this appendix. This material is included because these results are used many times in the main body of the text.

There is a theory of integration for functions of several variables closely patterned on the one -variable case. A rectangle in $\mathbb{R}^n$ is a product of bounded intervals

$$R = [a_1, b_1) \times \cdots \times [a_n, b_n).$$

The n-dimensional volume of R is defined to be

$$|R| = \prod_{j=1}^{n} (b_j - a_j).$$

Suppose that f is a bounded function with bounded support in $\mathbb{R}^n$. A partition of the support of f is a collection of disjoint rectangles $\{R_1, \ldots, R_N\}$ such that

$$\operatorname{supp} f \subset \bigcup_{j=1}^{N} R_j.$$

To each partition P of $\operatorname{supp} f$ we associate an upper and lower Riemann sum:

$$U(f, P) = \sum_{j=1}^{N} \sup_{x \in R_j} f(\boldsymbol{x}) |R_j|, \quad L(f, P) = \sum_{j=1}^{N} \inf_{x \in R_j} f(\boldsymbol{x}) |R_j|.$$

As before, f is integrable if

$$\inf_P U(f, P) = \sup_P L(f, P).$$

In this case the integral of f over $\mathbb{R}^n$ is denoted by

$$\int_{\mathbb{R}^n} f(\boldsymbol{x})\, d\boldsymbol{x}.$$

Let B_r denote the ball centered at zero of radius r. If $\chi_{B_r}|f|$ is integrable for every r and

$$\lim_{r\to\infty}\int_{\mathbb{R}^n}\chi_{B_r}|f|(x)\,dx$$

exists, then we say that f is absolutely integrable on $\mathbb{R}^n$. It is not difficult to extend the definition of absolute integrability to unbounded functions. Let f be a function defined on $\mathbb{R}^n$ and set

$$E_R = f^{-1}([-R, R]).$$

Suppose that for every positive number R the function $\chi_{E_R} f$ is absolutely integrable on $\mathbb{R}^n$. If the limit

$$\lim_{R\to\infty}\int_{\mathbb{R}^n}\chi_{E_R}|f|(x)\,dx$$

exists, then we say that f is absolutely integrable and let

$$\int_{\mathbb{R}^n} f(x)\,dx = \lim_{R\to\infty}\int_{\mathbb{R}^n}\chi_{E_R} f(x)\,dx.$$

Suppose that $n = k + l$ for two positive integers k and l. Then $\mathbb{R}^n = \mathbb{R}^k \times \mathbb{R}^l$. Let w be coordinates for $\mathbb{R}^k$ and y coordinates for $\mathbb{R}^l$. Assume that for each w in $\mathbb{R}^k$ the function $f(w, \cdot)$ on $\mathbb{R}^l$ is absolutely integrable and the function

$$g(w) = \int_{\mathbb{R}^l} f(w, y)\,dy$$

is an integrable function on $\mathbb{R}^k$. The integral of g over $\mathbb{R}^k$ is an *iterated integral* of f; it usually expressed as

$$\int_{\mathbb{R}^n} g(x)\,dx = \int_{\mathbb{R}^k}\int_{\mathbb{R}^l} f(w, y)\,dw\,dy.$$

It is reasonable to enquire how is the integral of g over $\mathbb{R}^k$ is related to the integral of f over $\mathbb{R}^n$. Fubini's theorem provides a comprehensive answer to this question.

Theorem B.8.1 (Fubini's theorem). *Let f be a function defined on $\mathbb{R}^n$ and let $n = k + l$ for positive integers k and l. If either of the iterated integrals*

$$\int_{\mathbb{R}^k}\int_{\mathbb{R}^l} |f(w, y)|\,dw\,dy \text{ or } \int_{\mathbb{R}^l}\int_{\mathbb{R}^k} |f(w, y)|\,dy\,dw$$

is finite, then the other is as well. In this case f is integrable over $\mathbb{R}^n$ and

$$\int_{\mathbb{R}^k}\int_{\mathbb{R}^l} f(w, y)\,dw\,dy = \int_{\mathbb{R}^n} f(x)\,dx = \int_{\mathbb{R}^l}\int_{\mathbb{R}^k} f(w, y)\,dy\,dw \tag{B.16}$$

Informally, the order of the integrations can be interchanged. Note that we assume that f is *absolutely* integrable in order to conclude that the order of integrations of f can be interchanged. There are examples of functions defined on $\mathbb{R}^2$ so that both iterated integrals,

$$\int_{-\infty}^{\infty}\int_{-\infty}^{\infty} f(x,y)\,dx\,dy \text{ and } \int_{-\infty}^{\infty}\int_{-\infty}^{\infty} f(x,y)\,dy\,dx$$

exist but are unequal, and f is not integrable on $\mathbb{R}^2$. A proof of Fubini's theorem can be found in [121] or [43].

The second problem we need to consider is that of differentiation under the integral sign. For a positive number ϵ, let f be a function defined on $\mathbb{R}^n \times (a-\epsilon, a+\epsilon)$. Suppose that for each y in $(a-\epsilon, a+\epsilon)$, the function $f(\cdot, y)$ is absolutely integrable on $\mathbb{R}^n$, and for each $\boldsymbol{x}$ in $\mathbb{R}^n$, the function $f(\boldsymbol{x}, \cdot)$ is differentiable at a. Is the function defined by the integral

$$g(y) = \int_{\mathbb{R}^n} f(\boldsymbol{x}, y)\,d\boldsymbol{x}$$

differentiable at a? In order for this to be true, we need to assume that the difference quotients

$$\frac{f(\boldsymbol{x}, a+h) - f(\boldsymbol{x}, a)}{h}$$

satisfy some sort of uniform bound. The following theorem is sufficient for our applications.

Theorem B.8.2. *With f as before, if there exists an absolutely integrable function F so that for every h with $|h| < \epsilon$, we have the estimate*

$$\left|\frac{f(\boldsymbol{x}, a+h) - f(\boldsymbol{x}, a)}{h}\right| \le F(\boldsymbol{x}),$$

then

$$g(y) = \int_{\mathbb{R}^n} f(\boldsymbol{x}, y)\,d\boldsymbol{x}$$

is differentiable at a and

$$g' = \int_{\mathbb{R}^n} \partial_y f(\boldsymbol{x}, y)\,d\boldsymbol{x}.$$

This theorem is a consequence of the Lebesgue dominated convergence theorem.

Bibliography

[1] A. Abragam, *Principles of nuclear magnetism*, Clarendon Press, Oxford, 1983.

[2] Lars V. Ahlfors, *Complex Analysis*, McGraw-Hill, New York, 1979.

[3] R. E. Alvarez and A. Macovski, *Energy selective reconstructions in x-ray computerized tomography*, Phys. Med. Biol. **21** (1976), 733–744.

[4] R. J. Barlow, *Statistics, A Guide to the Use of Statistical Methods in the Physical Sciences*, The Manchester Physics Series, John Wiley & Sons, 1989.

[5] Harrision H. Barrett and K. J. Myers, *Foundations of Image Science*, John Wiley and Sons, Hoboken, 2004.

[6] Harrison H. Barrett and William Swindell, *Radiological Imaging*, Academic Press, New York, 1981.

[7] R. Beals, *Advanced Mathematical Analysis*, Graduate Texts in Mathematics, vol. 119, Springer-Verlag, 1988.

[8] George B. Benedek and Felix M.H. Villars, *Physics with Illustrative Examples from Medicine and Biology, Electricity and Magnetism*, 2nd ed., AIP Press and Springer-Verlag, New York, 2000.

[9] Matt A. Bernstein, Kevin F. King, and Xiaohong Joe Zhou, *Handbook of MRI pulse sequences*, Elsevier Academic Press, London, 2004.

[10] M. Bertero and P. Boccacci, *Introduction to inverse problems in imaging*, Institute of Physics Publishing, Bristol, UK, 1998.

[11] Felix Bloch, *Nuclear induction*, Physical Review **70** (1946), 460–474.

[12] William E. Boyce and Richard C. DiPrima, *Elementary Differential Equations*, 6th ed., John Wiley & Sons, New York, 1997.

[13] D. P. Boyd and C. Haugland, *Recent progress in electron beam tomography*, Med. Imag. Tech. **11** (1993), 578–585.

[14] D. P. Boyd and M. J. Lipton, *Cardiac computed tomography*, Proceedings of the IEEE **71** (1983), 298–307.

[15] Ronald N. Bracewell, *The Fourier Transform and Its Applications,* 2nd revised edition, McGraw-Hill, New York, 1986.

[16] Robert Grover Brown, *Introduction to Random Signal Analysis and Kalman Filtering*, John Wiley & Sons, New York, 1983.

[17] Paul T. Callaghan, *Principles of nuclear magnetic resonance microscopy*, Clarendon Press, Oxford, 1993.

[18] Emmanuel J. Candes, Justin Romberg, and Terence Tao, *Robust uncertainty principles: exact signal reconstruction from highly incomplete frequency information*, IEEE Trans. Inform. Theory **52** (2006), 489–509.

[19] ______, *Stable signal recovery from incomplete and inaccurate measurements*, Comm. Pure Appl. Math. **59** (2006), 1207–1223.

[20] Yair Censor and Gabor T. Herman, *On some optimization techniques in image reconstruction from projections*, Appl. Numer. Math. **3** (1987), 365–391.

[21] C.-N. Chen and D. I. Hoult, *Biomedical magnetic resonance technology*, Adam Hilger, Bristol, 1989.

[22] Zang-Hee Cho, Joie P. Jones, and Manbir Singh, *Foundations of Medical Imaging*, John Wiley & Sons, New York, 1993.

[23] Committee on the Mathematics and Physics of Emerging Dynamic Biomedical Imaging, Washington, D.C., *Mathematics and Physics of Emerging Biomedical Imaging*, 1996, National Academy Press.

[24] A. M. J. Cormack, *Representation of a function by its line integrals, with some radiological applications I., II.*, J. Applied Physics **34,35** (1963,1964), 2722–2727, 195–207.

[25] Richard Courant and David Hilbert, *Methods of Mathematical Physics, I and II*, Wiley, New York, 1953.

[26] Carl R. Crawford and Kevin F. King, *Computed tomography scanning with simultaneous patient translation*, Med. Phys. **17** (1990), 967–982.

[27] John D'Angelo and Douglas West, *Mathematical Thinking, Problem-Solving and Proofs,* 2nd ed., Prentice Hall, Upper Saddle River, NJ, 2000.

[28] Wilbur B. Davenport, Jr. and William L. Root, *An Introduction to the Theory of Random Signals and Noise*, Lincoln Laboratory Publications, McGraw-Hill Co., New York, 1958.

[29] J. L. Doob, *Stochastic Processes*, John Wiley & Sons, Inc., New York, 1953.

[30] Edward R. Dougherty, *Random Processes for Image and Signal Processing*, SPIE/IEEE series on imaging science and engineering, IEEE press, Piscataway, NJ, 1999.

[31] W. A. Edelstein, G. H. Glover, C. Hardy, and R. Redington, *The intrinsic signal-to-noise ratio in NMR imaging*, Magn. Reson. Med. **3** (1986), 604–618.

[32] W. A. Edelstein, J. M. Hutchinson, J. M. Johnson, and T. Redpath, *Spin warp NMR imaging and applications to human whole-body imaging*, Phys. Med. Biol. **25** (1980), 751–756.

[33] Charles L. Epstein, *Introduction to magnetic resonance imaging for mathematicians*, Annales des L'Institut Fourier **54** (2004), 1697–1716.

[34] ______, *Minimum power pulse synthesis via the inverse scattering transform*, Jour. Mag. Res. **167** (2004), 185–210.

[35] ______, *How well does the finite Fourier transform approximate the Fourier transform?*, Comm. in Pure and Appl. Math. **58** (2005), 1421–1435.

[36] Charles L. Epstein and Bruce Kleiner, *Spherical means in annular regions*, CPAM **44** (1993), 441–451.

[37] Charles L. Epstein and Felix W. Wehrli, *Magnetic resonance imaging*, Encyclopedia of Mathematical Physics, Elsevier, 2006, pp. 367–375.

[38] Richard Ernst, Geoffrey Bodenhausen, and Alexander Wokaun, *Principles of nuclear magnetic resonance in one and two dimensions*, Clarendon, Oxford, 1987.

[39] Yu. L. Ershov, S. S. Goncharov, A. Nerode, J. B. Remmel, and V. W. Marek, *Handbook of Recursive Mathematics*, vol. 2 *of Recursive Algebra, Analysis and Combinatorics*, Studies in Logic and the Foundations of Mathematics, vol. 138, North-Holland, Amsterdam, 1998.

[40] L. A. Feldkamp, L. C. Davis, and J. W. Kress, *Practical cone-beam algorithm*, J. Opt. Soc. Am. **1(A)** (1984), 612–619.

[41] W. Feller, *Introduction to Probability Theory and its Applications, I and II*, John Wiley & Sons, New York, 1968, 1971.

[42] G.B. Folland, *Real Analysis, Modern Techniques and their Applications*, John Wiley & Sons, New York, 1984.

[43] ______, *Introduction to Partial Differential Equation*, 2nd ed., Princeton University Press, Princeton, NJ, 1995.

[44] I. M. Gelfand, M.I. Graev, and N. Ya. Vilenkin, *Generalized Functions. vol. 5. Integral Geometry and Representation Theory. translated from the Russian by Eugene Saletan*, Academic Press, New York-London, 1970.

[45] I. S. Gradshteyn and I. M. Ryzhik, *Table of Integrals, Series and Products*, Academic Press, New York, 1980.

[46] Pierre Grangeat, Pascal Sire, Règis Guillemaud, and Valérie La, *Indirect cone-beam three-dimensional image reconstruction*, Contemporary Perspectives in Three-dimensional Biomedical Imaging (C. Roux and J.-L. Coatrieux, eds.), IOS Press, 1997, pp. 29–52.

[47] Leslie Greengard and June-Yub Lee, *Accelerating the nonuniform fast Fourier transform*, SIAM Review **46** (2004), 443–454.

[48] Leslie Greengard, June-Yub Lee, and Souheil Inati, *The fast sinc transform and image reconstruction from nonuniform samples in k-space*, Commun. Appl. Math. Comput. Sci. **1** (2006), 121–131.

[49] Charles W. Groetsch, *Regularization and stabilization of inverse problems*, Handbook of Analytic-Computational Methods in Applied Mathematics (Boca Raton, FL) (George Anastassiou, ed.), Chapman and Hall/CRC, 2000, pp. 31–64.

[50] E. Mark Haacke, Robert W. Brown, Michael R. Thompson, and Ramesh Venkatesan, *Magnetic Resonance Imaging*, Wiley-Liss, New York, 1999.

[51] S. Helgason, *The Radon Transform*, 2nd ed., Birkhäuser, Boston, 1999.

[52] Gabor T. Herman, *Image Reconstruction from Projections*, Academic Press, New York, 1980.

[53] Gabor T. Herman, Arnold Lent, and Stuart Rowland, *Art: Mathematics and applications*, J. Theo. Bio. **42** (1973), 1–32.

[54] Gabor T. Herman and Dewey Odhner, *Performance evaluation of an iterative image reconstruction algorithm for positron emmision tomography*, IEEE Trans. on Med. Im. **10** (1991), 336–346.

[55] Gabor T. Herman, A.V. Lakshminarayanan, and A. Naparstek, *Reconstruction using divergent ray shadowgraphs*, in Ter-Pergossian [125], pp. 105–117.

[56] F. B. Hildebrand, *Introduction to Numerical Analysis*, McGraw-Hill, New York, 1956.

[57] Paul G. Hoel, Sidney C. Port, and Charles J. Stone, *Introduction to Stochastic Processes*, Houghton-Mifflin, Boston, Ma, 1972.

[58] M. Holschneider, *Wavelets, An Analysis Tool*, Clarendon Press, Oxford, 1995.

[59] L. Hörmander, *The Analysis of Linear Partial Differential Operators*, vol. 1, Springer-Verlag, Berlin, Heidelberg, New York, Tokyo, 1983.

[60] ———, *The Analysis of Linear Partial Differential Operators*, vol. 3, Springer-Verlag, Berlin, Heidelberg, New York, Tokyo, 1985.

[61] D. I. Hoult and P. C. Lauterbur, *The sensitivity of the zeugmatographic experiment involving human samples*, JMR **34** (1979), 425–433.

[62] D.I. Hoult, *The principle of reciprocity in signal strength calculations—A mathematical guide*, Concepts in Mag. Res. **12** (2000), 173–187.

[63] ———, *Sensitivity and power deposition in a high field imaging experiment*, JMRI **12** (2000), 46–67.

[64] G. N. Hounsfield, *Computerised transverse axial scanning tomography I. Description of system*, Br. J. Radiology **46** (1973), 1016–1022.

[65] Hui Hu, *Multi-slice helical scan CT: Scan and reconstruction*, Med. Phys. **26** (1999), 5–17.

[66] Robert J. Marks II (ed.), *Advanced Topics in Shannon Sampling and Interpolation Theory*, New York, Berlin, Heidelberg, Springer-Verlag, 1993.

[67] International Commission on Radiation Units and Measurement, Bethesda, MA, *Tissue substitute in radiation dosimetry and measurement Report 44*, 1998, Available at http://physics.nist.gov/PhysRefData/XrayMassCoef/cover.html.

[68] Bernd Jähne, *Digital Image Processing, Concepts, Algorithms and Scientific Applications, third ed.*, Springer-Verlag, Berlin, Heidelberg, 1995.

[69] Peter M. Joseph, *Image noise and smoothing in computed tomography (CT) scanners*, SPIE—Optical Instrumentation in Medicine VI **127** (1977), 43–49.

[70] ———, *The influence of gantry geometry on aliasing and other geometry dependent errors*, IEEE Transactions on Nuclear Science **NS-27** (1980), 1104–1111.

[71] ———, *Artifacts in computed tomography*, Technical Aspects of Computed Tomgraphy, vol. 5 (St. Louis) (M. D . Thomas H. Newton and M. D . D. Gordon Potts, eds.), The C.V. Mosby Company, 1981, pp. 3956–3992.

[72] Peter M. Joseph and Raymond A. Schulz, *View sampling requirements in fan beam computed tomography*, Med. Phys. **7** (1980), 692–702.

[73] Peter M. Joseph and Robin D. Spital, *A method for corrrecting bone induced artifacts in computer tomography scanners*, Journal of Computer Assisted Tomography **2** (1978), 100–108.

[74] Peter M. Joseph, Robin D. Spital, and Charles D. Stockham, *The effects of sampling on CT-images*, Computerized Tomography **4** (1980), 189–206.

[75] Peter M. Joseph and Charles D. Stockham, *The influence of modulation transfer function shape on computer tomographic image quality*, Radiology **145** (1982), 179–185.

[76] Avinash Kak and Malcolm Slaney, *Principles of Computerized Tomographic Imaging*, Classics Appl. Math. **33**, SIAM, Philadelphia, 2001.

[77] W. A. Kalender, *Computed tomography: Fundamentals, system technology, image quality, applications*, John Wiley & Sons, New York, 2000.

[78] A. Katsevich, *An improved exact filtered backprojection algorithm for spiral computed tomography*, Advances in Applied Mathematics **32** (2004), 681–697.

[79] Yitzhak Katznelson, *An Introduction to Harmonic Analysis*, Dover, New York, 1976.

[80] Joseph B. Keller, *Inverse problems*, American Math. Monthly **83** (1976), 107–118.

[81] Reinhard Klette and Piero Zamperoni, *Handbook of Image Processing Operators*, John Wiley & Sons, Chichester, 1996.

[82] Peter D. Lax, *Linear Algebra*, John Wiley & Sons, New York, 1997.

[83] Peter D. Lax and Ralph S. Phillips, *The Paley-Wiener theorem for the Radon transform*, CPAM **23** (1970), 409–424.

[84] M. H. Levitt, *Spin dynamics, basics of nuclear magnetic resonance*, John Wiley & Sons, Chichester, 2001.

[85] R. M. Lewitt, *Reconstruction algorithms: Transform methods*, Proceedings of the IEEE **71** (1983), no. 3, 390–408.

[86] R.M. Lewitt and Samuel Matej, *Overview of methods for image reconstruction in emission computed tomography*, Proceeding of the IEEE **91** (2003), 1588–1611.

[87] Peter Linz, *Theoretical Numerical Analysis*, John Wiley & Sons, New York, 1979.

[88] B. F. Logan, *The uncertainty principle in reconstructing functions from projections*, Duke Math. Journal **42** (1975), 661–706.

[89] B. F. Logan and L. A. Shepp, *Optimal reconstruction of a function from its projections*, Duke Math. Journal **42** (1975), 645–659.

[90] Donald Ludwig, *The Radon transform on Euclidean space*, CPAM **19** (1966), 49–81.

[91] Albert Macovski, *Medical Imaging Systems*, Prentice Hall, Englewood Cliffs, NJ, 1983.

[92] Jeremy Magland and Charles L. Epstein, *Practial pulse synthesis via the discrete inverse scattering transform*, Jour. Magn. Res. **172** (2004), 63–78.

[93] Wilhelm Magnus and Fritz Oberhettinger, *Formulas and Theorems for the Special Functions of Mathematical Physics*, Chelsea, New York, 1949, translated from the German by John Wermer.

[94] Sandy Napel, *Basic principles of spiral CT*, Spiral CT Principles, Techniques and Clinical Applications (Philadelphia, PA) (Elliot K. Fishman and R. Brooke Jeffrey, Jr., eds.), Lippincott-Raven, 1998, pp. 3–15.

[95] Frank Natterer, *The Mathematics of Computerized Tomography*, 2nd ed., Classics Appl. Math. **32**, SIAM, Philadelphia, 2001.

[96] Frank Natterer and Frank Wübbeling, *Mathematical Methods in Image Reconstruction*, SIAM Monogr. Math. Model. Comput. **5**, SIAM, Philadelphia, 2001.

[97] Zeev Nehari, *Conformal Mapping*, Dover, New York, 1952.

[98] R.G. Novikov, *An inversion formula for the attenuated X-ray transformation*, Ark. Mat. **40** (2002), 145–167.

[99] ______, *On the range characterization for the two-dimensional attenuated x-ray transformation*, Inverse Problems **18** (2002), 677–700.

[100] Alan V. Oppenheim and Ronald W. Schafer, *Digital Signal Processing*, Prentice Hall, 1975.

[101] John Pauly, Patrick Le Roux, Dwight Nishimura, and Albert Macovski, *Parameter relations for the Shinnar-Le Roux selective excitation pulse design algorithm*, IEEE Trans. on Med. Imaging **10** (1991), 53–65.

[102] Isaac Pesenson, *A sampling theorem on homogeneous manifolds*, Trans. of the Amer. Math. Soc. **352** (2000), 4257–4269.

[103] Mark A. Pinsky and Michael E. Taylor, *Pointwise Fourier inversion: a wave equation approach*, The Journal of Fourier Analysis and Applications **3** (1997), 647–703.

[104] G. Pólya and G. Szegő, *Problems and Theorems in Analysis, I*, Springer-Verlag, New York, 1972.

[105] Johan Radon, *Über die Bestimmung von Funktionen durch ihre Integralwerte längs gewisser Mannigfaltigkeiten*, Ber. Sachs. Akad. Wiss., Leipzig **69** (1917), 262–267.

[106] F. Riesz and B. Sz.-Nagy, *Functional Analysis*, Fredrick Ungar, New York, 1955.

[107] Theodore J. Rivlin, *The Chebyshev Polynomials*, New York, John Wiley & Sons, 1974.

[108] ______, *An Introduction to the Approximation Theory of Functions*, Dover, New York, 1981.

[109] H.L. Royden, *Real Analysis*, 2nd ed., Macmillan, New York, 1968.

[110] Walter Rudin, *Real and Complex Analysis*, 2nd ed., McGraw-Hill, New York, 1974.

[111] ______, *Principles of Mathematical Analysis*, 3rd ed., McGraw-Hill, New York, 1976.

[112] Hermann Schomberg and Jan Timmer, *The gridding method for image reconstruction by Fourier transformation*, IEEE Journal on Medical Imaging **14** (1995), no. 3, 596–607.

[113] L. A. Shepp and J. B. Kruskal, *Computerized tomography: The new medical X-ray technology*, Amer. Math. Monthly (1978), 421–439.

[114] L. A. Shepp and B. F. Logan, *The Fourier reconstruction of a head section*, IEEE Trans. Nuc. Sci. **NS-21** (1990), 21–43.

[115] L. A. Shepp and J. A. Stein, *Simulated reconstruction artifacts in computerized X-ray tomography*, in Ter-Pergossian [125], pp. 33–48.

[116] L. A. Shepp and Y. Vardi, *Maximum likelihood reconstruction for emission tomography*, IEEE Trans. on Med. Imaging **MI-1** (1982), 113–122.

[117] M. Shinnar and J.S. Leigh, *The application of spinors to pulse synthesis and analysis*, Mag. Res. in Med. **12** (1989), 93–98.

[118] D. L. Snyder, J. T. Lewis, and M. M. Ter-Pogossian, *A mathemtical model for positron-emission tomography systems having time-of-flight measurements*, IEEE Trans. on Nucl. Sci. **NS-28** (1981), 3575–3583.

[119] Michael Spivak, *Calculus on Manifolds*, Benjamin/Cummings, 1965.

[120] Elias M. Stein and Guido Weiss, *Introduction to Fourier Analysis on Euclidean Spaces*, Princeton University Press, Princeton, NJ, 1971.

[121] Robert Strichartz, *The Way of Analysis*, Jones and Bartlett, Boston, MA, 1995.

[122] Kunio Tanabe, *Projection method for solving a singular system of linear equations and its applications*, Num. Math. **17** (1971), 203–214.

[123] M. E. Taylor, *Pseudodifferential Operators*, Princeton Mathematical Series, vol. 34, Princeton University Press, Princeton, NJ, 1981.

[124] ______, *Partial Differential Equations*, vol. 2, Applied Mathematical Sciences, vol. 116, Springer-Verlag, New York, 1996.

[125] M. M. Ter-Pergossian (ed.), *Reconstruction Tomography in Diagnostic Radiology and Nuclear Medicine*, University Park Press, Baltimore, 1977.

[126] H. C. Torrey, *Bloch equations with diffusion terms*, Physical Review **104** (1956), 563–565.

[127] Lloyd N. Trefethen and David Bau III, *Numerical Linear Algebra*, SIAM, Philadelphia, 1997.

[128] S. R. S. Varadhan, *Stochastic Processes*, Courant Institute of Mathematical Sciences, New York, 1968.

[129] G. N. Watson, *A Treatise on the Theory of Bessel Functions,* 2nd ed., Cambridge University Press, Cambridge, 1948.

[130] E. T. Whittaker and G. N. Watson, *A Course of Modern Analysis,* 4th ed., Cambridge University Press, London, 1935.

[131] Harold Widom, *Lectures on Integral Equations*, Van Nostrand-Reinhold Co., New York, 1969.

[132] J. E. Wilting, *Technical aspects of spiral CT*, Medica Mundi **43** (1999), 34–43.

Index